Simons Symposia

Series Editor

Yuri Tschinkel, Courant Institute of Mathematical Sciences and Simons Foundation, New York University, New York, NY, USA

Working to foster communication and enable interactions between experts, volumes in the Simons Symposia series bring together leading researchers to demonstrate the powerful connection of ideas, methods, and goals shared by mathematicians, theoretical physicists, and theoretical computer scientists.

Symposia volumes feature a blend of original research papers and comprehensive surveys from international teams of leading researchers in thriving fields of study. This blend of approaches helps to ensure each volume will serve not only as an introduction for graduate students and researchers interested in entering the field, but also as the premier reference for experts working on related problems.

The Simons Foundation at its core exists to support basic, discovery-driven research in mathematics and the basic sciences, undertaken in pursuit of understanding the phenomena of our world without specific applications in mind. The foundation seeks to advance the frontiers of research in mathematics and the basic sciences by creating strong collaborations and fostering cross-pollination of ideas between investigators, leading to unexpected breakthroughs and a deeper understanding of the world around us.

Yuri Tschinkel
Editor

Arithmetic and Algebraic Geometry

A Mathematical Tribute to Yuri Manin

Springer

Editor
Yuri Tschinkel
Courant Institute of Mathematical Sciences
New York University
New York, NY, USA

ISSN 2365-9564 ISSN 2365-9572 (electronic)
Simons Symposia
ISBN 978-3-031-74133-3 ISBN 978-3-031-74134-0 (eBook)
https://doi.org/10.1007/978-3-031-74134-0

This Springer imprint is published by the registered company Springer Nature Switzerland AG
The registered company address is: Gewerbestrasse 11, 6330 Cham, Switzerland

Preface

This book is a tribute to the memory of Yuri Ivanovich Manin, who passed away on January 7, 2023. Manin was one of the giants of modern mathematics. His work covered a wide range of fields, including logic, number theory, geometry, mathematical physics, theoretical computer science, and linguistics. The papers collected here are on topics close to his life-long passion: arithmetic and algebraic geometry.

New York, NY, USA
March 2024

Yuri Tschinkel

Acknowledgements

The help of Dr. S. Wiedmann (University of Göttingen) in coordinating and typesetting this volume was invaluable. As always, it is a pleasure to work with the Springer team, and in particular, with Nicholas DiBenedetto and Elizabeth Loew.

Contents

Projecting Lattice Polytopes According to the Minimal Model Program

Victor V. Batyrev

Contents

1 Introduction

Let $P \subset \mathbb{R}^d$ be an arbitrary d-dimensional convex polytope.

Definition 1.1 For a nonzero lattice vector $\mathbf{a} = (a_1, \ldots, a_d) \in \mathbb{Z}^d \setminus \{0\}$ consider the **integral supporting hyperplane** of P:

$$H_{\mathbf{a},P} := \left\{ \mathbf{x} = (x_1, \ldots, x_d) \in \mathbb{R}^d \mid \sum_{i=1}^{d} a_i x_i = \mathrm{Min}_P(\mathbf{a}) := \min_{\mathbf{y} \in P} \left(\sum_{i=1}^{n} a_i y_i \right) \right\}.$$

If $\gcd(a_1, \ldots, a_d) = 1$ we call the number

$$\mathrm{dist}_{\mathbb{Z}}(\mathbf{y}, H_{\mathbf{a},P}) := \sum_{i=1}^{n} a_i y_i \; - \; \mathrm{Min}_P(\mathbf{a}) \geq 0$$

the **integral distance** between a point $\mathbf{y} = (y_1, \ldots, y_d) \in P$ and integral supporting hyperplane $H_{\mathbf{a},P}$.

V. V. Batyrev (✉)
Fachbereich Mathematik, Universität Tübingen, Tübingen, Germany
e-mail: batyrev@math.uni-tuebingen.de

Y. Tschinkel (ed.), *Arithmetic and Algebraic Geometry*, Simons Symposia,
https://doi.org/10.1007/978-3-031-74134-0_1

Definition 1.2 ([15, 25]) Let $P \subset \mathbb{R}^d$ be an arbitrary d-dimensional convex polytope. The set $F(P)$ of all points in P having integral distance at least 1 to any integral supporting hyperplane $H_{\mathbf{a},P}$ is called the **Fine interior of** P, i.e.,

$$F(P) := \{\mathbf{y} \in P \mid \mathrm{dist}_{\mathbb{Z}}(\mathbf{y}, H_{\mathbf{a},P}) \geq 1, \ \ \forall \mathbf{a} \in \mathbb{Z}^d \setminus \{0\}\}.$$

Remark 1.3 If the affine span of every facet of P is an integral supporting hyperplane, then Gordan's lemma shows that among countably many inequalities $\mathrm{dist}_{\mathbb{Z}}(\mathbf{y}, H_{\mathbf{a},P}) \geq 1$ defining $F(P)$ only finitely many $\mathbf{a} \in \mathbb{Z}^d$ are necessary. In particular, $F(P) \subset P$ is a rational polytope (or empty set) if all vertices of P belong to $\mathbb{Q}^d$. We explain more details concerning this fact in Proposition 3.1. **Lattice polytopes** P, i.e., polytopes having vertices in $\mathbb{Z}^d$, are main objects of our study. However, for some technical reasons it will be convenient to consider the Fine interior $F(P)$ of rational polytopes P and the Fine interior $F(\lambda P)$ of their arbitrary real positive multiples λP ($\lambda \in \mathbb{R}_{>0}$).

Remark 1.4 If $P \subset \mathbb{R}^d$ is a d-dimensional lattice polytope, then every interior lattice point $\mathbf{m} \in Int(P) \cap \mathbb{Z}^d$ necessarily belongs to $F(P)$, and we obtain the inclusion

$$\mathrm{Conv}\,(Int(P) \cap \mathbb{Z}^d) \subset F(P)$$

which is in fact equality for lattice polytopes P of dimension $d \in \{1, 2\}$ [9].

Recall some standard definitions.

Definition 1.5 Two lattice polytopes $P_1, P_2 \subset \mathbb{R}^d$ are called **unimodular equivalent** if there exists a lattice-preserving affine isomorphism $\varphi : \mathbb{R}^d \to \mathbb{R}^d$ such that $\varphi(P_1) = P_2$.

Definition 1.6 A k-dimensional lattice polytope $P' \subset \mathbb{R}^k$ ($1 \leq k < d$) is called **lattice projection**, or **$\mathbb{Z}$-projection**, of a d-dimensional lattice polytope $P \subset \mathbb{R}^d$ if there exists an affine map $\pi : \mathbb{R}^d \to \mathbb{R}^k$ inducing a surjective map of lattices $\pi : \mathbb{Z}^d \to \mathbb{Z}^k$ and $\pi(P) = P'$.

Remark 1.7 Since every epimorphism $\pi : \mathbb{Z}^d \to \mathbb{Z}^k$ splits, we can choose a splitting $\mathbb{Z}^d \cong \mathbb{Z}^k \oplus \mathbb{Z}^{n-k}$ such that the $\mathbb{Z}$-projection π has the standard form:

$$\pi(x_1, \ldots, x_d) = (x_1, \ldots, x_k) \in \mathbb{R}^k \ \ \forall (x_1, \ldots, x_d) \in \mathbb{R}^d.$$

Definition 1.8 We call a d-dimensional polytope $P \subset \mathbb{R}^d$ **F-hollow** if $F(P) = \emptyset$.

Remark 1.9 Let $P' \subset \mathbb{R}^k$ be a lattice projection of a d-dimensional polytope $P \subset \mathbb{R}^d$, then the lattice epimorphism $\pi : \mathbb{Z}^d \to \mathbb{Z}^k$ allows to lift every integral supporting hyperplane of P' to an integral supporting hyperplane of P. In particular, the condition $F(P') = \emptyset$ implies $F(P) = \emptyset$. This fact allows to construct infinitely many pairwise unimodular distinct d-dimensional F-hollow lattice polytopes P

whose lattice projections are equal to a given lower-dimensional F-hollow lattice polytope P'.

Definition 1.10 We call a d-dimensional F-hollow lattice polytope P **sporadic**, if P does not admit any lattice projection $\pi : P \to P'$ onto a k-dimensional F-hollow lattice polytope P' ($1 \leq k \leq d-1$).

The present paper shows that in any fixed dimension d, apart from finitely many unimodular equivalence classes of d-dimensional sporadic F-hollow lattice polytopes, every d-dimensional F-hollow lattice polytope P admits a $\mathbb{Z}$-projection to some lower-dimensional F-hollow lattice polytope P':

Theorem 1.11 *In each dimension d there exist up to unimodular transformations only finitely many d-dimensional sporadic F-hollow lattice polytopes P.*

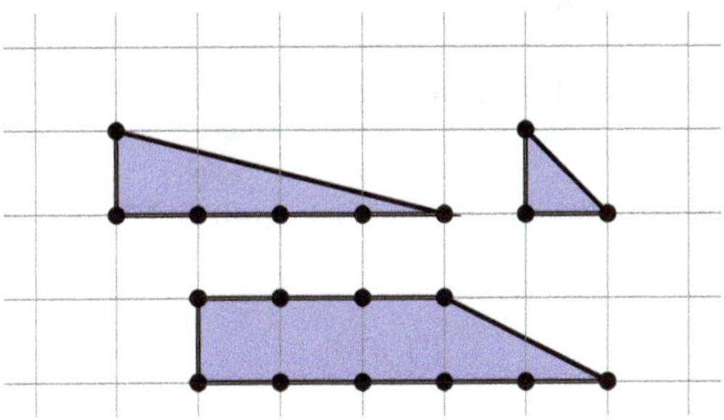

countably many F-hollow polygons

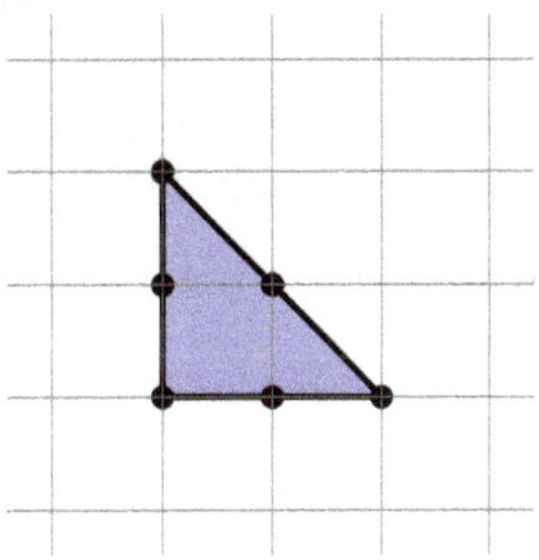

a sporadic polygon

Remark 1.12 Theorem 1.11 is a purely combinatorial contribution to the theory of lattice polytopes but its motivation comes from birational algebraic geometry [9, 27]. Recall that a lattice P is called **hollow** if P contains no lattice points in its interior. By Remark 1.4, any F-hollow lattice polytope P is hollow. In dimension $d \in \{1, 2\}$ the inverse statement is also true. Therefore, for $d \leq 3$, Theorem 1.11 follows from a result of Treutlein [27] which was later generalized by Nill and Ziegler for hollow lattice polytopes in arbitrary dimension d [24]. In case $d \geq 4$, Theorem 1.11 does not follow from Theorem of Nill and Ziegler because hollow lattice polytopes P of dimension ≥ 3 need not be F-hollow. For instance, there exist up to unimodular equivalence exactly 9 examples of 3-dimensional hollow lattice polytopes which are not F-hollow [10, Appendix B].

2 The Proof of Theorem 1.11

The proof of Theorem 1.11 uses standard combinatorial notions from the theory of toric varieties [12]. We consider two dual to each other lattices $M \cong \mathbb{Z}^d$ and $N := Hom(M, \mathbb{Z})$ together with their scalar extensions $M_{\mathbb{R}} := M \otimes \mathbb{R}$, $N_{\mathbb{R}} := N \otimes \mathbb{R}$ and the natural pairing

$$\langle *, * \rangle \; : \; M_{\mathbb{R}} \times N_{\mathbb{R}} \to \mathbb{R}.$$

Every d-dimensional rational polytope P defines a convex piecewise linear function

$$\mathrm{Min}_P \; : \; N_{\mathbb{R}} \to \mathbb{R}, \quad y \mapsto \min_{u \in P} \langle u, y \rangle$$

whose domains of linearity form a complete rational polyhedral fan Σ_P, a collection of rational polyhedral cones σ in $N_{\mathbb{R}}$, which is called the **normal fan of** P.

Using the above notations, the Fine interior $F(P)$ of a polytope $P \subset M_{\mathbb{R}}$ can be equivalently reformulated as follows:

Definition 2.1

$$F(P) := \{x \in M_{\mathbb{R}} \mid \langle x, \nu \rangle \geq \mathrm{Min}_P(\nu) + 1, \; \forall \nu \in N \setminus \{0\}\}.$$

Our first idea in the proof of Theorem 1.11 is to consider a positive number $\mu = \mu(P)$ attached to P:

Definition 2.2 Let $P \subset M_{\mathbb{R}}$ be a d-dimensional rational polytope. We call the number

$$\mu(P) := \inf \{\lambda \in \mathbb{R}_{>0} \mid F(\lambda P) \neq \emptyset\}$$

the **minimal multiplier of** P.

Remark 2.3 If $P \subset M_{\mathbb{R}}$ is a d-dimensional lattice polytope, then it follows from standard properties of Ehrhart polynomials that the lattice polytope $(d+1)P$ always contains at least one interior lattice point and hence $F((d+1)P) \neq \emptyset$. This implies the inequality

$$\mu(P) \leq d + 1$$

for all d-dimensional lattice polytopes P. Note that the inequality is sharp, because $\mu(P) = \dim P + 1$ if P is the d-dimensional lattice simplex spanned by $0 \in \mathbb{Z}^d$ and by the standard lattice basis $e_1, \ldots, e_d$ of $\mathbb{Z}^d$.

Remark 2.4 It is clear that $\mu(P) \leq 1$ if and only if P is not F-hollow.

We will use the following property of the minimal multiplier $\mu(P)$ which will be proved later in Propositions 3.2 and 3.4:

Proposition 2.5 *Let $P \subset M_{\mathbb{R}}$ be a d-dimensional rational polytope. Then the following statements hold:*

(i) *The number $\mu(P)$ is rational.*
(ii) *For a positive rational number λ, one has $\lambda = \mu(P)$ if and only if*

$$0 \leq \dim F(\lambda P) \leq d - 1.$$

(iii) *If* $\dim F(\mu P) = 0$, *then the convex hull* $Q := \mathrm{Conv}(S_F(\mu P))$ *of the set*

$$S_F(\mu P) := \{\nu \in N \mid \mathrm{Min}_{F(\mu P)}(\nu) = \mathrm{Min}_{\mu P}(\nu) + 1\} \subset N$$

is a d*-dimensional lattice polytope in* $N_{\mathbb{R}}$ *having a unique interior lattice point* $0 \in N$.

Finally, we need Theorem of Hensley [20] and some of its generalizations due to Lagarias and Ziegler [23].

Theorem 2.6 ([20]) *For any given positive integers* k, d, *there exists a constant* $C(k, d)$ *depending only on* k *and* d *such that the volume* $vol(P)$ *of any* d*-dimensional lattice polytope having exactly* k *interior lattice points is bounded from above by* $C(k, d)$.

Theorem 2.7 ([23, Thm. 2]) *If a* d*-dimensional lattice polytope* P *has volume* $vol(P) \leq V$, *then* P *is unimodular equivalent to a lattice polytope in the* d*-dimensional lattice cube*

$$\{\mathbf{x} \in \mathbb{R}^d \mid 0 \leq x_i \leq n \cdot n!V,\ i = 1, \ldots, d\}.$$

In particular, a family of d*-dimensional lattice polytopes* P_i $(i \in I)$ *contains only finitely many unimodular equivalence classes if and only if there exists a constant* $C > 0$ *such that*

$$vol(P_i) < C \ \ \forall i \in I.$$

Proof of Theorem 1.11 Let $P \subset M_{\mathbb{R}}$ be a d-dimensional F-hollow lattice polytope. We put $\mu := \mu(P)$. By Proposition 2.5(ii), we have $0 \leq \dim F(\mu P) \leq d - 1$. Consider two cases (two pictures below illustrate the case $d = 2$).

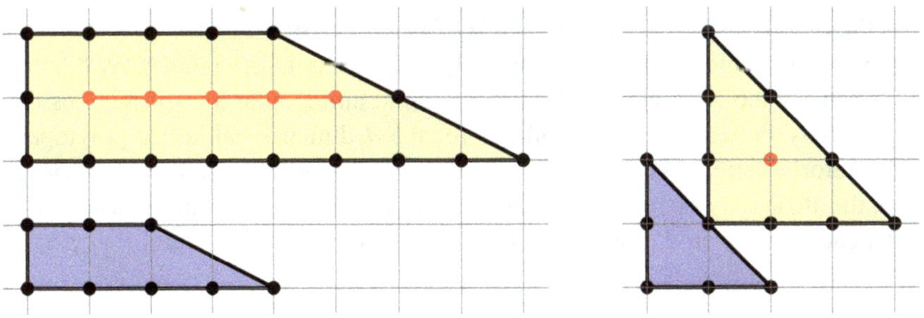

$\mu = 2$ $\mu = \frac{3}{2}$

Case 1. $1 \leq \dim F(\mu P) \leq d - 1$. We define the sublattice $N' \subset N$ consisting of all $\nu \in N$ such that $\langle x, \nu \rangle = \langle x', \nu \rangle$ for any two points $x, x' \in F(\mu P)$. Then N/N' has no torsion elements and N'. Indeed, if $l\nu \in N'$ for some positive integer l, then $\langle x, l\nu \rangle = \langle x', l\nu \rangle$ for any two points $x, x' \in F(\mu P)$, hence $\langle x, \nu \rangle = \langle x', \nu \rangle$, i.e., $\nu \in N'$. Therefore N' is a direct summand of N, and the embedding $N' \hookrightarrow N$ defines a lattice projection $\pi : M \to M' := Hom(N', \mathbb{Z})$, where

$$1 \le \text{rk}(M') = d - \dim F(\mu P) \le d - 1.$$

Consider the lattice polytope $P' := \pi(P)$. It remains to show that $F(P') = \emptyset$. By definition of N', $\pi(F(\mu P))$ is some rational point $q \in M'_{\mathbb{Q}}$. Moreover, one has $q = F(\mu\pi(P)) = F(\mu P')$. Since $\mu > 1$, by monotonicity of Fine interior [9, Remark 3.7], the polytope $F(P')$ must be strictly smaller than the point $q = F(\mu P')$. Hence $F(P') = \emptyset$, i.e., P' is F-hollow. In fact we have shown that $\mu(P') = \mu(P) = \mu$ and $\dim F(\mu P') = 0$.

Case 2. $\dim F(\mu P) = 0$. By Proposition 2.5(i), $F(\mu P)$ is a rational point $p \in M_{\mathbb{Q}}$ and for any $\nu \in S_F(\mu P)$ we have

$$\text{Min}_{F(\mu P)}(\nu) = \langle p, \nu \rangle = 1 + \text{Min}_{\mu P}(\nu) = 1 + \min_{x \in \mu P} \langle x, \nu \rangle.$$

Equivalently, we have $\langle x, \nu \rangle \ge -1 + \langle p, \nu \rangle$ for all $x \in \mu P$ and for all $\nu \in S_F(\mu P)$, or

$$\langle x, \nu \rangle \ge -1 \;\; \forall x \in \mu P - p, \; \forall \nu \in S_F(\mu P).$$

The last conditions imply the inequalities

$$\langle x, y \rangle \ge -1 \;\; \forall x \in \mu P - p, \; \forall y \in Q := \text{Conv}(S_F(\mu P)).$$

and we obtain the inclusion $\mu P - p \subseteq Q^*$, where

$$Q^* := \{x \in M_{\mathbb{R}} \mid \langle x, y \rangle \ge -1, \;\; \forall y \in Q\}$$

is the rational polar dual polytope of Q. By Proposition 2.5(iii), $Q \subset N_{\mathbb{R}}$ is a d-dimensional lattice polytope having only one interior lattice point 0. By theorems of Hensley 2.6 and Lagarias-Ziegler 2.7, up to unimodular transformations there exist only finitely many possibilities for the d-dimensional lattice polytope Q. Therefore, there exist only finitely many possible values for volumes $vol(Q^*)$ of the dual polytope of Q, i.e., $vol(Q^*)$ is bounded by some constant $C(d)$ depending only on d. Since $\mu > 1$, it follows from the inclusion $\mu P - p \subseteq Q^*$ that

$$C(d) \ge vol(Q^*) \ge vol(\mu P - p) = vol(\mu P) = \mu^d vol(P) > vol(P).$$

Hence, by Theorem 2.7 of Lagarias and Ziegler, up to unimodular equivalence we obtain only finitely many of d-dimensional lattice polytopes P such that $\dim F(\mu P) = 0$.

□

Definition 2.8 We call a d-dimensional F-hollow lattice polytope $P \subset \mathbb{R}^d$ **weakly sporadic**, if $\dim F(\mu P) = 0$, where μ is the minimal multiplier $\mu(P)$ of P.

Corollary 2.9 *In each dimension d there exist up to unimodular equivalence only finitely many weakly sporadic d-dimensional F-hollow lattice polytopes. Moreover, if a F-hollow lattice polytope P with minimal multiplier μ is not sporadic, then P admits a canonical lattice projection $\pi : P \to P'$, where P' is a k-dimensional weakly sporadic F-hollow lattice polytope* $(1 \le k \le d-1)$ *with the same minimal multiplier* $\mu = \mu(P) = \mu(P')$.

Proof The statement immediately follows from Cases 1 and 2 in the proof of Theorem 1.11. □

Remark 2.10 If $d \ge 2$, one can easily find examples of d-dimensional weakly sporadic F-hollow lattice polytopes P which are not sporadic F-hollow polytopes, i.e., P admitting lattice projections onto a k-dimensional F-hollow polytope P' $(1 \le k \le d-1)$.

Example 2.11 Up to unimodular equivalence there exist exactly three weakly sporadic F-hollow lattice polygons P which are not sporadic F-hollow lattice polygons:

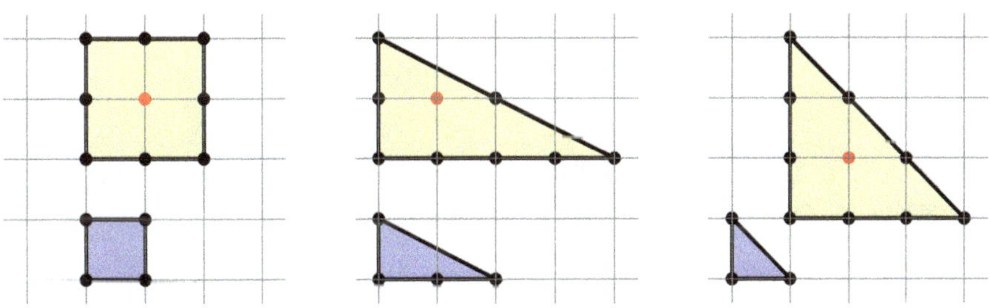

$\mu = 2$ $\mu = 2$ $\mu = 3$

Example 2.12 It follows from the combinatorial classification of all maximal hollow 3-dimensional lattice polytopes obtained by Averkov et al. [2, 3] that the following three 3-dimensional weakly sporadic F-hollow lattice polytopes with the minimal multipliers $\mu \in \{\frac{7}{6}, \frac{5}{4}, \frac{4}{3}\}$:

$$\Delta_1 := \text{Conv}\{(0,0,0), (2,0,0), (0,3,0), (0,0,6)\}, \ \mu = \frac{7}{6};$$

$$\Delta_2 := \text{Conv}\{(0,0,0), (2,0,0), (0,4,0), (0,0,4)\}, \ \mu = \frac{5}{4};$$

$$\Delta_3 := \text{Conv}\{(0,0,0), (3,0,0), (0,3,0), (0,0,3)\}, \ \mu = \frac{4}{3}$$

are in fact sporadic F-hollow polytopes.

Remark 2.13 We will see in the last section that every d-dimensional weakly sporadic F-hollow polytope P defines a family $\mathcal{F}(P)$ of non-degenerate $(d-1)$-dimensional $\mathbb{Q}$-Fano toric hypersurfaces with at worst canonical singularities. For example, three lattice tetrahedra Δ_i, $i \in \{1, 2, 3\}$ from Example 2.12 are Newton polytopes of smooth Del Pezzo surfaces of the anticanonical degree $i \in \{1, 2, 3\}$ naturally imbedded into 3-dimensional toric weighted projective spaces $\mathbb{P}(1, 1, 2, 3)$, $\mathbb{P}(1, 1, 1, 2)$, and $\mathbb{P}(1, 1, 1, 1)$, respectively. We expect that the complete list of all unimodular classes of 3-dimensional weakly sporadic F-hollow lattice polytopes P must have reasonable length. We draw attention to the fact that this list includes not only $\Delta_1, \Delta_2, \Delta_3$ but also 31 more 3-dimensional weakly sporadic F-hollow polytopes P with $\mu(P) = 2$ arising from 3-dimensional Gorenstein polytopes of index 2 which were classified in [8].

3 The Minimal Multiplier $\mu(P)$

Let $P \subset M_{\mathbb{R}}$ be a d-dimensional rational polytope. Theory of toric varieties associates with the normal fan Σ_P a d-dimensional projective toric variety X_P together with the ample $\mathbb{Q}$-Cartier divisor

$$L_P := \sum_{\nu \in \Sigma_P[1]} -\mathrm{Min}_P(\nu) D_\nu,$$

where D_ν ($\nu \in \Sigma_P[1]$) are torus invariant divisors on X_P corresponding to primitive lattice generators ν of 1-dimensional cones in Σ_P, and we have

$$P = \{x \in M_{\mathbb{Q}} \mid \langle x, \nu \rangle \geq \mathrm{Min}_P(\nu), \ \forall \nu \in \Sigma_P[1]\}.$$

Theory of toric varieties allows to describe the Fine interior $F(P)$ as rational polytope associated with the adjoint divisor on some smooth projective toric variety Y_Σ obtained by a regular refinement Σ of the normal fan Σ_P. More precisely, one has

Proposition 3.1 *Let $\rho : Y = Y_\Sigma \to X_P$ be a projective desingularization of the toric variety X_P corresponding to a regular simplicial refinement Σ of the normal fan Σ_P. Denote by $D_{\nu_1}, \ldots, D_{\nu_s}$ the torus invariant divisors corresponding to primitive lattice vectors in $\Sigma[1] = \{\nu_1, \ldots, \nu_s\}$. Then the Fine interior $F(P)$ is the rational polytope defined by s inequalities*

$$\langle x, \nu_i \rangle \geq \mathrm{Min}_P(\nu_i) + 1, \quad i \in \{1, \ldots, s\}$$

corresponding to the adjoint divisor on Y:

$$K_Y + \rho^*(L_P) = \sum_{\nu \in \Sigma[1]} (-1 - \mathrm{Min}_P(\nu))\, D_\nu.$$

Proof Let $\nu \in N$ an arbitrary nonzero lattice vector. Then there exists a minimal regular simplicial cone $\sigma \in \Sigma$ containing ν. Without loss of generality, we can assume that $\nu_1, \ldots, \nu_r \in \Sigma[1]$ $(r \leq d < s)$ are generators of σ. Then $\nu = \sum_{i=1}^r l_i \nu_i$ for some positive integer coefficients $l_1, \ldots, l_r$. Since Σ is a refinement of Σ_P, there exists a minimal r'-dimensional cone $\sigma' \in \Sigma_P$ containing σ $(r \leq r' \leq d)$. The cone σ' is dual to some $(d-r')$-dimensional face $\Theta' \prec P$ and we have

$$\mathrm{Min}_P(\nu_i) = \langle x, \nu_i \rangle, \quad \forall x \in \Theta', \ \forall i = 1, \ldots, r.$$

Since $\mathrm{Min}_P(\cdot)$ is linear on $\sigma \subset \sigma'$, we obtain

$$\mathrm{Min}_P(\nu) = \mathrm{Min}_P\left(\sum_{i=1}^r l_i \nu_i\right) = \sum_{i=1}^r l_i \mathrm{Min}_P(\nu_i),$$

and r inequalities appearing in definition of $F(P)$

$$\langle x, \nu_i \rangle \geq \mathrm{Min}_P(\nu_i) + 1, \quad i \in \{1, \ldots, r\}$$

imply that

$$\langle x, \nu \rangle = \sum_{i=1}^r l_i \langle x, \nu_i \rangle \geq \sum_{i=1}^r l_i \mathrm{Min}_P(\nu_i) + \sum_{i=1}^r l_i = \mathrm{Min}_P(\nu) + \sum_{i=1}^r l_i \geq \mathrm{Min}_P(\nu) + 1,$$

and the equality $\langle x, \nu \rangle = \mathrm{Min}_P(\nu) + 1$ for some $x \in F(P)$ can happen only if $r = 1$ and $\nu = \nu_1 \in \Sigma[1]$. Therefore, s inequalities

$$\langle x, \nu_l \rangle \geq \mathrm{Min}_P(\nu_i) + 1, \quad i \in \{1, \ldots, s\}$$

are already sufficient to obtain the Fine interior $F(P)$. Finally, we note that the canonical divisor of toric variety $Y = Y_\Sigma$ equals $-\sum_{i=1}^s D_{\nu_i}$ and $\mathbb{Q}$-Cartier divisor $\rho^* L_P$ equals $-\sum_{i=1}^s \mathrm{Min}_P(\nu_i) D_{\nu_i}$. Therefore, the rational polytope $F(P)$ corresponds to the adjoint divisor $K_Y + \rho^* L_P$ of Y. □

Proposition 3.2 *Let X_P be projective toric variety corresponding to a d-dimensional rational polytope P. Consider any projective toric desingularization $\rho : Y = Y_\Sigma \to X_P$ as in Proposition 3.1. Denote by $\Lambda_{\mathrm{eff}}(Y) \subset \mathrm{Pic}(Y)_{\mathbb{R}}$ the closed cone of effective divisors of the smooth projective toric variety Y. Let $L := \rho^* L_P \in \mathrm{Pic}(Y)_{\mathbb{Q}}$ be the pullback of the ample $\mathbb{Q}$-Cartier divisor L_P. Then*

$$\mu(P) = \inf\{\lambda \in \mathbb{R}_{>0} \mid [K_Y] + \lambda[L] \in \Lambda_{\mathrm{eff}}(Y)\},$$

is a rational number and for $\mu := \mu(P)$ one has

$$0 \leq \dim F(\mu P) < d.$$

Proof Using Cox coordinates on Y one easily obtains that the cone of effective divisors $\Lambda_{\text{eff}}(Y) \subset \text{Pic}(Y)_{\mathbb{R}}$ is a rational polyhedral cone generated by the classes $[D_\nu]$ $(\nu \in \Sigma[1])$ of torus invariant divisors. The class $[L]$ represents the class of a semi-ample big $\mathbb{Q}$-Cartier divisor $\rho^*(L_P)$ on Y which defines a rational point $[\rho^*(L_P)] \in \text{Pic}(Y)_{\mathbb{R}}$ in the interior of the cone $\Lambda_{\text{eff}}(Y)$. On the other hand, the canonical class $[K_Y] \in \text{Pic}(Y)_{\mathbb{R}}$ of toric variety Y does not belong to the cone $C_{\text{eff}}(Y)$, because Y is a rational toric variety. Therefore, the ray

$$\{[K_Y] + \lambda[\rho^* L_P] \mid \lambda \in \mathbb{R}_{\geq 0}\} \subset \text{Pic}(Y)_{\mathbb{R}}$$

with the rational origin $[K_Y]$ having the rational direction $[\rho^* L_P]$ must hit the rational polyhedral cone $\Lambda_{\text{eff}}(Y)$ in some rational point $[K_Y] + \mu[\rho^* L_P] \in \Lambda_{\text{eff}}(Y)$ located at the polyhedral boundary $\partial\Lambda_{\text{eff}}(Y)$ contained in some proper rational polyhedral facet $\Gamma \prec \Lambda_{\text{eff}}(X)$. Since the intersection point $[K_Y] + \mu[\rho^* L_P]$ of the rational ray $[K_Y] + \lambda[\rho^* L_P]$ with the facet Γ has rational coordinates, the number μ must be rational. Now we use the fact that the class $[D]$ of a $\mathbb{Q}$-divisor

$$D = \sum_{i=1} b_i[D_{\nu_i}] \in \text{Pic}(Y)_{\mathbb{Q}}, \quad b_i \in \mathbb{Q},$$

represents a point in $\Lambda_{\text{eff}}(Y)$ if and only if the rational polytope

$$P_D := \{x \in M_{\mathbb{R}} \mid \langle x, \nu_i \rangle \geq -b_i, \quad \forall i \in \{1, \ldots, s\}\}$$

is not empty. Moreover, $[D]$ represents an interior point in $\Lambda_{\text{eff}}(Y)$ if and only if the rational polytope P_D has maximal dimension d. Applying Proposition 3.1 to the adjoint $\mathbb{Q}$-divisor

$$D := K_Y + \mu\rho^* L_P = K_Y + \rho^* \mu L_P = K_Y + \rho^* L_{\mu P},$$

we obtain that $F(\mu P)$ is not empty and $\dim F(\mu P) < d$. □

Definition 3.3 Let $P \subset M_{\mathbb{R}}$ be an arbitrary d-dimensional rational polytope. Assume $F(P) \neq \emptyset$. Then we call the set

$$S_F(P) := \{\nu \in N \mid \text{Min}_{F(P)}(\nu) = \text{Min}_P(\nu) + 1\}$$

the **support of the Fine interior** of P. It follows from Proposition 3.1 that $S_F(P)$ is always a finite set whose positive convex span $\mathbb{R}_{\geq 0} S_F(P)$ equals $N_{\mathbb{R}}$ since

$$F(P) = \{x \in M_{\mathbb{R}} \mid \langle x, \nu \rangle \geq \text{Min}_P(\nu) + 1 \;\; \forall \nu \in S_F(P)\}$$

is compact.

Proposition 3.4 *Let P be a d-dimensional rational polytope with* $\dim F(P) = 0$. *Then the convex hull* $Q := \mathrm{conv}(S_F(P)) \subset N_{\mathbb{R}}$ *is a d-dimensional lattice polytope containing in its interior only one lattice point* $0 \in N$.

Proof Let $p := F(P) \in M_{\mathbb{Q}}$. Note that the shifted polytope $P_0 := P - p$ has the Fine interior $0 \in M$, and the sets $S_F(P)$ and $S_F(P_0)$ are the same, since $F(P_0) = F(P) - p$. Hence, we can assume $F(P) = p = \{0\} \subset Int(P)$. The lattice polytope $Q = \mathrm{conv}(S_F(P))$ is d-dimensional, since

$$0 = F(P) = \{x \in M_{\mathbb{R}} \mid \langle x, \nu \rangle \geq 0 \ \forall \nu \in S_F(P)\}.$$

Using the upper convex piecewise linear function $\mathrm{Min}_P : N_{\mathbb{R}} \to \mathbb{R}$, we obtain the dual to P rational polytope

$$P^* := \{y \in N_{\mathbb{R}} \mid \mathrm{Min}_P(y) \geq -1\}.$$

Since $\mathrm{Min}_{F(P)}(\nu) = \langle 0, \nu \rangle = 0$ for all $\nu \in N$, we obtain $S_F(P) = \partial P^* \cap N$, where $\partial P^* := \{y \in N_{\mathbb{R}} \mid \mathrm{Min}_P(y) = -1\}$ is the boundary of P^*. The d-dimensional rational polytope P^* has only $0 \in N$ as interior lattice point, because $\mathrm{Min}_P(\nu) \leq -1$ for all $\nu \in N \setminus \{0\}$. Hence, the d-dimensional lattice subpolytope $Q \subseteq P^*$ has also only $0 \in N$ as its interior lattice point. □

Remark 3.5 We note that the minimal multiplier $\mu(P)$ has naturally appeared in the arithmetical problem of counting rational points of bounded height on algebraic varieties considered in the author's joint paper with Yu. I. Manin [4]. The close relation between the boundary point $[K_Y]+\mu[L] \in \Lambda_{\mathrm{eff}}(Y)$ and the Minimal Model Program in dimension 3 was observed in [5].

Independently, for an arbitrary normal L-polarized projective variety V the critical number $\mu = \mu(V)$ with the opposite sign

$$\kappa\varepsilon(V) = -\inf\{t \mid [K_V] + t[L] \in \Lambda_{\mathrm{eff}}(V)\}$$

was introduced by Fujita and called the **Kodaira energy** of V [16, 17, 18] (see also [26]). The negative sign in Fujita's definition was chosen to maintain consistency with the sign of the Kodaira dimension $\kappa(V)$ of V. Fano fibrations of toric varieties Y associated with adjoint divisors $K_Y + \mu L$ were first considered by the author in joint papers with Yu. Tschinkel [6, 7]. In the language of smooth lattice polytopes such toric fibrations turned out to be very useful in the classification of smooth lattice polytopes P without interior lattice points obtained by Dickenstein, Di Rocco, and Piene [13]. These ideas were further elaborated by Di Rocco, Haase, Nill, and Paffenholz for lattice polytopes containing interior lattice points in the **polyhedral adjunction theory** [14]. A close relation between the combinatorial finiteness theorem of Nill-Ziegler and the finiteness of singular toric Fano fibrations was recently independently discovered by Ambro [1].

A recent paper of Garzón Mora and Haase [19] introduces a modified version of the polyhedral adjunction theory which uses a scaled version of the Fine interior

$F(P)$ and considers for a fixed positive rational number s interior points of P having integral distance at least s from all integral supporting hyperplanes of P. This approach attaches to a lattice polytope P the reciprocal critical number $s_0 = \mu(P)^{-1}$. I am very grateful to the anonymous reviewer who pointed out me that a proof of Theorem 1.11 can be independently given using the results of the paper by Garzón Mora and Haase. Comparing two polyhedral adjunction theories one sees that the Fine version is more suitable for obtaining finiteness results because it uses all integral supporting hyperplanes of P instead of only those spanned by facets of P. Note that in case of smooth lattice polytopes P these two polyhedral adjunction theories coincide.

4 Toric Hypersurfaces with Newton Polytope P

Let $A \subset M \cong \mathbb{Z}^d$ be a finite subset such that the convex hull $P := \text{Conv}(A) \subset M_{\mathbb{R}} \cong \mathbb{R}^d$ is a d-dimensional lattice polytope. Take an arbitrary field K and consider P as **Newton polytope** of a Laurent polynomial

$$f(\mathbf{t}) = \sum_{\mathbf{m} \in A} c_{\mathbf{m}} \mathbf{t}^{\mathbf{m}} \in K[t_1^{\pm 1}, \ldots, t_d^{\pm 1}],$$

that is, $c_{\mathbf{m}} \neq 0$ for all vertices $\mathbf{m} \in P$. The zero locus

$$Z := \{f(\mathbf{t}) = 0\} \subset \mathbb{T}_K^d := \text{Spec}\, K[t_1^{\pm 1}, \ldots, t_d^{\pm 1}]$$

we call **affine toric hypersurface**.

Remark 4.1 Note that the affine toric hypersurface $Z := \{f(\mathbf{t}) = 0\} \subset \mathbb{T}^d$ determines its defining non-constant Laurent polynomial f uniquely up to multiplication by a nonzero monomial $a\mathbf{t}^{\mathbf{m}} (a \neq 0)$ which shifts the Newton polytope of f by lattice vector $\mathbf{m} \in M$. So it will be convenient to refer to P as **Newton polytope** of the toric hypersurface $Z \subset \mathbb{T}^d$. Moreover, it is natural to consider Newton polytopes P of toric hypersurfaces $Z \subset \mathbb{T}^d$ up unimodular equivalence, since the group of affine linear transformations $\text{Aff}(\mathbb{Z}^d) = GL(d, \mathbb{Z}) \rtimes \mathbb{Z}^d$ acts on Laurent polynomials f via automorphisms $\text{Aut}(\mathbb{T}^d) \cong GL(n, \mathbb{Z})$ of the algebraic torus $\mathbb{T}^d$ and via multiplication by monomials $\mathbf{t}^{\mathbf{m}} = t_1^{m_1} \ldots t_d^{m_d}$. A unimodular isomorphism $\varphi \in \text{Aff}(\mathbb{Z}^d) = GL(d, \mathbb{Z}) \rtimes \mathbb{Z}^d$ transforms an affine toric hypersurface $Z_1 \subset \mathbb{T}^d$ with the Newton polytope P_1 into the isomorphic affine hypersurface $Z_2 \subset \mathbb{T}^d$ with the Newton polytope $P_2 = \varphi(P_1)$.

Now let us consider the geometric meaning of lattice projections $\pi : P \to P'$ from the viewpoint of toric hypersurfaces $Z \subset \mathbb{T}^d$ with the Newton polytope P.

Remark 4.2 Let P be the Newton polytope of a Laurent polynomial f. Assume that P admits a standard lattice projection onto a k-dimensional lattice polytope

$P' \subset \mathbb{R}^k$ $(0 < k < d)$. Then we can view P' as Newton polytope of the Laurent polynomial f considered as element of the Laurent polynomial ring $R[t_1^{\pm}, \ldots, t_k^{\pm}]$ whose coefficients ring is another Laurent polynomial ring $R := \mathbb{C}[t_{k+1}^{\pm 1}, \ldots, t_d^{\pm 1}]$. Using the splitting $\mathbb{T}^d \cong \mathbb{T}^k \times \mathbb{T}^{d-k}$ and the ring embedding $R \hookrightarrow R[t_1^{\pm}, \ldots, t_k^{\pm}]$, we obtain the surjective morphism $\mathbb{T}^d \to \mathbb{T}^{d-k} = \mathrm{Spec}(R)$ whose restriction to Z is a dominant morphism $Z \to \mathbb{T}^{d-k}$ such that general fibers are affine toric hypersurfaces in k-dimensional torus $\mathbb{T}^k$ having P' as Newton polytope.

Consider some examples of lattice projections.

Example 4.3 Assume that a d-dimensional lattice polytope P has **width 1**, that is, P has a lattice projection on the unique hollow (also F-hollow) lattice segment $[0, 1] \subset \mathbb{R}$. The lattice projection $\pi : P \to P' = [0, 1]$ means that the d-dimensional lattice polytope $P \subset \mathbb{R}^d$ is unimodular equivalent to a lattice polytope in $\mathbb{R}^d$ contained between two parallel integral affine hyperplanes $\{x_d = 0\}$ and $\{x_d = 1\}$. Up to this unimodular isomorphism, we obtain the corresponding Laurent polynomial $f(t_1, \ldots, t_d) \in K[t_1^{\pm 1}, \ldots, t_d^{\pm 1}]$ in the form:

$$f(t_1, \ldots, t_d) = g_0(t_1, \ldots, t_{d-1}) + t_d g_1(t_1, \ldots, t_{d-1}),$$

for some Laurent polynomials $g_0, g_1 \in K[t_1^{\pm 1}, \ldots, t_{d-1}^{\pm 1}]$. Since the polynomial $f(\mathbf{t})$ defining $Z \subset \mathbb{T}^d$ is linear with respect to the last variable t_d, we can rationally eliminate t_d from this equation $f = 0$ by the formula

$$t_d = -\frac{g_0(t_1, \ldots, t_{d-1})}{g_1(t_1, \ldots, t_{d-1})}$$

and obtain a birational isomorphism $Z \overset{bir}{\sim} \mathbb{A}_K^{d-1}$ over K, i.e., Z is an irreducible K-rational algebraic variety.

The following conjecture proposes a natural "inverse statement" to last example.

Conjecture 4.4 Let $P \subset \mathbb{R}^d$ be a d-dimensional lattice polytope. Assume that for any field K any toric hypersurface $Z \subset (K^*)^d$ with the Newton polytope P is irreducible and birational to $\mathbb{A}_K^{d-1}$ over K. Then the Newton polytope P admits a lattice projection onto $[0, 1]$.

Example 4.5 In case $d = 2$, the conjecture can be easily verified. Indeed, the rationality of general curve $Z \subset \mathbb{T}^2$ with Newton polygon P implies that P has no interior lattice points, i.e., P is hollow. Up to unimodular isomorphisms, the unique sporadic hollow lattice polygon is the triangle

$$Q := \mathrm{Conv}((0, 0), (2, 0), (0, 2)) \subset \mathbb{R}^2,$$

which is the Newton polytope of a general conic $C \subset \mathbb{A}_K^2$ defined by a quadratic equation

$$a_{0,0} + a_{1,0}t_1 + a_{0,1}t_2 + a_{2,0}t_1^2 + a_{1,1}t_1t_2 + a_{0,2}t_2^2 = 0, \quad (a_{i,j} \in K).$$

If $K = \mathbb{C}$, then C is birational to $\mathbb{A}^1_{\mathbb{C}}$. However, if $K = \mathbb{R}$ the conic

$$1 + t_1^2 + t_2^2 = 0$$

is not birational to $\mathbb{A}^1_{\mathbb{R}}$.

More generally, if a d-dimensional Newton polytope P hypersurface $Z \subset \mathbb{T}^d$ has a lattice projection onto Q, then the toric hypersurface Z becomes birational to a conic bundle over $(d-2)$-dimensional algebraic torus. Note that the corresponding toric hypersurface $Z \subset \mathbb{T}^d$ might be non-rational variety even over the algebraically closed field $\mathbb{C}$ (see Example 4.6).

Example 4.6 Let $\mathcal{K}_d \subset \mathbb{P}^{d+1}$ be smooth projective d-dimensional Klein cubic given by the homogeneous equation

$$z_0z_1^2 + z_1z_2^2 + \cdots + z_dz_{d+1}^2 + z_{d+1}z_0^2 = 0, \quad d \geq 2,$$

which is invariant under the cyclic permutation of the homogeneous coordinates $z_0, z_1, \ldots, z_{d+2}$. The Newton polytope P_{d+1} of $\mathcal{K}_d$ is a weakly sporadic $(d+1)$-dimensional F-hollow lattice simplex with the minimal multiplier $\mu = (d+2)/3$ having no other lattice points besides its vertices. The affine equation of $\mathcal{K}_d \cap \mathbb{A}^{d+1} \subset \mathbb{A}^{d+1} = \{z_0 = 1\}$:

$$t_1^2 + t_1t_2^2 + t_2t_3^2 + t_3t_4^2 + \cdots + t_{d+1} = 0$$

considered with respect to the pair of variables x_1 and x_3 defines a conic C from Example 4.5. Note that the lattice $(d+1)$-simplex P_{d+1} admits several different lattice projections onto 2-dimensional simplex $Q := \mathrm{Conv}((0,0),(2,0),(0,2))$ defining corresponding different conic bundle structures on the Klein cubic $\mathcal{K}_d$. If the dimension $d = 2k$ is even, then P_{2k+1} admits a lattice projection onto $[0,1]$ which sends $k+1$ lattice vertices of P_{2k+1} corresponding to cubic monomials $z_{2i}z_{2i+1}^2$ $(0 \leq i \leq k)$ to the lattice point 0 and the remaining $k+1$ lattice vertices of P_{2k+1} to the lattice point 1. By 4.3, we see that any even-dimensional Klein cubic $\mathcal{K}_{2k}$ is rational over any field K. Note that it is rather nontrivial to show that the 3-dimensional Klein cubic $\mathcal{K}_3$ is not rational over $\mathbb{C}$ (see [11]). Moreover, a still open conjecture claims that the Klein cubic $\mathcal{K}_{2k+1}$ is not rational over $\mathbb{C}$ in all odd dimensions $d = 2k+1 \geq 3$ (see [22]). This conjecture is supported by the fact that the $2k$-dimensional empty lattice simplex P_{2k} has no a lattice projection onto $[0,1]$ for all integers $k \geq 2$. We note that any d-dimensional toric hypersurface $Z \subset \mathbb{T}^{d+1}$ with the Newton polytope P_{d+1} is irreducible and smooth for $d \geq 3$.

5 Non-degenerate Toric Hypersurfaces and $\mathbb{Q}$-Fano Fibrations

Definition 5.1 Let $Z \subset \mathbb{T}_K^d$ be an affine toric hypersurface given as zero locus of a Laurent polynomial

$$f(\mathbf{t}) = \sum_{\mathbf{m}\in A} c_{\mathbf{m}}\mathbf{t}^{\mathbf{m}} \in K[t_1^{\pm 1}, \dots, t_d^{\pm 1}]$$

with the Newton polytope $P = \mathrm{Conv}(A)$. The toric hypersurface Z is called **non-degenerate** if Z is smooth over the algebraic closure $\overline{K}$, and for any face $\Theta \prec P$ $(0 < \dim \Theta < d)$ the affine toric hypersurface $Z_\Theta \subset \mathbb{T}^d$ defined as zero locus of the Laurent polynomial

$$f_\Theta(\mathbf{t}) := \sum_{\mathbf{m}\in A\cap\Theta} c_{\mathbf{m}}\mathbf{t}^{\mathbf{m}}$$

is reduced and smooth over $\overline{K}$ as well.

Example 5.2 Let P be a d-dimensional lattice simplex. Denote by A the set of its $d+1$ vertices. Assume that char $K = 0$. Then the non-degeneracy of $Z \subset \mathbb{T}^d$ given by $\sum_{\mathbf{m}\in A} c_{\mathbf{m}}\mathbf{t}^{\mathbf{m}}$ is equivalent to the nonvanishing condition

$$\prod_{\mathbf{m}\in A} c_{\mathbf{m}} \neq 0.$$

Now we want to explain why Theorem 1.11 is inspired by the Minimal Model Program for non-degenerate toric hypersurfaces $Z \subset \mathbb{T}^d$ with Newton polytope P.

It was proved in [9] that if the Fine interior $F(P)$ of a d-dimensional lattice polytope P is not empty, then every non-degenerate toric hypersurface $Z \subset \mathbb{T}^d$ with the Newton polytope P has a minimal model, i.e., a projective model $\widehat{Z}$ with at worst $\mathbb{Q}$-factorial terminal singularities and semi-ample canonical $\mathbb{Q}$-divisor. Moreover, the Fine interior $F(P)$ allows to compute the Kodaira dimension $\kappa(\widehat{Z})$ by the formula:

$$\kappa(\widehat{Z}) = \min\{d-1, \dim F(P)\}.$$

Now we are interested in birational geometry of non-degenerate toric hypersurfaces $Z \subset \mathbb{T}^d$ with F-hollow Newton polytope P.

Our main result is the following.

Theorem 5.3 *Let P be a d-dimensional weakly sporadic F-hollow lattice polytope P with the minimal multiplier $\mu = \mu(P) > 1$. Denote by $\mathbb{P}_{Q^*}$ the canonical toric $\mathbb{Q}$-Fano variety whose defining fan is spanned by faces of the canonical Fano polytope*

$$Q := \mathrm{Conv}(S_F(\mu P)).$$

We denote by $\widetilde{Z}$ the Zariski closures in $\mathbb{P}_{Q^*}$ of a non-degenerate toric hypersurface $Z \subset \mathbb{T}^d$ with the Newton polytope P. Then $\widetilde{Z}$ is a normal variety and the following adjunction formula holds:

$$K_{\widetilde{Z}} = (K_{\mathbb{P}_{Q^*}} + \widetilde{Z})|_{\widetilde{Z}} = \left(\frac{\mu - 1}{\mu}\right) K_{\mathbb{P}_{Q^*}}|_{\widetilde{Z}}.$$

Furthermore, $\widetilde{Z}$ has at worst canonical singularities and ample anticanonical $\mathbb{Q}$-divisor, i.e., $\widetilde{Z}$ is a $(d-1)$-dimensional canonical $\mathbb{Q}$-Fano hypersurface in $\mathbb{P}_{Q^*}$.

In the proof we need the following technical statement.

Lemma 5.4 Let P be as in Theorem 5.3. Denote by $L(P) \subset \mathbb{C}[t_1^{\pm 1}, \ldots, t_d^{\pm 1}]$ the linear system on the canonical toric $\mathbb{Q}$-Fano variety $\mathbb{P}_{Q^*}$ spanned by all monomials $\mathbf{t}^{\mathbf{m}}$ corresponding to lattice points $\mathbf{m} \in P \subseteq Q^* \cap M$. Then no codimension-2 torus orbit in $\mathbb{P}_{Q^*}$ is contained in the base locus of $L(P)$.

Proof of Lemma 5.4 Take an arbitrary 1-dimensional edge $E \prec Q$ of the canonical Fano polytope $Q \subset N_R$. Let $\sigma_E \subset N_\mathbb{R}$ be the 2-dimensional cone over E in the fan Σ_{Q^*} defining the canonical toric $\mathbb{Q}$-Fano variety $\mathbb{P}_{Q^*}$. Since the lattice polytope Q has only one interior lattice point $0 \in N$, the lattice polygon $\mathrm{Conv}(0, E) \subset Q$ has no lattice points other than 0 and $E \cap N$. This means that σ_E is unimodular equivalent to the 2-dimensional cone $\sigma_k \subset \mathbb{R}^2$ spanned by lattice vectors $(1, 0)$ and $(1, k)$ for some $k \geq 1$. Let $\{e_0, e_1, \ldots, e_k\} := E \cap S_F(\mu P)$ be the lattice vectors corresponding under the above equivalence to lattice vectors

$$(1, 0), (1, 1), \ldots, (1, k) \in \sigma_k.$$

Denote by $N' \subset N$ the sublattice in N spanned by $e_0, e_1, \ldots, e_k$. Then N' is a direct summand of N since each pair e_i, e_{i+1} forms a part of $\mathbb{Z}$-basis of N for every $i \in \{0, 1, \ldots, k-1\}$. Take a $\mathbb{Z}$-basis $v_1, v_2, \ldots, v_d$ of N such that $v_1 = e_0$, $v_2 = e_1$ and consider the standard lattice projection

$$\pi_E : M \to M' := Hom(N', \mathbb{Z}) \cong \mathbb{Z}^2$$

together with the identification of lattice vectors $e_i \in N'$ with $(1, i) \in \mathbb{Z}^2$ for all $i \in \{0, 1, \ldots, k\}$. Denote by $P'_E \subset \mathbb{R}^2$ the lattice polygon $\pi_E(P)$. Using a shift by an appropriate lattice vector $\mathbf{m} \in M$, we can assume without loss of generality that $\mathrm{Min}_P(e_0) = \mathrm{Min}_P(e_1) = 0$, that is, both hyperplanes $\langle x, e_0 \rangle = 0$ and $\langle x, e_1 \rangle = 0$ are integral supporting hyperplanes for the lattice polytope P and for the rational polytope μP. Since $F(\mu P)$ is just a rational point $p \in M_\mathbb{Q}$, we obtain

$$\langle p, e_0 \rangle = \mathrm{Min}_{F(\mu P)}(e_0) = \mathrm{Min}_{F(\mu P)}(e_1) = \langle p, e_1 \rangle = 1,$$

and by linearity of $\langle p, * \rangle$ we obtain

$$\langle p, e_i \rangle = \mathrm{Min}_{F(\mu P)}(e_i) = 1, \quad \forall i \in \{0, 1, \ldots, k\}.$$

Since all lattice points $e_0, e_1, \ldots, e_k \in E \prec Q$ are contained in $S_F(\mu P)$, we obtain

$$\mathrm{Min}_{\mu P}(e_i) = 0 \ \ \forall i \in \{0, 1, \ldots, k\},$$

i.e., the linear equations $\langle x, e_i \rangle = 0$ define integral supporting hyperplanes for μP and for P.

In order to show that the codimension-2 torus orbit corresponding to σ_E is not contained in the base locus of $L(P)$, we have to show that the $(d-2)$-dimensional linear subspace $l_E \subset M_{\mathbb{R}}$,

$$l_E := \langle x, e_i \rangle = 0, \quad \forall i \in \{0, 1, \ldots, k\},$$

contains at least one lattice vertex of P, or equivalently, the lattice π_E-projection $P'_E = \pi_E(P) \subset \mathbb{R}^2$ contains the origin $(0,0) = \pi_E(l_E) \in \mathbb{R}^2$. We consider two cases.

Case 1: $k \geq 2$. Then π_E-projection of P is a lattice polygon $P'_E \subset \mathbb{R}^2$ contained in the cone

$$C_k := \{(x_1, x_2) \in \mathbb{R}^2 \mid x_1 \geq 0, x_1 + kx_2 \geq 0\}.$$

The supporting integral hyperplane $\langle x, e_1 \rangle = 0$ for P must contain at least one vertex of P. On the other hand, π_E-projection of this hyperplane is the line in $\mathbb{R}^2$ with the equation $x_1 + x_2 = 0$, and it has only the origin $(0,0)$ as a common point with the cone C_k.

Case 2: $k = 1$. Then 2-dimensional cone $\sigma_1 \subset \mathbb{R}^2$ is regular, i.e., its generators $e_0, e_1 \in S_F(\mu P)$ form a $\mathbb{Z}$-basis of N'. Assume that $(0,0) \not\in P'_E = \pi_E(P)$. Then one of two generators $(0,1)$ and $(1,-1)$ of σ_1 must belong to P'_E, since otherwise the lattice polytope P'_E would be contained in the convex set

$$B := \mathrm{Conv}(\{\sigma_1 \cap \mathbb{Z}^2\} \setminus \{(0,0), (0,1), (1,-1)\})$$

and the lattice point $(1,0) = \pi_E(F(\mu P))$ would not be in the interior of $\mu P'_E = \pi_E(\mu P) \subset \mu B$ for $\mu > 1$. So we obtain $\mathrm{Min}_P(e_0 + e_1) = 1$ and $\mathrm{Min}_{\mu P}(e_0 + e_1) = \mu > 1$. However, we have $\mathrm{Min}_{F(\mu P)}(e_0 + e_1) = \mathrm{Min}_{F(\mu P)}(e_0) + \mathrm{Min}_{F(\mu P)}(e_1) = 1 + 1 = 2$. This implies

$$\mathrm{Min}_{F(\mu P)}(e_1 + e_2) - \mathrm{Min}_{\mu P}(e_1 + e_2) = 2 - \mu < 1.$$

The latter contradicts the definition of Fine interior $F(\mu P)$.

□

Proof of Theorem 5.3 Let $\mathbb{P}^{(2)}_{Q^*} \subset \mathbb{P}_{Q^*}$ be the Zariski open toric subvariety obtained from $\mathbb{P}_{Q^*}$ by deleting all torus orbits of codimension at least 3 in $\mathbb{P}_{Q^*}$. The Gorenstein toric variety $\mathbb{P}^{(2)}_{Q^*}$ has a minimal crepant toric desingularization ρ : $\widehat{\mathbb{P}}^{(2)}_{Q^*} \to \mathbb{P}^{(2)}_{Q^*}$ by resolving of possible A_{k-1}-singularities along codimension-2 strata in $\mathbb{P}^{(2)}_{Q^*}$. Applying Lemma 5.4, we obtain that the Zariski closure $\widehat{Z}^{(2)}$ of the non-degenerate toric hypersurface Z in $\widehat{\mathbb{P}}^{(2)}_{Q^*}$ is smooth, and the Zariski closure $\widetilde{Z}^{(2)}$ of Z in $\mathbb{P}^{(2)}_{Q^*}$ is Gorenstein, and it has at worst A_{k-1}-singularities along the transversal intersections of the non-degenerate quasi-projective toric hypersurface $\widetilde{Z}^{(2)} \subset \mathbb{P}^{(2)}_{Q^*}$ with codimension-2 torus orbits in $\mathbb{P}^{(2)}_{Q^*}$. Thus, we obtain that the singular locus of $(d-1)$-dimensional projective hypersurfaces $\widetilde{Z} \subset \mathbb{P}_{Q^*}$ has codimension at least 2, and, by Serre's criterion for normality, $\widetilde{Z}$ is normal. Now we can relate the canonical classes of $\widetilde{Z}$ and $\mathbb{P}_{Q^*}$ by the adjunction formula on the Gorenstein quasi-projective toric variety $\mathbb{P}^{(2)}_{Q^*}$.

Let $\dim F(\mu P) = 0$ and $p = F(\mu P)$. We consider the shifted rational polytope $P_0 := \mu P - p$ such that $F(P_0) = \{0\}$. Let $\Sigma_{Q^*}[1]$ be the set of generators of 1-dimensional cones in the fan defining the canonical toric $\mathbb{Q}$-Fano variety $\mathbb{P}_{Q^*}$. Then $\Sigma_{Q^*}[1] \subseteq S_F(\mu P)$ and we have $\mathrm{Min}_{P_0}(\nu) = -1$ for all $\nu \in \Sigma_{Q^*}[1]$.

The lattice polytope P defines on $\mathbb{P}_{Q^*}$ a toric Weil divisor

$$L_P := \sum_{\nu \in \Sigma_{Q^*}[1]} (-\mathrm{Min}_P(\nu))D_\nu$$

such that $\mathbb{Q}$-Cartier divisor μL_P is rationally equivalent to the anticanonical class $-K_{\mathbb{P}_{Q^*}}$ of $\mathbb{P}_{Q^*}$, since

$$K_{\mathbb{P}_{Q^*}} = \sum_{\nu \in \Sigma_{Q^*}[1]} -D_\nu = \sum_{\nu \in \Sigma_{Q^*}[1]} \mathrm{Min}_{P_0}(\nu) D_\nu = \sum_{\nu \in \Sigma_{Q^*}[1]} (\mu \mathrm{Min}_P(\nu) - \langle p, \nu \rangle)\, D_\nu,$$

and for any $p \in M_{\mathbb{Q}}$ the $\mathbb{Q}$-divisor

$$\sum_{\nu \in \Sigma_{Q^*}[1]} \langle p, \nu \rangle D_\nu$$

is a principal $\mathbb{Q}$-divisor of $\mathbb{P}_{Q^*}$. We write the adjunction formula on the Gorenstein toric variety $\mathbb{P}^{(2)}_{Q^*}$ in the form

$$K_{\widetilde{Z}} = (K_{\mathbb{P}_{Q^*}} + L_P)|_{\widetilde{Z}} = (-\mu L_P + L_P)\,|_{\widetilde{Z}} = \left(\frac{\mu - 1}{\mu}\right) K_{\mathbb{P}_{Q^*}}|_{\widetilde{Z}}.$$

In particular, the canonical divisor of $\widetilde{Z}$ is a $\mathbb{Q}$-Cartier divisor. It order to show that singularities of $\widetilde{Z}$ are at worst canonical, we use a result of Khovanskiî that a non-degenerate toric hypersurface with Newton polytope P always admits smooth projective birational model W in a smooth toric variety defined by a regular refinement Σ of the normal fan Σ_P [21]. We can always choose a regular refinement Σ of Σ_P which is simultaneously a regular refinement of Σ_{Q^*}, so that we obtain for smooth toric hypersurface $W \subset \mathbb{P}_\Sigma$ with a birational morphism $\rho : W \to \widetilde{Z}$ together with a formula

$$K_W = \rho^* K_{\widetilde{Z}} + \sum_{\nu \in \Sigma[1]} a_\nu (D_\nu \cap W),$$

where $a_\nu = 0$ if $\nu \in S_F(\mu P)$, or if $D_\nu \cap W = \emptyset$, and $a_\nu = (-1 - \mathrm{Min}_{P_0}(\nu)) > 0$ for all $\nu \notin S_F(\mu P)$. Thus, $\widetilde{Z}$ has at worst canonical singularities. □

Corollary 5.5 *Let P be a d-dimensional F-hollow lattice polytope and $\dim F(\mu P) = k \geq 1$. Denote by $\pi : P \to P'$ the lattice projection onto $(d-k)$-dimensional weakly sporadic F-hollow lattice polytope P' with $\mu(P') = \mu(P) = \mu$ constructed in the proof of Theorem 1.11. Then π defines a dominant morphism $\varphi : Z \to \mathbb{T}^k$ whose general fibers are non-degenerate hypersurfaces with the $(n-k)$-dimensional weakly sporadic F-hollow lattice polytope P' admitting $\mathbb{Q}$-Fano projective compactifications.*

Proof We use Theorem 5.3 in relative situation by considering the $(n-k)$-dimensional fan $\Sigma' \subset N'_{\mathbb{R}}$, where $N' \subset N$ is the $(n-k)$-dimensional sublattice from Case 1 in the proof of Theorem 1.11. The fan Σ' is spanned by faces of $(n-k)$-dimensional canonical Fano polytope Q', where $Q' = \mathrm{Conv}(S_F(\mu P)) \cap N')$. We can identify the finite set $S_F(\mu P)) \cap N'$ with the support $S_F(\mu P')$ of the $(n-k)$-dimensional weakly sporadic F-hollow lattice polytope $P' = \pi(P) \subset M_{\mathbb{R}}$ with the minimal multiplier $\mu = \mu(P') = \mu(P)$. By Theorem 5.3, each general fiber of the dominant morphism $\varphi Z \to \mathbb{T}^k$ admits a natural $\mathbb{Q}$-Fano projective compactification which can be obtained in the following way.

Using [9, Theorem 6.3], we embed the non-degenerate toric hypersurface $Z \subset \mathbb{T}^d$ into d-dimensional $\mathbb{Q}$-Gorenstein toric variety $\widetilde{\mathbb{P}}$ associated with the Minkowski sum $F(\mu P) + C(\mu P)$, where $C(\mu P)$ is the d-dimensional rational polytope containing μP defined by the inequalities:

$$C(\mu P) := \{x \in M_{\mathbb{R}} \mid \langle x, \nu \rangle \geq \mathrm{Min}_{\mu P}(\nu), \ \ \forall \nu \in S_F(\mu P)\}.$$

Since $F(\mu P)$ is a Minkowski summand of $\widetilde{P}$, the normal fan $\Sigma_{\widetilde{P}}$ contains the $(n-k)$-dimensional fan Σ' as a subfan describing toric $\mathbb{Q}$-Fano fibers of the toric morphism $\alpha : \widetilde{\mathbb{P}} \to \mathbb{P}_{F(\mu P)}$, where $\mathbb{P}_{F(\mu P)}$ is k-dimensional toric variety corresponding to k-dimensional rational polytope $F(\mu P)$. By restricting α to the Zariski closure $\widetilde{Z}$ of Z in $\widetilde{\mathbb{P}}$, we obtain the $\mathbb{Q}$-Fano fibration $\widetilde{\varphi} : \widetilde{Z} \to \mathbb{P}_{F(\mu P)}$ which extends the dominant morphism $\varphi : Z \to \mathbb{T}^k \subset \mathbb{P}_{F(\mu P)}$ to the hypersurface $\widetilde{Z} \subset \widetilde{\mathbb{P}}$. □

Corollary 5.6 *A non-degenerate hypersurface $Z \subset \mathbb{T}^d$ with Newton polytope P has negative Kodaira dimension if and only if P is F-hollow.*

Proof If $F(P) \neq \emptyset$, then $Z \subset \mathbb{T}^d$ has a minimal model which can be constructed as Zariski closure $\widehat{Z}$ in some simplicial torus embedding $\mathbb{T}^d \subset \widehat{V}$. In particular, the Kodaira dimension of Z is non-negative [9].

If $F(P) = \emptyset$, then, by Theorems 1.11 and 3.4, P is weakly sporadic and $Z \subset \mathbb{T}^d$ is birational to a $\mathbb{Q}$-Fano hypersurface in the canonical $\mathbb{Q}$-Fano variety $\mathbb{P}_{Q^*}$, or, by Corollary 5.5, Z is birational to a $\mathbb{Q}$-Fano fibration $\widetilde{Z} \to \mathbb{P}_{\widetilde{F(\mu P)}}$ over a k-dimensional toric variety. Therefore, the projective toric hypersurface $\widetilde{Z}$ has negative Kodaira dimension. □

References

1. F. Ambro, *On toric Fano fibrations*. https://doi.org/10.48550/arXiv.2212.13862
2. G. Averkov, C. Wagner, R. Weismantel, Maximal lattice-free polyhedra: finiteness and an explicit description in dimension three. Math. Oper. Res. **36**(4), 721–742 (2011)
3. G. Averkov, J. Krümpelmann, S. Weltge, Notions of maximality for integral lattice-free polyhedra: the case of dimension three. Math. Oper. Res. **42**(4), 1035–1062 (2017)
4. V.V. Batyrev, Yu. I. Manin, Sur le nombre des points rationnels de hauteur bornée des variétés algébriques. Math. Ann. **286**, 27–43 (1990)
5. V.V. Batyrev, *The Cone of Effective Divisors of Threefolds*, Proc. Intern. Conf. on Algebra, Novosibirsk (1989), vol. 131, Part 3. (Amer. Math. Soc., Contemp. Math., Providence (RI), 1992), pp. 337–352
6. V.V. Batyrev, Yu. Tschinkel, Height zeta functions of toric varieties. Algebraic geometry. J. Math. Sci. **82**, 3220–3239 (1996)
7. V.V. Batyrev, Yu. Tschinkel, Tamagawa numbers of polarized algebraic varieties. Astérisque **251**, 299–340 (1998)
8. V.V. Batyrev, D. Juni, Classification of toric Del Pezzo varieties in arbitrary dimension. Moscow Math. J. **10**(2), 285–316 (2010)
9. V.V. Batyrev, Canonical models of toric hypersurfaces. Algebraic Geometry **10**(4), 394–431 (2023)
10. V.V. Batyrev, A.M. Kasprzyk, K. Schaller, *On the Fine Interior of Three-Dimensional Canonical Fano Polytopes*, in *Interactions with Lattice Polytopes, ILP 2017*, ed. by A.M. Kasprzyk, B. Nill, Springer Proceedings in Mathematics & Statistics, vol. 386 (Springer, 2022)
11. C.H. Clemens, P.A. Griffiths, The intermediate Jacobian of the cubic threefold. Ann. Math. **95**, 281–356 (1972)
12. D.A. Cox, J.B. Little, H.K. Schenck, *Toric Varieties*, Graduate Studies in Mathematics, vol. 124 (Amer. Math. Soc., Providence, RI, 2011)
13. A. Dickenstein, S. Di Rocco, R. Piene, *Classifying smooth lattice polytopes via toric fibrations*. Adv. Math. **222**(1), 240–254 (2009)
14. S. Di Rocco, C. Haase, B. Nill, A. Paffenholz, Polyhedral adjunction theory. Algebra Number Theory **7**(10), 2417–2446 (2014)
15. J. Fine, *Resolution and Completion of Algebraic Varieties*, Ph.D. University of Warwick 1983. http://wrap.warwick.ac.uk/114676/, Accessed: 18/08/2022

16. T. Fujita, *On Polarized Manifolds Whose Adjoint Bundles Are Not Semipositive*, in *Algebraic Geometry, Proc. Symp., Sendai/Jap. 1985*, vol. 10 of *Adv. Stud. Pure Math.* (Kinokuniya Company Ltd., Tokyo; North-Holland, Amsterdam-New York-Oxford, 1987), pp. 167–178
17. T. Fujita, On Kodaira energy and adjoint reduction of polarized manifolds. Manuscr. Math. **76**, 59–84 (1992)
18. T. Fujita, On Kodaira energy of polarized log varieties. J. Math. Soc. Jpn. **48**, 1–12 (1996)
19. S. Garzón Mora, C. Haase, *Fine Polyhedral Adjunction Theory*. arXiv:2302.04074
20. D. Hensley, Lattice vertex polytopes with interior lattice points. Pac. J. Math. **105**(1), 183–191 (1983)
21. A.G. Khovanskiĭ, *Newton Polyhedra (Resolution of Singularities)*, Itogi Nauki i Tekhniki. Ser. Sovrem. Probl. Mat., vol. 22 (VINITI, Moscow, 1983), pp. 207–239; J. Soviet Math., **27**(3), 2811–2830 (1984)
22. J. Kollár, Algebraic hypersurfaces. Bull. Amer. Math. Soc. **56**, 543–568 (2019)
23. J.C. Lagarias, G.M. Ziegler, Bounds for lattice polytopes containing a fixed number of interior points in a sublattice. Can. J. Math., **43**, 1022–1035 (1991)
24. B. Nill, G.M. Ziegler, Projecting lattice polytopes without interior lattice points. Math. Oper. Res. **36**, 462–467 (2011)
25. M. Reid, *Young Person's Guide to Canonical Singularities*, Algebraic geometry, Bowdoin, 1985 (Brunswick, Maine, 1985), Proc. Sympos. Pure Math., vol. 46 (Amer. Math. Soc., Providence, RI, 1987), pp. 345–414
26. A.J. Sommese, *On the Adjunction Theoretic Structure of Projective Varieties*, in *Complex Analysis and Algebraic Geometry* (Göttingen, 1985), vol. 1194 of Lecture Notes in Math. (Springer, Berlin, 1986), pp. 175–213
27. J. Treutlein, *Birationale Eigenschaften generischer Hyperflächen in algebraischen Tori*, Ph.D. thesis, Universität Tübingen, 2010, Germany. https://d-nb.info/1003786367/34

Zeta-Polynomials, Superpolynomials, DAHA, and Plane Curve Singularities

Ivan Cherednik

Dedicated to the memory of Yuri Ivanovich Manin

Contents

1 Modular Form Periods

1.1 Manin, My Teacher

In 1965–1967, Yuri Ivanovich Manin and Ernest Borisovich Vinberg delivered special courses at Moscow School no. 2 for senior students. This is when I met Yu.I. With some stretch, I can say that Manin and Vinberg were my high school teachers and Vasilii Iskovskikh and Victor Kac were our tutors (teaching assistants of Yu.I. and E.B).

Our regular relations with Yu.I. began about 1968, when he took me as his student at Moscow State University. My first assignment was reading Serre's "Corps locaux"; I learned Herbrand theory (the higher ramification), but cannot say the same about French.

I. Cherednik (✉)
Department of Mathematics, University of North Carolina, Chapel Hill, NC, USA
e-mail: chered@email.unc.edu

Y. Tschinkel (ed.), *Arithmetic and Algebraic Geometry*, Simons Symposia,
https://doi.org/10.1007/978-3-031-74134-0_2

Let me omit 50 years and go to 2017, the Arbeitstagung devoted to his 80th birthday. It was a great meeting! Yu.I. and Ksenia Glebovna were terrific hosts, many people were around, a perfect view of the Rhein from their apartment, etc.

Mostly we discussed anything but mathematics, though something came up: *zeta-polynomials*, certain combinations of modular form periods satisfying Riemann hypothesis, predicted by Manin.

My talk was mostly about DAHA *superpolynomials* $\mathcal{H}(q,t,a)$ for *double affine Hecke algebras*. Superpolynomials have several interpretations; the main one is via *Khovanov-Rozansky triply graded homology*.

This direction is in progress. Conjecturally, topological superpolynomials, those from the BPS states (SCFT), DAHA superpolynomials, motivic ones, and L-functions of plane curve singularities coincide (when these theories overlap). I will focus on the latter three below; this note is introductory, with very few names and references.

1.2 Modular Periods

Let us begin with Manin's well-known paper "Periods of parabolic forms and p-adic Hecke series" (1973). Basically, you consider a parabolic (cusp) form $\Phi(z)$ of even weight w and calculate its *periods*

$$r_k(\Phi) = \int_0^{\imath\infty} \Phi(z)z^k dz, \quad \text{for } 0 \le k \le w-2.$$

Then the ratios of r_k for even k or those for odd k are rational numbers, which can be calculated (Manin's theorem). For instance, such Φ are proportional to

$$\Delta = e^{2\pi\imath z} \prod_{n=1}^{\infty}(1-e^{2\pi\imath nz})^{24},$$

for $w=12$. Then $r_2/r_0 = -\frac{2^2 3^4 5}{691}$, $r_3/r_1 = -\frac{2^4 3}{5^2}$, etc. We note a relation to the Ramanujan's $\tau(n)=\sigma_{11} \mod 691$ (1916); Manin reproved it.

The periods are essentially the values $L_\Phi(s)$ of the corresponding L-function for integer s inside the critical strip. This can be extended to

$$L_\Phi(s,\chi) = \sum_{n=1}^{\infty} \chi(n)\lambda_n n^{-s},$$

for suitable Dirichlet characters χ if Φ is an eigenfunction of the Hecke operators T_n with eigenvalues λ_n.

The second part of his paper was on the p-adic extrapolations of the ratios of the periods, which is closely related to the *Kubota-Leopoldt p-adic zeta-function* and *eigenvarieties*. Concerning the origins of this direction, let me mention at least Barry Mazur and Nicholas Katz.

The periods are generally for any paths $\gamma[0, \imath\infty]$ for $\gamma \in SL(2, \mathbb{Z})$, but $[0, \imath\infty]$ is sufficient due to the modularity of Φ. However, more general paths do occur in Manin's paper in the process of calculations.

The p-adic extrapolations and the Kubota-Leopoldt zeta (1964) are closely related to the *Iwasawa invariants* of Γ-extensions due to Mazur and Wiles in full generality. The examples of Γ-extensions are some towers of cyclotomic fields, where we monitor the class numbers. Following Mazur, the Iwasawa invariants are parallel to the *Alexander polynomials*. The covers of the $S^3 \setminus K$ for knots K in their theory are similar to Γ-extensions; those of $\mathbb{P}^2$ minus the corresponding plane curve singularity are sufficient for algebraic K due to Libgober and others.

The Alexander polynomials are $\mathcal{H}(q = t, t, a = -1)$ for the DAHA superpolynomials $\mathcal{H}$, and there is a relation to $\rho(q, t)$, refined *quasi-ρ-invariants* introduced and discussed below (for algebraic knots).

1.3 Using DAHA

The modular periods and Double Affine Hecke Algebras (DAHA), are quite different theories, but there is a clear common denominator: the action of $SL(2, \mathbb{Z})$. The main feature of DAHA is that it provides a universal formalization of *Fourier transforms* and the action of (projective) $SL(2, \mathbb{Z})$ in algebra, harmonic analysis, and physics.

To be more exact, DAHA serves the theories with the Fourier transform and the Gaussian, where the latter is an eigenfunction of the former. The classical Fourier transform, its q-counterparts, the Hankel transform, and the Verlinde S, T-operators are basic examples. DAHA is a universal (flat) deformation of the Heisenberg and Weyl algebras, so its role in Fourier analysis is not surprising.

Moreover, it appeared that DAHA provides invariants of iterated torus links. This is not very surprising because the *Verlinde algebras* are closely connected with the invariants of links and 3-folds. In DAHA theory, these algebras become *perfect quotients* of the polynomial representations. Hopefully, DAHA and the theory of modular forms and L-functions can eventually merge into one, but this will require efforts.

Number theory already provided some framework for quite a bunch of similar directions, which is hardly accidental. Let me quote from Manin's paper: "…any points of contact with concrete number-theoretical facts, whether old or new, take on a special significance. They discipline the imagination, and they provide a breathing space and the opportunity to evaluate the stunning beauty of past discoveries." This is very much applicable now to the relations between number theory and physics (in both directions).

1.4 Zeta-Polynomials

Next, in his "Local zeta factors and geometries under Spec $\mathbb{Z}$" (2014), Yu.I. conjectured that a certain combination of $L_\Phi(1), \ldots, L_\Phi(w-1)$ is a *zeta-polynomial*: satisfies the functional equation $s \mapsto 1-s$ and the Riemann hypothesis. This was fully confirmed for $w \geq 4$ by Ken Ono, Larry Rolen, and Florian Sprung in their paper "Zeta-polynomials for modular form periods" (2016).

Let

$$M_\Phi(m) \stackrel{\text{def}}{=\!=} \sum_{j=0}^{w-2} \left(\frac{\sqrt{N}}{2\pi}\right)^{j+1} \frac{L_\Phi(j+1)}{(w-2-j)!} j^m$$

for a $\Gamma_0(N)$-modular Φ. Then their zeta-polynomial $Z_\Phi(s)$ is a linear combination of $M_\Phi(m)$, for $m = 0, \ldots, w-2$ with the coefficients given in terms of Stirling polynomials of the 1st kind and the Fernando Rodriguez-Villegas transform. Manin used the latter too. There is a relation of zeta-polynomials to the Bloch-Kato Conjecture, which is a Galois cohomological interpretation of the periods, and related advanced number-theoretical problems.

What is important for us is that there is some "canonical" way to combine the modular periods in a zeta-polynomial, which resembles very much the theory of Witten-Reshetikhin-Turaev invariants of 3-folds and their relations to knot invariants. DAHA invariants are "canonical" combinations of basic coinvariants in a similar way.

There are various connections of the *WRT invariants* with modular forms. Let us at least mention "Quantum invariants, modular forms, and lattice points II" (K. Hikami, 2006). A challenge is to connect the inequalities $0 \leq k \leq w-2$ with those in *Verlinde algebras*, more exactly with the range of Macdonald polynomials in *prefect DAHA modules* at roots of unity, but this is fully open at the moment.

2 Basic DAHA Theory

2.1 Main Definitions

DAHA, denoted by $\mathcal{H}$, were initially introduced to complete the theory of Knizhnik-Zamolodchikov equations and quantum many-body problem. It is a universal flat deformation of $\mathcal{W}$, the Weyl algebra extended by the Weyl group W; this is for any reducible reduced root systems R (in this note). The projective $SL(2,\mathbb{Z})$ due to Steinberg acts in $\mathcal{H}$. This is actually the braid group B_3; the notation will be $\widetilde{SL(2,\mathbb{Z})}$.

For A_1, $\mathcal{H\!H}$ is generated by $X^{\pm 1}, Y^{\pm 1}, T$ subject to group relations

$$TXTX = 1 = TY^{-1}TY^{-1},\, Y^{-1}X^{-1}YXT^2 = q^{-1/2}$$

and the quadratic one

$$(T - t^{1/2})(T + t^{-1/2}) = 1.$$

The action of the $\widetilde{SL(2,\mathbb{Z})}$ is:

$$\tau_+ : Y \mapsto q^{-\frac{1}{4}}XY,\ X \mapsto X,\ T \mapsto T,\ \tau_- : X \mapsto q^{\frac{1}{4}}YX,\ Y \mapsto Y,\ T \mapsto T,$$

exactly as for $\mathcal{W}$ (there is no q here). The automorphisms $\tau_\pm$, the generators of $\widetilde{SL(2,\mathbb{Z})}$, are the preimages of $\begin{pmatrix}1 & 1\\ 0 & 1\end{pmatrix}$ and $\begin{pmatrix}1 & 0\\ 1 & 1\end{pmatrix}$, the standard generators of $SL(2,\mathbb{Z})$. The defining relation of $\widetilde{SL(2,\mathbb{Z})}$ is simple: $\tau_+\tau_-^{-1}\tau_+ = \sigma = \tau_-^{-1}\tau_+\tau_-^{-1}$, which formally gives that σ^2 is central. The element σ^{-1} is the (operator) *DAHA-Fourier transform*.

When $t^{1/2} = 1$, T becomes the inversion s of X and Y, and we arrive at $\mathcal{W}$. Upon $t = q$, DAHA is closely related to quantum groups. The case $t = q^k$ as $q \to 1$ serves the Harish-Chandra theory and its k-generalization, called Heckman-Opdam theory (in mathematics), including spherical functions and Jack polynomials. Also, $q \to 0$ is the p-adic limit and $t \to 0$ is the Kac-Moody limit. The action of $\widetilde{SL(2,\mathbb{Z})}$ generally collapses in the limits. However it survives in the important limit to rational DAHA and Hankel transforms, which is when $X = q^x$, $Y = q^{-y}$, $t = q^k$ and $q \to 1$ (for A_1).

Also, $\widetilde{SL(2,\mathbb{Z})}$ generally acts in finite-dimensional *rigid* $\mathcal{H\!H}$-modules. *Perfect representations* are such: canonical irreducible quotients of polynomial representations at the roots of unity q or for any q and *singular* k. The classical *Verlinde algebras* are symmetric (spherical) parts of perfect representations when $t = q$ and q is a root of unity.

We omit the general definition of $\mathcal{H\!H}$. For an arbitrary reduced irreducible root system R of rank n, it is generated by pairwise commutative X_λ for λ in the *the weight lattice* P, and pairwise commutative Y_λ and T_i for $1 \le i \le n$ such that

$$(T_i - t^{1/2})(T_i + t^{-1/2}) = 0,$$

where t_{sht} and t_{lng} can be generally used when α_i is short and long. *Spherical DAHA* is defined as $\mathcal{P}_+\mathcal{H\!H}\mathcal{P}_+$ for the t-symmetrizer

$$\mathcal{P}_+ = \sum_{w\in W} t^{l(w)/2}T_w \Big/ \sum_{w\in W} t^{l(w)}.$$

2.2 Polynomial Representation

The key property of DAHA is the *PBW theorem*: any element $H \in \mathcal{H}$ can be uniquely represented as

$$H = \sum c_{\lambda,w,\mu} X_\lambda T_w Y_\mu$$

for $\lambda, \mu \in P$ and $w \in W$ (the non-affine Weyl group). Equivalently, there is a faithful action of $\mathcal{H}$ in the *polynomial representation* $\mathscr{X} = \mathbb{C}[X_\lambda]$: the one induced from the character

$$Y_\lambda \mapsto t^{(\rho,\lambda)}, \quad T_i \mapsto t_i^{1/2}.$$

It is a deformation of the classical *Fock representation* of the corresponding Heisenberg-Weyl algebra. We will need below the *DAHA coinvariant*:

$$\{H\} = \sum c_{\lambda,w,\mu} t^{(\mu-\lambda,\rho)+\text{length}(w)/2},$$

which is $H(1)(X_\lambda \mapsto t^{-(\rho,\lambda)})$, where $H(1)$ is the action of $H \in \mathcal{H}$ at $1 \in \mathscr{X}$.

Technically, the simplest definition of $\mathcal{H}$ is via the usage of T_0 instead of Y_λ. This approach is directly related to the $T \times \mathbb{C}^*-$ equivariant K-theory of affine flag varieties. Then Y_λ are defined as some products in terms of T_i for $i \geq 0$ and their commutativity is some proposition. The construction of Y_λ via $\{T_i\}$ is essentially due to Bernstein-Zelevinsky and Lusztig. Finding explicit defining relations between X_λ and Y_μ is generally involved unless for A_n and in small ranks.

The exact connection with affine flag varieties was fully clarified by Garland-Grojnowski and Ginzburg-Kapranov-Vasserot (1995). We note that the action of $\widetilde{SL(2,\mathbb{Z})}$ is far from obvious from the K-theoretical viewpoint in spite of various attempts in this direction. The key property of any Fourier transforms is that they send polynomials to delta-functions, which is not simple to incorporate geometrically.

Generally, there are three fundamentally different definitions (interpretations) of DAHA, more than three for A_n. The 2nd definition is via the *orbifold* fundamental group of the so-called *elliptic configuration space*:

$$\pi_1^{orb}\Big(\{x \in E^n \mid \prod_\alpha (x,\alpha) \neq 0\}\}/W\Big).$$

Here E is an elliptic curve and W is the non-affine Weyl group acting in E^n. This action is well-defined because the roots $\alpha \in R$ are with integer coefficients in terms of the fundamental weights, which are considered as coordinates of E^n.

Then we take group algebra of this π_1^{orb} and impose the quadratic relations for T_i as above, the elements of which are half-turns corresponding to simple reflections.

The element σ becomes basically the transposition of the periods of E. The elliptic configuration space is the "big cell" in $Bun_G(E)$ for the corresponding simple Lie group G.

The existence of the action of $\widetilde{SL(2,\mathbb{Z})}$ is straightforward from the topological definition. However, the polynomial representation is far from immediate in this interpretation, which is almost by construction via the K-theory of affine flag varieties. I constantly use both approaches and the one via harmonic analysis, the 3rd major approach.

The action of $\widetilde{SL(2,\mathbb{Z})}$ becomes "natural" if DAHA is interpreted via the q-version of harmonic analysis on symmetric spaces. Namely, σ becomes the DAHA-Fourier transform, and τ_+ is associated with the multiplication by the Gaussian, $q^{x^2/2}$. Basically, this projective action is equivalent to the fundamental theorem of Fourier analysis: the Gaussian is an eigenfunction of the corresponding Fourier transform.

Spherical DAHA for GL_n can be interpreted as *elliptic Hall algebras* and are related to quantum groups due to Schiffmann-Vasserot and others. In spite of the usage of elliptic curves, the action of $\widetilde{SL(2,\mathbb{Z})}$ is not "immediate" in this approach; further work is needed. Also, it can be interpreted via certain *shuffle algebras*.

Given a reduced irreducible root system R, the *nonsymmetric Macdonald polynomials* E_λ for $\lambda \in P$ generalize the monomials X_λ for the corresponding Weyl algebra and form a basis of $\mathscr{X}$. They are eigenfunctions of Y_μ normalized by the conditions $E_\lambda = X_\lambda + \text{(lower terms)}$. In this approach, *Macdonald polynomials* P_λ are the t-symmetrizations of E_λ for $\lambda \in P_+$; the normalization is $P_\lambda = X_\lambda + \text{(lower terms)}$.

The action of $\mathcal{H}\!\!\mathcal{H}$ in $\mathscr{X}$ is sufficiently explicit. For A_1:

$$T \mapsto t^{1/2}s + \frac{t^{1/2} - t^{-1/2}}{X^2-1}(s-1), \quad X \mapsto X, \quad Y \mapsto spT,$$

where $s(X) = X^{-1},\ p(X) = q^{1/2}X$. The divided differences here and for any root systems R are very standard in the theory of affine Hecke algebras and are quite common in related geometry and combinatorics.

For GL_n, the corresponding $\mathcal{H}\!\!\mathcal{H}$ is generated by $X_i^{\pm 1}, Y_j^{\pm 1}, T_k$, where $1 \le i, j \le n$ and $1 \le k \le n-1$ for pairwise commutative $\{X_i\}$ and $\{Y_j\}$. One has:

$$\tau_+(Y_1) = q^{-1/2}X_1Y_1, \quad \tau_-(X_1) = q^{+1/2}Y_1X_1$$

and so on. The action of Y_1 in $\mathscr{X}$ is via the formula $Y_1 = \pi T_{n-1} \dots T_1$, where

$$\pi : X_1 \mapsto X_2, \quad X_2 \mapsto X_3, \dots, X_n \mapsto q^{-1}X_1.$$

These formulas are quite similar for any Y_i. We will use them below when calculating the DAHA-superpolynomial of trefoil (as an example).

Going back to the modular periods, the evaluation map $X \mapsto t^{-\rho}$, *the DAHA coinvariant*, plays a role of integration $\int_0^{\imath\infty}\{\cdot\}\Phi dz$, E_λ replace z^k, and the action of $\widetilde{\gamma}$ corresponds to the change of the integration path to $\gamma[0, \imath\infty]$. The main deviation is that the action of γ plays a much more significant role for DAHA superpolynomials versus that for the periods. In contrast to $\int_0^{\imath\infty}\{\cdot\}\Phi dz$, the DAHA coinvariant is not stable in any way with respect to the action of $\widetilde{SL(2,Z)}$.

When switching to the zeta-polynomials $Z_\Phi(s)$, special linear combinations of z^k-momenta must be considered, which resembles our usage of E_λ. Both constructions are "canonical" in a sense; the restriction $0 \le k \le w-2$ seems somewhat similar to those in Verlinde algebras.

The next topic, *DAHA-Verlinde algebras*, gives a direct link of DAHA at roots of unity to number theory. They are some counterparts of *Tate modules* for the towers of covers of an elliptic curve ramified at one point; the *absolute Galois group* acts there. The Verlinde algebras are one of the key ingredients of the *Witten-Reshetikhin-Turaev invariants*, generalize K_0 of the *reduced category* in representation theory of *quantum groups* at roots of unity, and that of integrable modules of *Kac-Moody algebras* (Kazhdan-Lusztig, Finkelberg).

2.3 Refined Verlinde Algebras

These algebras in the nonsymmetric variant are *perfect* finite-dimensional quotients of $\mathscr{X}$, those with the action of $\widetilde{SL(2,\mathbb{Z})}$ and DAHA-invariant non-degenerate quadratic forms. They are commutative algebras, but can be non-semisimple, those related to *logarithmic* CFT in examples. Technically, we divide $\mathscr{X}$ by the radical of the evaluation pairing. Such modules exist either when q is a root of unity or for *singular* k, where $t = q^k$ as above.

In the case of A_1, let $q = \exp(\frac{2\pi\imath}{N})$, $k \in \frac{\mathbb{Z}_+}{2}$ and $k < N/2$. The map $X(z) = q^z$ can be naturally extended to an $\mathcal{H}$-homomorphism $\mathbb{C}[X^{\pm 1}] \to V_{2N-4k}$. The latter is the *nonsymmetric Verlinde algebra*. It is the space of functions

$$f : \{-\frac{N+k+1}{2}, \ldots, -\frac{k+1}{2}, -\frac{k}{2}, \frac{k+1}{2}, \ldots, \frac{N-k}{2}\} \to \mathbb{C}$$

with pointwise multiplication and the action of X, T, Y induced from that in $\mathscr{X}$. These modules are *rigid*, which readily gives an action of $\widetilde{PSL_2(\mathbb{Z})}$ there, and in

$$V^{sym}_{N-2k+1} \overset{\mathbf{def}}{=\!=} \{v \in V \mid Tv = t^{\frac{1}{2}} v\}.$$

The low index is the dimension of the corresponding V: dim $V_{2N-4k} = 2N-4k$, and dim $(V_{2N-4k})^{sym} = N-2k+1$. The operators X, Y, T become unitary in V_{2N-4k} if the $q = e^{\frac{2\pi\imath}{N}}$, the "minimal" primitive N_{th} root.

We classified *rigid* modules for A_1 in "On Galois action in rigid DAHA modules" (2017). They are:

($\boldsymbol{\alpha}$) V_{2N-4k} as above.
($\boldsymbol{\beta}$) Non-semisimple $V_{2N+4|k|}$ for $k \in -\mathbb{Z}_+$ such that $-N/2 < k < 0$.
($\boldsymbol{\gamma}$) $V_{2|k|}$ for $k = -\frac{1}{2} - m > -N/2$, where $m \in \mathbb{Z}_+$.

There is a similar list for the *little* DAHA $\mathcal{HH}' = \langle X^{\pm 2}, Y^{\pm 2}, T\rangle \subset \mathcal{HH}$. Importantly, families ($\boldsymbol{\alpha}, \boldsymbol{\gamma}$) have *flat* q-deformations from roots of unity to arbitrary q. The unimodular q such that $\arg q \le \frac{2\pi}{N}$ result in the positivity of the invariant form in type ($\boldsymbol{\alpha}$). Such a deformation leads to some relations between V, V' defined for $N \mid N'$, generalizing those for the *Tate modules.*

The usual *Verlinde algebra* is V^{sym}_{N-1} of type ($\boldsymbol{\alpha}$), which is for $k = 1$, i.e., for $t = q$. Then τ_+ becomes the T-operator, and $\sigma = \tau_+\tau_-^{-1}\tau_+$ becomes the Verlinde S-operator. The *reduced characters* in Verlinde algebras generally become the images of the corresponding Macdonald E-polynomials for V and symmetric ones for V^{sym}.

Perfect representations are quotients of the ones obtained from $\mathscr{X}$ by fixing the corresponding central characters, which are of dimension $4N$. For $k = 1$, the symmetrizations V^{sym} of the latter are connected with the category of representations of *small quantum group.* For instance, V^{sym}_{N-1} is the Grothendieck ring K_0 of the so-called reduced category for A_1. The perfect representations for $\mathbb{Z}/2 \ni k \neq 1$ are generally beyond quantum groups, though the ones of type ($\boldsymbol{\beta}$) are connected with *logarithmic* conformal field theories and there are other links.

2.4 The Galois Action

The rigidity provides that the *absolute Galois group* acts in the modules above (including the usual Verlinde algebras). We use that *elliptic braid group* $\mathcal{B}_q$ generated by X, T, Y subject to the group relations in the definition of $\mathcal{HH}$ of type A_1 is a renormalization of the orbifold fundamental group $\pi_1^{orb}(E/\{1, s\})$, where E is an elliptic curve, $s : x \mapsto -x$. If E and its origin o are defined over some field $\mathbb{Q}[q^{1/4}] \subset K \subset \overline{\mathbb{Q}}$, then $\mathrm{Gal}(\overline{\mathbb{Q}}/K)$ acts projectively in these modules.

More exactly, setting $A = XT,\ B = XTY,\ C = T^{-1}Y$, the relations of $\mathcal{B}_q$ and the action of $\tau_\pm$ there become as follows:

$$A^2 = 1 = C^2 = q^{1/2}B^2, \ \text{ where } ABC = A^2YT^{-1}Y = YY^{-1}T = T,$$

$$\tau_+ : A \mapsto A,\ B \mapsto q^{-1/4}C,\ C \mapsto q^{1/4}C^{-1}BC,$$

$$\tau_- : A \mapsto q^{1/4}ABA^{-1},\ B \mapsto q^{-1/4}A,\ C \mapsto C.$$

The classification of $\mathcal{H}$-modules at roots of unity q, t becomes the corresponding multiplicative *Deligne-Simpson problem* with specific quadratic relations for $A, B, C, D = T^{-1}$. They can be arbitrary quadratic for DAHA of type $C^\vee C_1$ (Sahi, Noumi-Stokman); let me also mention Oblomkov-Stoica (2009). This algebra is generated by A, B, C, D such that $ABCD = 1$ satisfying any quadratic relations: those for the monodromy of the *Heun equation.* There are links to SCFT.

The images of $\mathcal{B}_q$ in type (α) rigid modules with positive-definite invariant forms are finite and we obtain finite covers of $\mathbb{P}^1$ ramified at $0, 1, \infty$, and $o \in E(K)$, where A, B, C, D are the corresponding monodromies. When $t = 1$, we arrive at unramified covers of E.

The case of the Hermitian invariant forms *with one minus* is interesting. Then the images of $\mathcal{B}_q$ are discrete groups. The smallest nontrivial such V is for *little* $\mathcal{H}$; $\dim V = 3$. In particular, we obtain then *all Livné lattices* in $PU(2, 1)$, which are examples of the *Mostow groups.* Livné used a branched 2-cover of degree 2 of the *universal elliptic curve.*

More generally (for the same V), there is a direct connection with the theory of *equilateral triangle groups* in $PU(2, 1)$; for instance, see "Complex hyperbolic triangle groups" (R.E. Schwartz, 2002) and "Cone metrics on the sphere and Livné's lattices" (Parker, 2006).

We mention here that the (regular) *Inverse Galois Problem* is based on *rigid triples*, which are $\{a, b, c\}$ generating a group G and satisfying $abc = 1$. They are assumed from given conjugacy classes in G and the rigidity means essentially the uniqueness of such $\{a, b, c\}$ up to (simultaneous) conjugations in G. We need $\{a, b, c, d\}$ here, 4 points in $\mathbb{P}^1$ and the *linear rigidity* (in matrices) based on Katz' theory of rigid systems (in the variant due to M. Dettweiler and others).

Such covers extend the Belyi's theorem and Grothendieck's program of *dessins d'enfants* to E; let us mention Beilinson-Levin (1991). We deal only with very "small" covers: those from DAHA modules. Our towers are similar to those from Tate modules, though our (ramified) version of $T_p(E) = \varprojlim(E/E_{p^n})$ is *not* a module over p-adic numbers. Here and above see "On Galois action in rigid DAHA modules".

3 Knot Invariants via DAHA

3.1 DAHA-Jones Polynomials

They can be defined for any reduced root system R; in the non-reduced case, see author's "Jones polynomials of torus knots via DAHA" and "DAHA-Jones polynomials of torus knots". The system $C^\vee C_1$ is discussed in the latter. Algebraic torus knots are for $r, s > 0$ such that $\gcd(r, s) = 1$. They can be represented as $T(r, s) = \{x^r = y^s\} \cap S^3_\epsilon$ for a small sphere S^3_ϵ centered at 0. The formula for the corresponding *DAHA-Jones invariant* is:

$$J^{\lambda}_{r,s}(q,t)=\{\widetilde{\gamma}\big(\frac{E_\lambda}{E_\lambda(t^{-\rho})}\big)(1)\},\ \{F(X)\}=F(X\mapsto t^{-\rho}),\ \lambda\in P_+.$$

where $(r,s)^{tr}$ is the 1st column of $\gamma\in SL(2,\mathbb{Z})$, $\widetilde{\gamma}$ is its action in $\mathcal{HH}$, and the Laurent polynomial $\widetilde{\gamma}(E_\lambda)(1)$ is $\widetilde{\gamma}(E_\lambda)\in\mathcal{HH}$ applied to $1\in\mathscr{X}$. It is not necessary to assume that $r,s>0$ in this definition; the corresponding torus knots will be non-algebraic for $rs<0$.

A remarkable theorem is that it is always a q,t-polynomial (up to some fractional power $q^\bullet t^\bullet$) in spite of the q,t-singularities of $E/E_\lambda(t^{-\rho})$, which exist only for some λ at roots of unity q and for singular t. The product formula for $E_\lambda(t^{-\rho})$ is the Macdonald evaluation conjecture in the nonsymmetric variant (now a relatively simple theorem).

The formulas for $J^{\lambda}_{r,s}(q,t)$ and their generalizations below for iterated torus knots will give exactly the same invariants if E_λ is replaced by the *symmetric* Macdonald polynomials P_λ for $\lambda\in P_+$. Note that $P_\lambda(t^{-\rho})=P_\lambda(t^{\rho})$. We employ the t-symmetrization inside the coinvariant $\{\cdot\}$.

The P-polynomials become the Schur-Weyl characters for R when $t=q$; they do not depend on q,t in this case; then $P_\lambda(t^\rho)$ is the corresponding q-dimension. Colored *Jones polynomials* for $T(r,s)$ are J^λ for A_1, $q=t$ and $\lambda=m\omega_1$, which is up to some power $t^\bullet$.

For instance, $E_{\omega_i}=X_{\omega_i}$ in the A_n-case, and $J^{\omega_i}_{r,s}(q,t)$ are closely related to the left-hand side of the *Shuffle Conjecture*, $\nabla e_n[X]$, proved by Carlsson-Mellit in their "A proof of the shuffle conjecture" (JAMS, 2018). The right-hand side of this conjecture (in the most general setting) is the corresponding *motivic superpolynomial* (below). In the uncolored case $\lambda=\omega_1$; *we will omit* λ *in* J^λ *and other formulas in this case.*

A_n*-stabilization.* For any $G=1+q\mathbb{C}[[q,t]]+t\mathbb{C}[[q,t]]\in C[q,t]$ and rational u,v, let $\big(q^u t^v G\big)^\circ \overset{\text{def}}{=\!=} G$. If G also depends on $a^{\pm1}$ (below), then

$$\big(q^\bullet t^\bullet a^\bullet\, G\big)^\circ\in 1+q\mathbb{C}[q,t]+t\mathbb{C}[q,t]+a\mathbb{C}[q^{\pm1},t^{\pm1}].$$

Given a Young diagram $\lambda=(\lambda_1\ge\lambda_2\ge\cdots\ge\lambda_m>0)$, $\lambda=\sum_{i=1}^m c_i\omega_i$ is considered a weight for A_n, where $n\ge m-1$ and c_i is the number of columns of size i in λ.

The claim is that given r,s as above, there exists a unique polynomial $\mathcal{H}_{r,s}(q,t,a)$ in $\mathbb{C}[q,t^{\pm1},a]$ such that

$$J^{\lambda}_{r,s}(q,t)^\circ=\mathcal{H}^{\lambda}_{r,s}(q,t,a=-t^{n+1})$$

for $J^\lambda_{r,s}$ of type A_n for $n\ge m-1$ Automatically, $\mathcal{H}_{r,s}(q,t,a)^\circ=\mathcal{H}_{r,s}(q,t,a)$.

Comments. The starting point of this theory was due to Aganagic-Shakirov (2011) and the author (2011). Concerning the related physics, let me mention at least the paper by Gukov, Iqbal, Kozcaz, and Vafa (2010). The stabilization of J^λ for A_n is based on a DAHA theorem due to Schiffmann-Vasserot (2012). Let me mention here the proof of the DAHA superduality for torus knots by Gorsky-Negut (2013).

It was conjectured by the author that the a-stabilization holds for B, C, D too; topologically, we generalize *Kauffman polynomials*. Moreover, superpolynomials were calculated for a couple of knots by the author and R. Elliot (2016) for the exceptional series from (Deligne-Gross, 2002): $A_0 \subset A_1 \subset A_2 \subset D_2 \subset F_4 \subset D_4 \subset F_4 \subset E_6 \subset E_7 \subset E_8$. Our paper was mostly on the superpolynomials in the case of the *annulus*.

Let us state a version of the stabilization conjecture for C_n (from my paper). One has $R_+ = \{\epsilon_i \pm \epsilon_j, 2\epsilon_i\}$ for $1 \le i, j \le n$ such that $i \le j$,

$$P_+ = \{\lambda = \sum_{i=1}^{n} \lambda_i \epsilon_i, \lambda_1 \ge \lambda_2 \ge \cdots \ge \lambda_n \ge 0\}.$$

Assuming that $\lambda_m > 0$, they become dominant weights for C_n for any $n \ge m$. We will treat λ as Young diagrams. DAHA invariants depend now on $q, t = t_{sht} = q^{k_{sht}}$ and the additional parameter $u = t_{lng} = q^{k_{lng}}$. We set $\rho_k = \frac{1}{2}\sum_{\alpha>0} k_\alpha \alpha$ and replace $t^\rho \mapsto q^{\rho_k}$ in all formulas.

Given a torus knot $T_{r,s}$ and $\lambda = (\lambda_i)$ with $\lambda_m > 0$, conjecturally there exists a polynomial $\mathcal{H}^C_{r,s}(\lambda\,;\, q,t,u,a)$ with integral coefficients in terms of positive powers of a, q, u and $t^{\pm 1}$ such that for any $n \ge m$

$$\mathcal{H}^C_{r,s}(\lambda\,;\, q,t,u,a = -t^{n-1}) = J^{C_n}_{r,s}(\lambda\,;\, q,t,u)^\circ.$$

The uncolored invariants are for $\omega_1 = \epsilon_1$, which is minuscule for C_n.

3.2 The Case of Trefoil

Let us calculate $\mathcal{H}_{3,2}$ for uncolored trefoil; we begin with A_1. As above,

$$\{H\} \stackrel{\mathbf{def}}{=\!=} H(1)(X \mapsto t^{-\rho}),$$

where $t^{-\rho} = t^{-\frac{1}{2}}$ for A_1. By $\sim$, we mean "up to $q^\bullet t^\bullet$." One has:

$$J_{3,2} = \{\tau_+ \tau_-^2(X)\} \sim \{(XY)(XY)X(1)\} \sim \{Y(X^2)\}$$
$$= t^{-\frac{1}{2}} q^{-1} X^2 - t^{\frac{1}{2}} + t^{-\frac{1}{2}} |_{X^2 \mapsto t^{-1}} \sim 1 + qt - qt^2.$$

When $q = t$, we obtain the *Jones polynomial*: $J_{3,2}(q \mapsto t)^\circ = 1 + t^2 - t^3$.

We use that $E_1 = X$: $Y(X) = (qt)^{-\frac{1}{2}} X$. Using the formula for Y_1 for A_n above and the action of $\tau_\pm$ on X_1, Y_1, we obtain that $J^\circ_{3,2} = 1 + qt - qt^{n+1}$, which gives that $\mathcal{H}_{3,2} = 1 + qt + aq$ for $a = -t^{n+1}$. The relations $\mathcal{H}_{3,2}(a \mapsto -t) = 1$ and $\mathcal{H}_{3,2}(a \mapsto -t^2) = 1 + qt - qt^2$ are sufficient to fix it uniquely if it is known that $deg_a \mathcal{H} = 1$. Generally, $deg_a \mathcal{H}^{\lambda}_{r,s} = |\lambda|\big(\mathrm{Min}(r,s)-1\big)$. A remarkable simplicity of $\mathcal{H}_{3,2}$ is fully clarified in the approach via motivic superpolynomials (below).

The case of $T(2p+1,2)$ is quite similar. For $p = 1, 2, \ldots,$ one has:

$$\mathcal{H}_{2p+1,2} = 1 + qt + q^2t^2 + \cdots + q^p t^p + aq(1 + qt + \cdots + (qt)^{p-1}).$$

Similarly, the Khovanov-Rozansky polynomials are the simplest for these knots.

The formula becomes significantly more involved with colors. Let $\lambda = m\omega_1 = \square\square\cdots\square$ (m boxes) for $m = 1, 2, \ldots$. Then:

$$\mathcal{H}^{\lambda}_{2p+1,2} = \frac{(q;q)_m}{(-a;q)_m(1-t)} \sum_{k=0}^{m} (-1)^{m-k}$$

$$(qt)^{\frac{m-k}{2}} \Big((q^{\frac{m(m+1)}{2}} - q^{\frac{k(k+1)}{2}})(t/q)^{\frac{m-k}{2}} \Big)^{2p+1}$$

$$\frac{(t;q)_k(-a;q)_{m+k}(-a/t;q)_{m-k}(1-q^{2k}t)}{(q;q)_k(qt;q)_{m+k}(q;q)_{m-k}},$$

where $(a;q)_n = (1-a)\cdots(1-aq^{n-1})$. This formula was proposed by Dunin-Barkowski, Mironov, Morozov, Sleptsov, and Smirnov (2011–2012) and, independently, by Fuji, Gukov, and Sulkowsky (2012). A somewhat different formula is

$$\mathcal{H}^{\lambda}_{3,2} = \sum_{k=0}^{m} q^{mk} t^k \frac{(q;q)_m(-a/t;q)_k}{(q;q)_k(q;q)_{m-k}}$$

(only for trefoil). The justifications were obtained via DAHA, i.e., for the DAHA superpolynomials. The Habiro's formula (2000) is for $p = 1, a = -t^2, t = q$.

Let us mention here "Torus knots and quantum modular forms" devoted to color Jones polynomials for $T(2p+1,2)$ (K.Hikami-Lovejoy, 2014), and the Kontsevich-Zagier series from "Vassiliev invariants and a strange identity related to the Dedekind eta-function" (Zagier, 2001). Presumably, our refined formulas above can be used in a similar way. See also Example 5 from "Quantum modular forms" (Zagier, 2010).

In the case of uncolored trefoil for the system C_n, $\mathcal{H}^{C}_{3,2}(\omega_1\,;\,q,t,u,a) = 1 + qt + a(qt - qu) + a^2\big(-qu + q^2u - q^2tu\big) + a^3\big(-q^2tu + q^2u^2\big)$.

The superpolynomial for the *exceptional series* for $T(3,2)$ and for the adjoint representation is due to Cherednik-Elliot (2016):
$\mathcal{H}^{E}_{3,2}(adj;\,q,t,a) = 1 + q(t - ta + a^2 - a^4 + t^{-1}a^5 - t^{-1}a^6) + q^2(t^2a^2 - ta^3 + a^4 + ta^5 + t^{-1}a^6 - 3a^6 + t^{-1}a^7 + a^7 - t^{-1}a^8 - t^{-1}a^9 + t^{-1}a^{10} - t^{-2}a^{11}) + q^3(t^{-1}a^6 - a^7 + ta^7 + t^{-1}a^8 - a^8 - ta^8 + a^9 - 2t^{-1}a^{10} + a^{10} + t^{-2}a^{11} - t^{-1}a^{11} - a^{11} - t^{-2}a^{12} + 2t^{-1}a^{12} - t^{-2}a^{13} + t^{-2}a^{15}) + q^4(t^{-2}a^{12} - t^{-1}a^{12} + t^{-1}a^{13} - a^{13} - t^{-2}a^{14} + t^{-1}a^{14} + t^{-2}a^{16} - t^{-1}a^{16} - t^{-3}a^{17} + t^{-2}a^{17}) + q^5(-t^{-3}a^{18} + t^{-2}a^{18})$.

Here $a = -t^{\frac{h}{6}}$ for the Coxeter number h for the A, D, E there.

3.3 Iterated Links

We will begin with the DAHA construction for iterated torus knots. For any sequence $\gamma_1, \gamma_2, \ldots, \gamma_\ell \in SL(2, \mathbb{Z})$, we set

$$J^\lambda = \Bigl(\cdots \widetilde{\gamma}_{\ell-1}\Bigl(\widetilde{\gamma}_\ell\bigl(\frac{E_\lambda}{E_\lambda(t^{-\rho})}\bigr)(1)\Bigr)(1)\cdots\Bigr)(t^{-\rho}).$$

This is due to Cherednik-Danilenko; the invariant J^λ depends only on the isotopy type of the corresponding *iterated torus knot*; see an example below.

This is somewhat similar to Manin's work "Iterated integrals of modular forms and noncommutative modular symbols" (2005). Basically, $\int_0^\infty$ is replaced by $\int_p^q$ for rational p, q in this paper and multiple zeta values occur. When the coinvariant is replaced by the corresponding integral formula (a DAHA theorem), the similarity becomes less speculative.

For instance, one obtains for $K = Cab(53, 2)Cab(13, 2)Cab(2, 3)$:

$$J^\lambda_{\mathcal{L}} = \{\mathcal{P}_\lambda\},\ \mathcal{P}_\lambda = \Downarrow \begin{pmatrix} 3 & * \\ 2 & * \end{pmatrix} \Downarrow \begin{pmatrix} 2 & * \\ 1 & * \end{pmatrix} \Downarrow \begin{pmatrix} 2 & * \\ 1 & * \end{pmatrix} \bigl(\frac{E_\lambda(X)}{E_\lambda(t^{-\rho})}\bigr),$$

where the γ-matrices act via their lifts to Aut($\mathcal{HH}$), $\Downarrow H \stackrel{\mathbf{def}}{=\!=} H(1)$, $\{H\} \stackrel{\text{def}}{=\!=} H(1)(t^{-\rho})$ is the *coinvariant*, and E_λ is the E-polynomial for dominant λ. Here $Cab(a, b)K$ is $T(b, a)$ plotted at the boundary of the solid torus around a given knot K for the Seifert zero-framing.

Generally, given a sequence $\begin{pmatrix} r_1 & r_2 & r_3 & \cdots \\ s_1 & s_2 & s_3 & \cdots \end{pmatrix}$ of the 1st columns of $\gamma_1, \gamma_2, \gamma_3, \ldots$, which is $\begin{pmatrix} 3 & 2 & 2 \\ 2 & 1 & 1 \end{pmatrix}$ in the example above, the corresponding cable is

$$\cdots Cab(a_3, r_3)Cab(a_2, r_2)Cab(a_1, r_1)$$

for $a_1 = s_1$, $a_2 = r_1 s_1 r_2 + s_2$, $a_3 = a_2 r_2 r_3 + s_3$, and so on. Note that

$$Cab(s_1, r_1) = T(r_1, s_1) = Cab(r_1, s_1);$$

the transposition of r_i, a_i for $i > 1$ changes the knot.

The a-stabilization theorem is the same as for torus knots. Generally, $\deg_a(\mathcal{H}^\lambda) = |\lambda|(mult - 1)$ for $mult = \mathrm{Min}(r_1, s_1) \cdot r_2 \cdot r_3 \cdots$, which is the multiplicity of singularity. When $a = 0$, $\deg_t(\mathcal{H}^\lambda) = \delta \sum_i m_i^2$ for $\lambda = (m_1 \geq m_2 \geq \cdots)$, and $\deg_q(\mathcal{H}^\lambda) = \delta \sum_i (m'_i)^2$ for the transposition λ' of

λ; δ is the arithmetic genus of the singularity (below). The justification is under minor assumptions; the superduality and the specialization $q = 1$ provide the reduction to pure columns.

Topological invariance. For torus knots, the isotopy invariance means that $T(r,s)$, $T(s,r)$ and $T(-s,-r)$ must have coinciding $\mathcal{H}^\lambda$, and that $\mathcal{H}^\lambda = 1$ for any $T(1,s)$. Also, the superpolynomial of $T(-s,r)$, which knot is the mirror image of $T(s,r)$, must be the "conjugation" $q,t,a \mapsto q^{-1},t^{-1},a^{-1}$ of $\mathcal{H}^\lambda$ for $T(s,r)$ up to a factor $q^\bullet t^\bullet a^\bullet$. This requires the usage of the DAHA automorphism η, which we omit.

We note that the symmetry $\mathcal{H}^\lambda_{r,s} = \mathcal{H}^\lambda_{s,r}$ can be a challenge for (other) algebraic and algebraic-geometric approaches even in the case of HOMFLY-PT polynomials. It is a simple DAHA lemma for us.

Generally, the theorem is that DAHA superpolynomials depend only the isotopy type of the corresponding iterated torus link. It is interesting, all fundamental properties of the *coinvariant* are needed, but not a difficult one. The case of iterated *links* is more involved, but this is mostly because *splice diagrams* are used, which are not too simple. The additional DAHA fact is the integral formula for action of τ^{-1}, which we will provide below when discussing *DAHA vertex.*

From knots to links. The construction becomes more ramified for iterated torus *links.* We will provide here the procedure from in "DAHA approach to iterated torus links" (Cherednik-Danilenko, 2015).

First, we switch from P_λ to the so-called J-polynomials. For A_n:

$$\widetilde{P}_\lambda \stackrel{\text{def}}{=\!=} h_\lambda P_\lambda, \qquad \text{for} \quad h_\lambda = \prod_{\Box\in\lambda}(1 - q^{arm(\Box)}t^{leg(\Box)+1}),$$

for the Macdonald polynomials P_λ. Here $arm(\Box)$ is the number of boxes in the same row as $\Box$ strictly after it; $leg(\Box)$ is the number of boxes in the column of $\Box$ strictly below it. The J-polynomials and the construction below are for any root systems R.

We now set

$$\widetilde{\mathcal{P}}^\lambda(X) = \Big(\cdots\widetilde{\gamma}_{\ell-1}\Big(\widetilde{\gamma}_\ell(\widetilde{P}_\lambda)(1)\Big)(1)\cdots\Big),$$

called *basic prepolynomials*. General *prepolynomials* are defined inductively as follows. Given two prepolynomials $\widetilde{\mathcal{P}}_1, \widetilde{\mathcal{P}}_2$ and $\gamma \in PSL(2,\mathbb{Z})$ (can be id), we define a new *prepolynomial* $\Big(\widetilde{\gamma}(\widetilde{\mathcal{P}}_1\widetilde{\mathcal{P}}_2)\Big)(1)$.

Combinatorially, we obtain a union of trees with marked *last vertices* and the corresponding ends "colored" by the diagrams $\{\lambda\} = \{\lambda^1, \cdots, \lambda^\kappa\}$, where κ is the number of connected components of a link. The prepolynomials for the (maximal) roads in this union are those for the connected components. A union of several trees is for a disconnected union of the corresponding links.

This is not the end. Given two prepolynomials, $\widetilde{\mathcal{P}}(X)$ and $\widetilde{\mathcal{Q}}(X)$, for the components colored by the sequences $\{\lambda^i\}$ and $\{\mu^j\}$ of diagrams,

$$J^{\{\lambda\},\{\mu\}} \stackrel{\mathbf{def}}{=\!=} \{\widetilde{\mathcal{Q}}(Y)\widetilde{\mathcal{P}}(X)\}/LCM(\widetilde{P}_{\lambda^i}(t^\rho), \widetilde{Q}_{\mu^j}(t^\rho) \text{ for all } i, j).$$

The division by the LCM provides that $J^{\{\lambda\},\{\mu\}}$ are polynomials in terms of q, t up to some (possibly fractional) powers of q and t. This can be extended to any root systems R. The a-stabilization theorem in type A is as for knots: we arrive at $\mathcal{H}^{\{\lambda\},\{\mu\}}(q, t, a)$.

Topologically, we consider the 1st link in the horizontal solid torus and the 2nd in the complementary vertical one. Applying the coinvariant $\{\cdot\}$, we obtain a topological invariant of the resulting link in S^3. Changing Y by Y^{-1} in this definition corresponds to changing the orientation of the 2nd link versus the 1st. Algebraic links are when $\tilde{\gamma}$ are products of positive powers of $\tau_\pm$ and the linking numbers between the components of the link for $\tilde{\mathscr{P}}$ and that for $\tilde{\mathscr{Q}}$ are positive.

For instance,

$$J = \{\tau_-(\tilde{P}_{\lambda^1}\tilde{P}_{\lambda^2})\}/LCM(\tilde{P}_{\lambda^1}(t^\rho), \tilde{P}_{\lambda^2}(t^\rho))$$

for the Hopf 2-link L with the linking number $+1$ (algebraic), colored by the Young diagrams λ^1 and λ^2. When $\lambda^1 = \square = \lambda^2$, the corresponding $\mathcal{H}$ becomes $1 - t + qt + aq$. Using the presentation of L with 2 solid tori, we obtain the alternative formula:

$$J = \{\tilde{P}_{\lambda^1}(Y)\tilde{P}_{\lambda^2}(X)\}/LCM(\tilde{P}_{\lambda^1}(t^\rho), \tilde{P}_{\lambda^2}(t^\rho)).$$

Their coincidence is one of the key DAHA identities in the proof of the topological invariance of J for links.

3.4 Superduality and RH

The DAHA superduality is based on the $q \leftrightarrow t$-symmetry of type-A *stable* Macdonald polynomials and some properties of the action of projective $PSL_2(\mathbb{Z})$.

In physics, the *superduality* is related to the S-*duality* in SCFT for the BPS states and to the *CPT symmetry*. We note that various formulas and properties of superpolynomials were obtained and/or conjectured by physicists. Their works are mostly experimental, though the BPS states can be defined rigorously.

In topology, the superduality for the Khovanov-Rozansky polynomials, *KhR-polynomials*, is a difficult matter. As far as we know, it was justified only for positive iterated torus links when one of their components is colored by a row or a column (uncolored otherwise). A related problem is the definition of the *reduced* KhR-polynomials, which was resolved only partially; see "Khovanov-Rozansky homology of two-bridge knots and links" (Rasmussen, 2005). Our superpolynomials are counterparts of *reduced* KhR-polynomials. Also, considering *links* is generally involved in the KhR theory. Let us mention here *Soergel modules*, an important tool in this theory.

The DAHA construction is for *any* iterated torus links and *arbitrary* colors. Moreover, $J^{\{\lambda\},\{\mu\}}$ can be defined for any reduced root systems and $C^\vee C_1$. When the theories overlap, it is expected that topological superpolynomials, the DAHA ones, motivic superpolynomials, and those from physics coincide up to renormalizations. Also, there are combinatorial conjectures and connections with the *Heegaard-Floer cohomology*. The Alexander polynomials and the ρ_{ab}-invariants, discussed below, are related to the latter.

The DAHA superduality, conjectured by the author, was justified by Gorsky-Negut for torus knots and Cherednik-Danilenko for iterated torus links. In terms of the standard DAHA parameters: $\mathcal{H}^{\lambda}(q,t,a) = q^{\bullet}t^{\bullet}\mathcal{H}^{\lambda'}(\frac{1}{t},\frac{1}{q},a)$, where by $q^{\bullet}t^{\bullet}$, we mean "up to some power of q, t"; λ' is the transposition of the diagram λ.

The conjectural coincidence of the DAHA superpolynomials $\mathcal{H}$ with motivic ones will be stated below for algebraic links and "rows" (weights $m\omega_1$). The motivic superpolynomials are defined by now only in this generality. They are conjectured to coincide with the corresponding *flagged L-functions* (below). The superduality for L-functions is the *functional equation*, not very difficult to check (for $m = 1$).

The a-stabilization and superduality are expected to hold for B, C, D. For the C-*hyperpolynomials* (above) and the transposition $\lambda \mapsto \lambda'$, the conjecture in "Jones polynomials of torus knots via DAHA" was:

$$\mathcal{H}^C(\lambda;\, q,t,u,a) = q^{\bullet}t^{\bullet}u^{\bullet}\mathcal{H}^C(\lambda';\, t^{-1}, q^{-1}, u^{-1}t/q, -aqu).$$

We conjecture in the case of $u = t$ that $\mathcal{H}^C$ are in terms of $q, t^{\pm 1}, \mathsf{a} \stackrel{\textbf{def}}{=} qta^2$ (odd powers of a vanish) for iterated torus knots; then the superduality is that for A: it fixes a and sends $q \leftrightarrow t^{-1}$. Let us provide the hyperpolynomial for $T(4,3)$ for $u = t$. One has: $\mathcal{H}^C_{4,3}(\Box; q,t,t,\mathsf{a})$=

$$1+qt+q^2t+q^2t^2+q^3t^3+\mathsf{a}(-1+q^3-2qt-2q^2t^2-q^3t^3)+\mathsf{a}^2(qt-q^2t+q^2t^2).$$

We can take here $\mathsf{a} = a^2t^2$, which is super-invariant too. Then the restriction to C_n will be $\mathsf{a} = t^{2n} = t^h$ for the Coxeter number h.

HOMFLY-PT polynomials. The a-stabilization of our J for $q = t$ corresponds to the relation between the HOMFLY-PT polynomials, $HOM(t,\mathsf{a};\lambda)$, and the quantum group invariants for A_n (the WRT invariants). Namely, the latter are essentially $HOM(t,\mathsf{a} = t^{n+1};\lambda)$. The stabilization of the QG invariants is connected with the *Deligne category* $Rep(GL(t))$. See *"New realizations of deformed double current algebras and Deligne categories"* (Etingof, Kalinov, Rains, 2020). We note that the symmetry $k \mapsto 1/k, \upsilon \mapsto \upsilon k$ is discussed there in Sect. 4.3; with $\upsilon \mapsto -\upsilon k$, it occurs below in the context of Riemann's zeta (not because of deformed current algebras).

The definition of $HOM(t,\mathsf{a};\lambda)$ is especially simple in the uncolored case, which is for $\lambda = \Box$ (i.e., for $\lambda = \omega_1$ for A_n). The following *skein relation* is sufficient to define them (the reduced ones):

$$\mathsf{a}^{1/2}HOM(\text{⤫})-\mathsf{a}^{-1/2}HOM(\text{⤫})=(t^{1/2}-t^{-1/2})HOM(\uparrow\uparrow),\ \ HOM(\bigcirc)=1.$$

Given λ (type A), $HOM^{\circ}(t,\mathsf{a};\lambda)=\mathcal{H}^{\lambda}(q=t,t,\mathsf{a}=-a)$ for *iterated torus knots*, where $\circ$ is as above (up to $t^{\bullet}\mathsf{a}^{\bullet}$) and HOM is reduced: 1 for the unknot. The coincidence is due to Cherednik (torus knots), Morton-Samuelson (iterated torus knots), and Cherednik-Danilenko (iterated torus links).

We note that the LCM-normalization of our $\mathcal{H}^{\{\lambda\},\{\mu\}}$ must be replaced by the division by *one* $\widetilde{P}_{\lambda^i}(t^{\rho})$ (or that for one μ^j) to match *reduced* "HOMFLY-PT" for links. The reduced HOM is defined with respect to one "distinguished" components of a link. Both normalizations coincide for links with uncolored non-distinguished components or if all colors coincide. Even in these cases, the passage from non-reduced KhR polynomials of links to reduced ones is known only partially. In full generality, the LCM normalization is "non-topological."

For the uncolored trefoil, i.e., for $T(3,2)$ and when $\lambda=\omega_1=\square$:

$$HOM=\mathsf{a}(t+t^{-1}-\mathsf{a}),\ \ HOM^{\circ}=1+t^2-t\mathsf{a};$$

recall that $\mathcal{H}=1+qt+qa$. The Alexander polynomials $Al(t)$ are generally $HOM(t,\mathsf{a}=1)(1-t)^{\kappa-1}$ for links with κ components; in particular, $Al=t^{-1}-1+t,\ Al^{\circ}=1-t+t^2$ for trefoil. The simplest link is the Hopf 2-plus-link, 2 unknots with the linking number $+1$. Then:

$$HOM=\mathsf{a}^{1/2}\,\frac{1+\mathsf{a}-t-t^{-1}}{t^{1/2}-t^{-1/2}},\ \text{ and } Al^{\circ}=1.$$

The superduality becomes $t^{\frac{1}{2}}\to -t^{-\frac{1}{2}},\mathsf{a}^{\frac{1}{2}}\to\mathsf{a}^{-\frac{1}{2}}$ for HOM; it is obviously compatible with the skein relation above (in the uncolored case). Generally, the Young diagram λ goes to its transpose. The symmetry $t^{\frac{1}{2}}\to -t^{-\frac{1}{2}}$ holds for Al. However, it does not hold for the Jones polynomials and for the (quantum group) A_n-invariants. The latter are basically $HOM(t,\mathsf{a}=t^{n+1};\lambda)$, where the substitution $\mathsf{a}=t^{n+1}$ is obviously incompatible with the superduality.

RH for superpolynomials. After our talks with Yu.I. in 2017, I focused on RH for DAHA superpolynomials. We need to adjust the parameters:

$$\mathbf{H}(q,t,a)\stackrel{\mathbf{def}}{=\!=}\mathcal{H}(qt,t,a),$$

i.e., we switch to $q_{new}=q/t$. Importantly, q_{new} is fixed under the superduality. Then $\mathbf{H}(q,1/(qt),a)=q^{\bullet}t^{\bullet}\mathbf{H}(q,t,a)$ and the "weak" (qualitative) RH is the claim that $|\xi|=1/\sqrt{q}$ for the t-zeros ξ of $\mathbf{H}(q,t,a)$ for *sufficiently small* q. We consider RH only for uncolored algebraic knots; otherwise the transposition $\lambda\mapsto\lambda'$ is necessary. There is a variant with colors: for *rectangle* Young diagrams.

Such *weak RH* can be justified for (uncolored) *motivic* superpolynomials, conjecturally coinciding with the DAHA ones. *Strong RH* for $a=0$ states:

$|\xi| = 1/\sqrt{q}$ holds for $0 < q \le 1/2$ for *any* algebraic knot. This is the exact bound conjecturally.

Let us mention that strong RH holds for any $q > 0$ in the case of the family of uncolored $T(2p+1, 2)$. In this case

$$\mathbf{H}(q, t, a=0) = \frac{1-(qt^2)^{p-1}}{1-qt^2},$$

where q is "new," i.e., after the substitution $q \mapsto qt$. Experimentally, RH holds for any $q < 1$ only for this family.

We note that the value $q = 1$ is special for torus knots. Then $\mathbf{H}(q=1, t, a=0)$ becomes then a product of cyclotomic polynomials due to the shuffle conjecture (now a theorem). However, we are looking for the minimal q_0 such that RH holds for *any* $q < q_0$; this bound q_0 is smaller than 1 generally, including sufficiently large torus knots.

Numerically, the bound q_0 tends to $\frac{1}{2}$ for $Cab(13+2m, 2)Cab(2, 3)$ as $m \to \infty$, which is not proven rigorously, but probably a very difficult to check. This is the only such family we found. These cables correspond to the singularity rings $\mathcal{R} = \mathbb{C}[[z^4, z^6+z^{7+2m}]]$ (see below). Interestingly, the bound q_0 frequently become greater (better!) for multiple cables or if the cables begin with torus knots different from $T(3, 2)$. For instance, it is somewhat better for $Cab(53, 2)Cab(13, 2)Cab(2, 3)$ versus $Cab(13, 2)Cab(2, 3)$, numerically 0.6816 versus 0.6686 for $a = 0$.

This is from my paper "Riemann hypothesis for DAHA superpolynomials and plane curve singularities" (2018). There are many examples of superpolynomials there, including colored ones and links. Weak RH can be stated for algebraic links too; namely, the conjectural claim is that for $a = 0$ sufficiently small the number of pairs of exceptional (non-RH) zeros is $\kappa - 1$, where κ is the number of components of an algebraic uncolored link. See the paper concerning rectangle diagrams.

Generally, RH totally fails for *non-algebraic* knots/links and beyond rectangle diagrams taken as colors for algebraic ones. It seems a really algebraic phenomenon. Another special feature of algebraic knots is the positivity of the coefficients of $\mathcal{H}$ for algebraic knots colored by rectangles. The positivity conjecture for rectangles is the last unresolved problem from my initial paper. There is a version for the algebraic links (our papers with Danilenko).

The substitution $q \mapsto qt$ in the passage from $\mathcal{H}$ to $\mathbf{H}$ occurs above as a technicality: the DAHA superduality $q \leftrightarrow t^{-1}$ then becomes $q \mapsto q, t \mapsto 1/(qt)$. However, the latter is exactly the Hasse-Weil symmetry from the functional equation for curves over finite fields. There is some connection with q-deformations of Riemann's zeta and the Dirichlet L-functions (below); it is based on my "RH paper." Generally, it is the passage from the superpolynomials of links to those for Seifert 3-folds and their special infinite sums.

4 Plane Curve Singularities

This section provides a conjectural formula for superpolynomials of algebraic links colored by "rows" in terms of the corresponding plane curve singularities. It corresponds to the most general case of affine Springer fibers of type A and matches well the DAHA formulas.

4.1 Basic Facts

Algebraic links are intersections of plane curve singularities at $(0,0) \in \mathbb{C}^2$ with small $S^3 \subset \mathbb{C}^2$ centered at $(0,0)$; they are *knots* for irreducible (unibranch) singularities. For such knots, the corresponding (local) singularity rings can be considered inside $\mathbb{C}[[z]]$, where z is the uniformizing parameter. They are any local rings $\mathcal{R} \subset \mathbb{C}[[z]]$ with 2 generators in $(z) = z\mathbb{C}[[z]]$ and the localization $\mathbb{C}((z))$. Such rings are always *Gorenstein*.

The simplest topological invariants of a singularity are its multiplicity $\dim \mathbb{C}[[z]]/\mathbb{C}[[z]]\mathfrak{m}$ for the maximal ideal $\mathfrak{m} \subset \mathcal{R}$, and the arithmetic genus $\delta = \dim \mathbb{C}[[z]]/\mathcal{R}$, the *Serre number*.

The rings $\mathcal{R} = \mathbb{C}[[x=z^r, y=z^s]]$ for $r, s \in \mathbb{N}$ such that $\gcd(r,s)=1$ correspond to unibranch *quasi-homogeneous singularities* $x^s = y^r$ and torus knots $T(r,s)$. The multiplicity is $\mathrm{Min}(r,s)$ and $\delta = \frac{(r-1)(s-1)}{2}$, which is actually due to Sylvester (the Frobenius coin problem). The simplest "non-torus" family is $\mathcal{R} = \mathbb{C}[[z^4, z^6 + z^{7+2m}]]$ for $m \in \mathbb{Z}_+$, which are of multiplicity 4 and with $\delta_m = 8+m$.

From families to towers. The simplest *family* is $\mathcal{R} = \mathbb{C}[[z^r, z^{s+mr}]]$ for $m \in \mathbb{Z}_+$. Our families can be naturally interpreted as towers of extensions of $\mathbb{C}[[x,y]]$ via the Puiseux theory. This is related to the theory of *Drinfeld-Vléduts bound* (1983) and the paper by Manin-Vléduts "Linear codes and modular curves" (1985). It is for growth of the arithmetic genus in some towers of curves X, which can be singular. The *Artin-Schreier towers* provide some important examples here.

We do plane curve singularities, when the "curve" is 1 point, and the problems become about finding some formulas-bounds for $|\mathcal{J}_m(\mathbb{F}_q)|$ for the corresponding compactified Jacobians (below). These numbers are the values of motivic superpolynomials as $t = 1, a = 0$. For smooth projective curves X over $\mathbb{F}_q$, the Hasse-Weil-Deligne formula can be used in terms of the eigenvalues of "Frobenius." We use different tools, but the functional equation and even some form of Riemann hypothesis work for plane curve singularities.

The formulas for $|\mathcal{J}_m(\mathbb{F}_q)|$ and our superpolynomials can be viewed as counterparts of *Iwasawa polynomials* for class numbers in Γ-extensions. According to Barry Mazur: Γ-extensions can be considered as counterparts of abelian coverings of S^3 ramified at a given link, where the Iwasawa polynomials can be seen as counterparts of Alexander polynomials ($q = t, a = -1$ for us). This is for any links. For algebraic links, cyclic (algebraic) coverings of P^2 branched over certain singular curves (can be assumed rational) are sufficient to consider (Libgober (1980) and others). This is similar to our towers.

Valuation semigroup. It is one of the key in the theory of curve singularities. The definition of this semigroup is as follows:

$$\Gamma \stackrel{\text{def}}{=\!=} \{ \nu_z(f), 0 \neq f \in \mathcal{R} \subset O \stackrel{\text{def}}{=\!=} \mathbb{C}[[z]] \},$$

where ν_z is the valuation, the order of z. We readily obtain that $\delta = |\mathbb{Z}_+ \setminus \Gamma|$. Importantly, Γ gives the topological type of the corresponding algebraic knot (considered up to isotopy), which is due to Zariski and others. Thus, topological invariants of rings $\mathcal{R}$ are exactly those expressed in terms of Γ.

For instance, the Alexander polynomial is immediate via Γ. Namely, Al° is $(1-t)\sum_{\nu\in\Gamma} t^\nu$ for any $\mathcal{R} \subset O$ (for its $\circ$-normalization). For instance, it is

$$(1-t)(\frac{1}{1-t} - t) = 1 - t + t^2$$

for $T(3, 2)$. The theory of topological equivalence of algebraic *links* is significantly more ramified; *splice diagrams* are very helpful. Generally, the coincidence of semigroups for the components and the corresponding pairwise linking numbers (all must be positive) is sufficient due to the Reeve theorem. The pairwise linking numbers can be algebraically calculated via the ring of singularity, which is not too involved.

For $Cab(53, 2)Cab(13, 2)Cab(3, 2)$ above (note the change $(2, 3) \mapsto (3, 2)$), the ring is $\mathcal{R} = \mathbb{C}[[x = z^8, y = z^{12} + z^{14} + z^{15}]]$. The Newton's pairs are generally $\{r_1, s_1\}, \{r_2, s_2\}, \cdots$, and the Puiseux-type equation is

$$y = x^{\frac{s_1}{r_1}}\Big(1 + c_1 x^{\frac{s_2}{r_1 r_2}}\big(1 + c_2 x^{\frac{s_3}{r_1 r_2 r_3}}(\cdots)\big)\Big),$$

for generic c_i. *We will assume that* $r_1 < s_1$, *which can be always imposed.*

The arithmetic genus is $\delta = 42$, and the valuation semigroup $\Gamma = \langle 8, 12, 26, 53\rangle$. Generally, $\Gamma = \langle r_1 r_2 r_3, a_1 r_2 r_3, a_2 r_3, a_3\rangle$ for the cable parameters (a_i, r_i) above (here $r_1 < s_1$ is used). Recall that $a_1 = s_1$, $a_2 = r_1 s_1 r_2 + s_2$, $a_3 = a_2 r_2 r_3 + s_3$, and so on for any number of $\{r_i, s_i\}$. In this example, the Newton's pairs are $\{(2, 3), (2, 1), (2, 1)\}$.

The passage from the base field $\mathbb{C}$ to finite fields $\mathbb{F}_q$ for $q = p^k$ and prime p is sufficiently straightforward; it will be needed below. We begin with $\mathcal{R}$ over $\mathbb{C}$, define it over $\mathbb{Z}$, which is always doable within a given isotopy type, and then consider $\mathcal{R} \otimes_{\mathbb{Z}} \mathbb{F}_p$. A prime number p is called a *prime of good reduction* if Γ remains unchanged over $\mathbb{F}_p$ upon this procedure. This definition is adjusted to the topological invariance.

All primes p are good for the rings $\mathbb{C}[[x = z^r, y = t^s]]$ as above. Presumably, there are no prime p of bad reduction in this sense within a given topological type for *any* algebraic knots: given any p, there exists $\mathcal{R}$ representing a given knot where this p is good.

To give an example, let $\mathcal{R}=\mathbb{Z}[[x=t^4,\ y=t^6+t^7]]$. Then $\delta=8$ and

$$\Gamma=\{0,4,6,8,10,13,14,16,17,18,\ldots\}.$$

This $\mathcal{R}$ has bad reduction only at $p=2$. Indeed, $\nu_z(y^2-x^3)=14$ in $\mathbb{F}_2$, which is 13 for $p\neq 2$. However, this singularity is equivalent over $\mathbb{C}$ (analytically, not only topologically) to the one for $\mathbb{Z}[[t^4+t^5, t^6]]$, where bad p is 3. We obtain that the corresponding cable has no primes of bad reduction.

4.2 Compactified Jacobians

Let $\mathcal{R}\subset \mathcal{O} \stackrel{\text{def}}{=\!=} \mathbb{F}[[z]]$ be the ring of an irreducible plane curve singularity over any field $\mathbb{F}$. The corresponding *flagged compactified Jacobian* $\mathcal{J}_\ell$, considered as a set of $\mathbb{F}$-points by now, is formed by *standard flags*

$$\overrightarrow{M}=M_0\subset M_1\subset\cdots\subset M_\ell\subset\mathcal{O}=\mathbb{F}[[z]]$$

of $\mathcal{R}$-submodules M_i of $\mathcal{O}$ such that:

(a) $M_0\ni\phi=1+z(\cdot)$ (where $(\cdot)\in\mathcal{O}$),
(b) $\dim M_i/M_{i-1}=1$ and $M_i=M_{i-1}\oplus\mathbb{C}\,z^{g_i}(1+z(\cdot))$,
(c) (important) $g_i<g_{i+1}$, where $i\geq 1$.

We will call them ℓ-flags.

When $\ell=0$ (0-flags), there is only one condition: $\mathcal{O}\supset M\ni\phi=1+z(\cdot)$. Equivalently, $\Delta(M)\ni 0$, where $\Delta(M)\stackrel{\text{def}}{=\!=}\{\nu_z(v)\mid 0\neq v\in M\}$.

Generally, $\Delta(M)$ are Γ-modules for any $\mathcal{R}$-modules M, i.e., $\Gamma+\Delta\subset\Delta$. *Standard* Δ are those in $\mathbb{Z}_+$ containing 0 and, therefore, containing the whole Γ. Thus, standard M are those with standard $\Delta(M)$.

For quasi-homogeneous singularities $\mathcal{R}=\mathbb{F}[[x=z^r, y=z^s]]$, where $\gcd(r,s)=1$, $r,s>1$, all standard Γ-modules Δ come from some standard M. There are several ways to see this. Piontkowski used the method of syzygies, which also gives that the corresponding cells are affine spaces and result in combinatorial formulas for their dimensions. Also, the $\mathbb{C}^*$-action and the *Bialynicki-Birula theorem* can be used.

This is a special feature of quasi-homogeneous (plane curve, unibranch) singularities; generally, not all Δ are present in the decomposition of $\mathcal{J}_0$. For instance, for $\mathbb{F}[[z^4, z^6+z^7]]$, two from 25 such Δ are not in the form $\Delta(M)$ for any standard M, the phenomenon of which is due to Piontkowski. It seems that this is always the case unless for quasi-homogeneous singularities. Also, generally, not all Piontkowski cells are affine spaces.

Let us supply $\mathcal{J}_0$ with a structure of a *projective* variety. We will describe the corresponding reduced scheme. By construction, this set is naturally a *disjoint* unions of quasi-projective varieties, those for different values of the deviations of M (below). Importantly, they can be combined in one projective variety. The main steps are as follows.

First, any standard M contains the ideal $(z^{2\delta}) = z^{2\delta}\mathcal{O}$. Indeed, the latter is the *conductor* of $\mathcal{R}$ for any Gorenstein $\mathcal{R}$, the greatest ideal in $\mathcal{O}$ that belongs to $\mathcal{R}$. Using this, $\phi = 1 + z(\cdot) \in M$ (it is standard) implies that $\phi \cdot (z^{2\delta}) = (z^{2\delta}) \subset M$.

Second, let $dev(M) \stackrel{\textbf{def}}{=} \delta - \dim(\mathcal{O}/M)$, its *deviation* from $\mathcal{R}$; this is for any $\mathcal{R}$-modules in $\mathcal{O}$. Then, $dev(M) \geq 0$ for standard M and it is 0 if and only if $M = \phi\mathcal{R}$ for some ϕ as above. The latter modules are called *invertible*. They form the *generalized Jacobian variety* of this singularity, which is an algebraic group. The third step (the key) is based on the fact that $z^{dev(M)}M \supset (z^{2\delta})$ for standard M due to Pfister-Steenbrink. Equivalently, $dev(M) + \Delta(M) \supset 2\delta + \mathbb{Z}_+$.

Finally, let $M \mapsto M' \stackrel{\textbf{def}}{=} z^{dev(M)}M$. Then $dev(M') = dev(M) - dev(M) = 0$. It establishes an identification of standard M with $\mathcal{R}$-modules $(z^{2\delta}) \subset M' \subset \mathcal{O}$ such that $dev(M') = 0$. The inverse map is $M' \mapsto z^{-d}M'$ for $d = \mathrm{Min}\{\nu_z(m) \mid 0 \neq m \in M'\}$. Then $\{M'\}$, the *compactified Jacobian*, becomes a projective subvariety of the Grassmannian of the subspaces of the middle dimension in $\mathcal{O}/(z^{2\delta})$. It is irreducible (Rego), which holds only for plane curve singularities among all Gorenstein ones. Then $\mathcal{J}_\ell$ become natural fiber spaces over $\mathcal{J}_0$; the fibers are not too difficult to describe, which will be used below.

Affine Springer fibers. The definition requires the equation $F(x, y) = 0$ for the generators x, y of $\mathcal{R}$. Let n and m be the top x-degree and y-degree of this equation, which we assume irreducible. Then our $\mathcal{J}_0$ can be interpreted as a (parahoric) affine Springer fiber $\mathcal{X}_\gamma$ defined either for GL_n or for GL_m; the equation connecting x and y becomes the corresponding *characteristic equation.*

The case of *arbitrary* $F(x, y)$, not irreducible and not square-free, will be addressed below; topologically, this is the case of algebraic links colored by any rows.

Generally, ASF are due to Kazhdan-Lusztig (1988). Their description entirely in terms of $\mathcal{R}$ is a remarkable feature of type A. The definition is via GL_n or via GL_m; but the corresponding ASF are isomorphic. This is not immediate from their definition (below). The standard modules M and the definition of $\mathcal{J}_0$ given above do *not* require the equation $F(x, y) = 0$; only $\mathcal{R} \subset \mathcal{O}$ is needed.

For semisimple Lie algebra $\mathfrak{g}$ and any field $\mathbb{F}$, let $\mathfrak{g}[[x]] = \mathfrak{g} \otimes_{\mathbb{F}} \mathbb{F}[[x]]$ and $\mathfrak{g}((x)) = \mathfrak{g} \otimes_{\mathbb{F}} \mathbb{F}((x))$. Accordingly, we define $G[[x]]$ and $G((x))$ for simply connected G with $\mathrm{Lie}(G) = \mathfrak{g}$. Given $\gamma \in \mathfrak{g}[[x]]$,

$$\mathcal{X}_\gamma \stackrel{\textbf{def}}{=} \{g \in G((x))/G[[x]] \mid g^{-1}\gamma g \in \mathfrak{g}[[x]]\},$$

where we assume that the centralizer of γ in $G((x))$ is *anisotropic* (the nil-elliptic case). Then $\mathcal{X}_\gamma \cong \mathcal{J}_0$ in type A, where the singularity is $P(x, y) = 0$ for the characteristic polynomial $P(x, y) = \det(\mathbf{1}y - \gamma)$. The corresponding *orbital integral* will become $\mathcal{H}^{mot}(q, t = 1, a = 0)$ for the motivic superpolynomials defined below, where $\mathbb{F} = \mathbb{F}_q$. It is conjectured to be a topological invariant, which implies that so is the orbital integral. For instance, $\mathcal{H}^{mot}(q=1,t=1,a=0)$ is conjecturally the Euler characteristic $e(\mathcal{X}_\gamma)$.

We note that our compactified Jacobians occur as *Jacobian factors* if projective *rational* singular curves are considered; they are basically *Hitchin fibers* over $\mathbb{P}^1$. However, *factorizable* Lie groups and algebras (below) are, generally, beyond Hitchin fibers. Given a *factorizable Lie algebras* $\mathfrak{G}$, we considered the families of subtori $\mathcal{T}\subset\mathcal{G}$ with fixed characteristic polynomials in the corresponding factorizable Lie group (Cherednik, 1983). The definitions are as follows.

The factorizable Lie algebras $\mathfrak{G}$ are vector bundles over a smooth projective curve E, with the structure of *relative* Lie algebra over E. The generic fiber must be $\mathfrak{g}$, but some fibers can be non-semisimple Lie algebras (all are of the same dimension). The factorization conditions are $H^0(E,\mathfrak{G})=\{0\}=H^1(E,\mathfrak{G})$ for Čech cohomology, which readily implies that genus$(E)\le 1$. If E is singular then $\mathfrak{G}$ must be assumed torsion-free. Main applications are for $E=\mathbb{P}^1$.

Such $\mathfrak{G}$ are in 1–1 correspondence with *not necessarily unitary classical r-matrices* $r(u,v)\in\mathfrak{g}^{\otimes 2}$: those satisfying the identity $[r^{12},r^{13}+r^{23}]=[r^{13},r^{32}]$, where r^{ij} is r^{ij} for $u=u_i, v=u_j$ considered with values in $\mathfrak{g}^{\otimes 2}$ embedded in the components i,j of $\mathfrak{g}^{\otimes 3}$. The parameters u_i are *local*: near 0. Additionally, we assume that $r-\Omega/(u-v)$ is regular at 0 for the "permutation matrix" $\Omega\in\mathfrak{g}^{\otimes 2}$, the Casimir element.

The link to ASF is basically as follows. Let $\mathcal{G}$ be the group scheme over E with the Lie algebra $\mathfrak{G}$. We obtain that $H^0(E,\mathcal{G})$ and $H^1(E,\mathcal{G})$ are trivial. The starting point is a subscheme $\mathcal{T}\subset\mathcal{G}$, which is assumed a maximal subtorus at the generic point of E. Since $H^1(E,\mathcal{G})=\{0\}$, any cocycle ϕ in the *generalized Jacobian*, which is $H^1(E,\mathcal{T})$, becomes the boundary $\{\phi_i\phi_j^{-1}\}$ for an open cover $E=\cup_i U_i$ and $\phi_i\in H^0(U_i,\mathcal{G})$. Then $\mathcal{T}_\phi=\phi_i^{-1}\mathcal{T}\phi_i\subset\mathcal{G}$ is another toric subscheme with the characteristic polynomial coinciding with that of $\mathcal{T}$.

Generally, any $\mathcal{T}$ can be represented as $\mathcal{T}=\mathbb{G}_m(C)$ for a projective curve C covering E, *possibly singular*. Let $\overline{Jac}(C)$ be the compactification of the generalized Jacobian of C. Then the *Jacobian factors* will be the contributions of singular points of C to $\overline{Jac}(C)$. One can take here $E=\mathbb{P}^1$ and consider rational curves C with only one singularity. Then it will give our $\mathcal{J}_0$ for the corresponding $\mathcal{R}$.

4.3 Motivic Superpolynomials

The rings $\mathcal{R}$ and $\mathcal{R}\subset\mathcal{O}$ will be now over $\mathbb{F}=\mathbb{F}_q$. Following "DAHA and plane curve singularities" (Cherednik-Philipp, 2017), the *motivic superpolynomial* of $\mathcal{R}$ is:

$$\mathcal{H}^{mot}\overset{\text{def}}{=\!=}\sum_{\{M_0\subset\cdots\subset M_\ell\}\in\mathcal{J}_\ell(\mathbb{F})} t^{\dim(\mathcal{O}/M_\ell)}a^\ell,$$

where the summation is over all ℓ-flags $\overrightarrow{M} \subset O$ and $\ell \geq 0$.

The flags are actually not necessary in this definition due to the following theorem. Let $rk_q(M) \stackrel{\mathbf{def}}{=} \dim_{\mathbb{F}_q} M/\mathfrak{m}M$ for the maximal ideal $\mathfrak{m}$ of $\mathcal{R}$. Then

$$\mathcal{H}^{mot} = \sum_{M} t^{\dim(O/M)}(1+aq)\cdots(1+aq^{rk_q(M)-1}),$$

where the summation is over all *standard* $M \subset O$. The justification uses Proposition 2.3 from the paper mentioned above.

We conjectured there that $\mathcal{H}^{mot} = \mathcal{H}$, i.e., the motivic one for $\mathcal{R}$ coincides with the uncolored DAHA superpolynomial $\mathcal{H}(q,t,a)$ associated with the link of the singularity associated with $\mathcal{R}$. The definition of $\mathcal{H}^{mot}$ and this conjecture were extended later (with Philipp) to torsion-free sheaves of any rank$= m$ over irreducible plane curve singularities, corresponding to the DAHA superpolynomials $\mathcal{H}^{\lambda}$ for $\lambda = m\omega_1$. The latest development is the generalization to non-unibranch singularities to be considered below. As we mentioned above, this corresponds to affine Springer fibers of type A with the most general characteristic polynomials. So it is a natural setting here.

The DAHA superpolynomials depend on q polynomially by construction and are topological invariants (a theorem). Thus, this conjecture includes the claims that $\mathcal{H}^{mot}$ polynomially depend on q and that these polynomials are *topological* invariants of the corresponding plane curve singularities. This was justified by the author for some families: when Γ has *two* generators (the case of torus knots) or *three* generators; the latter was with restrictions. Generally, algebraic/analytic types of plane curve singularities depend on "continuous" parameters; the classification is essentially known and we use its elements when considering Γ with 2–3 generators mentioned above.

We note that counterparts of the motivic superpolynomials can be defined in characteristic 0: for any p-adic integral domains O instead of $\mathbb{F}_q[[z]]$ and its *orders* $\mathcal{R}$, subrings with the same localization field. They count *standard* $\mathcal{R}$-modules $M \subset O$, those containing a unit in O, with the weights $t^{deg}a^{rk}$. Here $|O/M| = q^{\deg}$ for $\mathcal{R}/\mathfrak{m}_{\mathcal{R}} = \mathbb{F}_q$ and $rk = \dim_{\mathbb{F}_q} M/\mathfrak{m}_{\mathcal{R}}M$ for the maximal ideal $\mathfrak{m}_{\mathcal{R}} \subset \mathcal{R}$.

There will be no quasi-projective varieties and *Witt vectors* will be used, but the procedure is similar.

For instance, let $O = \mathbb{Z}_p[[\pi]]$ for the p-adic $\mathbb{Z}_p$, $\pi^s = p$ and $\mathcal{R} = \mathbb{Z}_p[[x = p, y = \pi^r]] \subset O$, where $r, s > 0$ and $\gcd(r,s) = 1 = \gcd(p,s)$ (the tamely ramified case). The corresponding superpolynomial will be then the same as the one for $\mathcal{R} = \mathbb{F}_p[[z^s, z^r]] \subset O = \mathbb{F}_p[[z]]$ in relatively simple examples. Generally, there are many possible domains O in the p-adic case. Counterparts of plane curve singularities are complete subrings in O with 1 and one generator (and the same field of rationals). Generally, the action of $\mathrm{Gal}(\overline{\mathbb{Q}}_p/\mathbb{Q}_p)$ becomes significantly more involved in the p-adic theory.

Piontkowski cells. We set $\Delta(\overrightarrow{M}) = \{\Delta(M_i)\}$. It is standard for standard M in the following sense. An abstract sequence of Γ-modules $\overrightarrow{\Delta} = \{\Delta_0 \subset \cdots \subset \Delta_\ell \subset$

$\mathbb{Z}_+\}$ is called ***standard*** if Δ_0 contains Γ, $\Delta_i = \Delta_{i-1} \cup \{g_i\}$, and $g_i < g_{i+1}$ for $1 \le i \le \ell$. Given a standard $\overrightarrow{\Delta}$, the corresponding ***Piontkowski cells*** is

$$\mathcal{J}_\ell(\overrightarrow{\Delta}) \stackrel{\mathbf{def}}{=\!=} \{\overrightarrow{M} \in \mathcal{J}_\ell \mid \Delta(\overrightarrow{M}) = \overrightarrow{\Delta}\}.$$

These cells are subsets in $\mathcal{J}_\ell$ and $\mathcal{J}_\ell = \cup \mathcal{J}_\ell(\overrightarrow{\Delta})$, where the union is disjoint. These cells are not always affine spaces $\mathbb{A}^m$ and some can be empty. Empty cells always occur (in examples) unless for quasi-homogeneous singularities $x^s = y^r$; all cells are non-empty affine spaces for them. Beyond them, all cells can be affine spaces for some "non-torus" exceptional "small" families. Then the lists of empty cells and $\dim \mathcal{J}_\ell(\overrightarrow{\Delta})$ for the other cells are sufficient to know.

The varieties $\mathcal{J}_\ell(\overrightarrow{\Delta})$ are conjectured to be ***configurations of affine spaces***, i.e., unions, and differences of affine spaces $\mathbb{A}^m$ in a bigger $\mathbb{A}^N$, not always equidimensional and connected. This conjecture includes the non-unibranch generalization considered below. Then $\mathcal{H}^{mot}$ becomes with the coefficients in terms of $[\mathbb{A}^1] \in K_0(Var/\mathbb{F})$ instead of q, i.e., motivic indeed; $K_0(Var/\mathbb{F}_q) \supset [X] \mapsto |X(\mathbb{F}_q)|$ is an important ***motivic measure***.

We mention that the connection between the dimensions of these cells in $\mathcal{J}_0$ and the **deviations** in the case of $\mathbb{F}_q[[z^r, z^s]]$ was observed by Lusztig-Smelt (1995). This is a special case of superduality. The deviations are readily given in terms of $\Delta(M)$.

The coincidence of $\mathcal{H}^{mot}$ with the DAHA superpolynomials $\mathcal{H}^{daha}$ is checked in many examples, including the cases when some Piontkowski cells are ***not*** affine spaces. If all of them are such, the method of ***syzygies*** provides formulas for their dimensions, and the calculation of $\mathcal{H}^{mot}$ is mostly reduced to the combinatorics of $\Delta(M)$; Dyck paths occur for $\mathcal{H}^{mot}_{r,s}$, etc. Thus, $\mathcal{H}^{daha} = \mathcal{H}^{mot}$ is actually an advanced version of the ***Shuffle Conjecture***.

Let us mention that motivic superpolynomials are always ***significantly*** faster to calculate than ***flagged L-functions*** defined below. This is especially true when explicit formulas for $\dim \mathcal{J}_\ell(\overrightarrow{\Delta})$ are known (for $T(r,s)$ and for several exceptional families of cables). Otherwise, DAHA calculations are, generally, faster than motivic ones, especially with colors.

If ***some*** covering of $\mathcal{J}_0$ by affine cells exists, then the coefficient of q^i in $\mathcal{H}^{mot}$ for $t=1, a=0$ is the ***Betti number*** $b_{2i} = \mathrm{rk}\, H_{2i}(\mathcal{J}_0; \mathbb{R})$ and $b_{2i+1}=0$. In particular, $\mathcal{H}^{mot}(q=1, t=1, a=0)$ is the Euler number $e(\mathcal{J}_0)$. The latter is the rational Catalan number $\frac{1}{r+s}\binom{r+s}{r}$ for $\mathcal{R} = \mathbb{F}[[z^r, z^s]]$ (Beauville), where $\gcd(r,s) = 1$ as above. This is the number of all standard Δ for such $\mathcal{R}$, which are 1-1 with Dyck paths in the rectangles "$r \times s$"; the approach to $e(\mathcal{J}_0)$ via the count of standard Γ-modules Δ is due to Piontkowski. However, this number is bigger than $e(\mathcal{J}_0)$ unless for torus knots (quasi-homogeneous singularities).

We conjectured with Ivan Danilenko that the relation to Betti numbers of $\mathcal{J}_0$ always holds for the corresponding DAHA superpolynomials. More generally, the conjecture is that the ***geometric superpolynomials*** defined in terms of ***Borel-Moore homology*** of $\mathcal{J}_\ell$ coincide with the DAHA superpolynomials for any algebraic knots. The geometric superpolynomials coincide with motivic ones if $\mathcal{J}_\ell$ can be covered by affine spaces, which is by the definition of the Borel-Moore homology.

From knots to links. The consideration of non-unibranch plane singularities colored by $m\omega_1$ (pure rows) is necessary for the theory of ASF of type A with arbitrary (not only irreducible) characteristic polynomials. Also, they occur in the inductive formulas for the superpolynomials, similar to the Rosso-Jones formula in topology, even if we begin with unibranch uncolored plane singularities.

The ring will be now $\mathcal{R} \subset \mathcal{O} \stackrel{\textbf{def}}{=\!=} \oplus_{i=1}^{\kappa} e_i \mathcal{O}_i$, where $\mathcal{O}_i = \mathbb{F}[[z_i]]$ and $e_i e_j = \delta_{ij} e_i$. We set $z \stackrel{\textbf{def}}{=\!=} \sum_{i=1}^{\kappa} z_i$ and $e \stackrel{\textbf{def}}{=\!=} \sum_{i=1}^{\kappa} e_i$; then $z_i = z e_i$ and e is the unit element 1 in the ring $\mathcal{O}$. Generally, f_i will be the projection $f e_i$ for any $f \in \mathcal{O}$. Here, $\mathcal{R}$ must contains $1 = e$ and have *two* generators: $x = \sum_{i=1}^{\kappa} x_i$ and $y = \sum_{i=1}^{\kappa} y_i$ in $\mathfrak{m}_{\mathcal{O}} = z\mathcal{O}$. Also, the localizations of the projection $\mathcal{R}_i$ of $\mathcal{R}$ onto $\mathcal{O}_i$ must be full $\mathbb{F}((z_i))$.

By construction, $\prod_{i=1}^{\kappa} F_i(x, y) = 0$, where $F_i(x_i, y_i) = 0$ for the corresponding (monic) irreducible polynomials F_i for $\mathcal{R}_i$, which will be assumed pairwise non-proportional. The assumption that $F(u, v) = \prod_{i=1}^{\kappa} F_i(u, v)$ is *square-free* is standard for curve singularities.

For $\mathbb{F} = \mathbb{C}$, the equation $F(u, v) = 0$ gives the corresponding singularity (with κ branches). The corresponding link is $\{F(u, v) = 0\} \cap S^3_\epsilon$ in $\mathbb{C}^2$ with the coordinates u, v; it has κ components. Its isotopy type gives the topological type of the singularity.

The passage from $\mathbb{C}$ to $\mathbb{F}_q$ is the same as in the unibranch case. Namely, we pick $x, y \in \mathbb{Z}[[z]]$ within a given topological type and then switch to $\mathbb{F}_q$ for $q = p^m$ provided that p is a prime of *good reduction*. By definition, good p are such that the corresponding F_i remain irreducible and pairwise non-proportional over $\mathbb{F}_q$. The semigroups Γ_i for $\mathcal{R}_i$ and the pairwise linking numbers must remain unchanged. The latter conditions are entirely algebraic: the linking numbers are the corresponding intersection numbers, which can be defined via $\mathcal{R}$.

The notion of good reduction is necessary for the conjectural coincidence of motivic superpolynomials with the DAHA superpolynomials and topology; the corresponding p must be good. The coincidence conjecture can be extended to $F(u, v) = \prod_{i=1}^{\kappa} F_i(u, v)^{c_i}$, i.e., to arbitrary F, not only square-free. The algebraic links colored by the sequence of weights $\sigma = \{c_i\omega_1, 1 \le i \le \kappa\}$ occur on the DAHA side in this case. The sequence $\sigma = \{c_i\}$ will be assumed ordered: $c_1 \ge c_2 \ge \cdots c_\kappa > 0$. These inequalities can be achieved by permuting $\{F_i\}$.

We extend the sequence of *uniformizing parameters* $\{z_i, 1 \le i \le \kappa\}$. It will be now $\{\zeta_i, 1 \le i \le \tau\}$, where $\tau \stackrel{\textbf{def}}{=\!=} \sum_{i=1}^{\kappa} c_i$. The connection is as follows:

$$z_1 = \zeta_1 + \cdots + \zeta_{c_1}, z_2 = \zeta_{c_1+1} + \cdots + \zeta_{c_1+c_2}, \ \ldots, \ z_\kappa = \zeta_{\tau-c_\kappa+1} + \cdots + \zeta_\tau.$$

Accordingly, $\{\epsilon_i, 1 \le i \le \tau\}$ will be the extended sequence of idempotents: $e_1 = \epsilon_1 + \cdots + \epsilon_{c_1}$, and so on. We set $\Omega = \mathbb{F}[[\zeta_1, \ldots, \zeta_\tau]] = \sum_{i=1}^{\tau} \Omega_i$ for $\Omega_i = \epsilon_i \Omega$, where $1 \le i \le \tau$, and $\mathcal{O}_i = e_i \Omega$ for $1 \le i \le \kappa$. The prior $\mathcal{R}$ can be naturally embedded in Ω: $\mathcal{R} \subset O = \sum_{i=1}^{\kappa} O_i \subset \Omega$, i.e., we embed diagonally for the segments of $\{\zeta_i\}$ associated with the multiplicities c_i.

Standard modules $\mathcal{M}$ are by definition $\mathcal{R}$-invariant $\mathbb{F}$-subspaces $\mathcal{M} \subset \Omega$ such that $\Omega\mathcal{M} = \Omega$. As above, $rk_q(\mathcal{M}) \stackrel{\text{def}}{=\!=} \dim_{\mathbb{F}_q} \mathcal{M}/\mathfrak{m}_{\mathcal{R}}\mathcal{M}$, where $\mathfrak{m}_{\mathcal{R}} = \mathcal{R} \cap zO$. The minimal q-rank is then $rk_{min} = c_1$ and the maximal q-rank, rk_{max}, is that for $\mathcal{M} = \Omega$, which is $\tau - 1 + \delta_\sigma$ for $\delta_\sigma = \dim_{\mathbb{F}_q} \mathfrak{m}_\Omega/\mathfrak{m}_{\mathcal{R}}$, where $\mathfrak{m}_\Omega = \oplus_{i=1}^{\tau} \zeta_i \Omega$. It equals $\sum_{i=1}^{\kappa} m_i c_i$, where m_i are multiplicities of the singularities associated with $F_i(x, y) = 0$, which are $m_i = \min\{\Gamma_i \setminus \{0\}\}$ for the corresponding Γ_i.

The *motivic superpolynomial* of $\mathcal{R}$, σ is defined as follows:

$$\mathcal{H}_\sigma^{mot} = \sum_{\mathcal{M}} t^{\dim(\Omega/\mathcal{M})} \prod_{j=rk_{min}}^{rk_q(\mathcal{M})-1} (1 + aq^j) \text{ summed over standard } \mathcal{M}.$$

The first product Π here (for $rk_q(\mathcal{M}) = rk_{min} = c_1$) is 1. The conjecture is that $\mathcal{H}^{mot}$ depend polynomially on q, t, a and coincide with the DAHA superpolynomials $\mathcal{H}^{\lambda}(q, t, a)$ for the corresponding links colored by the sequences $\{\lambda\} = \{c_1\omega_1, c_2\omega_1, \ldots, c_\kappa\omega_1\}$ (only pure rows). In particular, they are topological invariants. The conjecture is well-checked.

4.4 Some Examples

Let us begin with $\mathcal{J}_0$ for $\mathcal{R} = \mathbb{F}[[z^4, z^6 + z^7]]$ with $K = Cab(13, 2)Cab(2, 3)$ discussed above. All cells are affine spaces and we show only dim =dim $J_0(\Delta)$ in the table below for the corresponding sets of gaps $D \stackrel{\text{def}}{=\!=} \Delta \setminus \Gamma$ for standard Δ. One has $dev(D) = |D|$ and $\dim O/M = \delta - |D|$, which gives the power of t. Two standard Δ from 25 have no standard M, namely, for $D = [2, 15]$ and $D = [2, 11, 15]$. The table of D and the corresponding dimensions of the cells $\mathcal{J}_0(\Delta)$ is:

D-sets	dim	D-sets	dim
∅	8	1,3,5,7,9,11,15	2
15	7	2,7,11,15	6
11,15	6	2,9,15	7
7,11,15	6	2,9,11,15	6
9,15	7	2,7,9,11,15	5
9,11,15	5	2,3,7,9,11,15	4
7,9,11,15	4	2,5,9,11,15	5
3,7,9,11,15	4	2,5,7,9,11,15	3
5,9,11,15	5	2,3,5,7,9,11,15	1
5,7,9,11,15	3	1,2,5,7,9,11,15	3
3,5,7,9,11,15	2	1,2,3,5,7,9,11,15	0
1,5,7,9,11,15	4	2,15 and 2,11,15	∅

For instance, there are *three* cells of dimensions 7 in $\mathcal{J}_0$ (for $a = 0$). Namely, those with $D = [15], [9, 15], [2, 9, 15]$ and t^7, t^6, t^5. Generally, the number of cells of $\dim = \delta - 1$ is the multiplicity of singularity; it equals the coefficient of t for $q = 1, a = 0$ due to the superduality, which is for $\mathbb{Z}_+ \setminus \Delta = \{1\}, \{2\}, \{3\}$ in this example. Only $\{1\}$ results in $\dim = 1$ and qt in $\mathcal{H}^{mot}$.

We note the *reciprocity involution* of standard modules $\Delta \mapsto \Delta^\vee - \min\{\Delta^\vee\}$ for $\Delta^\vee = \Gamma \setminus \{(2\delta - 1) - D\}$, which preserves their *dim*. For instance, $[15] \mapsto [2, 9, 15]$ $[2, 9] \mapsto [2, 9]$.

In the case of $Cab(53, 2)Cab(13, 2)Cab(3, 2)$ discussed above and the corresponding ring $\mathcal{R} = \mathbb{F}[[z^8, z^{12} + z^{14} + z^{15}]]$, one has: $\mathcal{H}(q, t = 1, a = 0) = q^{42} + 7q^{41} + 24q^{40} + 56q^{39} + 104q^{38} + 166q^{37} + 236q^{36} + 306q^{35} + 370q^{34} + 424q^{33} + 465q^{32} + 492q^{31} + 507q^{30} + 510q^{29} + 504q^{28} + 488q^{27} + 466q^{26} + 437q^{25} + 406q^{24} + 370q^{23} + 335q^{22} + 298q^{21} + 264q^{20} + 230q^{19} + 199q^{18} + 168q^{17} + 143q^{16} + 118q^{15} + 97q^{14} + 78q^{13} + 63q^{12} + 48q^{11} + 38q^{10} + 28q^9 + 21q^8 + 15q^7 + 11q^6 + 7q^5 + 5q^4 + 3q^3 + 2q^2 + q + 1.$

Here $\delta = 42$, which corresponds to q^{42} (invertible modules). The coefficients of q^i are the Betty number b_{2i} (the odd ones vanish), and the Euler number $e(\mathcal{J}_0)$ is 8512 (which is for $a=0, t=1, q=1$).

The simplest $\mathcal{H}^{mot}$ is for trefoil $T(3, 2)$. Its singularity ring is $\mathcal{R} = \mathbb{F}_q[[z^2, z^3]]$ with $\Gamma = \mathbb{Z}_+ \setminus \{1\}$. There are no primes of bad reduction for this and any torus knots. The standard modules are $M_\lambda = (1 + \lambda z)$ (invertible ones) of $\dim \mathcal{O}/M = 1$ and $M = \mathcal{O}$, where there are *two* generators (dim=0). The standard flags for $\ell = 1$ are $\{M_0 = M_\lambda \subset M_1 = \mathcal{O}\}$; the dimension is $\dim \mathcal{O}/M_1 = 0$ for them. Thus, $\mathcal{H}^{mot} = 1$ (for $\mathcal{O}$) $+ qt$ (counting invertible modules) $+ aq$ (counting flags). This calculation is almost equally simple for $T(2p+1, 2)$.

Links. The simplest example is the uncolored Hopf (algebraic) 2-link, the one for $F(u, v) = uv$. We take $\mathcal{O} = \mathbb{F}_q[[e_1, e_2, z_1, z_2]]$ for the idempotents e_i such that $z_i e_j = \delta_{ij} z_i$, and $\mathcal{R} = \mathbb{F}_q[[e_1 + e_2, x = z_1, y = z_2]]$, where $e_1 + e_2$ is the unit 1 in $\mathcal{O}$. The standard modules are $\mathcal{M}_\circ = \mathcal{O}$ of q-rank 2 and $\dim(\mathcal{O}/\mathcal{M}) = 0$, and $\mathcal{M}_\alpha = (e_1 + \alpha e_2)\mathcal{R}$ for $0 \neq \alpha \in \mathbb{F}_q$ of q-rank 1 (invertibles) and $\dim(\mathcal{O}/\mathcal{M}) = 2$. The latter contain z_1, z_2: $(e_1 + \alpha e_2)x = \alpha z_1, (e_1 + \alpha e_2)y = \alpha z_2$. Thus, $\mathcal{H}^{mot} = (q - 1)t + (1 + aq)$.

Hopf links are quite interesting. Let us provide $\mathcal{H}^{mot}_{2,1,1}$ for the Hopf 3-link with the linking numbers $\{+1, +1, +1\}$ and the 1st component colored by $2\omega_1 = \square\square$. We take $\mathcal{R} = \mathbb{F}_q[[1, x, y]]$ for $x = \zeta_1 + \zeta_2 + \zeta_3 + \zeta_4, y = \zeta_1 + \zeta_2 - \zeta_3 + c\zeta_4$, where $c \neq \pm 1$ and $p \neq 2$. The corresponding embeddings are:

$$\mathcal{R} \subset \mathcal{O} = \mathbb{F}_q[\epsilon_1 + \epsilon_2, \epsilon_3, \epsilon_4, \zeta_1 + \zeta_2, \zeta_3, \zeta_4]] \subset \Omega = \mathbb{F}_q[[\epsilon_i, \zeta_i, 1 \le i \le 4]].$$

One has: $\mathcal{H}^{mot}_{2,1,1} = 1 + a^2q^5 - 2t + q^2t + q^3t + t^2 - 2q^2t^2 + q^4t^2 + q^2t^3 - q^3t^3 - q^4t^3 + q^5t^3 + a\big(q^2 + q^3 - 2q^2t + q^4t + q^5t + q^2t^2 - q^3t^2 - q^4t^2 + q^5t^2\big) = (q - 1)^2(1 + q)qt^2(1 + qt) + \big((q - 1)t(2 + 2q + q^2 + (q^2 - 1)t)\big)(1 + aq^2) + (1 + aq^2)(1 + aq^3)$.

5 Zetas for Singularities

5.1 Generalizing Galkin's Zeta

V.M. Galkin studied in 1973 zeta- and L-functions for Gorenstein rings in dimension one. Plane curve singularities are an important particular case. Let us consider the unibranch case: when $\mathcal{R} \subset \mathcal{O} = \mathbb{F}_q[[z]]$ with 2 generators and the localization $\mathbb{F}_q((z))$. Recall that $\delta = \dim_{\mathbb{F}_q} \mathcal{O}/\mathcal{R} = |\mathbb{Z}_+ \setminus \Gamma|$. He and Stöhr (in 1998) considered links; adding a to their definition and the case of arbitrary ranks closely follow what we did for the superpolynomials.

The *admissible flags* of ideals in $\mathcal{R}$ are $\overrightarrow{M} = \{M_0 \subset M_1 \subset \cdots \subset M_\ell \subset \mathcal{R}\}$ such that $\{z^{-m_0} M_i \subset \mathcal{O}\}$ for $m_0 = \mathrm{Min}(\Delta_0)$ are *standard flags* in $\mathcal{O}$ as in Sect. 4.2. The *flagged zeta-function* is:

$$Z(q,t,a) \stackrel{\mathbf{def}}{=\!=} \sum_{\overrightarrow{M}} a^\ell t^{\dim(\mathcal{R}/M_\ell)} = \sum_{M} t^{\dim(\mathcal{R}/M)} (1+aq) \cdots (1+aq^{rk_q(M)-1}),$$

where the summation is over all admissible flags $\overrightarrow{M} \subset \mathcal{R}$ and over all ideals $M \subset \mathcal{R}$ in the 2nd formula; dim=$\dim_{\mathbb{F}_q}$, $rk_q(M) = \dim M/\mathfrak{m}M$.

The 2nd formula becomes with $\prod_{i=rk_{min}}^{rk_q(M)-1}(1+aq^i)$ in the non-unibranch case, similar to that for the motivic superpolynomials; rk_{min} is the maximal multiplicity of irreducible factors in the factorization of the corresponding $F(u,v)$. The interpretation in terms of the standard flags, which is the 1st formula, becomes somewhat technical in the non-unibranch case.

The *flagged L-function* is then $L(q,t,a) \stackrel{\mathbf{def}}{=\!=} (1-t)Z(q,t,a)$; it is a polynomial in terms of t, a, and $t^{-\delta} L(q,t,a)$ is invariant under $t \mapsto 1/(qt)$, which is the *functional equation*; we note that $Z(q,t,a)$ does not satisfy the latter. In contrast to the smooth case, the Riemann hypothesis holds only for sufficiently small q (if we know that the dependence on q is polynomial).

The definition of the Galkin zeta, which is $Z(q,t,a)$ for $a = 0$ (no flags), is sufficiently standard: a Dirichlet series. The functional equation for L is actually surprising because there is no Poincaré duality for singular varieties (unless intersection cohomology is used or so). Stöhr found a short entirely combinatorial proof of this fact, a significant simplification of that from the John Tate's thesis. Tate's p-adic proof works well for curve singularities.

The key here is the following property of Γ, which is actually the defining property of *Gorenstein rings*: the map $g \mapsto g' = 2\delta - 1 - g$ identifies $\{g \in \mathbb{Z}_+ \setminus \Gamma\}$ (the set of "gaps") with $\{g' \in \Gamma \setminus \{2\delta + \mathbb{Z}_+\}\}$. For instance, the last gap, which is $2\delta - 1$, maps to $g' = 0$.

Let $\mathbf{H}^{mot}(q,t,a) \stackrel{\mathbf{def}}{=\!=} \mathcal{H}^{mot}(qt,t,a)$ for motivic $\mathcal{H}^{mot}$ above: we switch to $q_{new} = q/t$ as we did for the DAHA superpolynomials when defining $\mathbf{H}(q,t,a)$. The $\mathbf{H}^{mot} \leftrightarrow L$ *coincidence conjecture* reads:

$$\mathbf{H}^{mot}(q,t,a)=L(q,t,a) \text{ for any plane curve rings } \mathcal{R}\subset\mathcal{O},$$

$$\mathbf{H}^{mot}(q,t,a=-1/q)=L_{\text{prncpl}}(q,t) \text{ for Gorenstein } \mathcal{R}\subset\mathcal{O}.$$

The latter is the *Zúñiga zeta-function*: for $a = 0$ and when the summation is only over *principle* $M \subset \mathcal{R}$. It coincides with $L(q,t,a = -1/q)$, indeed; see the alternative formula for flagged L provided above. The conjecture can be extended to the non-unibranch case and to any ranks (in the DAHA setting, any colors $m\omega_1$).

When $q \to 1$ (for the "field" with 1 element): $Z_{\text{prncpl}} = \sum_{\nu\in\Gamma} t^{\nu}$ and $L_{\text{prncpl}} = (1-t)Z_{\text{prncpl}}$ is the Alexander polynomial $Al^{\circ}(t)$. Namely,

$$\lim_{q\to 1} L_{\text{prncpl}} = (1-t)\Big(\sum_{i=1}^{\delta} t^{g_i}\Big) + t^{2\delta} \text{ for } \{g_i\} = \Gamma\setminus(2\delta+\mathbb{Z}_+).$$

We obtain that $\big(L_{\text{prncpl}} - t^{2\delta}\big)/(1-t)$ becomes δ when $q \to 1$ and $t = 1$.

The conjectural coincidence of $\mathbf{H}(q,t,a)$ (DAHA), $\mathbf{H}^{mot}(q,t,a)$ and $L(q,t,a)$ identifies the superduality for the former with the functional equation for the later. Recall that the conjectural upper RH-bound for $\mathbf{H}(q,t,a=0)$ is $q \le 1/2$ in the unibranch uncolored case, which is far from "arithmetic" $q = p^m$ for $L(q,t,0)$.

The coincidence $\mathbf{H}^{mot}(q,t,a)$ and $L(q,t,a)$ is simple for $t = 1$ (for any rings $\mathcal{R}\subset\mathcal{O}$, not only Gorenstein ones). Indeed, any admissible flag of ideals $\overrightarrow{M}'\subset\mathcal{R}$ is $z^{m'}\overrightarrow{M}$ for standard $\overrightarrow{M}\subset\mathcal{O}$, where m', $\overrightarrow{M}$ are uniquely determined by $\overrightarrow{M}'$. Vice versa, given a standard flag $\overrightarrow{M}$, let:

$$\{m \mid z^m\overrightarrow{M}\subset\mathcal{R}\} = \{0\le m_1 < m_2 < \cdots < m_k < 2\delta\}\cup\{2\delta+\mathbb{Z}_+\}$$

for some $k = k_M$ and $\{m_i\}$. The contribution of $z^m\overrightarrow{M} \subset \mathcal{R}$ for such m to $L(q,t,a) = (1-t)Z(q,t,a)$ is $(1-t)t^{m-dev(M_\ell)}$. Recall: $dev(M) = \delta - \dim\mathcal{O}/M$. Thus, all such $z^m\overrightarrow{M}$ contribute $(1-t)t^{-dev}\big(\sum_{i=1}^{k_M} t^{m_i} + t^{2\delta}/(1-t)\big)$. This will be 1 for $t \to 1$ (and any q). We obtain that

$$L(q,t=1,a) = \sum_{\ell=0}^{2\delta-1} |\mathcal{J}_\ell(\mathbb{F}_q)|a^{\ell},$$

which is $\mathbf{H}^{mot}(q,t=1,a)$.

Using Hilbert schemes. For a *rational* projective curve $C\subset\mathbb{P}^2$, the following identity is a natural object for physicists and mathematicians (Gopakumar-Vafa and Pandharipande-Thomas):

$$\sum_{n\ge 0} q^{n+1-\delta} e(C^{[n]}) = \sum_{0\le i\le\delta} n_C(i)\Big(\frac{q}{(1-q)^2}\Big)^{i+1-\delta},$$

for the Euler numbers of *Hilbert schemes* $C^{[n]}$. The points of the latter are zero cycles of C, collections of ideals at any points, of the (total) colength n. Here δ is the arithmetic genus of C; $n_C(i)$ are some numbers. The passage from a series to a polynomial is far from obvious even in this relatively simple case. It is significantly more subtle to prove that $n_C(i)\in\mathbb{Z}_+$ (Göttsche and then Shende for all i); the usage of versal deformations of singularities appeared necessary in the Shende's proof.

Switching to local rings $\mathcal{R}$ of singularities, the following conjecture is for *nested* Hilbert schemes $Hilb^{[l\le l+m]}$ formed by pairs of ideals $\mathfrak{m}I' \subset I \subset I' \subset \mathcal{R}$ of colengths $l, l+m$ for the maximal ideal $\mathfrak{m} \subset \mathcal{R}$. One needs the *weight t-polynomial* $\mathfrak{w}(Hilb^{[l\le l+m]})$ defined for the weight filtration of $Hilb^{[l\le l+m]}$ due to Serre and Deligne. The *Oblomkov-Rasmussen-Shende conjecture* (2012) states that

$$\sum_{l,m\ge 0} q^{2l}\, a^{2m}\, t^{m^2}\, \mathfrak{w}(Hilb^{[l\le l+m]})$$

is proportional to the Poincaré series of the HOMFLY-PT triply graded homology of the corresponding link. The connection with the perverse filtration of $\mathcal{J}_0$ is due to Maulik-Yun and Migliorini-Shende. The ORS series is a geometric variant of $Z(q,t,a)$.

The ORS conjecture adds t to the *Oblomkov-Shende conjecture*, which was extended by adding colors λ and then proved by Maulik.

The passage from series to polynomials required superpolynomials. The corresponding topological ones are *reduced* KhR polynomials. The Cherednik-Danilenko conjecture was that they coincide with the uncolored DAHA $\mathcal{H}(q,t,a)$. The passage to polynomials is a nontrivial step. This is manifested for the DAHA and motivic superpolynomials, though it is a conjecture that the motivic ones for plane curve singularity depend on q polynomially.

5.2 The Passage to Links

Let us provide the definition of L-functions for *non-unibranch* plane curve singularities colored by rows. Without the colors and for $a=0$, such L are due to Galkin (1973) and Stöhr (1998). The colored case, considered below, is preliminary; further work is required.

Recall that $\mathcal{H}^{mot}_{\sigma}$ was defined for $\sigma=(c_1\ge c_2\ge\cdots\ge c_\kappa>0)$ and

$$\mathcal{R}=\mathbb{F}_q[[x,y]]\subset \mathcal{O}=\mathbb{F}_q[[z,e_1,\ldots,e_\kappa]]\subset \Omega=\mathbb{F}_q[[z,\epsilon_1,\ldots,\epsilon_\tau]],$$

for $\tau=\sum_{i=1}^{\kappa} c_i$, $\epsilon_i\epsilon_j=\epsilon_i\delta_{ij}$, $e_1=\epsilon_1+\cdots+\epsilon_{c_1}$, $e_2=\epsilon_{c_1+1}+\cdots+\epsilon_{c_1+c_2}$, etc. Accordingly, $z_i=ze_i(1\le i\le\kappa)$ and $\zeta_i=z\epsilon_i(1\le i\le\tau)$. Let $\varpi_{\mathsf{m}}=\sum_{i=1}^{\kappa}\epsilon_{m_i}$ for all sequences $\mathsf{m}=(m_1,m_2,\ldots,m_\kappa)$ such that $0\le m_1<c_1$, $0\le m_2<c_2$, $\tilde{\mathcal{R}}=\mathbb{F}_q[[x\varpi_{\mathsf{m}},y\varpi_{\mathsf{m}}]]$, $\tilde{\mathfrak{m}}=(x\varpi_{\mathsf{m}},y\varpi_{\mathsf{m}})$, and $\tilde{\Omega}=\oplus_{\mathsf{m}}\mathcal{R}\varpi_{\mathsf{m}}=\tilde{\mathfrak{m}}\oplus\sum_{\mathsf{m}}\mathbb{F}_q\varpi_{\mathsf{m}}$.

Note that $\tilde{\Omega} = \tilde{\mathcal{R}} = \mathcal{R}$ in the uncolored case. For $rk_q(M) = \dim(M/\tilde{\mathfrak{m}}M)$, $\dim = \dim_{\mathbb{F}_q}$, we set:

$$Z_\sigma(q,t,a) = \sum_M t^{\dim(\tilde{\Omega}/M)} \prod_{j=c_1}^{rk_q(M)-1} (1+aq^j),\ L_\sigma(q,t,a) = (1-t)^\tau Z(q,t,a),$$

where $M \subset \tilde{\Omega}$ are $\tilde{\mathcal{R}}$-invariant such that $rk_q(M_i) \geq c_i$ for $M_i = M\,e_i$.

Setting $\mathbf{H}_\sigma(q,t,a) = \mathcal{H}_\sigma^{mot}(qt,t,a)$ (as above), we conjecture that

$$\mathbf{H}_\sigma(q,t,a) = L_\sigma(q,t,a),$$

at least in the uncolored case, when $\sigma = (1,\ldots,1)$.

Let $\sigma = (2,1)$ for the Hopf 2-link. Then, $\mathcal{R} = \mathbb{F}_q[[x = \zeta_1 + \zeta_2, y = \zeta_3]]$, $\tilde{\mathcal{R}} = \mathbb{F}_q[[\zeta_j]]$ and $\tilde{\Omega} = \mathbb{F}_q(\epsilon_1+\epsilon_3) \oplus \mathbb{F}_q(\epsilon_2+\epsilon_3) \oplus (\zeta_j)$. It is easy to see that

$$\mathcal{H}_\sigma^{mot} = (1+aq^2) + (q^2-1)t;$$

the corresponding L is as follows.

The modules $M \subset \tilde{\Omega}$ of $rk_q(M) = 2$ are those generated over $\tilde{\mathcal{R}}$ by $\{\epsilon_1+\epsilon_3, \epsilon_2+\epsilon_3\}$ (containing $\tilde{\mathfrak{m}}$), $\{\epsilon_1+\epsilon_3, \zeta_2^u\}$, $\{\epsilon_2+\epsilon_3, \zeta_1^u\}$, and $\{\zeta_1^u + \alpha\zeta_3^w, \zeta_2^v + \beta\zeta_3^w\}$, where $u, v, w \geq 1$, $\alpha, \beta \in \mathbb{F}_q$ such that $\alpha \neq 0$ or $\beta \neq 0$.

Their contribution to $Z_\sigma(q,t,a)$ will be $\frac{(q^2-1)t^3}{(1-t)^3} + \frac{1+t}{(1-t)}$. The submodules of q-rank 3 are those generated by $\{\zeta_1^u, \zeta_2^v, \zeta_3^w\}$, which contribute $(1+aq^2)\frac{t^2}{(1-t)^3}$. Combining, $L_\sigma(q,t,a) = (q^2-1)t^3 + (1+t)(1-t)^2 + t^2(1+aq^2) = q^2t^3 + 1 - t + t^2aq^2 = (1+aq^2t^2) + (q^2t^2-1)t$.

More generally, let $\sigma = (m,1)$. Then

$$\mathcal{H}_\sigma^{mot} = (1+q^m a) + (q^m-1)t, \quad \mathbf{H}_\sigma^{mot} = (1+q^m t^m a) + (q^m t^m - 1)t,$$

which coincides with L_σ.

Double trefoil. Let us provide the motivic uncolored superpolynomial for $T(6,4)$ and comment on it and its L-counterpart. One has: $\mathcal{H}_{6,4}^{mot}(q,t,a) = q^8t^8 - q^7t^8 + q^3t^2(q + t - 2qt + q^2t - 2q^2t^2 + 2q^3t^2 - q^2t^3 + q^4t^3 - q^2t^4 + q^4t^4 - q^3t^5 + q^4t^5)(1+aq) + q^2t(1 + q - t + q^2t - t^2 - qt^2 + q^2t^2 + q^3t^2 - qt^3 + q^3t^3 - q^2t^4 + q^3t^4)(1+aq)(1+aq^2) + (1 - t + qt - qt^2 + q^2t^2)(1+aq)(1+aq^2)(1+aq^3)$.

It coincides with $\mathcal{H}_{6,4}^{daha}$ in formula (6.4) of Sect. 6.2 of "DAHA approach to iterated torus links" (Ch, Danilenko, 2015); the notation there was $\hat{\mathcal{H}}^{min}$. Also, it coincides with the reduced (uncolored) Khovanov-Rozansky polynomial defined and obtained via *Soergel modules*, which is from Example 1.3 in "Torus link homology" (Hogancamp-Mellit, 2019). Their expression is $\mathcal{H}_{6,4}^{daha}(q \mapsto 1/t, t \mapsto q, a)$ multiplied by $\frac{1+a}{(1-q)^2}$.

Here

$$\mathcal{R} = \mathbb{F}_q[[x = z_1^2 - z_2^2, y = z_1^3 + z_2^3]] \subset \mathcal{O} = \mathbb{F}_q[[e_1, e_2, z_1, z_2]];$$

recall that all our rings contain $1 = \sum_i e_i$. A natural basis in $\mathcal{O}/\mathcal{R}$ is formed by (the images of) $\{e_1, z_1, z_2, z_1^2, z_1^3, z_1^4, z_1^5, z_1^7\}$. We use different signs in x and y to have $z_{1,2}^6$ in $\mathcal{R}$; so are all $z_{1,2}^m$ for $m \geq 8$. Thus, $\dim_{\mathbb{F}_q} \mathcal{O}/\mathcal{R} = 8$ and the contribution of any *invertible modules* M (those with $rk_q = 1$) to $\mathcal{H}^{mot}$ is $(q^8 - q^7)t^8$. The difference $(q^8 - q^7)$ here is because the cyclic generator of M must contain $e_1 + \alpha e_2 + \ldots$ for $\alpha \neq 0$ (it must be standard).

Concerning large modules, $rk_q(\mathcal{O}) = 4$. There are *two* more families of standard modules with $rk_q = 4$: M'_α generated by $e_1 + \alpha e_2, z_1, z_2, z_1^2$ and $M''_{\alpha,\beta}$ generated by $e_1 + \alpha e_2 + \beta z_1, z_1 - \alpha z_2, z_1^2, z_2^2$, where $\alpha \neq 0$. They give

$$\big(1 + (q-1)t + q(q-1)t^2\big)(1 + aq)(1 + aq^2)(1 + aq^3) \in \mathcal{H}^{mot}.$$

Let us discuss a bit the corresponding contributions to the L-function. The *conductor* of $\mathcal{R}$ is $\mathcal{C} = z^8\mathcal{O}$, the greatest $\mathcal{O}$-submodule in $\mathcal{R}$. It has *four* generators, $z_{1,2}^8$ and $z_{1,2}^9$, over $\mathcal{R}$ and $\dim_{\mathbb{F}_q} \mathcal{R}/\mathcal{C} = 6$. Thus, the contribution of ideals $M_{i,j} = z_1^i\mathcal{C} + z_2^j\mathcal{C}$ for $i, j \geq 0$ to Z is $t^6(1+aq)(1+aq^2)(1+aq^3)/(1-t)^2$. The other two families are based on z^8M' and z^8M'' of q-rank 4 with the generators. They contribute $t^6(1 + t(q-1) + t^2(q^2-q)(1+aq)(1+aq^2)(1+aq^3)$ to L. The expected general formula is: $L_{6,4}(q,t,a) = \mathcal{H}_{6,4}^{mot}(qt,t,a) = 1 - 2t + t^2 + q^3t^4 - 2q^3t^5 + q^3t^6 + q^4t^6 - 2q^4t^7 + q^4t^8 + q^5t^8 - 2q^5t^9 + q^5t^{10} + q^6t^{10} - 2q^6t^{11} + q^7t^{12} + q^6t^{14} - 2q^7t^{15} + q^8t^{16} + t\big(1 - t + qt - t^2 - qt^2 + q^2t^2 + t^3 - qt^3 - 2q^2t^4 + 2q^3t^4 + 2qt^5 - q^3t^5 + q^4t^5 - qt^6 - 3q^3t^6 + q^4t^6 + 2q^2t^7 + q^3t^7 - q^4t^7 + q^5t^7 - q^2t^8 - 3q^4t^8 + q^5t^8 + 2q^3t^9 - q^5t^9 + q^6t^9 - 2q^5t^{10} + 2q^6t^{10} + q^4t^{11} - q^5t^{11} - q^5t^{12} - q^6t^{12} + q^7t^{12} - q^6t^{13} + q^7t^{13} + q^7t^{14}\big)(1+aq) + t^3\big(1 - t + qt + q^2t^2 - t^3 - 2qt^3 + t^4 + 2q^3t^4 - aq^4t^4 + qt^5 - 3q^2t^5 - q^3t^5 + 2aq^4t^5 - aq^5t^5 + q^2t^6 + 2q^4t^6 - aq^4t^6 - 2q^3t^7 + aq^5t^7 + q^5t^8 - q^4t^9 + q^5t^9 + q^5t^{10}\big)(1+aq)(1+aq^2) + t^6\big(1 - t + qt - qt^2 + q^2t^2\big)(1+aq)(1+aq^2)(1+aq^3)$.

The coincidence of $L_{6,4}$ with $\mathbf{H}_{6,4}$ was verified only partially. Recall that $L(q,t,-1/q)$ is conjecturally the *Zúñiga* L-function (principle ideals only).

5.3 *Quasi-Rho-Invariants*

The $\rho_{ab}-$ invariant is the von Neumann invariant defined for the abelianization representation $\pi_1(S^3 \setminus K) \to \mathbb{Z}$. We will define a superpolynomial for a certain *integer* variant of ρ_{ab} for algebraic knots K. The superduality

$$q^\delta t^{2\delta}\mathbf{H}_K(q, \frac{1}{qt}, a) = \mathbf{H}_K(q,t,a)$$

and the connection conjecture for $a \to -1/q$ will be used; the dependence on the knot K will be shown. We set:

$$R_K(q,t,a) \stackrel{\text{def}}{=} \big(\mathbf{H}_K(q,t,a) - t^{\delta}\mathbf{H}_K(q,t=1,a)\big)/\big((1-qt)(1-t)\big),$$

$$\rho_K(q,t) \stackrel{\text{def}}{=} R_K(q,t,a=-1/q), \text{ where } \mathbf{H}_K(q,1,a=-1/q) = q^{\delta}.$$

Switching to L_{prncpl}, our $\rho_K(q,t)$ is a sum of monic $q^i t^j$ (of multiplicity one); see the next section for the exact formula. The superduality holds

$$q^{\delta-1}t^{2\delta-2}R_K(q,\frac{1}{qt},a) = R_K(q,t,a);$$

the same holds for ρ_K.

Let us express $\rho_K(1,1)=\rho_K(q=1,t=1)$ in terms of $\Gamma=\nu_z(\mathcal{R}\setminus\{0\})$. We set

$$G \stackrel{\text{def}}{=} \mathbb{Z}_+ \setminus \Gamma = \mathbb{Z}_+ \cap S,$$

for $S = \cup_{i=1}^{\varpi}[g_i, g_i'+1]$, a disjoint union of segments, where $g_i \le g_i' \in G \not\ni g_i'+1$. Then $\delta = \sum_{i=1}^{\varpi} m_i$ for $m_i \stackrel{\text{def}}{=} g_i' - g_i + 1$. This is actually for any Gorenstein $\mathcal{R} \subset \mathbb{C}[[z]]$. Setting $\varsigma(x) = x$ for $x \in S$ and 0 otherwise,

$$\rho_K(1,1) = \textstyle\sum_{i=1}^{\varpi} m_i(g_i'+1-\frac{m_i}{2}) - \frac{\delta^2}{2} = \int_0^{\infty}\varsigma(x)dx - \frac{\delta^2}{2}.$$

Geometrically, $\rho_{ab} = \int_0^1 \sigma_K(e^{2\pi i x})dx$ for the *Tristram-Levine signature* σ_K; see, e.g., "Signatures of iterated torus knots" (Litherland, 1979). Our ς is some variant of σ_K; taking $\int_0^1$ in the formula for ρ_{ab}, makes ρ_{ab} "additive" for iterated knots (see below).

Quasi-rho for cables. For $r, s > 0$ such that $\gcd(r,s) = 1$, one has: $\rho_{r,s}(1,1) = \frac{(r^2-1)(s^2-1)}{24}$. We note that this is the maximal size of the (r,s)-core partitions. The classical ρ_{ab} is $-\frac{1}{3}\frac{(r^2-1)(s^2-1)}{rs}$. Thus, we basically obtain the same formula up to some renormalization. The values of our ρ are always natural numbers; so it can be used for *categorification*. This is not the case with ρ_{ab} due to its geometric origin. Our ρ is expected to have some natural geometric interpretation too.

The classical ρ_{ab} for cables is known to be additive. For instance let $K = Cab(m,n)Cab(s,r)$. Then

$$\rho_{ab} = -\frac{1}{3}\Big(\frac{(m^2-1)(n^2-1)}{mn} + \frac{(r^2-1)(s^2-1)}{rs}\Big).$$

This deviates from our ρ, which is additive with some weights.

Let $\mathcal{R} = \mathbb{F}[[z^{\upsilon r}, z^{\upsilon s} + z^{\upsilon s+p}]]$, where, $\gcd(r,s) = 1$ as above, $\upsilon > 1$ and $\gcd(\upsilon, p) = 1$ for $p \ge 1$. Then

$$\Gamma = \langle \upsilon r,\ \upsilon s,\ \upsilon rs+p\rangle 2\delta = \upsilon^2 rs - \upsilon(r+s) + (\upsilon-1)p + 1$$

and $K = Cab(m\!=\!vrs\!+\!p,\ n\!=\!v)\,Cab(s,r)$. One obtains:

$$\rho_K(1,1) = \frac{1}{24}\big((m^2-1)(n^2-1) + v^2(r^2-1)(s^2-1)\big).$$

More generally,

$$\rho_K(1,1) = \frac{1}{24}\sum_{i=1}^{k} v_i^2(a_i^2-1)(r_i^2-1)$$

for the cable

$$K = Cab(a_k, r_k)\cdots Cab(a_2, r_2)Cab(a_1, r_1),$$

where $1 \le i \le k$, $v_i = r_k \cdots r_{i+1}$ and $v_k = 1$. We will post the details elsewhere. Here $v_i = \gcd(u_1, \cdots, u_{i+1})$ for $\Gamma = \langle u_1, \cdots, u_{k+1}\rangle$, where $u_i < u_{i+1}$ and $v_{i+1}|v_i$; it is known that $\delta = \frac{1}{2}\sum_{i=1}^{k} v_i(a_i-1)(r_i-1)$.

We note the following natural embedding for $K' = Cab(a,r)K$ (if both are algebraic knots). If δ is that for the ring $\mathcal{R}$ of K, then $\rho_K(q,t)$ is the sum of monomials $q^i t^j$ in $\rho_{K'}(q,t)$ such that $j < 2\delta - 1$.

The case of Cab(13,2)Cab(2,3). Here $\mathcal{R}\!=\!\mathbb{C}[[z^4, z^6\!+\!z^7]]$, $r\!=\!3$, $s\!=\!2$, $v\!=\!2$, $\delta\!=\!8$. Then $\rho(1,1) = 25$ and its refined version is $\rho(q,t) = 1 + qt + q^2t^2 + q^3t^3 + q^3t^4 + q^4t^4 + q^4t^5 + q^5t^5 + q^4t^6 + q^5t^6 + q^6t^6 + q^5t^7 + q^6t^7 + q^7t^7 + q^5t^8 + q^6t^8 + q^7t^8 + q^6t^9 + q^7t^9 + q^6t^{10} + q^7t^{10} + q^7t^{11} + q^7t^{12} + q^7t^{13} + q^7t^{14}$.

RH holds for $\rho(q,t)$ when $q < q_{sup} \approx 0.802$. Presumably, $\lim_{p\to\infty} q_{sup} = 1$ for $Cab(2p\!+\!13, 2)Cab(2,3)$; for instance, $q_{sup}\approx 0.996$ for $p\!=\!2000$.

Let us provide now full $R(q,t,a)$ for $\mathbb{C}[[z^4, z^6\!+\!z^7]]$. It is:

$1 + t + qt + t^2 + 2qt^2 + q^2t^2 + t^3 + 2qt^3 + 3q^2t^3 + q^3t^3 + t^4 + 2qt^4 + 4q^2t^4 + 4q^3t^4 + q^4t^4 + t^5 + 2qt^5 + 4q^2t^5 + 6q^3t^5 + 4q^4t^5 + q^5t^5 + t^6 + 2qt^6 + 4q^2t^6 + 7q^3t^6 + 8q^4t^6 + 4q^5t^6 + q^6t^6 + t^7 + 2qt^7 + 4q^2t^7 + 7q^3t^7 + 10q^4t^7 + 8q^5t^7 + 4q^6t^7 + q^7t^7 + qt^8 + 2q^2t^8 + 4q^3t^8 + 7q^4t^8 + 8q^5t^8 + 4q^6t^8 + q^7t^8 + q^2t^9 + 2q^3t^9 + 4q^4t^9 + 6q^5t^9 + 4q^6t^9 + q^7t^9 + q^3t^{10} + 2q^4t^{10} + 4q^5t^{10} + 4q^6t^{10} + q^7t^{10} + q^4t^{11} + 2q^5t^{11} + 3q^6t^{11} + q^7t^{11} + q^5t^{12} + 2q^6t^{12} + q^7t^{12} + q^6t^{13} + q^7t^{13} + q^7t^{14} + a\Big(qt + qt^2 + 2q^2t^2 + qt^3 + 3q^2t^3 + 3q^3t^3 + qt^4 + 3q^2t^4 + 6q^3t^4 + 3q^4t^4 + qt^5 + 3q^2t^5 + 7q^3t^5 + 9q^4t^5 + 3q^5t^5 + qt^6 + 3q^2t^6 + 7q^3t^6 + 12q^4t^6 + 10q^5t^6 + 3q^6t^6 + qt^7 + 3q^2t^7 + 7q^3t^7 + 13q^4t^7 + 17q^5t^7 + 10q^6t^7 + 3q^7t^7 + q^2t^8 + 3q^3t^8 + 7q^4t^8 + 12q^5t^8 + 10q^6t^8 + 3q^7t^8 + q^3t^9 + 3q^4t^9 + 7q^5t^9 + 9q^6t^9 + 3q^7t^9 + q^4t^{10} + 3q^5t^{10} + 6q^6t^{10} + 3q^7t^{10} + q^5t^{11} + 3q^6t^{11} + 3q^7t^{11} + q^6t^{12} + 2q^7t^{12} + q^7t^{13}\Big) + a^2\Big(q^3t^3 + q^3t^4 + 2q^4t^4 + q^3t^5 + 3q^4t^5 + 3q^5t^5 + q^3t^6 + 3q^4t^6 + 6q^5t^6 + 3q^6t^6 + q^3t^7 + 3q^4t^7 + 7q^5t^7 + 8q^6t^7 + 3q^7t^7 + q^4t^8 + 3q^5t^8 + 6q^6t^8 + 3q^7t^8 + q^5t^9 + 3q^6t^9 + 3q^7t^9 + q^6t^{10} + 2q^7t^{10} + q^7t^{11}\Big) + a^3\Big(q^6t^6 + q^6t^7 + q^7t^7 + q^7t^8\Big)$.

Recall that $R(q,t,a) \mapsto \rho(q,t)$ upon the substitution $a \mapsto -\frac{1}{q}$ in the parameters of $\mathbf{H}(q,t,a)$, which is generally the passage to the Heegaard-Floer homology and Alexander polynomials (when $q=1, a=-1$).

6 On Physics Connections

Generally, a challenge is to associate the Riemann and Lindelöf hypotheses with some physics phenomena in SCFT or similar theories. SCFT is connected with quite a few recent mathematical developments. DAHA can be considered as its part; their origin was in the Knizhnik-Zamolodchikov equations. DAHA superpolynomials can be interpreted as some physics partition functions, those for knot operators.

The p-*adic strings* due to Witten and others must be mentioned in this context. The starting point of this theory was an adelic product formula for the *Veneziano amplitude*. In mathematics, we have a long history of understanding "geometry" via p-adic constructions and those over $\mathbb{F}_q$, including recent Peter Scholze theory (say, in his lectures at IHES, 2023–2024). The figures below are about such and similar connections. For instance, we expect that L-functions of plane curve singularities over $\mathbb{F}_q$ can be related to some Dirac operators; this can be connected with p-adic strings.

The *Lee-Yang circle theorems* provide a different perspective. The Ising model with an external magnetic field is the key example.

6.1 Lee-Yang Theorem

For any lattice (of any dimension) with N vertices and the connected pairs of vertices denoted by $\langle n, n'\rangle$, let $\mathcal{Z} = \lim_{N\to\infty} \frac{\log(Z_N)}{N}$ for the partition function $Z_N = \sum_{\{\sigma_n\}} e^{-\beta\mathcal{H}}$, where the Hamiltonian is

$$\mathcal{H} = -\sum_{\langle n,n'\rangle} J_{n,n'}\sigma_n\sigma_{n'} - H\sum_n \sigma_n$$

and $\sigma = \pm 1$. This is the Ising model with an external magnetic field H. Here $\beta = (k_B T)^{-1}$ is the inverse temperature for the Boltzmann constant k_B. Assuming that $J_{n,n'} \geq 0$ (the ferromagnetic case) and $\beta > 0$, Lee-Yang proved that the zeros of Z_N in terms of the "complex fugacity" $\mu = e^{-2\beta H}$ belong to the unit circle $|\mu| = 1$; the corresponding symmetry of Z_N is simply $\sigma \mapsto -\sigma$. For the square lattice with $J = const > 0$, Z_N is a polynomial in terms of μ and $0 < u \stackrel{\textbf{def}}{=\!=} e^{-4\beta J} < 1$. There is a q-version of this theorem and other physics-statistical variants.

The Lee-Yang-Fisher zeros are when u is considered as a free parameter for complex T. Numerical experiments showed that $|\mu| = 1$ for the μ-zeros can hold for some $u < 0$. The physics calculations are mostly when μ is fixed and the u-zeros are considered, but they can be used for the μ-zeros too. This phenomenon resembles the behavior of the t-zeros of our $\mathbf{H}(q,t,a)$. Actually, DAHA is directly related to the XXZ-model, which is somewhat similar to the Ising model with H as above, though all attempts to "integrate" the latter failed.

Only $\mathcal{Z}$ is physical; its phase transitions are positive *real* limits as $N \to \infty$ of (complex) μ-zeros of Z_N. Thus, these zeros can result in a phase transition only at $\mu = 1$ due to RH for Z_N, the point of which is the intersection of the unit μ-circle with $\mathbb{R}_+$. The relation between the failure of RH and "unwanted" phase transitions seems sufficiently general. Given $u < 0$, the μ-zeros of Z_N quickly become wild near the real line when N goes beyond N_u, the last N when RH still holds for $u < 0$. So do the corresponding points of phase transition. This can be clearly seen in the 1D Ising model.

One-dimensional case. The zeros of Z_N can be found explicitly in this case. One has:

$$-\beta\mathcal{H} = \beta\sum_{n=1}^{N} J\sigma_n\sigma_{n+1} + \frac{\beta}{2}H(\sigma_n + \sigma_{n+1}),$$

where the periodicity $\sigma_{N+1}=\sigma_1$ is assumed. Then $Z_N = \sum_{\{\sigma_n\}} e^{-\beta\mathcal{H}}$ can be calculated using the eigenvalues λ_1, λ_2 of the *transfer matrix* $\mathcal{T} = \begin{pmatrix} e^{\beta(J+H)} & e^{-\beta J} \\ e^{-\beta J} & e^{\beta(J-H)} \end{pmatrix}$. Namely, $Z_N = \operatorname{tr}(\mathcal{T}^N) = \lambda_1^N + \lambda_2^N$ for

$$\lambda_{1,2} = e^{\beta J}\cosh(\beta H) \pm \sqrt{e^{2\beta J}\sinh^2(\beta H) + e^{-2\beta J}}.$$

Upon some algebraic manipulations, the μ-zeros of Z_N are for $H = \imath\,\theta_n/\beta$, where $\cos(\theta_n) = \sqrt{1-u}\,\cos\frac{(2n-1)\pi}{2N}$ for $n = 1, \ldots, N$ and $u = e^{-4\beta J}$ as above. We obtain that RH, which is the condition $\theta_n \in \mathbb{R}$ for any n, holds if $\sqrt{1-u}\,\cos\frac{\pi}{2N} \le 1$, i.e., for $u \ge -\tan^2\left(\frac{\pi}{2N}\right)$. Thus, RH can hold for negative u, but the latter bound tends to 0 when $N \to \infty$.

This example provides some physical insight. A counterpart of N is our $\deg_t \mathcal{H}^\lambda$, which is $\delta\sum_i m_i^2$ for $\lambda = (m_i)$. For instance, $\deg_t \mathcal{H}^\lambda = \delta n$ as $a = 0$ for n-columns λ. We note that the distribution of μ-zeros above resembles our one for $\mathbf{H}(q,t,a=0)$ in the case of uncolored $T(2,2p+1)$.

6.2 Landau-Ginzburg Models

Another physics approach to singularities is presented in paper "Catastrophes and the classification of conformal theories" (Vafa-Warner, 1989). The authors consider LGSM, Landau-Ginzburg sigma models, for *superpotentials* $W(x, y)$ corresponding to isolated singularities. They can be with several variables, and more than one superpotential W can be considered. Let us mention here two publications by Alexander Zamolodchikov in 1986. We mention that there are many other classes of superpotentials, for instance those for quiver varieties and KZ. The correspondence between SCFT and LGSM is one of the key in string theory.

A lot of information can be obtained directly in terms of $W(x, y)$ and the corresponding singularities. If only the topological class of singularity matters, then this is some "topological LGSM." We note that analytic parameters of singularities can be interpreted as *modes* in physics, but motivic theory captures (by now) only topological types.

For instance, the *Milnor number* $\mu = 2\delta$, which is $(r-1)(s-1)$ for $W_{r,s}(x, y) = x^s - y^r$, coincides with the number of (independent) chiral operators or superfields. It is the *Witten index* for plane curve singularities: the number of zero energy bosonic vacuum states minus the number of zero energy fermionic vacuum states. The dimensions of superfields for $W_{r,s}(x, y)$ are proportional to the corresponding quasi-homogeneity weights, which are $1/s$ for x, $1/r$ for y and so on.

Another example is the central charge, which is $c = 6\beta$ for $\beta = (\frac{1}{2} - \frac{1}{r})(\frac{1}{2} - \frac{1}{s})$ for $W_{r,s}(x, y)$. Generally, β is obtained from the asymptotic formula

$$\int e^{\imath\lambda W(x_1,\ldots,x_m)} \prod_{i=1}^{m} \lambda^{1/2} dx_i \sim O(\lambda^{\beta}),$$

for large λ. Also, the adjacency of singularities plays an important role in this approach.

6.3 Refined Witten Index

Refined Witten and BPS indices were studied in the literature; see, e.g., Gaiotto-Moore-Neitzke (Adv. Theor. Math. Phys. 2013). Generally, the challenge is to "split" the vacuum states counted by these indices using some additional parameters.

Let us split $\mu = 2\delta$ using $\mathcal{H}^{mot}(q, t, a = -t/q)$, though we will actually use L_{prncpl} in the following calculation. We begin with

$$\delta_{q,t} \stackrel{\text{def}}{=\!=} \frac{\mathcal{H}^{mot}(q, t, a = -t/q) - (qt)^{\delta}}{1 - t} = \frac{L_{pncpl}(\frac{q}{t}, t) - (qt)^{\delta}}{1 - t}.$$

This formula is for any Gorenstein $\mathcal{R} \subset \mathbb{C}[[z]]$. The DAHA parameters q, t from $\mathcal{H}$ are used (not q_{new} from the definition of $\mathbf{H}$ and L).

As above: $G = \mathbb{Z}_+ \setminus \Gamma = \cup_{i=1}^{\varpi} \{g_i \le x \le g_i'\}$, where $g_i'+1 \in \Gamma$, and $m_i = g_i' - g_i + 1$. Then

$$\delta_{q,t} = \frac{1-t^{g_1}}{1-t} + \sum_{i=1}^{\varpi-1} \frac{t^{g_i'+1} - t^{g_{i+1}}}{1-t} \Big(\frac{q}{t}\Big)^{m_1+\cdots+m_i},$$

and $\delta_{1,1} = \delta$. This formula was actually used above in the definition of the refined quasi-rho invariant $\rho_K(q,t)$; also, see below.

Let

$$\mu_{q,t} \overset{\mathbf{def}}{=\!=} \delta_{q,t} + (qt)^{\delta-1}\delta_{t^{-1},q^{-1}} = \sum_{x=0}^{2\delta-1} t^{v(x)-1} q^{g(x)},$$

where $v(x) = |\{v \in \Gamma \mid 0 \le v \le x\}|$ and $g(x) = |\{g \in G \mid 0 \le g < x\}|$ for G as above. Then $\mu_{1,1} = \mu$ and this definition ensures the superduality:

$$(qt)^{\delta-1}\mu(1/t, 1/q) = \mu(q,t).$$

Not all monomials are monic in $\mu(q,t)$: ϖ of them are with coefficient 2, which correspond to $x \in \Gamma \not\ni x+1$.

For example,

$$\mu_{q,t} = 2 + q + q^2 + 2q^3 t + 2q^4 t^2 + 2q^5 t^3 + 2q^6 t^4 + q^7 t^5 + q^7 t^6 + 2q^7 t^7$$

for $\mathcal{R} = \mathbb{C}[[z^4, z^6+z^7]]$. Upon $q \mapsto qt$, it satisfies RH for $0 < q < 0.919090$. For $\mathcal{R} = \mathbb{C}[[z^6, z^9 + z^{460}]]$, this range becomes $0 < q < 0.852561$. For $\mathcal{R} = \mathbb{C}[[z^6, z^8 + z^{649}]]$, it is $0 < q < 0.846566$ and 0.848063 for $z^8 + z^{3003}$.

Compare with the formula for

$$\varrho(q,t) \overset{\mathbf{def}}{=\!=} \frac{\mathcal{H}(q,t,a=-t/q) - q^\delta}{(1-t)(1-q)} = \rho\Big(\frac{q}{t}, t\Big):$$

$$\varrho(q,t) = \sum_{x \in G} q^{g(x)} \frac{1-t^{v(x)}}{1-t} = \sum_{G \ni x > y \in \Gamma} q^{g(x)} t^{v(y)-1} = (qt)^{\delta-1} \varrho(t^{-1}, q^{-1}).$$

Recall that we use the DAHA parameters q, t from $\mathcal{H}$ in this section.

Adding colors. The substitution $a \mapsto -\frac{t}{q}$ has remarkable properties for $\mathcal{H}^\lambda(q,t,a)$ for partitions λ more general than $\square$ (the uncolored case). Let n be the number of rows of λ and m the number of its columns.

For *hooks* λ, we expect that

$$\mathcal{H}^{\lambda}(q,t,a\mapsto -\frac{t}{q}) = (qt^{n-1})^{\delta(n-1)} r^{\lambda}(q,t)$$

for $r^{\lambda}(q,t) = r(q \mapsto t^{n-1}q^m, t \mapsto q^{m-1}t^n)$. Here $r(q,t) = \mathcal{H}(q,t,a\mapsto -\frac{t}{q})$ is for $\lambda = \square$ considered above; its constant term is 1. For instance, $1 - r^{\lambda}(q,t)$ is divisible by $(1-q^m)$ for $n=1$. This was checked for pure columns/rows and several hooks with $m=2$ or $n=2$. For example,

$$(1-r^{\square\square}(q,t))/(1-q^2) = qt(1+q^2+q^4+q^7t+q^{10}t^2+q^{13}t^3+q^{16}t^4+q^{21}t^7),$$

for $K = Cab(13,2)Cab(2,3)$.

This is far from being that simple beyond the hooks. For instance, for $T(3,2)$ and $\lambda = 2\times 2 = \boxplus$: $1 - r^{2\times 2}_{3,2}(q,t) = q(1-qt)(1+q-q^2+t-q^2t+q^4t+q^3t^2-q^5t^2+q^2t^3-q^4t^3+q^5t^4+q^6t^4+q^4t^5-q^6t^5-q^6t^6+q^6t^7)$.

The definition of the $\varrho^{\lambda}(q,t)$ for *symmetric* λ can follow the uncolored case, but this is preliminary. For instance, one can set:

$$\varrho^{2\times 2}_K(q,t) \stackrel{\mathbf{def}}{=\!=} \big(r^{2\times 2}_K(q,t) - q^{4\delta}t^{4\delta}\big)/(1-qt)^2.$$

For the example above: $\varrho^{2\times 2}_{3,2}(q,t) = 1-q-q^2+q^3+qt-q^2t+q^4t-q^5t+2q^2t^2-q^3t^2-q^4t^2+q^5t^2+2q^3t^3-q^4t^3+q^6t^3+2q^4t^4-q^5t^4-q^6t^4+q^5t^5-q^6t^5+q^6t^6$, which satisfies the superduality $q \leftrightarrow t^{-1}$ with the multiplier q^6t^6. Presumably, $\varrho^{2\times 2}_K(1,1) = 4\varrho_K(1,1)$ for algebraic knots K, but there are negative terms in $\varrho^{2\times 2}_{3,2}(q,t)$. Recall that we omit λ in uncolored $\mathcal{H}_K, \varrho_K$ and so on, i.e., for $\lambda = \square$.

6.4 S-Duality

The relation SCFT $\leftrightsquigarrow$ LGSM suggests that the S-*duality* in the former can be seen via the superpotential $W(x,y)$. The superduality of physics superpolynomials can be connected with that in M-*theory* and the symmetry $\epsilon_1 \leftrightarrow \epsilon_2$ in *Nekrasov's instanton sums*. The general physics superduality (with λ) for superpolynomials was considered by Gukov-Stosic (2012) (and in some prior works). For us, this correspondence is between DAHA, a "representative" of SCFT, and L-functions of singularities, which presumably "represent" LGSM.

More specifically, we focus on the DAHA superpolynomials, which are certain partition functions of knot operators and are related to the BPS states. They can be expected to coincide with motivic superpolynomials (a solid mathematical conjecture), which are presumably some partition functions of properly defined *motivic* LGSM for plane curve singularities. Then, switching to the L-functions of the latter, the S-duality from SCFT becomes the Hasse-Weil functional equation, with some potential toward various generalizations.

This link may be not too much surprising. The S-duality and mirror symmetry (CPT) are very universal in physics. The functional equation is certainly of the same caliber in mathematics. There are more than 20 different zeta theories; RH does not always hold. By analogy with the Lee-Yang theorem, one can speculate that (topological) LGSM associated with plane curve singularities is "stable" when the "coupling constant" q is small enough to ensure RH for $\mathbf{H}(q, t, a)$. More generally, can the failure of RH be somehow connected with the presence of unwanted phase transitions? Something like this: "elementary particles" for $W_{2,2p+1} = x^{2p+1} - y^2$ can be "observed" for any $q > 0$, but this requires $q < 1/2$ or so for arbitrary superpotentials $W(x, y)$.

We took the bound $1/2$ from *strong RH* for $a = 0$; generally, we can make a a constant or any quantity invariant under the superduality. For instance, let $a = -t/q$, which is super-invariant. Then we arrive at

$$\varrho_{2,2p+1}(q,t) = \frac{q^p - tq^{p+1} - t(qt)^p + (qt)^{p+1} + t - 1}{(q-1)(t-1)(qt-1)} = \sum_{0 \le j \le i < p} q^i t^j,$$

and RH holds for $\rho_{2,2p+1}(q,t) = \varrho_{2,2p+1}(qt,t)$ only for sufficiently small q (not all).

Let me finish this section with little something on Manin's "Mathematics as metaphor". Yu.I. obviously expected number theory to play a major role in the alliance of physics and mathematics. If RH has something to do with the absence of unwanted phase transitions in physics theories or their stability of any kind, then number theory will not be just a "metaphor." Technically, DAHA accumulated quite a few integrable models and the fact that it appeared very "motivic" can be meaningful physically. I thank my friends-physicists for various talks on these matters (though they are not responsible for what I wrote).

7 Zeta-Functions as Invariants

7.1 The First Figure

Modern mathematics and physics very much rely (more than ever) on the progress in geometry. Any new geometric approaches to the classical zeta and L-functions can open new avenues toward the justification of "grand conjectures." This of course does not diminish the role of analytic methods, including classical Fourier analysis and (more recently) the p-adic methods. The Dwork proof of the rationality of zeta-functions is an example of the latter.

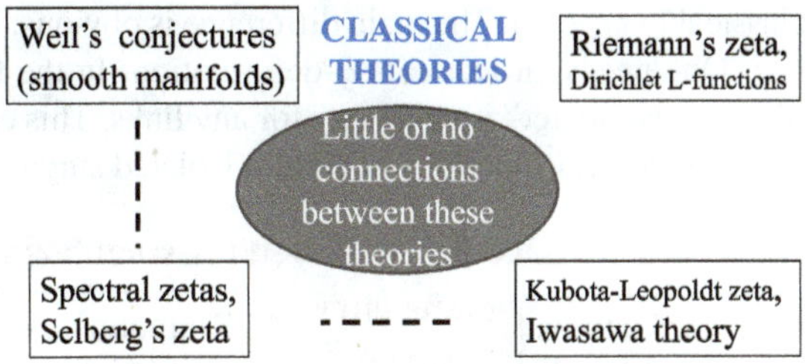

For instance, Dirichlet L-functions have no counterparts among Weil's L-functions (and they have no q) : two different universes. Also, ***zeta-equivalence*** of algebraic varieties over C (N. Katz) generally results only in the coincidence of their Hodge numbers.

In contrast to the Weil conjectures (proved by Deligne), the distribution of the zeros of Riemann's $\zeta(s)$ in the critical strip does not seem to reflect any "geometry." Riemann's zeta does occur in some geometric and physics considerations, but the interpretation of the grand conjectures geometrically or physically is (totally?) missing. We note here that the zeros of Selberg's zeta-functions are much more "geometric" and have many analytic applications. The zeros of Hasse-Weil zetas are very geometric too, which was the key for the justification of the corresponding RH in special cases and in general (Deligne).

Nicholas Katz proved that the zeta-equivalence of algebraic varieties X results in the coincidence of their virtual Hodge numbers. One needs to add the coefficients of the equations of X to the corresponding rings of functions and then consider the zeta-function ζ_X of the resulting scheme over $\mathbb{Z}$. If $\zeta_X = \zeta_Y$, then we call X and Y zeta-equivalent. The coincidence of Hodge numbers is of course very far from the existence of any kind of isomorphism between X and Y.

Generally, *Kapranov's (motivic) zetas* are with the coefficients in the Grothendieck ring $K_0(Var/\mathbb{F})$ of varieties over $\mathbb{F}$. The map $X \mapsto |X(\mathbb{F}_q)|$ is a *motivic measures*. Conjecturally, our motivic superpolynomials can be lifted to $K_0(Var/\mathbb{F})$. Recall that we conjecture that the Piontkowski cells are configurations of affine spaces. Let me mention Manin's "Lectures on zeta functions and motives" (1995).

In the first figure, the following four theories are presented as disconnected blocks:

(a) the Hasse-Weil zetas (over $\mathbb{F}_q$),
(b) the Selberg's zetas (via the Laplace operators),
(c) the Kubota-Leopoldt zetas,
(d) Riemann's $\zeta(s)$ and the Dirichlet L-functions.

They are really different, but not totally disconnected; there are some deep relations. For instance, the connection (not in this figure) of the Selberg's *"1/4 conjecture"* for arithmetic subgroups $\Gamma \subset SL(2, \mathbb{R})$ to the Riemann's zeta is such.

The upper-left block is connected with the upper-right one: global fields and the product formulas are here and there. However, this does not help much to understand the zeros of the classical $\zeta(s)$, $L(s)$. The main difference is obvious: Riemann's $\zeta(s)$ does not contain q. One needs, at least, its q-deformation. In the theory over $\mathbb{F}_q$, making q a "continuous parameter" is necessary for any links. This can be hopefully achieved via the passage to zeta-functions of certain isolated singularities.

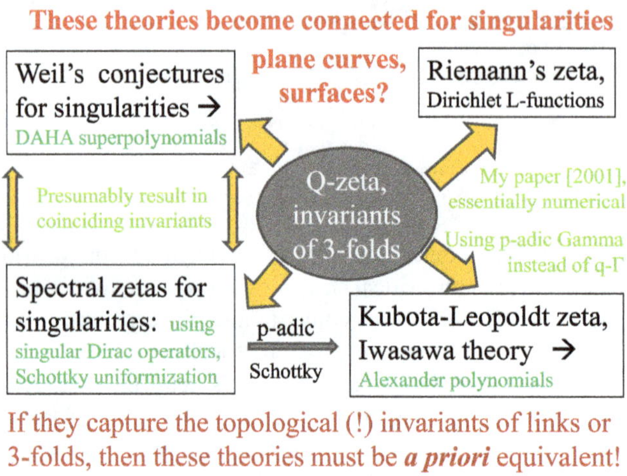

7.2 The Second Figure

Upon the switch to isolated singularities $\mathcal{X}$, the corresponding zetas can be expected to capture the topological type for some "good" $\mathcal{X}$, which conjecturally holds for *any* plane curve singularities. The "Hasse-Weil block" of the 2nd figure, the upper-left one, is discussed in this note for plane curve singularities. Let us comment a bit on the "spectral block."

Algebraically, we study the covers of $\mathbb{P}^2$ ramified at $\mathcal{X}$. This is closely related to the *Schottky uniformization* in the variant due to Tate-Mumford. Generally, it is for "relative curves" over $\mathbb{C}[[z]]$ or $\mathbb{Z}_p$ (2-dimensional schemes), with a singular rational curve as the closed fiber. Basically, the monoidal transformations are performed making the closed fiber a divisor with normal crossings. We need this for plane curve singularities, as closed fibers, and their deformations.

More analytically, one can consider the *versal deformations* of $\mathcal{X}$ and the invariants of the zeta-function of the *Dirac operator* upon the corresponding *Schottky uniformization*. Let us provide one reference: "Zeta functions that hear the shape of a Riemann surface" (Cornelissen-Marcolli, 2008). This is in the smooth case, but Schottky uniformization is compatible with the passage to singularities. For plane curve singularities $\mathcal{X}$, the corresponding Selberg-type zetas are expected to be related to the superpolynomials of $\mathcal{X}$.

Let me mention the p-adic uniformization of modular curves associated with division algebras over totally real number fields (my PhD thesis). It was used in the ε*-conjecture* and Ribet's theorem (in the proof of Last Fermat Theorem). See, for instance, recent "On the p-adic uniformization of quaternionic Shimura curves" (Boutot-Zink, 2022). This is an example of the role the p-adic block can play, which also shows the potential of the Schottky uniformization.

The middle oval. The DAHA approach provides a natural way of adding q to the Riemann's $\zeta(s)$ and Dirichlet L-functions, though without the symmetry $s \leftrightarrow 1-s$. The corresponding q-deformed zeros seem more regular at $0<q<1$. If true, then the stochastic behavior of the classical ones can be due to $q \to 1$.

I defined such q-analogs in "On q-analogues of Riemann's zeta function" (2001) for A_1. The starting point was the q-Mehta-Macdonald formula, which is a product formula in terms of q-Gamma functions for $\int_{\imath\mathbb{R}^n} \gamma(x)\,\mu(q^x)dx$ for the Gaussian $\gamma(x) = q^{-x^2/2}$ and the measure-function

$$\mu(q^x) = \prod_{\alpha, j\geq 0} \frac{(1-q^{(x,\alpha)+\nu_\alpha j})\,(1-q^{-(x,\alpha)+\nu_\alpha (j+1)})}{(1-t_\alpha q^{(x,\alpha)+\nu_\alpha j})(1-t_\alpha q^{(x,\alpha)+\nu_\alpha (j+1)})},$$

making the E-polynomials pairwise orthogonal. Here α are positive roots of a given reduced irreducible root system $R \subset \mathbb{R}^n$, normalized by the condition $(\alpha_{sht}, \alpha_{sht}) = 2$; we set $\nu_\alpha = 1$ for short α, $= 3$ for G_2, and 2 otherwise. Let $t_{sht} = q^{k_{sht}}$ and $t_{lng} = q^{k_{lng}}$; t_α depends only on $|\alpha|$. Also, let $\rho_k = \frac{1}{2}\sum_{\alpha>0} k_\alpha \alpha$ and $\mu_1 = \mu/\mathrm{CT}(\mu)$ for the constant term functional CT for Laurent series in terms of X_b (μ_1 is such). From now on, we assume that $0<q<1$ and $t_\alpha > 1$.

The link to the DAHA superpolynomials is due to the theorem that the integral $\int_{\imath\mathbb{R}^n} f(x)\gamma(x)\,\mu(q^x)dx$ is proportional to the *coinvariant* of $f(x)$, which is $f(x = -\rho_k)$ in suitable spaces of functions. Here the integration can be replaced by taking CT; see below. We note that the integration can be over $\mathbb{R}^n$ in this formula: γ must be replaced by $\gamma^{-1} = q^{x^2/2}$ and the proportionality factor changes (significantly).

Let

$$Z_n^+(q,t) = \frac{\int_{\varepsilon+\imath R^n} \gamma(x)/\,(1+\gamma(x))\;\mu(q^x)dx}{\int_{\varepsilon+\imath R^n} \gamma(x)\,\mu(q^x)dx}.$$

Due to the *Stirling-Moak formula* (Moak, 1984):

$$\lim_{q\to 1_-} Z_n^+(q,t) = \eta(s) \stackrel{\mathbf{def}}{=\!=} (1-2^{1-s})\zeta(s)$$

for the Riemann's $\zeta(s)$, where $s = k_{sht}|R_+^{sht}| + k_{lng}|R_+^{lng}| + \frac{n}{2}$. The usage of ε improves the range of k_ν where Z_n^+ is analytic. The basic range is $\Re k_\nu > 0$, which is for $\varepsilon = 0$. If $k_{lng} = k = k_{sht}$ and $\varepsilon = \rho/h$ for the Coxeter number, then $Z_n^+(q,t)$ is analytic for $\Re k > -1/h$, which corresponds to $\Re s > 0$, i.e., we cover the critical strip $0 < s < 1$ in the limit. By the analytic continuation, the convergence to $\eta(s)$ holds for any $s \in \mathbb{C}$. The analytic continuation is essentially by the procedure of "picking up residues" due to Weyl-Arthur-Heckman-Opdam.

The case of A_n. Setting $\upsilon = n+1$ and $\upsilon^\circ = -k\upsilon$, we obtain:

$$s = k\frac{\upsilon(\upsilon-1)}{2} + \frac{\upsilon-1}{2} = \frac{1}{2}(\upsilon-1)(k\upsilon^\circ+1) = -\frac{1}{2}(\upsilon-1)(\upsilon^\circ-1).$$

The integral

$$\mathcal{I}_n^+ = \int_{\varepsilon+\imath R^n} \frac{\gamma(x)}{1+\gamma(x)}\,\mu(q^x)dx,$$

for $\varepsilon = \rho/\upsilon$, is an analytic function for $\Re k > -1/\upsilon$ and, accordingly, for $s > -\frac{1}{\upsilon}\frac{\upsilon(\upsilon-1)}{2} + \frac{\upsilon-1}{2} = 0$. Generally, $\mathcal{I}_R^+$ defined by the same formula for a root system R is analytic if:

(a) $k_{sht} > \max\{-(\varepsilon,\alpha_i), (\varepsilon,\theta_{sht})-1\}$, where α_i are short simple roots and θ_{sht} is the maximal short root in R_+,
(b) k_{lng} satisfies the analogous inequalities for long instead of short.

We conjecture in type A that there exists a meromorphic function $\mathcal{Z}(q,t,\mathsf{a})$ in terms of q,t,a such that $\mathcal{Z}(t^{-1},q^{-1},\mathsf{a}) = \mathcal{Z}(q,t,\mathsf{a})$ and $\eta(s)$ is the limit $q \to 1_-$ of $\mathcal{Z}(q,t=q^k,\mathsf{a}=t^\upsilon)$. Note the usage of $\mathsf{a}=-a$ instead of a in superpolynomials.

The superduality becomes $k \mapsto 1/k$, $\upsilon \to -k\upsilon = \upsilon^\circ$ in terms of k,υ. The corresponding s remains fixed under this symmetry (it must!). However, we have a nontrivial connection between the values of $\mathcal{Z}$ at k and $1/k$ in the q-theory. For instance, the a-coefficients of $\mathcal{Z}$ are (conditionally) bounded as $|k|\to\infty$ and $\Re s > 0$, or its values for super-invariant a, which is a variant of the *Lindelöf hypothesis*.

One can expect similar features for classical root systems modulo the stabilization and superduality conjectures. For C_n in the case $t_{\rm lng} = t_{sht}$, briefly discussed above, the *hyperpolynomials* are conjectured to depend polynomially on $q, t^{\pm1}, \mathsf{a}$, where the passage to C_n is $\mathsf{a}=t^{2n}$ and the superduality is $q \leftrightarrow t^{-1}, \mathsf{a}\mapsto \mathsf{a}$, as for the A-series.

In this case: $k_{sht} = k = 2k_{lng}$, $\upsilon = n$, and $s = k\big(\upsilon(\upsilon-1)+\upsilon/2\big)+\upsilon/2$. The superduality becomes $k \mapsto 1/k$, $\upsilon \mapsto -k\upsilon$, i.e., the same as for A; however, s is different for C. As for A, this s is fixed under the superduality:

$$s \mapsto \frac{1}{k}(k\upsilon)(\frac{1}{2}+k\upsilon) - k\frac{\upsilon}{2} = k(\upsilon^2-\frac{\upsilon}{2}) + \frac{\upsilon}{2} = s.$$

DAHA vertex. The rationale for the existence of $\mathcal{Z}(q,t,\mathsf{a})$ is the following theorem. Let $\mathbb{J}_m = \mathrm{CT}\big(\Theta^m(q^x)\,\mu_1\big)$ for

$$\Theta(q^x) \stackrel{\mathbf{def}}{=\!=} \sum_{b\in P} q^{b^2/2+(x,b)}.$$

The latter is $\gamma(x)$ "presented" as a Laurent series in terms of $X_b = q^{(x,b)}$; its defining property is the P-periodicity of $\gamma(x)^{-1}\Theta(q^x)$. This series naturally occurs when we switch from $\int_{\varepsilon+\imath\mathbb{R}^n}\{\cdots\}\mu(q^x)dx$ to the integration over the periods of

$\mu(q^x)$. For instance, $\int_{\varepsilon+\imath R^n}\gamma(x)\,\mu_1(q^x)dx$ coincides with $\mathbb{J}_1 = \mathrm{CT}\big(\Theta\mu_1(q^x)\big)$ up to a simple factor. To see this, replace $\gamma(x)$ by $\sum_{b\in Q}\gamma(x+2\pi\imath\,\log(q)\,b)$ for the *root lattice* Q, use the functional equation for Θ, and then switch to CT.

Given $m \geq 1$, the claim is that $\mathbb{H}_m(q,t,\mathsf{a})$ exists such that

$$\mathbb{H}_m(q,t,\mathsf{a}=t^{n+1}) = \mathbb{J}_m/\mathbb{J}_1^m,$$

for any A_n, and the superduality holds:

$$\mathbb{H}_m(t^{-1},q^{-1},\mathsf{a}) = \mathbb{H}_m(q,t,\mathsf{a})$$

(without any q,t-factors).

Technically, this theorem follows from the a-stabilization and super-invariance of $\frac{q^{b^2/2+k(\rho,b)}}{\langle\mathcal{P}_b,\mathcal{P}_b\rangle_1}$ and those for $\mathcal{P}_b(q^{-c-k\rho})$ for any $b,c\in P_+$, where we set:

$$\mathcal{P}_b(X) \stackrel{\mathbf{def}}{=\!=} P_b(X)/P_b(t^\rho)$$

for the Macdonald polynomials P_b, and

$$\langle f,g\rangle_1 \stackrel{\mathbf{def}}{=\!=} \mathrm{CT}\big(f(X)g(X^{-1})\mu_1(q^x)\big).$$

For instance, let $b = \begin{smallmatrix}\square\square\\ \square\end{smallmatrix} = \omega_1+\omega_2,\ c = \square = \omega_1$. Then

$$\mathcal{P}_b(q^{-c-k\rho}) = 1 + \frac{(q^{-1}-1)(1-t)}{1-\mathsf{a}}\big(1+q^{-1}+t\big),$$

which is, indeed, super-invariant, as well as:

$$\frac{q^{b^2/2+(b,\rho_k)}}{\langle\mathcal{P}_b,\mathcal{P}_b\rangle_1} = \Big(\frac{q}{t}\Big)^{5/2}\mathsf{a}^{-3/2}(1-\mathsf{a})\frac{(1-t^{-1}\mathsf{a})(1-q\mathsf{a})(1-t^{-1}q\mathsf{a})(1-t^{-1}q^2\mathsf{a})(1-t^2q\mathsf{a})}{(1-q)^2(1-t^{-1})^2(1-qt^{-2})(1-q^2t^{-1})}.$$

These quantities are the key in the theory of ***DAHA vertex*** due to the author and Danilenko. Namely, the series $\mathrm{CT}\big(\Theta^m(q^x)\mu_1(q^x)\big)$ can be expressed in terms of products of them using the expansion

$$\Theta(q^x) = \sum_{b\in P_+}\frac{q^{b^2/2+k(b,\rho)}}{\langle\mathcal{P}_b,\mathcal{P}_b\rangle_1}\mathcal{P}_b(X)\mathrm{CT}\big(\Theta\mu_1\big).$$

Here the formula for the proportionality coefficient is

$$\mathrm{CT}\big(\Theta\mu_1\big) = \prod_{\alpha\in R_+}\prod_{j=1}^{\infty}\Big(\frac{1-t_\alpha^{-1}q_\alpha^{(\rho_k,\alpha^\vee)+j}}{1-q_\alpha^{(\rho_k,\alpha^\vee)+j}}\Big),$$

which is the q-Mehta-Macdonald identity. It follows from the theorem that the symmetric form $\langle f, g\,\Theta(q^x)\,\rangle_1$ corresponds to a *Shapovalov-type* anti-involution of $\mathcal{HH}$ and is unique such up to proportionality (for this anti-involution).

When $t = 0$, our $\mathbb{J}_m$ become generalized *Rogers-Ramanujan* series. Actually, Θ-functions "with characteristics" are needed for them, where $X_b \mapsto \xi(b)X_b$ for characters $\xi : P \to \mathbb{C}^*$, which are via $P/Q = \mathbb{Z}_{n+1}$ for A_n. See our paper with Boris Feigin (2012). The case $t = 0$ is incompatible with the super-invariance, but we obtain "instead" the modularity of the resulting q-series.

Without going int detail, $\mathbb{H}_m$ can be interpreted as invariants of the *Lens spaces* $L(m, 1)$; they are sums of colored superpolynomials for the m-chains of consecutive unknots with linking numbers -1 between the neighboring ones. Not much is published in this direction. The necessary DAHA theory is still in progress. Let me mention "BPS spectra and 3-manifold invariants" (Gukov-Du Pei-Putrov-Vafa, 2017).

The next level is the passage to q-zeta: when we consider special generating functions for the family $\{\mathbb{H}_m\}$ and more general ones. Any "geometric" interpretation of this passage can be valuable. There are quite a few challenges, including the following one.

Higher theta-functions. Given m, the passage from the imaginary integration in

$$\int_{\varepsilon+\imath R^n} \frac{\gamma(x)}{1+\gamma(x)}\,\mu(q^x)dx = \int_{\varepsilon+\imath R^n} \sum_{m=1}^{\infty}(-1)^{m-1}\gamma^m(x)\,\mu(q^x)dx$$

to CT results in $\gamma^m(x) \mapsto \Theta(q^{mx})$. The latter function has the same multiplicator as $\Theta^m(q^x)$ upon the action of P and behaves as $\sim q^{-mx^2/2}$ when $q \to 1_-$ with a proper proportionality factor. It is quite different from $\Theta^m(q^x)$ for $m \geq 2$. To employ the theorem on super-invariance of the quantities above, one needs either a counterpart of $\mathbb{Z}(q, t, a)$ defined in terms of $\{\Theta^m(q^x)\}$, which seems doable, or the theory of *DAHA vertex* based on $\Theta(q^{mx})$ instead of $\Theta^m(q^x)$.

More generally, theta-functions of level m, those with the same multiplicator as for Θ^m, are in 1-1 correspondence with the *DAHA coinvariants* of level m. The latter are defined algebraically using the action of τ_+^m in DAHA. We note that finding relations between theta-functions of arbitrary levels m is a subtle algebraic problem even for A_1.

Further comments. The convergence of $Z_n^+(q,t)$ above will become $\zeta(s)$ instead of $\eta(s)$ when we switch to $Z_n^-(q,t)$ with $(1-\gamma(x))$ instead of $(1+\gamma(x))$. However, this will hold only for $\Re s > n$. It will diverge otherwise, which can be fixed upon multiplication by proper $\omega^\bullet$ for $q \stackrel{\mathbf{def}}{=\!=} e^{-1/\omega}$; the limit becomes then some Γ-type function (no zeta!). For instance, let

$$\mathcal{I}^-(k) = \frac{1}{i}\int_{1/2+\imath\mathbb{R}} \left(q^{x^2}-1\right)^{-1}\mu(q^x)dx,$$

for $n = 1$. Then $\left(\frac{\omega}{4}\right)^{k-1/2}\mathcal{I}^-(k)$ converges to $\Gamma(s)\zeta(s)$ for $\Re k > \frac{1}{2}$, where $s = k+\frac{1}{2}$. However, the limit of $\omega^{2k-1}\mathcal{I}^-(k)$ becomes $\tan(\pi k)\Gamma(k)^2$ when $\Re k < \frac{1}{2}$. Generally, $\omega^{s-n}\mathcal{I}^{\pm}(k)$ converges to $\zeta(s)$ or $\eta(s)$ times proper products of Γ-factors when $\Re s > n$ in the case of $\mathcal{I}^-$ and $\Re s > 0$ for $\mathcal{I}^+$. Using the analytic continuation, s can be arbitrary in the second case.

Interestingly, the q-zeros of $\mathcal{I}^-(k)$ with $\Re k \sim 0$ (not in the range $\Re k > \frac{1}{2}$) approach the classical ones for q sufficiently close to 1_- (but not too close!), and then $\mathcal{I}$ slowly begin to "switch" to the Gamma-limit as q continues to approach 1_-. This can be clearly seen numerically.

This can be potentially related to the *Gram law*. We change $\mathcal{I}^-(k)$ to $\widetilde{\mathcal{I}}^-(k)$ with the limit to the classical

$$\widetilde{\zeta}(s) = \pi^{-s/2}\Gamma(\frac{s}{2})\zeta(s),$$

invariant under $s \mapsto 1 - s$; this limit is only for $\Re s > 1/2$ and upon $\omega^\bullet$. Then the zeros of proper linear combination of $\mathcal{I}^-(k)$ and its complex conjugation approach those of $\zeta(k + 1/2)$ for $\Re k \sim 0$ and large ω (but not too large) and then eventually tend to properly adjusted *Gram points* as ω increases. This argument is of course qualitative. So is the Gram law: the zeros of $\zeta(1/2 + k)$ "mostly" alternate with the Gram points.

We note (again) that the invariance of $\widetilde{\zeta}(s)$ under $s \mapsto 1 - s$ fails for the q-deformations. However, a variant of "q-RH" can hold for A_1. This was discussed in my paper mostly for the *sharp q-zeta* introduced below. A qualitative version is that the tendency is strong for the *sharp q-deformations* of the classical zeros to stay in the half-plane $\Re s > 1/2$ (in certain horizontal strips depending on $q < 1$). The corresponding half-plane becomes $\Re s < 1/2$ for the imaginary integration.

A direct counterpart of RH can hold too. This requires the consideration of a proper linear combination of $\mathcal{I}^+(k)$ and $\mathcal{I}^+(-k)$. The q-zeros of the latter we were able to find (not too many) satisfy $\Re k = 0$. The case of the simplest *sharp* q-L-function is touched upon below.

Peter Sarnak noted once that many applications are based on the absence of zeta-zeros with $\frac{1}{2} < \Re s < 1$. This is, basically, what we see for "small" q-zeros. However, there was (and there is) uncertainty when the corresponding neighboring zeros of the classical $\zeta(s)$ are getting "too close." Namely, the linear approximations of sharp q-deformations of such "unusual" zeros of $\zeta(s)$ can be with $\Re s > 1/2$ (in the opposite half-plane). We think that the linear approximations can be irrelevant for such q-zeros; theoretically, the convergence of the corresponding Taylor expansions is not known. Let me quote Harold Edwards: "the existence of nearly coincident zeros must give pause to even the most convinced believer" (his "Riemann's Zeta Function").

7.3 Sharp q-Zeta

As above, $q = \exp(-1/\omega)$ for $\omega > 0$; let $\sigma \stackrel{\mathbf{def}}{=\!=} \sqrt{\pi\omega/2}$. The integration path will be now $\circlearrowleft_{\infty-\sigma i}^{\infty+\sigma i}$ around zero. For A_1 and

$$\delta_k(x;q) \stackrel{\mathbf{def}}{=\!=} \prod_{j=0}^{\infty} \frac{(1-q^{j+2x})(1-q^{j-2x})}{(1-q^{j+k+2x})(1-q^{j+k-2x})},$$

the symmetric variant of μ, the function

$$\mathfrak{Z}_q^{\sqsubset}(k) \stackrel{\mathbf{def}}{=\!=} \frac{1}{2i}\oint_{\infty-\sigma i}^{\infty+\sigma i} \frac{\delta_k(x;q)}{1+q^{-x^2}}\,dx$$

is analytic in the horizontal strip $K^\sharp = \{-2\sigma < \Im k < +2\sigma\}$ as $\Re k > -1/2$. Its meromorphic continuation to *all* $k \in \mathbb{C}$ via Cauchy's theorem, the *sharp q-zeta*, is:

$$\mathfrak{Z}_q^{\sharp}(k) = -\frac{\omega\pi}{2}\prod_{j=0}^{\infty}\frac{(1-q^{j+k})(1-q^{j-k})}{(1-q^{j+2k})(1-q^{j+1})}$$

$$\times\sum_{j=0}^{\infty}\frac{(1-q^{j+k})q^{-kj}}{(1-q^k)(q^{-\frac{(k+j)^2}{4}}+1)}\prod_{l=1}^{j}\frac{1-q^{l+2k-1}}{1-q^l}.$$

It has poles at $\{-\frac{1}{2}-\mathbb{Z}_+\}$ in $K^\sharp$. This strip is between the first zeros of $1+q^{-\frac{k^2}{4}}$. For all k apart from the poles,

$$\lim_{\omega\to\infty}\left(\frac{\omega}{4}\right)^{k-1/2}\mathfrak{Z}_q^{\sharp}(k) = \sin(\pi k)(1-2^{\frac{1}{2}-k})\Gamma(k+\frac{1}{2})\zeta(k+\frac{1}{2}).$$

Given a classical zero $k=z$ of $\zeta(1/2+k)$, let us assume that its $\sharp$-deformation $z^\sharp(\omega)$ exists and is differentiable with respect to $\varpi = 1/\omega$. Then the formula for its linear approximation $\tilde{z}^\sharp(\omega)$ is as follows:

$$\tilde{z}^\sharp(\omega) = z\Big(1-\frac{4(z+\frac{1}{2})\zeta_+(z+\frac{3}{2})-(z-1)\zeta_+(z-\frac{1}{2})}{12\omega\zeta'(z+\frac{1}{2})(1-2^{\frac{1}{2}-z})}\Big).$$

Thus, the linear ϖ-approximations of all classical zeros z exist if and only if all of them are simple, an interesting interpretation of the classical conjecture.

A similar approach can be employed for $\mathfrak{Z}_q^\sharp(k;d)$ with for q^{-dx^2} instead of q^{-x^2} in $\mathfrak{Z}_q^\sharp$ and for the *sharp L-functions* $\mathfrak{L}_q^\sharp(k;d)$ that are for

$$\frac{q^{x^2/2}-q^{-x^2/2}}{q^{(d+1)x^2/2}-q^{-(d+1)x^2/2}}.$$

The L-functions are somewhat simpler to analyze numerically; their first zeros are smaller than those of $\zeta(s)$. The usage of d has some practical advantages too.

Taking the classical $z = 14.1347i$ and $\omega = 750$ for $\mathfrak{Z}_q^\sharp(k;2)$:

$$z^\sharp = 0.1304 + 14.1450i,\ \tilde{z}^\sharp = 0.1302 + 14.1465i.$$

Other zeros in $K^\sharp$ for $\omega = 750$, $d = 2$ are:

zeta	*sharp − zeta*	*linear approx.*
$21.0220i$	$0.3514 + 21.0702i$	$0.3504 + 21.0771i$
$25.0109i$	$0.5641 + 24.9586i$	$0.5745 + 24.9643i$
$30.4249i$	$0.9046 + 30.4014i$	$0.9134 + 30.4077i$
$32.9351i$	$1.1051 + 33.0341i$	$1.0998 + 33.0854i$
$37.5862i$	$1.6449 + 37.9660i$	$1.7675 + 38.1895i$
$40.9187i$	$1.9080 + 40.8119i$	$1.9141 + 40.7816i$
$43.3271i$	$2.2860 + 43.2485i$	$2.4497 + 43.3138i$
$48.0052i$	$2.9259 + 47.8424i$	$3.1103 + 47.5578i$.

There is a clear tendency for $z^\sharp$ to move to the right. If true, this would give the classical RH. A direct q-counterpart of RH was the observation at the end of paper "On q-analogs ..." that several (not too many) "small" zeros of $\mathfrak{L}_q^\sharp(k;d) - \mathfrak{L}_q^\sharp(-k;d)$ were all with $\Re k = 0$. This is within the corresponding strips. The convergence is very good, including large ω, but the calculations become involved for large $\Im(k)$; this restricted our simulations. The passage from A_1 to A_n and the stabilization add *superduality* to this approach.

Our calculations indicate that the zeros of the q-deformed zeta-functions and L-functions become more "regular" for $q < 1$ than the corresponding classical zeros. It is expected that the zeros of Riemann zeta-function in the critical strip are totally random subject to the distribution for the eigenvalues of random Hermitian matrices (Dyson, Montgomery, Odlyzko, and others). This can be due to the limit $q \to 1_-$ of "relatively regular" q-zeros. Similar quasi-classical limits, in physics and mathematics, are known to create chaotic behavior.

Two bottom blocks. Concerning the "p-adic block," it is expected that there is a p-adic DAHA theory, where q-Gamma functions are replaced by their p-adic counterparts. This theory is important and doable. Technically, the cyclotomic q-Gauss-Selberg sums in the DAHA theory, where q are roots of unity, will be replaced by the classical *modular* Gauss-Selberg sums, those over $\mathbb{F}_q$. The q-

Mehta-Macdonald formulas is expected then to become in terms of the p-adic Gamma.

The "spectral block" has been partially discussed. The ρ_{ab}-invariants provide important links. Generally, $\rho_\xi(M)$ is the value of at $s = 0$ of $\tilde\eta_M(s) - n\,\eta_M(s)$ for odd-dimensional closed oriented Riemannian manifolds M, where $\eta_M(s)$ is due to Atiyah-Patodi-Singer and "twisted" $\tilde\eta_M(s)$ is defined for the flat bundle over M associated with the representation $\xi : \pi_1(M) \to U_n$. We take $M = S^3 \setminus K$ for a knot K and the abelianization of $\pi_1(M)$ as ξ. Thus, our $\rho_{q,t} = R_K(q,t,a{=}{-1}/q)$ from Sect. 5.3 can be considered as a q-version of $\tilde\eta_M(s)$. One can expect a "triply-graded" categorification of $R(q,t,a)$. Recall that they are defined in terms of $\mathcal{H}^{mot}$, which conjecturally coincide with superpolynomials in any other theories and, also, with the generalized L-functions of plane curve singularities. Thus, *quasi-rho invariants* connect the "Weil block" with the "spectral block" and then, potentially, with the "p-adic block" via the p-adic Schottky uniformization.

Further perspectives. Assuming that zeta-functions of singularities $\mathcal{X}$ and their corresponding a,q,t-versions are topological invariants/moduli of some sort, one can expect *a priori* links between the Hasse-Weil zetas, Selberg's zetas, and p-adic zetas for such $\mathcal{X}$. If this is true, then these zetas are different invariants of the same singularity and must be connected as such. This is expected in the case of plane curve singularities.

A program is to switch from plane curve singularities to *surface* singularities serving Seifert 3-folds and more general *plumbed manifolds*. An example is our q,t,a-deformation of the Dirichlet $\eta(s)$, essentially, $\sum_{m=1}^{\infty}(-1)^{m-1}\mathbb{H}_m(q,t,\mathsf{a})$, an alternating sum of certain invariants of Lens spaces $L(m,1)$. Thus, the invariants of the latter "replace" $\frac{1}{m^s}$ in the classical $\eta(s) = \sum_{m=1}^{\infty}(-1)^{m-1}\frac{1}{m^s}$ in our approach.

One can try to interpret such deformed sums geometrically: by looking for a manifold M such that its "triply-graded" homology reduces to the sums of those over its special submanifolds, Lens or Seifert spaces. In geometry, this can be due to the *localization* in certain (co)homology of M or via the count of *closed* totally geodesic submanifolds in M (with proper weights). Also, it is not rare when some invariants of fibered spaces M reduce to sums over proper *special* fibers. However, here we have *infinite* sums and quite involved homology theories.

We note that the superduality holds for any (convergent) series in the form $\sum_{m=1}^{\infty} c_m(q,t)\mathbb{H}_m(q,t,\mathsf{a})$ provided the super-invariance of $c_m(q,t)$. The superduality alone does not require specific $\lim_{q\to 1_-} c_m(q,t) = \pm 1$ from $\eta(s)$ and those for the L-functions. However, the fact that we "lift" the symmetry $k \mapsto 1/k,\ \upsilon \mapsto -\upsilon k$ of s in $\eta(s)$ to the fundamental superduality $q \leftrightarrow t^{-1}$ in the theory of superpolynomials is of interest.

7.4 Strongly-Polynomial Count

There are significant restrictions for the types of singularities and their usage in the 2nd figure.

First, the motivic zetas are supposed to be *topological* invariants. This is not granted in their definition, which is in terms of the corresponding singularity ring. They are of course of "discrete nature," which makes them potentially topological; however, the topological invariance is a conjecture even for general plane curve singularities.

Second, we need to check that $\mathcal{X}$ can be defined over $\mathbb{Z}$ within its topological type and has good reductions for almost all prime p.

Third, the varieties of "standard modules" and ideals of finite colength in the corresponding local rings must be of *strong polynomial count*: the number of their points over $\mathbb{F}_q$ must depend polynomially on q.

The latter condition is very restrictive. It holds if a variety is paved by *configurations of affine spaces*, their unions and differences in a bigger affine space. This is conjectured for the *Piontkowski cells* in our varieties of standard modules $\mathcal{J}_\ell$ and their multi-rank generalizations. We mention that affine Springer fibers can be *not* of strong polynomial count (unless in type A). There is an example of Bernstein-Kazhdan where the counting their $\mathbb{F}_q$-points involves zeta-functions of elliptic curves over $\mathbb{F}_q$, certainly not q-polynomial. See Appendix to "Fixed point varieties on affine flag manifolds" by Kazhdan-Lusztig (1988).

Under these conditions, the corresponding $\zeta_{\mathcal{X}}(q,t,a)$ or $L_{\mathcal{X}}(q,t,a)$ can be expected powerful topological invariants of $\mathcal{X}$. Presumably, they can capture the topological types of $\mathcal{X}$ in some cases. They certainly do this for plane curve singularities $\mathcal{X}$; however, the topological invariance of the motivic superpolynomials $\mathcal{H}^{mot}$ is a conjecture.

Plane curve singularities provide the main example by now. There is a bunch of constructions, theorems, and conjectures in this case. One of the purposes of this work is to show that the theory of their superpolynomials can be developed in various directions, which, presumably, includes isolated surface singularities of toric type and the corresponding Seifert-type manifolds.

Needless to say that isolated singularities are (and always were) among the key objects of algebraic geometry. Smooth projective manifolds proved to be very helpful in their study, but they are not really necessary for many aspects of singularity theory. We try to do as much as we can directly in terms of the singularity rings.

Knörrer's periodicity. In topology, there is a fundamental connection between knots/links and 3-folds. Its certain algebraic counterpart is the Knörrer's periodicity for singularities: a connection between the plane curve singularities $W(x,y)=0$ and the ones given by the equations $u^2 = W(x,y)$ is its important part. Actually, space singularities and 5-folds fit this picture too, but only "good" ones. For instance, the singularities in the form $uv = W(x,y)$ naturally occur here; such Calabi-Yau threefolds were considered by Vafa-Dijkgraaf.

The expected connection between the *superpolynomials* of algebraic links and the *superseries* of the corresponding Seifert spaces resembles that between zeta-polynomials and the corresponding L-functions $L_\Phi(s,\chi)$ we began with. This is very far from being exact. Our superpolynomials are much simpler than the zeta-polynomials, and our superduality is that from the Hasse-Weil functional equation, very different from that for L-functions. The passage from the *superseries* of Lens spaces to q, t, a-deformations of Dirichlet L-functions can be viewed as an attempt to unify these two theories.

Let me mention (again) that this note is very incomplete concerning the names and contributions; only very few papers are mentioned. It is focused mostly on superpolynomials and some perspectives of their theory. We tried to outline some number-theoretical perspectives of this direction and possible physics connections. The exposition is sketchy and speculative in several places. There are various omissions; for instance, we do not discuss much recent developments, even those directly related to the topics we touched upon.

To conclude, Manin's works and his vision of the role of number theory greatly influenced a lot of people, certainly all his students. We thank very much Yuri Tschinkel, Michael Finkelberg, and the referee.

Rational Points over C_1 Fields

Hélène Esnault

Contents

1 Introduction

Early in his work, Yuri Manin established a conjectural relation between the geometry of certain smooth projective varieties and the existence of rational points on them on some specific fields. We shall focus on the first such example, published in 1966 in [26]. He was 29.

In his PhD thesis [23], Lang defines the notion of a quasi-algebraically closed field, a concept he attributes to E. Artin; see also the preface [25, p.x] to Artin's Collected Papers by Lang and Tate: this is a field k for which every hypersurface of degree $d \leq n$ in $\mathbb{P}^n$ over k admits a rational point. He defines more generally a field to be C_i for some natural number i if every hypersurface of degree d with $d^i \leq n$ in $\mathbb{P}^n$ over k admits a rational point. So quasi-algebraically closed fields are precisely the C_1 fields. He lists fields which are C_1: finite fields, according to the theorem of Chevalley [4, Théorème], and its refinement by Warning (see [32, Satz 1a] in which he shows a congruence modulo the characteristic p of k for the number of rational points), and function fields over an algebraically closed field, according to the theorem of Tsen [31, Satz 5] (who had also developed a variant of the C_i-notion). Lang proves in *loc. cit.* that complete discretely valued fields with an algebraically closed residue field are C_1. In equal characteristic, it includes the

H. Esnault (✉)
Mathematik, Freie Universität Berlin, Berlin, Germany
e-mail: esnault@math.fu-berlin.de

Y. Tschinkel (ed.), *Arithmetic and Algebraic Geometry*, Simons Symposia,
https://doi.org/10.1007/978-3-031-74134-0_3

henselization at a finite place of a function field over an algebraically closed field (as then rational points are dense in rational points over the completion); in unequal characteristic it includes finite extensions of the maximal unramified extension of $\mathbb{Q}_p$.

Else there are conjectures. For example E. Artin conjectured that the maximal abelian extension $\mathbb{Q}^{ab}$ of $\mathbb{Q}$ is C_1 (see the preface by Lang and Tate in *loc. cit.*). More recently, Fargues conjectured in [16, Conjecture 3.10] that the field of functions $\bar{\mathbb{Q}}_p(FF)$ of the Fargues-Fontaine curve is C_1; in analogy with Lang's conjecture on the function field $\mathbb{R}(C_\circ)$ of the conic, $C_\circ = \mathrm{Spec}\big(R[x, y]/(x^2 + y^2 + 1)\big)$ over $\mathbb{R}$ without a point [24, p. 379].

Manin in *loc. cit.* Theorem 3.12 (b) proves that del Pezzo surfaces X over a finite field $\mathbb{F}_q$ with very ample anticanonical bundle, i.e., of degree ≥ 3, have a rational point. He uses the Weil conjectures for del Pezzo surfaces: the second ℓ-adic cohomology $H^2(X_{\bar{\mathbb{F}}_p}, \mathbb{Q}_\ell)$ is spanned by algebraic cycles, which are all defined over a finite extension of $\mathbb{F}_q$, and the odd cohomology vanishes. So the Grothendieck-Lefschetz trace formula [18, (25)] expresses the number of $\mathbb{F}_q$-points $|X(\mathbb{F}_q)|$ of X in the form

$$(\star)\quad |X(\mathbb{F}_q)| = 1 + q\sum_{j=1}^{b_2} \xi_j + q^2$$

where the ξ_i are roots of unity. This implies that $|X(\mathbb{F}_q)| \neq \emptyset$. In fact, more is true: as $(\sum_{i=1}^{b_2} \xi_j)$ is an algebraic integer, it implies that $|X(\mathbb{F}_q)|$ is congruent to 1 modulo q. Manin refers to Weil's article [33, p.557] for the shape of the trace formula in ℓ-adic cohomology. Weil did not have ℓ-adic cohomology at disposal, he was conjecturing its existence and purity, proved later by Deligne in [11] and in [12]. In *loc. cit.* Weil mentions explicitly the formula $(\star)$ for rational surfaces as a consequence of his conjecture. He argues that on those, cohomological correspondences on $X \times_{\mathbb{F}_q} X$ are coming from algebraic cycles defined over some finite extension of $\mathbb{F}_q$. Even if he does not quote them, Manin likely had [18] at disposal and [30] in which Tate formulates the Tate conjecture and details the relation between the zeta function and the cohomology as a Galois module.

Manin then formulates in *loc. cit.* Conjecture 4.1 to the effect that every rational surface X over a C_1 field k admits a rational point. He proves it by purely geometric methods for rational surfaces with an extra geometric property, for example, when the surface is fibered $f : X \to C$ in conics. Then C itself is a conic, that is, a curve of degree ≤ 2 in $\mathbb{P}^2$, so has a rational point by the C_1 definition and by the same argument its fiber as well.

Manin's conjecture was really *programmatic*, which made arithmetic and geometric properties intertwined. It is interesting to mention that Deligne, to quote himself, coined the term *Arithmetic Geometry*, influenced by the title of the conference "Arithmetic Algebraic Geometry" held in Purdue in December 1963, in which Tate exposed his conjectures, but also after Manin's article *loc. cit.*. Of course, the Weil conjecture, the Tate conjecture, etc. postulate properties shared by all varieties over certain fields, while Manin initiates the study of the relation between some specific fields and some specific varieties.

Campana in [3] in complex geometry and Kollár, Miyaoka, and Mori in [21] in more generality develop the notion of separably rationally connected varieties, a vast generalization of rational varieties on one hand and of hypersurfaces of degree $d \leq n$ in $\mathbb{P}^n$ on the other. In dimensions 1 and 2, rationally connected varieties are precisely geometrically rational varieties. In any dimension, a smooth projective connected variety X over a field k is separably rationally connected if over the algebraic closure $\bar{k}$ it admits a free rational curve, that is, a morphism $\mathbb{P}^1 \to X_{\bar{k}}$, such that the pull-back of the tangent bundle is ample on the rational curve. If k has characteristic 0 this is equivalent to X being rationally chain connected, that is to say the property that any two geometric points can be linked by a finite connected chain of rational curves. If k has characteristic $p > 0$, if X is separably rationally connected, it is rationally chain connected. The positivity of the pull-back of the tangent bundle forces those rational curves to be separable.

Kollár published his book [22] in 1996 in which he studied the general properties of rationally connected varieties. Manin's conjecture by this time has become the *Lang-Manin conjecture*: separably rationally connected varieties should have a rational point over a C_1 field. In the sequel we call it *the C_1-conjecture.* We mention also [22, IV. 6.1.1] in which Kollár poses the conjecture as a problem over a function field and [22, IV. 6.1.2] in which he asks for all fields over which separably rationally connected varieties have a rational point.

In this short note, we report on some progress on the C_1-conjecture. There are two main directions.

The first one relies on a classification of the varieties considered; see Sect. 5.1. The proof by Colliot-Thélène [5, Proposition 2] of Manin's initial conjecture, that is, for rational surfaces, relies on the classification of rational surfaces over any field (by Iskovskikh [19] for the last stone). They are fibered in conics, the case already essentially treated by Manin or del Pezzo surfaces; see [35, Théorème 1.11].

The second one consists in going through all the known C_1 fields one by one and trying to apply methods specific to those fields. For example, if k is a field of functions, do global geometry and degeneration on the special fibers; see Sect. 3.1. If k is the henselization at a finite place of a field of functions, reduce to the previous case; see Sect. 5.2. If k is finite, make sure that even if ℓ-adic cohomology is not spanned by algebraic cycles, the "level" of the cohomology is at least one so as to be able to apply q-divisibility in the Grothendieck-Lefschetz trace formula; see Sect. 4. If k is a complete discretely valued field with an algebraically closed residue field in unequal characteristic, we know very little; see Sect. 5.3 for a few modest facts on the index and on a "dynamical" result.

On the other hand, we do not know a complete list of C_1 fields. This shows that we are very far from understanding the conjecture in all generality. At the same time, there are classical varieties which are separably rationally connected and not defined by equations, but, e.g., by a moduli functor, like in characteristic 0 the moduli of stable bundles of coprime rank and degree on a curve (see [20] for first steps in this direction). It would be appealing if over a field from which one conjectures it is C_1, one could find a rational point on a separably rationally connected variety which is not defined by equations. Or perhaps no point: in the latter case, we would have to revise the hierarchy of problems!

This small text is by no means exhaustive and selects a few spots which are closer to the author's interests and understanding. For background and general studies, we refer to [7, 35] and [6].

2 General Facts

2.1 *The Brauer Group*

The Brauer group of a C_1 field is zero; see [23, Corollary of Theorem 14], and see [35, Proposition 1.15] for a self-contained proof, where the C_1 property is applied to the reduced norm of the skew field to its center.

In particular, to check whether a field k is C_1, one might first compute that its Brauer group $\mathrm{Br}(k)$ is equal to zero.

Witt proves in [34, Satz 4] that the norm map is surjective on finite extensions of $\mathbb{R}(C_\circ)$, which shows that $\mathrm{Br}(\mathbb{R}(C_\circ)) = 0$ (see [29, X. Proposition 11, 3)]).

For $\mathbb{Q}^{\mathrm{ab}}$, one proof is suggested in [29, p. 162, d)]: $\mathbb{Q}^{\mathrm{ab}}$ contains the maximal cyclotomic extension of $\mathbb{Q}$ (in fact it is equal to it by the Kronecker-Weber theorem). The Brauer group of a p-adic field is trivialized by the maximal cyclotomic extension which is defined over $\mathbb{Q}$. The local-global principle applied to the tower of cyclotomic extensions of $\mathbb{Q}$ allows one to conclude.

Fargues proves $\mathrm{Br}(\mathbb{Q}_p(FF)) = 0$ in [16, Théorème 2.2, Corollaire 2.5].

2.2 *Severi-Brauer Varieties*

As the Brauer group of a C_1 field k is trivial, Severi-Brauer varieties over k, that is, smooth projective varieties over k which are isomorphic to $\mathbb{P}^d$ over an algebraic closure $\bar{k}$, have a rational point.

3 Geometry

In this section, we follow the Bourbaki talk by Debarre [7].

3.1 *The Theorem of Graber-Harris-Starr*

In [17, Theorem 1.2] the authors prove the C_1 conjecture for $k = \mathbb{C}(B)$, the field of functions of a curve B over the field of complex numbers $\mathbb{C}$. This is equivalent to saying that if $f : X \to B$ is a projective morphism over a smooth curve B over $\mathbb{C}$, with rationally connected generic fiber $X \times_B \mathbb{C}(B)$, then f has a section.

By replacing f by its Weil restriction with respect to a Noether normalization $B \to \mathbb{P}^1$, we may assume that $B = \mathbb{P}^1$.

A closed point of $X \times_{\mathbb{P}^1} \mathbb{C}(\mathbb{P}^1)$ corresponds to a diagram

$$\begin{array}{ccc} C & \xrightarrow{g} & X \\ {\scriptstyle u}\downarrow & \swarrow {\scriptstyle f} & \\ \mathbb{P}^1 & & \end{array}$$

where C is a curve and $u : C \to \mathbb{P}^1$ is finite. If g, viewed as a point in the moduli space of maps to $\mathbb{P}^1$ with certain properties at the ramification of $C \to g(C)$, was smooth, then it would deform to a section of f [7, Theorem 4.1]. This is the first place where the authors use transcendental methods. The moduli space $M(X, g_*[C])$ is the space of such maps for a fixed Betti class $g_*[C] \in H_2(X, \mathbb{Z})$. They need its compactification for the proof. Then they have to put themselves in this situation. To this aim, they use topological arguments combined with the abundance of rational curves to prove that once they have g, they can attach to C many rational curves so the union deforms well.

3.2 *The Theorem of de Jong-Starr*

In [9, Theorem] the authors prove the C_1 conjecture for $k = F(B)$ where F is any algebraically closed field. Strongly using the freeness of the rational curves, they deform directly algebraically the starting g bypassing the use of $M(X, g_*[C])$ and its compactification and the topological arguments.

4 Motives and Cohomology

4.1 *Grothendieck-Lefschetz Trace Formula*

As we saw in the introduction, if the whole ℓ-adic cohomology of a smooth geometrically connected projective variety X is spanned by the cycle classes of algebraic cycles, then the Grothendieck-Lefschetz trace formula reads

$$|X(\mathbb{F}_q)| = 1 + \sum_{i=1}^{2d} (-1)^i \operatorname{Tr} \operatorname{Fr}|H^i(X_{\bar{\mathbb{F}}_p}, \mathbb{Q}_\ell) = 1 + \sum_{i=1}^{d} q^i \sum_{j_i=1}^{b_{2i}} \xi_{j_i}$$

where the ξ_{j_i} are roots of unity, the b_{2i} are the Betti numbers of the even weighted ℓ-adic cohomology of $X_{\bar{\mathbb{F}}_p}$, and Fr is the geometric Frobenius acting on ℓ-adic cohomology. (In addition, the coefficient of q^d is equal to 1). As $\sum_{j=1}^{b_{2i}} \xi_j$ is an algebraic integer, from the formula follows the congruence

$$|X(\mathbb{F}_q)| \equiv 1 \bmod q.$$

More generally, if the trace formula reads

$$|X(\mathbb{F}_q)| = 1 + q\xi$$

where ξ is an algebraic integer, then we can conclude that $|X(\mathbb{F}_q)| \equiv 1 \bmod q$.

In [13], we prove this property for rationally chain connected varieties, which in particular proves the C_1 conjecture (without the extra separability assumption) over finite fields.

We explain how we reach this congruence in the sequel. In *loc. cit.* we used crystalline cohomology. The same argument is transposed here on the ℓ-adic side.

4.2 Integrality: Deligne's Theorem

In [10, Corollary 5.5.3] Deligne proves that the eigenvalues of Fr on $H^i(X_{\bar{\mathbb{F}}_p}, \mathbb{Q}_\ell)$ are algebraic integers. This is an important property in our context as it shows:

> the C_1 conjecture over finite fields follows from the statement that the eigenvalues of Fr on $H^i(X_{\bar{\mathbb{F}}_p}, \mathbb{Q}_\ell)$ for $i > 0$ are all divisible by q as algebraic integers.

4.3 Motivic Analogy

In [14, Introduction], we raise the question of

> the analogue in ℓ-adic cohomology of the Hodge level for Betti cohomology.

If X is a smooth complex projective variety, if $H^i(X, \mathbb{Q})$ is supported in codimension ≥ 1, then the Hodge level of $H^i(X, \mathbb{Q})$ is ≥ 1 as well. The converse is an example of Grothendieck's generalized Hodge conjecture.

Using purity, due to Gabber, we compute in [13, Lemma 2.1] that if X is a smooth projective variety defined over $\mathbb{F}_q$, and if $H^i(X_{\bar{\mathbb{F}}_p}, \mathbb{Q}_\ell)$ is supported in codimension ≥ 1, then the eigenvalues of Fr on $H^i(X_{\bar{\mathbb{F}}_p}, \mathbb{Q}_\ell)$ are divisible by q as algebraic integers.

Thus, for proving the C_1 conjecture over finite fields, we are reduced to proving that $H^i(X_{\bar{\mathbb{F}}_p}, \mathbb{Q}_\ell)$ is supported in codimension ≥ 1 for all $i \geq 1$.

In [1, Theorem 1.3] we relate the Hodge level in characteristic 0 to the analogue in crystalline cohomology of the q-divisibility of the eigenvalues of Fr in ℓ-adic cohomology: the former on a smooth projective variety in characteristic 0 implies that the slopes in rigid cohomology H^i of a specialization of a regular model all are ≥ 1 for $i \geq 1$. This makes in this context the philosophical analogy a concrete theorem.

4.4 Trivial Chow Group of 0-Cycles After Base Change to Any Algebraic Closed Field

By the very definition of rationally chain connected varieties, any two geometric points are linked by a finite connected chain of rational curves. In particular, the group $CH_0(X_K)$ of algebraic 0-cycles is equal to $\mathbb{Z}$ for any embedding $k \hookrightarrow K$ of k into an algebraically closed field K. Taking K to contain $k(X)$, the resulting decomposition of the diagonal initiated by Bloch in his book [2, Appendix to Lecture 1], a method which later became the basis of manifold developments, proves that $H^i(X_{\bar{\mathbb{F}}_p}, \mathbb{Q}_\ell)$ is supported in codimension ≥ 1 for all $i \geq 1$. This finishes the proof of the C_1 conjecture over finite fields.

5 Remarks

5.1 C_1 Conjecture and Minimal Model Program in Characteristic 0

A generalization of the study case by case of rational surfaces by Colliot-Thélène *loc. cit.* is performed in characteristic 0 by Pieropan in [27]: over a given field of characteristic 0, the minimal model program implies that all rationally connected varieties of dimension $\leq d$ have a rational point if and only if $\mathbb{Q}$-factorial terminal Fano varieties of dimension $\leq d$ and Picard rank 1 do. It is roughly a generalization of the reduction from all rational surfaces to the del Pezzo ones.

5.2 Henselization of Function Fields at Finite Places

In [6, Théorème 7.5], Colliot-Thélène proves that the C_1 conjecture holds true for k being the field of fractions of a henselian discrete valuation ring with algebraically closed residue field F in equal characteristic. Indeed, if X is defined over k, its rational points $X(k)$ are dense in $X(\hat{k})$ for the topology defined by the discrete valuation, where $\hat{k} = F((t))$ is the completion with respect to the valuation. If X was defined over $k(t)$, we could then apply directly the theorems in Sects. 3.1 and 3.2. In general X descends to X_A where $A = F[T_1, \ldots, T_n]/I \subset F[[t]]$ is an affine and smooth algebra over F, where the embedding is defined by a power series expansion in t of the T_i. For $n \in \mathbb{N}_{\geq 1}$ large enough, depending on the multiplicity of $\mathrm{Spec}F \in S = \mathrm{Spec}(A)$, the $F[[t]]/(t^n)$-point of S integrates to a finite type normal curve $C_n \to S$ over F, generically embedded in S, thus over which X_{C_n} has a C_n-point. In particular, $X(F[[t]]/(t^n)) \neq \emptyset$. As X is proper, a compactness argument allows one to conclude that $X(F[[t]]) \neq \emptyset$.

5.3 *Some Remarks on Complete Discrete Valuation Rings $\mathcal{O}$ with Algebraically Closed Residue Field F in Unequal Characteristic*

5.3.1 A Small Dream

We mention now where the proof from Sect. 5.2 cannot be generalized to the unequal characteristic situation. By the same argument as in Sect. 3.1, we may assume that $\mathcal{O}$ is the ring of Witt vectors $W(F)$ over the field F, an algebraically closed residue field of characteristic $p > 0$. So we have $A \subset W$, which is now defined by the p-adic expansion of the T_i, we have the resulting $W_n = W/p^n$ points of S, and we have X_A. But we do not have the integration C_n as in the equal characteristic case. At the ICM 2014 in Seoul, we asked Scholze, who had just proved Deligne's weight-monodromy conjecture [28, Theorem 1.15] for complete intersections by applying his tilting method, whether he would think there is a way to tilt X_W to, say, $Y_{F[[t]]}$, keeping the rational connectivity property for $Y_{F((t))}$, and if so if one could tilt back $Y_{F[[t]]}(F[[t]])$ constructed in Sect. 5.2 to $X_W(W)$. It is a difficult and badly posed question for many reasons, two immediate ones being that the tilt of the field of fractions of W contains also all the p-roots of t, and that over this large field, Y is not rationally connected. Nonetheless we can dream of a more evolved analogue of the classical proof in Sect. 5.2 in this context.

5.3.2 Back to the Hodge Level

In [15, Corollary 3] a weak form of the C_1 conjecture is proved over the field of fractions of $\mathcal{O}$: the index of a separably rationally connected variety X, that is, the g.c.d. of the degrees of its closed points, is equal to 1 if $\dim X + 1 < p$. It is really a very weak form. Indeed, the index 1 statement is true as soon as the Hodge level of X is ≥ 1. It is in the spirit of the theorem [1, Theorem 1.3] mentioned in Sect. 4.3.

5.3.3 Dynamic

As explained previously, fixing a field K, complete with respect to a discrete valuation with algebraically closed residue field in unequal characteristic, the C_1 conjecture is not understood. In [8, Theorem 1.3], Duesler and Knecht prove a "dynamical" version of the conjecture which is close to the theorem of Ax-Knochen according to which, except for finitely many p, any hypersurface in $\mathbb{P}^n$ over $\mathbb{Q}_p$ of degree d with $d^2 \leq n$ has a rational point. They fix a Hilbert polynomial P and show using similar methods that, except for finitely many p, any rationally connected variety with Hilbert polynomial P and defined over the maximal unramified extension of $\mathbb{Q}_p$ has a rational point. In [27, Corollary 1.4] this statement is improved replacing the Hilbert polynomial by the dimension of the variety.

Acknowledgements We thank Pierre Deligne for a discussion on [33] and on the origins of the terminology "Arithmetic Geometry" which is reflected in the Introduction. We thank Lars Hesselholt for discussions on $\bar{\mathbb{Q}}_p(FF)$; Olivier Benoist and Kay Rülling for discussions on $\mathbb{R}(C_\circ)$; Marco D'Addezio for discussions on $\mathbb{Q}^{ab}$, which are reflected in Sect. 2.1; Olivier Wittenberg for general discussions on the topic when it came; János Kollár for precise references in his book; and Akhil Mathew for his interest. We thank the referee for a careful and friendly reading.

References

1. P. Berthelot, H. Esnault, K. Rülling, Rational points over finite fields for regular models of algebraic varieties of Hodge type greater or equal to 1. Ann. Math. **176**, 1–96 (2012)
2. S. Bloch, *Lectures on Algebraic Cycles*, Duke University Mathematics Series, vol. IV (1980)
3. F. Campana, On twistor spaces of the class C. J. Diff. Geom. **33**, 541–549 (1991)
4. C. Chevalley, Démonstration d'une hypothèse de M. Artin. Hamburg Abh. **11**, 73–75 (1935)
5. J.-L. Colliot-Thélène, *Arithmétique des variétés rationnelles et problèmes birationnels*, in *Proceedings ICM 1986*, vol. 1 (Amer. Math. Soc., Berkeley, 1987), pp. 641–653
6. J.-L. Colliot-Thélène, *Variétés presque rationnelles, leurs points rationnels et leurs dégénérescences*, in *Arithmetic Geometry*, Lecture Notes in Math., vol. 2009 (Springer, 2011), pp. 1–44
7. O. Debarre, *Variétés rationnellement connexes*, Exposé, vol. 905, Séminaire Bourbaki 2001/2002, Astérisque, vol. 290, pp. 243–266 (2003)
8. B. Duesler, A. Knecht, Rationally connected varieties over the maximally unramified extension of p-adic fields. Rocky Mountain J. Math. **47**(8), 2605–2617 (2017)
9. J. de Jong, J. Starr, Every rationally connected variety over the function field of a curve has a rational point. Amer. J. Math. **125**(3), 567–580 (2003)
10. P. Deligne, *Théorème d'intégralité*, Appendix to "Le eniveau de la cohomologie des intersections complètes" by N. Katz, Exposé XXI in SGA 7, Lecture Notes in Math., vol. 340 (Springer, New York, 1973), pp. 363–400
11. P. Deligne, La conjecture de Weil I. Publ. Math. IHÉS **43**, 273–308 (1974)
12. P. Deligne, La conjecture de Weil II. Publ. Math. IHÉS **52**, 137–252 (1974)
13. H. Esnault, Varieties over a finite field with trivial Chow group of 0-cycles have a rational point. Invent. math. **151**, 187–191 (2003)
14. H. Esnault, Deligne's integrality theorem in unequal characteristic and rational points over finite fields. Ann. Math. **164**, 715–730 (2006)
15. H. Esnault, M. Levine, O. Wittenberg, Index of varieties over Henselian fields and Euler characteristic of coherent sheaves. J. Algebraic Geometry **24**, 693–718 (2015)
16. L. Fargues, G-torseurs en théorie de Hodge p-adique. Compositio Math. **156**, 2076–2110 (2020)
17. T. Graber, J. Harris, J. Starr, Families of rationally connected varieties. J. AMS **16**(1), 57–67 (2003)

18. A. Grothendieck, *Formule de Lefschetz et rationalité des fonctions L*. Séminaire Bourbaki, vol. 279 (Soc. Math. France, Paris (1964/1965), 1995), pp. 41–55
19. V. Iskovskikh, Minimal models of rational surfaces over arbitrary fields. Izv. Akad. Nauk SSSR Ser. Mat. **43**(1), 19–43 (1979) (in russian). Math. USSR Izv. **14**(1), (1980). (in english)
20. I. Kaur, *A Pathological Case of the C_1 Conjecture in Mixed Characteristic*, in *Mathematical Proceedings of the Cambridge Philosophical Society*, vol. 167, 61–64 (2019)
21. J. Kollár, Y. Miyaokoa, S. Mori, Rational connectedness and boundedness of Fano manifolds. J. Diff. Geom. **36**, 765–779 (1992)
22. J. Kollár, *Rational Curves on Algebraic Varieties*, Ergebnisse der Mathematik und ihrer Grenzgebiete, vol. 32 (Springer, 1996)
23. S. Lang, On quasi-algebraic closure. Ann. Math. **55**, 373–390 (1952)
24. S. Lang, The theory of real places. Ann. Math. **57**(2), 378–391 (1957)
25. S. Lang, J. Tate, *The Collected Papers of Emil Artin* (Addison-Wesley Publishing Co., 1965), xvi+560 pp.
26. Yu. Manin, Rational surfaces over perfect fields. Publ. Math. IHÉS **30**, 55–113 (1966)
27. M. Pieropan, *On Rationally Connected Varieties over C_1 Fields of Characteristic* 0, Algebra Number Theory, vol. 16(8), 1811–1844 (2022)
28. P. Scholze, Perfectoid spaces. Publ. Math. IHÉS **116**(1), 245–313 (2012)
29. J.-P. Serre, *Local Fields*, Graduate Texts in Mathematics, vol. 67 (Springer, 1979)
30. J. Tate, *Algebraic Cycles and Poles of Zeta Functions*, in *Arithmetical Algebraic Geometry*. Proc. Conf. Purdue Univ., (1963) (Harper and Row, New York, 1965), pp. 93–110
31. Ch.C. Tsen, *Algebren über Funktionenkörpern*, Ph.D. dissertation, Göttingen University (1934)
32. E. Warning, Bemerkung zur vorstehender Arbeit von Herrn Chevalley. Hamburg Abh. **11**, 76–83 (1935)
33. A. Weil, Abstract versus classical algebraic geometry. Matematika **4**(2), 59–66 (1958)
34. E. Witt, Zerlegung reeller algebraischer Funktionen in Quadrate. Schiefkörper über reellem Funktionenkörper. J. Reine Angew. Math. **171**, 4–11 (1934)
35. O. Wittenberg, *La connexité rationnelle en arithmétique*, in *Variétés rationnellement connexes: aspects géométriques et arithmétiques*, Panorama et Synthèses, vol. 31 (Société mathématique de France, 2010)

On Isomorphisms of Ind-Varieties of Generalized Flags

Lucas Fresse and Ivan Penkov

Contents

1 Introduction

If X and Y are two finite-dimensional flag varieties, possibly one or both being varieties of isotropic flags, the problem of whether X and Y are isomorphic is easily solvable. One can approach it in different ways, one of which is to look at the automorphism groups of X and Y. This yields an elegant proof of the following theorem, for whose statement we need to introduce some notation. If $X = \mathrm{Fl}(a_1, \ldots, a_i, V)$ is the variety of flags with dimension sequence $(a_1, \ldots, a_i, a_{i+1} = \dim V)$ in a finite-dimensional vector space V with $\dim V \geq 2$, we say that X is of *general type*. If $X = \mathrm{FlO}(a_1, \ldots, a_i, V)$ is a connected variety of isotropic flags with dimension sequence $(a_1, \ldots, a_i, a_{i+1} = \dim V)$ in an orthogonal space V with $\dim V \geq 5$, we say that X is *of orthogonal type*. If $X = \mathrm{FlS}(a_1, \ldots, a_i, V)$ is a (automatically connected) variety of isotropic flags with dimension sequence

Dedication: To the memory of Yuri Manin, an extraordinary human, mathematician, and mentor.

L. Fresse (✉)
Université de Lorraine, CNRS, Institut Élie Cartan de Lorraine, UMR, Vandoeuvre-lès-Nancy, France
e-mail: lucas.fresse@univ-lorraine.fr

I. Penkov
Constructor University, Bremen, Germany
e-mail: ipenkov@constructor.university

Y. Tschinkel (ed.), *Arithmetic and Algebraic Geometry*, Simons Symposia,
https://doi.org/10.1007/978-3-031-74134-0_4

$(a_1, \ldots, a_i, a_{i+1} = \dim V)$ in a symplectic space V with $\dim V \geq 6$, we say that X is *of symplectic type*.

Theorem 1.1 Let X and Y be two flag varieties of the same type as above. Then X and Y are isomorphic if and only if their dimension sequences coincide, or both X and Y are of general type and their respective dimension sequences $(a_1, \ldots, a_i, a_{i+1})$ and $(b_1, \ldots, b_j, b_{j+1})$ satisfy $i = j$, $a_{i+1} = b_{i+1}$, and $a_k = a_{i+1} - b_k$ for $k \in \{1, \ldots, i\}$.

If X and Y are of different types, then the only possible isomorphisms are as follows:

- $\mathrm{Fl}(1, \mathbb{C}^{2n}) \cong \mathrm{FlS}(1, \mathbb{C}^{2n})$.
- $\mathrm{FlO}(n-1, \mathbb{C}^{2n-1}) \cong \mathrm{FlO}(n, \mathbb{C}^{2n})$.

This theorem can be considered a corollary of Onishchik's result [5] claiming that the connected component of unity of the automorphism group of a flag variety X, or a variety of isotropic flags, is the centerless adjoint group corresponding to the variety, except when X is isomorphic to $\mathrm{FlS}(1, \mathbb{C}^{2n})$ or $\mathrm{FlO}(n-1, \mathbb{C}^{2n-1})$. Indeed, since the algebraic groups $\mathrm{SL}(n)$ for $n \geq 2$, $\mathrm{SO}(m)$ for $m \geq 4$, $\mathrm{Sp}(r)$ for even $r \geq 4$, are pairwise non-isomorphic, Onishchik's result reduces the problem to comparing two flag varieties X and Y of the same type for the same vector space V. The proof of Theorem 1.1 gets then easily completed by comparing the grassmannians, or isotropic grassmannians, to which X and Y project.

In the present paper we prove an exact analogue of Theorem 1.1 for ind-varieties of, possibly isotropic, generalized flags. These ind-varieties are homogeneous ind-spaces for the groups $\mathrm{GL}(\infty)$, $\mathrm{SO}(\infty)$, $\mathrm{Sp}(\infty)$ and have been studied quite extensively in the last twenty years [1, 2, 3, 6, 7, 8]. Nevertheless, a precise statement of when two such ind-varieties are isomorphic has been missing in the literature.

First, let us note that Theorem 1.1 does not imply directly any statement of isomorphism or non-isomorphism of ind-varieties of generalized flags, since two non-isomorphic ind-varieties may admit exhaustions with pairwise isomorphic finite-dimensional varieties, and conversely, an ind-variety may admit two exhaustions by pairwise non-isomorphic finite-dimensional varieties. Next, we recall that the automorphism groups of ind-varieties of, possibly isotropic, generalized flags have been computed in [3]. However, since the question of when two such groups are isomorphic as abstract groups has not yet been addressed (and may be quite hard), we are unable to produce an argument as direct as in the outline of proof of Theorem 1.1 given above. Instead, we rely on some basic information about automorphism groups of ind-varieties of generalized flags and on a technique developed in the papers [7, 8]. This technique turns out to be very useful also in the problem of isomorphisms.

The precise statement of our main result is Theorem 2.5. In Sect. 3 we have collected preliminaries on finite-dimensional flag varieties. Sections 4 and 5 are devoted to the proof of Theorem 2.5.

2 Statement of Result

2.1 *Short Review of Ind-Varieties of Generalized Flags*

The base field is the field of complex numbers $\mathbb{C}$. Let V be a countable-dimensional vector space, possibly equipped with an orthogonal (i.e., non-degenerate, symmetric) or symplectic (i.e., non-degenerate, antisymmetric) bilinear form ω. By E we denote a basis of V. In the presence of a form ω, we make the following definition.

Definition 2.1 Assume that V is equipped with an orthogonal or symplectic form ω. A basis E is said to be *isotropic* if it is equipped with an involution $i_E : E \to E$ with at most one fixed point, such that $\omega(e, e') \neq 0$ if and only if $e' = i_E(e)$. Then:

- If ω is orthogonal and i_E has one fixed point, then the basis E is said to be of type B.
- If ω is orthogonal and i_E has no fixed point, then E is said to be of type D.
- If ω is symplectic, then i_E cannot have a fixed point, and the basis E is said to be of type C.

In [1], the homogeneous spaces of the form $\mathbf{G}/\mathbf{P}$ have been described, where $\mathbf{G}$ is one of the classical ind-groups $\mathrm{SL}(\infty)$ (or $\mathrm{GL}(\infty)$), $\mathrm{SO}(\infty)$, $\mathrm{Sp}(\infty)$, and $\mathbf{P} \subset \mathbf{G}$ is a splitting parabolic subgroup. The term "splitting" means that $\mathbf{P}$ contains the Cartan subgroup of transformations inside $\mathbf{G}$ which are diagonal in some basis E (isotropic in the case of $\mathrm{SO}(\infty)$ and $\mathrm{Sp}(\infty)$) of the underlying space V.

The description is by means of the notion of generalized flag.

Definition 2.2

(a) A *generalized flag* of V is a collection $\mathcal{F}$ of subspaces of V which is totally ordered by inclusion and such that

- Every $F \in \mathcal{F}$ has an immediate predecessor F' or an immediate successor F'' in $\mathcal{F}$.
- Every vector $v \in V \setminus \{0\}$ belongs to $F'' \setminus F'$ for a unique pair of consecutive subspaces (F', F'') of $\mathcal{F}$.

(b) In the case where V is equipped with an orthogonal of symplectic form ω, we say that $\mathcal{F}$ is *isotropic* if there is an involution $i_{\mathcal{F}} : \mathcal{F} \to \mathcal{F}$ such that $i_{\mathcal{F}}(F) = F^{\perp}$ for all $F \in \mathcal{F}$, where $F^{\perp}$ stands for the orthogonal subspace to F with respect to ω.

(c) If E is a basis of V, then $\mathcal{F}$ is said to be *compatible with* E if each subspace $F \in \mathcal{F}$ has a basis formed by elements of E. We say that $\mathcal{F}$ is *weakly compatible with* E if it is compatible with some basis E' which differs from E by finitely many vectors, i.e., $\#E \setminus (E \cap E') = \#E' \setminus (E \cap E') < +\infty$.

In [1], an equivalence relation called *E-commensurability* is introduced on generalized flags. Then, given a generalized flag $\mathcal{F}$ compatible with a basis E, one defines $\mathrm{Fl}(\mathcal{F}, E, V)$ as the set of all generalized flags in V which are E-commensurable with $\mathcal{F}$. If $\mathcal{F}$ and E are isotropic, one defines instead $\mathrm{Fl}_\omega(\mathcal{F}, E, V)$ as the set of all isotropic generalized flags in V which are E-commensurable with $\mathcal{F}$. It is known that $\mathrm{Fl}(\mathcal{F}, E, V)$ and $\mathrm{Fl}_\omega(\mathcal{F}, E, V)$ have natural structures of ind-varieties. We will recall these structures later on. In what follows, whenever we write $\mathrm{Fl}(\mathcal{F}, E, V)$ or $\mathrm{Fl}_\omega(\mathcal{F}, E, V)$, we assume that the generalized flag $\mathcal{F}$ is compatible with the basis E.

We will adopt the following notation:

- If ω is an orthogonal form on V, then E is of type B or D, and we set in both cases $\mathrm{FlO}(\mathcal{F}, E, V) := \mathrm{Fl}_\omega(\mathcal{F}, E, V)$, with the following exception. If $\mathcal{F}$ contains a subspace F such that $F^\perp = F$ or a subspace F' such that $\dim F'^\perp/F' = 2$ (in both cases E has to be of type D), then $\mathrm{Fl}_\omega(\mathcal{F}, E, V)$ consists of two isomorphic connected components, and we define $\mathrm{FlO}(\mathcal{F}, E, V)$ as either one. In addition, if there is $F' \in \mathcal{F}$ with $\dim F'^\perp/F' = 2$, we assume that $\mathcal{F}$ contains also one of the two maximal isotropic subspaces F containing F'.
- If ω is a symplectic form on V, then E is of type C and we set $\mathrm{FlS}(\mathcal{F}, E, V) = \mathrm{Fl}_\omega(\mathcal{F}, E, V)$.

Ind-grassmannians correspond to quotients $\mathbf{G}/\mathbf{P}$ with $\mathbf{P}$ maximal:

- If $\mathcal{F} = \{\{0\} \subset F \subset V\}$ then we set $\mathrm{Gr}(F, E, V) = \mathrm{Fl}(\mathcal{F}, E, V)$.
- In the case where V is equipped with an orthogonal or symplectic form ω, a minimal isotropic generalized flag has the form $\mathcal{F} = \{\{0\} \subset F \subset F^\perp \subset V\}$ (where F and $F^\perp$ may coincide), and we set $\mathrm{GrO}(F, E, V) = \mathrm{FlO}(\mathcal{F}, E, V)$ or $\mathrm{GrS}(F, E, V) = \mathrm{FlS}(\mathcal{F}, E, V)$ depending on whether ω is orthogonal or symplectic.
- In accordance with our convention above, when using the notation $\mathrm{GrO}(F, E, V)$, we exclude the case when $\dim F^\perp/F = 2$. Instead, we consider $\mathrm{FlO}(\mathcal{F}, E, V)$ where $\mathcal{F} = \{\{0\} \subset F \subset \tilde{F} \subset F^\perp \subset V\}$, $\tilde{F}$ being one of the two maximal isotropic spaces containing F.

Remark 2.3 If V is an orthogonal space, then V admits isotropic bases E_1 and E_2 of respective types B and D. Accordingly, maximal isotropic subspaces F of V are of two types: either $\dim F^\perp/F = 1$ or $F^\perp = F$. As we will see below, the corresponding ind-grassmannians $\mathrm{GrO}(F_1, E_1, V)$ for $\dim F_1^\perp/F_1 = 1$ and $\mathrm{GrO}(F_2, E_2, V)$ for $F_2^\perp = F_2$ are isomorphic as ind-varieties. This property is an infinite-dimensional analogue of the isomorphism stated in the second bullet point of Theorem 1.1.

2.2 Main Result

Definition 2.4

(a) Let $\mathcal{F}$ and $\mathcal{G}$ be generalized flags of countable dimensional spaces V and W, respectively. Without further assumption, we say that $\mathcal{F}$ and $\mathcal{G}$ are *isomorphic* if there exists a linear isomorphism $\phi : V \to W$ such that $\mathcal{G} = \{\phi(F) : F \in \mathcal{F}\}$.
(b) In the case where V and W are equipped with symplectic forms (resp., orthogonal forms) ω and ω', we assume that $\mathcal{F}$ and $\mathcal{G}$ are isotropic generalized flags and say that they are *isomorphic* if the isomorphism ϕ preserves the forms: $\omega'(\phi(x), \phi(y)) = \omega(x, y)$ for all $(x, y) \in V \times V$.

If $\mathcal{F} = \{F_\theta : \theta \in \Theta\}$ is a generalized flag in a countable-dimensional vector space V, compatible with a basis E of V, then we define its orthogonal as the chain $\mathcal{F}^\perp = \{F_\theta^\perp : \theta \in \Theta\}$ where $F_\theta^\perp$ is the annihilator of F_θ in the space $\langle E^* \rangle$, and E^* denotes the system of linear functionals on V dual to the basis E. If V is equipped with an orthogonal or a symplectic form ω and the basis E is isotropic, then we use this form to identify V and $\langle E^* \rangle$. Moreover, the above definition of an isotropic generalized flag $\mathcal{F}$ is equivalent to the requirement $\mathcal{F} = \mathcal{F}^\perp$.

Theorem 2.5 Let X and Y be ind-varieties of, possibly isotropic, generalized flags as above. In other words, $X = \mathrm{Fl}(\mathcal{F}, E, V)$, or $X = \mathrm{FlO}(\mathcal{F}, E, V)$, or $X = \mathrm{FlS}(\mathcal{F}, E, V)$, and similarly $Y = \mathrm{Fl}(\mathcal{G}, E', W)$, or $Y = \mathrm{FlO}(\mathcal{G}, E', W)$, or $Y = \mathrm{FlS}(\mathcal{G}, E', W)$. Then the ind-varieties X and Y are isomorphic whenever $\mathcal{F}$ and $\mathcal{G}$, or $\mathcal{F}$ and $\mathcal{G}^\perp$, are isomorphic, possibly as isotropic flags. The only additional isomorphisms $X \cong Y$ are the following:

- $X = \mathrm{Gr}(F, E, V)$, $Y = \mathrm{GrS}(G, E', W)$, where $\dim F = \dim G = 1$.
- $X = \mathrm{GrO}(F, E, V)$, $Y = \mathrm{GrO}(G, E', W)$, where $\dim F^\perp/F = 1$ and $G^\perp = G$.

Remark 2.6 Note that two ind-varieties $\mathrm{FlO}(\mathcal{F}_1, E_1, V)$ and $\mathrm{FlO}(\mathcal{F}_2, E_2, V)$, where E_1 is an isotropic basis of type B and E_2 is an isotropic basis of type D, may be isomorphic also beyond the special case of Remark 2.3. This is a consequence of the observation that a given isotropic generalized flag $\mathcal{F}$ may be compatible with two different isotropic bases E_1 and E_2 of respective types B and D, as illustrated by the following example.

Example 2.7 Consider an isotropic flag $\mathcal{F}$ in an orthogonal space V, with the property that $W_\mathcal{F} := \sum_{\substack{F \in \mathcal{F} \\ F \subset F^\perp}} F$ has infinite codimension in its orthogonal. Then there exist two isotropic bases E_1 and E_2 of respective types B and D with which $\mathcal{F}$ is compatible and $E_1 \cap W_\mathcal{F} = E_2 \cap W_\mathcal{F}$. Consequently, $\mathrm{FlO}(\mathcal{F}, E_1, V) = \mathrm{FlO}(\mathcal{F}, E_2, V)$.

3 A Review on Embeddings of Flag Varieties

Throughout this section, V, V', W, W' are finite-dimensional vector spaces.

3.1 *Short Review of Picard Groups for Flag Varieties*

For an integer $0 < p < \dim V$, we denote by $\mathrm{Gr}(p; V)$ the grassmannian of p-dimensional subspaces in V. It can be realized as a projective variety via the Plücker embedding $\pi : \mathrm{Gr}(p; V) \hookrightarrow \mathbb{P}(\bigwedge^p V)$. Moreover, the Picard group $\mathrm{Pic}(\mathrm{Gr}(p; V))$ of $\mathrm{Gr}(p; V)$ is isomorphic to $(\mathbb{Z}, +)$, and its generators are $\mathcal{O}_{\mathrm{Gr}(p;V)}(1) := \pi^*\mathcal{O}_{\mathbb{P}(\bigwedge^p V)}(1)$ and $\mathcal{O}_{\mathrm{Gr}(p;V)}(-1) := \pi^*\mathcal{O}_{\mathbb{P}(\bigwedge^p V)}(-1)$. Here $\mathcal{O}_{\mathbb{P}(\bigwedge^p V)}(-1)$ stands for the tautological bundle of $\mathbb{P}(\bigwedge^p V)$ and $\mathcal{O}_{\mathbb{P}(\bigwedge^p V)}(1)$ stands for its dual.

For a sequence of integers $0 < p_1 < \ldots < p_k < \dim V$, we denote by $\mathrm{Fl}(p_1, \ldots, p_k; V)$ the variety of (partial) flags

$$\mathrm{Fl}(p_1, \ldots, p_k; V) = \{(V_1, \ldots, V_k) \in \mathrm{Gr}(p_1; V) \times \cdots \times \mathrm{Gr}(p_k; V) : V_1 \subset \ldots \subset V_k\}.$$

We have

$$\mathrm{Pic}(\mathrm{Fl}(p_1, \ldots, p_k; V)) \cong \mathbb{Z}^k.$$

More precisely, if we denote by L_i the pull-back

$$L_i = \mathrm{pr}_i^* \mathcal{O}_{\mathrm{Gr}(p_i;V)}(1)$$

along the projection

$$\mathrm{pr}_i : \mathrm{Fl}(p_1, \ldots, p_k; V) \to \mathrm{Gr}(p_i; V)$$

(for $i = 1, \ldots, k$), then $[L_1], \ldots, [L_k]$ is a set of generators of the Picard group, which we will refer to as the set of *preferred generators*.

If V is a vector space endowed with an orthogonal or symplectic form ω, we assume that the sequence $(p_1, \ldots, p_k)$ satisfies

$$p_i + p_{k-i+1} = \dim V \quad \text{for all } i = 1, \ldots, k.$$

Orthogonal Case Here we assume that the form ω is orthogonal. If $\frac{\dim V}{2}, \frac{\dim V}{2} - 1 \notin \{p_1, \ldots, p_k\}$ (which is automatic when $\dim V$ is odd), we define

$$\mathrm{FlO}(p_1, \ldots, p_k; V) \subset \mathrm{Fl}(p_1, \ldots, p_k; V)$$

as the subvariety of *isotropic flags*, i.e., flags $(F_1 \subset \ldots \subset F_k)$ such that $F_i^\perp = F_{k-i+1}$ for all i. If $\frac{\dim V}{2} \in \{p_1, \ldots, p_k\}$ or $\frac{\dim V}{2} - 1 \in \{p_1, \ldots, p_k\}$, the subvariety of $\mathrm{Fl}(p_1, \ldots, p_k; V)$ of isotropic flags consists of two irreducible components, and we define $\mathrm{FlO}(p_1, \ldots, p_k; V)$ as either of these two components.

Moreover, as it is well known every isotropic subspace of dimension $\frac{\dim V}{2} - 1$ is contained in exactly two Lagrangian subspaces, so we lose no generality in considering only sequences $(p_1, \ldots, p_k)$ which satisfy the condition

$$\frac{\dim V}{2} - 1 \in \{p_1, \ldots, p_k\} \quad \Rightarrow \quad \frac{\dim V}{2} \in \{p_1, \ldots, p_k\}.$$

We denote $\mathrm{GrO}(p; V) := \mathrm{FlO}(p, \dim V - p; V)$ if $p \notin \{\frac{\dim V}{2}, \frac{\dim V}{2} - 1\}$ and $\mathrm{GrO}(\frac{\dim V}{2}; V) := \mathrm{FlO}(\frac{\dim V}{2}; V)$, assuming that $\dim V$ is even in the latter case. We do not define an orthogonal grassmannian for $p = \frac{\dim V}{2} - 1$ as we consider instead $\mathrm{FlO}(\frac{\dim V}{2} - 1, \frac{\dim V}{2}, \frac{\dim V}{2} + 1; V)$.

Let $\ell = \lfloor \frac{k+1}{2} \rfloor$. If $\frac{\dim V}{2} - 1 \notin \{p_1, \ldots, p_k\}$, then the pull-backs $L_i := \mathrm{pro}_i^* \mathcal{O}_{\mathrm{GrO}(p_i; V)}(1)$ by the projections $\mathrm{pro}_i : \mathrm{FlO}(p_1, \ldots, p_k; V) \to \mathrm{GrO}(p_i; V)$, for $i \in \{1, \ldots, \ell\}$, are a set of generators of the Picard group $\mathrm{Pic}\,\mathrm{FlO}(p_1, \ldots, p_k; V)$, which we call *preferred generators*. If $\frac{\dim V}{2} - 1 \in \{p_1, \ldots, p_k\}$, that is, $\frac{\dim V}{2} - 1 = p_{\ell-1}$, then the preferred generators L_i of $\mathrm{Pic}\,\mathrm{FlO}(p_1, \ldots, p_k; V)$ are as above except for $i = \ell - 1$, and the $(\ell - 1)$-th preferred generator is by definition $(\bigwedge^{\frac{\dim V}{2} - 1} S_{\ell-1})^*$ where $S_{\ell-1}$ is the tautological bundle of rank $\frac{\dim V}{2} - 1$ on $\mathrm{FlO}(p_1, \ldots, p_k; V)$.

Symplectic Case If the form ω is symplectic, we denote by $\mathrm{FlS}(p_1, \ldots, p_k; V) \subset \mathrm{Fl}(p_1, \ldots, p_k; V)$ the subvariety of isotropic flags. Moreover, we set $\mathrm{GrS}(p; V) := \mathrm{FlS}(p, \dim V - p; V)$ if $\dim V \neq 2p$, and $\mathrm{GrS}(\frac{\dim V}{2}; V) := \mathrm{FlS}(\frac{\dim V}{2}; V)$.

Let $\ell = \lfloor \frac{k+1}{2} \rfloor$. Then $\mathrm{Pic}\,\mathrm{FlS}(p_1, \ldots, p_k; V) \cong \mathbb{Z}^\ell$, and the pull-backs $L_i := \mathrm{prs}_i^* \mathcal{O}_{\mathrm{GrS}(p_i; V)}(1)$ by the projections $\mathrm{prs}_i : \mathrm{FlS}(p_1, \ldots, p_k; V) \to \mathrm{GrS}(p_i; V)$ for $i \in \{1, \ldots, \ell\}$ yield a set of generators $[L_1], \ldots, [L_\ell]$ of $\mathrm{Pic}\,\mathrm{FlS}(p_1, \ldots, p_k; V)$, which again we refer to as *preferred generators*.

We close this subsection with the following well-known fact.

Lemma 3.1 Let $\mathcal{M}$ be a line bundle on $\mathrm{Fl}(p_1, \ldots, p_k; V)$, $\mathrm{FlO}(p_1, \ldots, p_k; V)$, or $\mathrm{FlS}(p_1, \ldots, p_k; V)$, and assume that the equality

$$[\mathcal{M}] = n_1[L_1] + \ldots + n_\ell[L_\ell] \quad \text{with } n_1, \ldots, n_\ell \in \mathbb{Z}$$

holds in the Picard group, where $[L_1], \ldots, [L_\ell]$ is the respective set of preferred generators. Then the following conditions are equivalent:

(i) $\mathcal{M}$ is very ample.
(ii) $\mathcal{M}$ is ample.
(iii) $n_i > 0$ for all $i \in \{1, \ldots, \ell\}$.

3.2 Embeddings of Flag Varieties

In this section, we denote by X one of the flag varieties

$$\mathrm{Fl}(p_1, \ldots, p_K; V), \quad \mathrm{FlS}(p_1, \ldots, p_K; V), \quad \mathrm{FlO}(p_1, \ldots, p_K; V),$$

and by Y a respective flag variety

$$\mathrm{Fl}(q_1, \ldots, q_L; W), \quad \mathrm{FlS}(q_1, \ldots, q_L; W), \quad \mathrm{FlO}(q_1, \ldots, q_L; W)$$

of the same type as X. Consider an embedding (i.e., closed immersion) of flag varieties

$$\varphi : X \hookrightarrow Y,$$

together with the group homomorphism on Picard groups

$$\varphi^* : \mathrm{Pic}\, Y \to \mathrm{Pic}\, X$$

which it induces. Let $[L_1], \ldots, [L_k]$ and $[M_1], \ldots, [M_\ell]$ be the respective sets of preferred generators of Pic X and Pic Y (in the sense of the previous subsection), where $k = K$ and $\ell = L$, or $k = \lfloor\frac{K+1}{2}\rfloor$ and $\ell = \lfloor\frac{L+1}{2}\rfloor$, depending on whether a flag variety of general type or a variety of isotropic flags is considered.

Lemma 3.2 For all $j \in \{1, \ldots, \ell\}$, we have $\varphi^*([M_j]) \in \mathbb{Z}_{\geq 0}[L_1]+\ldots+\mathbb{Z}_{\geq 0}[L_k]$.

Proof Since φ is an embedding, if $\mathcal{M}$ is an ample line bundle on Y then $\varphi^*\mathcal{M}$ should be an ample line bundle on X. In view of Lemma 3.1, we must have

$$\varphi^*(\mathbb{Z}_{>0}[M_1] + \ldots + \mathbb{Z}_{>0}[M_\ell]) \subset \mathbb{Z}_{>0}[L_1] + \ldots + \mathbb{Z}_{>0}[L_k].$$

The claim of the lemma follows. □

We now recall from [8] the notion of linear embedding, standard extension, and factorization through direct product.

Definition 3.3 Let $\varphi : X \hookrightarrow Y$ be an embedding of flag varieties as above.

(a) We say that φ is *linear* if

$$\varphi^*[M_j] = 0 \quad \text{or} \quad \varphi^*[M_j] \in \{[L_1], \ldots, [L_k]\}$$

for all $j \in \{1, \ldots, \ell\}$.

(b.1) We say that φ is a *strict standard extension* if there are:

- a linear monomorphism $\alpha : V \hookrightarrow W$ and a decomposition $W = \mathrm{Im}\,\alpha \oplus K$;
- a nondecreasing sequence of subspaces $K_0 = \{0\} \subset K_1 \subset K_2 \subset \ldots \subset K_\ell = K$;

- a surjective, nondecreasing map $\kappa : \{0, 1, \dots, \ell\} \to \{0, 1, \dots, k\}$ such that, for all $i \in \{1, \dots, \ell\}$, $K_{i-1} = K_i \Rightarrow \kappa(i-1) < \kappa(i)$;
- in the case where V and W are equipped with nondegenerate symmetric or antisymmetric forms ω and ϕ, respectively, then the monomorphism α is compatible with the forms in the sense that $\phi(\alpha(v_1), \alpha(v_2)) = \omega(v_1, v_2)$ and the decomposition $W = \operatorname{Im} \alpha \oplus K$ is orthogonal;

so that φ can be expressed as

$$\varphi : (F_0 = \{0\}, F_1, \dots, F_k) \mapsto (\alpha(F_{\kappa(1)}) \oplus K_1, \dots, \alpha(F_{\kappa(\ell)}) \oplus K_\ell).$$

(b.2) When $X = \mathrm{Fl}(p_1, \dots, p_k; V)$ and $Y = \mathrm{Fl}(q_1, \dots, q_\ell; W)$, we say that φ is a *modified standard extension* if φ equals the composition $\delta \circ \tilde{\varphi}$ of a strict standard extension $\tilde{\varphi} : X \hookrightarrow Y^\vee := \mathrm{Fl}(\dim W - q_\ell, \dots, \dim W - q_1; W^*)$ with the isomorphism

$$\delta : Y^\vee \to Y,\ (Z_1, \dots, Z_\ell) \mapsto (Z_\ell^\perp, \dots, Z_1^\perp).$$

(b.3) We say that φ is a *standard extension* if φ is a strict or a modified standard extension.

(c) We say that φ *factors through a direct product* if there are $s \geq 2$, a decomposition $\{p_1, \dots, p_k\} = R_1 \sqcup \dots \sqcup R_s$ into nonempty subsets, and exponents $t_1, \dots, t_s \geq 1$ such that φ factors as the composition

$$X \overset{\psi_R}{\hookrightarrow} \prod_{i=1}^{s} \mathrm{Fl}'(R_i; V)^{t_i} \overset{\psi}{\hookrightarrow} Y$$

where ψ_R is the canonical embedding and ψ is an embedding, and the notation Fl' means Fl or Fl_ω depending on whether X consists of general or isotropic flags.

(d.1) Assume that W is endowed with an orthogonal or, respectively, symplectic form so that V is an isotropic subspace of W. Then, there are natural embeddings

$$X = \mathrm{Fl}(p_1, \dots, p_k; V) \hookrightarrow \mathrm{FlO}(p_1, \dots, p_k; W) \text{ and } X^\vee \hookrightarrow \mathrm{FlO}(p_1, \dots, p_k; W),$$

respectively,

$$X = \mathrm{Fl}(p_1, \dots, p_k; V) \hookrightarrow \mathrm{FlS}(p_1, \dots, p_k; W) \text{ and } X^\vee \hookrightarrow \mathrm{FlS}(p_1, \dots, p_k; W),$$

which we call *isotropic extensions*.

(d.2) A *combination of standard and isotropic extensions* is an embedding of the form

$$\mathrm{FlO}(p_1,\ldots,p_k;V)\stackrel{t}{\hookrightarrow}\mathrm{Fl}(p_1,\ldots,p_k;V)\stackrel{\zeta}{\hookrightarrow}\mathrm{Fl}(q_1,\ldots,q_\ell;V')$$
$$\stackrel{\chi}{\hookrightarrow}\mathrm{FlO}(q_1,\ldots,q_\ell;W)\stackrel{\xi}{\hookrightarrow}\mathrm{FlO}(r_1,\ldots,r_m;W'),$$

respectively,

$$\mathrm{FlS}(p_1,\ldots,p_k;V)\stackrel{t}{\hookrightarrow}\mathrm{Fl}(p_1,\ldots,p_k;V)\stackrel{\zeta}{\hookrightarrow}\mathrm{Fl}(q_1,\ldots,q_\ell;V')$$
$$\stackrel{\chi}{\hookrightarrow}\mathrm{FlS}(q_1,\ldots,q_\ell;W)\stackrel{\xi}{\hookrightarrow}\mathrm{FlS}(r_1,\ldots,r_m;W'),$$

where t is the tautological embedding, ζ,ξ are standard extensions, and χ is an isotropic extension.

Remark 3.4 Given a standard extension $\varphi : X \to Y$, the splitting $W = \mathrm{Im}\,\alpha \oplus K$ is in general not unique if X and Y are of general type, and is unique if X and Y are both of orthogonal or symplectic type, since in the latter cases this splitting is assumed orthogonal.

The following proposition is based on [8, Proposition 2.3].

Proposition 3.5 Let $\varphi : X = \mathrm{Fl}(p_1,\ldots,p_k;V) \hookrightarrow Y = \mathrm{Fl}(q_1,\ldots,q_\ell;W)$ be an embedding of flag varieties.

The following conditions are equivalent:

(i) The embedding φ is linear.
(ii) There are

- A partition $\{1,\ldots,\ell\} = I_0 \sqcup I_1 \sqcup \ldots \sqcup I_k$, with $I_i \neq \emptyset$ for $i \neq 0$.
- A sequence of linear embeddings

$$\varphi[i] = (\varphi_{i,j})_{j\in I_i} : \mathrm{Gr}(p_i;V) \hookrightarrow \prod_{j\in I_i} \mathrm{Gr}(q_j;W) \text{ for } 0 \le i \le k$$

and if $I_0 \neq \emptyset$ a constant map $X_0 := \{\mathrm{pt}\} \hookrightarrow \prod_{j\in I_0} \mathrm{Gr}(q_j;W)$ such that the following diagram commutes:

$$\begin{array}{ccc} X = \mathrm{Fl}(p_1,\ldots,p_k;V) & \xrightarrow{\varphi} & Y = \mathrm{Fl}(q_1,\ldots,q_\ell;W) \\ \downarrow \mu & & \downarrow \pi \\ X_0 \times \prod_{i=1}^k \mathrm{Gr}(p_i;V) & \xrightarrow{\prod \varphi[i]} & \prod_{j=1}^\ell \mathrm{Gr}(q_j;W), \end{array}$$

where the vertical arrows are the natural embeddings.

A similar result holds in the symplectic and orthogonal cases.

Proof (i)⇒(ii) is shown in [8, Proposition 2.3]. (ii)⇒(i): for every $j \in \{1, \dots, \ell\}$, assuming that $j \in I_i$ with $i \neq 0$, we have

$$\begin{aligned}[(\pi \circ \varphi)^* \mathrm{pr}_j^* \mathcal{O}_{\mathrm{Gr}(q_j;W)}(1)] &= [\mu^* \mathrm{pr}_i^* \varphi[i]^* \mathrm{pr}_j^* \mathcal{O}_{\mathrm{Gr}(q_j;W)}(1)] \\ &\in \{0, [\mu^* \mathrm{pr}_i^* \mathcal{O}_{\mathrm{Gr}(p_i;V)}(1)]\} = \{0, [L_i]\}\end{aligned}$$

by the assumption that $\varphi[i]$ is linear. If $j \in I_0$, then

$$[(\pi \circ \varphi)^* \mathrm{pr}_j^* \mathcal{O}_{\mathrm{Gr}(q_j;W)}(1)] = [\mu^* \mathrm{pr}_0^* \varphi[0]^* \mathrm{pr}_j^* \mathcal{O}_{\mathrm{Gr}(q_j;W)}(1)] = 0.$$

The conclusion follows. □

A key result is now the following:

Theorem 3.6 ([7, Theorem 1], [8, Theorem 4.2])

(a) Let $\varphi : X \hookrightarrow Y$ be an embedding of flag varieties of the same type. Assume that φ is linear and does not factor through a direct product, and all the maps $\varphi[i]$ of Proposition 3.5 are standard extensions. Then φ is a standard extension.
(b) Assume that X and Y are grassmannians of the same type. In addition, in the orthogonal case suppose that X and Y are of the form $\mathrm{GrO}(p; V)$ with $p \notin \{\frac{\dim V}{2} - 1, \frac{\dim V}{2}\}$.

(i) If X and Y are of general type, then $\varphi : X \hookrightarrow Y$ is a standard extension if and only if it is linear and does not factor through a projective space.
(ii) If X and Y are of orthogonal or symplectic type, then $\varphi : X \hookrightarrow Y$ is a standard extension or a combination of standard and isotropic extensions if and only if φ is linear and does not factor through a projective space and, in the orthogonal case, also not through a quadric.

3.3 Additional Lemmas

Lemma 3.7

(a) The composition of two standard extensions is a standard extension.
(b) The composition of two standard extensions φ_1 and φ_2 is strict if and only if φ_1 and φ_2 are both strict or are both modified.

Proof Straightforward. □

Lemma 3.8 Let

$$X = \mathrm{Fl}(n_1, \dots, n_k; V) \xhookrightarrow{\ \chi\ } \mathrm{Fl}(q_1, \dots, q_m; U) = Z$$
$$X \xhookrightarrow{\ \varphi\ } Y = \mathrm{Fl}(p_1, \dots, p_\ell; W) \xhookrightarrow{\ \psi\ } Z$$

be a commutative diagram of strict standard extensions. Assume that:

- φ corresponds to $\alpha : V \hookrightarrow W$, a decomposition $W = \operatorname{Im}\alpha \oplus K$, a nondecreasing sequence of subspaces $K_0 = \{0\} \subset K_1 \subset \ldots \subset K_\ell = K$, and a surjective map $\kappa : \{0, \ldots, \ell\} \to \{0, \ldots, k\}$, in the sense of Definition 3.3 (b.1).
- ψ corresponds similarly to $\beta : W \hookrightarrow U$, $U = \operatorname{Im}\beta \oplus L$, $L_0 \subset \ldots \subset L_m = L$, $\lambda : \{0, \ldots, m\} \to \{0, \ldots, \ell\}$.
- χ corresponds similarly to $\gamma : V \hookrightarrow U$, $U = \operatorname{Im}\gamma \oplus M$, $M_0 \subset \ldots \subset M_m = M$, $\mu : \{0, \ldots, m\} \to \{0, \ldots, k\}$.

Then we have $\mu = \kappa \circ \lambda$, $M_i = L_i \oplus \beta(K_i)$ for all $i \in \{1, \ldots, m\}$, and up to modifying β we can assume that $\gamma = \beta \circ \alpha$.

Similar statements hold in the symplectic and orthogonal cases, and the equality $\gamma = \beta \circ \alpha$ always holds in this case.

Proof Since $\chi = \psi \circ \varphi$, for all $\mathcal{F} = (F_1, \ldots, F_k) \in X$, all $i \in \{1, \ldots, m\}$, we have

$$\gamma(F_{\mu(i)}) \oplus M_i = \beta\alpha(F_{\kappa\lambda(i)}) \oplus \beta(K_{\lambda(i)}) \oplus L_i. \tag{1}$$

It follows from the equality $\bigcap_{\mathcal{F} \in X} F_{\mu(i)} = \{0\}$ that

$$M_i = \beta(K_{\lambda(i)}) \oplus L_i \quad \text{for all } i = 1, \ldots, m.$$

Then, since the dimensions of $F_1, \ldots, F_k$ are pairwise distinct, formula (1) implies for dimension reasons

$$\mu(i) = \kappa \circ \lambda(i) \quad \text{for all } i = 1, \ldots, m.$$

Take $i_0 \in \{1, \ldots, m\}$ minimal such that $\mu(i_0) \neq 0$. Then $\bigcup_{\mathcal{F} \in X} F_{\mu(i_0)} = V$ and we must have

$$\operatorname{Im}\gamma \oplus M_{i_0} = \operatorname{Im}\beta \circ \alpha \oplus M_{i_0}.$$

Up to replacing β by some $\tilde{\beta}$ such that $\tilde{\beta}(x) - \beta(x) \in M_{i_0}$ for all $x \in \operatorname{Im}\alpha$ (which will not affect the map ψ), we can assume that

$$\operatorname{Im}\gamma = \operatorname{Im}\beta \circ \alpha.$$

Then by projecting (1) to $\operatorname{Im}\gamma = \operatorname{Im}\beta \circ \alpha$, with respect to the decomposition $U = \operatorname{Im}\gamma \oplus M$, we get

$$\gamma(F_{\mu(i)}) = \beta \circ \alpha(F_{\mu(i)}) \quad \text{for all } i = 1, \ldots, m, \text{ all } \mathcal{F} \in X.$$

Up to multiplying β by a scalar, we can assume that the equality $\gamma = \beta \circ \alpha$ holds. □

4 Construction of Isomorphisms

In this and the next section we prove Theorem 2.5. Here we show that all pairs of ind-varieties that are claimed to be isomorphic in Theorem 2.5 are indeed isomorphic.

We start with the following known fact.

Lemma 4.1

(a) Let $\mathcal{F}$ be a generalized flag in V compatible with two bases E and E'. Then the ind-varieties $\mathrm{Fl}(\mathcal{F}, E, V)$ and $\mathrm{Fl}(\mathcal{F}, E', V)$ are isomorphic.

(b) Moreover, in the case where V is endowed with an orthogonal (respectively, a symplectic) form ω, $\mathcal{F}$ is isotropic, and E and E' are isotropic, then the ind-varieties $\mathrm{FlO}(\mathcal{F}, E, V)$ and $\mathrm{FlO}(\mathcal{F}, E', V)$ (respectively, $\mathrm{FlS}(\mathcal{F}, E, V)$ and $\mathrm{FlS}(\mathcal{F}, E', V)$) are isomorphic.

Proof It suffices to construct a linear automorphism $\alpha : V \to V$ such that

$$\alpha(E) = E', \qquad \forall F \in \mathcal{F},\ \alpha(F) = F,$$

and α preserves the form ω in the situation (b) of the lemma. Then α clearly induces an isomorphism $\mathcal{G} \mapsto \alpha(\mathcal{G})$ between the two considered ind-varieties.

(a) For $F \in \mathcal{F}$, we denote $E_F := \{e \in E : e \in F\}$ and $\hat{E}_F := E_F \setminus \bigcup_{\substack{F' \in \mathcal{F} \\ F' \subset F}} E_{F'}$. We define similarly E'_F and $\hat{E}'_F$. Since the generalized flag $\mathcal{F}$ is E- and E'-compatible, we have

$$F = \langle E_F \rangle = \langle \hat{E}_F \rangle \oplus \sum_{\substack{F' \in \mathcal{F} \\ F' \subset F}} F' \quad \text{for all } F \in \mathcal{F}.$$

This yields decompositions $E = \bigsqcup_{F \in \mathcal{F}} \hat{E}_F$, $E' = \bigsqcup_{F \in \mathcal{F}} \hat{E}'_F$, and, moreover, $|\hat{E}_F| = |\hat{E}'_F| = \dim F / (\sum_{\substack{F' \in \mathcal{F} \\ F' \subset F}} F')$ for all $F \in \mathcal{F}$. Next, for every $F \in \mathcal{F}$, we can choose a bijection $\alpha_F : \hat{E}_F \to \hat{E}'_F$. This defines a bijection $\bigsqcup_{F \in \mathcal{F}} \alpha_F : E \to E'$, whose corresponding automorphism $\alpha : V \to V$ stabilizes each subspace of $\mathcal{F}$.

(b) We adapt the construction made in (a) in the following way. The generalized flag $\mathcal{F}$ is equipped with the involution $F \mapsto F^{\perp}$, and the bases E and E' are equipped with involutions $i_E : E \to E$ and $i_{E'} : E' \to E'$ satisfying the conditions of Definition 2.1. For every $F \in \mathcal{F}$ such that $F \subset F^{\perp}$ we have

$$\langle E_F \rangle = F \subset F^{\perp} = \langle E_{F^{\perp}} \rangle \subset V = F^{\perp} \oplus \langle i_E(E_F) \rangle$$

and for all $F \in \mathcal{F}$ we have either $i_E(\hat{E}_F) \cap \hat{E}_F = \emptyset$ or $i_E(\hat{E}_F) = \hat{E}_F$; the latter equality holds for at most one F, namely, the one, if it exists, such that $F^{\perp} \subsetneq F$ are consecutive subspaces in $\mathcal{F}$. The same applies to E'. Then we have decompositions

$$E = \bigsqcup_{\substack{F \in \mathcal{F} \\ F \subset F^\perp}} i_E(\hat{E}_F) \cup \hat{E}_F, \qquad E' = \bigsqcup_{\substack{F \in \mathcal{F} \\ F \subset F^\perp}} i_{E'}(\hat{E}'_F) \cup \hat{E}'_F.$$

Now for all $F \in \mathcal{F}$ we can find a bijection $\alpha_F : i_E(\hat{E}_F) \cup \hat{E}_F \to i_{E'}(\hat{E}'_F) \cup \hat{E}'_F$ such that $\alpha_F(i_E(e)) = i_{E'}(\alpha_F(e))$ for all e. Whence we have a bijection $\alpha : E \to E'$ and up to replacing the elements in E' by suitable scalar multiples, we can assume that the corresponding automorphism $\alpha : V \to V$ preserves ω. In addition, $\alpha(F) = F$ for all $F \in \mathcal{F}$, and this concludes the proof of the lemma. □

Next we show that $\mathrm{Fl}(\mathcal{F}, E, V)$ and $\mathrm{Fl}(\mathcal{G}, E', W)$ are isomorphic whenever

(A) $\mathcal{F}$ and $\mathcal{G}$ are isomorphic in the sense of Definition 2.4 (a), or
(B) $\mathcal{F}$ is isomorphic to the dual generalized flag $\mathcal{G}^\perp$.

Assume first that $\mathcal{F}$ and $\mathcal{G}$ are isomorphic; hence, there is an isomorphism $\phi : V \to W$ such that $\phi(\mathcal{F}) = \mathcal{G}$. Then $E'' := \phi(E)$ is a basis of W; moreover, $\mathcal{G}$ is compatible with E'', and the map ϕ induces an isomorphism of ind-varieties

$$\mathrm{Fl}(\mathcal{F}, E, V) \xrightarrow{\sim} \mathrm{Fl}(\mathcal{G}, E'', W), \ \mathcal{F}' \mapsto \phi(\mathcal{F}').$$

Thanks to Lemma 4.1, we have an isomorphism $\mathrm{Fl}(\mathcal{G}, E'', W) \cong \mathrm{Fl}(\mathcal{G}, E', W)$. Altogether, we get an isomorphism $\mathrm{Fl}(\mathcal{F}, E, V) \cong \mathrm{Fl}(\mathcal{G}, E', W)$ as desired.

The case (B) is a consequence of (A), Lemma 4.1, and the fact that the map $\mathcal{G}' \mapsto \mathcal{G}'^\perp$ clearly defines an isomorphism of ind-varieties $\mathrm{Fl}(\mathcal{G}, E', W) \xrightarrow{\sim} \mathrm{Fl}(\mathcal{G}^\perp, E'^*, \langle E'^* \rangle)$ where E'^* is the dual family of the basis E'.

Similar reasoning shows that $\mathrm{FlO}(\mathcal{F}, E, V)$ and $\mathrm{FlO}(\mathcal{G}, E', W)$ (respectively, $\mathrm{FlS}(\mathcal{F}, E, V)$ and $\mathrm{FlS}(\mathcal{G}, E', W)$) are isomorphic ind-varieties whenever $\mathcal{F}$ and $\mathcal{G}$ are isomorphic in the sense of Definition 2.4 (b). Note that in this case we have $\mathcal{F}^\perp = \mathcal{F}$ and $\mathcal{G}^\perp = \mathcal{G}$.

We now turn our attention to the additional isomorphisms from Theorem 2.5. First, since in a symplectic space every line is isotropic, the isomorphism between $X = \mathrm{Gr}(F, E, V)$, $Y = \mathrm{GrS}(G, E', W)$, where $\dim F = \dim G = 1$, is obvious.

Finally, the isomorphism between $X = \mathrm{GrO}(F, E, V)$, $Y = \mathrm{GrO}(G, E', W)$, where $F^\perp = F$ and $\dim G^\perp/G = 1$, is also easy to verify. The key observation is that the well-known isomorphism $\mathrm{GrO}(n-1, \mathbb{C}^{2n-1}) \cong \mathrm{GrO}(n, \mathbb{C}^{2n})$ is compatible with standard extensions. More precisely, thanks to Lemma 4.1, we can assume without loss of generality that W is a subspace of codimension one in V. Let

$$E = \{e_1, \hat{e}_1, e_2, \hat{e}_2, \ldots, e_n, \hat{e}_n, \ldots\}$$

be a basis of type D in V, with involution $i_E : e_i \mapsto \hat{e}_i$. Let $F := \langle e_i : i \geq 1 \rangle$ so that $F = F^\perp$. Consider also

$$E' = \{e_1 + \hat{e}_1, e_2, \hat{e}_2, \ldots, e_n, \hat{e}_n, \ldots\}$$

which is a basis of type B in the subspace $W \subset V$, with involution $i_{E'} : e_i \mapsto \hat{e}_i$ for all $i \geq 2$ and $i_{E'}(e_1 + \hat{e}_1) = e_1 + \hat{e}_1$. Let $G := \langle e_i : i \geq 2 \rangle$; thus, $\dim G^\perp/G = 1$.

Set $V_n := \langle e_1, \hat{e}_1, \ldots, e_n, \hat{e}_n \rangle$ and $W_n := \langle e_1 + \hat{e}_1, e_2, \hat{e}_2, \ldots, e_n, \hat{e}_n \rangle$. We have exhaustions

$$\cdots \hookrightarrow \mathrm{GrO}(n, V_n) \overset{\alpha_n}{\hookrightarrow} \mathrm{GrO}(n+1, V_{n+1}) \hookrightarrow \cdots, \qquad X = \varinjlim \mathrm{GrO}(m, V_m)$$

and

$$\cdots \hookrightarrow \mathrm{GrO}(n-1, W_n) \overset{\beta_n}{\hookrightarrow} \mathrm{GrO}(n, W_{n+1}) \hookrightarrow \cdots, \qquad Y = \varinjlim \mathrm{GrO}(m-1, W_m),$$

where $\alpha_n : L \mapsto L \oplus \langle e_{n+1} \rangle$ and $\beta_n : M \mapsto M \oplus \langle e_{n+1} \rangle$.

For every n, there is an isomorphism

$$\phi_n : \mathrm{GrO}(n-1, W_n) \to \mathrm{GrO}(n, V_n), \; M \mapsto \begin{pmatrix} \text{The unique Lagrangian subspace} \\ \hat{M} \in \mathrm{GrO}(n, V_n) \text{ containing } M. \end{pmatrix}$$

Moreover, the diagram

$$\begin{array}{ccc} \mathrm{GrO}(n, V_n) & \overset{\alpha_n}{\hookrightarrow} & \mathrm{GrO}(n+1, V_{n+1}) \\ \uparrow \phi_n & & \uparrow \phi_{n+1} \\ \mathrm{GrO}(n-1, W_n) & \overset{\beta_n}{\hookrightarrow} & \mathrm{GrO}(n, W_{n+1}) \end{array}$$

is commutative. Indeed $\alpha_n \circ \phi_n(M)$ is a Lagrangian subspace in $\mathrm{GrO}(n+1, V_{n+1})$ containing M and e_{n+1}, thus containing $M \oplus \langle e_{n+1} \rangle = \beta_n(M)$, and therefore coinciding with $\phi_{n+1} \circ \beta_n(M)$. Hence, X and Y are isomorphic.

5 Non-existence of Other Isomorphisms

In this section we complete the proof of Theorem 2.5. This is done by proving the following two statements.

Theorem 5.1 Assume that X, Y is a pair of ind-varieties of generalized flags of the same type (general, orthogonal, or symplectic), different from the pair

$$\mathrm{GrO}(F, E, V), \mathrm{GrO}(G, E', W)$$
$$\text{with } \dim F^\perp/F = 0, \dim G^\perp/G = 1, \text{ or vice versa.}$$

Consider two arbitrary (possibly isotropic) generalized flags $\mathcal{F} \in X$ and $\mathcal{G} \in Y$. Then X is isomorphic to Y if and only if $\mathcal{F}$ is isomorphic to $\mathcal{G}$ or to $\mathcal{G}^\perp$.

Theorem 5.2 Assume X, Y are two ind-varieties of generalized flags of different types. Then X is isomorphic to Y if and only if X, Y (or Y, X) is the pair

$$\begin{aligned} &\mathrm{Gr}(F,E,V),\ \mathrm{GrS}(G,E',W) \quad \text{with } \dim F=\dim G=1 \text{ or} \\ &\dim V/F=\dim G=1. \end{aligned} \tag{2}$$

The direct implications in Theorems 5.1–5.2 are shown in Sect. 4. It remains to prove the reverse implications.

We start with some auxiliary results. By $\mathbb{P}^n$ we denote the n-dimensional projective space and by $\mathbb{P}^\infty$ we denote the infinite-dimensional projective ind-space: $\mathbb{P}^\infty = \varinjlim \mathbb{P}^n$. The following proposition extends Theorem 2 from [7].

Proposition 5.3 Let X, Y be two ind-grassmannians, so $X = \mathrm{Gr}(F,E,V)$, $X = \mathrm{GrO}(F,E,V)$, or $X = \mathrm{GrS}(F,E,V)$, and $Y = \mathrm{Gr}(G,E',W)$, $Y = \mathrm{GrO}(G,E',W)$, or $Y = \mathrm{GrS}(G,E',W)$. If $Y = \mathrm{GrO}(G,E',W)$, we assume that $\dim G^\perp/G \notin \{0,1\}$. Then X is isomorphic to Y if and only if one of the following condition holds.

(A) $X = \mathrm{Gr}(F,E,V)$ and $Y = \mathrm{GrS}(G,E',W)$ with $\dim F = \dim G = 1$ or $\dim V/F = \dim G = 1$ (or vice versa).
(B) X and Y are of the same type with $\dim F = \dim G$, or $X = \mathrm{Gr}(F,E,V)$ and $Y = \mathrm{Gr}(G,E',W)$ with $\dim V/F = \dim G < \infty$ (or vice versa).

Proof The case where X is different from $\mathrm{GrO}(F,E,V)$ with $\dim F^\perp/F \in \{0,1\}$ is treated in [7, Theorem 2]. Hence, it remains to show that $X \not\cong Y$ whenever $X = \mathrm{GrO}(F,E,V)$ with $\dim F^\perp/F \in \{0,1\}$.

We will do this by the same method used in [7]. Indeed, using results in [4, Section 4], it is not difficult to check that through any point $x \in X$, there is a family $\mathcal{P}^3$ consisting of maximal 3-dimensional linearly embedded projective subspaces of X, and a family $\mathcal{P}^\infty$ of maximal linearly embedded infinite-dimensional projective ind-spaces. Moreover, the intersection of any space in $\mathcal{P}^3$ with a space in $\mathcal{P}^\infty$ is isomorphic to $\mathbb{P}^2$ or equals the point x.

We claim that this type of configuration of maximal linearly embedded projective spaces passing through a point does not appear on any ind-grassmannian Y. Indeed, it is well known that Y admits a linear embedding into an ind-grassmannian of general type (this embedding being the identity of Y itself is of general type). Using an appropriate such embedding, it is easy to check that the complete list of ind-grassmannians Y having a family of maximal linearly embedded projective spaces $\mathbb{P}^3$ and a family of maximal linearly embedded projective spaces $\mathbb{P}^\infty$ passing through a fixed point $y \in Y$ is:

- $\mathrm{Gr}(F,E,V)$ where $\dim F = 3$ or $\dim V/F = 3$.
- $\mathrm{GrO}(F,E,V)$ where $\dim F = 3$.
- $\mathrm{GrO}(F,E,V)$ where $\dim F^\perp/F \in \{6,7\}$.
- $\mathrm{GrS}(F,E,V)$ where $\dim F = 3$.
- $\mathrm{GrS}(F,E,V)$ where $\dim F^\perp/F = 2$.

However, in all these cases, the intersection of a maximal linearly embedded space $\mathbb{P}^3$ and a maximal linearly embedded ind-space $\mathbb{P}^\infty$ passing through the same point y is isomorphic to $\mathbb{P}^1$ or is the point y itself. This proves our claim. □

Lemma 5.4 Let $X = \mathrm{Fl}(\mathcal{F}, E, V)$, $X = \mathrm{FlO}(\mathcal{F}, E, V)$, or $X = \mathrm{FlS}(\mathcal{F}, E, V)$, and let $\pi : X \to Y$ be an $\mathrm{Aut}\, X$-equivariant smooth surjective morphism, where Y is another ind-variety of generalized flags. Then Y is isomorphic respectively to $\mathrm{Fl}(\mathcal{F}', E, V)$, $\mathrm{FlO}(\mathcal{F}', E, V)$, or $\mathrm{FlS}(\mathcal{F}', E, V)$, where $\mathcal{F}'$ is a generalized subflag of $\mathcal{F}$.

Proof Consider an exhaustion of X by standard extensions

$$\cdots \hookrightarrow X_n \hookrightarrow X_{n+1} \hookrightarrow \cdots, \quad X = \varinjlim X_m$$

and a corresponding exhaustion of Y

$$\cdots \hookrightarrow Y_n := \pi(X_n) \hookrightarrow Y_{n+1} := \pi(X_{n+1}) \hookrightarrow \cdots, \qquad Y = \varinjlim Y_m. \tag{3}$$

Since X is a homogeneous space for the finitary ind-group $\varinjlim \mathrm{Aut}^0 X_n$, where Aut^0 stands for connected component of unity (see also [3] for a description of $\mathrm{Aut}\, X$), any automorphism in $\mathrm{Aut}^0 X_n$ extends to an automorphism of X. Therefore, each projection $\pi_n : X_n \to Y_n$ is $\mathrm{Aut}^0 X_n$-equivariant. Through the theory of finite-dimensional flag varieties, this implies that Y_n is isomorphic to a shorter flag variety X_n' of same type as X_n so that π_n corresponds to the natural projection $X_n \to X_n'$. The standard extensions $X_n \hookrightarrow X_{n+1}$ then induce an exhaustion by standard extensions

$$\cdots \hookrightarrow X_n' \hookrightarrow X_{n+1}' \hookrightarrow \cdots, \qquad X' := \varinjlim X_m'$$

which commutes with the exhaustion (3) via the isomorphisms $Y_n \cong X_n'$. Therefore, Y is isomorphic to the ind-variety of generalized flags $\varinjlim X_n'$ of the form indicated in the statement. □

We can now prove Theorems 5.1 and 5.2.

Proof of Theorem 5.2 First we suppose that $X = \mathrm{Fl}(\mathcal{F}, E, V)$ and $Y = \mathrm{FlS}(\mathcal{G}, E', W)$ are isomorphic ind-varieties. By Lemma 5.4, for every nonzero proper subspace $F \in \mathcal{F}$ there is an isotropic subspace $G \in \mathcal{G}$ such that $\mathrm{Gr}(F, E, V) \cong \mathrm{GrS}(G, E', W)$, and for every nontrivial isotropic subspace $G' \in \mathcal{G}$ there is a subspace $F' \in \mathcal{F}$ such that $\mathrm{GrS}(G', E', W) \cong \mathrm{Gr}(F', E, V)$. Since an isomorphism $\mathrm{Gr}(F, E, V) \cong \mathrm{GrS}(G, E', W)$ can exist only if $\dim G = 1$ and $\dim F = 1$ or $\dim V/F = 1$ (see Proposition 5.3), we infer that $\mathcal{G}$ must be of the form $\mathcal{G} = \{\{0\} \subset G \subset G^\perp \subset W\}$ with $\dim G = 1$, while $\mathcal{F}$ could be only of the form $\mathcal{F} = \{\{0\} \subset F \subset V\}$ with $\dim F = 1$ or $\dim V/F = 1$, or of the form $\mathcal{F} = \{\{0\} \subset F_1 \subset F_2 \subset V\}$ with $\dim F_1 = \dim V/F_2 = 1$. However, the latter is not an option since we would then have $\operatorname{Pic} X = \mathbb{Z}^2 \not\cong \mathbb{Z} = \operatorname{Pic} Y$. The conclusion of the theorem follows.

Next we suppose that $X = \mathrm{Fl}(\mathcal{F}, E, V)$, or $X = \mathrm{FlS}(\mathcal{F}, E, V)$, and $Y = \mathrm{FlO}(\mathcal{G}, E', W)$ are isomorphic. Arguing as in the first case, it suffices to note that an isotropic ind-grassmannian $\mathrm{GrO}(G, E', W)$ is never isomorphic to an ind-grassmannian $\mathrm{Gr}(F, E, V)$ or $\mathrm{GrS}(F, E, V)$. Thus, the claim follows again from Proposition 5.3. □

Proof of Theorem 5.1 Assume that two ind-varieties X and Y as in the statement of Theorem 5.1 are isomorphic. Without loss of generality we may suppose that $X = \mathrm{Fl}(\mathcal{F}, E, V)$, or $X = \mathrm{FlO}(\mathcal{F}, E, V)$, or $X = \mathrm{FlS}(\mathcal{F}, E, V)$, and $Y = \mathrm{Fl}(\mathcal{G}, E', V)$, or $Y = \mathrm{FlO}(\mathcal{G}, E', V)$, or $Y = \mathrm{FlS}(\mathcal{G}, E', V)$, respectively. Fix an isomorphism $\varphi : X \to Y$ with $\varphi(\mathcal{F}) = \mathcal{G}$. In the case of ind-varieties of orthogonal generalized flags, we first assume that $\mathcal{F}$ and $\mathcal{G}$ do not contain isotropic subspaces F with $\dim F^\perp/F \leq 2$. According to our convention from Sect. 2, this is equivalent to assuming that both $\mathcal{F}$ and $\mathcal{G}$ do not contain maximal isotropic subspaces.

The existence of the isomorphism φ implies the existence of a commutative diagram

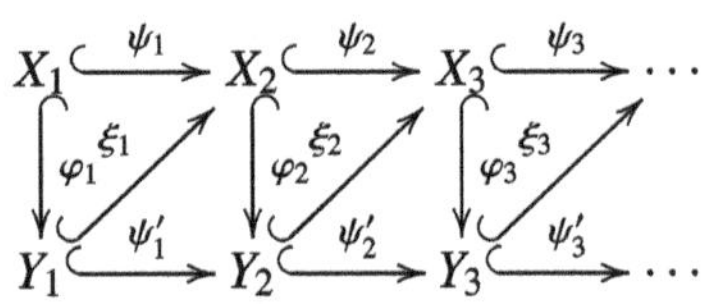
(4)

where all the maps are embeddings and the rows are exhaustions of X and Y, respectively, by standard extensions.

We claim that φ_i and ξ_i are standard extensions for all $i \geq 1$. First we check that φ_i and ξ_i are linear. To do this, we analyze the maps induced on Picard groups. Since $\xi_i^* \circ \varphi_{i+1}^* = \psi_i'^*$ and $\psi_i'^*$ is surjective, it follows that ξ_i^* is surjective. Letting $[M]$ be one of the preferred generators of $\operatorname{Pic} Y_i$, there is $a \in \operatorname{Pic} X_{i+1}$ with $\xi_i^* a = [M]$. Due to Lemma 3.2 we can choose $a = [L]$, where $[L]$ is one of the preferred generators of $\operatorname{Pic} X_{i+1}$. Then $\varphi_i^*[M] = \psi_i^*[L]$ should be equal to 0 or be a preferred generator of $\operatorname{Pic} X_i$ because ψ_i is linear. This shows that φ_i is linear, and a similar argument works for ξ_i. Next, since $\psi_i = \xi_i \circ \varphi_i$ and ψ_i is a standard extension, we deduce from Theorem 3.6 that φ_i and ξ_i do not factor through a direct product or a maximal quadric (in the orthogonal case). Now Theorem 3.6 implies that φ_i and ξ_i are standard extensions as long as all maps induced on grassmannians by φ_i and ξ_i, see Proposition 3.5, are standard extensions. However, if some of these maps induced on grassmannians were not a standard extension, then by Theorem 3.6 (b) it would be necessarily a combination of isotropic and standard extensions. Now the commutativity of diagram (4), together with [7, Lemma 3.9], implies that some map induced on grassmannians by φ_i or ξ_i would be a combination of isotropic and standard extensions. By [7, Remark 3.10], this contradicts the assumption that ψ_i are standard extensions. This establishes the claim that φ_i and ξ_i are standard extensions for all $i \geq 1$.

In the isotropic case, φ_i and ξ_i are strict standard extensions. Let us show that also for X and Y of general type we can reduce the problem to the case where φ_i and ξ_i are strict standard extensions. Indeed, note that there is also the diagram

$$\begin{array}{ccccccc} Y_1 & \overset{\psi_1'}{\hookrightarrow} & Y_2 & \overset{\psi_2'}{\hookrightarrow} & Y_3 & \overset{\psi_3'}{\hookrightarrow} & \cdots \\ {\scriptstyle\sim}\downarrow{\scriptstyle\delta_1} & & {\scriptstyle\sim}\downarrow{\scriptstyle\delta_2} & & {\scriptstyle\sim}\downarrow{\scriptstyle\delta_3} & & \\ Y_1^\vee & \xrightarrow{\psi_1'^\vee} & Y_2^\vee & \overset{\psi_2'^\vee}{\hookrightarrow} & Y_3^\vee & \overset{\psi_3'^\vee}{\hookrightarrow} & \cdots \end{array} \tag{5}$$

where $Y_i^\vee$ and δ_i are as Definition 3.3 (b.2); the bottom line of the diagram forms an exhaustion of $\mathrm{Fl}(\mathcal{G}^\perp, E'^*, \langle E'^*\rangle)$ (see the notation in Sect. 2.2). Up to composing the two diagrams (4) and (5), thus considering $\mathrm{Fl}(\mathcal{G}^\perp, E'^*, \langle E'^*\rangle) = \varinjlim Y_n^\vee$ instead of $\mathrm{Fl}(\mathcal{G}, E', V)$, we can assume that φ_1 is a strict standard extension. Then it follows from Lemma 3.7 (b) that every map φ_i, ξ_i is a strict standard extension.

Let now $V = \bigcup_{n\geq 1} V_n$ be an exhaustion of the vector space V so that X_n and Y_n are varieties of flags in V_n. By Definition 3.3 (b.1), the strict standard extensions in (4) are induced by a diagram of injective linear maps

$$\begin{array}{ccccccc} V_1 & \xrightarrow{\iota_1} & V_2 & \xrightarrow{\iota_2} & V_3 & \xrightarrow{\iota_3} & \cdots \\ \downarrow{\scriptstyle\alpha_1} & \overset{\beta_1}{\nearrow} & \downarrow{\scriptstyle\alpha_2} & \overset{\beta_2}{\nearrow} & \downarrow{\scriptstyle\alpha_3} & \overset{\beta_3}{\nearrow} & \\ V_1 & \xrightarrow{\iota_1'} & V_2 & \xrightarrow{\iota_2'} & V_3 & \xrightarrow{\iota_3'} & \cdots \end{array} \tag{6}$$

where the $\iota_n : V_n \to V_{n+1}$ are the inclusion maps, and the maps α_n are induced by the morphisms φ_n and make the diagram commutative if one disregards the diagonal arrows. Moreover, by Lemma 3.8, the maps β_n can be chosen so that the entire diagram (6) is commutative. The maps ι_n' are then simply the compositions $\iota_n' = \alpha_{n+1}\circ\beta_n$. In the orthogonal and symplectic cases the maps β_n are unique and $\iota_n' = \iota_n$ (see Remark 3.4).

Therefore, this diagram induces a linear isomorphism $\alpha : V \to V'$ where V' is the direct limit of the lower row of the diagram, such that the generalized flags $\alpha(\mathcal{F})$ in V' and $\mathcal{G}$ in V coincide as points of Y. In the orthogonal and symplectic cases, we have simply $\alpha(\mathcal{F}) = \mathcal{G}$, and hence $\mathcal{F}$ and $\mathcal{G}$ are isomorphic.

If X and Y are of general type, we need an additional argument. Note that in this case we have $Y = \mathrm{Fl}(\alpha(\mathcal{F}), \alpha(E), V')$. Since the generalized flags $\mathcal{F}$ and $\alpha(\mathcal{F})$ are isomorphic via the linear operator α, it remains to show that the generalized flags $\alpha(\mathcal{F})$ and $\mathcal{G}$ are isomorphic. This follows from the easy observation that if a generalized flag $\mathcal{H}$ in V and a generalized flag $\mathcal{H}'$ in V' represent the same point in Y, then $\mathcal{H}$ and $\mathcal{H}'$ are isomorphic. Indeed, one builds inductively an isomorphism of direct systems

$$\begin{array}{ccccccc} V_1 & \xrightarrow{\iota_1} & V_2 & \xrightarrow{\iota_2} & V_3 & \xrightarrow{\iota_3} & \cdots \\ \downarrow{\scriptstyle \eta_1} & & \downarrow{\scriptstyle \eta_2} & & \downarrow{\scriptstyle \eta_3} & & \\ V_1 & \xrightarrow{\iota_1'} & V_2 & \xrightarrow{\iota_2'} & V_3 & \xrightarrow{\iota_3'} & \cdots \end{array}$$

such that $\eta_i(\mathcal{H} \cap V_i) = \mathcal{H}' \cap V_i$. Then the direct limit map $\varinjlim \eta_n$ provides an isomorphism between $\mathcal{H}$ and $\mathcal{H}'$. This allows us to conclude that $\mathcal{F}$ and $\mathcal{G}$ are isomorphic generalized flags.

To finish the proof of the theorem, it remains to consider the orthogonal case where

$$X = \mathrm{FlO}(\mathcal{F}, E, V) \quad \text{and} \quad Y = \mathrm{FlO}(\mathcal{G}, E', V),$$

and where one of the generalized flags $\mathcal{F}$ or $\mathcal{G}$ contains a maximal isotropic subspace. There is no loss of generality in assuming that $X \neq \mathrm{GrO}(F, E, V)$ and $Y \neq \mathrm{GrO}(G, E', V)$.

Consider first the case where $\mathcal{F} = \{\{0\} \subset F \subset \tilde{F} \subset F^\perp \subset V\}$ with $\dim \tilde{F}/F = 1$ and $\tilde{F}^\perp = \tilde{F}$. Then $\mathrm{Pic}\, X \cong \mathbb{Z}^2$ and hence $\mathcal{G} = \{\{0\} \subset G \subset \tilde{G} \subseteq \tilde{G}^\perp \subset G^\perp \subset V\}$ with $\dim \tilde{G}^\perp/\tilde{G} \in \{0, 1\}$. If $\mathcal{G}$ is not isomorphic to $\mathcal{F}$, then $\dim \tilde{G}/G \geq 2$ or $\dim \tilde{G}^\perp/\tilde{G} = 1$. In both cases Y admits a proper smooth surjection to $\mathrm{GrO}(G, E', V)$, while the only orthogonal ind-grassmannian to which X admits a proper smooth surjection is $\mathrm{GrO}(\tilde{F}, E, V)$ with $\tilde{F}^\perp = \tilde{F}$. Since $\mathrm{GrO}(G, E', V)$ is not isomorphic to $\mathrm{GrO}(\tilde{F}, E, V)$ by Proposition 5.3, this case is settled.

Now we consider the case of arbitrary orthogonal generalized flags $\mathcal{F}$ and $\mathcal{G}$ containing respective maximal isotropic subspaces. Define projections as follows:

- $\pi_X : X \to \hat{X}$ where $\hat{X} := \mathrm{FlO}(\hat{\mathcal{F}}, E, V)$ is the ind-variety of generalized flags associated to $\hat{\mathcal{F}} := \mathcal{F} \setminus \{F, F^\perp : F \in \mathcal{F},\ \dim F^\perp/F \leq 2\}$,
- $\pi_Y : Y \to \hat{Y}$ where $\hat{Y} := \mathrm{FlO}(\hat{\mathcal{G}}, E', V)$ is the ind-variety of generalized flags associated to $\hat{\mathcal{G}} := \mathcal{G} \setminus \{G, G^\perp : G \in \mathcal{G},\ \dim G^\perp/G \leq 2\}$,

We can assume without loss of generality that $\hat{X}$ and $\hat{Y}$ are both proper ind-varieties of generalized flags (not points) because otherwise we land in the case already considered. (The case of $X = \mathrm{FlO}(\mathcal{F}, E, V)$ for $\mathcal{F} = \{\{0\} \subset F \subset \tilde{F} \subset F^\perp \subset V\}$ where $\dim \tilde{F}/F = 1$ and $\tilde{F}^\perp = \tilde{F}$, and $Y = \mathrm{GrO}(G, E', V)$ for $\mathcal{G} = \{\{0\} \subset G \subset G^\perp \subset V\}$ with $\dim \tilde{G}^\perp/\tilde{G} = 1$ is ruled out by the existence of the isomorphism φ because $\mathrm{Pic}\, X \cong \mathbb{Z}^2$ and $\mathrm{Pic}\, Y \cong \mathbb{Z}$.)

By Lemma 5.4 the isomorphism φ induces an isomorphism $\hat{\varphi} : \hat{X} \to \hat{Y}$ with $\hat{\varphi}(\hat{\mathcal{F}}) = \hat{\mathcal{G}}$. Now the first part of the proof allows us to conclude that there exists an automorphism

$$\hat{\varphi}_V : V \to V, \text{ preserving } \omega \text{ and such that } \hat{\mathcal{G}} = \hat{\varphi}_V(\hat{\mathcal{F}}).$$

We claim that $\hat{\varphi}_V(\mathcal{F}) = \mathcal{G}$, implying that the isotropic generalized flags $\mathcal{F}$ and $\mathcal{G}$ are isomorphic. Indeed, the maximal isotropic space $\tilde{F} \in \mathcal{F}$ is the union of all subspaces $F'' \subsetneq \tilde{F}$ with the property that F'' belongs to some point $\mathcal{F}'' \in X$ and has codimension 2 or more in $\tilde{F}$. A similar statement applies to the maximal isotropic space $\tilde{G} \in \mathcal{G} = \varphi(\mathcal{F})$. Therefore, $\tilde{G}$ equals the union of the spaces $\hat{\varphi}_V(F'')$ and hence coincides with $\hat{\varphi}_V(\tilde{F})$. The same argument applies to spaces $F \in \mathcal{F}$ of codimension 1 in $\tilde{F}$ and $G \in \mathcal{G}$ of codimension 1 in $\tilde{G}$, if they exist, i.e., $\hat{\varphi}_V(F) = G$. Therefore, $\hat{\varphi}_V(\mathcal{F}) = \mathcal{G}$. □

Remark 5.5 Note that the appearance of the space V' in the proof of Theorem 5.1 correlates well with the results in [3] on automorphism groups of ind-varieties of generalized flags. It is essential that V' does not have to coincide with V. Indeed, let $X = \mathrm{Gr}(F, E, V) = Y$ for $\dim V/F = 1$. Then $\mathrm{Aut}\, X = \mathrm{PGL}(\langle E^* \rangle)$ by [3]. Consider the automorphism $\varphi : X \to X$ defined with respect to the basis E^* by the infinite $\mathbb{Z}_{>0} \times \mathbb{Z}_{>0}$ matrix with entries equal to 1 in the entire first row and the diagonal, and equal to 0 elsewhere. Then one can compute that $V' = \langle e_1 - \sum_{i \geq 2} e_i, e_2, e_3, \ldots \rangle$ where the vectors in V are interpreted as linear functions on $\langle E^* \rangle$.

Acknowledgements We thank Valdemar Tsanov for providing with the reference [4] and explaining its relevance. The work of I. P. has been partially supported by DFG Grant PE 980/9-1.

References

1. I. Dimitrov, I. Penkov, Ind-varieties of generalized flags as homogeneous spaces for classical ind-groups. Int. Math. Res. Not. **2004**(55), 2935–2953 (2004)
2. I. Dimitrov, I. Penkov, J.A. Wolf, A Bott-Borel-Weil theory for direct limits of algebraic groups. Am. J. Math. **124**, 955–998 (2002)
3. M.V. Ignatyev, I. Penkov, Automorphism groups of ind-varieties of generalized flags. Transform. Groups (2022). https://doi.org/10.1007/s00031-022-09703-1
4. J.M. Landsberg, L. Manivel, On the projective geometry of rational homogeneous varieties. Comment. Math. Helv. **78**, 65–100 (2003)
5. A.L. Onishchik, Transitive compact transformation groups. Mat. Sb. (N.S.) **60**(102), 447–485 (1963)
6. I. Penkov, C. Hoyt, *Classical Lie Algebras at Infinity*. Springer Monographs in Mathematics (Springer, Cham, 2022)
7. I. Penkov, A.S. Tikhomirov, Linear ind-Grassmannians. Pure Appl. Math. Q. **10**, 289–323 (2014)
8. I. Penkov, A.S. Tikhomirov, An algebraic-geometric construction of ind-varieties of generalized flags. Ann. Mat. Pura Appl. (4) **201**, 2287–2314 (2022)

Spectral Description of Non-Commutative Local Systems on Surfaces and Non-Commutative Cluster Varieties

Alexander Goncharov and Maxim Kontsevich

To the memory of Yuri Ivanovich Manin

Contents

1 Introduction

1.1 Summary

We start from a self-contained exposition of several key results of the chapter, presented in Sect. 1.2. All of them but Proposition 1.2.7 are equipped with complete proofs. Computations proving Proposition 1.2.7 are presented in Sect. 6.3. They show how Gelfand-Retakh's quasideterminants [18] appear from the geometric considerations, making many of their properties obvious.

In particular, we prove in Sect. 1.2 the following. Take a surface S, possibly singular and non-orientable, obtained by gluing triangles and deleting their vertices.

A. Goncharov (✉)
Yale University, New Haven, CT, USA

M. Kontsevich
IHES, Bures sur Yvette, France
e-mail: maxim@ihes.fr

Y. Tschinkel (ed.), *Arithmetic and Algebraic Geometry*, Simons Symposia,
https://doi.org/10.1007/978-3-031-74134-0_5

Then generic local system of m-dimensional vector spaces over a skew field R on $\mathcal{S}$, equipped with a flat section of the associated local system of flags near each deleted vertex, can be described by a flat line bundle on a certain graph Γ.

One can glue a surface from triangles in various ways. Different gluing patterns lead to different graphs Γ. The main question not resolved in Sect. 1.2 is how to relate flat line bundles on different graphs Γ describing the same generic local system on the surface. To address this question we introduce in Sects. 2 and 4 non-commutative cluster varieties of two flavors, related to bipartite ribbon graphs. They carry, respectively, a non-commutative Poisson structure and a closed non-commutative 2-form. In Sects. 5 *and* 6 we define these cluster structures on moduli spaces of R-local systems on surfaces of two flavors, depending on the type of flags near the vertices. Bipartite graphs play a crucial role on the surface called GL_m-*graphs* [19]. A preview can be found in Sect. 1.3.

GL_m-graphs appear in the description of local systems on surfaces out of the blue. In Sect. 7 we present another approach, which explains the role of GL_m-graphs and proves the main results in a bit stronger form. We start from any collection $\mathcal{H}$ of smooth cooriented curves with disjoint Legendrians on an oriented surface. We assign to $\mathcal{H}$ a finite-type stack $\mathcal{M}(\mathcal{H}, \mathbf{1})$. We show explicitly that deformations of the collection $\mathcal{H}$ leaving its components smooth and their Legendrians disjoint lead to equivalent stacks.

Here is an example. Let $\mathcal{S}$ be a surface S with a finite set of punctures. Draw m small simple non-intersecting loops around each puncture s. Denote by $\mathcal{H}_m$ their union. Then $\mathcal{M}(\mathcal{H}_m, \mathbf{1})$ is the stack $\mathcal{X}_m(S)$ of m-dimensional local systems on S with a complete flag near each puncture, invariant under the monodromy around s. On the other hand, zig-zag paths of any bipartite graph Γ on S provide a collection $\mathcal{Z}_\Gamma$ of cooriented curves on S. We show that $\mathcal{M}(\mathcal{Z}_\Gamma, \mathbf{1})$ contains as an open subspace the space $\mathrm{Loc}_1(\Gamma)$ of flat line bundles on the graph Γ. By the very definition of a GL_m-graph, the collection of its zig-zag paths is isotopic to the collection of loops $\mathcal{H}_m$. Such an isotopy provides an open embedding:

$$\mathrm{Loc}_1(\Gamma) \subset \mathcal{M}(\mathcal{Z}_\Gamma, \mathbf{1}) \stackrel{\sim}{=} \mathcal{M}(\mathcal{H}_m, \mathbf{1}) = \mathcal{X}_m(S).$$

So we embed the non-commutative torus $\mathrm{Loc}_1(\Gamma)$ as an open part in $\mathcal{X}_m(S)$. The union of these tori for all GL_m-graphs on S is a cluster Poisson atlas in the sense of Sect. 2, equivariant under the mapping class group action. See the preview in Sect. 1.4.

In Sect. 8 we construct a wide supply of stacks with a similar cluster description. We start from a split reductive group G over $\mathbb{Q}$ and pick any collection $\{\beta_s\}$ of elements of the cyclic envelope of the Artin braid semigroup $\mathrm{Br}^+_{\mathrm{G}}$, assigned to the punctures s of S. We introduce a stack[1] $\mathcal{X}_{\mathrm{G}}(S, \{\beta_s\})$ of G-local systems on S with a pattern of flags near each puncture s determined by the β_s. When $\beta_s = e$ is the unit for all s, and $\mathrm{G} = \mathrm{GL}_m$, we get the stack $\mathcal{X}_m(S)$. When $S = S^2 - \{0, \infty\}$

[1] Here stack means traditional stack used in algebraic geometry.

and $\beta_0 = e$, we recover the cluster Poisson varieties introduced in [9]. Using this and [22] we show that, skipping $S^2 - \{0\}$, any stack $\mathcal{X}_{\rm G}(S, \{\beta_s\})$ carries a cluster Poisson structure equivariant under the action of the wild mapping class group for the pair $(S, \{\beta_s\})$. We get similar results for any surface $\mathcal{S}$ glued from triangles, decorated in addition to the $\{\beta_s\}$ at the punctures s by the elements of the braid semigroup ${\rm Br}^+_{\rm G}$ at the corners.

These stacks are of independent interest. For example, any stack of Stokes data for holomorphic G-bundles with meromorphic connections on a Riemann surface is holomorphically equivalent to one of them. However, unless G is of type A_1, many stacks $\mathcal{X}_{\rm G}(S, \{\beta_s\})$ do not appear this way.

We describe non-commutative analogs of stacks $\mathcal{X}_{{\rm GL}_m}(\mathcal{S}, \{\beta_s\})$ and prove that they carry a non-commutative cluster Poisson structure, equivariant under the wild mapping class group action.

Section 3 contains an application of the constructions of Sect. 2 to non-commutative cluster integrable systems. It is entirely independent of the rest of the chapter.

1.2 Spectral Description of Generic Triples of Flags

Let R be a *skew field*, i.e., a non-commutative division algebra. An m-dimensional R-vector space is a free rank m left R-module. If $m = 1$, we call it an R-line.

A *flag* $\mathcal{F}$ in an m-dimensional R-vector space V is a filtration by R vector spaces:

$$V = \mathcal{F}^0 \supset \mathcal{F}^1 \supset \mathcal{F}^2 \supset \ldots \supset \mathcal{F}^{m-1} \supset \mathcal{F}^m = 0, \quad {\rm codim}\, \mathcal{F}^i = i. \tag{1}$$

We associate to a flag $\mathcal{F}$ a collection of m lines ${\rm gr}^i\mathcal{F} := \mathcal{F}^i/\mathcal{F}^{i+1}, i = 0, \ldots, m-1$.

The group of automorphisms of the R-vector space V acts transitively on the set of all flags in V.

In Sect. 1.2 we describe generic triples of flags in an m-dimensional R-vector space via flat line bundles on the bipartite graph Γ_m, shown on Figs. 1 and 2. It is a self-contained part of the chapter, describing the simplest and at the same time crucial result with all details. Let us start with generic pairs of flags.

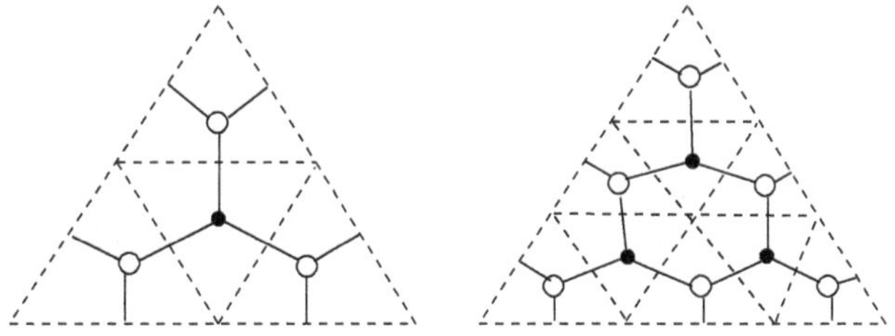

Fig. 1 The bipartite graphs Γ_2 and Γ_3, shown by solid lines

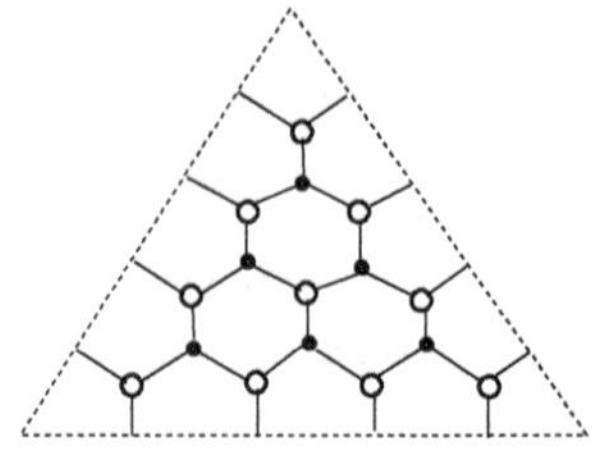

Fig. 2 A bipartite graph Γ_4 for a generic triple of flags $(\mathcal{A}, \mathcal{B}, \mathcal{C})$ in a 4-dimensional space

1. Generic Pairs of Flags
A pair of flags $(\mathcal{A}, \mathcal{B})$ in an m-dimensional R-vector space V is *generic* if for any integers $a, b \geq 0$ such that $a + b = m$ the projections $V \longrightarrow V/\mathcal{A}^a$ and $V \longrightarrow V/\mathcal{B}^b$ induce an isomorphism:

$$V \stackrel{\sim}{\longrightarrow} V/\mathcal{A}^a \oplus V/\mathcal{B}^b.$$

Given a generic pair of flags $(\mathcal{A}, \mathcal{B})$, let us set

$$\mathrm{L}^{b+1} := \mathcal{A}^a \cap \mathcal{B}^b, \quad a + b = m - 1, \;\; a, b \geq 0.$$

The embeddings $\mathrm{L}^a \hookrightarrow V, a = 1, \ldots, m$, give rise to a canonical isomorphism:

$$\mathrm{L}^1 \oplus \ldots \oplus \mathrm{L}^m \stackrel{\sim}{\longrightarrow} V.$$

Lemma 1.2.1 *Assigning to a generic pair of flags* $(\mathcal{A}, \mathcal{B})$ *in an* m*-dimensional* R*-vector space* V *the ordered collection of* m *lines* $(\mathrm{L}^1, \ldots, \mathrm{L}^m)$*, we get an equivalence of groupoids:*

$$\text{Groupoid of generic pairs of flags in } m\text{-dimensional spaces } (V, \mathcal{A}, \mathcal{B}) \stackrel{\sim}{\longrightarrow} \text{Groupoid of ordered collections of } m \text{ lines } (\mathrm{L}^1, \ldots, \mathrm{L}^m). \tag{2}$$

Proof The inverse functor $(\mathrm{L}^1, \ldots, \mathrm{L}^m) \longrightarrow (V, \mathcal{A}, \mathcal{B})$ is given by

$$V := \mathrm{L}^1 \oplus \ldots \oplus \mathrm{L}^m, \quad \mathcal{A}^a := \mathrm{L}^1 \oplus \ldots \oplus \mathrm{L}^{m-a}, \quad \mathcal{B}^b := \mathrm{L}^{b+1} \oplus \ldots \oplus \mathrm{L}^m.$$ □

2. Spectral Description of the Groupoid of Generic Triples of Flags
A triple of flags $(\mathcal{A}, \mathcal{B}, \mathcal{C})$ in an m-dimensional R-vector space V is *generic* if for any nonnegative integers (a, b, c) with $a + b + c = m$, the following map is an

isomorphism:[2]

$$V \longrightarrow V/\mathcal{A}^a \oplus V/\mathcal{B}^b \oplus V/C^c, \qquad a+b+c=m. \tag{3}$$

Let us describe generic triples of flags in an m-dimensional R-vector space.

Let us assign a triple of flags $(\mathcal{A}, \mathcal{B}, C)$ to the vertices of a triangle t, labeled by A, B, C.

We introduce a bipartite graph Γ_m, shown in Figs. 1, 2 for $m \leq 4$. Recall that a *bipartite graph* is a graph with vertices of two types, •-vertices and ∘-vertices, such that each edge has a vertex of each type.

To define the graph Γ_m, take an m-triangulation of the triangle t, shown by punctured lines on Fig. 1. It subdivides each side of the triangle into m equal segments and tessellates the triangle into little triangles of two types: the •-triangles and the ∘-triangles. The bipartite graph Γ_m is the dual graph for this tessellation.

Recall that a flat line bundle L on a graph Γ is given by the following data:

1. A line L_v for every vertex v of Γ
2. An isomorphism $t_{w,v} : L_v \to L_w$ for each edge of Γ with the ends v, w, with $t_{v,w} = t_{w,v}^{-1}$

Theorem 1.2.2 *There is a canonical equivalence between the following groupoids:*

- *Groupoid of generic triples of flags in an m-dimensional R-vector space*
- *Groupoid of R-line bundles with connections on the bipartite graph Γ_m (see Fig. 4)*

Proof

1. *Functor* $\mathcal{L}$: {Generic triples of flags} $\to$ {Flat line bundles on the graph Γ_m}. Given a generic triple of flags $(\mathcal{A}, \mathcal{B}, C)$, let us define a flat R-line bundle $\mathcal{L}(\mathcal{A}, \mathcal{B}, C)$ on the graph Γ_m.

1. The ∘-vertices of the graph Γ_m are parametrized by the triples of nonnegative integers (a, b, c) such that $a+b+c=m-1$. We assign to each ∘-vertex a one-dimensional subspace:

$$\mathrm{L}^{\circ}_{a,b,c} := \mathcal{A}^a \cap \mathcal{B}^b \cap C^c, \qquad a+b+c=m-1. \tag{4}$$

2. The •-vertices of the graph Γ_m are parametrized by the triples of nonnegative integers (a, b, c) such that $a+b+c=m-2$. We assign to each •-vertex a two-dimensional subspace:

$$\mathrm{P}_{a,b,c} := \mathcal{A}^a \cap \mathcal{B}^b \cap C^c, \qquad a+b+c=m-2. \tag{5}$$

[2] Note that a flag in V gives rise to the dual flag in V^*, but the dual to a generic triple of flags may not be generic.

The plane $P_{a,b,c}$ assigned to a $\bullet$-vertex b contains three lines assigned to the $\circ$-vertices incident to b:

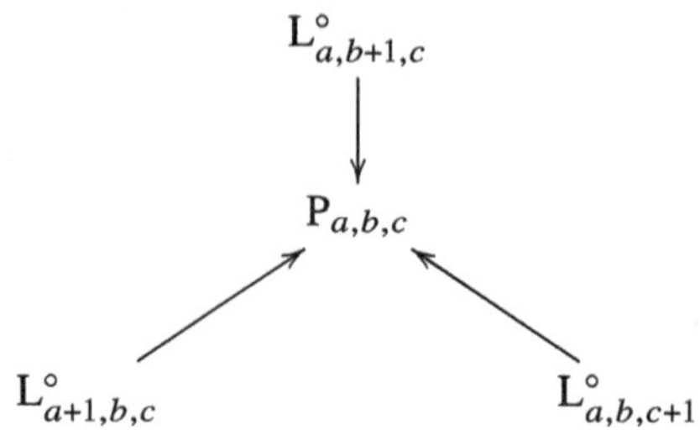

(6)

Indeed, the embedding $\mathcal{A}^{a+1} \hookrightarrow \mathcal{A}^a$ induces the embedding $\mathcal{A}^{a+1} \cap \mathcal{B}^b \cap \mathcal{C}^c \hookrightarrow \mathcal{A}^a \cap \mathcal{B}^b \cap \mathcal{C}^c$. Condition (3) just means that for each plane $P_{a,b,c}$, the three lines (6) are disjoint. We define a line $L^\bullet_{a,b,c}$ as the kernel of the natural map from the sum of the three lines in (6) to the plane:

$$L^\bullet_{a,b,c} := \mathrm{Ker}\Big(L_{a+1,b,c} \oplus L_{a,b+1,c} \oplus L_{a+1,b,c} \longrightarrow P_{a,b,c}\Big). \tag{7}$$

Then there are three maps, induced by projections of the subspace $L^\bullet_{a,b,c}$ onto the summands:

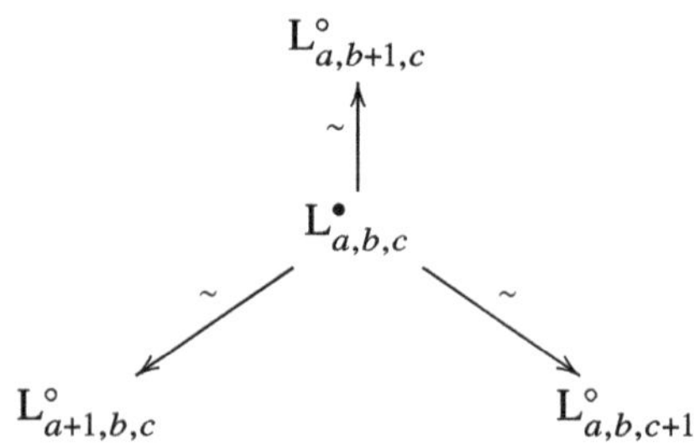

(8)

These three maps are isomorphisms if and only if the three lines in the plane $P_{a,b,c}$ are disjoint.

So we get a flat line bundle on the graph Γ_m: its fibers at the vertices are given by the lines L° and $L^\bullet$, and the parallel transport along an edge $\bullet \to \circ$ is given by a map $L^\bullet \longrightarrow L^\circ$ in (8).

Remark If $R = K$ is commutative, the moduli space of flat line bundles on a graph Γ is identified with $H^1(\Gamma, K^*)$. So flat line bundles on the graph Γ_m are described by their monodromies around the $\binom{m-1}{2}$ holes in the graph. These are literally the canonical coordinates from [7], called the *triple ratios*.

If R is non-commutative, we no longer have canonical coordinates describing generic triples. For example, the isomorphism classes of flat R-line bundles on the graph Γ_3 are described by the elements of R^* considered modulo the conjugation. For $m > 3$ we need to pick a base point and paths to get a collection of elements of R^* defined up to a common conjugation.

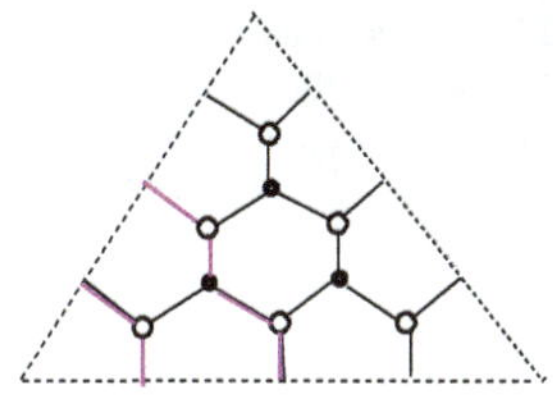

Fig. 3 Vertices of the graph Γ_3 on the distance ≤ 2 from the vertex A are on the two (purple) zig-zags

2. ***Reconstruction Functor*** $\mathcal{R}$: **{Flat line bundles on** Γ_m**}** $\rightarrow$ **{generic triples of flags}**
Given a flat line bundle $\mathcal{L}$ on the graph Γ_m, we get a map from the direct sum over all internal $\bullet$-vertices $\{\mathbf{b}\}$ of the graph Γ_m to the direct sum over all $\circ$-vertices $\{\mathbf{w}\}$ of the graph Γ_m:

$$\mathrm{K}_{\mathcal{L}}: \bigoplus_{\text{internal } \bullet\text{-vertices } \mathbf{b}} \mathcal{L}_{\mathbf{b}} \longrightarrow \bigoplus_{\circ-\text{vertices } \mathbf{w}} \mathcal{L}_{\mathbf{w}}. \tag{9}$$

We define a vector space $V_{\mathcal{L}}$ as the cokernel of this map:

$$V_{\mathcal{L}} := \mathrm{Coker}(\mathrm{K}_{\mathcal{L}}). \tag{10}$$

Let us define three flags in this vector space, assigned to the vertices of the triangle t. Given a vertex A, consider a function d_{A} : {vertices of the graph Γ_m} $\longrightarrow$ $\{1, \ldots, m\}$, given by the distance from a vertex to the A; see Fig. 3.[3] The function $m - d_{\mathrm{A}}$ induces a decreasing filtration on the complex (9).[4] So it induces a filtration on $\mathrm{Coker}(\mathrm{K}_{\mathcal{L}})$. We define the flag $\mathcal{A}^{\bullet}_{\mathcal{L}}$ as the space (10) with this filtration. Explicitly,

$$\mathcal{A}^{a}_{\mathcal{L}} := \mathrm{Coker}\Big(\bigoplus_{d_{\mathrm{A}}(\mathbf{b}) \leq m-a} \mathcal{L}_{\mathbf{b}} \longrightarrow \bigoplus_{d_{\mathrm{A}}(\mathbf{w}) \leq m-a} \mathcal{L}_{\mathbf{w}} \Big), \quad a = 0, \ldots, m-1. \tag{11}$$

This way we get a functor:

$$\mathcal{R} : \mathcal{L} \longrightarrow (\mathcal{A}^{\bullet}_{\mathcal{L}}, \mathcal{B}^{\bullet}_{\mathcal{L}}, C^{\bullet}_{\mathcal{L}}). \tag{12}$$

By the definition, the functors $\mathcal{L}$ and $\mathcal{R}$ provide the equivalence of categories. Theorem 1.2.2 is proved.

□

[3] That is the distance to A from the zig-zag strand containing the vertex on the graph Γ_m and parallel to the side BC.

[4] That is a filtration on each of the spaces, preserved by the operator $K_{\mathcal{L}}$.

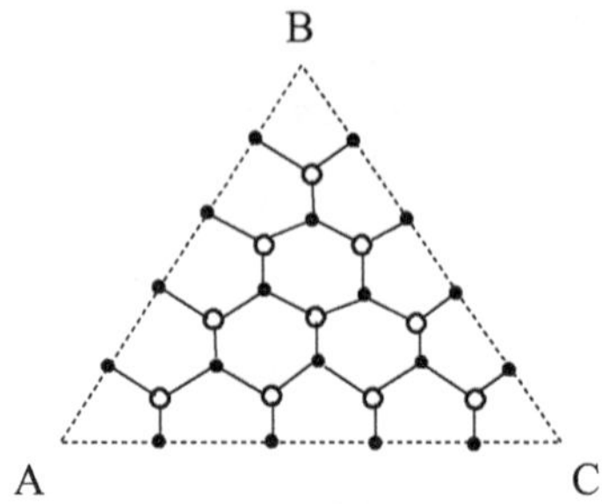

Fig. 4 The bipartite graph $\overline{\Gamma}_4$ includes the $\bullet$-vertices on the sides of the triangle

3. Reconstruction Functor Revisited

Denote by Γ'_m the graph obtained from the graph Γ_m by adding $\bullet$-vertices at the endpoints of external edges (see Fig. 4), pictured on the sides of the triangle.

A flat line bundle $\mathcal{L}$ on $\overline{\Gamma}_m$ provides three m-dimensional R-vector spaces assigned to the sides of the triangle as follows. For example, take the side AB. Let $V^\bullet_{\rm AB}$ be the direct sum of the fibers of $\mathcal{L}$ at the $\bullet$-vertices on the side AB and $V^\circ_{\rm AB}$ the direct sum of the fibers of $\mathcal{L}$ at the $\circ$-vertices closest to the side AB. The parallel transport along the edges from $\bullet$-vertices on the side AB to the closest $\circ$-vertices provides a canonical isomorphism $V^\bullet_{\rm AB} = V^\circ_{\rm AB}$. So we set

$$V_{\rm AB} := V^\bullet_{\rm AB} = V^\circ_{\rm AB}.$$

Denote by $V^\bullet_{\mathcal{L}}$ (respectively $V^\circ_{\mathcal{L}}$) the direct sum of the lines $\mathcal{L}_{\rm b}$ over all $\bullet$-vertices $\mathbf{b}$ (respectively the lines $\mathcal{L}_{\rm w}$ over the $\circ$-vertices $\mathbf{w}$) of the graph Γ_m on Fig. 2. Then there is a canonical map, induced by the embedding $V^\circ_{\rm AB} \hookrightarrow V^\circ_{\mathcal{L}}$, followed by the projection to the quotient $V_{\mathcal{L}}$, defined in (10):

$$\psi_{\rm AB} : V_{\rm AB} \longrightarrow V_{\mathcal{L}}. \tag{13}$$

Lemma 1.2.3 *One has $\psi_{\rm AB} = \psi_{\rm BA}$. The map $\psi_{\rm AB}$ is an isomorphism.*

Proof The first claim is the tautology. The second is equivalent to the statement that

$$V^\circ_{\rm AB} \cap {\rm K}_{\mathcal{L}}(V^\bullet_{\mathcal{L}}) = 0.$$

Let $\widetilde{V}^\circ_{\mathcal{L}}$ be the direct sum of $\mathcal{L}_{\rm w}$ over all but the ones closest to the side AB $\circ$-vertices. So $V^\circ_{\mathcal{L}} = V_{\rm AB} \oplus \widetilde{V}^\circ_{\mathcal{L}}$. Note that the map $K_{\mathcal{L},\rm C} : V^\bullet_{\mathcal{L}} \stackrel{\sim}{\longrightarrow} \widetilde{V}^\circ_{\mathcal{L}}$ given by the parallel transport from each $\bullet$-vertex along the edge towards the vertex C is an isomorphism. There are increasing filtrations of the spaces $\widetilde{V}^\circ_{\mathcal{L}}$ and $V^\bullet_{\mathcal{L}}$, given by the codistance from the $\circ/\bullet$-vertices to the vertex C of the triangle, such that the map ${\rm gr}K_{\mathcal{L}}$ of their associated graded induced by the map $K_{\mathcal{L}}$ is the map $K_{\mathcal{L},\rm C}$. This implies the claim. □

Therefore, we have three canonical isomorphisms $\psi_{\rm AB}, \psi_{\rm BC}, \psi_{\rm CA}$.

$$
\begin{array}{ccccc}
V_{\mathrm{AB}} & & & & V_{\mathrm{BC}} \\
& \searrow^{\psi_{\mathrm{AB}}} & & \swarrow^{\psi_{\mathrm{BC}}} & \\
& & V_{\mathcal{L}} & & \\
& & \uparrow{\scriptstyle \psi_{\mathrm{CA}}} & & \\
& & V_{\mathrm{CA}} & &
\end{array}
\qquad (14)
$$

Our next goal is to calculate the composition

$$\psi_{\mathrm{AC}}^{-1} \circ \psi_{\mathrm{AB}} : V_{\mathrm{AB}} \longrightarrow V_{\mathcal{L}} \longrightarrow V_{\mathrm{AC}}. \qquad (15)$$

To present the answer, we define explicitly an isomorphism:

$$\varphi_{\mathrm{A,B}\to\mathrm{C}} : V_{\mathrm{AB}} \longrightarrow V_{\mathrm{AC}}. \qquad (16)$$

Namely, we define an A-*path* as a path on the graph from a vertex on the side AB to a vertex on the side AC, which always goes towards AC. So given a vertex v_i on the side AB on the distance i from A and a vertex v'_j on the side AC on the distance j from A, there are $\binom{i-1}{j-1}$ A-paths $\gamma : v_i \longrightarrow v'_j$; see Fig. 5. Then we set

$$\varphi_{\mathrm{A,B}\to\mathrm{C}} := \sum_{\gamma} (-1)^{i-1} t_{\gamma}, \qquad (17)$$

where the sum is over all A-paths $\gamma : v_i \longrightarrow v'_j$ from AB to AC, and $t_\gamma : \mathcal{L}_{v_i} \longrightarrow \mathcal{L}_{v'_j}$ is the parallel transport in the local system $\mathcal{L}$ along the path γ.

Proposition 1.2.4 *Map (15) coincides with map (16):*

$$\psi_{\mathrm{AC}}^{-1} \circ \psi_{\mathrm{AB}} = \varphi_{\mathrm{A,B}\to\mathrm{C}}.$$

Proof Let f be a $\bullet$-vertex of the graph Γ_m. Denote by a, b, c the $\circ$-vertices next to f, so that they are in the natural bijection with the vertices of the triangle, e.g., the a is north of f and matches A. Pick a nonzero vector $f \in \mathcal{L}_f$. Denote by e_a, e_b, e_c its parallel transports to the fibers at the vertices a, b, c, respectively.

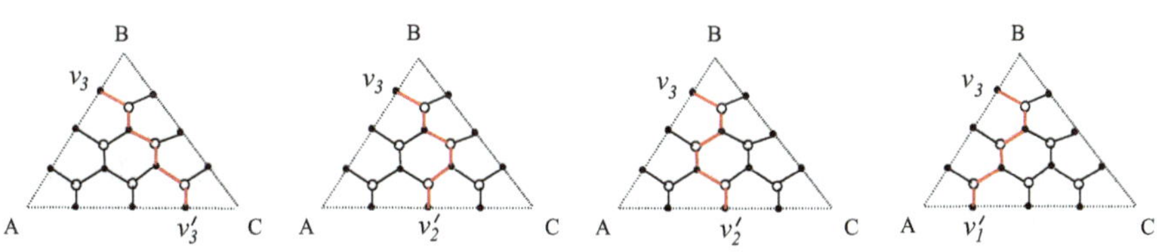

Fig. 5 There are four A-paths from the vertex v_3 on the side AB to the side AC

Denote by e'_a, e'_b, e'_c their projections to $V_{\mathcal{L}}$. Then $e'_a = -e'_b - e'_c$. Note that e_b and e_c are parallel transports of the vector e_a to the fibers of $\mathcal{L}$ at the vertices b and c, respectively. Then apply the same procedure to e_b and e_c and so on. In the end we express the vector e'_a as a linear combination of the vectors at the vertices on the side AC. Applying this to the ∘-vertices closest to the side AB we prove the formula. □

The space V_{AB} carries flags $\mathcal{F}^\bullet_{\text{AB}}$ and $\mathcal{F}^\bullet_{\text{BA}}$, assigned to the orientations of the side. Namely, let $\mathbf{b}_1, \ldots, \mathbf{b}_m$ be the •-vertices on the side AB, oriented A → B. Then the subspaces of the flag $\mathcal{F}^\bullet_{\text{AB}}$ are

$$\mathcal{F}^a_{\text{AB}} := \mathcal{L}_{\mathbf{b}_1} \oplus \ldots \oplus \mathcal{L}_{\mathbf{b}_{m-a}}. \tag{18}$$

It is clear from the definitions that the following three flags in the space $V_{\mathcal{L}}$ coincide:

$$\psi_{\text{AB}}(\mathcal{F}^\bullet_{\text{AB}}) = \psi_{\text{AC}}(\mathcal{F}^\bullet_{\text{AC}}) = \mathcal{A}^\bullet_{\mathcal{L}}. \tag{19}$$

So we can define a variant $\mathcal{R}'$ of the reconstruction functor $\mathcal{R}$ as follows.

Take the three spaces $V_{\text{AB}}, V_{\text{BC}}, V_{\text{CA}}$ assigned to the sides of the triangle. There are six explicitly defined isomorphisms between them: the isomorphism $\psi_{\text{A,B}\to\text{C}}$ given by (17) and the ones obtained by permuting $\{\text{A, B, C}\}$. Proposition 1.2.4 implies that

$$\begin{aligned} &\varphi_{\text{A,C}\to\text{B}} \circ \varphi_{\text{A,B}\to\text{C}} = \text{Id}. \\ &\varphi_{\text{B,C}\to\text{A}} \circ \varphi_{\text{C,A}\to\text{B}} \circ \varphi_{\text{A,B}\to\text{C}} = \text{Id}. \end{aligned} \tag{20}$$

Similar identities are obtained by permuting $\{\text{A, B, C}\}$. Therefore, we can identify the three vector spaces and transform the flags to one of them, getting a configuration of three flags. These flags are in generic position. Finally, it is clear from the construction that we have the following proposition.

Proposition 1.2.5 *The two reconstruction functors $\mathcal{R}$ and $\mathcal{R}'$ coincide.*

The functor $\mathcal{R}'$ is a variant of the snake algorithm [7, Section 9.7] in the commutative case. We use it to prove in Theorem 1.2.9 an analog of the Laurent phenomenon in the non-commutative case.

4. Spectral Description of Decorated Triples of Flags

A *decorated flag* is a flag $\mathcal{F}^\bullet$ as in (1) with a choice of a nonzero element $f_i \in \text{gr}^i\mathcal{F} := \mathcal{F}^{i-1}/\mathcal{F}^i$ in each successive quotient.

A *zig-zag path*, or simply a *zig-zag*, on a ribbon graph Γ is a path turning at every vertex either left or right, so that the left and right turns alternate. Zig-zags on a bipartite ribbon graph are oriented so that they turn right at ∘-vertices. The graph Γ_m carries $3m$ oriented zig-zags; see Fig. 6.

Theorem 1.2.6 *There is a canonical equivalence between the following groupoids:*

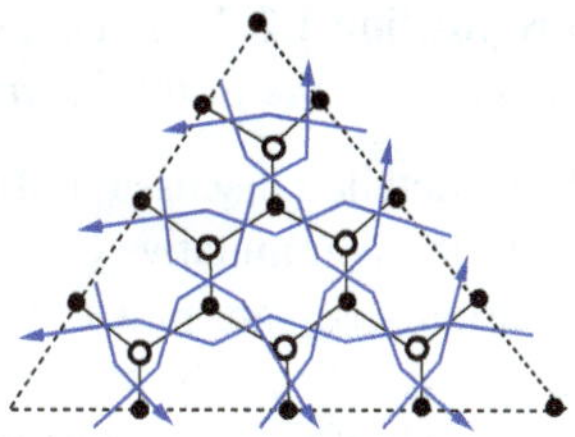

Fig. 6 Nine zig-zag paths on the bipartite graph Γ_3. We push zig-zags a bit out of the vertices

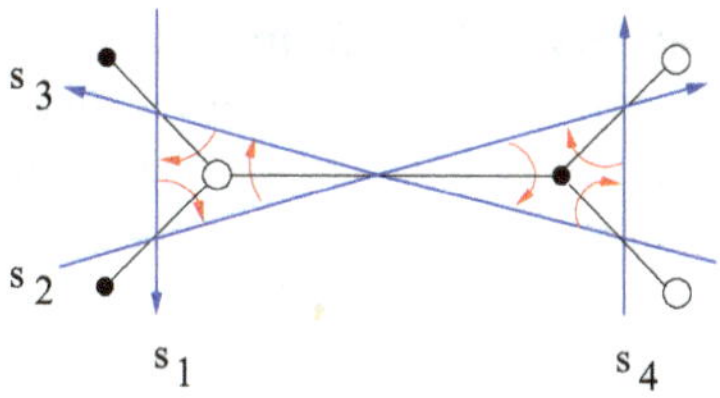

Fig. 7 Writing $\Delta_E = s_1/s_2$, etc. where s_i are flat sections on zig-zags, the counterclockwise products are $s_1/s_2 \cdot s_2/s_3 \cdot s_3/s_1 = 1$ near the ∘-vertex and $s_2/s_3 \cdot s_3/s_4 \cdot s_4/s_2 = 1$ near the •-vertex

1. *Groupoid of generic triples of decorated flags in an m-dimensional R-vector space*
2. *Groupoid of R-line bundles with connections on the bipartite graph Γ_m, trivialized on every zig-zag*

Theorem 1.2.6 is immediately deduced from Theorem 1.2.2 and the description of the reconstruction functor, which turns trivializations of the local system on zig-zags into decoration vectors of the flags.

The notable difference is that in the decorated setting there is a coordinate description. Indeed, for each edge E of the graph Γ_m there are two zig-zags γ_1, γ_2 containing E. They provide two trivializations s_{γ_1} and s_{γ_2} of the restriction of a flat line bundle to E. We assume that the oriented zig-zag γ_1 goes along the edge E in the direction ∘ → •. We define an edge coordinate Δ_E as their ratio:

$$s_{\gamma_1} = \Delta_E \cdot s_{\gamma_2}, \quad \Delta_E \in R^*. \tag{21}$$

Elements $\{\Delta_E\}$ satisfy the following monomial relations. For each internal vertex of Γ_m, the counterclockwise product of the elements Δ_E over the edges incident to the vertex is equal to 1; see Fig. 7. There are no other relations between elements $\{\Delta_E\}$.

Recall that, given an $n \times n$ matrix M over a non-commutative skew field R, Gelfand and Retakh [18] defined quasideterminants M_{ij}, $1 \leq i, j \leq n$. If R is commutative, then the quasideterminant M_{ij} is, up to a sign, the ratio of the determinant of M by a minor obtained by deleting the row i and the column j of M. See (139) for examples of quasideterminants of 2×2 matrices.

Proposition 1.2.7 *Given a generic triple of decorated flags, the invariants* $\{\Delta_E\}$ *at the edges of the graph* Γ_m *are ratios of two Gelfand-Retakh quasideterminants.*

See a detailed discussion of the simplest $m = 2$ case in Sect. 6.3.

In the commutative case, assuming that the space V carries a volume form, the invariants Δ_E are ratios of the $\mathcal{A}$-coordinates A_f [7], assigned to the faces of the graph Γ_m. Precisely, each edge E is shared by two faces f^+ and f^-, where f^+ is to the left of the $\circ \to \bullet$ oriented edge E, and $\Delta_E = A_{f^+}/A_{f^-}$.

5. A Generalization for Surfaces Glued from Triangles
Theorems 1.2.2 and 1.2.6 have the following immediate generalizations. Take a finite collection of triangles, glue them edge to edge in an arbitrary way, and delete the vertices of the glued triangles. The obtained "surface" $\mathcal{S}$ is *singular* along the edges where more than two triangles meet. It might be *non-orientable*. Its boundary is the union of the edges which are not glued to other edges. It has *corners*, also referred to as *special points*, and *punctures* provided by the deleted vertices. The punctures correspond to those deleted vertices whose neighborhoods are topologically nontrivial. We call the union of the corners and punctures *marked points*. These are precisely the points which have been deleted. It comes with a triangulation $\mathcal{T}$.

Let us consider the following groupoids:

1. Groupoid $\mathcal{X}_{m,\mathcal{S}}$ of local systems of m-dimensional R-vector spaces on $\mathcal{S}$, equipped with a flag in a fiber near each special point s, invariant under the monodromy around s, such that for each triangle of $\mathcal{T}$ the triple of flags obtained by moving the flags near the vertices to one point is generic
2. Groupoid $\mathcal{A}_{m,\mathcal{S}}$ of local systems of m-dimensional R-vector spaces on $\mathcal{S}$, equipped with a *decorated* flag in a fiber near each special point s, invariant under the monodromy s, such that for each triangle of $\mathcal{T}$ the triple of flags obtained by moving the flags near the vertices to one point is generic

We call the local systems with the data as in (1) *framed* local systems and as in (2) the *decorated* ones.

If $\mathcal{S} = t$ is just a triangle, $\mathcal{X}_{m,t}$ is the groupoid of triples of flags in generic position an m-dimensional R-vector space. Indeed, any local system on a triangle is trivial, so the triple of flags at the vertices is the only data left. Similarly, $\mathcal{A}_{m,t}$ is the groupoid of triples of decorated flags in generic position in an m-dimensional R-vector space.

The graphs Γ'_m are glued into a bipartite graph $\Gamma_m(\mathcal{T})$: gluing the triangles along the sides we glue the matching black vertices on the sides.

Theorem 1.2.8 *Denote by* $\mathcal{S}$ *such a possibly singular and possibly unorientable triangulated surface.*

(1) The groupoid $\mathcal{X}_{m,\mathcal{S}}$ *is equivalent to the groupoid of flat line bundles on the graph* $\Gamma_m(\mathcal{T})$.
(2) The groupoid $\mathcal{A}_{m,\mathcal{S}}$ *is equivalent to the groupoid of flat line bundles on the graph* $\Gamma_m(\mathcal{T})$*, trivialized on each zig-zag path.*

Proof

(1) Take triangles with the vertices (a, b, c) and (b', c', d) and glue them along the bc-sides, matching the b and c vertices, getting a rectangle $abcd$. The GL_m-graphs Γ_{abc} and $\Gamma_{b'c'd}$ assigned to these triangles are glued into a GL_m-graph Γ_{abcd} for the rectangle $abcd$. Given flat line bundles $\mathcal{L}_{abc}$ and $\mathcal{L}_{b'c'd}$ on the graphs Γ_{abc} and $\Gamma_{b'c'd}$, Theorem 1.2.2 provides generic triples of flags (A, B, C) and (B', C', D) in m-dimensional R-vector spaces V_{abc} and $V_{b'c'd}$, respectively. Each of the two spaces is decomposed into a direct sum of m lines, provided by the pair of flags BC and $B'C'$. The sets of these lines are identified. So there is a collection of isomorphisms of R-vector spaces $i_{bc} : V_{abc} \longrightarrow V_{b'c'd}$ which identifies the lines and the flags $\{B, C\}$ and $\{B', C'\}$. These isomorphisms form a right $(R^\times)^m$-torsor. Indeed, an isomorphism i_{bc} is defined uniquely up to the composition with an R-linear map φ' preserving the lines assigned to the flags B', C'. So we get an $(R^\times)^m$-torsor of configurations of four flags A, B, C, D. Given such an isomorphism i_{bc}, there is a canonical way to glue the flat line bundles $\mathcal{L}_{abc}$ and $\mathcal{L}_{b'c'd}$ into a flat line bundle $\mathcal{L}_{abcd}$ on the graph Γ_{abcd}. The obtained line bundles form a $(R^\times)^m$-torsor. Conversely, we can start from gluing the flat line bundles $\mathcal{L}_{abc}$ and $\mathcal{L}_{b'c'd}$. This way we get a canonical equivalence between the two torsors. Performing this equivalence for each pair of glued triangles, we prove the claim.

(2) The same arguments plus Theorem 1.2.6 provide flat line bundles on the graphs $\Gamma_m(\mathcal{T})$ trivialized on zig-zag paths. □

6. The Non-Commutative Laurent Phenomenon

Take a generic framed local system $\mathcal{V}$ of m-dimensional R-vector spaces on $\mathcal{S}$. Denote by $\mathcal{L}$ the associated flat line bundle on the graph $\Gamma_m(\mathcal{T})$. Take an edge e of the triangulation $\mathcal{T}$. Recall the reconstruction functor $\mathcal{R}'$. The local system $\mathrm{V} :=\mathcal{R}'(\mathcal{L})$ is isomorphic to $\mathcal{E}$ by Theorem 1.2.2 and Proposition 1.2.5. Its fiber V'_e over e is the direct sum of the fibers $\mathcal{L}_{\mathbf{b}_i}$ of the flat line bundle $\mathcal{L}$ over the $\bullet$-vertices $\mathbf{b}_i$ on the edge e:

$$\mathrm{V}'_e = \bigoplus_{\mathbf{b}_i} \mathcal{L}_{\mathbf{b}_i}.$$

So the monodromy of V along any closed path α on $\mathcal{S}$ intersecting the edge e provides a linear map:

$$\mathrm{M}_\alpha : \bigoplus_{\mathbf{b}_i} \mathcal{L}_{\mathbf{b}_i} \longrightarrow \bigoplus_{\mathbf{b}_i} \mathcal{L}_{\mathbf{b}_i}.$$

It is described by the matrix elements $\mu_{ij} : \mathcal{L}_{\mathbf{b}_i} \longrightarrow \mathcal{L}_{\mathbf{b}_j}$.

Given a decorated local system $\mathcal{F}$, the lines $\mathcal{L}_{\mathbf{b}_i}$ carry nonzero vectors. Therefore, the monodromy is described by a matrix μ_{ij} with the entries in $R^\times$. Recall that in this case there are elements $\Delta_E \in R^\times$ assigned to the edges E of the bipartite graph $\Gamma_m(\mathcal{T})$, called the $\mathcal{A}$-coordinates of $\mathcal{F}$.

Theorem 1.2.9 *The matrix elements μ_{ij} of the monodromy of a framed local system $\mathcal{V}$ on $\mathcal{S}$ along a loop α are finite signed sums of the monodromies of the associated flat line bundle $\mathcal{L}$ on the graph $\Gamma_m(\mathcal{T})$ along certain paths from $\mathbf{b}_i$ to $\mathbf{b}_j$ on the graph, homotopic to α.*

The matrix elements of the monodromy of a decorated local system $\mathcal{F}$ on $\mathcal{S}$ are finite signed sums of the products of the associated $\mathcal{A}$-coordinates or their inverses:

$$\mu_{ij} = \sum_{\gamma} \pm \Delta^{\varepsilon_1}_{E_{\gamma,1}} \cdots \Delta^{\varepsilon_k}_{E_{\gamma,k}}, \qquad \varepsilon_j \in \pm 1. \tag{22}$$

The sum is over certain paths γ on the graph from $\mathbf{b}_i$ to $\mathbf{b}_j$, and $E_{\gamma,j}$ are the consecutive edges of the γ. The sign ε in Δ^{ε}_{E} is determined by the direction of the edge E on the path γ: from $\circ \to \bullet$, or $\bullet \to \circ$.

Proof The loop α intersects edges $e_1, e_2, \ldots, e_n$ of $\mathcal{T}$, ordered by the intersection points $e_i \cap \alpha$ on the α. The reconstruction functor $\mathcal{R}'$ for the triangle provides explicit isomorphisms defined by formula (17):

$$\mathrm{V}_e \stackrel{\mathcal{R}'}{\longrightarrow} \mathrm{V}_{e_1} \stackrel{\mathcal{R}'}{\longrightarrow} \ldots \stackrel{\mathcal{R}'}{\longrightarrow} \mathrm{V}_{e_n} \stackrel{\mathcal{R}'}{\longrightarrow} \mathrm{V}_e.$$

The monodromy M_α is their composition. This proves the first claim. The second follows from this since for the decorated local system each map $\mathcal{R}'$ is described explicitly by a matrix, whose entries are given by the signed sums over paths, and each path provides a monomial in $\mathcal{A}$-coordinates or their inverses. □

Formula (22) is the non-commutative analog of the Laurent phenomenon for the monodromies.

When $\mathcal{S}$ is smooth and orientable, it is the surface glued from triangles into a decorated surface. Recall that a *decorated surface* $\mathbb{S}$ is a smooth oriented surface with a finite set of *punctures* inside and *special points* on the boundary, considered modulo isotopy. We assume that each boundary component has special points. A *marked point* is either a special point or a puncture.

We described generic framed/decorated local systems on a decorated surface $\mathbb{S}$ via flat line bundles on the graph $\Gamma_m(\mathcal{T})$ for an ideal triangulation $\mathcal{T}$ of $\mathbb{S}$. The mapping class group $\Gamma_{\mathbb{S}}$ of $\mathbb{S}$ acts by automorphisms of the space of local systems on $\mathbb{S}$, preserving the data at the marked points. We want to see how its action affects the above description. For this we need, for each $\gamma \in \Gamma_{\mathbb{S}}$, to relate the flat line bundles on the graphs $\Gamma_m(\mathcal{T})$ and $\Gamma_m(\gamma(\mathcal{T}))$ corresponding to a generic local system on $\mathbb{S}$. Recall that a flip of a triangulation $\mathcal{T}$ is a new triangulation $\mathcal{T}'$ obtained by flipping the diagonal in one of the rectangles of $\mathcal{T}$. Since any two ideal triangulations are related by a sequence of flips, we have to solve the problem for a single flip. However, if $m > 2$, a flip $\mathcal{T} \to \mathcal{T}'$ results in a complicated transformation of the bipartite graphs $\Gamma_m(\mathcal{T}) \to \Gamma_m(\mathcal{T}')$. In [19] this transformation was decomposed into a composition of $(m-1)^2$ elementary moves, known as *two by two moves*, of bipartite graphs of special kind, called GL_m-graphs. So to relate the flat line bundles on the graphs $\Gamma_m(\mathcal{T})$ and $\Gamma_m(\mathcal{T}')$ assigned to a generic framed/decorated local system on $\mathbb{S}$, we generalize the main constructions of Sect. 1.2 to GL_m-graphs. We discuss this in Sects. 1.3, 1.4.

1.3 Describing Non-Commutative Local Systems on Surfaces

1. Spectral Description of Local Systems on Decorated Surfaces via GL_m**-*graphs***

For a decorated surface $\mathbb{S}$ there are two more flexible variants of Theorem 1.2.8 described below.

Let us recall few more facts about bipartite ribbon graphs; see also Sect. 5.1. A bipartite ribbon graph Γ gives rise to a decorated surface $\mathbb{S}_\Gamma$ homotopy equivalent to Γ, obtained by gluing the ribbons. The graph Γ is embedded to the surface $\mathbb{S}_\Gamma$. A bipartite ribbon graph Γ gives rise to the conjugate bipartite ribbon graph Γ^*, obtained by altering the cyclic order of the edges at each $\bullet$-vertex. The surface Σ_Γ associated with the ribbon graph Γ^* is called the *spectral surface*. The collection $\mathcal{Z}_\Gamma$ of oriented zig-zags of Γ, being pushed to $\mathbb{S}_\Gamma$, is well defined up to an isotopy. It allows to reconstruct the graph Γ. There are elementary transformations of bipartite ribbon graphs, called the two by two move.

A special type of bipartite graphs on $\mathbb{S}$, called GL_m-*graphs*, $m \geq 2$, was introduced in [19]. They are characterized the way their zig-zags relate to the marked points. Precisely (on a finite cover of $\mathbb{S}$), each zig-zag path goes around a single marked point on $\mathbb{S}$, and for each marked point s on $\mathbb{S}$ there are exactly m zig-zag paths going around s. There is also a technical condition: no parallel bigons; see Fig. 30.

Any *ideal triangulation* $\mathcal{T}$ of $\mathbb{S}$, i.e., a triangulation with vertices at the marked points, gives rise to a GL_m-graph $\Gamma_m(\mathcal{T})$ on $\mathbb{S}$. The GL_m-graphs $\Gamma_m(\mathcal{T})$ and $\Gamma_m(\mathcal{T}')$ assigned to the triangulations $\mathcal{T}$ and $\mathcal{T}'$ related by a flip of an edge can be connected by a sequence of two by two moves [19, Section 1.2].

A flat connection on an R-vector bundle induces a flat connection on the bundle of flags in its fibers.

Definition 1.3.1 A framed R-local system on a decorated surface $\mathbb{S}$ is a local system of finite-dimensional R-vector spaces on $\mathbb{S}$ with a framing, given by flat sections of the restriction of the associated local system of flags to the neighborhoods of the marked points.

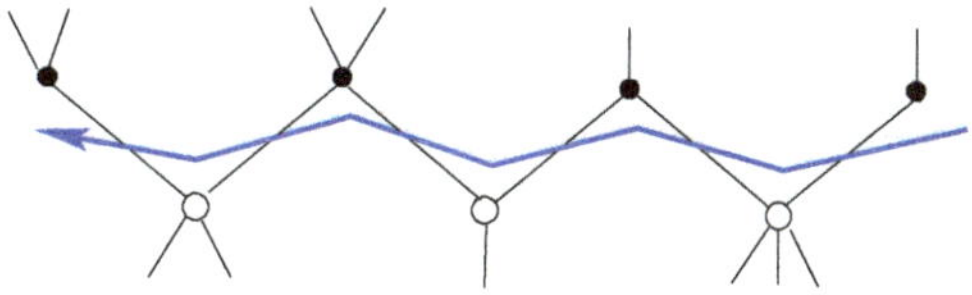

Fig. 8 A zig-zag on a bipartite graph, naturally oriented, and pushed out of the vertices

Theorem 1.3.2 *Let* Γ *be a* $\mathrm{GL_m}$*-graph on a decorated surface* $\mathbb{S}$*, homotopy equivalent to* $\mathbb{S}$*, and* Σ_Γ *the spectral surface. Consider the following two groupoids:*

1. *Groupoid* $\mathcal{X}_{m,\mathbb{S}}$ *of* m*-dimensional framed* R*-local systems on* $\mathbb{S}$
2. *Groupoid* $\mathrm{Loc}_1(\Sigma_\Gamma)$ *of flat* R*-line bundles with connection on the spectral surface* Σ_Γ

Then there is a canonical open embedding:[5]

$$\varphi_\Gamma : \mathrm{Loc}_1(\Sigma_\Gamma) \hookrightarrow \mathcal{X}_{m,\mathbb{S}}. \tag{23}$$

The spectral surface Σ_Γ is homotopy equivalent to the graph Γ^*. So local systems on Σ_Γ are the same thing as local systems on the graph Γ^*. Note also that $\Gamma = \Gamma^*$. So there are canonical identifications:

$$\mathrm{Loc}_1(\Sigma_\Gamma) = \mathrm{Loc}_1(\Gamma) = \mathrm{Loc}_1(\Gamma^*). \tag{24}$$

Theorem 1.3.2 is proved in Sect. 7. The main idea is outlined in Sect. 1.4. It is based on the interpretation of each of the moduli spaces as a moduli space of *admissible dg-sheaves*. We use only a combinatorial description of admissible dg-sheaves, and elementary linear algebra, valid over a skew field.[6]

An elementary proof of a slightly weaker version of Theorem 1.3.2 is given in Theorem 5.1.4. Namely, we restrict to bipartite graphs which can be obtained by compositions of two by two moves from the graphs $\Gamma_m(\mathcal{T})$ related to ideal triangulations and give a simple construction of a functor assigning to a generic local system on $\mathbb{S}$ a flat line bundle on the spectral surface Σ_Γ, providing a birational equivalence of the groupoids (1) and (2). Theorem 5.1.4 is all we need to describe explicitly the action of the mapping class group $\Gamma_\mathbb{S}$ on the space of all framed local systems, discussed in the end of Sect. 1.2.

We show in Sect. 2.1 that any two by two move of bipartite ribbon graphs $\mu : \Gamma \to \Gamma'$ gives rise to a non-commutative birational isomorphism μ intertwining φ_Γ and $\varphi_{\Gamma'}$:

$$\mu : \mathrm{Loc}_1(\Gamma) \stackrel{\sim}{\longrightarrow} \mathrm{Loc}_1(\Gamma'). \tag{25}$$

Therefore, the natural action of the mapping class group of $\mathbb{S}$ on the non-commutative moduli space $\mathcal{X}_{m,\mathbb{S}}$ can be presented by compositions of birational isomorphisms (25).

[5] In particular, they are birationally equivalent.

[6] We do not use constructions involving tensor products of vector spaces. The latter are not defined as R-vector spaces.

Theorems 1.3.2 *and* 5.1.4 are closely related to the Gaiotto-Moore-Neitzke abelianization of vector bundles with a flat connection on a Riemann surface via a flat line bundle on the associated spectral curve [14].[7]

When $R = K$ is commutative and Γ is a bipartite ribbon graph, $\mathrm{Loc}_1(\Gamma)(K) = H^1(\Gamma, K^*)$. Any loop γ on Γ defines a function on $\mathrm{Loc}_1(\Gamma)$ given by the monodromy around the loop. Given a $\mathrm{GL_m}$-graph $\Gamma \subset \mathbb{S}$, the birational isomorphism (23) is nothing else but the cluster Poisson coordinate system on the space $\mathcal{X}_{\mathrm{GL_m},\mathbb{S}}$ of framed $\mathrm{GL_m}$-local systems on $\mathbb{S}$ [19]. The PGL_m version of the story was done in [7].

Going back to a skew field R, the monodromy of a flat R-line bundle over a circle is described by an element of R^* modulo conjugation. So it does not give rise to a well-defined rational function on non-commutative space. The analogs of cluster Poisson coordinate systems are non-commutative birational isomorphisms (23). The analogs of cluster Poisson transformations are non-commutative birational isomorphisms (25) intertwining the ones (23).

Let us introduce the non-commutative analog of the Poisson structure in a more general setup of non-commutative cluster Poisson varieties assigned to arbitrary bipartite ribbon graphs. Here Theorem 5.1.4 is sufficient for the development of the theory and all its applications discussed in the chapter.

2. Non-Commutative Cluster Poisson Varieties

In the commutative case, a cluster Poisson variety is defined by starting from a quiver and considering a collection of quivers related by sequences of mutations. We assign to each quiver a cluster Poisson torus and glue them by using birational isomorphisms between the tori assigned to mutations.

A special class of quivers and mutations is provided by bipartite ribbon graphs and two by two moves between them [20]. A cluster Poisson torus assigned to a bipartite graph Γ is given by $\mathrm{H}^1(\Gamma, \mathbb{G}_m) = \mathrm{Loc}_1(\Gamma)$. It has the following Poisson structure. Recall that the graphs Γ and Γ^* coincide—only their ribbon structures are different. On the other hand, the graph Γ^* is homotopy equivalent to the spectral surface Σ_Γ. So the intersection pairing on $\mathrm{H}_1(\Sigma_\Gamma, \mathbb{Z})$ induces a Poisson structure on $\mathrm{H}^1(\Gamma, \mathbb{G}_m)$. Birational transformations (25) in the commutative case are cluster Poisson transformations.

[7] Following the request of the referee, let us elaborate this. Gaiotto-Moore-Neitzke [14] start from the spectral cover $\pi : \Sigma_{\mathrm{an}} \to C$ related to a point of the Hitchin base of a Riemann surface C and assign to it a spectral network on C. Given a flat line bundle $\mathcal{L}$ on Σ_{an}, they define, using the spectral network, an n-dimensional flat bundle $\mathcal{E}$ on C. The construction in general involves infinite formal sums. When $n > 2$, it is not clear (and probably not always true) that the topological spectral covers corresponding to ideal webs can be realized as the analytic ones Σ_{an}. However, when the ideal web comes from an ideal triangulation of S, [15] defines the spectral network bypassing the spectral curve, and the flat bundle $\mathcal{E}$ is given then by a finite sum construction. It is not clear how to do this for more general ideal webs. In particular, it is not clear how to prove using this approach that flips of triangulations give rise to non-commutative cluster transformations, or to describe the action of the mapping class group of $\mathbb{S}$ on the non-commutative moduli space $\mathcal{X}_{m,\mathbb{S}}$.

In the non-commutative case, the analogs of quivers are bipartite ribbon graphs. The analogs of mutations are two by two moves. The non-commutative cluster Poisson torus assigned to a bipartite graph Γ is the moduli space $\mathrm{Loc}_1(\Gamma)$. Consider a collection of bipartite ribbon graphs $\{\Gamma'\}$ obtained by a sequence of two by two moves from the original graph Γ. Then groupoids $\mathrm{Loc}_1(\Gamma')$ are related by birational transformations (25). We show in the end of Sect. 2.1 that they satisfy the pentagon relations. Therefore, we get a non-commutative analog of a cluster Poisson variety $\mathcal{X}_{|\Gamma|}$ by gluing them according to these birational isomorphisms.

To introduce a non-commutative analog of the Poisson structure, consider the free abelian group $L(\Gamma)$ generated by the loops on the graph Γ. It has a Lie algebra structure, provided by the bipartite ribbon structure on Γ. Since the graphs Γ and Γ^* coincide, there is another Lie bracket on the same vector space, provided by the bipartite ribbon structure of Γ^*. We denote by $L(\Gamma^*)$ the latter Lie algebra.

The commutative algebra given by the symmetric algebra of $L(\Gamma^*)$ has a unique commutative Poisson algebra structure extending the Lie bracket on $L(\Gamma^*)$. It is the non-commutative analog of the Poisson structure on $H^1(\Gamma, \mathbb{G}_m)$. We prove that it is preserved by non-commutative birational isomorphisms (25).

We conclude that the moduli spaces of m-dimensional R-local systems on decorated surfaces carry a non-commutative cluster Poisson variety structure.

Another class of interesting examples is non-commutative Grassmannians, parametrizing the m-dimensional subspaces in a standard N-dimensional R-vector space, and their strata.

Finally, a few words about "non-commutative spaces." We say that X is a non-commutative space if for any skew field R there is a set $X(R)$ of R-points, depending functorially on R. We say that X is a non-commutative stack if there is a non-commutative space $\mathcal{X}$ such that for each skew field R the group $\mathrm{GL}_n(R)$ acts on $\mathcal{X}(R)$, and $X(R) := \mathcal{X}(R)/\mathrm{GL}_n(R)$. For example, for the non-commutive space $\mathrm{Loc}'_m(S)$ of m-dimensional local systems on a surface S, trivialized at a point $x \in S$, the $\mathrm{Loc}'_m(S)(R)$ is the set of m-dimensional R-local systems on S, trivialized x. For the non-commutative stack $\mathrm{Loc}_m(S)$ we have $\mathrm{Loc}_m(S)(R) := \mathrm{Loc}'_m(S)(R)/\mathrm{GL}_m(R)$. Non-commutative spaces/stacks in the chapter, e.g., the ones $\mathcal{X}_m(\mathbb{S})/\mathcal{A}_m(R)$ of framed/decorated m-dimensional local systems on a decorated surface $\mathbb{S}$, are understood this way.

Quite often one can define points of a non-commutative space X with values in the $N \times N$ matrices $\mathrm{Mat}_N(K)$ over a field K, and there are commutative spaces X_N such that $X(\mathrm{Mat}_N(K)) = X_n(K)$. For example, $\mathrm{Loc}'_m(S)(\mathrm{Mat}_N(K)) := \mathrm{Loc}'_{mN}(S)(K)$. Then a non-commutative space gives rise to a collection of classical moduli spaces parametrized by positive integers N. The non-commutative differential forms and symplectic or Poisson structures on such a non-commutative space X are defined so that they provide the usual differential forms, symplectic or Poisson structures on $X(\mathrm{Mat}_N(K))$.

3. An Application: Non-Commutative Cluster Integrable Systems

In Sect. 3, in the special case when Γ is a bipartite graph on a torus, we define a non-commutative generalization of the dimer models and a non-commutative analog of the dimer cluster integrable system constructed in [20]. After the definition of cluster Poisson varieties is given, the construction of the Hamiltonians and the proof that they Poisson commute follows literally the commutative case. In the simplest example we recover the non-commutative integrable system from [26].

Since the dimer cluster integrable systems is equivalent to the relativistic Toda systems studied by Fock and Marshakov [11], this also gives a non-commutative analog of the latter for GL_m.

4. Spectral Description of *Twisted* Local Systems on Decorated Surfaces
Let $T'\mathbb{S}$ be the complement to the zero section of the tangent bundle on $\mathbb{S}$.

Definition 1.3.3 A *twisted local system on* $\mathbb{S}$ is a local system on $T'\mathbb{S}$ with the monodromy -1 around a loop generating $\pi_1(T'_s\mathbb{S})$.

We consider twisted framed local systems of R-modules of rank m on $\mathbb{S}$. Notice that the -1 monodromy around a loop generating $\pi_1(T'_s\mathbb{S})$ preserves the framing flags.

Theorem 1.3.4 *Let* Γ *be a* $\mathrm{GL_m}$*-graph on a decorated surface* $\mathbb{S}$*, homotopy equivalent to* $\mathbb{S}$*, and* Σ_Γ *the spectral surface. Consider the following two groupoids:*

1. *Groupoid of* m*-dimensional twisted framed* R*-local systems on* $\mathbb{S}$
2. *Groupoid of flat twisted* R*-line bundles with connection on the spectral surface* Σ_Γ

Then there is a canonical open embedding of the second groupoid to the first.

Theorem 1.3.2 is the "untwisted" version of Theorem 1.3.4, with the underlined words "twisted" deleted. We can identify untwisted and twisted local systems on Σ_Γ since the obstruction lies in $H^2(\Sigma_\Gamma, \mathbb{Z}/2\mathbb{Z}) = 0$.

5. Non-Commutative Cluster $\mathcal{A}$-Varieties
In the commutative setup, cluster varieties come in pairs, as cluster Poisson varieties and cluster K_2-varieties, also referred to as cluster $\mathcal{A}$-varieties [8]. Moreover, there are two dual variants of moduli spaces of local systems on decorated surfaces, which carry an equivariant cluster variety structure of either Poisson or K_2-flavor. We show that in the non-commutative setup the story is similar. Let us discuss now the non-commutative cluster $\mathcal{A}$-varieties.

A bipartite ribbon graph Γ gives rise to a non-commutative torus $\mathcal{A}_\Gamma$—the moduli space of twisted flat R-line bundles on the associated spectral surface Σ_Γ, trivialized at the boundary. It comes with a canonical collection of $\mathcal{A}$-coordinates $\{\Delta_E\}$, parametrized by the edges of the graph Γ and satisfying monomial relations parametrized by the vertices of Γ.[8] For a two by two move $\mu : \Gamma_1 \longrightarrow \Gamma_2$, we define in Sect. 4.1 a non-commutative birational isomorphism:

[8] Note that boundary components of the spectral surface Σ_Γ are in natural bijection with zig-zags on the original surface $\mathbb{S}_\Gamma$. So this description match the one in Subsection 3, except that now we can talk about the twisted flavor of the story.

$$\mu : \mathcal{A}_{\Gamma_1} \longrightarrow \mathcal{A}_{\Gamma_2}. \tag{26}$$

It serves the role of a non-commutative cluster $\mathcal{A}$-transformation. We show that transformations (26) satisfy the pentagon relations. Therefore, we can glue them into a non-commutative space $\mathcal{A}_{|\Gamma|}$, which we call a non-commutative cluster $\mathcal{A}$-variety.

We prove that a non-commutative cluster $\mathcal{A}$-variety carries a canonical non-commutative 2-form Ω. Precisely, given a bipartite ribbon graph Γ we introduce a non-commutative 2-form Ω_Γ on the torus $\mathcal{A}_\Gamma$. It has a simple expression in the cluster $\mathcal{A}$-coordinates. Birational isomorphisms (26) preserve them:

$$\mu^*\Omega_{\Gamma_2} = \Omega_{\Gamma_1}.$$

This feature of non-commutative cluster $\mathcal{A}$-varieties is quite nontrivial. Its proof Sect. 9 leads to a version of the Steinberg relation for non-commutative 2-forms; see Theorem 9.2.1. In the commutative case, we recover the 2-form associated with cluster algebras/cluster $\mathcal{A}$-varieties [17, 8].

Just as in the commutative case, there is a canonical map of non-commutative cluster varieties:

$$p : \mathcal{A} \longrightarrow \mathcal{X}.$$

It has Hamiltonian properties similar to the ones in the commutative case, where it is the projection along the null foliation of the canonical 2-form on Ω on $\mathcal{A}$.

We stress that in sharp contrast with the commutative case, non-commutative cluster Poisson varieties do not carry any cluster coordinates. Therefore, cluster Poisson considerations require new language.

On the other hand, the non-commutative cluster $\mathcal{A}$-varieties come with a canonical collection of functions. Although we refer to them as cluster $\mathcal{A}$-coordinates, unlike in the commutative case, cluster coordinates in each cluster are not independent; they satisfy certain monomial relations.

6. Spectral Description of the Non-Commutative *Decorated* Local Systems on $\mathbb{S}$ In the commutative case there is the dual moduli space $\mathcal{A}_{\mathrm{SL_m},\mathbb{S}}$ parametrized twisted $\mathrm{SL_m}$-local systems on $\mathbb{S}$ equipped with a *decoration*: a choice of an invariant decorated flag near every marked point of $\mathbb{S}$. It has a cluster $\mathcal{A}$-variety structure [7], which is a geometric incarnation of the Fomin-Zelevinsky upper cluster algebras [13]. There are cluster $\mathcal{A}$-coordinate systems on $\mathcal{A}_{\mathrm{SL_m},\mathbb{S}}$ parametrized by ideal triangulations [7] or, more generally, $\mathrm{SL_m}$-graphs [19].

The non-commutative analog is the moduli space $\mathcal{A}_{m,\mathbb{S}}$ parametrized twisted decorated rank m R-local systems on $\mathbb{S}$. Note that there is no non-commutative analog of $\mathrm{SL_m}$. Its generic points parametrize twisted flat line bundles on the spectral surface Σ which are trivialized at the boundary of Σ. The trivilalizations allow to construct rational functions Δ_E on the moduli space $\mathcal{A}_{m,\mathbb{S}}$ parametrized by the edges E of the graph Γ. These functions are analogs of the cluster $\mathcal{A}$-coordinates.

As we stressed above, the functions Δ_E do not form a coordinate system: they satisfy nontrivial monomial relations. There is no natural way to select a subset of the generators. A similar phenomenon occurs already for the commutative moduli space $\mathcal{A}_{\mathrm{GL_m},\mathbb{S}}$ [19].

7. Berenstein-Retakh Non-Commutative Cluster Algebras Related to Surfaces In Sect. 6 we show that in the case of two-dimensional twisted decorated non-commutative local systems, the obtained coordinate description recovers the non-commutative cluster functions described as ratios of 2×2 quasideterminants by A. Berenstein and V. Retakh in [1]. In our approach formulas for these functions were obtained by geometric considerations, which made their properties evident.

In general the functions Δ_E on the moduli spaces $\mathcal{A}_{m,\mathbb{S}}$ are ratios of the Gelfand-Retakh quasideterminants [18]. This way we get a geometric approach to quasideterminants.

1.4 Non-Commutative Cluster Varieties and Stacks of Admissible dg-Sheaves

Given a bipartite ribbon graph Γ, we take stacks $\mathrm{Loc}_1(\Gamma')$, where graphs Γ' are obtained from Γ by any sequences of two by two moves, and glue them via birational isomorphisms (25) into a space $\mathcal{X}_{|\Gamma|}$.

Then a question arises: can we assign to each bipartite ribbon graph Γ a finite stack $\mathcal{M}^{\circ}_{|\Gamma|}$, whose equivalence class is invariant under two by two moves and which contains $\mathrm{Loc}_1(\Gamma)$ as a Zariski open substack? Equivalently, we want to have a Zariski open embedding of $\mathcal{X}_{|\Gamma|}$ into an a priori defined stack:

$$\mathcal{X}_{|\Gamma|} \subset \mathcal{M}^{\circ}_{|\Gamma|}. \tag{27}$$

To approach this question, we introduce stacks of admissible dg-sheaves on manifolds. This allows to answer the question in many interesting situations, discussed below, although not in full generality yet.

1. Admissible dg-Sheaves
Given a cell complex C, we define in Sect. 7.1 a dg-category of *dg-sheaves*, a dg-version of the derived category of complexes of sheaves constructible for the stratification of C.

Recall that the microlocal support of a constructible complex of sheaves on a manifold X is a complex of sheaves on a conical Lagrangian in the cotangent bundle T^*X [25]. A cooriented hypersurface in a manifold X gives rise to a Lagrangian in T^*X and a Legendrian in its spherical bundle ST^*X. They are given, respectively, by the conormal vectors/rays in the direction specified by the coorientation.

Let $\mathcal{H}$ be a collection of *smooth* cooriented hypersurfaces in a manifold X with *disjoint* Legendrians.

We consider a subcategory of $\mathcal{H}$*-supported* dg-sheaves. These are dg-sheaves whose microlocal support is given by complexes supported at the zero section and Lagrangians assigned to the hypersurfaces of $\mathcal{H}$.

Then we consider a dg-subcategory of $\mathcal{H}$*-admissible* dg-sheaves. Its key features are the following:

1. Admissible dg-sheaves on a manifold X are concentrated in the degrees $[0, 1]$.
2. The microlocal support of an $\mathcal{H}$-admissible dg-sheaf is supported on the union of the zero section of T^*X and a Lagrangian $\mathrm{T}^*_{\mathcal{H}}X$, given by the union of conormal bundles to the cooriented hypersurfaces of $\mathcal{H}$. For each component C of $\mathcal{H}$, the microlocalization along $\mathrm{T}^*_C X$ can be identified with a local system on C.
3. Stacks of $\mathcal{H}$-admissible dg-sheaves are invariant under deformations of $\mathcal{H}$ on X, called below *admissible isotopies*, which keep each component of $\mathcal{H}$ smooth and their Legendrians disjoint.

We demonstrate property (3) directly in Sect. 7.2 in dimension 2. We stress that we allow in the process of deformation non-transversal intersections, e.g., three lines intersecting in a point. Note that a smooth Hamiltonian isotopy of a Legendrian curve can develop a cusp, which we do not allow.

Note that according to [24], the derived category of all constructible sheaves with a given microlocal support is invariant under Legendrian isotopies. However, this does not imply that admissible isotopies preserve $\mathcal{H}$-admissibility, and not all Legendrian isotopies preserve $\mathcal{H}$-admissibility.

Since local systems on the conormal bundle to a cooriented hypersurface are pullbacks of local systems on the hypersurface, we abuse terminology by referring to them as local systems on the hypersurface. Therefore, below we refer to the microlocal support on $\mathrm{T}^*_{\mathcal{H}}X$ as the *Legendrian microlocal support.*

Since $\mathcal{H}$-admissible dg-sheaves are concentrated in degrees $[0, 1]$, not all complexes of constructible sheaves with Legendrian microlocal support in $\mathrm{T}^*_{\mathcal{H}}X$ are $\mathcal{H}$-admissible. Stacks of $\mathcal{H}$-admissible dg-sheaves are of finite type, while stacks of complexes of constructible sheaves with the Legendrian microlocal support in $\mathrm{T}^*_{\mathcal{H}}X$ are not: the latter contain local systems on X placed in arbitrary degrees.

Let Γ be a bipartite ribbon graph. Denote by $\mathcal{Z}_\Gamma$ the collection of oriented zig-zags on the decorated surface $\mathbb{S}_\Gamma$ associated to Γ, well defined up to isotopy. Since the surface $\mathbb{S}_\Gamma$ is oriented, zig-zags are cooriented. The complement $\mathbb{S}_\Gamma - \mathcal{Z}_\Gamma$ is a union of domains of three types, depending on the orientation types of their boundary segments; see Fig. 9 *and* Sect. 7.3.

Lemma 1.4.1 *The stack* $\mathrm{Loc}_1(\Gamma)$ *of flat line bundles on a bipartite ribbon graph* Γ *is canonically equivalent to the stack of* $\mathcal{Z}_\Gamma$*-admissible dg-sheaves on* $\mathbb{S}_\Gamma$ *satisfying the following conditions:*

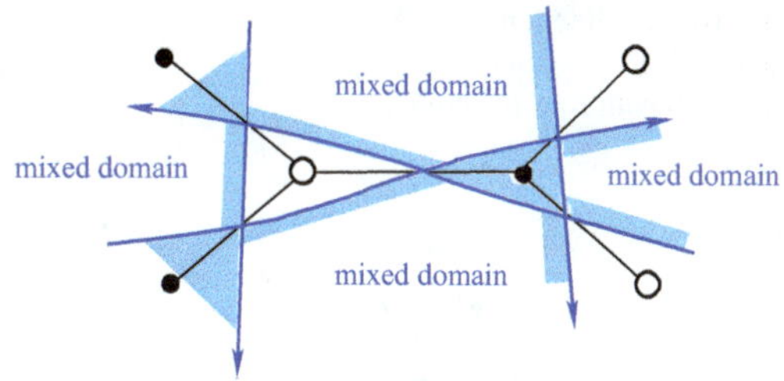

Fig. 9 Zig-zag paths for a bipartite graph Γ on a surface $\mathbb{S}$ cut the surface into •, ◦ and mixed domains. Each zig-zag path is coorientated by a shaded area to the right of the path

1. *They are acyclic on mixed domains—note that this is an open condition.*[9]
2. *Their Legendrian microlocal support is given by flat R-line bundles at every curve of* $\mathcal{Z}_\Gamma$.

Our construction of the functor from the sheaves on the surface to flat line bundles on the graph Γ uses dg-sheaves: the connection on the line bundle on Γ is given by "homotopies between morphisms," absent in the classical description of the derived category. In the commutative setting, another construction of the functor follows from [34].[10]

Denote by $\mathcal{M}(\mathcal{H}, \mathbf{1})$ the stack of $\mathcal{H}$-admissible dg-sheaves on a surface satisfying property 2) of Lemma 1.4.1 for every curve of $\mathcal{H}$. It is a finite-type stack. Then by Lemma 1.4.1 there is an embedding:

$$\mathrm{Loc}_1(\Gamma) \subset \mathcal{M}(\mathcal{Z}_\Gamma, \mathbf{1}). \tag{28}$$

We define $\mathcal{M}^\circ_{|\Gamma|}$ as the component of $\mathcal{M}(\mathcal{Z}_\Gamma, \mathbf{1})$ containing $\mathrm{Loc}_1(\Gamma)$. We show that in many interesting cases we get this way Zarisky open embeddings (27).

2. Examples of Stacks of $\mathcal{H}$-Admissible dg-Sheaves
They include the stacks of Stokes data, which are the Betti side of stacks of flat connections with irregular singularities on Riemann surfaces; see Sects. 8.1, 8.2. This way we get non-commutative stacks of Stokes data stacks as well.

Let us explain how to get the Betti stack of framed flat connections in the case of regular singularities.

For each marked point s of $\mathbb{S}$, in a small punctured at s disc/half disc, consider m non-intersecting and non-self-intersecting circles/arcs ending on the boundary,

[9] The condition that a complex does not have cohomology in two neighboring degrees is an open condition. In particular the conditions that a complex is acyclic, or has a single nontrivial cohomology group, are open conditions. The latter are precisely the conditions we impose in Lemma 1.4.1.

[10] Namely, as was pointed out by the referee, in [34] a Lagrangian L with the topology of the spectral surface of Γ is constructed; it fills the Legendrian of the zig-zag curves. Let $\widetilde{L}$ be its Legendrian lift to $J^1(S) \subset T^*S \times R^1$. By Guillermou's quantization theorem [23, 24], $\mathrm{Loc}_1(L)$ can be identified with the sheaves which are zero at $S \times -\infty$ and pure of microlocal rank one along the conormal to $\widetilde{L}$. Restricting the sheaves to $S \times \infty$ we get the functor.

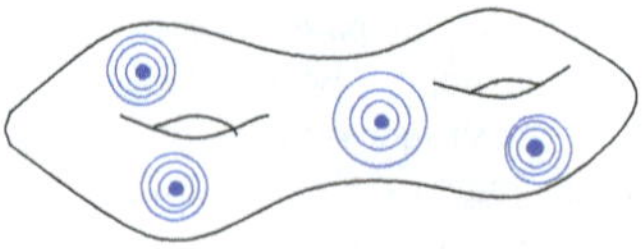

Fig. 10 A collection of curves $\mathcal{I}_{\mathbb{S},3}$ on a genus two surface $\mathbb{S}$ with four punctures

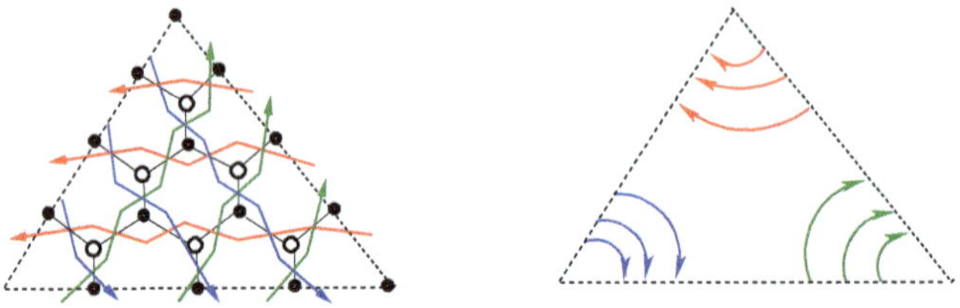

Fig. 11 Zig-zag paths on the bipartite graph Γ_3 on the left can be admissibly deformed to concentric arcs around the vertices, shown on the right

cooriented out of s. Denote by $\mathcal{I}_{\mathbb{S},m}$ their union; see Fig. 10. Its isotopy class is determined uniquely by the surface $\mathbb{S}$ and the integer m.

The following observation is an immediate corollary of the definition of $\mathcal{I}_{\mathbb{S},m}$-admissible dg-sheaves.

Lemma 1.4.2 *The stack $\mathcal{X}_{\mathbb{S},m}$ of m-dimensional R-local systems on a decorated surface $\mathbb{S}$ with complete filtrations near marked points is canonically equivalent to the stack of $\mathcal{I}_{\mathbb{S},m}$-admissible dg-sheaves on $\mathbb{S}$ satisfying the following two conditions:*

1. *They are acyclic near the marked points.*
2. *Their microlocal support is the union of the zero section and lat R-line bundles at every curve of $\mathcal{I}_{\mathbb{S},m}$.*

Lemmas 1.4.1, 1.4.2 provide a non-commutative cluster Poisson structure on the stack $\mathcal{X}_{\mathbb{S},m}$. Indeed, by Lemma 1.4.2, $\mathcal{X}_{\mathbb{S},m}$ is realized as the substack $\mathcal{M}_0(\mathcal{I}_{\mathbb{S},m}, \mathbf{1}) \subset \mathcal{M}(\mathcal{I}_{\mathbb{S},m}, \mathbf{1})$ given by condition (1):

$$\mathcal{X}_{\mathbb{S},m} = \mathcal{M}_0(\mathcal{I}_{\mathbb{S},m}, \mathbf{1}) \subset \mathcal{M}(\mathcal{I}_{\mathbb{S},m}, \mathbf{1}).$$

Consider a bipartite $\mathrm{GL_m}$-graph Γ on $\mathbb{S}$. By its very definition [19], there is an admissible isotopy of the collection $\mathcal{Z}_\Gamma$ of its zig-zags to the collection of curves $\mathcal{I}_{\mathbb{S},m}$; see Fig. 11:

$$i_\Gamma : \mathcal{Z}_\Gamma \longrightarrow \mathcal{I}_{\mathbb{S},m}. \tag{29}$$

By property (3), isotopy (29) provides an equivalence of the related stacks of $\mathcal{H}$-admissible dg-sheaves:

$$i_{\Gamma *} : \mathcal{M}(\mathcal{Z}_\Gamma, \mathbf{1}) \xrightarrow{\sim} \mathcal{M}(\mathcal{I}_{\mathbb{S},m}, \mathbf{1}).$$

By Lemma 1.4.1, the stack $\mathrm{Loc}_1(\Gamma)$ is realized as the Zariski open substack $\mathcal{M}_{\mathrm{mix}=0}(\mathcal{Z}_\Gamma, \mathbf{1})$ of $\mathcal{M}(\mathcal{Z}_\Gamma, \mathbf{1})$ defined by condition (1):

$$\mathrm{Loc}_1(\Gamma) = \mathcal{M}_{\mathrm{mix}=0}(\mathcal{Z}_\Gamma, \mathbf{1}) \subset \mathcal{M}(\mathcal{Z}_\Gamma, \mathbf{1}).$$

Indeed, the acyclicity on mixed domains in Lemma 1.4.1 is an open condition. Acyclicity on mixed domains implies acyclicity near marked points since marked points are contained in the mixed domains (see Fig. 11) when $\mathbb{S}$ is a triangle. The general case reduces to this since we can get a GL_m-graph on $\mathbb{S}$ by amalgamating the graphs Γ_m on triangles of an ideal triangulation of $\mathbb{S}$. Therefore, $i_{\Gamma *}$ induces an open embedding $j_\Gamma : \mathcal{M}_{\mathrm{mix}=0}(\mathcal{Z}_\Gamma, \mathbf{1}) \subset \mathcal{M}_0(\mathcal{I}_{\mathbb{S},m}, \mathbf{1})$. Summarizing, we get the commutative diagram:

$$\begin{array}{ccc} \mathrm{Loc}_1(\Gamma) = \mathcal{M}_{\mathrm{mix}=0}(\mathcal{Z}_\Gamma, \mathbf{1}) & \xrightarrow{\subset} & \mathcal{M}(\mathcal{Z}_\Gamma, \mathbf{1}) \\ {\scriptstyle j_\Gamma}\big\downarrow{\scriptstyle \cap} & & {\scriptstyle i_{\Gamma *}}\big\downarrow{\scriptstyle \sim} \\ \mathcal{X}_{\mathbb{S},m} = \mathcal{M}_0(\mathcal{I}_{\mathbb{S},m}, \mathbf{1}) & \xrightarrow{\subset} & \mathcal{M}(\mathcal{I}_{\mathbb{S},m}, \mathbf{1}). \end{array} \tag{30}$$

In particular, for any GL_m-graph Γ on $\mathbb{S}$, we have, as was promised in (27),

$$\mathcal{M}^0_{|\Gamma|} = \mathcal{X}_{\mathbb{S},m}. \tag{31}$$

For example, the bipartite graph Γ_m on Fig. 11 allows to recover the whole stack of all triples of flags as the stack $\mathcal{M}^0_{|\Gamma_m|}$, rather than describing its generic part, as Theorem 1.2.2 does.

The stack $\mathrm{Loc}_1(\Gamma)$ is a non-commutative Poisson torus. So open embeddings $j_\Gamma : \mathrm{Loc}_1(\Gamma) \subset \mathcal{X}_{\mathbb{S},m}$ for bipartite $\mathrm{GL_m}$-graphs Γ on $\mathbb{S}$ define a non-commutative cluster Poisson structure on the stack $\mathcal{X}_{\mathbb{S},m}$. The action of the mapping class group $\Gamma_\mathbb{S}$ of $\mathbb{S}$ preserves the class of $\mathrm{GL_m}$-graphs on $\mathbb{S}$. Any two of them are related by a sequence of two by two moves [19], providing admissible isotopies of their zig-zags. Therefore, we get a $\Gamma_\mathbb{S}$ equivariant non-commutative cluster Poisson structure on the stack $\mathcal{X}_{\mathbb{S},m}$.

In Sect. 8.3 we describe stacks of framed Stokes data via admissible dg-sheaves and show that

$$\textit{Stacks of (framed) Stokes data are shadows}$$
$$\textit{of non-commutative (framed) stacks of Stokesdata.} \tag{32}$$

Theorem 1.4.3 *For any stack of framed Stokes data $\mathcal{S}$ on a decorated surface $\mathbb{S}$, there is a non-commutative stack $\mathcal{S}^{\mathrm{nc}}$ over* $\mathrm{Spec}(\mathbb{Z})$, *whose restriction to commutative fields is the stack $\mathcal{S}$. Assume that $\mathbb{S} \neq S^2 - \{\infty\}$. Then the stack $\mathcal{S}^{\mathrm{nc}}$ carries a non-commutative cluster Poisson structure equivariant under the wild mapping class group.*

The argument is based on the observation that, similarly to Lemma 1.4.2, stacks of Stokes data can be realized as stacks of $\mathcal{L}$-admissible dg-sheaves on $\mathbb{S}$ vanishing near the punctures, where $\mathcal{L}$ is the union of certain local collection of curves $\mathcal{L}_p$ around the punctures p.

When $\mathbb{S} = S^2 - \{0, \infty\}$, with one regular and one irregular point, these stacks are identified birationally with cluster Poisson varieties assigned in [9] to the cyclic closures of elements of the braid semigroup. Thanks to [9, Section 3.10], their quivers are described by bipartite graphs. The general case is reduced to this one plus the regular triangle case. The case when $\mathbb{S} = S^2 - \{\infty\}$ requires a different technology and is not considered in the chapter. Yet the result is true in this case as well.

3. Stacks of Framed G-Stokes Data, Their Cluster Poisson Structure *and* **Quantization**
Stacks of G-Stokes data provide the Betti realization of stacks of G-bundles with meromorphic connections with arbitrary singularities on a Riemann surface Σ. We introduce stacks of *framed* G-Stokes data and prove that they carry a cluster Poisson structure, equivariant under the wild mapping class group action.

Theorem 1.4.4 *Assume that* $\mathbb{S} \neq S^2 - \{\infty\}$. *For any split reductive group* G *over* $\mathbb{Q}$ *with connected center, the stack of framed* G*-Stokes data on* $\mathbb{S}$ *has a cluster Poisson structure, equivariant under the action of the wild mapping class group. Stacks of plain* G*-Stokes data carry compatible Poisson structures.*

The Poisson nature of the moduli spaces of Stokes data was studied by P. Boalch [3].

Theorems 1.4.4– 1.4.5 are valid for a much larger class of *ideal* Legendrian links $\mathcal{L}$; see Sect. 8.2.

Theorem 1.4.4 is proved in Sect. 8.4 using the following key ingredients:

1. If $\mathbb{S} = S^2 - \{0, \infty\}$, with one regular and one irregular point, the stacks of G-Stokes data are identified birationally with the cluster Poisson varieties assigned in [9] to the cyclic closures of certain elements of the braid semigroup $\mathrm{Br}_{\mathrm{G}}^{+}$ related to G.
2. When $\mathbb{S} = t$ is a triangle, the cluster Poisson structure of the related space $\mathcal{P}_{\mathrm{G},t}$ of configurations of triples of flags with pinnings was described in [22]. The key point is its cyclic invariance.
3. If $\mathbb{S} = S^2 - \{\infty\}$, the stacks of Stokes data are given by certain cyclically ordered configurations of flags. The cluster Poisson structure for sufficiently generic cyclic configurations follows from [22].

The general case is built from these ones, except for single irregular point—that is $\mathbb{S} = S^2 - \{\infty\}$—of rather degenerate type, which requires a different technique, and is omitted.

Combining Theorem 1.4.4 with the cluster quantization machine [10], we get the following.

Theorem 1.4.5 *Under the same assumptions as in Theorem 1.4.4, the stacks of G-Stokes data, plain or framed, admit a cluster Poisson quantization, equivariant under the wild mapping class group action.*

Precisely, we apply the cluster quantization to the cluster Poisson space of framed G-Stokes data. Then we use the fact that forgetting the framing we get a map to the space of plane G-Stokes data, which is finite Galois map over a generic point. Its Galois group, given by a product of the Weyl groups of G, acts by cluster Poisson transformations. This allows us to pass to its invariants.

Theorem 1.4.5 completes the program of cluster quantization of the moduli spaces of flat connections on Riemann surfaces. In the regular case this was done in [7] for $G = PGL_m$ and in [22] in general.

4. Cluster $\mathcal{A}$-Varieties via Stacks of Admissible dg-Sheaves

The analogs of cluster $\mathcal{A}$-varieties in this setup are given by stacks of (twisted) $\mathcal{H}$-admissible dg-sheaves with the following extra condition:

Local systems on all components of the microlocal support but the zero section are trivialized.

In particular, given a bipartite ribbon graph Γ we recover non-commutative cluster $\mathcal{A}$-varieties discussed in Sect. 1.3. Indeed, zig-zags on the surface $\mathbb{S}_\Gamma$ match boundary curves on the spectral surface Σ_Γ. So the trivialization of the local systems on zig-zags on $\mathbb{S}_\Gamma$, describing the microlocal support of an admissible dg-sheaf, amounts to the trivialization of the local system on the spectral surface Σ_Γ on the boundary of Σ_Γ. The cluster $\mathcal{A}$-coordinates are assigned to the edges of Γ. They are obtained by comparing trivializations of the local systems on the two boundary components by using the parallel transport along the canonical up to isotopy path connecting two boundary components.

Remark The stacks of Stokes data are the geometric analogs of (families of) wildly ramified representations of the absolute Galois group; see [6]. So Theorem 1.4.5 should be useful in various flavors of the geometric quantum Langlands correspondence.

5. CY_3-Categories Assigned to Stacks of Stokes Data

Description of the cluster Poisson structure on stacks of Stokes data via webs of zig-zags for certain bipartite graphs on the surface S has another important application. Recall that any bipartite graph gives rise to a canonical quiver with potential, which determines a combinatorially defined 3d CY A_∞-category. The categories assigned to bipartite graphs related by two by two moves are equivalent. Therefore, we arrive to a combinatorially defined 3d Calabi-Yau A_∞-category assigned to any topological Stokes data $(\mathcal{L}, d)$. This category should describe the Fukaya category of the family of open complex CY threefold assigned to the same data $(\mathcal{L}, d)$. This story in the regular case is described in [19]. The $m = 2$ case is established in [4, 36].

6. Comments

Sections 7 and 8.3 are very closely related to the series of works of Shende-Treumann-Williams-Zaslow [33, 34, 35], where they studied extensively, in the commutative setting, moduli spaces of rank one microlocal sheaves, including the ones related to bipartite surface graphs. They usually work with all constructible complexes of sheaves with given microlocal support rather than with finite-type stacks of admissible ones. The restrictions on constructible sheaf categories to cut down to finite-type stacks appear in [33, Proposition 5.20], [34, Proposition 5.20]. The admissibility condition appears in [34, Proposition 5.3].[11] We use the very existence of the category of sheaves with given microlocal support [25]. Other than that, our exposition is elementary.

Our work also relates to some aspects of the Gao-Shen-Weng program [12]. In particular, non-commutative Plucker relations appear in their approach as well.

2 Non-Commutative Cluster Poisson Varieties

2.1 Two by Two Moves for Non-Commutative Flat Line Bundles

Definition 2.1.2 for the action of two by two moves on flat line bundles is dictated by the example provided by configurations of four lines in a two-dimensional space given in Sect. 1.2.

Recall that a ribbon graph is a graph with a cyclic order of the edges at every vertex. We allow ribbon graphs with external, that is, 1-valent, vertices. A ribbon graph Γ is the same thing as a graph on a decorated surface $\mathbb{S}$ which is homotopy equivalent to the graph.

A *face path* on Γ is a path turning right at every vertex. There is a bijection:

$$\{\text{face paths on } \Gamma\} \longleftrightarrow \{\text{marked points on } \mathbb{S}\}. \tag{33}$$

Face intervals correspond to special points on $\mathbb{S}$. Face loops correspond to punctures of $\mathbb{S}$.

A *zig-zag path* on Γ is a path turning at every vertex either left or right, so that the left and right turns alternate. A zig-zag path on a bipartite ribbon graph Γ is oriented so that it turns right at the $\circ$-vertices and turns left at the $\bullet$-vertices; see Fig. 29.

1. Two by Two Moves

Definition 2.1.1 Let Γ be a bipartite ribbon graph containing a subgraph α on the left of Fig. 12. A *two by two move* $\mu_\alpha : \Gamma \to \Gamma'$ *at* α produces a new bipartite graph Γ', obtained by replacing α by the subgraph α' on the right of Fig. 12.

[11] We are grateful to the referee for the last two comments.

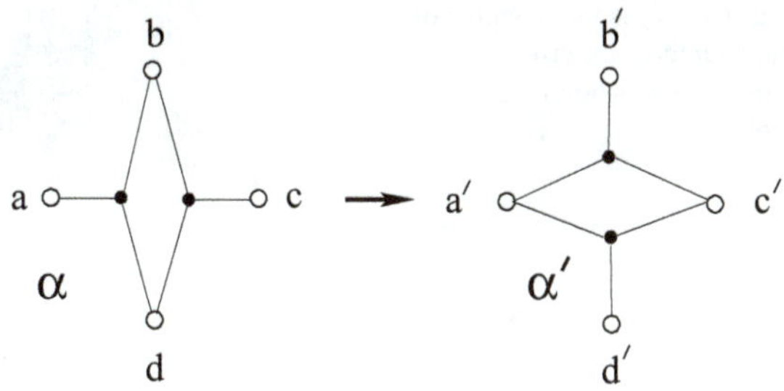

Fig. 12 A two by two move $\alpha \to \alpha'$

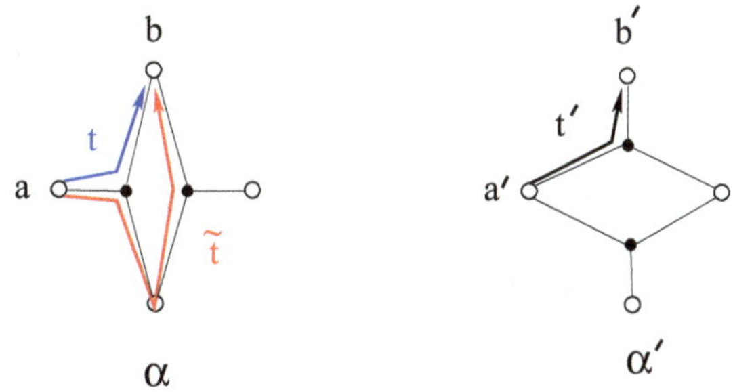

Fig. 13 The two by two move action on the parallel transform: $t' := t + \widetilde{t} = (1 + M_b)t$

Given a two by two move $\mu_\alpha : \Gamma \to \Gamma'$, let us define a birational isomorphism of the moduli spaces of non-commutative flat line bundles on the graphs:

$$\mu_\alpha : \mathrm{Loc}_1(\Gamma) \longrightarrow \mathrm{Loc}_1(\Gamma'). \tag{34}$$

Consider a two by two move at α shown on Fig. 12:

$$\mu_\alpha : \Gamma \to \Gamma'.$$

Let α_0 be the graph α without the $\circ$-vertices and similarly α_0'. By the definition, $\Gamma - \alpha_0 = \Gamma' - \alpha_0'$. Denote by $b \bullet a$ the arc t on the graph α going from the $\circ$-vertex a to the $\bullet$-vertex and then to the $\circ$-vertex b—see Fig. 13. We use a similar notation for the graph α'.

Starting from a flat line bundle L on Γ, we produce a new flat line bundle L' on Γ' as follows:

Let $M_v : L_v \to L_v$ be the *minus* of the operator of counterclockwise monodromy around the simple loop on L_α, acting at the fiber at v; see Fig. 14.

- The restriction of L' to $\Gamma' - \alpha_0'$ is defined as the restriction of L to $\Gamma - \alpha_0$.
- The restriction of L' to α', denoted by $L_{\alpha'}$, is defined in Definition 2.1.2.

Definition 2.1.2 Given a flat line bundle L_α on the graph α, we define a flat line bundle $L_{\alpha'}$ on the graph α' by setting the parallel transports along the arcs between the $\circ$-vertices on α' to be

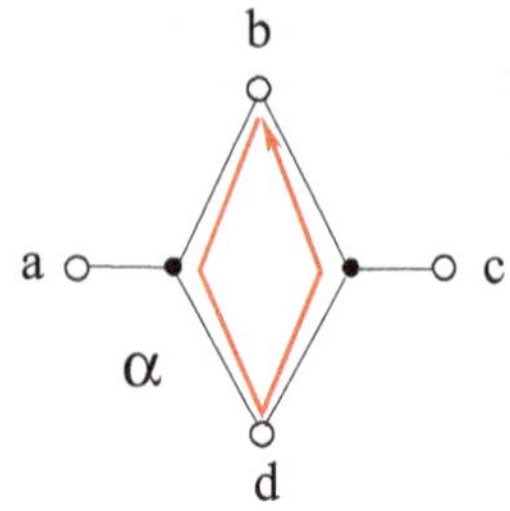

Fig. 14 M_b is the negative of the counterclockwise monodromy around the loop α

$$\begin{aligned} t_{b'\bullet a'} &:= (1+M_b)t_{b\bullet a}, \quad t_{c'\bullet b'} := t_{c\bullet b}(1+M_b)^{-1}, \\ t_{d'\bullet c'} &:= (1+M_d)t_{d\bullet c}, \quad t_{a'\bullet d'} := t_{a\bullet d}(1+M_d)^{-1}. \end{aligned} \tag{35}$$

We glue the flat line bundle $L_{\alpha'}$ to the restriction of the flat line bundle L to $\Gamma - \alpha_0$.

Let $\widetilde{t}_{b,a}$ be "the other way around α" parallel transport from a to b; see Fig. 13. Then

$$t_{b'\bullet a'} \overset{(35)}{=} (1+M_b)t_{b\bullet a} = t_{b\bullet a}(1+M_a) = t_{b\bullet a} + \widetilde{t}_{b,a}. \tag{36}$$

It is a sum of the two different simple paths from a to b on α.

Lemma 2.1.3 *Monodromies of the local systems L_α and $L_{\alpha'}$ at each of the four $\circ$-points coincide:*

$$M_a = M'_a, \quad M_b = M'_b, \quad M_c = M'_c, \quad M_d = M'_d.$$

Proof We have

$$\begin{aligned} &M'_a = t_{a'\bullet d'}t_{d'\bullet c'}t_{c'\bullet b'}t_{b'\bullet a'} = \\ &t_{a\bullet d}(1+M_d)^{-1}(1+M_d)t_{d\bullet c}t_{c\bullet b}(1+M_b)^{-1}(1+M_b)t_{b\bullet a} = \\ &t_{a\bullet d}t_{d\bullet c}t_{c\bullet b}t_{b\bullet a} = M_a. \end{aligned} \tag{37}$$

Next, using $M_a = M'_a$ we have

$$M'_b = t_{b'\bullet a'}M'_a t^{-1}_{b'\bullet a'} \overset{(35)}{=} t_{b\bullet a}(1+M_a)M_a t^{-1}_{b'\bullet a'} = t_{b\bullet a}M_a(1+M_a)t^{-1}_{b'\bullet a'}. \tag{38}$$

The first two equalities in (36) imply

$$t^{-1}_{b'\bullet a'} \overset{(36)}{=} (t_{b\bullet a}(1+M_a))^{-1} = (1+M_a)^{-1}t^{-1}_{b\bullet a}.$$

Substituting this to (38), we get

$$M_{b'} \stackrel{(38)}{=} t_{b\bullet a} M_a (1+M_a)(1+M_a)^{-1} t_{b\bullet a}^{-1} = t_{b\bullet a} M_a t_{b\bullet a}^{-1} = M_b.$$

The last two equalities follow from the cyclic shift by two symmetry of the picture. □

Corollary 2.1.4
(i) We can rewrite Definition 2.1.2 as follows:

$$\begin{aligned} t_{b'\bullet a'} &:= (1+M_b)t_{b\bullet a}, \quad & (1+M_{c'})t_{c'\bullet b'} &:= t_{c\bullet b}, \\ t_{d'\bullet c'} &:= (1+M_d)t_{d\bullet c}, \quad & (1+M_{a'})t_{a'\bullet d'} &:= t_{a\bullet d}. \end{aligned} \tag{39}$$

(ii) One has $\mu_{\alpha'} \circ \mu_\alpha = \mathrm{Id}$.

Proof
(i) Indeed, using Lemma 2.1.3,

$$\begin{aligned} t_{c'\bullet b'} = t_{c\bullet b}(1+M_b)^{-1} = t_{c\bullet b}(1+M_{b'})^{-1} \quad &\Longrightarrow \quad t_{c'\bullet b'}(1+M_{b'}) = t_{c\bullet b} \\ &\Longrightarrow \quad (1+M_{c'})t_{c'\bullet b'} = t_{c\bullet b}. \end{aligned}$$

The identity $(1+M_{a'})t_{a'\bullet d'} := t_{a\bullet d}$ follows by the cyclic shift by two symmetry. (ii) Since the cyclic shift by one of α' is isomorphic to α, this implies $\mu_{\alpha'} \circ \mu_\alpha = \mathrm{Id}$. □

2. Configurations of Lines and Two by Two Moves
To prove that transformations (34) satisfy the pentagon relations, we need the following comparison result. Our construction from Sect. 1.2 (see Theorems 1.2.2 and 1.2.8) provides a birational equivalence of groupoids (see Fig. 12 for the notation):

$$\{\text{Quadruples of lines in a 2-dimensional } R\text{-vector space } V\text{, at the } \circ\text{-vertices of } \alpha\}$$
$$\longleftrightarrow$$
$$\{\text{Flat } R\text{-line bundles on the graph } \alpha\}. \tag{40}$$

Proposition 2.1.5 *The transformation in Definition 2.1.2 is characterized by the condition that it commutes with equivalence (40), providing a commutative diagram*

$$\begin{array}{ccc} \{\text{Line bundles } \mathcal{L} \text{ on } \alpha\} & \xrightarrow{\text{Definition 2.1.2}} & \{\text{Line bundle } \mathcal{L}' \text{ on } \alpha'\} \\ & {\scriptstyle (40)}\searrow \quad \swarrow{\scriptstyle (40)} & \\ & (A,B,C,D) \subset V & \end{array} \tag{41}$$

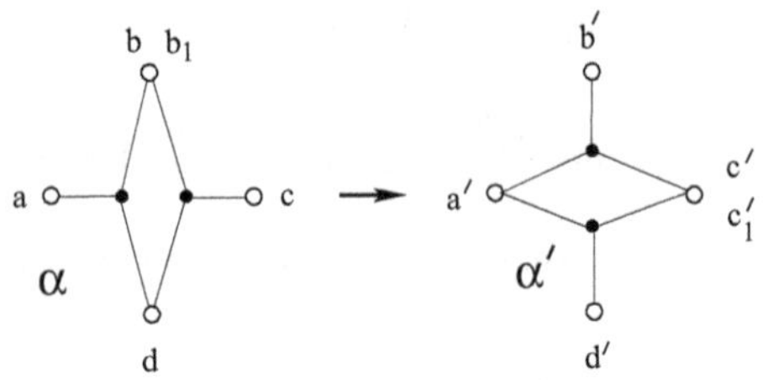

Fig. 15 Calculating the two by two move for twisted configurations of lines in V_2

Proof We start with a configuration of four generic lines (A, B, C, D) in V. We assign them to the ∘-vertices (a, b, c, d) of the graph α and to the ∘-vertices (a', b', c', d') of the graph α'. □

We abuse notation by denoting by the same letter a a vector in the line A, etc.

The two •-vertices provide us five vectors in the lines (A, B, C, D), defined uniquely up to a common left R^*-factor by the condition that the sum of the three vectors at the ∘-vertices incident to any •-vertex is zero. These are the vectors illustrated on the left in Fig. 15:

$$(a, b, b_1, c, d), \qquad \text{So } b_1 = Xb, \quad X \in R^*.$$

The counterclockwise monodromy around α acts on any of them by the left multiplication by X. Indeed,

$$t_{a \bullet b}(b) = a, \quad t_{d \bullet a}(a) = d, \quad t_{c \bullet d}(d) = c, \quad t_{b_1 \bullet c}(c) = b_1 = Xb. \tag{42}$$

There are similar five vectors in V related to the graph α', illustrated on the right in Fig. 15:

$$(a', b', b'_1, c', d'), \ \text{So } b'_1 = Yb', \quad Y \in R^*.$$

The counterclockwise monodromy around α' acts on them as the left multiplication by Y.

One summarizes the definition of these vectors by the following relations, one per each •-vertex:

$$a+b+d = 0, \quad d+c+b_1 = 0; \qquad a'+b'+c' = 0, \quad a'+d'+c'_1 = 0. \tag{43}$$

Notice that $b_1 = Xb$ and $c'_1 = Yc'$. So the equations read now as

$$a + b + d = 0, \quad d + c + Xb = 0; \qquad a' + b' + c' = 0, \quad a' + d' + Yc' = 0.$$

These vectors are related as follows.

Lemma 2.1.6 *One has*

$$a' = a, \qquad b' = (1 - X)b, \qquad c' = -c, \qquad d' = (1 - X^{-1})d. \tag{44}$$

Proof Each pair of the vectors $(a', a), \ldots, (d', d)$ is collinear, so $a' = \lambda_a a$, $b' = \lambda_b b$, $c' = \lambda_c c$, $d' = \lambda_d d$. Let us solve these equations. First, let us express c, d via a, b and c', d' via a', b', respectively:

$$d = -a - b, \quad c = (1 - X)b + a, \quad c' = -a' - b', \quad d' = -(1 - Y)a' + Yb'.$$

Next, the identities $\lambda_c c = c'$ and $\lambda_d d = d'$ give

$$\lambda_c((1 - X)b + a) = -\lambda_a a - \lambda_b b, \qquad \lambda_d(a + b) = (1 - Y)a - Y\lambda_b b.$$

Set $\lambda_a = 1$. Then we get the following equations:

$$\lambda_b = 1 - X, \quad \lambda_c = -1, \quad \lambda_d = 1 - Y, \quad -Y\lambda_b = \lambda_d.$$

From the first and last equations we see that $Y = X^{-1}$, and therefore, we finally get

$$\lambda_a = 1, \quad \lambda_b = 1 - X, \quad \lambda_c = -1, \quad \lambda_d = 1 - X^{-1}.$$

□

Lemma 2.1.6 together with (42) implies

$$\begin{aligned} t_{b'\bullet a'} &= (1 - X)t_{b\bullet a}, \quad t_{c'\bullet b'} = (1 - X^{-1})^{-1}t_{c\bullet b_1}, \\ t_{d'\bullet c'_1} &= (1 - X)t_{d\bullet c}, \quad t_{a'\bullet d'} = (1 - X^{-1})^{-1}t_{a\bullet d}. \end{aligned} \tag{45}$$

Recall that M_a is the *negative* of the counterclockwise monodromy around α.

Lemma 2.1.7 *Formulas (45) are equivalent to formulas (35).*

Proof The first line in (35) is equivalent to the second one by the cyclic shift by two arguments. The right formula in each line in (35) is equivalent to the left one, applied to the two by two move $\mu_{\alpha'} : \Gamma' \to \Gamma$. So we need to check only one of the four formulas in (35).

In the top left formula in (45) the variable X, by its definition, is the action of the counterclockwise monodromy along α on the vector b. So it can be written as $t_{b'\bullet a'} = (1 + M_b)t_{b\bullet a}$, which is just the top left formula in (35). □

Proposition 2.1.5 is proved

3. The Pentagon Relation

Theorem 2.1.8 *The fifth power of "two by two move followed by the nontrivial symmetry of the triangulated pentagon" for the two by two moves on Fig. 16 acts as the identity on flat line bundles.*

Proof Let $\{\Gamma_i\}$, where $i \in \mathbb{Z}/5\mathbb{Z}$, be the sequence of bipartite graphs on Fig. 16. They are assigned to triangulations of the pentagon. The two by two move $\Gamma_i \to$

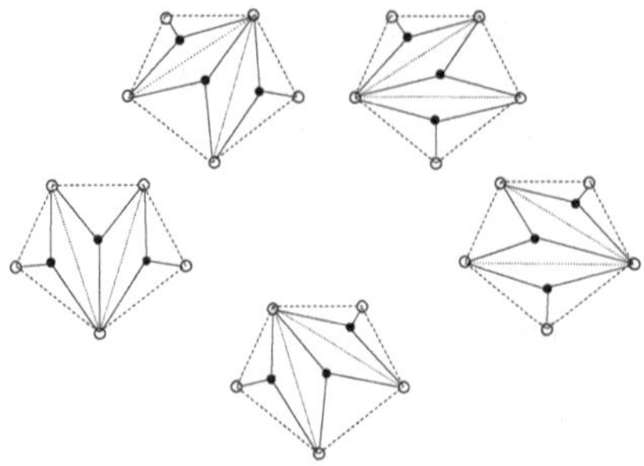

Fig. 16 The pentagon relation

Γ_{i+1} corresponds to a flip of the triangulation. For each of the graphs Γ_i there is an equivalence of groupoids:

$$\{\text{Generic flat line bundles on the graph } \Gamma_i\}$$
$$\longleftrightarrow$$
$$\{\text{Generic 5-tuples of lines in a two dimensional space } V, \text{ assigned to } \circ\text{-vertices of a pentagon.}\} \quad (46)$$

Thanks to Lemma 2.1.7 the mutation of flat line bundles corresponding to a two by two move intertwines these equivalences. Indeed, a two by two move corresponding to a flip of a triangulation of the pentagon keeps intact one of the triangles of the triangulation and does not affect the restriction of the flat line bundle to the graph inside of that triangle. □

2.2 *Non-Commutative Cluster Poisson Varieties*

In this section we define a non-commutative analog of the Poisson algebra assigned to a bipartite ribbon graph in [20]. The output is still a *commutative* Poisson algebra. We prove that two by two moves give rise to homomorphisms of Poisson algebras.

1. The Algebra $\mathcal{O}_\Gamma$
Let Γ be a graph. An oriented loop α on Γ is a continuous map $\alpha : S^1 \to \Gamma$ of an oriented circle. A vertex of the loop α is a point $v \in S^1$ which is mapped to a vertex of Γ.

Denote by l_Γ the set of all oriented loops on Γ, considered modulo homotopy. Let

$$L_\Gamma := \mathbb{Z}[l_\Gamma]$$

be the free abelian group generated by the set l_Γ. Let $\mathcal{O}_\Gamma$ be its symmetric algebra:

$$\mathcal{O}_\Gamma := S^*(L_\Gamma). \quad (47)$$

Let $\mathrm{Loc}_{N,\mathbb{C}}(\Gamma)$ be the moduli space of all N-dimensional complex vector bundles with connections on Γ. Denote by $\mathcal{O}(\mathrm{Loc}_{N,\mathbb{C}}(\Gamma))$ the algebra of regular functions on this space. There is a canonical map of commutative algebras:

$$r : \mathcal{O}_\Gamma \longrightarrow \mathcal{O}(\mathrm{Loc}_{N,\mathbb{C}}(\Gamma)). \tag{48}$$

Namely, a loop α on Γ gives rise to a function M_α on $\mathrm{Loc}_{N,\mathbb{C}}(\Gamma)$ given by the $1/N$ times the trace of the monodromy along the loop α. Homotopic loops give the same function. Since $\mathcal{O}_\Gamma$ is the free commutative algebra generated by the set l_Γ, the map $\alpha \longmapsto M_\alpha$ extends uniquely to a homomorphism (48). The trivial loop maps to the unit.

2. The Goldman Bracket for a Ribbon Graph Γ
Let us define a Lie bracket $\{*, *\}_{\mathrm{G}}$ on L_Γ.

A valency m vertex v of a ribbon graph Γ gives rise to a free abelian group $\mathbb{A}_v$ of rank $m-1$, given by $\mathbb{Z}$-linear combinations of the oriented out of v edges $\vec{E}_i$ at v, of total degree zero:

$$\mathbb{A}_v = \{\sum n_i \ \vec{E}_i \mid \sum n_i = 0, \ n_i \in \mathbb{Z}\}.$$

It is generated by the "oriented paths" $\vec{E}_i - \vec{E}_j$, where $-\vec{E}_j$ is the edge E_j oriented towards v.

Lemma 2.2.1 *Given a vertex v of Γ, there is a unique skew symmetric bilinear form:*

$$\delta_v : \mathbb{A}_v \wedge \mathbb{A}_v \longrightarrow \frac{1}{2}\mathbb{Z}$$

such that for any triple E_0, E_1, E_2 of edges as in Fig. 17 one has

$$\delta_v(\alpha_r \wedge \alpha_l) = \frac{1}{2}, \qquad \alpha_r := \vec{E}_1 - \vec{E}_0, \quad \alpha_l := \vec{E}_2 - \vec{E}_0 \,. \tag{49}$$

Take two loops α and β on a ribbon graph Γ. Let $\alpha_v \in \mathbb{A}_v$ be the element induced by the loop α at the vertex v. Set $\delta_v(\alpha, \beta) := \delta_v(\alpha_v, \beta_v)$. Given a pair (v, w), where v is a vertex of α, w is a vertex of β, and $\alpha(v) = \beta(w)$, consider a new loop $\alpha \circ_{v=w} \beta$

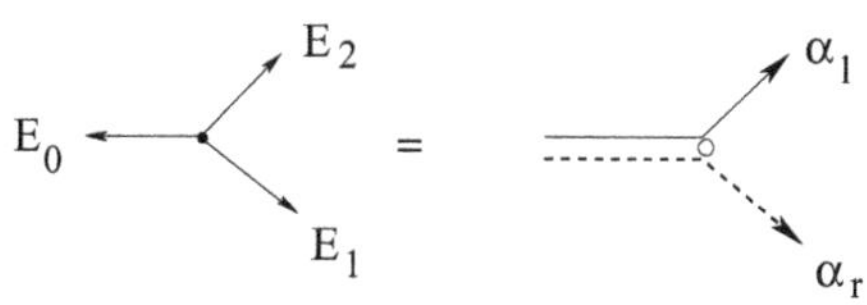

Fig. 17 One has $\delta_v(\alpha_r, \alpha_l) = \frac{1}{2}$

obtained by starting at the vertex $\beta(w)$, going around the loop β, and then around the loop α. We define a bracket $\{\alpha, \beta\}_{\mathrm{G}}$ by taking the sum over all such pairs of vertices (v, w):

$$\{\alpha, \beta\}_{\mathrm{G}} := \sum_{v=w} \delta_v(\alpha, \beta) \cdot \alpha \circ_{v=w} \beta.$$

This is a ribbon graph version of the Goldman bracket. It is easy to check the following.

Theorem 2.2.2 *The bracket $\{*, *\}_{\mathrm{G}}$ provides L_Γ with a Lie algebra structure. So it extends via the Leibniz rule to a Poisson bracket $\{*, *\}_{\mathrm{G}}$ on the commutative algebra $\mathcal{O}_\Gamma$. The map (48) is a map of Poisson algebras.*

3. The Lie Bracket $\{*, *\}$ on the Space L_Γ for a Bipartite Ribbon Graph Γ
Let Γ be a bipartite ribbon graph. We apply the previous construction to the conjugate graph Γ^*.

Definition 2.2.3
(i) The Lie bracket $\{*, *\}$ on L_Γ is induced, via the isomorphism $L_\Gamma = L_{\Gamma^*}$, by the Lie bracket on L_{Γ^*}, for the conjugate graph Γ^*.
(ii) It induces, via the Leibniz rule, a Poisson algebra structure $\{*, *\}$ on $\mathcal{O}_\Gamma$.

Let α, β be two oriented loops on a bipartite surface graph Γ. Then

$$\{\alpha, \beta\} = \sum_v \mathrm{sgn}(v) \delta_v(\alpha, \beta) \cdot \alpha \circ_v \beta.$$

The sum is over all vertices $v \in \alpha \cap \beta$ and $\mathrm{sgn}(v) = 1$ for the $\circ$-vertex and -1 for the $\bullet$-vertex.

To calculate the bracket $\{\alpha, \beta\}$ between two loops α, β on Γ, we have the following two cases:

- A vertex $v \in \alpha \cap \beta$ is an isolated intersection point. Then its contribution is $\pm 1, 0$ depending on the geometry of the intersection, as depicted on Fig. 18.
- A vertex $v \in \alpha \cap \beta$ lies on an edge $vw \subset \alpha \cap \beta$. Then its contribution is $\pm 1, 0$ depending on the geometry of the intersection, as depicted on Figs. 19 and 20.

Changing the orientation of the surface amounts to changing the sign of the Poisson bracket. Interchanging black and white we change the sign of the Poisson bracket.

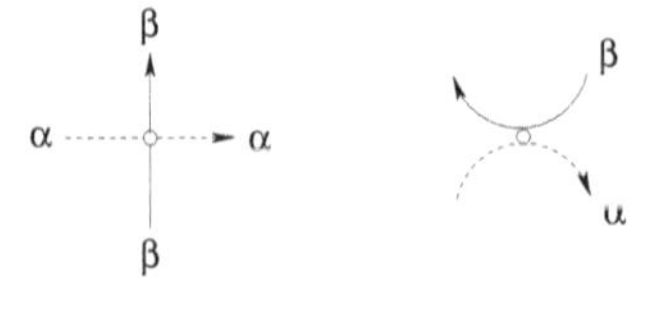

Fig. 18 On the left: $\delta_v(\alpha, \beta) = 1$. On the right: $\delta_v(\alpha, \beta) = 0$

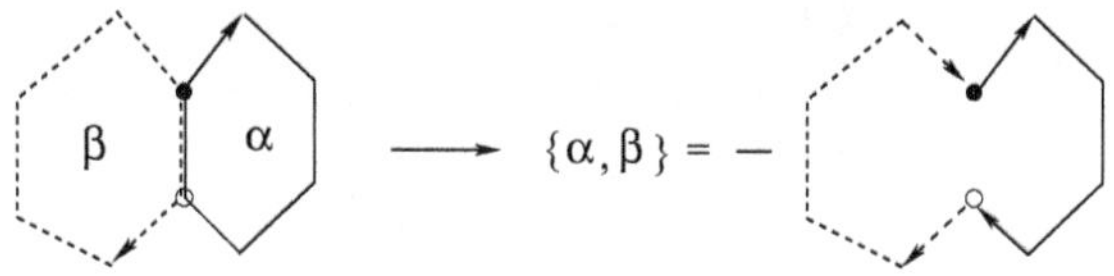

Fig. 19 One has $\{\alpha, \beta\} = -\alpha \circ \beta$, where $\alpha \circ \beta$ is the loop on the right

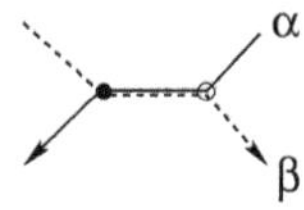

Fig. 20 The total contribution of the $\circ$ and $\bullet$-vertices to $\{\alpha, \beta\}$ is zero

Let us consider the formal completion $\widehat{L_\Gamma}$ of L_Γ by the length of the loop on the graph Γ, counted as the total number of edges in the loop. Then we make sense of $(1+M)^{-1}\gamma$ as $(1 - M + M^2 - M^3 + \ldots)\gamma$.

Theorem 2.2.4 *Let $\Gamma \to \Gamma'$ be a two by two move of bipartite ribbon graphs. Then the map induced by Definition 2.1.2 is a Lie algebra homomorphism of completed Lie algebras:*

$$\mu_{\Gamma\to\Gamma'} : \widehat{L_\Gamma} \longrightarrow \widehat{L_{\Gamma'}}.$$

Proof Denote by $\gamma_{b\bullet a}$ a loop on Γ containing the segment $t_{b\bullet a}$, etc . Consider the following three identities:

$$\begin{aligned}
\{\gamma_{b'\bullet a'}, \gamma_{d'\bullet c'}\} &= \{\gamma_{b\bullet a}, \gamma_{d\bullet c}\},\\
\{\gamma_{b'\bullet a'}, \gamma_{c'\bullet b'}\} &= \{\gamma_{b\bullet a}, \gamma_{c\bullet b}\},\\
\{\gamma_{b'\bullet a'}, \gamma_{c'\bullet b'}\} &= \{\gamma_{b\bullet a}, \gamma_{c\bullet b}\}.
\end{aligned} \tag{50}$$

Theorem 2.2.4 follows from them. Indeed, we have twelve identities like this to check. Using the skew symmetry of the Poisson bracket, we reduce their number to six. These six can be reduced to the three identities (50): two identities are reduced by the rotation by 180° of the graph and one by changing the plane orientation. Relation (3) in (50) is reduced to (1) by changing the plane orientation. Let us check identities (1) and (2).

(1) We have the following, where the first equality is Definition (35), and the second is explained on Fig. 21:

$$\{\gamma_{b'\bullet a'}, \gamma_{d'\bullet c'}\} = \{(1+M_b)\gamma_{b\bullet a}, (1+M_d)\gamma_{d\bullet c}\} \stackrel{\text{Fig. 21}}{=} \{\gamma_{b\bullet a}, \gamma_{d\bullet c}\}.$$

(2) The second boils down to

$$\{(1+M_b)\gamma_{b\bullet a}, (1+M_c)^{-1}\gamma_{c\bullet b}\} \stackrel{?}{=} \{\gamma_{b\bullet a}, \gamma_{c\bullet b}\}. \tag{51}$$

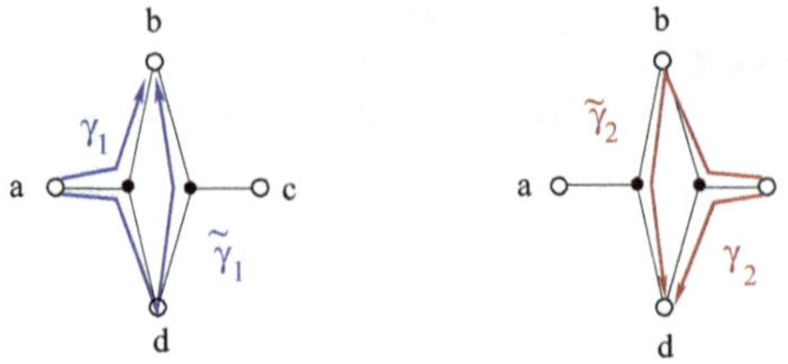

Fig. 21 The sum of the contributions of the edges $\bullet b$ and $\bullet d$ to $\{\gamma_1, \widetilde{\gamma}_2\} + \{\widetilde{\gamma}_1, \gamma_2\}$ is zero. Thus, $\{\gamma_1, \widetilde{\gamma}_2\} + \{\widetilde{\gamma}_1, \gamma_2\} = 0$. One has $\{\widetilde{\gamma}_1, \widetilde{\gamma}_2\} = 0$ since the paths are enter and exit at the common edges $b\bullet$ and $d\bullet$ in the same direction. So $\{\gamma_1 + \widetilde{\gamma}_1, \gamma_2 + \widetilde{\gamma}_2\} = \{\gamma_1, \gamma_2\}$

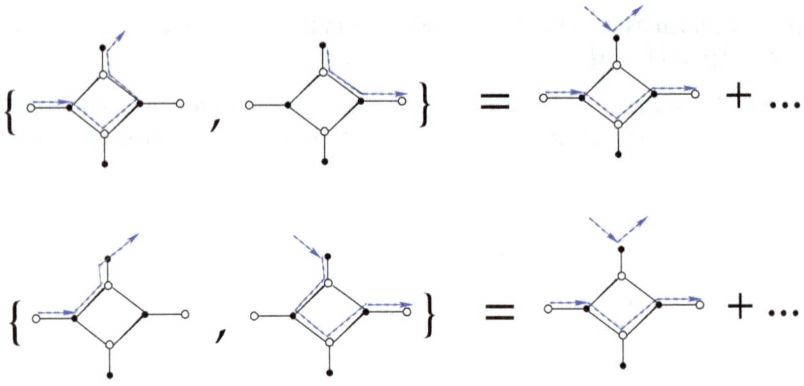

Fig. 22 Calculating the Poisson brackets $\{M_b\gamma_{b\bullet a}, X_{c\bullet b}\}$ and $\{\gamma_{b\bullet a}, M_cX_{c\bullet b}\}$ we get the same result

So we need to prove that

$$\{M_b\gamma_{b\bullet a}, (1 - M_c^{-1} + M_c^2 - \dots)\gamma_{c\bullet b}\} \stackrel{?}{=} \{\gamma_{b\bullet a}, (M_c^{-1} - M_c^2 + M_c^3 - \dots)\gamma_{c\bullet b}\}. \tag{52}$$

This is checked by using the following general identity, where $X_{c\bullet b}$ is a loop containing the segment $c \bullet b$:

$$\{M_b\gamma_{b\bullet a}, X_{c\bullet b}\} = \{\gamma_{b\bullet a}, M_cX_{c\bullet b}\}.$$

This identity proved by a calculation illustrated in Fig. 22. □

Corollary 2.2.5 *A two by two move gives rise to a birational isomorphism of Poisson algebras:*

$$\mu_{\Gamma\to\Gamma'} : \widehat{\mathcal{O}}_\Gamma \longrightarrow \widehat{\mathcal{O}}_{\Gamma'}.$$

Conclusion Groupoids $\mathrm{Loc}_1(\Gamma)$ assigned to bipartite graphs Γ and related by birational transformations (34) form a non-commutative analog of a cluster Poisson variety, with the Poisson bracket on the algebra (47) given by Definition 2.2.3.

3 Dimers and Non-Commutative Cluster Integrable Systems

In Sect. 3 we apply the general results about non-commutative cluster Poisson varieties from previous sections to produce non-commutative cluster integrable systems, which in the commutative case were studied in [20]. Section 3 will not be used anywhere in the rest of the chapter.

1. Dimer Covers
A *dimer cover* of a bipartite graph Γ is a subset of edges of Γ such that each vertex is the endpoint of a unique edge in the dimer cover. Given a dimer cover M, we assign to each edge E of Γ the weight $\omega_M(E) \in \{0, 1\}$ such that $\omega_M(E) = 1$ if and only if the edge E belongs to M. Let $[E]$ be the edge E oriented as $\bullet \to \circ$. Then we get a 1-chain:

$$[M] := \sum_E \omega_M(E)[E].$$

Here the sum is over all edges. We define $\mathrm{sgn}(v) := -1$ for the $\bullet$-vertex v and $\mathrm{sgn}(v) := 1$ for the $\circ$-vertex. The condition that M is a dimer cover just means that

$$d[M] = \sum_v \mathrm{sgn}(v)[v].$$

Here the sum is over all vertices v of Γ. So if M_1, M_2 are dimer covers, $[M_1]-[M_2]$ is a 1-cycle.

2. Zig-Zag Paths Revisited
Recall that a zig-zag γ on a bipartite ribbon graph Γ is oriented so that it turns right at the $\circ$-vertices and turns left at the $\bullet$-vertices. For every edge there are exactly two zig-zag paths containing the edge. They traverse the edge the opposite ways.

3. A Reference Dimer Cover Φ for a Bipartite Graph on a Torus [20]
We consider from now bipartite graphs Γ on a torus T. A bipartite graph Γ on a torus is *minimal* if it does not have parallel bigons; see the left picture on Fig. 30. Homology classes of oriented zig-zag paths on a bipartite graph $\Gamma \subset \mathrm{T}$ form a collection of vectors $e_1, \ldots, e_n$ of the rank two lattice $H_1(\mathrm{T}, \mathbb{Z})$. We order them cyclically by the angle they make with a given direction. Their sum is zero. So they are oriented sides of a convex polygon $N \subset H_1(\mathrm{T}, \mathbb{Z})$, called the *Newton polygon* of Γ. The set $\{e_1, \ldots, e_n\}$ is identified with the set of primitive edges of the boundary ∂N of N. Alternatively, N is the convex hull of $0, e_1, e_1 + e_2, \ldots, e_1 + \ldots + e_{n-1}$.

Let X be the set of circular-order-preserving maps:

$$\alpha : \{e_1, \ldots, e_n\} \rightarrow \mathbb{R}/\mathbb{Z}. \tag{53}$$

Let E be an edge of Γ. Denote by z_r (respectively z_l) the zig-zag path containing E in which the $\circ$-vertex of E precedes (respectively follows) the $\bullet$-vertex. Then a map α provides a function:

$$\varphi_\alpha = \varphi : \{\text{Edges of } \Gamma\} \longrightarrow \mathbb{R}, \qquad \varphi(E) := \alpha_r - \alpha_l. \tag{54}$$

Here we use the identification of zig-zag loops with their homology classes e_i. Then α_r (respectively α_l) is the evaluation of the map α on the zig-zag path z_r (respectively z_l) crossing the edge E, and $\alpha_r - \alpha_l$ denotes the length of the counterclockwise arc on $S^1 = \mathbb{R}/\mathbb{Z}$ from α_l to α_r.

Theorem 3.0.1 ([20, Theorem 3.3]) *For any map $\alpha \in X$, the function φ_α satisfies*

$$d\varphi_\alpha = \sum_v \text{sgn}(v)[v].$$

Take a map α which sends all e_i to 0 and assigns 1 to the arc between e_i and e_{i+1} and 0 to all other arcs. Then φ_α is a dimer cover of Γ. We denote it by Φ. It corresponds to a vertex of the Newton polygone, perhaps a degenerate one, i.e., located on a side.

4. The Commuting Hamiltonians

Since M and Φ are perfect matchings, the 1-cycle $[M] - [\Phi]$ is a union of non-intersecting and non-self-intersecting loops. Considered modulo homotopy on the graph Γ, it defines an element:

$$(M) := [M] - [\Phi] \in L(\Gamma).$$

Definition 3.0.2 Let Γ be a bipartite graph on a torus. The non-commutative partition function $\mathcal{P}_\Phi$ is an element of $L(\Gamma)$ given as the sum over all dimer covers M of Γ:

$$\mathcal{P}_\Phi := \sum_M \text{sgn}(M)(M). \tag{55}$$

The element $[M] - [\Phi]$ defines a class in $H_1(\mathrm{T}, \mathbb{Z})$, denoted by $h_{\mathrm{T}}(M)$. The set of homology classes for different dimer covers M of Γ coincides with the set of internal points of a shift of the Newton polygon N. Take the sum over the dimer covers M with the same homology class a:

$$H_a := \sum_{h_{\mathrm{T}}(M)=a} (M) \in L(\Gamma). \tag{56}$$

Furthermore, for any nonzero integer n, we define a new element:

$$H_{na} \in L(\Gamma).$$

Namely, if $n > 0$, it is obtained by travelling n times each loop of H_a. If $n < 0$, we reverse the orientations of the loops of H_a and then travel each obtained loop $|n|$ times.

Definition 3.0.3 The elements H_{na} are the *Hamiltonians of the dimer system.*

Remark A map α related to another vertex of the Newton polygon leads to another collection of Hamiltonians which differ from $\{H_a\}$ by a sum of zig-zag loops, which lies in the center of the Poisson algebra $\mathcal{O}(\Gamma)$. Therefore, the Hamiltonian flows do not depend on the choice of α, and on each symplectic leaf any two collections of Hamiltonians differ by a common factor.

Theorem 3.0.4 *Let Γ be a minimal bipartite graph on a torus* T. *Then the Hamiltonians H_{na} commute under the Lie bracket on $L(\Gamma)$. Therefore, they Poisson commute in $\mathcal{O}(\Gamma)$.*

Proof We prove that $\{H_a, H_b\} = 0$. The proof almost literally follows the proof of [20, Theorem 3.7]. For the convenience of the reader we indicate the main steps. The claim that $\{H_{na}, H_{mb}\} = 0$ follows from this.

Take a pair of matchings (M_1, M_2) on Γ. Let us assign to them another pair of matchings $(\widetilde{M}_1, \widetilde{M}_2)$ on Γ. Since $[M_1] - [M_2]$ is a 1-cycle, and M_1, M_2 are dimer covers, it is a disjoint union of non-intersecting and non-self-intersecting loops. So $[M_1] - [M_2]$ is a disjoint union of:

1. Homologically trivial on the torus loops
2. Homologically nontrivial loops
3. Edges shared by both matchings

□

For every edge E of each homologically trivial loop, we switch the label of E: if E belongs to a matching M_1 (respectively M_2), we declare that it will belong to a matching $\widetilde{M}_2$ (respectively $\widetilde{M}_1$). For all other edges we keep their labels intact. Switching the labels of a homologically trivial loop, we do not change the homology classes $[M_1] - [\Phi]$, $[M_2] - [\Phi]$.[12] Clearly $\widetilde{\widetilde{M}}_i = M_i$.

Lemma 3.0.5 ([20, Lemma 3.8]) *One has*

$$\{(M_1), (M_2)\} + \{(\widetilde{M}_1), (\widetilde{M}_2)\} = 0. \tag{57}$$

[12] Contrary to this, switching the labels of homologically nontrivial loops in $[M_1] - [M_2]$, we do change the homology classes $[M_i] - [\Phi]$.

Proof Let us write (57) as a sum of the local contributions corresponding to the vertices v of Γ and break it in three pieces as follows:

$$\varepsilon(\mu_1, \mu_2) + \varepsilon(\widetilde{\mu}_1, \widetilde{\mu}_2) = (\sum_{v\in\mathcal{E}_1} + \sum_{v\in\mathcal{E}_2} + \sum_{v\in\mathcal{E}_3})\Big(\delta_v(\mu_1, \mu_2) + \delta_v(\widetilde{\mu}_1, \widetilde{\mu}_2)\Big).$$

Here $\mathcal{E}_1$ (respectively $\mathcal{E}_2$ and $\mathcal{E}_3$) is the set of vertices which belongs to the homologically trivial loops (respectively homologically nontrivial loops, double edges).

For any $v \in \mathcal{E}_1$ we have $\delta_v(\mu_1, \mu_2)+\delta_v(\widetilde{\mu}_1, \widetilde{\mu}_2) = 0$ since the elements $l_v(\mu_1) \in \mathbb{A}_v$ provided by μ_i coincides with the $l_v(\widetilde{\mu}_2)$, and similarly $l_v(\mu_2) = l_v(\widetilde{\mu}_1)$. So the first sum is a sum of zeros.

The third sum is also a sum of zeros by the skew symmetry of the bracket.

It remains to prove that

$$\sum_{v\in\mathcal{E}_2} \Big(\delta_v(\mu_1, \mu_2) + \delta_v(\widetilde{\mu}_1, \widetilde{\mu}_2)\Big) = 0.$$

Let γ be an oriented loop on Γ. The *bending* $b_v(\gamma; \varphi)$ of the function φ at a vertex v of γ is

$$b_v(\gamma; \varphi) = \sum_{E\in R_v} \varphi(E) - \sum_{E\in L_v} \varphi(E) \in \mathbb{R}.$$

Here R_v (respectively L_v) is the set of all edges sharing the vertex v which are on the right (respectively left) of the oriented path γ. Lemma 3.0.5 now follows from Lemma 3.0.6 below. □

Lemma 3.0.6 ([20, Lemma 3.9]) *For any simple topologically nontrivial loop γ and for any $\alpha \in X$, see (53), the corresponding function φ_α satisfies*

$$\sum_{v\in\gamma} b_v(\gamma; \varphi_\alpha) = 0.$$

4 Non-Commutative Cluster $\mathcal{A}$-Varieties

4.1 Non-Commutative Cluster $\mathcal{A}$-Varieties from Bipartite Ribbon Graphs

Convention Dealing with *twisted* local systems on a decorated surface $\mathbb{S}$, we alter $\mathbb{S}$ by cutting out a little disc near each marked point m on $\mathbb{S}$. Abusing notation, we denote the obtained surface by $\mathbb{S}$. The tangent vectors to each boundary component

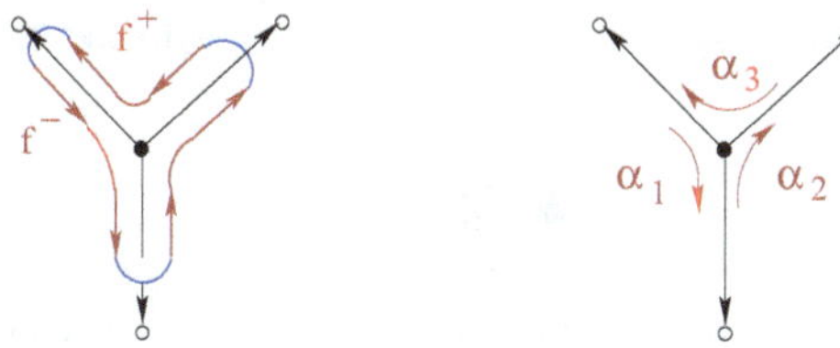

Fig. 23 An edge E of a ribbon graph Γ determines two faces of $\mathbb{S}_\Gamma$, denoted f_E^+ and f_E^-. There is a canonical homotopy class of path p_E, connecting boundary components f_E^+ and f_E^-, shown by a blue arc

of $\mathbb{S}$, positively oriented for the boundary orientation of $\mathbb{S}$, form a homotopy (semi)circle S_m^1 in the bundle of nonzero tangent vectors to $\mathbb{S}$. We perform a similar procedure with the marked points on the spectral curve Σ. Talking about a flat section of a twisted local system near a marked point m of $\mathbb{S}$ or Σ, we mean a flat section of its restriction to the (semi)circle S_m^1.

The definition of a framed local system on a decorated surface in Definition 1.3.1 easily extends to the case of a twisted framed local system on $\mathbb{S}$. So by the monodromy of a twisted local system near a puncture, we mean the monodromy of its restriction to the oriented homotopy circle near the puncture.

Pick a vector field tangent to the boundary of each face of a decorated surface, following the boundary orientation; see Fig. 23. A trivialization of a twisted flat line bundle $\mathcal{L}$ on the surface at the boundary is given by a choice of a nonzero section of the fiber of $\mathcal{L}$ over the tangent boundary vector field.

Recall the surface $\mathbb{S}_\Gamma$ for a ribbon graph Γ and the spectral surface Σ_Γ for a bipartite ribbon graph Γ.

Given a bipartite ribbon graph Γ, consider the groupoid, defined in either of the following two ways:

$$\begin{aligned}\mathcal{A}_\Gamma :=&\{\text{Twisted flat line bundles on spectral surface } \Sigma_\Gamma, \text{ trivialized at the boundary}\}\\ =&\{\text{Twisted flat line bundles on surface} \mathbb{S}_\Gamma, \text{ trivialized at the zig-zag strands.}\}\end{aligned} \tag{58}$$

These definitions are equivalent since there are canonical identifications:

$$\Gamma = \Gamma^*, \quad \Sigma_\Gamma = \mathbb{S}_{\Gamma^*}, \quad \{\text{zig-zags on } \mathbb{S}_\Gamma\} = \{\text{boundary components of } \Sigma_\Gamma\}.$$

Let us give a coordinate description of the non-commutative tori $\mathcal{A}_\Gamma$ in terms of the $\mathcal{A}$-coordinates.

1. $\mathcal{A}$-Coordinates on Ribbon Graphs

Definition 4.1.1 Let Γ be a ribbon graph. The *$\mathcal{A}$-coordinates* on Γ are the elements $\{\Delta_E \in R^*\}$ assigned to the *oriented* edges E of Γ, satisfying the following monomial relations:

- Let E be an oriented edge and $\overline{E}$ the same edge with the opposite orientation. Then

$$\Delta_E \Delta_{\overline{E}} = -1. \tag{59}$$

- Let $E_1, \ldots, E_n$ be the edges incident to any vertex v of Γ, oriented out of the vertex v, whose order is compatible with their cyclic order. Then

$$\Delta_{E_1} \Delta_{E_2} \ldots \Delta_{E_n} = -1. \tag{60}$$

Calling the elements Δ_E "coordinates" is an abuse of terminology since they are not independent. We use the name *$\mathcal{A}$-decorated ribbon graph* for a ribbon graph with a collection of $\mathcal{A}$-coordinates.

Lemma 4.1.2 *Given a ribbon graph Γ, the $\mathcal{A}$-coordinates on Γ parametrize the isomorphism classes of twisted flat line bundles on the surface $\mathbb{S}_\Gamma$, trivialized at the boundary.*

Proof We start from a twisted line bundle on the oriented decorated surface $\mathbb{S}_\Gamma$. An edge E of Γ determines two faces f_E^+ and f_E^- of $\mathbb{S}_\Gamma$. There is a canonical homotopy class of path p_E, connecting boundary components f_E^+ and f_E^-; see Fig. 23. The parallel transport along the p_E acts on the trivializations $s_E^\pm$ at the boundary components $f_E^\pm$ as follows, defining elements $\Delta_E \in R^*$:

$$\text{par}_{p_E} : s_E^+ \longrightarrow \Delta_E s_E^-, \quad \Delta_E \in R^*. \tag{61}$$

Going around any edge, or any vertex along the cyclic order at this vertex, as on Fig. 23, we rotate the tangent vector by 2π, getting both relations (59) and (60). For example, for a 3-valent vertex we get

$$\begin{aligned} s_{E_1}^+ &\longrightarrow \Delta_{E_1} s_{E_1}^- = \Delta_{E_1} s_{E_2}^+ \longrightarrow \Delta_{E_1} \Delta_{E_2} s_{E_2}^- = \Delta_{E_1} \Delta_{E_2} s_{E_3}^+ \\ &\longrightarrow \Delta_{E_1} \Delta_{E_2} \Delta_{E_3} s_{E_3}^- = -\Delta_{E_1} \Delta_{E_2} \Delta_{E_3} s_{E_1}^+. \end{aligned} \tag{62}$$

Here the arrows are induced by the parallel transports (61) $\text{par}_{p_{E_i}}$, $i \in \mathbb{Z}/3\mathbb{Z}$. The $=$ signs mean canonical identifications provided by the parallel transport along the arcs α_i between E_i to E_{i+1} (see Fig. 23), identifying $s_{E_i}^-$ with $s_{E_{i+1}}^+$. The $-$ sign in the end results from the fact that going around the circle amounts to the monodromy -1.

Conversely, picking a section s over a tangent vector to a boundary component and using elements Δ_E satisfying the monomial relations (59) and (60), we recover the twisted flat line bundle on $\mathbb{S}_\Gamma$, trivialized at the boundary. Changing s we get an isomorphic object. □

2. Conventions on $\mathcal{A}$-Coordinates on *Bipartite* Ribbon Graphs
Given a bipartite ribbon graph, its edges are orientated $\circ \longrightarrow \bullet$.

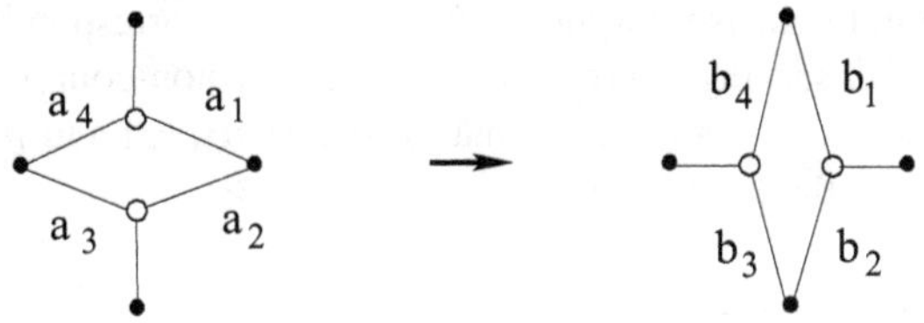

Fig. 24 The relations between the coordinates a_i and b_i are $a_i a_{i+1} = (b_i b_{i+1})^{-1}$, $\forall i \in \mathbb{Z}/4\mathbb{Z}$. The cases $i = 1, 3$ are checked using Fig. 25, and the cases $i = 2, 4$ are similar

Using this convention, the set of $\mathcal{A}$-coordinates on any bipartite ribbon graph can be given by a collection of elements $\{\Delta_E\}$ assigned to *nonoriented* edges E, assuming their default $\circ \longrightarrow \bullet$ orientation, subject to the following relations, where *anticyclic* means the order opposite to the cyclic order:

$$\begin{aligned} &\textit{Cyclic product of elements on the edges sharing any} \circ \textit{-vertex is} \\ &\qquad \textit{equal to } -1. \\ &\textit{Antycyclic product of elements on the edges sharing any} \bullet \\ &\qquad \textit{-vertex } v \textit{ is} (-1)^{\mathrm{val}(v)-1}. \end{aligned} \tag{63}$$

Indeed, given a $\bullet$-vertex v, we have, using $\Delta_{\overline{E}_i} = -\Delta_{E_i}^{-1}$,

$$-1 = \prod_{i=1}^{\mathrm{val}(v)} \Delta_{\overline{E}_i} = \prod_{i=1}^{\mathrm{val}(v)} (-\Delta_{E_i}^{-1}) = (-1)^{\mathrm{val}(v)} \Big(\prod_{\mathrm{val}(v)}^{i=1} \Delta_{E_i} \Big)^{-1}. \tag{64}$$

We apply this convention to the bipartite ribbon graph Γ^*, since the spectral surface where the twisted flat line bundles live is $\mathbb{S}_{\Gamma^*}$.

Note that the bipartite ribbon graphs Γ and Γ^* are identical as graphs. Thinking about the graph Γ, e.g., drawing figures, we depict the cyclic order on Γ as the counterclockwise one and use below and in Sect. 6.3 a different convention:

$$\begin{aligned} &\textit{Counterclockwise product of the el. on edges sharing a } \circ \textit{-vertex is equal} \\ &\qquad \textit{to } -1. \\ &\textit{Counterclockwise product of the el. on edges sharing a } \bullet \textit{-vertex } v \\ &\qquad \textit{is } (-1)^{\mathrm{val}(v)-1}. \end{aligned} \tag{65}$$

3. The Action of the Two by Two Moves on $\mathcal{A}$-Coordinates on Bipartite Ribbon Graphs

Recall the two by two moves of bipartite ribbon graphs, shown on Fig. 24; see also Definition 2.1.1 below.

We define a birational isomorphism $\mathcal{A}_\Gamma \longrightarrow \mathcal{A}_{\Gamma'}$ corresponding to a two by two move $\Gamma \longrightarrow \Gamma'$, serving the role of an elementary non-commutative $\mathcal{A}$-cluster transformation. Let us assign the coordinates $\{a_1, a_2, a_3, a_4\}$ and $\{b_1, b_2, b_3, b_4\}$ to the internal edges on the graphs on Fig. 24. Let us set

$$A_i = a_i a_{i+1} a_{i+2} a_{i+3}, \quad B_i = b_i b_{i+1} b_{i+2} b_{i+3}, \quad i \in \mathbb{Z}/4\mathbb{Z}. \tag{66}$$

Consider the following transformation of the coordinates illustrated in Fig. 24:

$$\begin{aligned} b_1 = (1 + A_3^{-1})a_3 &= a_3(1 + A_4^{-1}); \\ b_2 = (1 + A_4)^{-1}a_4 &= a_4(1 + A_1)^{-1}; \\ b_3 = (1 + A_1^{-1})a_1 &= a_1(1 + A_2^{-1}). \\ b_4 = (1 + A_2)^{-1}a_2 &= a_2(1 + A_3)^{-1}; \end{aligned} \tag{67}$$

To check the second equality in each line, note that $a_i a_{i+1} a_{i+2} a_{i+3} a_{i+4}$ can be written in two ways:

$$a_i A_{i+1} = A_i a_i, \quad \forall i \in \mathbb{Z}/4\mathbb{Z}; \tag{68}$$

Lemma 4.1.3 *The square of the two by two transformation (67) is the identity transformation.*

Proof This can be proved, by using the calculation of the coordinate transformations for the space $\mathrm{Conf}_4^{\mathrm{df}}(V_2)$ of four decorated flags in Sect. 6.3. Since the direct check is easy, let us elaborate it.

Going from bs to as, the analog of the first formula in (67) is the following formula: $a_2 = b_4(1+B_1^{-1})$. Since $B_1^{-1} = A_3$, it is equivalent to $b_4 = a_2(1+A_3)^{-1}$, which is indeed the case by (67).

Similarly, the analog of the second formula in (67) is the following: $a_3 = b_1(1 + B_2)^{-1}$. Since $B_2^{-1} = A_4$, it is equivalent to $b_1 = a_3(1 + A_4^{-1})$, which is indeed the case by (67). □

Lemma 4.1.4 *We have for any $i \in \mathbb{Z}/4\mathbb{Z}$,*

$$\begin{aligned} b_i b_{i+1} &= (a_i a_{i+1})^{-1}, \\ A_i &= B_{i+2}^{-1}. \end{aligned} \tag{69}$$

Proof We have

$$\begin{aligned} b_3 b_4 a_3 a_4 &= a_1(1 + A_2^{-1})(1 + A_2)^{-1} a_2 a_3 a_4 = a_1 A_2^{-1} a_2 a_3 a_4 = A_2^{-1} A_2 = 1. \\ b_4 b_1 a_4 a_1 &= a_2(1 + A_3)^{-1}(1 + A_3^{-1}) a_3 a_4 a_1 = a_2 A_3^{-1} a_3 a_4 a_1 = A_3^{-1} A_3 = 1. \end{aligned} \tag{70}$$

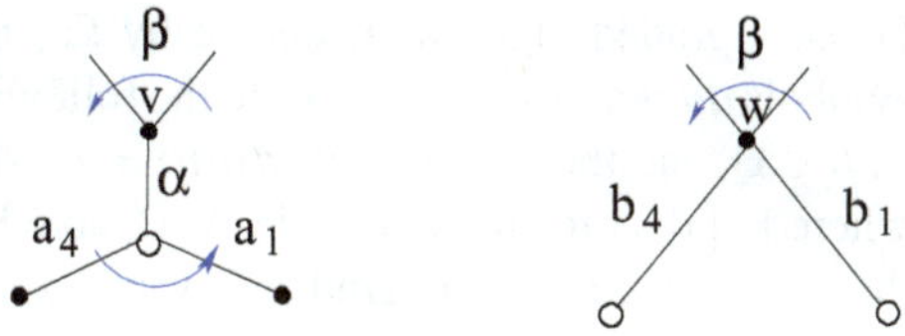

Fig. 25 Using conventions (65), we have $a_4a_1\alpha = -1$, $\alpha\beta = (-1)^{\text{val}(v)-1}$, $b_4b_1\beta = (-1)^{\text{val}(v)}$. Therefore, $a_4a_1 = (b_4b_1)^{-1}$

Since transformation formulas (67) are invariant under the cyclic shift $i \longmapsto i+2$, the first formula in (69) follows. The second formula follows from the first. □

Figure 25 shows that formula (69) is consistent with the condition that the counterclockwise product of the coordinates around a ∘-vertex is 1 and around a •-vertex is 1.

In Sect. 6.3 we define $\mathcal{A}$-coordinates for GL_2-graphs on a decorated surface $\mathbb{S}$, parametrizing the moduli space $\mathcal{A}_{2,\mathbb{S}}$. The change of these $\mathcal{A}$-coordinates under a flip of a triangulation, studied in Sect. 6.3, motivated the definition of the $\mathcal{A}$-cluster coordinate transformations in (67).

Theorem 4.1.5 *The coordinate transformation (67) for a two by two move satisfies the pentagon relation.*

The proof can be obtained by a long calculation. Instead, we give a simple proof based on Theorem 6.3.3 in the end of Sect. 6.3.

Theorem 4.1.5 allows to define the non-commutative cluster variety $\mathcal{A} = \mathcal{A}_{|\Gamma|}$ assigned to a bipartite ribbon graph Γ. Its clusters are parametrized by the bipartite ribbon graphs which can be obtained from Γ by two by two moves and the elementary transformations given by shrinking a two-valent vertex. We assign to each cluster the non-commutative torus $\mathcal{A}_\Gamma$ and glue them using the coordinate transformations (67) corresponding to the two by two moves. In Sect. 4.2 we will show that there is a canonical non-commutative 2-form $\Omega_{\mathcal{A}}$ on the cluster variety $\mathcal{A}$, preserved by the two by two moves of the graphs.

4.2 *The Canonical 2-Form on a Non-Commutative Cluster $\mathcal{A}$-Variety*

1. The Non-Commutative Analog of $d\log(a_1) \wedge \ldots \wedge d\log(a_n)$
Recall the cyclic envelope of the tensor algebra $\mathrm{T}(V)$ of a graded vector space V. It is spanned by the elements $(v_1 \otimes v_2 \otimes \ldots \otimes v_n)_C$, which are cyclically invariant:

$$(v_1 \otimes v_2 \otimes \ldots \otimes v_n)_C = (-1)^{|v_1|\cdot(|v_2|+\ldots+|v_n|)}(v_2 \otimes v_3 \otimes \ldots \otimes v_n \otimes v_1)_C.$$

Below, if it cannot lead to a misunderstanding, we use a shorthand:

$$v_1v_2\ldots v_n := (v_1 \otimes v_2 \otimes \ldots \otimes v_n)_C.$$

Given a graded algebra A, denote by Ω_A the $\mathbb{Q}$-vector space[13] generated by the symbols da where $a \in A$, modulo the following relations: $d(\alpha a+\beta b) = \alpha da+\beta db$, $\alpha, \beta \in \mathbb{Q}$, and the Leibniz rule $d(ab) = da \cdot b + (-1)^{|a|} a \cdot db$. Here $|da| = |a| + 1$, where $|\cdot|$ denotes the degree in A. Denote by $CT(A \oplus \Omega_A)$ the cyclic envelope of the tensor algebra of the graded $\mathbb{Q}$-vector space $\mathrm{T}(A \oplus \Omega_A)$.

Definition 4.2.1 Given any graded algebra A and any invertible elements $a_i \in A^\times$, the cyclic product

$$\{a_1, \ldots, a_n\} := da_1 \ldots da_n a_n^{-1} \ldots a_1^{-1} \in CT(A \oplus \Omega_A) \tag{71}$$

is the non-commutative analog of $d \log(a_1) \wedge \ldots \wedge d \log(a_n)$.

For example, the non-commutative 2-form generalizing $d \log(a) \wedge d \log(b)$ is given by

$$\{a, b\} = da\, db\, b^{-1} a^{-1}. \tag{72}$$

Note that $\{\alpha a, \beta b\} = \{a, b\}$ for any $\alpha, \beta \in \mathbb{Q}^\times$. In particular, we often use below that $\{\pm a, \pm b\} = \{a, b\}$.

Theorem 4.2.2 *For any elements $a_0, \ldots, a_n$ we have the Hochschild cocycle property:*

$$\begin{aligned} \{a_1, \ldots, a_n\} - \sum_{i=0}^{n-1} (-1)^i \{a_0, \ldots, a_{i-1}, a_i a_{i+1}, a_{i+2}, \ldots, a_n\} \\ + (-1)^n \{a_0, \ldots, a_{n-1}\} = 0. \end{aligned} \tag{73}$$

For example,

$$\begin{aligned} &\{a\} - \{ab\} + \{b\} = 0, \\ &\{b, c\} - \{ab, c\} + \{a, bc\} - \{a, b\} = 0. \end{aligned} \tag{74}$$

Proof Each of the following terms appears in the sum (73) twice, with the opposite signs:

$$da_0 \ldots da_{i-1} da_i\, a_{i+1} da_{i+2} \ldots da_n a_n^{-1} \ldots a_0^{-1}.$$

□

Lemma 4.2.3
(i) One has

$$\{a, b\} = -\{b^{-1}, a^{-1}\}. \tag{75}$$

[13] Characteristic p would be fine as well, but we do not need this in the chapter.

(ii) Suppose that $abc = \pm 1$. Then (76) implies that $\{a, b\}$ is cyclic invariant:

$$\{a, b\} + \{b, c\} + \{c, a\} = 3 \cdot \{a, b\}. \tag{76}$$

(iii) We have $\{x, 1 + x\} = \{1 + x, x\}$.

Proof
(i) Recall the identity $c\, dc^{-1} = -dc\, c^{-1}$. Using this, we have

$$\begin{aligned}\{a, b\} &= da\, db\, b^{-1}a^{-1} = a^{-1}da\, db\, b^{-1} = da^{-1}\, a\, b\, db^{-1}\\ &= -db^{-1}da^{-1}\, ab = -\{b^{-1}, a^{-1}\}.\end{aligned}$$

(ii) Since $\{a, b\} = \{a, -b\}$, we may assume that $abc = 1$. Then we have to show that

$$\{a, b\} + \{b, (ab)^{-1}\} + \{(ab)^{-1}, a\} = 3\{a, b\}. \tag{77}$$

Let us write the second identity in (74) for $c = (ab)^{-1}$:

$$\{a, b\} - \{a, a^{-1}\} + \{ab, (ab)^{-1}\} - \{b, (ab)^{-1}\} = 0. \tag{78}$$

By (75), we have $\{a, a^{-1}\} = -\{a, a^{-1}\}$. So $\{a, a^{-1}\} = 0$. So

$$\{a, b\} = \{b, (ab)^{-1}\}. \tag{79}$$

Using (75) we have $\{a, b\} = -\{b^{-1}, a^{-1}\} \overset{(79)}{=} -\{a^{-1}, ab\} = \{(ab)^{-1}, a\}$. This and (79) imply the claim.
(iii) One has $\{x, 1 + x\} = dxdx(1 + x)^{-1}x^{-1} = dxdxx^{-1}(1 + x)^{-1} = \{1 + x, x\}$. □

2. The Non-Commutative 2-Form Ω_Γ for an $\mathcal{A}$-Decorated Bipartite Ribbon Graph Γ

Recall that an edge of a graph Γ is called an *external edge* if one of its vertices is univalent.

For simplicity, we do not consider graphs which have a component given by a single external edge.

We start from a trivalent ribbon graph Γ_v with a single-vertex v, decorated by a_1, a_2, a_3 in the order compatible with the cyclic structure at v, such that $a_1a_2a_3 = \pm 1$. Then we set, using notation (72),

$$\Omega_v := \{a_1, a_2\}. \tag{80}$$

It depends only on the cyclic order of the edges. Indeed, $\{a_1, a_2\} = \{a_2, a_3\} = \{a_3, a_1\}$ since $a_1a_2a_3 = \pm 1$.

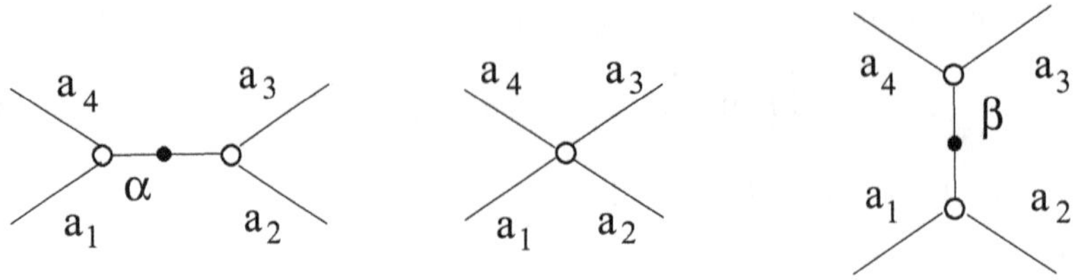

Fig. 26 The cocycle condition guarantees that the 2-form Ω_Γ is well defined

Given a ribbon graph Γ_v with a single vertex v of valency > 3, we expand this vertex by adding two-valent vertices of the opposite color, as shown on Fig. 26, producing a bipartite ribbon graph Γ'_v with trivalent vertices of the original color and two-valent vertices of the opposite color. We set

$$\Omega_v := \sum_{x \in \Gamma'_v} \Omega_x. \tag{81}$$

By default, the contribution of the 2-valent vertex is zero.[14] The sum is over trivalent vertices of the graph Γ'_v. To see that it does not depend on the choice of Γ'_v, it is sufficient to check this for the graphs on Fig. 26. This boils down to the cocycle condition for $\{a, b\}$:

$$\begin{aligned} &\{a_2, a_3\} + \{\alpha, a_4\} = \{a_1, a_2\} + \{\beta, a_3\} \longleftrightarrow \\ &\{a_2, a_3\} + \{a_1, a_2a_3\} = \{a_1, a_2\} + \{a_1a_2, a_3\}. \end{aligned} \tag{82}$$

Now we introduce the non-commutative 2-form assigned to an any decorated bipartite ribbon graph Γ.

Recall the $\mathcal{A}$-decorated ribbon graph; see Definition 4.1.1.

Definition 4.2.4 Given an $\mathcal{A}$-decorated bipartite ribbon graph Γ, we set

$$\Omega_\Gamma := \sum_{w \in \Gamma} \Omega_w - \sum_{b \in \Gamma} \Omega_b. \tag{83}$$

Here the first sum is over all $\circ$-vertices of Γ and the second over all $\bullet$-vertices.

We assign to each edge E of an $\mathcal{A}$-decorated ribbon graph Γ, decorated by an element a_E, the following 3-form:

$$\omega_E := (a_E^{-1} d a_E)^3. \tag{84}$$

[14] This is consistent with $\{a, a^{-1}\} = 0$ since the decorations of a 2-valent vertex are $a, \pm a^{-1}$.

Theorem 4.2.5 *For an $\mathcal{A}$-decorated bipartite ribbon graph Γ, the $d\Omega_\Gamma$ is a sum is over all external edges E of Γ:*

$$d\Omega_\Gamma = \sum_E \omega_E. \tag{85}$$

The 2-form Ω_Γ assigned to an $\mathcal{A}$-decorated ribbon graph Γ without external edges is closed: $d\Omega_\Gamma = 0$.

Proof Clearly the first claim implies the second. □

Lemma 4.2.6 *Let $abc = 1$. Then*

$$3d\{a, b\} = -(a^{-1}da)^3 - (b^{-1}db)^3 - (c^{-1}dc)^3. \tag{86}$$

Proof A straightforward calculation, using $c^{-1}dc = -d(c^{-1})c$, shows that the right-hand side is

$$-(a^{-1}da)^3-(b^{-1}db)^3+\Big((ab)^{-1}d(ab)\Big)^3=3dadbd(b^{-1})a^{-1} + 3d(a^{-1})dadbb^{-1}$$
$$= 3d\{a, b\}.$$
□

Therefore, for each ∘-vertex w of Γ the differential of the 2-form Ω_w assigned to w is

$$d\Omega_w = -\sum_{w\in E_i} \omega_{E_i}. \tag{87}$$

Here the sum is over all edges E_i sharing the vertex w.

On the other hand, for each •-vertex b of Γ the differential of the 2-form Ω_b assigned to b is

$$d\Omega_b = \sum_{b\in E_i} \omega_{E_i}. \tag{88}$$

Here the sum is over all edges sharing the vertex b. Then evidently

$$d\Omega_\Gamma = \sum_{v\in\Gamma} d\Omega_v = -\sum_{E\subset\Gamma} \omega_E + \sum_{E\subset\Gamma} \omega_E = 0. \tag{89}$$

For the convenience of the reader, we reproduce the picture illustrating the change of the coordinates under the two by two move, which we refer to in Theorem 4.2.7.

Theorem 4.2.7 *The 2-form Ω_Γ assigned to a decorated bipartite ribbon graph Γ is invariant under the two by two moves. This just means that the following identity holds, using the notation (67):*

Fig. 27 The right picture is obtained from the left one by the clockwise rotation by 90°, followed by the cyclic shift $a_i \longmapsto b_{i+1}, i \in \mathbb{Z}/4\mathbb{Z}$

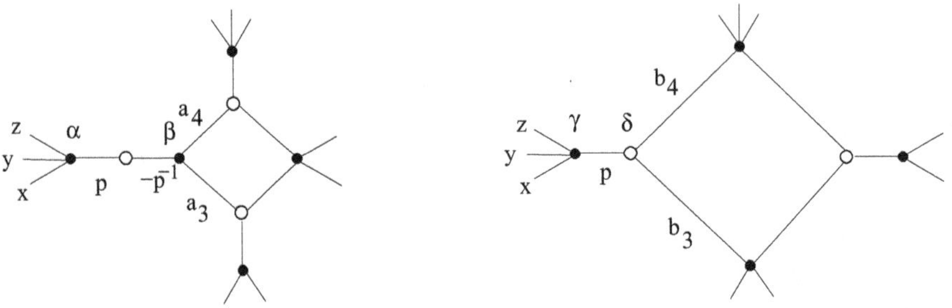

Fig. 28 Contributions of vertices α and γ to Ω_Γ and $\Omega_{\Gamma'}$, respectively, coincide: $\Omega_\alpha = \Omega_\gamma$. We use the identity $a_3a_4 = (b_3b_4)^{-1}$ to see that the decorations of right edges of vertices α and γ are the same: p. We use sign convention (65). The signs of $\mathcal{A}$-decorations do not affect the 2-forms since $\{\pm a, \pm b\} = \{a, b\}$

$$\{a_4, a_1\} - \{a_1, a_2\} + \{a_2, a_3\} - \{a_3, a_4\} = \{b_1, b_2\} - \{b_2, b_3\} + \{b_3, b_4\} - \{b_4, b_1\}. \tag{90}$$

Proof Denote by Γ and Γ' the two ribbon graphs related by a two by two move. Then we claim that

$$\begin{aligned} \Omega_\Gamma - \Omega_{\Gamma'} = &\Big(\{a_4, a_1\} - \{a_1, a_2\} + \{a_2, a_3\} - \{a_3, a_4\}\Big) \\ &- \Big(\{b_1, b_2\} - \{b_2, b_3\} + \{b_3, b_4\} - \{b_4, b_1\}\Big). \end{aligned} \tag{91}$$

Indeed, the contribution of the left $\bullet$-vertex on Fig. 27 is equal to $-\Omega_\alpha - \Omega_\beta$, where α and β are the two $\bullet$-vertices on the left graph Γ on Fig. 28. The contribution of the $\bullet$- and $\circ$-vertices γ and δ on the right graph Γ' on Fig. 28 is equal to $-\Omega_\gamma + \Omega_\delta$. Since, as Fig. 28 shows, $\Omega_\alpha = \Omega_\gamma$, the total contribution of the left vertices on Fig. 27 to $\Omega_\Gamma - \Omega_{\Gamma'}$ is equal to $-\Omega_\beta - \Omega_\delta$. Note that by default the contribution of the two-valent vertices are zero. A similar analysis for the top, right, and bottom vertices on Fig. 27 shows that their total contribution to the left-hand side of (91) is equal to the right-hand side of (91). On the other hand, the contributions of all other matching vertices of the graphs Γ and Γ' evidently coincide.

Therefore, it remains to prove the identity (90). This is done by a tedious calculation in Sect. 9.

Note that, as we show there, the identity follows from a more elementary identity from Theorem 9.2.1. The latter is the non-commutative analog of the Steinberg relation $d\log(1 + X) \wedge d\log(X)$. □

Remarks

1. The right picture on Fig. 27 can be obtained from the left one by the clockwise rotation by 90°, followed by the cyclic shift $a_i \longmapsto b_{i+1}$. The right-hand side of relation (90) is obtained from the left-hand side by the same cyclic shift.
2. Figure 28 shows that the identity $a_3a_4 = (b_3b_4)^{-1}$ is crucial to check that the invariance of the 2-form Ω_Γ under two by two moves is equivalent to the identity (90).

A Variant Take any ribbon graph Γ. Assign to its oriented edges $\mathbf{E}$ elements $a_\mathbf{E}$ such that denoting by $\overline{\mathbf{E}}$ the edge $\mathbf{E}$ with the opposite orientation, we have $a_\mathbf{E}a_{\overline{\mathbf{E}}} = 1$. For each vertex $v \in \Gamma$ we define Ω_v using the edges that are oriented out of the vertex and take the sum over all vertices $v \in \Gamma$ of valency > 1:

$$\Omega_\Gamma := \sum_{v \in \Gamma} \Omega_v. \tag{92}$$

5 Spectral Description of Framed Local Systems on Surfaces

In Sect. 5.1 we start from recalling some material from [19]: a class of bipartite ribbon graphs called GL_m-graphs and their spectral surfaces. Then we describe non-commutative framed local systems on a decorated surface $\mathbb{S}$ via framed flat line bundles on the spectral surface Σ of a bipartite graph on $\mathbb{S}$. Namely, we show that the main construction of Sect. 1.2 can be generalized to any GL_m-graph.

In Sects. 5.2 and 5.3 we consider the *twisted* variant: describe non-commutative *twisted* framed local systems on $\mathbb{S}$ via *twisted* framed flat line bundles on the spectral surface Σ.

In Sect. 5.2 we consider the key example, when $\mathbb{S}$ is a triangle t.

In Sect. 5.3 we consider the general case.

5.1 Spectral Description of Non-Commutative Framed Local Systems

1. The GL_m-Graphs and Spectral Surfaces [19]
Given a bipartite ribbon graph Γ, the *conjugate ribbon graph* Γ^* introduced in [20], is obtained by changing the cyclic order of the edges at every $\bullet$-vertex. There is a bijection:

$$\{\text{zig-zag paths on } \Gamma\} \longleftrightarrow \{\text{face paths on } \Gamma^*\}. \tag{93}$$

The conjugate ribbon graph Γ^* provides a topological surface Σ with holes, whose boundaries are the face paths on Γ^*, described as follows. Let us add to

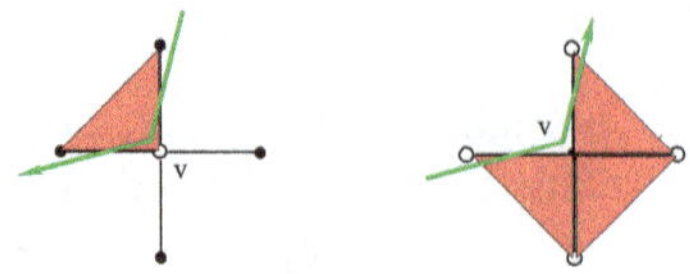

Fig. 29 A (green) oriented zig-zag strand γ near a vertex v, deformed slightly off the vertex

the ribbon graph Γ^* its punctured faces F_{γ_i} bounding the face paths γ_i. Precisely, if γ_i is a face loop, then F_{γ_i} is a disc bounded by this loop, punctured inside. If γ_i is a face interval, then F_{γ_i} is a disc; the half of its boundary is identified with the path γ_i, and the rest becomes a boundary of Σ. It is handy to puncture it at the marked point. We arrive at a topological surface, called the *spectral surface* [19]:

$$\Sigma = \Gamma^* \cup \cup_{\gamma_i} F_{\gamma_i}. \tag{94}$$

Let $\mathbb{S}$ be a decorated surface with marked points $\{s_1, \dots, s_n\}$. A *strand* on $\mathbb{S}$ is either an oriented loop or an oriented path on $\mathbb{S} - \{s_1, \dots, s_n\}$ connecting boundary points. Let γ be a strand such that $\mathbb{S} - \gamma$ has two connected components:

$$\mathbb{S} - \gamma = S^\circ_\gamma \cup S^{\circ\prime}_\gamma. \tag{95}$$

They inherit orientatations from $\mathbb{S}$. The domain S°_γ is the one whose orientation induces the original (clockwise) orientation of γ. We denote by S_γ the closure of S°_γ.

Let $\mathbb{S}$ be a decorated surface associated with a bipartite ribbon graph Γ, so that $\Gamma \subset \mathbb{S}$. Given a zig-zag path γ on Γ, the shape of the domain S_γ near a vertex v of Γ depends on the vertex type. Namely, the intersection $S_\gamma \cap U_v$ with a neighborhood U_v of v is a single sector if v is a $\circ$-vertex and a union of $\text{val}(v) - 1$ sectors if v is a $\bullet$-vertex; see Fig. 29.

Pushing a bit zig-zag paths out of the vertices as on Fig. 29, we get *zig-zag strands* on $\mathbb{S}$.

Let $\{\gamma\}$ be the collection of zig-zag strands on S. We assume that for each zig-zag strand γ:

- $\mathbb{S} - \gamma$ has two connected components, and the domains S_γ are discs.

Using bijection (93), we construct the spectral surface Σ as follows. Denote by F_γ a copy of the disc S_γ. We attach the disc F_γ to every zig-zag path γ on Γ. Since the F_γ are discs, we get a surface homeomorphic to surface (94). Canonical maps $F_\gamma \longrightarrow S_\gamma$ provide a ramified map:

$$\pi : \Sigma \longrightarrow \mathbb{S}. \tag{96}$$

Definition 5.1.1 The degree m of the map π is called the *rank* of the bipartite ribbon graph Γ.

The map π has the following properties:

Fig. 30 No parallel bigons (left) and half-bigons (right; the boundary of $\mathbb{S}$ is punctured)

- The ramification points of π are the $\bullet$-vertices of Γ of valency ≥ 3: the ramification index of a $\bullet$-vertex equals its valency minus 2—this is clear from Fig. 29.
- For every marked point $s \in \mathbb{S}$ there is a bijection:

$$\varphi_s : \{1, \dots, m\} \longrightarrow \pi^{-1}(s). \tag{97}$$

Indeed, the set $\pi^{-1}(s)$ is identified canonically with the set of zig-zag loops/arcs going around the point s. Each of the latter is the boundary of a disc/angle containing s, which are ordered by the inclusion.

Definition 5.1.2 Let $\Gamma \subset \mathbb{S}$ be a bipartite graph associated with a decorated surface $\mathbb{S}$.

(i) A zig-zag strand γ on $\mathbb{S}$ is *ideal* if the domain S_γ° in (95) contains a single marked point s. We say that γ is associated to s.
(ii) Γ is a *strict* GL_m*-graph* if its rank is $m > 1$, and all zig-zag strands are ideal.

- There are m zig-zag strands associated with each marked point.
- There are no parallel bigons and parallel half-bigons; see Fig. 30.

(iii) Γ is a GL_m*-graph* if for some finite unramified cover $\rho : \widetilde{\mathbb{S}} \to \mathbb{S}$, $\rho^{-1}(\Gamma)$ is a strict GL_m-graph.

For example, the bipartite ribbon graph $\Gamma_m(\mathcal{T})$ assigned to an ideal triangulation $\mathcal{T}$ of S just before Theorem 1.2.8 is a GL_m-graph.

For a strict GL_m-graph Γ on $\mathbb{S}$, the *codistance* $\langle x, s\rangle$ from a point $x \in \mathbb{S}$ to a special point s is the number of zig-zag strands γ associated with s such that x belongs to the *open* disc S_γ°. For example, the codistance to a special point s from a point x very close to s is m.

Lemma 5.1.3 *Let Γ be a strict GL_m-graph on $\mathbb{S}$. Then zig-zags associated with a marked point s form m concentric (half-) circles, going counterclockwise around s.*

For any point $x \in \mathbb{S} - \Gamma$, the sum of codistances from x to the marked points is equal to $\mathrm{rk}(\Gamma)$*:*

$$\sum_s \langle x, s\rangle = \mathrm{rk}(\Gamma). \tag{98}$$

Proof Formula (98) is equivalent to the definition of rank as the degree of the covering π.

Alternatively, formula (98) is evident for a point x close to a marked point: in this case all codistances but one are 0, and the nonzero one is m. When x moves, crossing an edge E in an internal point, we alter codistances as follows: one of them increases by 1, another decreases by 1, and the rest stay intact. □

2. Spectral Description of Non-Commutative Framed Local Systems
Given a decorated surface $\mathbb{S}$, recall the notion of a *framing* on a local system of finite-dimensional vector spaces over a skew field R; see Definition 1.3.1. A framing is a choice of a flat section of the associated local system of flags near each marked point. A local system on $\mathbb{S}$ is framed if it is equipped with a framing.

The monodromy of an m-dimensional framed R-local system $\mathcal{V}$ on $\mathbb{S}$ around a puncture p is an operator N in the fiber $\mathcal{V}_x$ over a nearby point x preserving the framing flag $\mathcal{F}$ in $\mathcal{V}_x$. The triple $(\mathcal{V}_x, \mathcal{F}, N)$ determines m operators, called the *monodromy eigenvalue operators* at p:

$$N_i : \mathrm{gr}^i\mathcal{F} \longrightarrow \mathrm{gr}^i\mathcal{F}, \quad i = 0, \ldots, m-1.$$

If R is commutative, these are the eigenvalues of N, which determine the semi-simple part of N.

The points of Σ_Γ projecting to the marked points of $\mathbb{S}$ are the *marked points of* Σ_Γ.

Theorem 5.1.4 *Let Γ be a* $\mathrm{GL_m}$*-graph on a decorated surface $\mathbb{S}$, homotopy equivalent to $\mathbb{S}$. We assume that it can be related by a sequence of two by two moves to a* $\mathrm{GL_m}$*-graph $\Gamma_m(\mathcal{T})$ assigned to an ideal triangulation of $\mathbb{S}$.*[15] *Then the following two groupoids are canonically birationally equivalent:*

1. *Groupoid of m-dimensional framed R-local systems on $\mathbb{S}$*
2. *Groupoid of flat R-line bundles with connection on the spectral surface Σ_Γ*

The equivalence identifies the monodromy eigenvalue operators of the connection on $\mathcal{V}$ around a puncture p with the monodromy operators of $\mathcal{L}$ around the punctures on Σ_Γ over p, so that the eigenvalue operator N_i is the monodromy of $\mathcal{L}$ around the point $\varphi_p(i) \in \pi^{-1}(p)$; see (97).

Proof Flat line bundles on the spectral surface Σ_Γ are the same thing as flat line bundles on the graph Γ. Follow the proof of Theorem 1.2.2, adjusted to the graph Γ; let us define a functor:

$$\mathcal{L} : \{\text{Generic } m\text{-dimensional framed local systems } \mathcal{V} \text{ on } \mathbb{S}\} \longrightarrow \{\text{Flat line bundles on } \Gamma\}. \quad (99)$$

[15] We do not know whether any $\mathrm{GL_m}$-graph on a decorated surface $\mathbb{S}$ has this property. This is the case when $\mathbb{S}$ is a polygon by [19, Section 3]. Since the collection of $\mathrm{GL_m}$-graphs with this property is equivariant under the action of the mapping class group of $\mathbb{S}$, Theorem 5.1.4 is sufficient for all our purposes.

Given a vertex v of Γ and a marked point s on $\mathbb{S}$ with the codistance $\langle v, s\rangle > 0$, there is a unique zig-zag γ containing v such that $s \in S_\gamma$. Since S_γ is a punctured disc, and the flag $\mathcal{F}^\bullet(s)$ near s is monodromy invariant, there is a canonical parallel transport of the flag $\mathcal{F}^\bullet(s)$ to a flag $\mathcal{F}^\bullet_{v,s}$ in the fiber $\mathcal{V}_v$ of $\mathcal{V}$ at v.

1. Let $\mathbf{w}$ be a $\circ$-vertex of the graph Γ. Denote by $a_1, \dots, a_n$ the codistances from the $\mathbf{w}$ to the marked points $s_1, \dots, s_n$ on $\mathbb{S}$. Then

$$a_1 + \ldots + a_n = m - 1, \qquad a_i \geq 0.$$

We assign to the $\circ$-vertex $\mathbf{w}$ a one-dimensional subspace in the vector space $\mathcal{V}_{\mathbf{w}}$:

$$\mathrm{L}^\circ_{\mathbf{w}} := \mathcal{F}^{a_1}_{\mathbf{w},s_1} \cap \ldots \cap \mathcal{F}^{a_n}_{\mathbf{w},s_n}. \tag{100}$$

2. Let $\mathbf{b}$ be a k-valent $\bullet$-vertex of the graph Γ. Denote by $b_1, \dots, b_n$ the codistances from the $\mathbf{b}$ to the marked points $s_1, \dots, s_n$ on $\mathbb{S}$. Then

$$b_1 + \ldots + b_n = m - k + 1, \qquad b_i \geq 0.$$

We assign to the k-valent $\bullet$-vertex $\mathbf{b}$ a $(k-1)$-dimensional subspace in the vector space $\mathcal{V}_{\mathbf{b}}$:

$$\mathrm{P}_{\mathbf{b}} := \mathcal{F}^{b_1}_{\mathbf{b},s_1} \cap \ldots \cap \mathcal{F}^{b_n}_{\mathbf{b},s_n}. \tag{101}$$

The condition that for a configuration of flags the intersections (100) and (101) have given dimensions (1 or 2) is an open condition. To check that the collection of flags satisfying this condition is non-empty, we note that this follows by Theorem 1.2.2 for the bipartite graphs $\Gamma_m(\mathcal{T})$. Since two by two moves provide birational isomorphisms, the claim follows.

The plane $\mathrm{P}_{\mathbf{b}}$ contains the k lines $\mathrm{L}^\circ_{\mathbf{w}}$ assigned to the $\circ$-vertices $\mathbf{w}$ incident to $\mathbf{b}$. Indeed, the codistances $(a_1, \dots, a_n)$ and $(b_1, \dots, b_n)$ from the vertices $\mathbf{w}$ and $\mathbf{b}$ are related as follows: $0 \leq a_i - b_i \leq 1$, and $a_j = b_j + 1$ for exactly k indices j. For any such an index j the embedding $\mathcal{F}^{b_j+1} \hookrightarrow \mathcal{F}^{b_j}$ induces the embedding $i_{\mathbf{w}\to\mathbf{b}} : \mathrm{L}^\circ_{\mathbf{w}} \longrightarrow \mathrm{P}_{\mathbf{b}}$.

We define a line $\mathrm{L}^\bullet_{\mathbf{b}}$ as the kernel of the sum of these maps over all $\circ$-vertices $\mathbf{w}$ incident to the $\mathbf{b}$:

$$\mathrm{L}^\bullet_{\mathbf{b}} := \mathrm{Ker}\Big(\bigoplus_{\mathbf{w}-\mathbf{b}} \mathrm{L}^\circ_{\mathbf{w}} \stackrel{i_{\mathbf{w}\to\mathbf{b}}}{\longrightarrow} \mathrm{P}_{\mathbf{b}}\Big). \tag{102}$$

Then there is a canonical map, induced by projections of the subspace $\mathrm{L}^\bullet_{\mathbf{b}}$ onto the summands:

$$\mathrm{L}^\bullet_{\mathbf{b}} \longrightarrow \bigoplus_{\mathbf{w}} \mathrm{L}^\circ_{\mathbf{w}}. \tag{103}$$

So we get a flat line bundle $\mathcal{L}_{\mathcal{V}}$ on the graph Γ. Its fibers at the vertices are given by the lines L° and $L^\bullet$, and the parallel transport along an edge $\bullet \to \circ$ is given by a map $L^\bullet \longrightarrow L^\circ$ in (103).

Remark If $R = K$ is commutative, the moduli space of flat line bundles on a graph Γ is identified with $H^1(\Gamma, K^*)$. The latter includes the monodromies around the holes in the graph. These are the coordinates from [19], generalizing the coordinates [7] to ideal bipartite graphs on $\mathbb{S}$.

If R is non-commutative, we no longer have canonical coordinates.

Let us show that the functor $\mathcal{L}$ is an equivalence. Pick an ideal triangulation $\mathcal{T}$ of $\mathbb{S}$. Recall the GL_m-graph $\Gamma_m(\mathcal{T})$ related to $\mathcal{T}$. By Theorem 1.2.8, applied to the surface $\mathcal{S}$ glued from the triangles of $\mathcal{T}$, which is just the decorated surface $\mathbb{S}$, the functor (99) is an equivalence of groupoids, where "generic" means that the triple of flags for each triangle of $\mathcal{T}$ is generic in the sense of Sect. 1.2. Then the claim for a graph Γ related to $\Gamma_m(\mathcal{T})$ by a sequence of two by two moves follows from results of Sect. 2.1. □

5.2 Spectral Description of Generic Twisted Triples of Flags

Let us replace a triangle t by an oriented circle S^1 with three marked points A, B, C. Twisted framed local systems on $(S^1; \mathrm{A}, \mathrm{B}, \mathrm{C})$ are the same thing as the flat line bundle on S^1 with the monodromy -1 and flags $(\mathcal{A}, \mathcal{B}, \mathcal{C})$ in the fibers over the tangent vectors $v_{\mathrm{A}}, v_{\mathrm{B}}, v_{\mathrm{C}}$ to S^1 at the points A, B, C, following the orientation of S^1. We call such data *twisted triples of flags*. There is a canonical one-dimensional local system ε on S^1 with the monodromy -1. Multiplying by ε, we identify the usual and twisted triples of flags. So there is almost no difference between the two notions. The distinction between the two will be important when we move to *decorated* twisted configurations of flags.

Take a generic triple of flags $(\mathcal{A}, \mathcal{B}, \mathcal{C})$ in an m-dimensional R-vector space. We assign the flags to the vertices of a triangle t. The bipartite ribbon graph Γ_m on Fig. 4 assigned to t gives rise to the spectral surface Σ_m. Namely, each zig-zag γ on Γ_m determines a smaller triangle t_γ, given by the component of $t - \gamma$ containing a single vertex of t. Take the disjoint union of all triangles t_γ. Then, for each edge E of the graph Γ_m, glue the triangles $t_{\gamma'}$ and $t_{\gamma''}$ assigned to the zig-zags γ' and γ'' containing E over the edge E. We get the surface Σ_m and the projection $\Sigma_m \longrightarrow t$, ramified over the $\bullet$-vertices. To avoid confusion, we denote by F_γ the triangle t_γ when it sits on the spectral surface.

Theorem 5.2.1 *There is a canonical equivalence between the following groupoids:*

1. *Groupoid of generic triples of flags in an m-dimensional R-vector space*
2. *Groupoid of flat twisted R-line bundles with connection on the spectral surface Σ_m of the graph Γ_m*

Proof 1 $\longrightarrow$ 2. A triple of flags $(\mathcal{A}, \mathcal{B}, C)$ in an m-dimensional R-vector space V gives rise to $3m$ lines:

$$\mathrm{gr}^a\mathcal{A}, \;\; \mathrm{gr}^b\mathcal{B}, \;\; \mathrm{gr}^c C, \;\; a, b, c = 0, \ldots, m-1. \tag{104}$$

Given a zig-zag γ on the graph Γ_m going around of the vertex A, we assign to the face F_γ of the spectral surface the line $\mathrm{gr}^{a_\gamma - 1}\mathcal{A}$, where $a_\gamma \in \{1, \ldots, m\}$ is the codistance from γ to A. Abusing notation, we denote by $\mathrm{gr}^{a_\gamma-1}\mathcal{A}$ the twisted flat line bundle on F_γ whose restriction to oriented tangent vectors to the oriented side BC is given by the line $\mathrm{gr}^{a_\gamma-1}\mathcal{A}$. We do the same for the vertices B, C. Let us glue these twisted flat line bundles on the $3m$ faces F_γ to a twisted line bundle on the spectral surface.

Note that the edges of the graph Γ_m are parametrized by the following data:

(i) A triple of nonnegative integers (a, b, c) such that $a + b + c = m - 1$
(ii) A choice of two of the three numbers (a, b, c)

The faces of the graph Γ_m are labeled by the triples of nonnegative integers (a, b, c) with $a + b + c = m$.

Given a data (a, b, c) assigned to an edge E, where the chosen pair in (ii) is the pair (a, b), the two faces $F_{\gamma'}$ and $F_{\gamma''}$ of the spectral surface sharing the edge E are labeled by the triples $(a, b+1, c)$ and $(a+1, b, c)$, accordantly. Then the faces $F_{\gamma'}$ and $F_{\gamma''}$ carry the line bundles:

$$\mathrm{L}_{\gamma'} := \mathrm{gr}^a\mathcal{A}, \qquad \mathrm{L}_{\gamma''} := \mathrm{gr}^b\mathcal{B}.$$

We assign to the edge E the line:

$$\mathrm{L}_E := \mathcal{A}^a \cap \mathcal{B}^b \cap C^c.$$

Then there are canonical isomorphisms, obtained by projecting along $\mathcal{A}^{a+1}$ and $\mathcal{B}^{b+1}$:

$$\mathrm{gr}^a\mathcal{A} \stackrel{\sim}{\longleftarrow} \mathcal{A}^a \cap \mathcal{B}^b \cap C^c \stackrel{\sim}{\longrightarrow} \mathrm{gr}^b\mathcal{B}. \tag{105}$$

They allow to glue the twisted flat line bundles over the faces $F_{\gamma'}$ and $F_{\gamma''}$ along their common edge E.

Let us extend the obtained twisted flat line bundle on $\Sigma_m - \{\text{the vertices of } \Gamma_m^*\}$ to the spectral surface. In order to do this, we make the following observations.

For any triple (a, b, c) such that $a + b + c = m - 1$, there are three canonical isomorphisms:

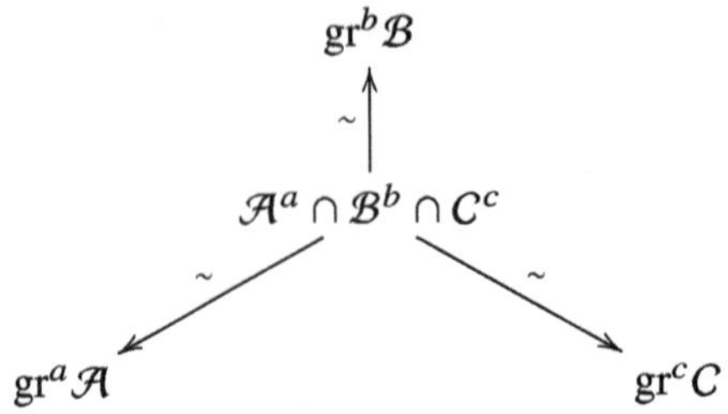

(106)

The first is the composition $\mathcal{A}^a \cap \mathcal{B}^b \cap \mathcal{C}^c \hookrightarrow \mathcal{A}^a \to \mathcal{A}^a/\mathcal{A}^{a+1}$. The others are similar. □

Lemma 5.2.2 *There are the following relations between the isomorphisms (106):*

1. *For any integers $a, b, c \geq 0$ such that $a+b+c = m-1$ there is a commutative triangle, i.e., the composition of the three isomorphism is the identity:*

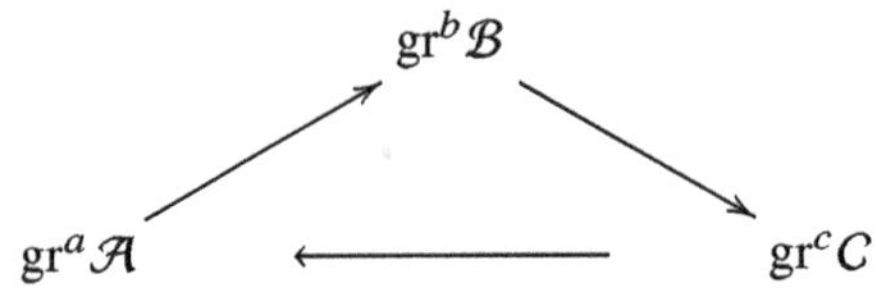

(107)

2. *For any integers $a, b, c \geq 0$ such that $a+b+c = m-2$ there is a sign-commutative hexagon, i.e., the composition of the following six isomorphisms/their inverses is equal to -1:*

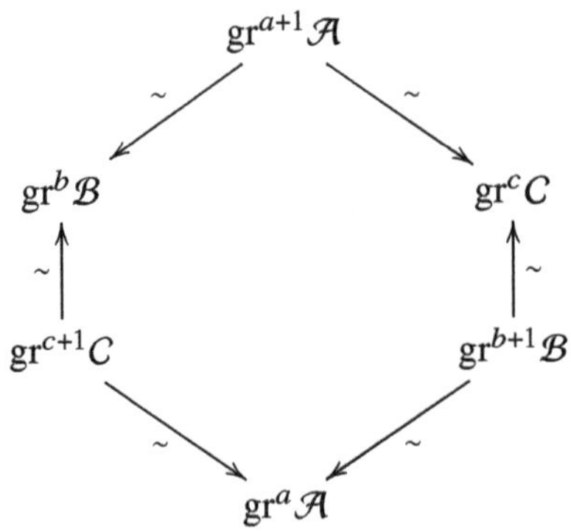

(108)

Proof
(1) Starting with a pair of isomorphisms $\mathrm{gr}^a\mathcal{A} \xleftarrow{\sim} \mathcal{A}^a \cap \mathcal{B}^b \cap \mathcal{C}^c \xrightarrow{\sim} \mathrm{gr}^b\mathcal{B}$ and composing the inverse of the first one with the second, we get an isomorphism $\mathrm{gr}^a\mathcal{A} \xrightarrow{\sim} \mathrm{gr}^b\mathcal{B}$. The other two are similar. This makes the first claim is evident.
(2) Intersecting with the plane $\mathcal{A}^a \cap \mathcal{B}^b \cap \mathcal{C}^c$, we reduce the claim to the following. Let A, B, C be three different lines in a plane P. Consider the quotient lines:

$$\overline{A} := \mathrm{P}/A, \quad \overline{B} := \mathrm{P}/B, \quad \overline{C} := \mathrm{P}/C.$$

Projecting the line A to the quotient line $\overline{B}$ we get an isomorphism $A \longrightarrow \overline{B}$. There are six such line isomorphisms:

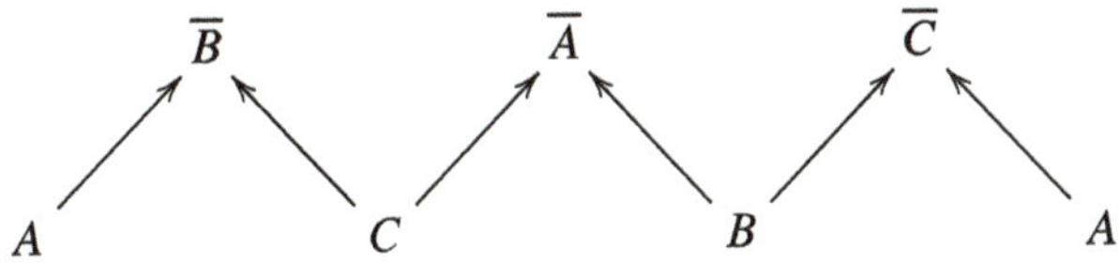

(109)

We compose them to a map $A \to A$, where going against an arrow means inverting the isomorphism. We claim that the composition is the minus identity map. Indeed, choose a triple of nonzero vectors $x \in A, y \in B, z \in C$ whose sum is zero:

$$x + y + z = 0. \tag{110}$$

Such a triple exists and is defined uniquely up to a multiplication from the left by a nonzero scalar. Condition (110) just means that $x + z$ projects to zero in $\overline{B}$. So the map $A \longrightarrow C$ provided by the diagram $A \overset{\sim}{\longrightarrow} \overline{B} \overset{\sim}{\longleftarrow} C$ is given by $x \longmapsto -z$. Therefore, the composition $A \longrightarrow C \longrightarrow B \longrightarrow A$ is given by $x \longmapsto -z \longmapsto y \longmapsto -x$. □

Part (i) of Lemma 5.2.2 allows to extend the twisted flat line bundle to the ∘-vertices. Indeed, although the composition of the three maps on Fig. 31 is the identity map, the tangent vectors are rotated by 2π, thus resulting the $-\mathrm{Id}$ map.

Similarly, part (ii) of Lemma 5.2.2 allows to extend the twisted flat line bundle to the •-vertices. Indeed, although the composition of the six maps (108) is $-\mathrm{Id}$, the tangent vectors are rotated by 4π, thus still resulting the $-\mathrm{Id}$ map.

$2 \longrightarrow 1$. *The functor* $\mathcal{R}$: {Twisted flat line bundles on Σ_m} $\longrightarrow$ {generic twisted triples of flags}. Take a twisted flat line bundle $\mathcal{L}$ on the spectral surface Σ_m of the bipartite ribbon graph Γ_m. It contains the graph Γ_m^*. For each vertex $v \in \Gamma_m^*$, restrict $\mathcal{L}$ to the positively oriented tangent vectors to an oriented circle $S_v^1 \subset T_v\Sigma - \{0\}$, getting a flat line bundle $\mathcal{L}_{S_v^1}$ on S_v^1 with the monodromy -1.

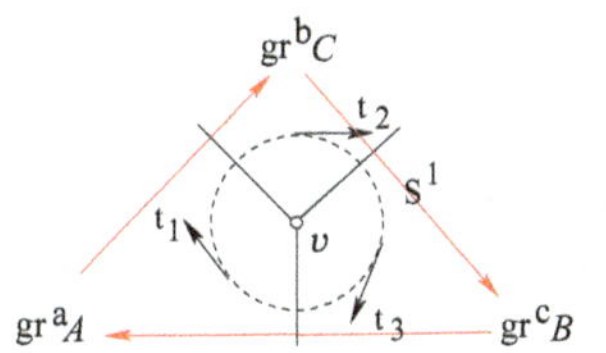

Fig. 31 Let $a+b+c = m-1$, and the line $\mathrm{gr}^a\mathcal{A}$ sits over the tangent vector t_1, etc. By (107), the composition $\mathrm{gr}^a\mathcal{A} \to \mathrm{gr}^b\mathcal{B} \to \mathrm{gr}^c\mathcal{C} \to \mathrm{gr}^a\mathcal{A}$ is the identity map. The tangent vector t_1 rotates by $t_1 \to t_2 \to t_3 \to t_1$ by 2π

Then the flat connection on $\mathcal{L}$ provides a natural map:

$$\mathrm{K}_{\mathcal{L}}: \bigoplus_{\text{internal } \bullet\text{-vertices } \mathrm{b}} \mathcal{L}_{S^1_{\mathrm{b}}} \longrightarrow \bigoplus_{\circ\text{-vertices } \mathrm{w}} \mathcal{L}_{S^1_{\mathrm{w}}}. \tag{111}$$

We identify all circles S^1_{w} with a circle S^1 and define the vector space $V_{\mathcal{L}}$ as the cokernel of this map:

$$V_{\mathcal{L}} := \mathrm{Coker}(\mathrm{K}_{\mathcal{L}}). \tag{112}$$

It is a flat bundle on the circle S^1 with the monodromy $-\mathrm{Id}$. Let us define three flags in the fibers of this local system over the points A, B, C. The distance functions $d_{\mathrm{A}}, d_{\mathrm{B}}, d_{\mathrm{C}}$ (see the proof of Theorem 1.2.2) induce three filtrations in the fibers of complex (111) over the points A, B, C. They induce filtrations on $\mathrm{Coker}(\mathrm{K}_{\mathcal{L}})$, providing a twisted triple of flags. The functor $\mathcal{R}$ is the inverse to the one defined before.

Remark Let us compare constructions of the connection in the proofs of Theorems 1.2.2 and 5.2.1. The key point is that given three vectors x, y, z in a two-dimensional vector space with $x+y+z=0$; we can define a connection either by $x \to y \to z \to x$, as we did proving Theorem 1.2.2, or by $x \to -y \to z \to -x$, getting a twisted connection.

5.3 Spectral Description Twisted Local Systems on Surfaces

Below we use the notation Σ for the spectral surface Σ_Γ since the bipartite graph Γ is fixed.

Theorem 5.3.1 *Let Γ be a GL_m-graph on a decorated surface $\mathbb{S}$ and Σ the corresponding spectral surface. Then the following two groupoids are canonically birationally equivalent:*

1. *Groupoid of m-dimensional twisted framed R-vector bundles with flat connections $\mathcal{V}$ on $\mathbb{S}$*
2. *Groupoid of flat twisted R-line bundles with connection $\mathcal{L}$ on the spectral surface Σ*

The equivalence identifies the monodromy eigenvalue operators of the connection on $\mathcal{V}$ around a puncture p with the monodromy operators of $\mathcal{L}$ around the punctures on Σ over p just as in Theorem 5.1.4.

Proof Let $\mathcal{V}$ be an m-dimensional twisted framed R-local system on $\mathbb{S}$. Let us construct a twisted flat line bundle $\mathcal{L}$ on Σ. Let $\mathcal{F}(s)$ be a flat section of the local system of flags in $\mathcal{V}$ defining a framing near a marked point $s \in \mathbb{S}$. For each zig-zag γ going around s, it provides a flat section over the punctured disc S_γ. Recall the punctured face F_γ of the spectral surface Σ assigned to γ. By the very construction of the spectral surface, there is a canonical identification:

$$F_\gamma = S_\gamma.$$

We use it to transport the flat section of the local system of flags on S_γ to F_γ and denote it by $\mathcal{F}(s)$. We assign to a zig-zag γ a number $a_\gamma \in \{1, \ldots, m\}$—the codistance from a point of the open disc S_γ° close to γ to the marked point associated with γ. Then there is a twisted flat line bundle on the face F_γ of Σ:

$$\mathcal{L}_{F_\gamma} := \mathrm{gr}^{a_\gamma - 1}\mathcal{F}(s) \tag{113}$$

Let us glue them into a twisted flat line bundle $\mathcal{L}$ on Σ.

□

Lemma 5.3.2 *Let $s_1, \ldots, s_n$ be the marked points on $\mathbb{S}$. Let y be a point of Σ, and $x := \pi(y) \in \mathbb{S}$.*

Let $0 \leq a_j \leq m$ be the codistance from x to s_j. Then we have:

1. Let $y \in \Sigma - \Gamma^*$. Then

$$a_1 + \ldots + a_n = m. \tag{114}$$

2. Let $y \in \Gamma^* - \{\bullet\text{-vertices}\}$. Then

$$a_1 + \ldots + a_n = m - 1. \tag{115}$$

3. Let y be a k-valent $\bullet$-vertex of Γ^*. Then

$$a_1 + \ldots + a_n = m - k + 1. \tag{116}$$

Proof

1. When a point x is inside of a face, the claim follows from (98).
2. When a point x' moves and hits an inside point x of an edge E, just one of the codistances a_i decreases by 1. Namely, it is the one corresponding to the marked point s_γ surrounded by the zig-zag γ containing E, such that the open domain S_γ° contains x'. The same thing happens when x is a $\circ$-vertex. Indeed, take a small disc U_x containing x. Let $\gamma_1, \ldots, \gamma_k$ be the zig-zags containing x. The domains $S_\gamma^\circ \cap U_x$ are disjoint, and their closures cover U_x, as shown on the left in Fig. 29.
3. When a point x' moves and hits a $\bullet$-vertex x, the $k - 1$ of the codistances $\{a_i\}$ decrease by 1. They are the ones corresponding to the zig-zags γ containing the vertex x such that $x' \in S_\gamma^\circ$. □

Gluing over an edge E of Γ. Let γ_1 and γ_2 be the two zig-zags passing through E and s_1 and s_2 the related marked points of $\mathbb{S}$. Faces F_{γ_1} and F_{γ_2} on Σ intersect along the edge (identified with) E. We denote by $\mathcal{F}^{a_i}$ the codimension a_i subspace of the flag $\mathcal{F}(s_i)$. Then, thanks to (115), $a_1 + a_2 + \ldots + a_n = m - 1$, and so there is a twisted flat line bundle near the edge E on $\mathbb{S}$:

$$\mathcal{L}_E := \mathcal{F}^{a_1} \cap \mathcal{F}^{a_2} \cap \mathcal{F}^{a_3} \cap \ldots \cap \mathcal{F}^{a_n}, \qquad a_1 + a_2 + \ldots + a_n = m - 1. \tag{117}$$

Note that there are canonical projections:

$$\mathcal{F}^{a_1} \longrightarrow \mathrm{gr}^{a_1}\mathcal{F}(s_1), \qquad \mathcal{F}^{a_2} \longrightarrow \mathrm{gr}^{a_2}\mathcal{F}(s_2).$$

Restricting them to intersection (117) we get isomorphisms of twisted flat line bundles near the edge E:

$$\mathcal{L}_{F_{\gamma_1}} \stackrel{(113)}{=} \mathrm{gr}^{a_1}\mathcal{F}(s_1) \stackrel{\sim}{\longleftarrow} \mathcal{L}_E \stackrel{\sim}{\longrightarrow} \mathrm{gr}^{a_2}\mathcal{F}(s_2) \stackrel{(113)}{=} \mathcal{L}_{F_{\gamma_2}}. \tag{118}$$

Here the very left and right isomorphisms come from the fact that the codistances from a generic point of $S^{\circ}_{\gamma_1}$ close to the zig-zag γ_1 are given by $(a_1 + 1, a_2, \ldots, a_n)$, and for $S^{\circ}_{\gamma_2}$ they are $(a_1, a_2+1, \ldots, a_n)$. Using isomorphisms (118), we glue twisted flat line bundles (113) on faces F_γ into a twisted flat line bundle:

$$\mathcal{L}^{\times} \text{on } \Sigma - \{\text{vertices of } \Gamma^*\}.$$

Proposition 5.3.3 *The twisted flat line bundle $\mathcal{L}^{\times}$ extends to a twisted flat line bundle $\mathcal{L}$ on Σ.*

Proof Take a small circle S^1_v around a vertex $v \in \Gamma^*$ on the spectral surface Σ. Restrict the $\mathcal{L}^{\times}$ to a section of the bundle of nonzero tangent vectors on S^1_v. We get a flat line bundle $\mathcal{L}^{\times}_v$ over the circle S^1_v. □

Lemma 5.3.4 *The monodromy of the flat line bundle $\mathcal{L}^{\times}_v$ around every $\circ$-vertex v is equal to -1.*

Proof Denote by $E_{i,i+1}$ the edge of Γ^* shared by the faces F_i and F_{i+1}; see Fig. 32. Take a nonzero tangent vector t_i at a point of $S^1_v \cap F_i$ in the clockwise direction.

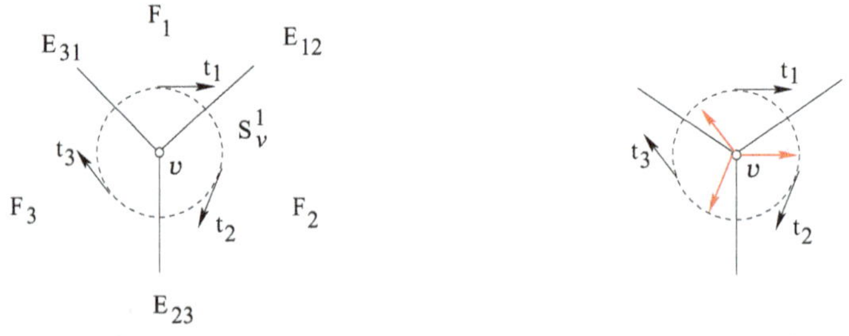

Fig. 32 On the left: the edges $E_{i,i+1}$ and faces F_i of Γ^* near a $\circ$-vertex v

Fig. 33 The shaded domain is the cosector $C_{v_4 \bullet v_1}$. The red ones are domains C_{v_i}

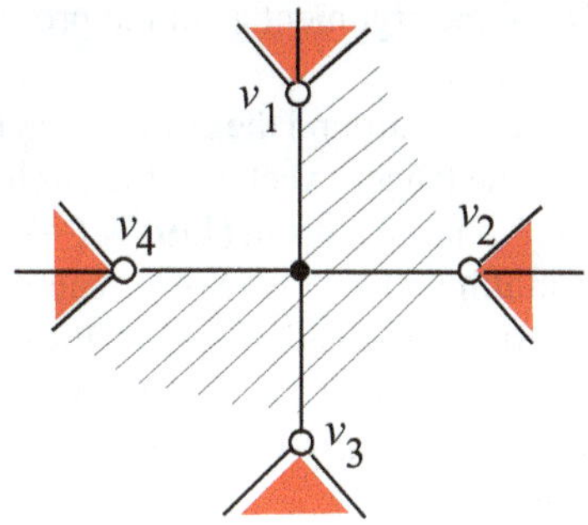

Denote by $\mathcal{L}_{t_i}$ the fiber of the twisted local system $\mathcal{L}_{F_i}$ over t_i. To define the parallel transform $\mathcal{L}_{t_i} \to \mathcal{L}_{t_{i+1}}$, we move tangent vectors t_i and t_{i+1} towards each other to the tangent vector $t_{i,i+1}$ at the point $S^1_v \cap E_{i,i+1}$ and use isomorphisms (118). Denote by $\widetilde{t_i}$ the tangent vectors at the vertex v parallel to the tangent vectors t_i; see the red vectors on the right picture on Fig. 32. We have canonical isomorphisms $\mathcal{L}_{t_i} = \mathcal{L}_{\widetilde{t_i}}$ provided by the parallel transform of tangent vectors $t_i \to \widetilde{t_i}$. After these identifications, the parallel transform of the vector t_i around the circle S^1_v amounts to the parallel transform $\mathcal{L}_{\widetilde{t_i}}$ for the rotation of the tangent vector $\widetilde{t_i}$ by 360°, resulting the -1 monodromy. □

Lemma 5.3.4 implies Proposition 5.3.3 a ∘-vertex.

Now let • be a •-vertex of Γ. Let $v_1, \ldots, v_k$ be the ∘-vertices incident to •, whose order is compatible with the cyclic structure at •. A zig-zag $\ldots v_i \bullet v_{i+1} \ldots$, where $i \in \mathbb{Z}/k\mathbb{Z}$, passing via the vertex • determines a •-*cosector* (see Fig. 33):

$$C_{v_i \bullet v_{i+1}} := U_\bullet \cap S_\gamma.$$

The flat line bundle over this •-cosector is denoted by $\mathcal{L}_{C_{v_i \bullet v_{i+1}}}$. Next, for each ∘-vertex v_i consider the domain C_{v_i} given by the intersection of U_{v_i} with an angle formed by the two edges at v_i on the left and on the right of the edge $\bullet v_i$; see Fig. 33. Denote the twisted flat line bundle over this domain by $\mathcal{L}_{C_{v_i}}$.

For each $i = 1, \ldots, k$ the twisted flat structure of $\mathcal{L}$ provides a pair of isomorphisms:

$$\mathcal{L}_{C_{v_i}} \longrightarrow \mathcal{L}_{C_{v_i \bullet v_{i+1}}} \longleftarrow \mathcal{L}_{C_{v_{i+1}}}.$$

They combine into a diagram of $2k$ line isomorphisms. For example, for $k = 3$ we get

$$\begin{array}{ccccccccccccc} & & \mathcal{L}_{C_{v_1 \bullet v_2}} & & & & \mathcal{L}_{C_{v_2 \bullet v_3}} & & & & \mathcal{L}_{C_{v_3 \bullet v_1}} & & \\ & \nearrow & & \nwarrow & & \nearrow & & \nwarrow & & \nearrow & & \searrow & \\ \mathcal{L}_{C_{v_1}} & & & & \mathcal{L}_{C_{v_2}} & & & & \mathcal{L}_{C_{v_3}} & & & & \mathcal{L}_{C_{v_1}} \end{array} \tag{119}$$

The same argument as in the proof of Lemma 5.2.2(ii) shows that the composition is $(-1)^k$.

Moving around the circle $S^1_\bullet$ on the spectral surface Σ, we rotate the projection to $\mathbb{S}$ of the tangent vector to the circle by $(\text{val}(\bullet) - 1)2\pi$. The composition of $2\text{val}(\bullet)$ line isomorphisms in (119) is $(-1)^{\text{val}(\bullet)}$. Since $\text{val}(\bullet) - 1 + \text{val}(\bullet)$ is odd, the result is multiplied by -1. So we can extend the twisted flat line bundle $\mathcal{L}^\times$ across the $\bullet$-vertex. Proposition 5.3.3 is proved.

To prove that the constructed map is a birational equivalence of groupoids, we proceed similarly to the proof of equivalence in Theorem 5.2.1. Theorem 5.3.1 is proved.

6 Spectral Description of Twisted Decorated Local Systems

6.1 The Spectral Description

We use the same conventions about twisted local systems on $\mathbb{S}$ as in Sect. 5.3. We recall decorated flags, defined in Sect. 1.3.2.

Definition 6.1.1 A twisted decorated R-local system on a decorated surface $\mathbb{S}$ is a twisted local system of finite-dimensional R-vector spaces on $\mathbb{S}$ with a flat section of the associated local system of decorated flags over each marked boundary component.

Let Γ be a GL_m-graph on $\mathbb{S}$, homotopy equivalent to $\mathbb{S}$. Denote by Σ its spectral surface.

Theorem 6.1.2 *Given a* GL_m*-graph* Γ *on* $\mathbb{S}$*, homotopy equivalent to* $\mathbb{S}$*, there is a birational equivalence of groupoids of non-commutative local systems:*

$$\begin{aligned} &\{m\text{-}dimensionaltwisted\ decoratedlocal\ systems\ \mathcal{V}\ on\ \mathbb{S}\} \longleftrightarrow \\ &\{Flat\ twisted\ line\ bundles\ \mathcal{L}\ on\ \Sigma,\ trivialized\ on\ the\ boundary\}. \end{aligned} \tag{120}$$

Proof Equivalence (120) is obtained by applying the equivalence from Theorem 5.3.1 to twisted framed local systems on $\mathbb{S}$, equipped with an extra data: a decoration.[16] The description of the eigenvalue monodromy operators in Theorem 5.3.1 implies that the twisted flat line bundle $\mathcal{L}$ on Σ assigned to $\mathcal{V}$ has the following property:

- The restrictions of $\mathcal{L}$ to punctured discs around the marked points in $\pi^{-1}(s)$ are identified with the twisted flat line bundles $\text{gr}^a\mathcal{A}$ associated to the decoration of $\mathcal{V}$ near s.

[16] Note that a decorated local system is unipotent.

So a choice of a decoration on $\mathcal{V}$ near a marked point s is equivalent to a choice of trivializations of the restriction of $\mathcal{L}$ to the homotopy circles near the marked points $\pi^{-1}(s)$. □

A coordinate description. Recall that $\Gamma = \Gamma^*$ as graphs and faces of Γ^* match zig-zags on Γ. By Theorem 4.1.2, the isomorphism classes of *twisted* flat line bundles on the spectral surface Σ, trivialized on the marked boundary, are described by a collection of invariants $\Delta_{\mathbf{E}} \in R^{\times}$ at the oriented edges $\mathbf{E}$ of Γ, subject to the monomial relations (60).

Since the edges have canonical orientation $\circ \to \bullet$, given a GL_m-graph Γ on $\mathbb{S}$, we get a collection of elements $\Delta_E \in R^{\times}$, assigned to the edges E of Γ. They describe generic twisted decorated local systems of m-dimensional R-vector spaces on $\mathbb{S}$. They are non-commutative counterparts of the ratios of the cluster $\mathcal{A}$-coordinates [7], introduced in the context of GL_m-graphs Γ in [19].

Precisely, in the commutative case, for the twisted SL_m-local systems, the $\mathcal{A}$-coordinates $\{A_{\mathrm{F}}\}$ are assigned to the faces F of the graph Γ. Then $\Delta_E = A_{\mathrm{F}_+}/A_{\mathrm{F}_-}$, where F_+ and F_- are the two faces sharing the edge E, with the face F_+ on the right of the oriented $\circ \to \bullet$ edge. However, the functions $\{\Delta_E\}$ do not form a coordinate system: they satisfy monomial relations (60).

In Sect. 6.2 we elaborate the $m = 2$ example.

6.2 Gluing Spectral Surfaces for GL_2-Graphs from Hexagons

The spectral surface Σ of a GL_m-graph Γ was discussed in Sect. 1. Below we revisit the case of a GL_2-graph from a more elementary perspective and use this to describe non-commutative twisted decorated GL_2-local systems on a decorated surface $\mathbb{S}$ via flat line bundles on the spectral surface. Then in Sect. 6.3 we elaborate this description in detail in coordinates, relating to the Gelfand-Retakh quasideterminants and algebraic relations between them.

1. The Spectral Surface for the Triangle

Let Γ be the GL_2-graph associated to the triangle t; see Fig. 36. The spectral surface of Γ is a hexagon h_t providing a $2:1$ cover of the triangle t, ramified at the central $\bullet$-vertex. The map $h_t \to t$ can be realized by the map $z \to z^2$ sending a regular hexagon in the complex plane z with the vertices at the sixth roots of unity onto a triangle with vertices at the cubic roots of unity. The vertices of the hexagon h_t are labeled by elements of the set $\{1, 2\}$, so that going around the hexagon the labels alternate. For each vertex v of t there is a bijection:

$$\{\text{The vertices of the hexagon } h_t \text{ over } v\} \stackrel{\sim}{\longrightarrow} \{1, 2\}.$$

The sides of the hexagon inherit a "$1 \to 2$" orientation.

2. Gluing the Spectral Surface from Hexagons

Let $\mathcal{T}$ be an ideal triangulation of $\mathbb{S}$. For each triangle t of $\mathcal{T}$ choose a hexagon h_t over t with the vertices labeled as above. Let us glue the hexagons h_t into a complete smooth topological surface $\mathbf{\Sigma}$. Given an edge E of $\mathcal{T}$, there are two triangles t and t' sharing E. Let h and h' be the hexagons covering these triangles. We identify the vertices of the hexagons with the same labels, projecting onto the same vertex of $\mathcal{T}$. Then we glue two edges whose vertices are identified. For example, gluing two hexagons assigned to a triangulation of a rectangle we get an annulus. Denote by $\mathbf{S}$ a complete surface obtained by filling the punctures on $\mathbb{S}$. Gluing hexagons h_t we get a surface $\mathbf{\Sigma}$ with the following properties:

- There is a $2:1$ map $\pi : \mathbf{\Sigma} \to \mathbf{S}$ with one ramification point in every triangle of $\mathcal{T}$.
- For every marked point $s \in \mathbf{S}$ there is a bijection $\pi^{-1}(s) \longrightarrow \{1, 2\}$. Moving around a triangle of $\mathcal{T}$ we get a permutation $\{1, 2\} \longrightarrow \{2, 1\}$.

The spectral surface Σ is obtained by deleting the vertices of the triangles from $\mathbf{\Sigma}$.

3. The Twisted Decorated Flat Line Bundle on Σ

Let $\mathcal{V}$ be a twisted decorated local system of two-dimensional R-vector spaces over $\mathbb{S}$. So there is an invariant decorated flag $\mathcal{A}_s$ near each marked point s. For every ideal triangle t there are decorated flags $\mathcal{A}$, $\mathcal{B}$, $\mathcal{C}$ assigned to the vertices of t. We assign to the vertices of the hexagon h_t the following six trivialized lines:

$$\begin{aligned} A &:= \mathcal{A}^1, \quad & B &:= \mathcal{B}^1, \quad & C &:= \mathcal{C}^1, \\ \overline{A} &:= \mathcal{A}/\mathcal{A}^1, \quad & \overline{B} &:= \mathcal{B}/\mathcal{B}^1, \quad & \overline{C} &:= \mathcal{C}/\mathcal{C}^1, \end{aligned} \tag{121}$$

so that going around the hexagon we get the lines $(A, \overline{B}, C, \overline{A}, B, \overline{C})$. We denote by a_1 the vector trivializing the line A, by a_2 the vector trivializing the line $\overline{\text{A}}$, etc.

The map $h_t \to t$ sends the vertices labeled by $A, \overline{A}$ to the vertex a, and so on; see Fig. 34. We get a twisted flat line bundle $\mathcal{L}_{h_t}$ over the hexagon, trivialized at the tangent vectors to the boundary near the vertices by the vectors provided by the decorations. For example, the line bundle near the vertex A of the hexagon is trivialized by the vector $a_1 \in$ A. The parallel transports along the oriented edges are defined by the following maps/their inverses:

$$\varphi_{a_1,b_2} : A \longrightarrow \overline{B}. \tag{122}$$

Given an ideal triangulation of $\mathbb{S}$ and gluing the hexagons h_t into the surface Σ, the twisted flat line bundles $\mathcal{L}_{h_t}$ are glued into a twisted flat line bundle $\mathcal{L}$ on Σ, trivialized near the marked points.

We construct $\mathcal{A}$-coordinates by assigning to each edge E of the hexagonal tessellation of Σ an $\Delta_E \subset R^\times$ as follows. Let A and $\overline{B}$ be the lines over the vertices of E, trivialized by nonzero vectors $a_1 \in A$ and $b_2 \in \overline{B}$, provided by the decorations.

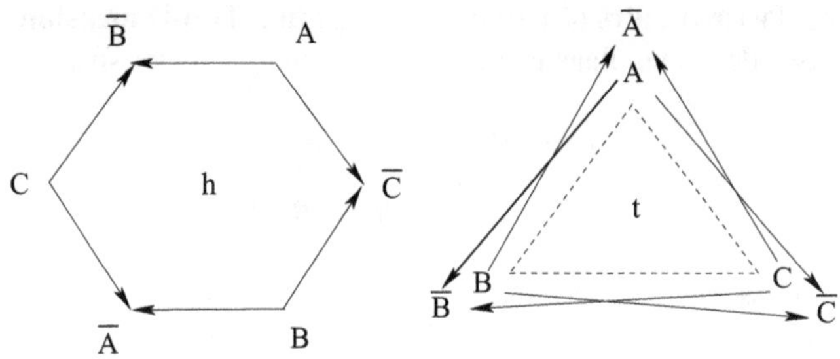

Fig. 34 The twisted flat line bundle on a hexagon assigned to a triple of flags

The parallel transport $\varphi_E : A \to \overline{B}$ along the E acts by $\varphi_E(a_1) = \Delta_E b_2$. For every hexagon h_t, the alternated product of the six functions Δ_E assigned to its consecutive edges is equal to -1. There are no more relations between the functions $\{\Delta_E\}$.

Let us elaborate this, to see what happens when we alter a GL_2-graph Γ by a two by two move.

6.3 *Coordinates for Twisted Decorated Non-Commutative 2-Dimensional Local Systems on* $\mathbb{S}$

1. Twisted Triples of Decorated Flags

We define them as collections of decorated flags:

$$\{\mathcal{F}_i\},\ i \in \mathbb{Z}/6\mathbb{Z},\quad \text{such that } \mathcal{F}_{i+3} = -\mathcal{F}_i\ \ \forall i \in \mathbb{Z}/6\mathbb{Z}.$$

Lemma 6.3.1 *A twisted triple of decorated flags* $\{\mathcal{F}_i\}$ *is the same thing as a twisted decorated local system on the triangle* t.

Proof We consider t as a disc with three marked points x_1, x_2, x_3. Pick a nonzero tangent vector v_i at the point x_i, which follow the boundary orientation. There is the unique twisted local system $\mathcal{L}$ on the disc. We identify the fibers $\mathcal{L}_{v_i}$ of $\mathcal{L}$ at the vectors v_i by the parallel transport from v_1 to v_2 and to v_3 along the boundary circle, following the circle orientation, depicted as counterclockwise on the pictures.

Set $V := \mathcal{L}_{v_1}$. We assign to a triple of decorated flags $\{\mathcal{F}_i\}$ in V a pair $(\mathcal{L}, \alpha)$, where the decoration α is given by the flag $\mathcal{F}_i$ at the fibers $\mathcal{L}_{v_i}$ for $i = 1, 2, 3$, identified with V as above. □

A triple of decorated flags $(\mathcal{A}, \mathcal{B}, C)$ gives rise to a twisted triple, setting

$$(\mathcal{A}, \mathcal{B}, C) \longmapsto (\mathcal{A}, \mathcal{B}, C, \mathcal{A}', \mathcal{B}', C'),\ \ \mathcal{A}' = -\mathcal{A},\ \mathcal{B}' = -\mathcal{B},\ C' = -C. \quad (123)$$

2. Describing Twisted Pairs of Decorated Flags in a Two-Dimensional Space Let $\mathcal{A}$, $\mathcal{B}$ be two decorated flags in a two-dimensional R-vector space V_2. Let

$$\begin{aligned} A &:= \mathcal{A}^1, \qquad B := \mathcal{B}^1, \\ \overline{A} &:= \mathcal{A}/\mathcal{A}^1, \quad \overline{B} := \mathcal{B}/\mathcal{B}^1 \end{aligned} \tag{124}$$

be the pairs of associate graded lines of these flags. Each of the lines is equipped with a nonzero vector. We can describe decorated flags by these vectors, using the notation

$$\mathcal{A} = (a_1, a_2), \quad \mathcal{B} = (b_1, b_2). \tag{125}$$

Projecting the line A to the quotient line $\overline{B}$ we get an isomorphism of lines:

$$\varphi_{a_1,b_2} : A \longrightarrow \overline{B}. \tag{126}$$

Since the lines are equipped with nonzero vectors, it is described by an element $\Delta(a_1, b_2) \in R^*$:

$$\varphi_{a_1,b_2} : a_1 \longmapsto \Delta(a_1, b_2)b_2. \tag{127}$$

The inverse map $\varphi_{b_2,a_1} : \overline{B} \longrightarrow$ A, $b_2 \longmapsto \Delta(b_2, a_1)a_1$ is described by the invariant:

$$\Delta(b_2, a_1) := \Delta(a_1, b_2)^{-1}. \tag{128}$$

Generic pairs of flags $(\mathcal{A}, \mathcal{B})$ are described up to an isomorphism by the invariants $\{\Delta(a_1, b_2), \Delta(b_1, a_2)\}$.

Let us consider now *twisted* pairs of decorated flags $(\mathcal{A}, \mathcal{B}, -\mathcal{A}, -\mathcal{B})$. Then the natural invariants are

$$\{\Delta(a_1, b_2), \Delta(b_1, a_2')\}, \quad a_2' := -a_2.$$

They are invariant under the twisted cyclic shift:

$$(\mathcal{A}, \mathcal{B}, -\mathcal{A}, -\mathcal{B}) \to (\mathcal{B}, -\mathcal{A}, -\mathcal{B}, -\mathcal{A}).$$

Note that on the bipartite graph on Fig. 35, consistently with (65), there are two more invariants:

$$\Delta(a_2, b_1)^{-1} = -\Delta(b_1, a_2'), \quad \Delta(b_2, a_1') := -\Delta(a_1, b_2)^{-1}. \tag{129}$$

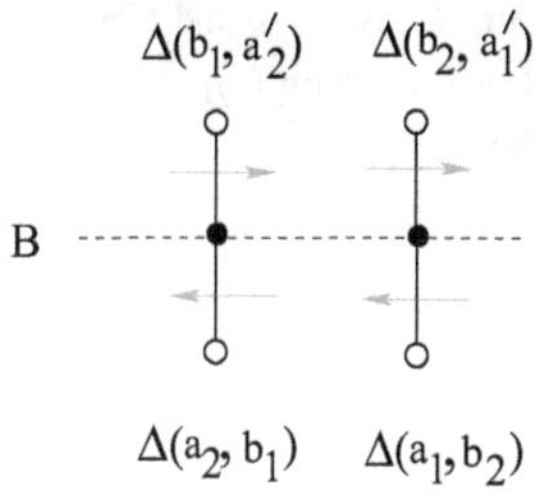

Fig. 35 Invariants of a twisted pair of decorated flags $(\mathcal{A}, \mathcal{B}, -\mathcal{A}, -\mathcal{B})$ at the edges of a (disconnected) bipartite ribbon graph

3. Describing Twisted Triples of Decorated Flags in a Two-Dimensional Space Let $\mathcal{A}$, $\mathcal{B}$, $\mathcal{C}$ be three decorated flags in a two-dimensional R-vector space V_2. Similar to (124), we set

$$\begin{aligned} A &:= \mathcal{A}^1, & B &:= \mathcal{B}^1, & C &:= \mathcal{C}^1, \\ \overline{A} &:= \mathcal{A}/\mathcal{A}^1, & \overline{B} &:= \mathcal{B}/\mathcal{B}^1, & \overline{C} &:= \mathcal{C}/\mathcal{C}^1. \end{aligned} \tag{130}$$

We describe decorated flags by the vectors in these lines, using the notation

$$\mathcal{A} = (a_1, a_2), \quad \mathcal{B} = (b_1, b_2), \quad \mathcal{C} = (c_1, c_2). \tag{131}$$

The three flags provide six such line isomorphisms, which we organize into a diagram:

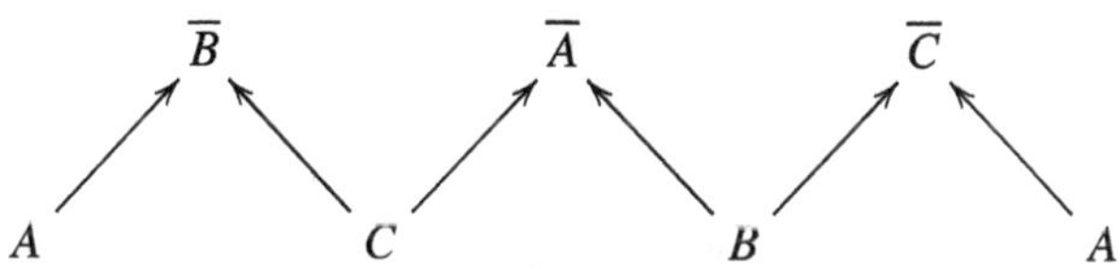

(132)

We compose the six maps to a map $A \to A$, where going against an arrow means inverting the isomorphism. By Lemma 5.2.2 the composition is equal to $-\mathrm{Id}$. This just means that

$$\Delta(a_1, b_2)\Delta(c_1, b_2)^{-1}\Delta(c_1, a_2)\Delta(b_1, a_2)^{-1}\Delta(b_1, c_2)\Delta(a_1, c_2)^{-1} = -1. \tag{133}$$

Using the twisted triples of decorated flags (123), setting $a_1' := -a_1$, etc., and using (128), we write this as

$$\Delta(a_1, b_2)\Delta(b_2, c_1)\Delta(c_1, a_2')\Delta(a_2', b_1')\Delta(b_1', c_2')\Delta(c_2', a_1') = 1. \tag{134}$$

In fact this is just equivalent to

$$\Delta(a_1, b_2)\Delta(b_2, c_1)\Delta(c_1, a_2)\Delta(a_2, b_1)\Delta(b_1, c_2)\Delta(c_2, a_1') = 1. \tag{135}$$

Fig. 36 Sectors $C_{\mathcal{A}}$ and C_C (pink) and cosector $\widehat{C}_{\mathcal{B}}$ (shaded)

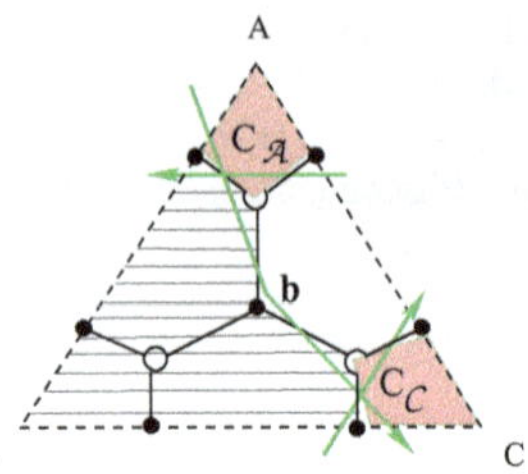

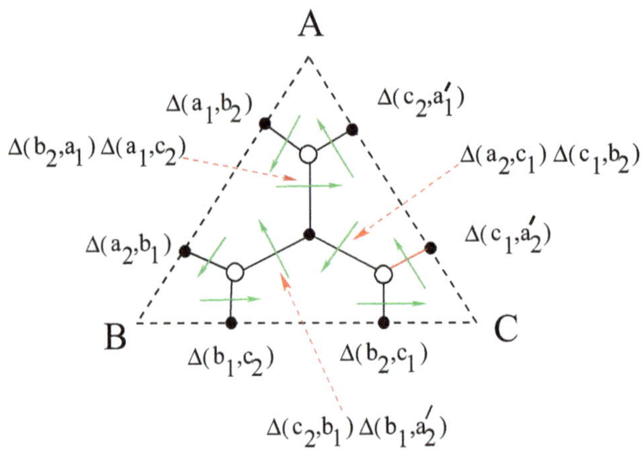

Fig. 37 The $\mathcal{A}$-coordinates on the bipartite GL_2-graph assigned to a triangle. The coordinates at the side CA are assigned to the pair $(C, \mathcal{A}')$ where $\mathcal{A}' := -\mathcal{A}$, and $(a_1', a_2') = (-a_1, -a_2)$. Identity (134) means that the counterclockwise product of the coordinates at the edges at the central •-vertex is 1

Let Γ be the GL_2-graph associated to the triangle t; see Fig. 36. The lines A, B, C are assigned to the sectors $C_{\mathcal{A}}, C_{\mathcal{B}}, C_C$, respectively. The lines $\overline{A}, \overline{B}, \overline{C}$ are assigned to the cosectors $\widehat{C}_{\mathcal{A}}, \widehat{C}_{\mathcal{B}}, \widehat{C}_C$, respectively. The spectral surface of Γ is a hexagon h_t providing a $2:1$ cover of the triangle t, ramified at the central •-vertex.

Using this and notation (128), we assign the coordinates to the edges of the bipartite GL_2-graph related to the counterclockwise oriented triangle on Fig. 37. Then we assign coordinates to the edges of a bipartite GL_2-graph associated with an ideal triangulation of the surface, as shown on Fig. 38.

The cyclic product of elements at the edges incident to a ∘-vertex is -1 and for the •-vertices $(-1)^{\mathrm{val}(v)-1}$. For a •-vertex this is the six-term identity (134), or, equivalently, (135).

The product of the two coordinates at the edges sharing a two-valent •-vertex is -1. We shrink the edge containing a two-valent •-vertex into a ∘-vertex. Then the cyclic product of the resulted coordinates at the edges of the new ∘-vertex is -1, as needed. The resulting coordinates for the two triangulations of a rectangle are shown on Figs. 39, 40.

Fig. 38 The $\mathcal{A}$-coordinates on the bipartite GL_2-graph associated with a triangulated rectangle

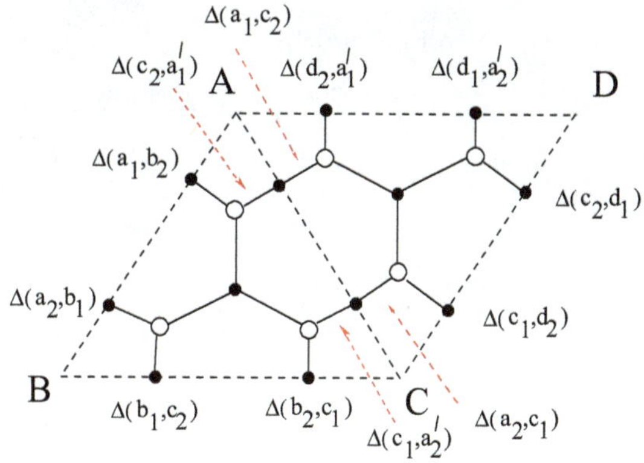

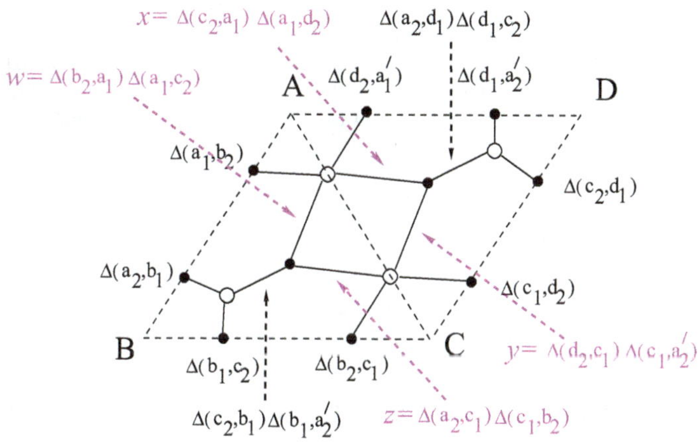

Fig. 39 The $\mathcal{A}$-coordinates for the triangulation AC of counterclockwise oriented rectangle $ABCD$

4. In Coordinates
Choose a basis (e_1, e_2) in the two-dimensional R-vector space V_2. Then the three pairs of vectors $(a_1, a_2; b_1, b_2; c_1, c_2)$ are described by a 2×6 matrix with the columns given by decompositions:

$$\begin{aligned} a_1 &= a_{11}e_1 + a_{12}e_2, \quad a_2 = a_{21}e_1 + a_{22}e_2; \\ b_1 &= b_{11}e_1 + b_{12}e_2, \quad b_2 = b_{21}e_1 + b_{22}e_2; \\ c_1 &= c_{11}e_1 + c_{12}e_2, \quad c_2 = c_{21}e_1 + c_{22}e_2. \end{aligned} \tag{136}$$

The condition that $a_1 - \Delta(a_1, b_2)b_2 = \lambda b_1$ for some $\lambda \in R^*$ just means that

$$(a_{11}e_1 + a_{12}e_2) - \Delta(a_1, b_2)(b_{21}e_1 + b_{22}e_2) = \lambda(b_{11}e_1 + b_{12}e_2).$$

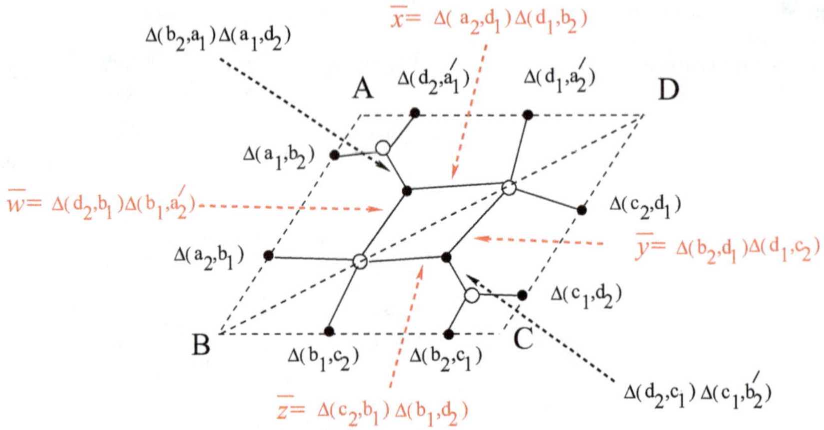

Fig. 40 The $\mathcal{A}$-coordinates for the triangulation BD of counterclockwise oriented rectangle $ABCD$. We start with a configuration of 4 decorated flags $(\mathcal{A}, \mathcal{B}, \mathcal{C}, \mathcal{D})$. It induces triples $(\mathcal{A}, \mathcal{B}, \mathcal{C})$ and $(\mathcal{B}, \mathcal{C}, \mathcal{D})$, which we promote to twisted configurations, and introduce the coordinates as on Fig. 37

So we get a system of equations:

$$\begin{cases} a_{11} - \Delta(a_1, b_2)b_{21} = \lambda b_{11}, \\ a_{12} - \Delta(a_1, b_2)b_{22} = \lambda b_{12}. \end{cases} \tag{137}$$

To solve the system, multiply the first equation from the right by b_{11}^{-1}, the second by b_{12}^{-1}, and subtract. We get the unique solution, which can be written in two slightly different ways:

$$\Delta(a_1, b_2) = (a_{11} - a_{12}b_{12}^{-1}b_{11})(b_{21} - b_{22}b_{12}^{-1}b_{11})^{-1} = (a_1, b_1)_{11}(b_1, b_2)_{21}^{-1}; \tag{138}$$

$$\Delta(a_1, b_2) = (a_{11}b_{11}^{-1}b_{12} - a_{12})(b_{21}b_{11}^{-1}b_{12} - b_{22})^{-1} = (a_1, b_1)_{12}(b_1, b_2)_{22}^{-1}.$$

Here $(x_1, x_2)_{ij}$ is the quasideterminant [18] assigned to the (ij) entry of a 2×2 matrix (x_{ij}):

$$(x_1, x_2)_{11} := x_{11} - x_{12}x_{22}^{-1}x_{21}, \quad (x_1, x_2)_{12} := x_{12} - x_{11}x_{21}^{-1}x_{22}. \tag{139}$$

$$(x_1, x_2)_{21} := x_{21} - x_{22}x_{12}^{-1}x_{11}, \quad (x_1, x_2)_{22} := x_{22} - x_{21}x_{11}^{-1}x_{12}.$$

Remarks

1. Equations (137) reflect the parallel transform in the *twisted* flat line bundle; see remark at the end of Sect. 5.2.

2. If R is commutative, formulas (138) look as follows. Pick a symplectic form $\omega \in \det(V_2^*)$. Then for any two vectors $v, w \in V_2$ we have an invariant $\omega(v, w)$, and formula (138) can be written as

$$\begin{aligned}\Delta(a_1, b_2) =& -\omega(a_1, b_1)\omega(b_1, b_2)^{-1} \\ =&\omega(b_1, a_1)\omega(b_1, b_2)^{-1}.\end{aligned} \tag{140}$$

5. Flips of Triangulations and Non-Commutative Plücker Relations
Given a basis (e_1, e_2) in the R-vector space V_2, the four pairs of vectors

$$(a_1, a_2; b_1, b_2; c_1, c_2; d_1, d_2)$$

are described by a 2×8 matrix:

$$\begin{aligned}
a_1 &= a_{11}e_1 + a_{12}e_2, \quad a_2 = a_{21}e_1 + a_{22}e_2;\\
b_1 &= b_{11}e_1 + b_{12}e_2, \quad b_2 = b_{21}e_1 + b_{22}e_2;\\
c_1 &= c_{11}e_1 + c_{12}e_2, \quad c_2 = c_{21}e_1 + c_{22}e_2;\\
d_1 &= d_{11}e_1 + d_{12}e_2, \quad d_2 = d_{21}e_1 + d_{22}e_2.
\end{aligned} \tag{141}$$

Recall the Plücker identity for quasideterminants:

$$(a_1, c_1)_{11} =(a_1, b_1)_{11}(d_1, b_1)_{11}^{-1}(d_1, c_1)_{11} + (a_1, d_1)_{11}(b_1, d_1)_{11}^{-1}(b_1, c_1)_{11}. \tag{142}$$

Proposition 6.3.2 *There is a non-commutative Plücker relation for the elements $\Delta(*, *)$ from (138):*

$$\Delta(a_1, c_2) =\Delta(a_1, b_2)\Delta(d_1, b_2)^{-1}\Delta(d_1, c_2) + \Delta(a_1, d_2)\Delta(b_1, d_2)^{-1}\Delta(b_1, c_2). \tag{143}$$

It is equivalent to the Plücker identity for quasideterminants.

Proof We use formula $\Delta(a_1, b_2) = (a_1, b_1)_{11}(b_1, b_2)_{21}^{-1}$ from (138). Then we have to check that

$$\begin{aligned}
&(a_1, c_1)_{11}(c_1, c_2)_{21}^{-1} \stackrel{?}{=}\\
&\quad (a_1, b_1)_{11}(b_1, b_2)_{21}^{-1} \cdot (b_1, b_2)_{21}(d_1, b_1)_{11}^{-1} \cdot (d_1, c_1)_{11}(c_1, c_2)_{21}^{-1}+\\
&\quad (a_1, d_1)_{11}(d_1, d_2)_{21}^{-1} \cdot (d_1, d_2)_{21}(b_1, d_1)_{11}^{-1} \cdot (b_1, c_1)_{11}(c_1, c_2)_{21}^{-1}\\
&\qquad = (a_1, b_1)_{11}(d_1, b_1)_{11}^{-1}(d_1, c_1)_{11}(c_1, c_2)_{21}^{-1}+\\
&\qquad\quad (a_1, d_1)_{11}(b_1, d_1)_{11}^{-1}(b_1, c_1)_{11}(c_1, c_2)_{21}^{-1}.
\end{aligned} \tag{144}$$

Multiplying this from the right by $(c_1, c_2)_{21}$ we get precisely (142). □

Preliminary Comments Recall from Lemma 4.1.2 that $\mathcal{A}$-decorated ribbon graphs Γ are in bijection with twisted one-dimensional local systems on the corresponding spectral surface $\Sigma = \mathbb{S}_\Gamma$, trivialized at the boundary. By Theorem 1.2.8, proved in Sect. 1.2, the one-dimensional local systems on the spectral surface inject into twisted decorated local systems on the base surface $\mathbb{S} = \mathbb{S}_\Gamma$. In the case considered in Theorem 6.3.3, the latter are just twisted configurations of quadruples of flags in V_2. In Theorem 6.3.3 below we consider the case when the ribbon graphs Γ and Γ' are dual to the two triangulations of the rectangle $ABCD$. If a certain quadruple of flags is in the image of both injections, for the graphs Γ and Γ', then Theorem 6.3.3 tells how the corresponding $\mathcal{A}$-decorations of the graphs Γ and Γ' are related. Note that a quadruple of flags is described by $12 = 8 + 4$ $\mathcal{A}$-coordinates. We stress that they are not coordinates, since there are monomial relations between them.[17] The eight $\mathcal{A}$-coordinates at the external edges (see Fig. 39) remain intact. So we have to explain how the $\mathcal{A}$-coordinates at the four rhombus edges change under a flip. The answer is given by formulas (145).[18]

Theorem 6.3.3 *The flip $AC \to BD$ of the triangulated rectangle $ABCD$ amounts to the following transformation of the $\mathcal{A}$-coordinates, as explained in the comment above, depicted on Figs. 39, 40:*

$$\begin{aligned}
\overline{x} &= (1 + (zwxy)^{-1})z = z(1 + (wxyz)^{-1});\\
\overline{y} &= (1 + wxyz)^{-1}w = w(1 + xyzw)^{-1};\\
\overline{z} &= (1 + (xyzw)^{-1})x = x(1 + (yzwx)^{-1});\\
\overline{w} &= (1 + yzwx)^{-1}y = y(1 + zwxy)^{-1}.
\end{aligned} \tag{145}$$

Each of these formulas is equivalent to the non-commutative Plücker identity (143).

Proof The clockwise products of the rhombus edge coordinates on Fig. 39, starting from a •-vertex, are given by

$$\begin{aligned}
yzwx &= -\Delta(d_2, c_1)\Delta(c_1, b_2)\Delta(b_2, a_1)\Delta(a_1, d_2),\\
wxyz &= -\Delta(b_2, a_1)\Delta(a_1, d_2)\Delta(d_2, c_1)\Delta(c_1, b_2).
\end{aligned} \tag{146}$$

Clockwise products of rhombus edge coordinates on Fig. 40, starting from a •-vertex, are given by

$$\begin{aligned}
\overline{x}\,\overline{y}\,\overline{z}\,\overline{w} &= -\Delta(a_2, d_1)\Delta(d_1, c_2)\Delta(c_2, b_1)\Delta(b_1, a_2),\\
\overline{z}\,\overline{w}\,\overline{x}\,\overline{y} &= -\Delta(c_2, b_1)\Delta(b_1, a_2)\Delta(a_2, d_1)\Delta(d_1, c_2).
\end{aligned} \tag{147}$$

[17] The coordinate at the internal non-rhombus edge in each triangle is recovered by the monomial relation at its ∘-vertex.

[18] Decorated flags in V_2 depend on 3 parameters. So the space of isomorphism classes of quadruples of decorated flags in V_2 depends on $8 = 3 \cdot 4 - 4$ parameters. There are 4 monomial relations between the twelve "$\mathcal{A}$-coordinates," one for each rhombus vertex.

Now the proof is deduced easily from the following lemma. □

Lemma 6.3.4 *We have:*

$$\begin{aligned} y\overline{w}^{-1} &= 1 + \Delta(d_2, c_1)\Delta(c_1, b_2)\Delta(b_2, a_1)\Delta(a_1, d_2) = 1 + yzwx; \\ w\overline{y}^{-1} &= 1 + \Delta(b_2, a_1)\Delta(a_1, d_2)\Delta(d_2, c_1)\Delta(c_1, b_2) = 1 + wxyz; \\ z^{-1}\overline{x} &= 1 + \Delta(b_2, c_1)\Delta(c_1, d_2)\Delta(d_2, a_1)\Delta(a_1, b_2) = 1 + (wxyz)^{-1}; \\ x^{-1}\overline{z} &= 1 + \Delta(d_2, a_1)\Delta(a_1, b_2)\Delta(b_2, c_1)\Delta(c_1, d_2) = 1 + (yzwx)^{-1}. \end{aligned} \tag{148}$$

Proof It is easy to see that we have the following:

$$\begin{aligned} w\overline{y}^{-1} &= \Delta(b_2, a_1)\Delta(a_1, c_2)\Delta(c_2, d_1)\Delta(d_1, b_2), \\ z^{-1}\overline{x} &= \Delta(b_2, c_1)\Delta(c_1, a_2)\Delta(a_2, d_1)\Delta(d_1, b_2), \end{aligned} \tag{149}$$

and similar two identities obtained by the cyclic shift by two:

$$(a, b, c, d) \longmapsto (c, d, a, b).$$

Substituting the non-commutative Plücker identity (143) to formula (149) for $w\overline{y}^{-1}$ we get

$$\begin{aligned} w\overline{y}^{-1} &= \Delta(b_2, a_1)\Big(\Delta(a_1, b_2)\Delta(b_2, d_1)\Delta(d_1, c_2) + \Delta(a_1, d_2)\Delta(d_2, b_1)\Delta(b_1, c_2)\Big) \\ &\quad \Delta(c_2, d_1)\Delta(d_1, b_2) \\ &= 1 + \Delta(b_2, a_1)\Delta(a_1, d_2)\Delta(d_2, b_1) \cdot \Delta(b_1, c_2)\Delta(c_2, d_1)\Delta(d_1, b_2) \\ &= 1 + \Delta(b_2, a_1)\Delta(a_1, d_2)\Delta(d_2, b_1) \cdot \Delta(b_1, d_2)\Delta(d_2, c_1)\Delta(c_1, b_2) \\ &= 1 - \Delta(b_2, a_1)\Delta(a_1, d_2)\Delta(d_2, c_1)\Delta(c_1, b_2) \\ &= 1 + wxyz. \end{aligned} \tag{150}$$

Here to get the third equality we use the following identity, equivalent to (133), combined with (128):

$$\Delta(b_1, c_2)\Delta(c_2, d_1)\Delta(d_1, b_2) = -\Delta(b_1, d_2)\Delta(d_2, c_1)\Delta(c_1, b_2). \tag{151}$$

The identity for $y\overline{w}^{-1}$ is obtained from this by the cyclic shift by two.

Similarly, to get formula (149) for $z^{-1}\overline{x}$ we need the non-commutative Plücker identity:

$$\Delta(c_1, a_2) = \Delta(c_1, d_2)\Delta(d_2, b_1)\Delta(b_1, a_2) + \Delta(c_1, b_2)\Delta(b_2, d_1)\Delta(d_1, a_2). \tag{152}$$

Substituting identity (152) to formula (149) for $z^{-1}\overline{x}$, we get

$$\begin{aligned} z^{-1}\overline{x} &= \Delta(b_2, c_1)\Delta(c_1, a_2)\Delta(a_2, d_1)\Delta(d_1, b_2) \\ &\overset{(152)}{=} \Delta(b_2, c_1)\Big(\Delta(c_1, d_2)\Delta(d_2, b_1)\Delta(b_1, a_2)+\Delta(c_1, b_2)\Delta(b_2, d_1) \\ &\qquad \Delta(d_1, a_2)\Big)\Delta(a_2, d_1)\Delta(d_1, b_2) \\ &= \Delta(b_2, c_1)\Delta(c_1, d_2)\Delta(d_2, b_1)\cdot\Delta(b_1, a_2)\Delta(a_2, d_1)\Delta(d_1, b_2) + 1 \\ &= 1 - \Delta(b_2, c_1)\Delta(c_1, d_2)\Delta(d_2, b_1)\cdot\Delta(b_1, d_2)\Delta(d_2, a_1)\Delta(a_1, b_2) \\ &= 1 - \Delta(b_2, c_1)\Delta(c_1, d_2)\Delta(d_2, a_1)\Delta(a_1, b_2) \\ &\overset{(146)}{=} 1 + (wxyz)^{-1}. \end{aligned} \tag{153}$$

Here to get the fourth equality we use the identity similar to (151). □

Remark Formulas (145) imply that the monodromies starting at the •-vertex are related as follows:

$$\overline{x}\,\overline{y}\,\overline{z}\,\overline{w} = (zwxy)^{-1}. \tag{154}$$

Indeed, the left-hand side is equal to

$$\begin{aligned} &z(1 + (wxyz)^{-1})(1 + wxyz)^{-1}wx(1 + (yzwx)^{-1})(1 + yzwx)^{-1}y \\ =&z\cdot(wxyz)^{-1}\cdot wx\cdot(yzwx)^{-1}\cdot y = (zwxy)^{-1}. \end{aligned} \tag{155}$$

It is also clear from Figs. 39, 40 that each side of (154) is equal to

$$\Delta(a_2, d_1)\Delta(d_1, c_2)\Delta(c_2, b_1)\Delta(b_1, a_2).$$

We say that a twisted decorated two-dimensional R-local system on $\mathbb{S}$ is *generic*, if for any triangle t of any ideal triangulation of $\mathbb{S}$ the decorations at the vertices of t are in generic position.

Corollary 6.3.5 *The isomorphism classes of generic twisted decorated two-dimensional local systems on $\mathbb{S}$ can be described by a collection of $-3\chi(\mathbb{S}) - 1$ elements $a_i \in R^\times$ which satisfy a countable collection of inequalities $F_j(a_i) \neq 0$, where F_j are non-commutative rational functions.*

For example, generic quadruples of decorated flags in a two-dimensional R-vector space V_2 are parametrized by $f_1, \dots, f_8, x, y, z, w \in R^\times$ where $xyzw \neq -1$ satisfying four monomial relations.

6. Proof of Theorem 4.1.5

Proof There are two ways to assign the non-commutative cluster $\mathcal{A}$-coordinates to generic quadruples of decorated flags in a two-dimensional R-vector space V_2. They correspond to the two bipartite GL_2-graphs assigned to the two triangulations of the rectangle. Formula (145) proved in Theorem 6.3.3 calculates the related coordinate transformation. Formula (145) is the same as the one (67).

Denote by $\mathrm{Conf}_n^{\mathrm{df}}(V_2)^\circ$ the isomorphism classes of n generic decorated flags in V_2. Consider the cluster $\mathcal{A}$-coordinates on $\mathrm{Conf}_5^{\mathrm{df}}(V_2)^\circ$, assigned to the bipartite GL_2-graphs provided by the five different triangulations of the pentagon. By Theorem 6.3.3, each flip of a triangulation of the pentagon is described by the transformation (67) of $\mathcal{A}$-coordinates for the corresponding bipartite ribbon graphs. By the construction, the composition of the five consecutive transformations describes the identity map on $\mathrm{Conf}_5^{\mathrm{df}}(V_2)^\circ$. Since the cluster $\mathcal{A}$-coordinates for a given triangulation of the pentagon provide an open embedding of $\mathrm{Conf}_5^{\mathrm{df}}(V_2)$ to the cluster torus, this implies the pentagon relation for the $\mathcal{A}$-coordinates. □

7 Admissible dg-Sheaves on Surfaces and Non-Commutative Cluster Varieties

7.1 Admissible dg-Sheaves

1. DG-Category of Constructible dg-Sheaves on a Cell Complex

Let X be a stratified topological space, such that the closure of each stratum is a ball. The strata form a partially ordered set. We write $\sigma_1 < \sigma_2$ if $\sigma_1 \subset \overline{\sigma}_2$ and $\sigma_1 \neq \sigma_2$. It determines a quiver Q, whose vertices are the strata σ, and vertices σ_1, σ_2 are related by an arrow $\sigma_1 \to \sigma_2$ if and only if $\sigma_1 < \sigma_2$.

The quiver Q determines a nonunital A_∞-category $\mathcal{Q}$ whose objects are the strata σ, and

$$\mathrm{Hom}_{\mathcal{Q}}(\sigma_1, \sigma_2) := \begin{cases} \mathbb{Z} & \text{if } \sigma_1 < \sigma_2, \\ 0 & \text{otherwise.} \end{cases} \tag{156}$$

The composition of Hom's is defined by the composition of arrows.

Definition 7.1.1 A constructible dg-sheaf on X is an A_∞-functor from the A_∞-category $\mathcal{Q}$ to the dg-category $\mathrm{Vect}^\bullet$ of complexes of R-vector spaces. Constructible dg-sheaves form a dg-category:

$$\mathcal{C}_X := \mathrm{Fun}_{A_\infty}(\mathcal{Q}, \mathrm{Vect}^\bullet).$$

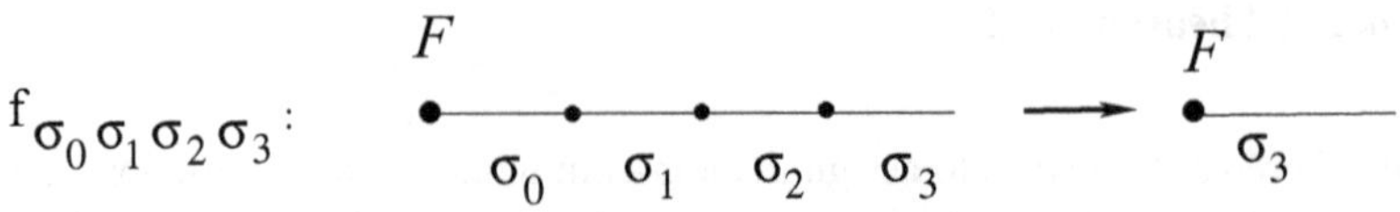

Fig. 41 A map $\mathcal{F}(\sigma_0) \otimes \mathrm{Hom}(\sigma_0, \sigma_1) \otimes \mathrm{Hom}(\sigma_1, \sigma_2) \otimes \mathrm{Hom}(\sigma_2, \sigma_3) \longrightarrow \mathcal{F}(\sigma_3)[-2]$

We consider it as a category linear over the underlying ground field—that is, $\mathbb{Q}$ if R is of characteristic zero, since Hom_R and $\otimes_R$ do not have a natural R-linear structure.

Elaborating this definition, an A_∞-functor $\mathcal{F} \in \mathrm{Ob}(C_X)$ is given by the following data:

- For each nested collection of strata $\sigma_0 < \sigma_1 < \ldots < \sigma_n$, a map

$$f_{\sigma_0,\ldots,\sigma_n} : \mathcal{F}(\sigma_0) \otimes \mathrm{Hom}_{\mathcal{Q}}(\sigma_0, \sigma_1) \otimes \ldots \otimes \mathrm{Hom}_{\mathcal{Q}}(\sigma_{n-1}, \sigma_n) \longrightarrow \mathcal{F}(\sigma_n)[1-n]. \tag{157}$$

These maps satisfy the well-known quadratic equations.

Since the spaces $\mathrm{Hom}_{\mathcal{Q}}(\sigma_0, \sigma_1)$ are canonically identified with $\mathbb{Z}$, one can write (157) just as

$$f_{\sigma_0,\ldots,\sigma_n} : \mathcal{F}(\sigma_0) \longrightarrow \mathcal{F}(\sigma_n)[1-n]. \tag{158}$$

One pictures a map (157) by a collection of points on the half-line, with the objects σ_i at the intervals between the points and the object $\mathcal{F}$ at the endpoint; see Fig. 41. Then the terms of the quadratic equations match all the ways to collapse a collection of consecutive points.

Elaborating further, an object of the category C_X is given by the following data:
A complex C_σ assigned to each cell σ.
For each pair of strata $\sigma_0 < \sigma_1$, a map of complexes $f_{\sigma_0,\sigma_1} : C_{\sigma_0} \longrightarrow C_{\sigma_1}$.
For each triple of strata $\sigma_0 < \sigma_1 < \sigma_2$, a homotopy $f_{\sigma_0,\sigma_1,\sigma_2} : C_{\sigma_0} \to C_{\sigma_2}[-1]$:

$$[d, f_{\sigma_0,\sigma_1,\sigma_2}] = f_{\sigma_1,\sigma_2} \circ f_{\sigma_0,\sigma_1} - f_{\sigma_0,\sigma_2}.$$

And so on.

Given two A_∞-functors $\mathcal{F}$ and $\mathcal{G}$, one defines a graded vector space:

$$\begin{aligned} &\mathrm{Hom}^\bullet(\mathcal{F}, \mathcal{G}) := \\ &\prod_{n \geq 0} \prod_{\sigma_0 < \ldots < \sigma_n} \mathrm{Hom}^\bullet\Big(\mathcal{F}(\sigma_0) \otimes \mathrm{Hom}_{\mathcal{Q}}(\sigma_0, \sigma_1) \otimes \ldots \\ &\qquad \otimes \mathrm{Hom}_{\mathcal{Q}}(\sigma_{n-1}, \sigma_n), \mathcal{G}(\sigma_n)\Big)[-n] = \\ &\prod_{n \geq 0} \prod_{\sigma_0 < \ldots < \sigma_n} \mathrm{Hom}^\bullet\Big(\mathcal{F}(\sigma_0), \mathcal{G}(\sigma_n)\Big)[-n]. \end{aligned} \tag{159}$$

One equips it with a natural differential, getting a complex $\mathrm{Hom}^\bullet(\mathcal{F}, \mathcal{G})$.

2. $\mathcal{H}$-Supported Complexes of Constructible Sheaves on a Manifold
Let X be an oriented n-dimensional manifold. A *coorientation* of a hypersurface H in X is a choice of a connected component of the conormal bundle to H minus the zero section. This component is called the conormal bundle to the cooriented hypersurface H.

Recall that the *microlocal support* of a complex of constructible sheaves $\mathcal{F}$ on X [25]. It is a closed conical Lagrangian subvariety of T^*X, defined as follows. Take a nonzero covector $\eta \in T_x^*X$ and a little ball passing through the point x and tangent to the hyperplane $\eta = 0$ in T_xX. Move it slightly, getting a ball B_η containing x; move it the other way, getting a ball B'_η which does not contain x. Let $B''_\eta := B'_\eta \cap B_\eta$. Consider the cone of the restriction map from B_η to B''_η:

$$\mathrm{Cone}\Big(\mathcal{F}(B_\eta) \stackrel{\mathrm{Res}}{\longrightarrow} \mathcal{F}(B''_\eta)\Big). \tag{160}$$

Then η does not enter the microlocal support of $\mathcal{F}$ if and only if the complex (160) is acyclic. Finally, if an open neighborhood U of x is contained in the support of $\mathcal{F}$, then the zero section of T^*U belongs to the microlocal support of $\mathcal{F}$.

Definition 7.1.2 Let $\mathcal{H}$ be a collection of cooriented hypersurfaces in a manifold X. A complex of constructible sheaves on X is $\mathcal{H}$-supported if its microlocal support is contained in the union of the zero section of T^*X and the conormal bundles to the cooriented hypersurfaces in $\mathcal{H}$.

The triangulated categories of complexes of constructible sheaves with microlocal support on a given collection of hypersurfaces has been studied before; see [33]. Our goal is to describe a simple dg-model of its certain subcategory, that is, a dg-category canonically equivalent to this subcategory, which can be studied by elementary means, which also work in the non-commutative setup.

3. Generic Cooriented Stratification $\mathcal{H}$ of a Manifold
Take a collection of cooriented hypersurfaces in generic position, that is, locally isomorphic to a collection of coordinate hyperplanes. It provides a stratification $\mathcal{H}$ of X. The 0-cells are the intersection points of n hypersurfaces. The 1-cells are components of the intersections of $n - 1$ hypersurfaces minus 0-cells, and so on. Let S be a codimension k stratum of the stratification $\mathcal{H}$. We can identify an open neighborhood of S with $S \times \mathbb{R}^k$, where $\mathbb{R}^k$ is the vector space with coordinates $\{x_i\}$, so that the codimension 1 strata containing S are given by equations $x_i = 0$, cooriented by $dx_i > 0$. The conormal bundle T_S^*X contains 2^k octants. Using an isomorphism $T_S^*X = \mathbb{R}^k \times S$, they are given by inequalities $\pm x_1 \geq 0, \ldots, \pm x_k \geq 0$. The octant $x_1 \geq 0, \ldots, x_k \geq 0$ is called the *positive octant* and denoted by T_S^+X. The dual stratification is described by an oriented cube $\mathcal{K}_S$. Its vertices match the octants in $\mathbb{R}^k$. The oriented edges match the cooriented codimension 1 strata, etc. The cubes $\mathcal{K}_S$ for different strata S are assembled into a cubical complex $\mathcal{K}$ dual to the stratification $\mathcal{H}$; see Fig. 42.

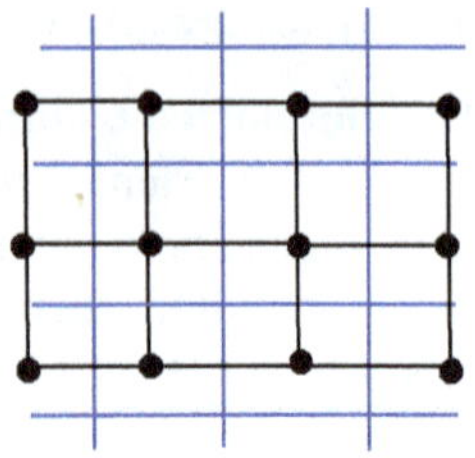

Fig. 42 A cubical complex dual to a generic stratification

In particular, for each face K of $\mathcal{K}$ there is a vertex $v_-(K)$ assigned to the negative octant and the opposite vertex $v_+(K)$ assigned to the positive octant. They coincide if $\dim(K) = 0$.

Lemma 7.1.3 *The microlocal support of an $\mathcal{H}$-supported constructible complex of sheaves $\mathcal{F}$ on X is contained in the union of the positive octants $T^+_S X$, where S are the strata of $\mathcal{H}$.*

Proof The microlocal support of $\mathcal{F}$ is a union of certain octants in $T^*_S X$. Indeed, varying the convector η inside of one octant, we do not change the cohomology groups of complex (160). If the microlocal support intersects one of the nonpositive open octants, the closure of this octant intersects a negative component of the conormal bundle to one of the codimension 1 strata, which contradicts to the condition that $\mathcal{F}$ is $\mathcal{H}$-supported. □

4. $\mathcal{H}$-Supported dg-Sheaves on a Manifold

A dg-enhancement of the derived category of $\mathcal{H}$-supported complexes of constructible sheaves is given by the $\mathcal{H}$-supported dg-sheaves. Let us define a more economic model for it for stratifications by generic cooriented hypersurfaces.

Proposition 7.1.4 *Let $\mathcal{H}$ be a stratification of a manifold X by generic cooriented hypersurfaces, such that the closures of all strata are balls. Then $\mathcal{H}$-supported dg-sheaves can be described as follows, using the dual cubical complex $\mathcal{K}$:*

Data:

- *A graded vector space $C^\bullet_v$ assigned to each vertex v of $\mathcal{K}$*
- *A map $\delta_K : C^\bullet_{v_-(K)} \longrightarrow C^\bullet_{v_+(K)}[1-k]$ for each k-dimensional face K of $\mathcal{K}$, where $k \geq 0$*

Condition: *For each face K of $\mathcal{K}$, take the direct sum of the graded spaces $C^\bullet_v$ at the vertices v of K, shifted by the distance d_v from v to the vertex $v_-(K)$:*[19]

$$C^\bullet_K := \oplus_{v\in K} C^\bullet_v[-d_v].$$

[19] If one parametrizes the vertices of a cube K by $(\varepsilon_1, \ldots, \varepsilon_k)$, where $\varepsilon_i \in \{0, 1\}$, then the distance is $\varepsilon_1 + \ldots + \varepsilon_k$.

- *Then the signed sum d_K of the maps δ_F assigned to the faces F of K in $\mathcal{K}$ satisfies $d_K^2 = 0$.*

Example Let us elaborate the cubical description of an $\mathcal{H}$-supported dg-sheaf in the dimension 3:

For each vertex v we get a complex $(C_v^\bullet, \delta_v)$.

For each oriented edge $E = (0, 1)$ we get a morphism of complexes:

$$\delta_E : C_0^\bullet \longrightarrow C_1^\bullet.$$

For each square K with the vertices $(00, 01, 10, 11)$ we get a homotopy:

$$h_K : C_{00}^\bullet \longrightarrow C_{11}^\bullet[-1], \quad \delta_{11}h_K + h_K\delta_{00} = \delta_{01\to 11} \circ \delta_{00\to 01} - \delta_{10\to 11} \circ \delta_{00\to 10}.$$

For each cube C with vertices $\{\varepsilon_1, \varepsilon_2, \varepsilon_3\}$ we get a higher homotopy $H_C : C_{000}^\bullet \longrightarrow C_{111}^\bullet[-2]$:

$$\delta H_C - H_C\delta = \delta_{110\to 111} \circ h_{000\to 110} + \delta_{101\to 111} \circ h_{000\to 101} + \delta_{011\to 111} \circ h_{000\to 011}$$
$$- h_{100\to 111} \circ \delta_{000\to 100} - h_{010\to 111} \circ \delta_{000\to 010} - h_{001\to 111} \circ \delta_{000\to 001}. \tag{161}$$

Proof Let us show how a data $(C_v^\bullet, \delta_K)$ determines a dg-sheaf on X. We start with a simple observation. Let H be a component of a cooriented hypersurface strata. Denote by $\mathcal{D}^+$ and $\mathcal{D}^-$ the connected domains of $X - \mathcal{H}$ sharing H, so that H is cooriented towards $\mathcal{D}^+$. Let $\mathcal{F}$ be an $\mathcal{H}$-supported dg-sheaf on X. Then the natural map

$$\mathcal{F}_{|H} \longrightarrow \mathcal{F}_{|\mathcal{D}^-} \tag{162}$$

is a quasi-isomorphism. Indeed, its cone detects the microlocal support of $\mathcal{F}$ in the direction opposite to the coorientation of H and hence is acyclic. We can alter the dg-sheaf data on the component H so that the map (162) is an isomorphism, getting a quasi-isomorphic dg-sheaf.

Generalizing this, let us construct an A_∞-functor $\mathcal{F}$. Given a stratum σ of $\mathcal{H}$, let us define $\mathcal{F}(\sigma)$. Let us move a bit the point $p \in \sigma$ inside of the unique n-dimensional stratum S_p in the negative direction from p, whose closure contains p. The stratum S_p is encoded by a vertex v of $\mathcal{K}$. We set $\mathcal{F}(\sigma) := C_v^\bullet$. The maps $f_{\sigma_1,\ldots,\sigma_n}$ are defined naturally.

Let us describe a dg-sheaf $C^\bullet$ near the intersection point v of two cooriented lines in $\mathbb{R}^2$ on Fig. 43. The cubical description of the dg-sheaf $C^\bullet$ near v is given by the following data, where $h : C_{00}^\bullet \longrightarrow C_{11}^\bullet[-1]$ (see also Fig. 43):

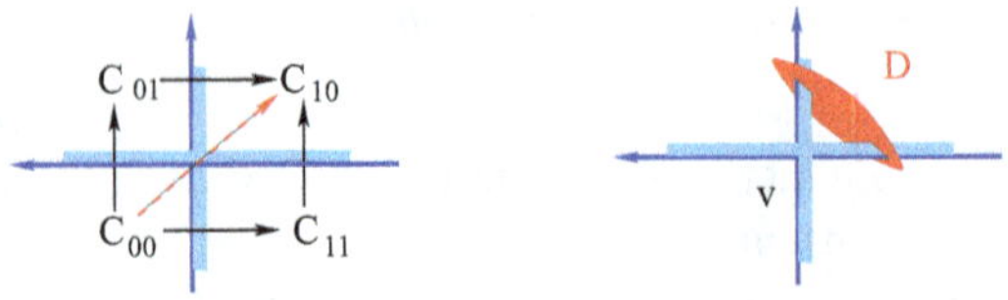

Fig. 43 The domain $\mathcal{D}$ intersects the two lines, but does not contain their intersection v

$$\begin{array}{ccc} C^\bullet_{01} & \xrightarrow{g_1} & C^\bullet_{11} \\ f_0 \uparrow & \overset{h}{\nearrow} & \uparrow f_1 \\ C^\bullet_{00} & \xrightarrow[g_0]{} & C^\bullet_{10} \end{array} \tag{163}$$

The restriction of $C^\bullet$ to a ball $\mathcal{B}$ around v is given, by the definition, by the complex $C^\bullet_{00}$. The restriction of $C^\bullet$ to the domain $\mathcal{D}$ on Fig. 43 is given by the following complex, where C^0_{ij} is in the degree $i + j - 1$:

$$\begin{array}{ccc} C^\bullet_{01} & \xrightarrow{g_1} & C^\bullet_{11} \\ & & \uparrow f_1 \\ & & C^\bullet_{10} \end{array} \tag{164}$$

Indeed, to calculate $C^\bullet_{|\mathcal{D}}$ we cover $\mathcal{D}$ by open subsets $\mathcal{U}$ and $\mathcal{V}$, where $\mathcal{U}$ is disjoint with the horizontal line and $\mathcal{V}$ with the vertical. Then $C^\bullet_{|\mathcal{U}} = C^\bullet_{01}$, $C^\bullet_{|\mathcal{V}} = C^\bullet_{10}$. Their restrictions to $\mathcal{V} \cap \mathcal{U}$ are given by the maps g_1, f_1.

Assume that $\mathcal{D} \subset \mathcal{B}$. Then, using (164), the restriction of $C^\bullet_{|\mathcal{B}}$ to $\mathcal{D}$ is given by the map of complexes:

$$(f_0, g_0, h) : C^\bullet_{00} \longrightarrow \text{Cone}\Big(C^\bullet_{01} \oplus C^\bullet_{10} \overset{(g_1, f_1)}{\longrightarrow} C^\bullet_{11}\Big)[-1]. \tag{165}$$

The shifted cone of this map is the total complex associated with (163). □

Lemma 7.1.5 *Given an $\mathcal{H}$-supported complex of sheaves $\mathcal{F}$ on X, the complex (160) for any convector η inside the positive octant $T^+_S X$ is quasi-isomorphic to the complex $(C^\bullet_K, d_K)$, where K is the face of $\mathcal{K}$ dual to the strata S.*

Proof Let us elaborate the microlocal support at the vertex v on Fig. 43. Then the map (160) is given by the map (165). So the claim boils down to the last sentence of the proof of Proposition 7.1.4. □

5. $\mathcal{H}$-Admissible dg-Sheaves

Among all $\mathcal{H}$-supported dg-sheaves we distinguish a crucial for us subcategory of *$\mathcal{H}$-admissible dg-sheaves*.

Definition 7.1.6 Given a stratification $\mathcal{H}$ provided by a collection of cooriented hypersurfaces in a manifold X whose Legendrians are disjoint, an $\mathcal{H}$-supported dg-sheaf as in Proposition 7.1.4 is $\mathcal{H}$-admissible if:

- The complexes $(C_K^\bullet, d_K)$ satisfy the following conditions:

$$
\begin{aligned}
&i)\ \dim\ \mathrm{K} = 0: \quad H^p(C_K^\bullet) = 0 \ \text{ if } p \notin \{0, 1\}.\\
&ii)\ \dim\ \mathrm{K} = 1: \quad H^p(C_K^\bullet) = 0 \ \text{ if } p \notin \{0\}.\\
&iii)\ \dim\ \mathrm{K} > 1: \quad H^p(C_K^\bullet) = 0 \ \text{ for all } p.
\end{aligned}
\tag{166}
$$

Lemma 7.1.7 *The restriction of an $\mathcal{H}$-admissible dg-sheaf to each hypersurface of $\mathcal{H}$ is a local system.*

Proof Follow the same idea as the proof of Lemma 7.3.1. □

Remark Below we need subcategories of $\mathcal{H}$-admissible dg-sheaves with microlocal support on the (conormal bundles to) hypersurfaces of an a priori given collection of hypersurfaces $\mathcal{H}$. However, we do not want to assume that the induced stratification is topologically trivial. A typical example is a punctured surface S with a collection of small loops near each puncture. So we consider in addition to $\mathcal{H}$ a collection of dummy hypersurfaces and require that the microlocal support of dg-sheaves on them is trivial, i.e., the dg-sheaves are non-singular along them. We denote by $\widetilde{\mathcal{H}}$ the bigger collection of hypersurfaces and choose it so that the induced stratification is topologically trivial. There is a functor from the category of $\widetilde{\mathcal{H}}$-admissible dg-sheaves to the triangulated category of complexes of constructible sheaves with the microlocal support in $\mathcal{H}$, where the latter was defined in [25], which is an equivalence. Using these functors one sees that the categories of $\widetilde{\mathcal{H}}$-admissible dg-sheaves are equivalent, and they are all equivalent to the original category of $\mathcal{H}$-admissible dg-sheaves. See also [33, Section 3] where, as the referee pointed out, the related issues are discussed.

7.2 *The Bigon and Triangle Moves*

We consider moves of the hypersurfaces supporting the microlocal support of an admissible dg-sheaf such that the corresponding Legendrians remain disjoint. We call them admissible moves. Focusing on the dimension 2, we prove that admissible moves preserve the category of $\mathcal{H}$-admissible dg-sheaves.

The Bigon Moves

There are two different admissible moves of a pair of cooriented lines in the plane, called bigon moves and shown on Fig. 44. Complexes assigned to the faces and arrows between them corresponding to the cooriented lines separating them are described by the vertices and oriented edges of a square, shown at the bottom of each move.

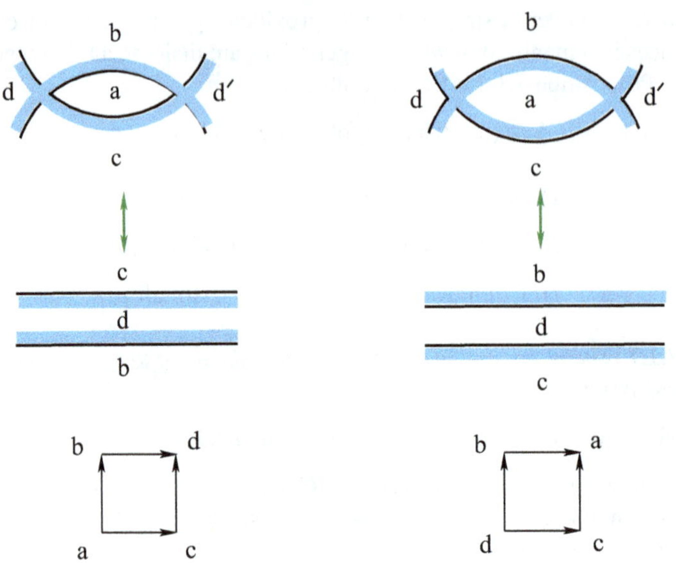

Fig. 44 There are two different admissible bigon moves

Recall that any object of the triangulated category of representations of a Dynkin quiver is a direct sum of the shifts $R_i[n_i]$, $n_i \in \mathbb{Z}$ of indecomposable representations R_i. The isomorphism classes of the latter are parametrized by positive roots of the root system.

The two arrows from the bottom left vertex of the square form an A_3-quiver. It has six indecomposable representations. Taking into account a $\mathbb{Z}/2\mathbb{Z}$-symmetry of the A_3-quiver, there are just four of them to consider, depicted on Fig. 45. Each of them determines uniquely an acyclic homotopy square, as shown on Fig. 45.[20]

[20] Below we use the fact that the classification of the indecomposable representations of quivers of type $A_m, m \leq 3$, and D_4 over a skew field is exactly the same as over a field. Following the request of the referee, we present the proof in the D_4 case when all arrows to a vertex v. In the chapter we also need the case when they go out of v. Over a field one reduces to the other by dualization. Over a skew field dualization does not work, but one proceeds by the analogy. So take a D_4-quiver with vertices a, b, c, v and all arrows going to v. The only fact we use is that a proper linear subspace of an R-vector space has a complement. So we reduce to the case when all maps are injective: otherwise we split off a representation with $V_v = 0$. So from now on $V_a, V_b, V_c \subset V_v$. Similarly we can assume $V_a + V_b + V_c = V_d$. If $W_{abc} := V_a \cap V_b \cap V_c \neq 0$, we get a non-decomposable if $V_a = V_b = V_c$ is of rank 1, which we can split off otherwise. So we can assume $W_{abc} = 0$. If $V_c = 0$, we get a representation of the quiver A_3. If also $V_b = 0$, we split off an indecomposable with the dimension vector $(1, 0, 0, 1)$. Otherwise we get an indecomposable only if $V_a = V_b$ is one-dimensional. So we can assume $V_a, V_b, V_c \neq 0$. Next, we can assume that any two of V_a, V_b, V_c intersect by 0: otherwise we split off a representation of an A_3 quiver with the dimension vector $(1, 1, 1)$. If $\dim(V_a) > 1$, we can split off a subspace from V_a. So we can assume V_a, V_b, V_c are one-dimensional. We get the last indecomposable with the dimension vector $(1, 1, 1, 2)$,

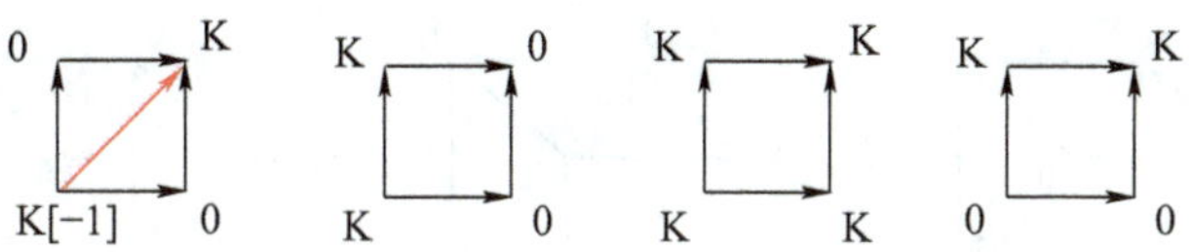

Fig. 45 The squares for bigon moves. The nontrivial homotopy is shown by a red arrow. Each arrow $K \to K$ is an isomorphism

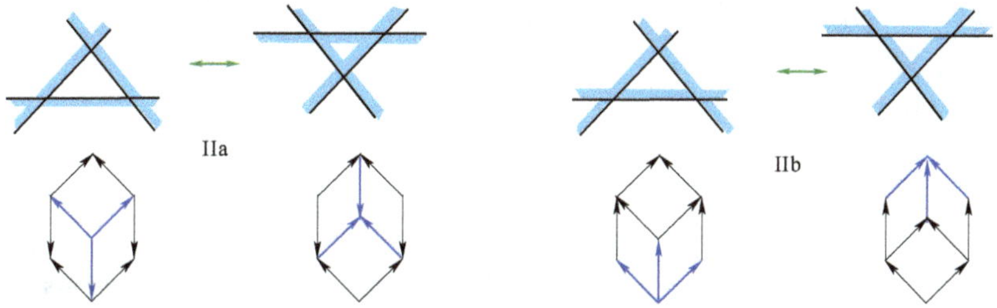

Fig. 46 The triangle moves. The D_4-quivers are shown by blue arrows

The quiver representation obtained by forgetting the vertex a of the square and the arrows from this vertex describe a dg-sheaf obtained by the elementary move of the original dg-sheaf.

Theorem 7.2.1 *The bigon moves transform admissible dg-sheaf to admissible ones.*

Proof We use the remark in the end of Sect. 7.1 to extend to a bigger collection $\widetilde{\mathcal{H}}$ with topologically trivial strata of the induced stratification, to be able to restrict considerations to a small ball where we will perform the isotopy. Then the claim is clear from Fig. 45. □

The Triangle Moves

There are two moves of a triple of cooriented lines in the plane, called IIa and IIb moves, or just triangle moves, shown on Fig. 46. The complexes assigned to the faces and the arrows between them are described by the quivers shown on the bottom.

The three arrows from the central vertex form an D_4-quiver. A D_4-quiver has 12 indecomposable representations. Pick an orientation of the edges of the D_4-quiver such that the three arrows go out of the central vertex. The symmetry group of this quiver is the group S_3. Modulo the S_3-symmetry, there are 6 indecomposable representations. They are depicted on Fig. 47 for the D_4-quiver whose central vertex is the bottom left vertex of the cube.

A triangle move of three cooriented lines in a 2d space is described in the 3d space-time by the three cooriented coordinate planes. The $\mathcal{H}$-supported dg-sheaves for the union of cooriented coordinate planes are described by data on a cube shown on Fig. 47 and their shifts.

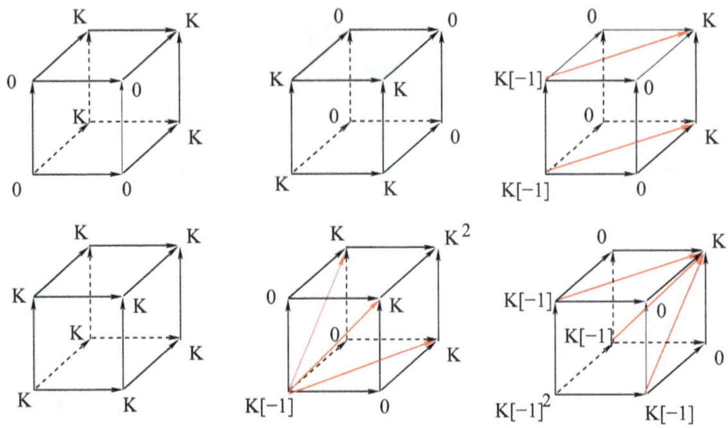

Fig. 47 The cubes for elementary moves IIa and IIb. Homotopies are shown by red arrows

Forgetting the top right vertex of the cube diagram we describe the original admissible dg-sheaf. Forgetting the bottom left vertex we describe the resulting one. This shows how indecomposable dg-sheaves transforms under type IIa moves. Type IIb moves are treated similarly, with the comment that this time we remove bottom right vertex instead of the top right.

Theorem 7.2.2 *Triangle moves transform admissible dg-sheafs to admissible ones.*

Proof This is clear from Fig. 47. Indeed, the dg-sheaf in $\mathbb{R}^3$ assigned to the cube is admissible, so forgetting any vertex we still get an admissible dg-sheaf. □

Triangle moves provide a bijection between the isomorphism classes of indecomposable objects for two differently oriented A_3-quivers: the one oriented out of the central vertex to the one oriented towards the central vertex for the IIa move and to the one with two edges oriented towards the central vertex for the IIb move.

7.3 Local Systems on Ideal Bipartite Graphs and Decorated Surfaces

1. Bipartite Ribbon Graphs and Zig-Zag Paths

A bipartite graph Γ gives rise to a collection of zig-zag paths on the decorated surface $\mathbb{S}$ associated with Γ. Denote by $\mathcal{Z}$ their union. The complement $\mathbb{S} - \mathcal{Z}$ is a disjoint union of connected domains of three types: •, ∘, and mixed domains (see Fig. 48). The •-domains are the ones containing a single •-vertex of Γ. The ∘-domains contain a single ∘-vertex of Γ. The rest are mixed domains. The • and ∘-domains are contractible.

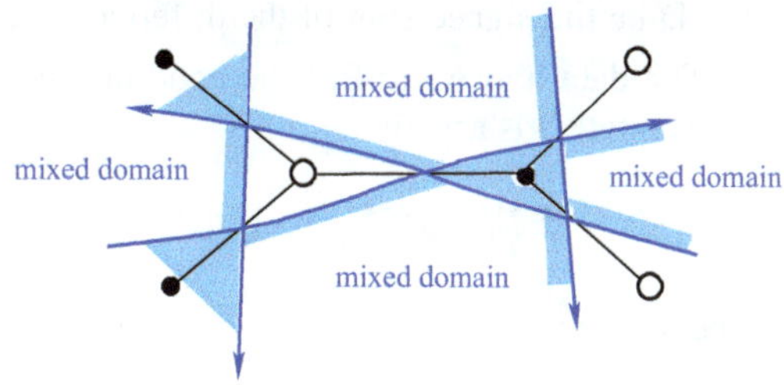

Fig. 48 Zig-zag paths for a bipartite graph Γ on a surface $\mathbb{S}$ cut the surface into $\bullet$, $\circ$, and mixed domains. Each zig-zag path is coorientated by a shaded area to the right of the path

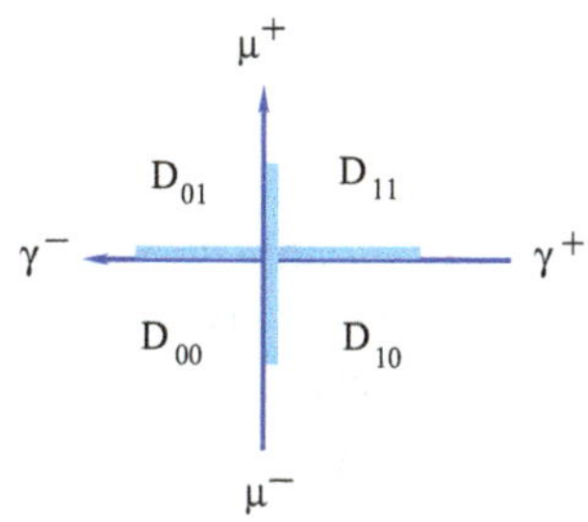

Fig. 49 The stratification near a crossing point of zig-zag paths γ and μ has one 0-dimensional stratum, four 1-dimensional strata $\gamma^{\pm}$, $\mu^{\pm}$ and four 2-dimensional strata $\mathcal{D}_{**}$

Each zig-zag path is orientated so that the $\bullet$-vertices are on the right. This plus the surface orientation provides coorientation of zig-zag paths: moving along a zig-zag path the coorientation points to the right. All sides of a $\bullet$-domain are cooriented inside, all sides of a $\circ$-domain are cooriented outside, and coorientations of the sides of a mixed domain alternate.

So the union of zig-zag paths give rise to a cooriented stratification $\mathcal{Z}$ of $\mathbb{S}$. Its zero stratum is the set $\mathcal{Z}^0$ of crossing points. So there is a category of $\mathcal{Z}$-admissible dg-sheaves on $\mathbb{S}$.

2. $\mathcal{Z}$-Admissible dg-Sheaves

A $\mathcal{Z}$-admissible dg-sheaf on $\mathbb{S}$ is given by the following data:

- A complex of local systems $\mathcal{F}^\bullet_{\mathcal{D}}$ on each domain $\mathcal{D}$ of $\mathbb{S} - \mathcal{Z}$, concentrated in degrees [0, 1].
- A map of complexes $\varphi_\gamma : \mathcal{F}^\bullet_{\mathcal{D}^-} \to \mathcal{F}^\bullet_{\mathcal{D}^+}$ for each component γ of $\mathcal{Z} - \mathcal{Z}^0$, such that

$$H^i\mathrm{Cone}(\varphi_\gamma) = 0 \ \text{ unless } i = 0. \tag{167}$$

- For each crossing point of zig-zags γ and μ, a homotopy (see Fig. 49):

$$h : \mathcal{F}^\bullet_{\mathcal{D}_{00}} \longrightarrow \mathcal{F}^\bullet_{\mathcal{D}_{11}}[-1], \quad dh + hd = \varphi_{\mu^+} \circ \varphi_{\gamma^-} - \varphi_{\gamma^+} \circ \varphi_{\mu^-}.$$

$$\begin{array}{ccc} \mathcal{F}^\bullet_{\mathcal{D}_{01}} & \xrightarrow{\varphi_{\mu^+}} & \mathcal{F}^\bullet_{\mathcal{D}_{11}} \\ {\scriptstyle\varphi_{\gamma^-}}\uparrow & \overset{h}{\nearrow} & \uparrow{\scriptstyle\varphi_{\gamma^+}} \\ \mathcal{F}^\bullet_{\mathcal{D}_{00}} & \xrightarrow[\varphi_{\mu^-}]{} & \mathcal{F}^\bullet_{\mathcal{D}_{10}} \end{array} \tag{168}$$

- Let D be the signed sum of the differentials on $\mathcal{F}^\bullet_{\mathcal{D}_{ij}}$, the maps $\varphi_{\mu^\pm}$, $\varphi_{\gamma^\pm}$, and h. The data above satisfies the condition that the following complex with the differential D is acyclic:

$$\mathcal{F}^\bullet_{\mathcal{D}_{00}} \oplus \mathcal{F}^\bullet_{\mathcal{D}_{01}}[-1] \oplus \mathcal{F}^\bullet_{\mathcal{D}_{10}}[-1] \oplus \mathcal{F}^\bullet_{\mathcal{D}_{11}}[-2]. \tag{169}$$

The last condition is equivalent to either of the following conditions:

1. The natural map of complexes $\mathrm{Cone}(\varphi_{\gamma^-}) \longrightarrow \mathrm{Cone}(\varphi_{\gamma^+})$ is a quasi-isomorphism.
2. The natural map of complexes $\mathrm{Cone}(\varphi_{\mu^-}) \longrightarrow \mathrm{Cone}(\varphi_{\mu^+})$ is a quasi-isomorphism.

(a) $\mathcal{Z}$-Admissible dg-Sheaves on $\mathbb{S} \longrightarrow$ Local Systems on Zig-Zag Paths
Condition (167) just means that on each component γ of $\mathcal{Z} - \mathcal{Z}^0$ there is a local system $\mathrm{Cone}(\varphi_\gamma)$.

Lemma 7.3.1 *Given a zig-zag path γ, local systems* $\mathrm{Cone}(\varphi_\gamma)$ *on $\gamma - \{crossing\ pts\}$ extend to a local system on γ, denoted by* $\mathrm{Cone}(\varphi_\gamma)$.

Proof Condition (169) implies that for each crossing point on γ separating two germs of zig-zags γ^- and γ^+, there is an isomorphism $\mathrm{Cone}(\varphi_{\gamma^-}) \stackrel{\sim}{\longrightarrow} \mathrm{Cone}(\varphi_{\gamma^+})$. It allows to extend local systems on components of γ − {crossing points} to a local system on γ. □

(b) Local Systems on $\Gamma \longrightarrow \mathcal{Z}$-Admissible dg-Sheaves on $\mathbb{S}$, Acyclic on Mixed Strata
A local system of vector spaces on a graph Γ is given by a collection of vector spaces V_v at the vertices v of Γ and linear maps $\varphi_{(v,w)} : V_v \to V_w$ for each edge (v, w) of Γ, so that $\varphi_{(w,v)} = \varphi^{-1}_{(v,w)}$.

Given a local system $\mathcal{V}$ on a bipartite graph Γ on $\mathbb{S}$, let us define a $\mathcal{Z}$-admissible dg-sheaf $\mathcal{F}_\mathcal{V}$ on $\mathbb{S}$. Since the $\bullet$ and $\circ$ domains of $\mathbb{S} - \mathcal{Z}$ are contractible, we can talk about the fibers of a complex of $\mathcal{Z}$-constructible sheaves over them. Then we define:

- The fiber of $\mathcal{F}_\mathcal{V}$ at the $\circ$-domain with a vertex $\circ$ of Γ is the shifted fiber $\mathcal{V}_\circ[-1]$ of $\mathcal{V}$ at $\circ$.
- The fiber of $\mathcal{F}_\mathcal{V}$ at the $\bullet$-domain containing a vertex $\bullet$ of Γ is the fiber $\mathcal{V}_\bullet$ of $\mathcal{V}$.
- The fibers of $\mathcal{F}_\mathcal{V}$ over mixed domains are zero. So the maps φ_γ for zig-zags γ are zero.
- The homotopy map for each edge $\circ - \bullet$ of Γ, as illustrated in Fig. 50:

$$h : \mathcal{V}_\circ[-1] \longrightarrow \mathcal{V}_\bullet[-1]. \tag{170}$$

This map is the shift of the parallel transport map along the edge $\circ - \bullet$.

So we get the following diagram near each crossing of two zig-zag paths:

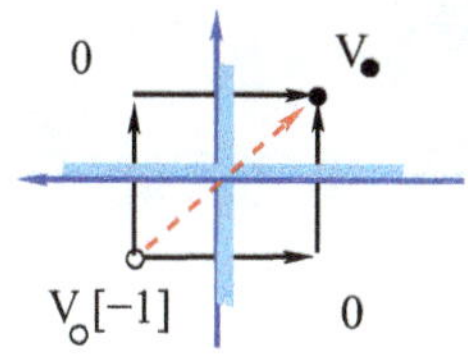

Fig. 50 The admissible dg-sheaf $\mathcal{F}_{\mathcal{V}}$ on $\mathbb{S}$ assigned to a local system $\mathcal{V}$ on Γ

$$\begin{array}{ccc} 0 & \longrightarrow & \mathcal{V}_\bullet \\ \uparrow & \overset{h}{\nearrow} & \uparrow \\ \mathcal{V}_\circ[-1] & \longrightarrow & 0 \end{array} \tag{171}$$

(c) $\mathcal{Z}$-Admissible dg-Sheaves on $\mathbb{S}$, Acyclic on Mixed Strata $\longrightarrow$ Local Systems on Γ

Construction (b) can be inverted. Namely, let $\mathcal{F}$ be a $\mathcal{Z}$-admissible dg-sheaf $\mathcal{F}$ on $\mathbb{S}$, acyclic on mixed strata. Then $\mathcal{Z}$-admissibility implies that

$$H^i(\mathcal{F}_{\mathcal{D}_\circ}) = 0 \text{ if } i \neq 1, \qquad H^i(\mathcal{F}_{\mathcal{D}_\bullet}) = 0 \text{ if } i \neq 0. \tag{172}$$

Furthermore, the homotopy h provides an isomorphism:

$$h : H^1(\mathcal{F}_{\mathcal{D}_\circ}) \xrightarrow{=} H^0(\mathcal{F}_{\mathcal{D}_\bullet})[-1].$$

So we get a local system $\mathcal{L}_{\mathcal{F}}$ on the bipartite graph Γ by assigning the spaces $H^1(\mathcal{F}_{\mathcal{D}_\circ})[1]$ to the $\circ$-vertices $\circ$ of Γ, the ones $H^0(\mathcal{F}_{\mathcal{D}_\bullet})$ to the $\bullet$-vertices $\bullet$ of Γ, and using the isomorphism h as the parallel transport along the edge $\circ \to \bullet$.

(d) Framed Local Systems on $\mathbb{S}$ as $\mathcal{Z}_m$-Admissible dg-Sheaves

Given a decorated surface $\mathbb{S}$, consider a collection of m nested simple clockwise loops around each puncture and a collection of m simple clockwise arcs around each special point, which do not intersect and do not self-intersect. They are cooriented "inside the surface"; see Fig. 51. Denote by $\mathcal{Z}_m$ their union.

Lemma 7.3.2 *A $\mathcal{Z}_m$-admissible dg-sheaf on a decorated surface $\mathbb{S}$, acyclic near the punctures and special points, is equivalent to an m-dimensional local system on $\mathbb{S}$ with a framing.*

Proof Take a punctured disc $\mathcal{D}^*$ with m concentric circles going clockwise around the puncture; see Fig. 51. Take a radius of the disc and restrict to it a $\mathcal{Z}_m$-admissible complex of sheaves on $\mathcal{D}^*$. The restriction is described by complexes $V_0^\bullet, \ldots V_m^\bullet$ and maps $\varphi_{i,i+1} : V_i^\bullet \longrightarrow V_{i+1}^\bullet$ between them. To proceed further we need the following simple observation.

Let $f : W_1^\bullet \longrightarrow W_2^\bullet$ be a map of complexes of vector spaces with the following properties:

1. Cone(f) is quasi-isomorphic to a one-dimensional vector space in the degree 0.

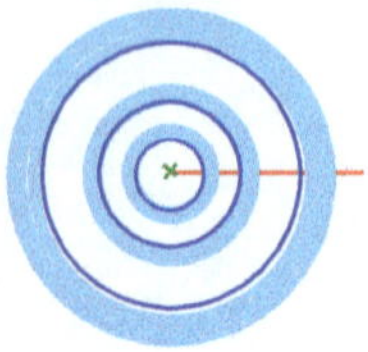

Fig. 51 An admissible dg-sheaf on a punctured disc stratified by m concentric outward cooriented circles, which is zero near the puncture, is a filtered vector space

2. $H^i(W_1^\bullet) = 0$ if $i \neq 0$.

Then $H^i(W_2^\bullet) = 0$ if $i \neq 0$, and the induced map $H^0(W_1^\bullet) \longrightarrow H^0(W_2^\bullet)$ is injective.

We apply this to the maps $\varphi_{i,i+1}$. Since $H^\bullet(V_0^\bullet) = 0$, we get a flag of vector spaces:

$$V_0 \hookrightarrow V_1 \hookrightarrow V_2 \hookrightarrow \ldots \hookrightarrow V_m, \quad \dim V_i = i, \quad V_i := H^0(V_i^\bullet).$$

The monodromy around the puncture provides an operator in the vector space V_m preserving the flag. This just means that we get a framed m-dimensional local system of vector spaces over the punctured disc $\mathcal{D}^*$. Applying this argument at all punctures and special points we get an m-dimensional local system on the surface $\mathbb{S}$ equipped with an invariant flag near each puncture and each special point, that is, a framed local system on the decorated surface $\mathbb{S}$. □

3. Framed Local Systems on $\mathbb{S} \longleftrightarrow$ Flat Line Bundles on an Ideal Bipartite Graph Γ

Let $\mathcal{L}$ be a one-dimensional local system on a GL_m-bipartite graph Γ on $\mathbb{S}$. It determines a $\mathcal{Z}$-admissible dg-sheaf $\mathcal{F}_{\mathcal{L}}$ on $\mathbb{S}$ via construction (b). By the very definition, the zig-zag paths of Γ can be deformed using the elementary moves to a collection $\mathcal{Z}_m$, formed by m nested loops/arcs near each marked points. By Lemma 7.3.2, the resulting $\mathcal{Z}_m$-admissible dg-sheaf is just a framed m-dimensional local system on $\mathbb{S}$. The elementary moves provide birational equivalences between the corresponding categories of $\mathcal{Z}$-admissible dg-sheaves.

Conversely, let $\mathcal{L}$ be a framed m-dimensional local system on $\mathbb{S}$. Construction (d) assigns to it a $\mathcal{Z}_m$-admissible dg-sheaf on $\mathbb{S}$. Deforming back using the elementary moves the collection $\mathcal{Z}_m$ of loops/arcs around the marked points to the collection $\mathcal{Z}_\Gamma$ of zig-zag paths of the ideal bipartite graph Γ, we get a $\mathcal{Z}_\Gamma$-admissible dg-sheaf $\mathcal{E}_{\mathcal{L}}$ on $\mathbb{S}$, which is zero near the marked points. As Fig. 52 shows, this admissible dg-sheaf may not be acyclic on all mixed domain. However, for a generic $\mathcal{Z}_m$-admissible dg-sheaf on $\mathbb{S}$ we get an admissible dg-sheaf acyclic on the mixed domain. The microlocal support of the obtained $\mathcal{Z}_\Gamma$-admissible dg-sheaf is described by flat line bundles on zig-zag paths γ of Γ. They are the same line

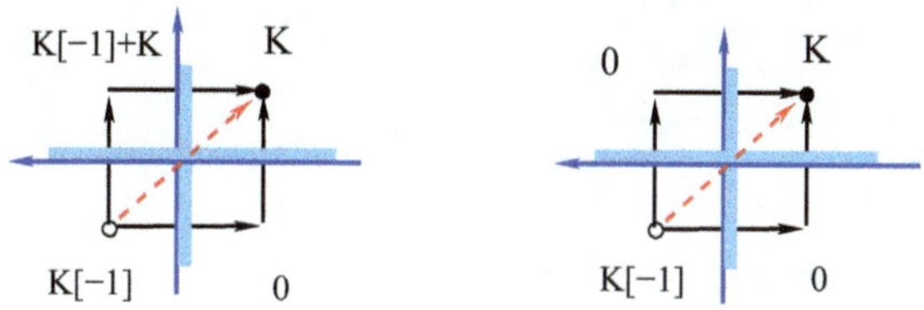

Fig. 52 Two admissible dg-sheaves near a crossing point. Each provides a rank one local system on the microlocal support

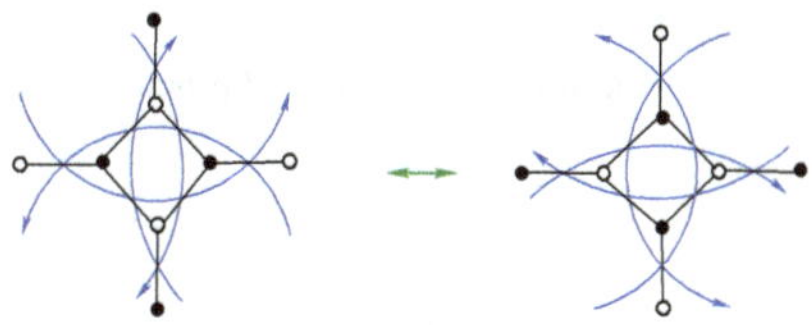

Fig. 53 A two by two move of bipartite graphs

bundles we had on γ before the deformation. Construction (c) gives us a flat line bundle on the graph Γ.

So we get a regular injective map from the moduli space $\mathrm{Loc}_1(\Gamma)$ of flat line bundles on the graph Γ to the moduli space $\mathcal{X}_{\mathrm{GL_m},\mathbb{S}}$ of framed m-dimensional local systems on $\mathbb{S}$, called the reconstruction map:

$$R_\Gamma : \mathrm{Loc}_1(\Gamma) \hookrightarrow \mathcal{X}_{\mathrm{GL}_m,\mathbb{S}}. \tag{173}$$

It provides an open domain in the moduli space $\mathcal{X}_{\mathrm{GL}_m,\mathbb{S}}$, which is a cluster Poisson torus. This is achieved by the following chain of constructions:

$$\begin{aligned} &\text{Flat line bundles on a rank } m \text{ ideal bipartite graph } \Gamma \text{ on } \mathbb{S} \stackrel{b)+c)}{\longleftrightarrow} \\ &\mathcal{Z}_\Gamma\text{-admissible dg-sheaves on } \mathbb{S}, \text{ acyclic on mixed domains} \stackrel{\text{Sec. 7.2}}{\subset} \\ &\mathcal{Z}_m\text{-admissible dg-sheaves, vanishing near the marked points} \stackrel{d)}{\longleftrightarrow} \\ &\text{Framed } m\text{-dimensional local systems on } \mathbb{S} \end{aligned} \tag{174}$$

The middle map in (174) is an open embedding. The others are isomorphisms of moduli spaces.

4. A Lagrangian Surface in $\mathcal{L}_\Gamma \subset T^*\mathbb{S}$
Given a bipartite graph Γ, note the following:

1. Each $\bullet$ or $\circ$-domain $\mathcal{D} \subset \mathbb{S}$ gives rise to a domain $\mathcal{D} \subset T^*\mathbb{S}$—the zero section on $\mathcal{D}$.
2. Each cooriented zig-zag path γ gives rise to the conormal bundle $T^*_\gamma\mathbb{S} \subset T^*\mathbb{S}$.

Definition 7.3.3 The surface $\mathcal{L}_\Gamma \subset T^*\mathbb{S}$ is the union of all these domains in $T^*\mathbb{S}$:

$$\mathcal{L}_\Gamma := \cup_\bullet \mathcal{D}_\bullet \cup_\circ \mathcal{D}_\circ \cup_\gamma T^*_\gamma \tag{175}$$

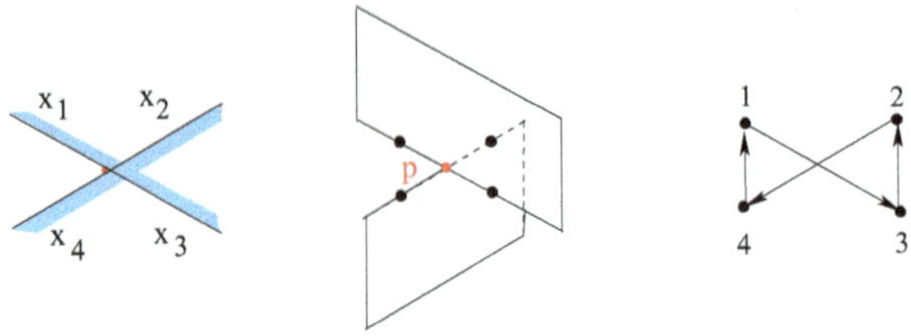

Fig. 54 The link of the surface $\mathcal{L}_\Gamma$ near a crossing point p is a circle $1 \to 3 \to 2 \to 4 \to 1$

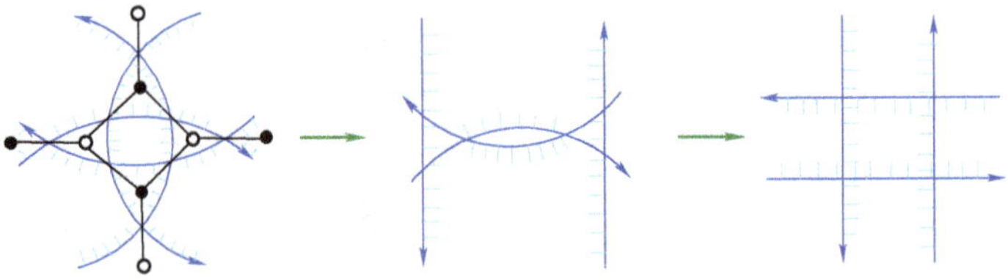

Fig. 55 From the left to the right: two type II triangle moves, followed by a bigon move

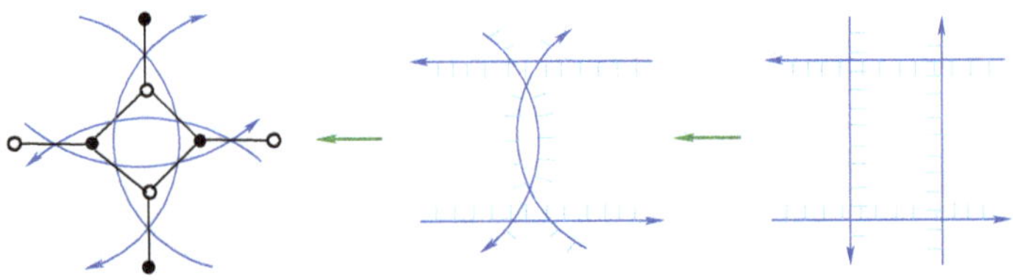

Fig. 56 From the right to the left: a bigon move, followed by two triangle moves

Lemma 7.3.4 *The surface $\mathcal{L}_\Gamma$ is homeomorphic to a smooth oriented surface, identified with the oriented spectral surface Σ_Γ.*

Proof The surface $\mathcal{L}_\Gamma$ is evidently homeomorphic to a smooth surface everywhere except the crossing point of the conormal bundles $p := T^*_{\gamma_1}\mathbb{S} \cap T^*_{\gamma_2}\mathbb{S}$ given by the zero cotangent vector at the crossing point $\gamma_1 \cap \gamma_2$ of two zig-zag paths. To prove that it is homeomorphic to a smooth surface at p, we calculate the intersection of a small sphere at p with $T^*_{\gamma_1}\mathbb{S} \cup T^*_{\gamma_2}\mathbb{S}$. We claim that it is homeomorphic to a circle. This is proved on Fig. 54, where we showed the link of a crossing point, observing at the same time that the projection $\mathcal{L}_\Gamma \longrightarrow \mathbb{S}$ reversing the orientation on one of the triangles. (Indeed, both arrows $3 \to 2$ and $4 \to 1$ look up on Fig. 54.) □

5. Two by Two Moves of Bipartite Graphs

Recall the two by two move $\Gamma_1 \to \Gamma_2$ of bipartite graphs on Fig. 53. It is decomposed into a sequence of bigon and triangle moves on Figs. 55, 56. Note that the collections of cooriented lines which we get in the process are no longer associated with bipartite graphs.

Starting from a local system $\mathcal{L}_1$ on the graph Γ_1, we interpret it as an admissible dg-sheaf and transform accordantly along the sequence of admissible moves above.

In the process we get admissible dg-sheaf not associated with any bipartite graphs. However, assuming that we started from a generic local system $\mathcal{L}_1$ on the graph Γ_1, the resulting admissible dg-sheaf in the end does correspond to a local system $\mathcal{L}_1$ on the graph Γ_2. So we get a birational isomorphism of stacks of local systems on the graphs Γ_1 and Γ_2. In particular, we get a birational isomorphism between the stacks of flat line bundles:

$$\mathrm{Loc}_1(\Gamma_1) \longrightarrow \mathrm{Loc}_1(\Gamma_2).$$

Our next goal is to get an explicit expression for this map for any skew field R.

Let B_1, B_2, B_3, B_4 be the fibers of the local system over the $\bullet$-domains and W_1, W_3 the fibers over the $\circ$-domains on the left on Fig. 55. Then in the domain on the figure the local system and hence the corresponding dg-sheaf are described by the following diagram of isomorphisms:

$$\begin{array}{ccccccc} & & & B_2 & & & \\ & & \overset{\sim}{\nearrow} & & \overset{\sim}{\nwarrow} & & \\ B_1 & \overset{\sim}{\longleftarrow} & W_1 & & W_3 & \overset{\sim}{\longrightarrow} & B_3 \\ & & \underset{\sim}{\searrow} & & \underset{\sim}{\swarrow} & & \\ & & & B_4 & & & \end{array} \tag{176}$$

Let us set

$$\mathcal{W}_1 := \frac{B_1 \oplus B_2 \oplus B_4}{W_1}, \quad \mathcal{W}_3 := \frac{B_2 \oplus B_3 \oplus B_4}{W_3}.$$

Here we identify W_1 with its image under the map $W_1 \hookrightarrow B_1 \oplus B_2 \oplus B_4$ and similarly for W_3. Then after the triangle moves at the two $\circ$-vertices we get a dg-sheaf with the microlocal support shown in the middle picture on Fig. 55. Since an admissible dg-sheaf related to a bipartite graph has zero fibers over the mixed domains, the only relevant picture describing a single type IIa triangle move is the fifth picture on Fig. 47. It follows that the resulting dg-sheaf is described by the following diagram:

$$\begin{array}{ccccccc} & & & B_2 & & & \\ & & \swarrow & & \searrow & & \\ B_1 & \longrightarrow & \mathcal{W}_1 & & \mathcal{W}_3 & \longleftarrow & B_3 \\ & & \nwarrow & & \nearrow & & \\ & & & B_4 & & & \end{array} \tag{177}$$

These maps are no longer isomorphisms since $\mathrm{rk}(\mathcal{W}_1) = \mathrm{rk}(\mathcal{W}_3) = 2\mathrm{rk}(B_i)$, and the diagram is no longer assigned to a bipartite graph. There are canonical isomorphisms:

$$\mathcal{W}_1 \stackrel{\sim}{\longleftarrow} B_2 \oplus B_4 \stackrel{\sim}{\longrightarrow} \mathcal{W}_3.$$

The left one is the embedding $B_2 \oplus B_4 \hookrightarrow B_1 \oplus B_2 \oplus B_4$, followed by the projection to $\mathcal{W}_1$. The right one is similar. Combining them we get an isomorphism $f : \mathcal{W}_1 \stackrel{\sim}{\longrightarrow} \mathcal{W}_3$. Setting $X := \mathcal{W}_1 \stackrel{f}{=} \mathcal{W}_3$, we get a diagram:

$$\begin{array}{ccccc} & & B_2 & & \\ & & \downarrow & & \\ B_1 & \longrightarrow & X & \longleftarrow & B_3 \\ & & \uparrow & & \\ & & B_4 & & \end{array} \tag{178}$$

It describes the dg-sheaf on the right of Fig. 55. The transition (176) $\rightarrow$ (178) describes the composition of the three moves on Fig. 55. Then we apply the three moves on Fig. 56. The corresponding transformations of quiver representations are recorded on diagram (179).

$$\begin{array}{ccccc} & & B_2 & & \\ & & \uparrow & & \\ & & W_2 & & \\ & \swarrow & & \searrow & \\ B_1 & & & & B_3 \\ & \nwarrow & & \nearrow & \\ & & W_4 & & \\ & & \downarrow & & \\ & & B_4 & & \end{array} \quad \longleftarrow \quad \begin{array}{ccccc} & & B_2 & & \\ & & \downarrow & & \\ & & \mathcal{W}_2 & & \\ & \nearrow & & \nwarrow & \\ B_1 & & & & B_3 \\ & \searrow & & \swarrow & \\ & & \mathcal{W}_4 & & \\ & & \uparrow & & \\ & & B_4 & & \end{array} \quad \longleftarrow \text{(178)}. \tag{179}$$

As the result we altered the dg-sheaf in the domain $\mathcal{D}$ inside of the four $\bullet$-vertices, but did not alter it outside. Let us calculate how the corresponding local system inside of the domain $\mathcal{D}$ changes. Consider the following diagram on the left, where all maps are isomorphisms:

$$
\begin{array}{ccccc}
 & & \mathrm{L}_2 & & \\
 & & \uparrow{\scriptstyle i_2} & & \\
\mathrm{L}_1 & \xleftarrow{i_1} & \mathrm{L} & \xrightarrow{i_3} & \mathrm{L}_3
\end{array}
\qquad\qquad
\begin{array}{ccccc}
 & & \mathrm{L}_2 & & \\
 & & \downarrow & & \\
\mathrm{L}_1 & \longrightarrow & \mathcal{L} & \longleftarrow & \mathrm{L}_3
\end{array}
\tag{180}
$$

Setting $\mathcal{L} := (\mathrm{L}_1 \oplus \mathrm{L}_2 \oplus \mathrm{L}_3)/\mathrm{L}$, it determines the diagram on the right. So there are three subspaces in $\mathcal{L}$. Then the third subspace L_3 provides a linear isomorphism:

$$\varphi_{\mathrm{L}_1} : \mathrm{L}_2 \longrightarrow \mathrm{L}_3.$$

We claim that $\varphi_{\mathrm{L}_1} = -i_3 \circ i_2^{-1}$. Indeed, for any $l \in \mathrm{L}$, let $l_k := i_k(l)$. Then there are three vectors in $\mathcal{L}$ which sum to zero:

$$(l_1, 0, 0) + (0, l_2, 0) + (0, 0, l_3) = 0 \quad \text{in } \mathcal{L}.$$

Now return to diagram (178), providing four subspaces in X. Then there are two isomorphisms:

$$\varphi_{\mathrm{B}_3} : \mathrm{B}_1 \longrightarrow \mathrm{B}_2, \quad \varphi_{\mathrm{B}_4} : \mathrm{B}_1 \longrightarrow \mathrm{B}_2.$$

Let $M_{\mathrm{B}_2} : \mathrm{B}_2 \longrightarrow \mathrm{B}_2$ be the *minus* counterclockwise monodromy in diagram (176), given by

$$\mathrm{B}_2 \longrightarrow \mathrm{W}_1 \longrightarrow \mathrm{B}_4 \longrightarrow \mathrm{W}_3 \longrightarrow \mathrm{B}_2.$$

Lemma 7.3.5 *One has*

$$\varphi_{\mathrm{B}_4} = (1 + M_{\mathrm{B}_2})\varphi_{\mathrm{B}_3}.$$

Proof If our local system is a rank one flat R-line bundle, it is proved by the calculation done in proof of Proposition 2.1.5. □

Changing {•-vertices} $\longleftrightarrow$ {∘-vertices} and reversing the arrows amounts to dualization of the diagrams. Precisely, let W_1, W_2, W_3, W_4 be the fibers of the local system over the black domains and B_1, B_3 the fibers over the white domains. Then the local system is described by the following diagram where all maps are isomorphisms:

$$
\begin{array}{ccccccc}
 & & & W_2 & & & \\
 & & \swarrow{\scriptstyle\sim} & & {\scriptstyle\sim}\searrow & & \\
W_1 & \xrightarrow{\sim} & B_1 & & B_3 & \xleftarrow{\sim} & W_3 \\
 & & \nwarrow{\scriptstyle\sim} & & {\scriptstyle\sim}\nearrow & & \\
 & & & W_4 & & &
\end{array}
\tag{181}
$$

Let us set

$$\mathcal{B}_1 := \mathrm{Ker}\Big(W_1 \oplus W_2 \oplus W_4 \longrightarrow B_1\Big), \quad \mathcal{B}_3 := \mathrm{Ker}\Big(W_2 \oplus W_3 \oplus W_4 \longrightarrow B_3\Big).$$

After the triangle moves at the two •-vertices we get a dg-sheaf described by the diagram:

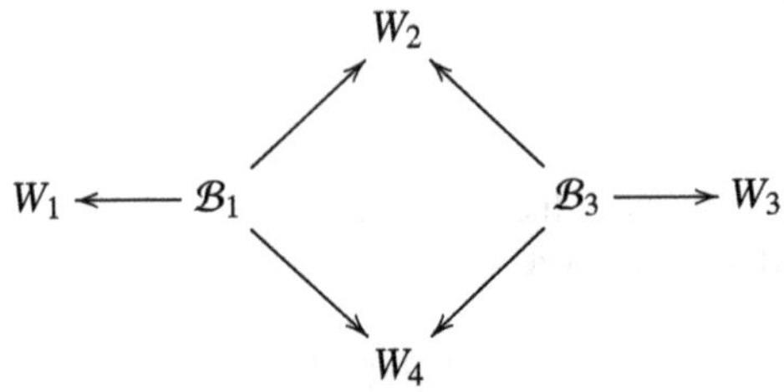

(182)

The composition of canonical isomorphisms $\mathcal{B}_1 \xrightarrow{\sim} W_2 \oplus W_4 \xleftarrow{\sim} \mathcal{B}_3$ provides an isomorphism $g : \mathcal{B}_1 \xrightarrow{\sim} \mathcal{B}_3$. Setting $Y := \mathcal{B}_1 = \mathcal{B}_3$, we get a diagram:

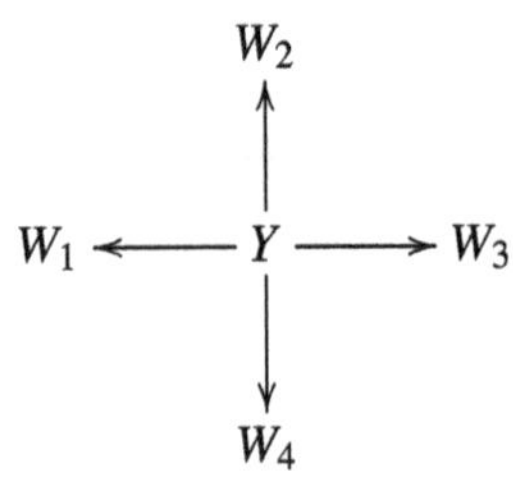

(183)

Conclusion Section 7.3 gives an alternative approach to main definitions *and* results of Sect. 2.1.

6. Twisted and Untwisted dg-Sheaves

Given a sheaf of categories C on a space X and a cocycle c_2 representing a class in $H^2(X, \mathrm{Cent}(C)^*)$, where $\mathrm{Cent}(C)^*$ is given by the invertible elements of C, one can make a new sheaf of categories $\widetilde{C}$, by twisting C by the 2-cocyle c_2. Precisely, take an open cover $\{U_i\}$ of X. Then an object of $\widetilde{C}$ over X is given by objects A_i of categories $C(U_i)$, related by isomorphisms $f_{ij} : A_i \longrightarrow A_j$ on $U_i \cap U_j$ which satisfy the following relations on the triple intersections:

$$f_{jk} \circ f_{ij} = c_2(U_i, U_j, U_k) f_{ik}.$$

Below we use twisting by the second Stiefel-Whitney class $\mathrm{sw}_2(\mathbb{S}) \in H^2(\mathbb{S}, \mathbb{Z}/2\mathbb{Z})$. We can twist both the Fukaya category and the category of constructible or dg-sheafs. The Fukaya category of objects with the microlocal support on $T^*_Z\mathbb{S}$ is naturally equivalent to the twisted constructible category on $\mathbb{S}$, and vice versa; see [16, Remark 5.2]. Note that $\mathrm{sw}_2(\mathbb{S}) = 0$ unless $\mathbb{S}$ is a compact non-oriented surface. So in our case the twisted category is equivalent to the original one. Yet the twisted version is somewhat better. It is the one providing positivity phenomenon in all formulas.

8 Non-Commutative Stacks of Framed Stokes Data

In Sect. 8 we adopt a faster pace of the exposition than in the rest of the chapter.

8.1 Stokes Data: Local and Global Pictures

1. Admissible Sheaves and Local Stokes Data
We start with two topological definitions.

Definition 8.1.1 A collection $\mathcal{L}$ of smooth oriented loops $L_1, \ldots, L_n$ in the punctured disc $\mathrm{D}^* \subset \mathbb{C}^*$ is called *ideal* if it enjoys the following properties:

1. The (self-)intersection $L_i \cap L_j$ of any two loops is transversal.
2. Each loop L_i rotates to the right, that is, the argument function φ on D^* strictly decreases as we go along the loop following its orientation. We coorient the loops out of the puncture.

Cooriented loops L_i are the connected components of the Legendrian link in $\mathrm{ST}(\mathrm{D}^*)$ provided by $\mathcal{L}$.

We refer to $\mathcal{L}$ as an *ideal Legendrian link*, although it is only a projection of a Legendrian.

Lemma 8.1.2 *An ideal Legendrian link on the punctured disc* D^* *is the same thing as an element of the cyclic envelope* $[\mathrm{Br}_n^+]$ *of the braid semigroup,* $n \geq 0$.

Proof Cut the collection of curves $\mathcal{L}$ by a generic ray r from the origin. Scanning the $\mathcal{L}$ by rotating the ray to the right, we get a sequence of permutations and hence an element of the braid semigroup Br_n^+, given by their product. The $n = 0$ case is the empty link. This construction can evidently be inverted. Scanning stating from a different ray produces the same element of the cyclic braid semigroup. □

Consider the subcategory $\mathcal{C}_0(\mathrm{D}^*, \mathcal{L})$ of the category $\mathcal{C}(\mathrm{D}^*, \mathcal{L})$ of admissible dg-sheaves on D^* for an ideal Legendrian $\mathcal{L}$ consisting of admissible dg-sheaves which are acyclic near the puncture. By Proposition 8.1.5, they are automatically concentrated in the degree zero, so we refer to them just as admissible sheaves. We denote by $[\mathcal{L}]$ the homotopy class of an ideal Legendrian $\mathcal{L}$ on D^* under Legendrian isotopies keeping Legendrians ideal and disjunct, called admissible isotopies. Categories assigned to admissibly isotopic Legendrians $\mathcal{L}$ are equivalent by Theorems 7.2.1, 7.2.2. So the category $\mathcal{C}_0(\mathrm{D}^*, \mathcal{L})$ is determined by the admissible isotopy class $[\mathcal{L}]$.

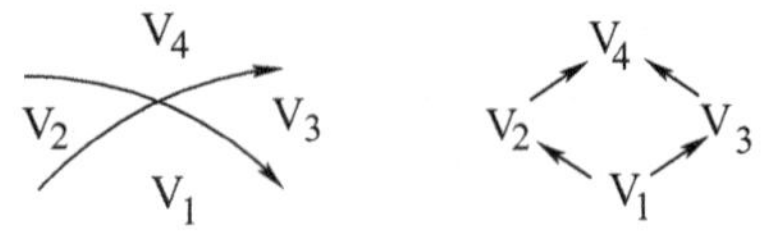

Fig. 57 An admissible sheaf near a crossing point

Definition 8.1.3 An ideal Legendrian $\mathcal{L}$ in the punctured disc $\mathrm{D}^* \subset \mathbb{C}^*$ is of *Stokes type* if:

- The intersection $L_i \cap L_j$ of any two different components of $\mathcal{L}$ is non-empty.

Definition 8.1.4 A local *Stokes data* is a local system $\mathcal{V}$ of vector spaces on the circle of rays in the tangent space at zero, with an increasing filtration $\mathcal{F}_\bullet$ everywhere except the fibers over a finite number of rays, called *Stokes rays*, related near them as follows.

Denote by $\mathcal{F}^{\pm}_\bullet(\mathcal{V})$ filtrations to the right and to the left of a Stokes ray r. Then $\mathcal{F}^+_i(\mathcal{V}) = \mathcal{F}^-_i(\mathcal{V})$ if $i \neq a$, and the subquotient $\mathrm{gr}_{[a+1,a]}\mathcal{F}(\mathcal{V})$ is the direct sum of its subspaces $\mathrm{gr}_a\mathcal{F}^{\pm}(\mathcal{V})$:

$$\mathrm{gr}_{[a+1,a]}\mathcal{F}(\mathcal{V}) := \mathcal{F}_{a+1}(\mathcal{V})/\mathcal{F}_{a-1}(\mathcal{V}) = \mathrm{gr}_a\mathcal{F}^+(\mathcal{V}) \oplus \mathrm{gr}_a\mathcal{F}^-(\mathcal{V}). \tag{184}$$

Proposition 8.1.5 *The following categories are canonically equivalent:*

1. *The category of local Stokes data from Definition 8.1.4*
2. *The direct sum of categories* $\bigoplus_{[\mathcal{L}]}\mathcal{C}_0(\mathrm{D}^*, \mathcal{L})$ *over all homotopy types* $[\mathcal{L}]$ *of Stokes Legendrians*

Proof Before we proceed to the proof, let us state the following simple Lemma.

Lemma 8.1.6 *An admissible sheaf near a simple crossing point is determined by a commutative diagram of four vector spaces* V_i *on the right of Fig. 57, where all arrows are injective, and the map*

$$V_2/V_1 \oplus V_3/V_1 \xrightarrow{\sim} V_4 \tag{185}$$

is an isomorphism.

Now let us prove the proposition.

2) $\longrightarrow$ 1). The circle S^1 of rays minus the Stokes rays is a union of a finite number of intervals:

$$S^1 - \{\text{Stokes rays}\} = \mathrm{I}_1 \cup \ldots \cup \mathrm{I}_m.$$

Since the sheaf is zero near the puncture, the admissibility just means that on each of the intervals I_k we get a local system of filtered vector spaces. It follows from Lemma 8.1.6 that the admissibility implies that this local system extends across the Stokes rays, while the filtration is altered: the filtration in $\mathrm{gr}_{[a+1,a]}$ flips, and the rest of the filtration data is preserved.

1) $\longrightarrow$ 2). We start from a local Stokes data on S^1. For each ray $r \in \mathrm{I}_k$, we assign the associate graded quotients $\mathrm{gr}_i\mathcal{F}^+(\mathcal{V})$ to some points of the ray so that the index i grows when we go along the ray, and when the ray moves in the interval I_k, we get a collection of curves together with local systems of vector spaces on them for each sector between the Stokes rays. We erase the curves with the zero local systems on them.

Let us glue these curves with local systems, which we have so far in the sectors of the punctured disc determined by the Stokes rays. For each $i \neq a, a+1$, at a Stokes ray r, we glue the curves which carry the local systems gr_i and then glue the local systems on them using the local system structure on $\mathcal{V}$ near the ray. Finally, we glue the curve which carries gr_a on the one side of r with the one with gr_{a+1} on the other side and glue local systems on them via the isomorphisms provided by condition (184):

$$\mathrm{gr}_a\mathcal{F}^-(\mathcal{V}) \stackrel{\sim}{\longrightarrow} \mathrm{gr}_{a+1}\mathcal{F}^+(\mathcal{V}), \qquad \mathrm{gr}_{a+1}\mathcal{F}^-(\mathcal{V}) \stackrel{\sim}{\longrightarrow} \mathrm{gr}_a\mathcal{F}^+(\mathcal{V}). \tag{186}$$

We get a collection $\mathcal{L}$ of loops on the disc, which never turn back, with simple crossings at the Stokes rays. Each of the loops carries a local system of vector spaces.

The claim that we get an admissible sheaf follows from Lemma 8.1.6. Proposition 8.1.5 is proved. □

Remark One can relax condition (1) in Definition 8.1.3 allowing $m > 2$ curves intersect at a point, with different tangents. Let us describe the corresponding analog of condition (184) in Definition 8.1.4. Recall that a pair of length m flags $\mathcal{F}_\bullet$ and $\mathcal{G}_\bullet$ in a vector space V is complimentary if $V = \mathcal{F}_a \oplus \mathcal{G}_b$ if $a + b = m$.

Lemma 8.1.7 *An admissible sheaf near an m-fold crossing point with different tangents (see Fig. 58) is determined by a pair of vector spaces $V_0 \subset V$ and two length m complimentary flags in V/V_0.*

Proof Given such an admissible sheaf, going from the bottom to the top on the left of the crossing points gives one flag, and going from the right gives the other flag. □

Then there is an analog of Proposition 8.1.5 if instead of (184) we have the following condition: *The left and right flags in the subquotient associated with the crossing point are opposite to each other.*

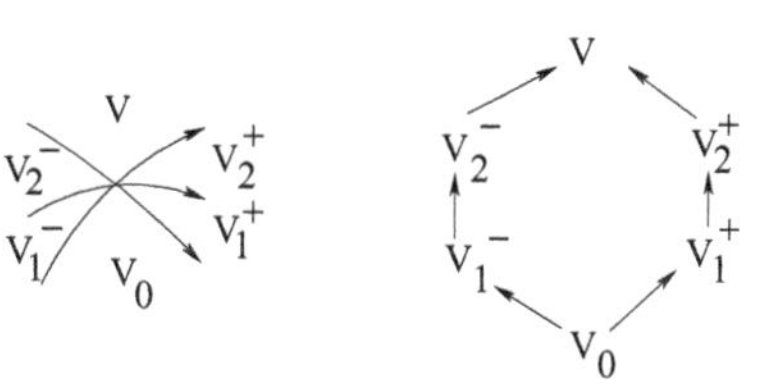

Fig. 58 An admissible sheaf near a triple crossing point: the following two flags in V/V_0 are complimentary to each other: $V_1^-/V_0 \subset V_2^-/V_0$ and $V_1^+/V_0 \subset V_2^+/V_0$

Our next goal is to explain how Stokes Legendrian links $\mathcal{L}$ and the categories $\mathcal{C}_0(\mathrm{D}^*, \mathcal{L})$ arise in complex geometry, as topological invariants of holomorphic vector bundles with connections on a punctured disc with any, possibly irregular, singularity at zero. For this we need to recall classification of connections on a formal punctured disc Spec $\mathbb{C}((z))$, due to Hukuhara et al. [30].

2. Formal Classification of Irregular Singularities

A Puiseux series is a Laurent series in $z^{1/d}$, $d \in \mathbb{Z}_{\geq 1}$. The field of all Puiseux series is the algebraic closure $\overline{\mathbb{C}((z))}$ of the field $\mathbb{C}((z))$ of Laurent series in z, with the Galois group $\widehat{\mathbb{Z}}$. The integral closure of the ring of Taylor series $\mathbb{C}[[z]]$ is the local ring $\mathcal{O}_{\overline{\mathbb{C}((z))}}$ of this field. It is given by Taylor series in $z^{1/d}$, $d \geq 1$.

Consider the quotient of the field of Puiseux series by its local ring:

$$\overline{\mathbb{C}((z))} \,/\, \mathcal{O}_{\overline{\mathbb{C}((z))}} = \lim_{d \geq 1} \mathbb{C}((z^{1/d})) \,/\, \mathbb{C}[[z^{1/d}]]. \tag{187}$$

Definition 8.1.8 The space $\mathcal{G}$ is the space of orbits of the Galois action on the elements $f(z)$ in (187).

So an element $f(z) \in \mathcal{G}$ is presented as a Puiseux polynomial in $z^{-1/d}$ for some positive integer d:

$$f(z) = \sum_{k \geq 1} \lambda_k z^{-k/d}, \quad \lambda_k \in \mathbb{C}, \tag{188}$$

where we identify $\{f(z)\} \sim \{f(e^{2\pi i/d}z)\}$ and assume d is the smallest such integer.

Consider the $d:1$ cover of the formal punctured disc:

$$\pi_d : \mathrm{Spec}\ \mathbb{C}((w)) \longrightarrow \mathrm{Spec}\ \mathbb{C}((z)), \quad z = w^d, \ d \geq 1.$$

Take a holonomic $\mathcal{D}$-module on this cover, given by the tensor product of the $\mathcal{D}$-module generated by $e^{f(z)}$, and a local system $\mathcal{V}_w$ on Spec $\mathbb{C}((w))$:

$$e^{f(z)} \otimes \mathcal{V}_w.$$

The $\mathcal{V}_w$ is determined by a local system on the circle of the rays in the tangent space to 0 in the w-plane. We consider the push-forward of this $\mathcal{D}$-module under the map π_d:

$$\pi_{d*}(e^{f(z)} \otimes \mathcal{V}_w).$$

Theorem 8.1.9 *The category of holonomic $\mathcal{D}$-modules on the formal punctured disc is decomposed into blocks, parametrized by the elements of $\mathcal{G}$. So any holonomic $\mathcal{D}$-modules on* Spec $\mathbb{C}((z))$ *is isomorphic to a direct sum of the standard $\mathcal{D}$-modules:*

$$\bigoplus_{j=1}^{m} \pi_{d_j*}(e^{f_j(z)} \otimes \mathcal{V}_{w_j}), \quad f_j(z) \in \mathcal{G}. \tag{189}$$

3. Legendrian Links $\mathcal{L}$ of Stokes Type and Local Stokes Data
Take a flat connection on a disc with a single singular point. By Theorem 8.1.9, its restriction to the formal punctured disc is a direct sum (189). So it determines a collection of different Puiseux polynomials with negative exponents:[21]

$$f_j(z) = \sum_{k \in \mathbb{Z}_{\geq 1}} \lambda_{j,k} z^{-k/d_j}, \quad \lambda_{j,k} \in \mathbb{C}, \;\; d_j \in \mathbb{Z}_{\geq 1}, \quad j = 1, \ldots, m. \tag{190}$$

They define multivalued functions $f_j(z)$. Restricting them to a certain concrete but sufficiently small radius ε circle $\{\varepsilon e^{i\theta}\}$ centered at 0, we get a collection of curves $\{L_j\}$ given parametrically by

$$Z_j(\theta) := R_j(\theta) \cdot e^{i\theta}, \quad R_j(\theta) := e^{\mathrm{Re} f_j(\varepsilon e^{i\theta})} \quad j = 1, \ldots, m. \tag{191}$$

Taking the 1-jet of the radial part $R_j(\theta)$ of the function $Z_j(\theta)$, we get a Legendrian knot L_j in the space $\mathcal{J}^1(S^1)$ of 1-jets of real functions on the circle. Explicitly,

$$\theta \longmapsto (\theta, R_j(\theta), dR_j(\theta)).$$

Their union is a Legendrian link $\mathcal{L}$ in $\mathcal{J}^1(S^1)$. We call it a Legendrian link of *Stokes type*. For a generic small ε it fits Definition 8.1.3.

The push-forward of the local systems $\mathcal{V}_{w_j}$ provides a local system $\mathcal{V}_j$ of vector spaces on the loop L_j.

The pairs $(L_j, \mathcal{V}_j)$ describe the microlocal support of an admissible sheaf.

The filtration by the order of growth provides the filtration at all but Stokes rays.

Therefore, we get a local Stokes data from Definition 8.1.4.

Legendrian links $\mathcal{L}$ assigned to different sufficiently small ε are isotopic, although considered as collections of curves they may look differently. Therefore, the corresponding categories are canonically equivalent by Theorem 7.2.2.

One can consider, as Deligne and Malgrange do [29, 30], the limiting case when $\varepsilon \to 0$. Then the corresponding Legendrian link may have non-transversal crossings, with different tangents. Note that Deligne-Malgrange Stokes rays are determined by the singularity, while our Stokes rays depend also on the choice of a small radius ε.

[21] An irreducible analytic flat connection on a punctured disc, restricted to the formal punctured disc, can be a sum of several $\mathcal{D}$-modules from different blocks.

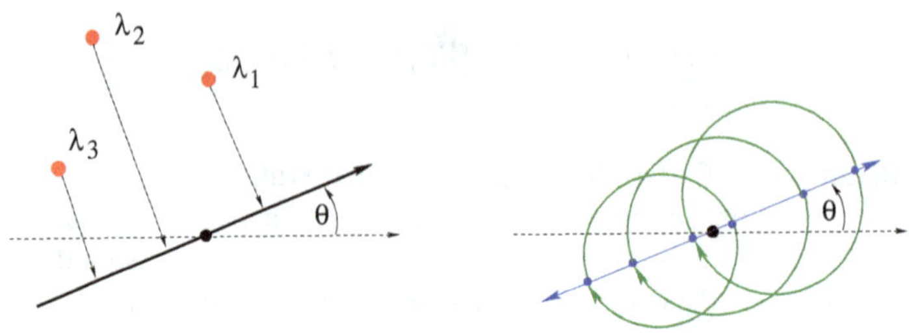

Fig. 59 On the left: projecting red complex points $\lambda_1, \lambda_2, \lambda_3$ onto a line. On the right: rotating their exponentials - blue points on the axis. The curve $\mathcal{L}$ is the union of three green loops obtained this way

4. Example
Given an $n \times n$ matrix A with distinct eigenvalues $\{\lambda_j\}$, consider a differential equation:

$$df = \frac{A}{z^2} f(z)dz. \tag{192}$$

Its solutions in the eigenbasis of A are linear combinations of functions $f_j(z) = e^{-\lambda_j/z}$. To draw the curve $\mathcal{L}$, we project orthogonally points $\{\lambda_j\}$ onto the oriented line obtained by rotating the real axis by an angle θ, exponentiate their coordinates on the line, and put them onto the positive ray in the direction θ. Rotating θ from 0 to 2π, these points move along the curve $\mathcal{L}$. Rotating $\theta \to \theta + \pi$, we change the order of the points on the ray to the opposite one; see Fig. 59. So any two of these loops intersect. Therefore, we get a collection of loops as in Definition 8.1.3.

The Stokes rays are the rays $\text{Re}(\lambda_i - \lambda_j) = 0$ in $\mathbb{C}$. The local Stokes data from Definition 8.1.4 is given by filtration of the space of solutions in a sector around a non-Stokes ray r by the growth of a solution. It satisfies the transversality assumption (184) if no three λ_is are on the same line.

In a more general situation, solutions of type $f_j(z) = z^\alpha e^{-\lambda_j/z}$ describe one-dimensional local systems on these circles with the monodromies $e^{2\pi i\alpha}$.

5. Legendrian Links of Analytic Stokes Type
Only some Stokes Legendrians describe the singular type of a connection on a disc. We call them Legendrian links of *analytic Stokes* type.

One can describe Legendrian links of analytic Stokes type as follows. Take a radius ε sphere $S^3 \subset \mathbb{C}^2$ and delete from it the circle in the $x = 0$ plane. The obtained contact threefold $S^3_{(0)}$ is identified with $\mathcal{J}^1(S^1)$. Let us consider a germ of an algebraic curve C in $\mathbb{C}^2$, reduced, but possibly reducible, with a singularity at $(0, 0)$. So it is the set of zeros of a product of different irreducible polynomials in two variables. Intersecting the curve C with $S^3_{(0)} \subset \mathbb{C}^2$ we get a Legendrian link. All Legendrian links of analytic Stokes type are obtained this way.

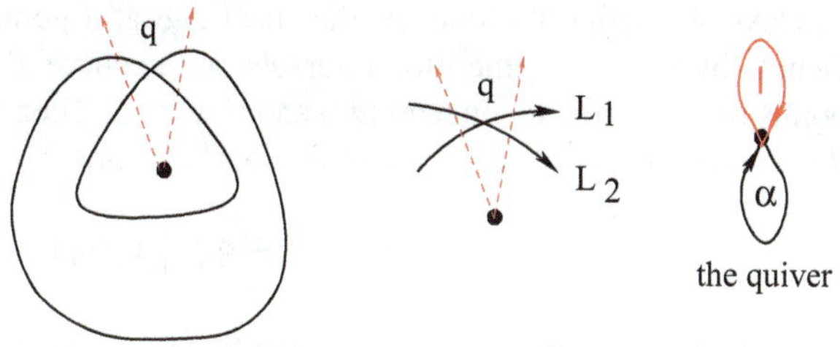

Fig. 60 At the crossing point q, the local arc L_1 is above the one L_2, producing a loop α of the quiver. The loop l on the diagram, shown red, corresponds to the loop itself

We now return back to arbitrary Stokes Legendrians and describe the corresponding category of admissible sheaves in terms of representations of a certain quiver.

6. Quiver Description

Consider an oriented quiver $Q_{\mathcal{L}}$ with the vertices v_i assigned to the oriented loops L_i of $\mathcal{L}$ and the following arrows assigned to the loops and crossing points of $\mathcal{L}$ (see Fig. 60):

1. For each vertex v, there is a simple loop $l_v : v \longrightarrow v$ at this vertex.
2. For each crossing $q \in L_i \cap L_j$, we have an arrow $\alpha_q : v_i \rightarrow v_j$ if L_i is above L_j to the right of q.

Theorem 8.1.10

1. *The category* $C(\mathrm{D}^*, \mathcal{L})$ *is abelian. It is equivalent to the category* $\mathcal{R}(Q^*_{\mathcal{L}})$ *of representations of the quiver* $Q_{\mathcal{L}}$*, with invertible arrows* l_v*.*
2. *If a Legendrian link* $\mathcal{L}$ *is of analytic Stokes type, that is, arises from a singular point of a holomorphic vector bundle with connection on* D^**, possibly irregular, then the stack* $\mathcal{M}_{\mathrm{D}^*}(\mathcal{L})$ *is equivalent to the Betti stack of given irregular type, also known as the stack of Stokes data.*

Remark Since the category of representations of the quiver $Q_{\mathcal{L}}$ is abelian, the category $C_{\mathrm{D}^*}(\mathcal{L})$ is abelian. The category $C(\mathrm{D}^*, \mathcal{L})$ assigned to a general ideal Legendrian is not abelian. Indeed, if $\mathcal{L}$ is isotopic to a collection of concentric circles, we get the category of filtered vector spaces.

Proof

(i) Pick a reference point p_i at each loop L_i of $\mathcal{L}$.

For each intersection/self-intersection point $q \in L_i \cap L_j$, let κ_1, κ_2 be little intervals on the loops L_i, L_j containing q. They are ordered so that the interval κ_1 goes above κ_2 to the right of q; see Fig. 60. Pick paths in the positive direction connecting the point q with reference points p_i and p_j.

Let us assign to a representation $\mathcal{R}$ of the quiver $Q_{\mathcal{L}}$ a sheaf $S_{\mathcal{R}}$ on the punctured disc D^*.

First, we define a local system $\mathcal{V}_i$ on each loop L_i. Recall the vertex v of the quiver assigned to the loop L_i. Put the vector space $\mathcal{R}(v)$ to the reference point p_i on L_i. Then the operator $\mathcal{R}(l_v) : \mathcal{R}(v) \longrightarrow \mathcal{R}(v)$ for the loop l_v describes the monodromy of the local system on the loop L_i in the positive direction.

Next, we define the fiber of the sheaf $\mathcal{S}_{\mathcal{R}}$ at a point $d \in \mathrm{D}^* - \{\text{Stokes rays}\}$. Denote by $\gamma_1, \ldots, \gamma_k$ the little intervals on the curve $\mathcal{L}$ containing the intersection points of the segment connecting 0 and d with $\mathcal{L}$. Then the fiber of the sheaf $\mathcal{S}_{\mathcal{R}}$ at d is the direct sum of the fibers of the local systems $\mathcal{V}_i$ at the intervals:

$$\mathcal{S}_{\mathcal{R}}(d) := \oplus_{i=1}^{k} \mathcal{V}_i(\gamma_i).$$

We order the intervals away from 0 and filter the fiber by subspaces $\oplus_{i=1}^{p} \mathcal{V}_i(\gamma_i)$.

We extend the sheaf to $\mathrm{D}^* - \{\text{crossing points}\}$ in the obvious way. Note that the filtration above is defined only away from the Stokes rays. Let us use the maps $\mathcal{R}(\alpha_q)$ assigned to the arrows α_q of the quiver to extend the sheaf to the crossing points.

Denote by $\mathcal{V}_{\kappa_i}$ the fiber of the local system on $\mathcal{L}$ over the interval κ_i, $i = 1, 2$. The stalk $\mathcal{S}_{\mathcal{R}}(q)$ of the sheaf at a crossing point q is mapped by the restriction map to the stalks $\mathcal{S}_{\mathcal{R}}(q_\pm)$ on the right/left of q. These maps are isomorphisms, since the sheaf does not have microlocal support in those directions. So an extension of the sheaf to the point q provides isomorphisms:

$$\mathcal{V}_{<q} \oplus \mathcal{V}_{\kappa_1} \oplus \mathcal{V}_{\kappa_2} =: \mathcal{S}_{\mathcal{R}}(q_-) \stackrel{\sim}{\longleftarrow} \mathcal{S}_{\mathcal{R}}(q) \stackrel{\sim}{\longrightarrow} \mathcal{S}_{\mathcal{R}}(q_+) := \mathcal{V}_{<q} \oplus \mathcal{V}_{\kappa_1} \oplus \mathcal{V}_{\kappa_2}.$$

So to extend the sheaf to the point q one just needs to define a linear automorphism of $\mathcal{V}_{<q} \oplus \mathcal{V}_{\kappa_1} \oplus \mathcal{V}_{\kappa_2}$, which is the identity on $\mathcal{V}_{<q}$ and preserves $\mathcal{V}_{\kappa_1} \oplus \mathcal{V}_{\kappa_2}$. We define the latter by setting

$$(v_1, v_2) \longrightarrow (v_1, \mathcal{R}(\alpha_q)(v_1) + v_2). \qquad \mathcal{R}(\alpha_q) : V_{\kappa_1} \longrightarrow V_{\kappa_2}. \tag{193}$$

Then it also preserving the filtration $V_{\kappa_1} \subset V_{\kappa_1} \oplus V_{\kappa_2}$, as required in our situation.

This functor is an equivalence. This is a nontrivial result, and we do not present here a proof.

(ii) This is a reformulation of the known description [6, 30] of the analytic Stokes data in terms of the filtration of the space of solutions given by the order of growth and the way it changes when we cross Stokes rays. □

7. Global Stokes Data

Below S is an oriented topological surface, possibly with punctures.

Definition 8.1.11 A Legendrian $\mathcal{L}$ on S is *ideal / Stokes type*, if it admissibly isotopic to a disjoint union of Legendrians $\mathcal{L}_p$ from Definition 8.1.1 / 8.1.3 near punctures p on S:

$$\mathcal{L} = \cup_p \mathcal{L}_p, \quad \mathcal{L}_p \cap \mathcal{L}_q = \emptyset.$$

Definition 8.1.12 A *Stokes data* on S is a local system $\mathcal{V}$ of vector spaces on S, plus a local Stokes data at the specialization of $\mathcal{V}$ to each puncture.

The following proposition is an immediate consequence of Proposition 8.1.5.

Proposition 8.1.13 *Given an ideal Legendrian $\mathcal{L}$ on S, the category of Stokes data on S of the topological type $\mathcal{L}$ is equivalent to the category $\mathcal{C}_0(S; \mathcal{L})$ of admissible sheaves on S with the microlocal support on $\mathcal{L}$ and the zero section of T^*S, which vanish near the marked points.*

8. Non-Commutative Stacks of Stokes Data
Denote by $\mathcal{S}(\mathcal{L}, d)$ the stack of Stokes data of a given topological type $(\mathcal{L}, d)$ on S. The topological types are described by the pairs $(\mathcal{L}, d)$, where $\mathcal{L}$ is an ideal Legendrian link on $\mathrm{ST}(S)$ from Definition 8.1.11, modulo admissible Legendrian isotopies, equipped with local systems of vector spaces on the connected components $\mathcal{L}_\alpha$ of the Legendrian $\mathcal{L}$. Dimensions of local systems are encoded by the dimension vector $d = \{d_\alpha\}$. When $\mathcal{L}$ is empty, we recover the stack $\mathrm{Loc}_m(S)$.

Definition 8.1.14 The non-commutative stack of Stokes data $\mathcal{S}^{\mathrm{nc}}(\mathcal{L}, d)$ of the topological type $(\mathcal{L}, d)$ is the stack $\mathcal{M}_0(\mathcal{L}, d)$ of $\mathcal{L}$-admissible dg-sheaves of R-vector spaces on S vanishing near the punctures of S, whose microlocal support on $\mathcal{L}_\alpha$ is a local system on the loop $\mathcal{L}_\alpha$ of dimension d_α:

$$\mathcal{S}^{\mathrm{nc}}(\mathcal{L}, d) := \mathcal{M}_0(\mathcal{L}, d).$$

8.2 *The Riemann-Hilbert Correspondence and the Wild Mapping Class Group*

In Sect. 8.2 we work in the commutative setup.

1. The Classical Riemann-Hilbert Correspondence
Given a Riemann surface Σ, recall the de Rham stack $\mathcal{M}_{\mathrm{DR}}^{\mathrm{reg}}(m, \Sigma)$ of rank m vector bundles with meromorphic connections with regular singularities on Σ. Its Betti variant is the stack $\mathrm{Loc}_m(S)$ of local systems of m-dimensional vector spaces on the topological surface S underlying Σ.

The classical Riemann-Hilbert correspondence [5] identifies the associated complex analytic stacks:

$$\mathcal{RH}_{\Sigma}^{\mathrm{reg}} : \mathcal{M}_{\mathrm{DR}}^{\mathrm{reg}}(m, \Sigma)(\mathbb{C}) \xrightarrow{\sim} \mathrm{Loc}_m(S)(\mathbb{C}).$$

However, the algebraic structures of these stacks are different. The algebraic structure of $\mathcal{M}_{\mathrm{DR}}^{\mathrm{reg}}(m, \Sigma)$ depends on the complex structure on Σ. When Σ varies, we get the algebraic de Rham stack $\mathcal{M}_{\mathrm{DR}}^{\mathrm{reg}}(m)$ over the moduli space $\mathcal{M}_{g,n}$. Contrary to this, the algebraic structure of the Betti stack depends only on the topology of Σ. When Σ varies, we get a local system of the Betti stacks over $\mathcal{M}_{g,n}$. Equivalently, the mapping class group $\Gamma_S = \pi_1(\mathcal{M}_{g,n})$ of S acts by automorphisms of the stack $\mathrm{Loc}_m(S)$.

Summarizing, we get the canonical universal Riemann-Hilbert equivalence:

$$\begin{array}{ccc} \mathcal{M}^{\mathrm{reg}}_{\mathrm{DR}}(m)(\mathbb{C}) & \xrightarrow{\mathcal{RH}} & \mathcal{M}^{\mathrm{reg}}_{\mathrm{B}}(m)(\mathbb{C}) \\ & \searrow \quad \swarrow & \\ & \mathcal{M}_{g,n} & \end{array} \tag{194}$$

To generalize it to irregular singularities, we need the Betti side, given by moduli spaces of Stokes data.

2. The Riemann-Hilbert Correspondence
We start with a pair $(S, \mathcal{L})$, where $\mathcal{L} = \cup \mathcal{L}_p$ is a Legendrian link of analytic Stokes type on a surface S, considered up to admissible isotopies.

Definition 8.2.1 The moduli space $\mathcal{P}_{S,\mathcal{L}}$ parametrizes data $(\Sigma, \{f_j\})$, where Σ is a Riemann surface of the topological type S, and $\{f_j \in \mathcal{G}_p\}$ is a collection of distinct Puiseux polynomials[22] from Definition 8.1.8 near the punctures p on Σ, with a given admissible isotopy class of the corresponding Legendrian $\mathcal{L}$.

Forgetting Puiseux polynomials, we get a projection, where S is a genus g surface with n punctures:

$$\mathcal{P}_{S,\mathcal{L}} \longrightarrow \mathcal{M}_{g,n}. \tag{195}$$

Given a Legendrian link $\mathcal{L}$ of Stokes type on S, there are the following categories and stacks:

1. The category of admissible sheaves $\mathcal{C}(S, \mathcal{L})$ and the stack of its objects $\mathcal{M}(S, \mathcal{L})$.
2. The DeRham stack $\mathcal{M}_{\mathrm{DR}}(S, \mathcal{L})$ of holomorphic vector bundles with connection over a Riemann surface Σ of topological type S, with the given homotopy class of the Legendrian link $\mathcal{L}$ assigned to its formal equivalence class. It is fibered over the stack $\mathcal{P}_{S,\mathcal{L}}$:

$$\mathcal{M}_{\mathrm{DR}}(S, \mathcal{L}) \longrightarrow \mathcal{P}_{S,\mathcal{L}}. \tag{196}$$

3. The Betti stack $\mathcal{M}_{\mathrm{B}}(S, \mathcal{L})$, which is a local system of stacks $\mathcal{M}_{\mathrm{B}}(S, \mathcal{L})$ over $\mathcal{P}_{S,\mathcal{L}}$.

   Its fiber over a point $(\Sigma, \{f_j\}) \in \mathcal{P}_{S,\mathcal{L}}$ is the stack $\mathcal{M}(S, \mathcal{L})$ for the Legendrian link $\mathcal{L}$ of $\{f_j\}$.

The fact that it is indeed a well-defined local system of stacks, e.g., does not depend on the choice of radius ε in the assignment $\{f_j\} \longmapsto \mathcal{L}$, and locally constant over the base, is guaranteed by Theorem 7.2.2.

[22] The space $\mathcal{G}_p$ is defined via the local ring $\mathcal{O}_p$ of functions at a point $p \in \Sigma$. It does not depend on a local parameter z.

The wild Riemann-Hilbert correspondence [30] is a complex analytic equivalence of stacks over $\mathcal{P}_{S,\mathcal{L}}$:

$$\begin{array}{ccc} \mathcal{M}_{\rm DR}(S,\mathcal{L}) & \xrightarrow{\mathcal{RH}} & \mathcal{M}_{\rm B}(S,\mathcal{L}) \\ & \searrow \quad \swarrow & \\ & \mathcal{P}_{S,\mathcal{L}} & \end{array} \tag{197}$$

The algebraic structure of the fibers of the De Rham stack over $\mathcal{P}_{S,\mathcal{L}}$ varies, while for the Betti stack it does not. The Betti stack forms a local system of stacks over $\mathcal{P}_{S,\mathcal{L}}$. In particular, we can define the analog of isomonodromic deformation for irregular holomorphic connections.

3. The Wild Mapping Class Group

Comparing the classical Riemann-Hilbert correspondence (194) with the one (197), we see that the right analog of $\mathcal{M}_{g,n}$ is the moduli space $\mathcal{P}_{S,\mathcal{L}}$ and arrive at:

Definition 8.2.2 The *wild mapping class group* is the fundamental group $\Gamma_{S,\mathcal{L}} := \pi_1(\mathcal{P}_{S,\mathcal{L}})$.

The wild mapping class group $\Gamma_{S,\mathcal{L}}$ acts by automorphisms of the stack $\mathcal{M}(S,\mathcal{L})$.

The map (195) provides a surjection $\Gamma_{S,\mathcal{L}} \longrightarrow \Gamma_S$. The group $\Gamma_{S,\mathcal{L}}$ is an extension of the mapping class group Γ_S by the product over the punctures p of S of the local wild mapping class groups $\Gamma_{\mathrm{D}^*,\mathcal{L}_p}$:

$$1 \longrightarrow \prod_p \Gamma_{\mathrm{D}^*,\mathcal{L}_p} \longrightarrow \Gamma_{S,\mathcal{L}} \longrightarrow \Gamma_S \longrightarrow 1. \tag{198}$$

Example In example (192) in Sect. 8.1, the space $\mathcal{P}_{\mathrm{D}^*,\mathcal{L}}$ for the punctured disc is given by collections of distinct complex numbers $\{\lambda_j\}$. The wild mapping class group $\Gamma_{S,\mathcal{L}}$ is thc braid group Br_n.

Below we will discuss topological ramifications of this definition.

4. Three Flavors of Legendrian Mapping Class Groups

There are three flavors of the space of Legendrian links $\mathcal{L} \subset \mathrm{ST}^*(S)$ and Hamiltonian isotopies between them. Therefore, there are three flavors of Legendrian mapping class group, defined as the fundamental group of the connected component of $\mathcal{L}$.

1. The space of *all* Legendrians $\mathcal{L} \subset \mathrm{ST}^*(S)$ and Hamiltonian isotopies. The *Legendrian mapping class group* $\Gamma_{\mathrm{leg},S,\mathcal{L}}$ is the fundamental group of the connected component containing $\mathcal{L}$. By [24], it acts by autoequivalences of the triangulated/dg category of constructible sheaves on S.

2. Legendrians given by *smooth* loops in S and *admissible* Hamiltonian isotopies between them. The *admissible Legendrian mapping class group* $\Gamma_{\text{adm},S,\mathcal{L}}$ is the fundamental group of the component of $\mathcal{L}$. By Theorems 7.2.1–7.2.2, it acts by autoequivalences of the category $C(S, \mathcal{L})$ of admissible dg-sheaves.
3. The *ideal* Legendrians $\mathcal{L}$ which are, by the definition, admissibly isotopic to a union of local ideal Legendrians $\mathcal{L}_p$ near the punctures p of S, modulo admissible Hamiltonian isotopies. The *wild mapping class group* $\Gamma_{S,\mathcal{L}}$ is the fundamental group of the connected component of $\mathcal{L}$. It acts by preserving cluster structures of the moduli spaces of Stokes data; see Sect. 8.3.

These spaces are nested: (3) $\subset$ (2) $\subset$ (1). Therefore, there are maps of the corresponding groups:

$$\Gamma_{S,\mathcal{L}} \longrightarrow \Gamma_{\text{adm},S,\mathcal{L}} \longrightarrow \Gamma_{\text{leg},S,\mathcal{L}}. \tag{199}$$

5. A Combinatorial Description of the Wild Mapping Class Group
Given a Weyl group W, consider the following stratified space C_W. Its strata $\mathcal{S}_\beta$ are given by collections $\beta = \{w_1, \ldots, w_n\}$ of the cyclically ordered elements of W assigned to distinct points on a circle, so that the cyclic order of the points on the circle coincides with the one of β. So $\mathcal{S}_\beta$ is homeomorphic to $S^1 \times \mathbb{R}^{n-1}$.

If $l(w_i w_{i+1}) = l(w_i) + l(w_{i+1})$, the stratum $\mathcal{S}_{w_1,\ldots,w_i w_{i+1},\ldots,w_n}$ lies on the boundary of $\mathcal{S}_{w_1,\ldots,w_i,w_{i+1},\ldots w_n}$. Denote by C_β the connected component of the cell complex C_W containing the strata $\mathcal{S}_w$. The space C_β is determined by the image $[\beta]$ of the element β in the cyclic envelope of the braid semigroup Br^+_W assigned to W, discussed in Sect. 8.4.

Let $W = S_N$. Then $[\beta]$ determines a cyclic braid. Denote the corresponding Legendrian by $\mathcal{L}_{[\beta]}$.

Proposition 8.2.3 *The local wild mapping class group $\Gamma_{\mathcal{L}_{[\beta]}}$ is the fundamental group of C_β:*

$$\Gamma_{\mathcal{L}_{[\beta]}} := \pi_1(C_\beta).$$

8.3 Non-Commutative Cluster Structure of Spaces of Stokes Data

By a non-commutative space $\mathcal{X}$ we mean a functor from a category of skew fields to the category of sets which assigns to a skew field R a set $\mathcal{X}(R)$. We say that $\mathcal{X}$ is defined over $\mathrm{Spec}(\mathbb{Q})$ if we consider all skew fields R of characteristic zero.

By a non-commutative stack $\mathcal{X}$ we just mean the quotient of a non-commutative space $\mathcal{Y}$, equipped with an action of the group $\mathrm{GL}_N(R)$, by the action of this group. So for any skew field R we have a set $\mathcal{X}(R) := \mathcal{Y}(R)/\mathrm{GL}_N(R)$.

We say, e.g., in Theorem 8.3.2, that a non-commutative space $\mathcal{X}$ admits a structure of a *non-commutative cluster Poisson variety* if it is birationally isomorphic to a cluster Poisson variety associated to a bipartite ribbon graph Γ, as discussed in paragraph 3 of Sect. 1.3 and in Sect. 2.2.

This means that there exists a non-empty collection of bipartite ribbon graphs $\{\Gamma\}$ such that any two of the graphs are related by a sequence of two by two moves, for each Γ from the collection there is a birational isomorphism

$$\alpha : \mathrm{Loc}_1(\Gamma) \longrightarrow \mathcal{X},$$

and for any pair of the graphs related by a two by two move $\mu : \Gamma \longrightarrow \Gamma'$ there is a commutative diagram:

$$\begin{array}{ccc} \mathrm{Loc}_1(\Gamma) & \xrightarrow{\alpha} & \mathcal{X} \\ \downarrow{\scriptstyle \mu_*} & & \downarrow{\scriptstyle =} \\ \mathrm{Loc}_1(\Gamma') & \xrightarrow{\alpha'} & \mathcal{X} \end{array} \tag{200}$$

where the birational isomorphism μ_* was described in Sect. 2.1.

Next, let $\mathrm{Mod}_{\mathcal{X}}$ be a discrete group acting by birational automorphisms of $\mathcal{X}$. We say that $\mathcal{X}$ admits a structure of a *non-commutative* $\mathrm{Mod}_{\mathcal{X}}$*-equivariant cluster Poisson variety* if for any $g \in \mathrm{Mod}_{\mathcal{X}}$ there is a bipartite ribbon graph Γ' together with a birational isomorphism:

$$\alpha' : \mathrm{Loc}_1(\Gamma') \longrightarrow \mathcal{X},$$

and a composition of two by two moves $\mu_g : \Gamma' \longrightarrow \Gamma'$, which induces a birational non-commutative cluster Poisson transformation $\mu_{g*} : \mathrm{Loc}_1(\Gamma) \longrightarrow \mathrm{Loc}_1(\Gamma')$ providing a commutative diagram:

$$\begin{array}{ccc} \mathrm{Loc}_1(\Gamma)(R) & \xrightarrow{\alpha} & \mathcal{X} \\ \downarrow{\scriptstyle \mu_{g*}} & & \downarrow{\scriptstyle g} \\ \mathrm{Loc}_1(\Gamma')(R) & \xrightarrow{\alpha'} & \mathcal{X} \end{array} \tag{201}$$

1. Non-Commutative Stacks of Framed Stokes Data
Just like in the regular case, to quantize the stack of Stokes data $\mathcal{S}(\mathcal{L}, d)$, we introduce a larger stack $\mathcal{X}(\mathcal{L}, d)$ of *framed* Stokes data. Namely, we add a framing, defined as a complete filtration of the local system $\mathcal{L}_\alpha$ on each loop α of the Legendrian $\mathcal{L}$, invariant under the monodromy around the loop.

We interpret a filtration on a d_j-dimensional local system on the loop α_j of the Legendrian $\mathcal{L}$ as follows. We replace the loop α_j by a collection of d_j loops $\alpha_j^{(1)}, \ldots, \alpha_j^{(d_j)}$ obtained by expanding the loop α_j out of the puncture by

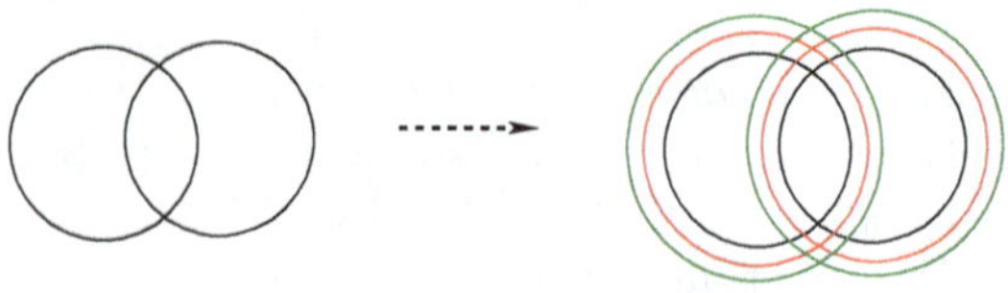

Fig. 61 Inflating a Legendrian link

$\varepsilon, 2\varepsilon, \ldots, d_j\varepsilon$; see Fig. 61. We equip each loop $\alpha_j^{(p)}$ with the local system given by gr^p of the filtration on the loop α_j. So get a new link $\mathcal{L}_d$ supporting one-dimensional local systems.

Set $\mathbf{1} := (1, \ldots, 1)$. Consider the stack $\mathcal{M}_0(\mathcal{L}_d, \mathbf{1})$ of $\mathcal{L}_d$-admissible dg-sheaves on S, vanishing near the punctures, whose microlocal support is described by the union of the zero section and arbitrary 1-dimensional local systems on the Legendrian $\mathcal{L}_d$. The next observation generalizes Lemma 1.4.2.

Lemma 8.3.1 *The non-commutative stack $\mathcal{M}_0(\mathcal{L}_d, \mathbf{1})$ is canonically equivalent to the one of framed Stokes data of type $(\mathcal{L}, d)$:*

$$\mathcal{M}_0(\mathcal{L}_d, \mathbf{1}) = \mathcal{X}(\mathcal{L}, d). \tag{202}$$

There is a projection given by assigning the microlocal support local system on the Legendrian $\mathcal{L}_d$:

$$\mu : \mathcal{M}_0(\mathcal{L}_d, \mathbf{1}) \longrightarrow \mathrm{Loc}_1(S^1)^{|d|}, \qquad |d| := \sum_\alpha d_\alpha. \tag{203}$$

The product of the symmetric groups $\prod_\alpha S_{d_\alpha}$ acts by birational automorphisms of $\mathcal{X}(\mathcal{L}, d)$. To define it, note that given a generic p-dimensional R-local system on a circle, there are $p!$ complete filtrations of the local system. Indeed, a generic linear automorphism L of an p-dimensional vector space V_p is decomposed uniquely into a direct sum of L-invariant one-dimensional subspaces. The filtrations are parametrized by orderings of these subspaces. Therefore, the symmetric group S_p acts on the filtrations.

This just means that the canonical projection, obtained by forgetting the filtrations

$$p : \mathcal{X}(\mathcal{L}, d) \longrightarrow \mathrm{Loc}(\mathcal{L}, d), \tag{204}$$

is a Galois cover at the generic point with the Galois group $\prod_\alpha S_{d_\alpha}$.

Theorem 8.3.2 *For any decorated surface $\mathbb{S}$ and any topological data $(\mathcal{L}, d)$ where $\mathcal{L}$ is an ideal Legendrian on $\mathbb{S}$, the stack $\mathcal{X}(\mathcal{L}, d)$ of framed non-commutative Stokes data has a non-commutative cluster Poisson structure equivariant under the wild cluster mapping class group $\Gamma_{\mathcal{L},d}$.*

In the commutative case, the subalgebra $\mu^*\mathcal{O}(\mathbb{G}_m^{|d|})$ (see (203)) is the Poisson center of $\mathcal{O}(\mathcal{X}(\mathcal{L}, d))$.

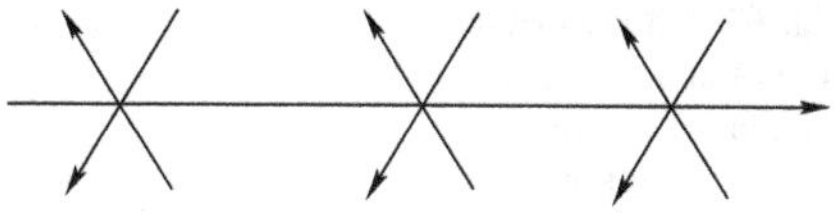

Fig. 62 A triple crossing diagram

We prove Theorem 8.3.2 assuming that in the case when $\mathbb{S} = S^2 - \{\infty\}$ the Legendrian $\mathcal{L}$ is not very degenerate; see Theorem 8.3.4 for the precise statement.

2. Idea of the Proof of Theorem 8.3.2

Lemma 8.3.1 tells that any stack of framed Stokes data has a canonical realization as a stack of admissible sheaves with the following two features:

(i) Admissible sheaves vanish near the punctures.
(ii) Microlocal support local systems on the Legendrian $\mathcal{L}_d$ are *one-dimensional.*

We show that the Legendrian link $\mathcal{L}_d$ has an admissible deformation to (a slight modification of) the web $\mathcal{Z}_\Gamma$ of zig-zag strands for a certain bipartite graph Γ on S. By the invariance of the stack of admissible dg-shaves under admissible deformations, $\mathcal{M}(\mathcal{L}_d, \mathbf{1}) = \mathcal{M}(\mathcal{Z}_\Gamma, \mathbf{1})$. By Lemma 1.4.1, the stack $\mathrm{Loc}_1(\Gamma)$ of 1-dimensional local systems on Γ is realized as the open substack defined by the condition of vanishing on mixed domains: $\mathrm{Loc}_1(\Gamma) \subset \mathcal{M}(\mathcal{Z}_\Gamma, \mathbf{1})$. Since vanishing on mixed domains implies vanishing near the punctures, we get an open embedding:

$$\mathrm{Loc}_1(\Gamma) \subset \mathcal{M}_0(\mathcal{L}_d, \mathbf{1}) = \mathcal{X}(\mathcal{L}, d).$$

The last equivalence is given by Lemma 1.4.1. This way we get an open cluster Poisson torus assigned to the graph Γ. The tori assigned to different graphs Γ provide the cluster atlas. The elements of the cluster wild mapping class group transform such a graph Γ to another one Γ', related to Γ by two by two moves. Therefore, we get a cluster Poisson atlas equivariant under the wild mapping class group.

Before we start the proof, let us recall the following.

3. Bipartite Graphs and Triple Crossing Diagrams

Let us recall triple crossing diagrams [37] (Fig. 62).

Definition 8.3.3 A *triple crossing diagram* is a collection $\mathcal{I}$ of smooth oriented strands on an oriented decorated surface $\mathbb{S}$ such that:

1. Each intersection point of $\mathcal{I}$ is a triple point.
2. Strand orientations are consistent with an orientation of each component of $\mathbb{S} - \mathcal{I}$.
3. Each strand either is a loop or ends on the boundary of $\mathbb{S}$. Endpoints of strands are disjunct.

See counterexamples to condition (2) on Fig. 63. Note that condition (2) implies the following:

- For each intersection point, passing strands are oriented as on Fig. 62.

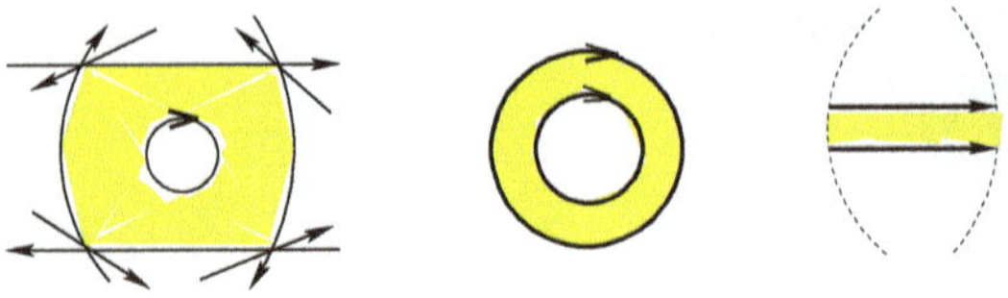

Fig. 63 Strand orientations are not consistent with any orientations of internal (yellow) components

- Going along a strand of $\mathcal{I}$, the orientations of the intersecting strands alternate, as on Fig. 62.

Any bipartite graph can be transformed to a bipartite graph with 3-valent •-vertices obtained by splitting the •-vertices of valency > 3 by adding edges and introducing 2-valent ∘-vertices in the middle of these edges. A bipartite ribbon graph Γ with 3-valent •-vertices gives rise to a triple crossing diagram on the associated surface S_Γ, provided by its zig-zag strands [20, Section 2.6]. Namely, we first push zig-zag strands out of the graph and then shrink the triangles near the •-vertices to points; see Fig. 48. Then all original double crossings collide into triple crossings corresponding to the •-vertices.

This construction can be inverted. Any triple crossing diagram $\mathcal{I}$ on $\mathbb{S}$ has a canonical up to isotopy resolution defined as follows [20, Section 2.3]. For each crossing, move slightly one of the strands, inflating the crossing into a little triangle, whose strands are oriented consistently with the orientation of $\mathbb{S}$—counterclockwise—getting a web $\mathcal{I}'$. There are three types of components of $\mathbb{S} - \mathcal{I}'$ (see Fig. 48):

(i) *Black* components, whose boundary strands are oriented clockwise
(ii) *White* components, whose boundary strands are oriented counterclockwise
(iii) *Mixed* components, whose boundary strands orientations alternate

We shrink all black (respectively white) components to black (respectively white) vertices and connect two vertices by an edge if they are separated by a crossing of the web $\mathcal{I}'$. We get a bipartite graph $\Gamma_{\mathcal{I}}$.

4. Ideal Legendrians → Triple Crossing Diagrams → Non-Commutative Cluster Poisson Varieties
Take an ideal Legendrian $\mathcal{L}$ on a punctured surface S, whose local components $\mathcal{L}_p$ are ideal Legendrians near punctures p. So each $\mathcal{L}_p$ satisfies conditions of Definition 8.1.1, different $\mathcal{L}_p$ are disjunct, and $\mathcal{L} = \cup_p \mathcal{L}_p$. The number of crossings of a generic ray r from p with $\mathcal{L}_p$ is the *local rank* of $\mathcal{L}_p$ at p. An ideal Legendrian $\mathcal{L}$ is of *rank m* if for each local Legendrian $\mathcal{L}_p$ its local rank at p is equal to m.

Theorem 8.3.4 *Any rank m ideal Legendrian $\mathcal{L}$ on a punctured surface S different from $S^2 - \{\infty\}$, or a rank m ideal Legendrian on $S^2 - \{\infty\}$ corresponding to a cyclic closure of the braid, $w_0 x w_0 y$, for any $x, y \in \mathrm{Br}_m$, can be admissibly isotoped to a triple crossing diagram with the following property:*

The black and white components are either contractible or homeomorphic to an annulus.

Fig. 64 Left: the link $\mathcal{L}_3$ describing the stack of rank 3 local systems on $\mathbb{C}^*$ with complete filtrations near the punctures. Right: admissible deformed link providing an open locus identified with $\mathbb{G}_m^3$

The union of all annuli is a disjoint union of punctured discs around (some of the) punctures.

Before we proceed to the proof of Theorem 8.3.4, let us discuss its main implications.

The triple crossing diagrams $\mathcal{L}$ which appear in our story have black and white components which are either contractible or annuli. Furthermore, let us assume that the union of all annuli is given by a union of discs around the punctures. Assuming these conditions, consider the category $C(S; \mathcal{L}, \mathbf{1})$ of $\mathcal{L}$-admissible dg-sheaves such that the local systems of the microlocal support supported on the components of $\mathcal{L}$ are 1-dimensional. Denote by $\mathcal{M}_0(S; \mathcal{L}, \mathbf{1})$ the substack of the stack of objects $\mathcal{M}(S; \mathcal{L}, \mathbf{1})$ of the category $C(S; \mathcal{L}, \mathbf{1})$ given by $\mathcal{L}$-admissible dg-sheaves which are zero on the mixed components.

Lemma 8.3.5 *The stack $\mathcal{M}_0(S; \mathcal{L}, \mathbf{1})$ is a torus.*

Proof According to our assumption, the annuli surrounding a given puncture: □

Corollary 8.3.6 *Any rank m ideal Legendrian $\mathcal{L}$ on S gives rise to a Zariski open torus $\mathcal{M}_0(S; \mathcal{L}, \mathbf{1})$ in the non-commutative moduli stack $\mathcal{M}(S; \mathcal{L}, \mathbf{1})$ of $\mathcal{L}$-admissible dg-sheaves with one-dimensional local systems on the components of $\mathcal{L}$.*

We will proceed to the proof of Theorem 8.3.4, after discussing two examples.

5. The Simplest Example
The stack of m-dimensional local systems on $S^2 - \{0, \infty\}$ with complete filtrations near the punctures is just the stack of admissible sheaves vanishing near the punctures for the oriented web $\mathcal{L}_m$, illustrated for $m = 3$ on the left of Fig. 64. An admissible deformation of $\mathcal{L}_m$ shown on the right describes the following constructible sheaves. Denote by $\mathcal{A}_1, \mathcal{A}_2, \ldots, \mathcal{A}_{2m+1}$ the open components of $S^2 - \mathcal{L}_m$, counted from 0. Then the sheaves are zero on $\mathcal{A}_{2k+1}$ and are 1-dimensional local systems on $\overline{\mathcal{A}}_{2k}$. Their moduli space is $(\mathbb{G}_m)^m$, parametrizing the monodromies.

6. Another Simple Example
Let $S^2 - \{0, \infty\}$ be a sphere a regular puncture at ∞ and an irregular one at 0. The Legendrian $\mathcal{L}_0$ goes clockwise around 0. It bounds a domain $\mathcal{D}_{\mathcal{L}_0}$. Precisely, take a point $x \in \mathrm{D}^*$. Let r_x be the ray from p to x. Then $x \in \mathcal{D}_{\mathcal{L}_0}$ if and only if it is in the convex hull of $\{r_x \cap \mathcal{L}_0\}$.

Fig. 65 Left: a triple crossing diagram from two intersecting clockwise-oriented loops around the puncture. Right: the corresponding bipartite graph

Fig. 66 Inflating an edge to a pair of bigons. Inflating edges of a triangle to pairs of colored bigons

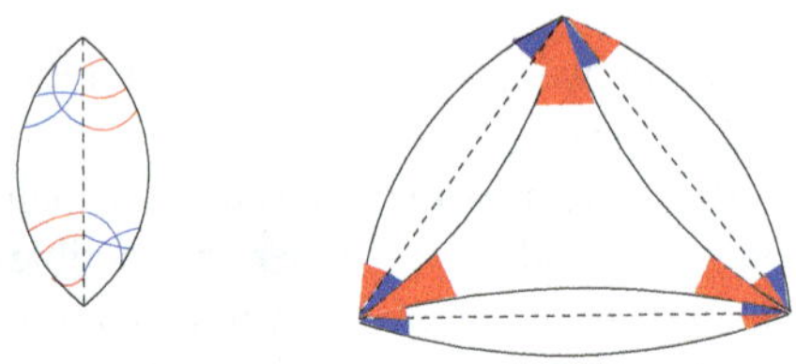

Now we follow [9]. Call the self-intersection points of $\mathcal{L}_0$ *vertices*. Take a component $\mathcal{V}$ of $\mathcal{D}_{\mathcal{L}_0} - \mathcal{L}_0$. Connect the rightmost vertex on the boundary of $\mathcal{V}$ with the leftmost one by a counterclockwise path $\gamma_{\mathcal{V}} \subset \mathcal{V}$, transversal to $\mathcal{L}_0$. Take the union of these paths, a counterclockwise arc β near ∞, and $\mathcal{L}_0$ (see Fig. 65):

$$\mathcal{I} := \cup_{\mathcal{V}} \gamma_{\mathcal{V}} \cup \beta \cup \mathcal{L}_0. \tag{205}$$

It is a triple point diagram. Its number of counterclockwise arcs is the same as the local rank of $\mathcal{L}_0$ at 0.

7. Proof of Theorem 8.3.4
Take a triangulation $\mathcal{T}$ of S with vertices at the punctures. Inflate each side of $\mathcal{T}$ to a pair of thin *bigons*; see Fig. 66. Then S - {the bigons} = a union of *internal triangles.* Performing admissible isotopies, we can assume that $\mathcal{L}$ crosses internal triangles t and bigons b as follows:

1. For each vertex v of t: the $\mathcal{L}$ induces m non-intersecting, clockwise (red) arcs around v.
2. For each bigon b: near one of its angles, called the *red angle*, $\mathcal{L}$ induces m non-intersecting clockwise arcs near its vertex. We put no conditions on the restriction of $\mathcal{L}$ to the other one, called the *blue angle*.
3. For each vertex v of each edge of $\mathcal{T}$: just one of the bigons at this edge has a red angle at v.

Let b be a bigon. Denote by $\mathcal{L}_b$ the Legendrian in the bigon.

Lemma 8.3.7 *One can admissibly isotope the Legendrian $\mathcal{L}_b$ to a Legendrian $\mathcal{L}'_b$ so that:*

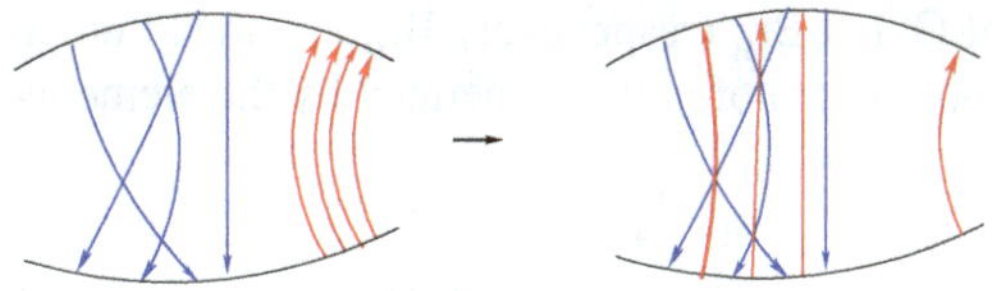

Fig. 67 From concentric red and blue arcs near the vertices of a bigon, to a triple crossing diagram

Fig. 68 A triple crossing diagram assigned to an m-triangulation of the triangle t, for $m = 1, 2$

1. *On each side of the bigon b, the endpoints of the red and blue strands of* $\mathcal{L}'_b$ *alternate.*
2. *The strands of* $\mathcal{L}'_b$ *form a triple crossing diagram.*

Proof Move the red arcs towards the blue angle of b so that condition (1) holds. Then there is a unique up to isotopy way to deform them to a triple crossing diagram, with the minimal number of crossings.

Namely, forget the red arcs. Then for each component of the complement to the blue arcs pick the leftmost vertex/boundary point, and the similar rightmost one, and connect them by the unique up to isotopy red arc inside the component; see Fig. 67. We get a triple crossing diagram. □

Lemma 8.3.7 provides triple crossing diagrams in the bigons. For each internal triangle t (see Fig. 66), there is a standard triple crossing diagram provided by the m-triangulation of the triangle, which induces on the boundary the given alternating sequence of intersection points of $\mathcal{L}'_h$ provided by Lemma 8.3.7. Indeed, take the m-triangulation of a triangle (see Fig. 68), and move each of its strands towards the parallel side of the triangle, so that the centers of looking down small triangles become the triple crossing points. Then the endpoints alternate, and we can match them with the ones on the bigons $\mathcal{L}'_b$. We denote this triple crossing diagram by $\mathcal{L}'_t$. Take the union of these triple point diagrams:

$$\mathcal{L}_{\mathcal{T}} := \cup_t \mathcal{L}'_t \bigcup \cup_b \mathcal{L}'_b. \tag{206}$$

Theorem 8.3.4 is proved.

8.4 Cluster Structure of Stacks of G-Stokes Data on Surfaces

1. The Cyclic Envelope $\mathrm{Br}^+_{\mathfrak{g}}$ of the Braid Semigroup

Let G be a split semi-simple algebraic group over $\mathbb{Q}$. Denote by I the set of positive simple roots and by C_{ij} the Cartan matrix of the root system of the Lie algebra $\mathfrak{g}$

of G. Let $\text{Br}_{\mathfrak{g}}$ (respectively $\text{Br}^+_{\mathfrak{g}}$) be the braid group (respectively semigroup) of the root system of $\mathfrak{g}$. It is generated by the elements s_i, $i \in \text{I}$, subject to the relations

$$\begin{array}{ll} s_i s_j = s_j s_i & \text{if } C_{ij} = C_{ji} = 0, \\ s_i s_j s_i = s_j s_i s_j & \text{if } C_{ij} = C_{ji} = -1, \\ s_i s_j s_i s_j = s_j s_i s_j s_i & \text{if } C_{ij} = 2C_{ji} = -2, \\ s_i s_j s_i s_j s_i s_j = s_j s_i s_j s_i s_j s_i & \text{if } C_{ij} = 3C_{ji} = -3. \end{array} \tag{207}$$

The Weyl group W is the quotient of the group $\text{Br}_{\mathfrak{g}}$ by the relations $s_i^2 = 1$. Denote by $l(*)$ the length function on W. There is a set theoretic section $\mu : W \to \text{Br}^+_{\mathfrak{g}}$, such that $\mu(s_i) = s_i$ and $\mu(ss') = \mu(s)\mu(s')$ if $l(ss') = l(s) + l(s')$. Abusing notation, we denote by s_i the elements $\mu(s_i) \in \text{Br}^+_{\mathfrak{g}}$.

Denote by $[\text{Br}^+_{\mathfrak{g}}]$ the set of coinvariants of the cyclic shift $s_{j_1} \dots s_{j_{n-1}} s_{j_n} \longmapsto s_{j_n} s_{j_1} \dots s_{j_{n-1}}$ on $\text{Br}^+_{\mathfrak{g}}$, called the *cyclic envelope* of the braid semigroup.

We denote by $[s_{j_1} \dots s_{j_n}] \in [\text{Br}^+_{\mathfrak{g}}]$ the cyclic element corresponding to $s_{j_1} \dots s_{j_n} \in \text{Br}^+_{\mathfrak{g}}$.

Given a pair of flags B, B′, we define $w(\text{B}, \text{B}') \in W$ via the isomorphism $\text{G}\backslash(\mathcal{B} \times \mathcal{B}) = W$.

Take a defining relation $s_{l_1} s_{l_2} \dots s_{l_m} = s_{r_1} s_{r_2} \dots s_{r_m}$ in (207). For any flags $(\text{B}_0, \dots, \text{B}_m)$ with $w(\text{B}_{a-1}, \text{B}_a) = s_{j_a}$, where $a = 1, \dots, m$, there is a unique collection of flags $(\text{B}'_0, \dots, \text{B}'_m)$ with $w(\text{B}'_{a-1}, \text{B}'_a) = s_{r_a}$.

Given a reduced decomposition $s_{j_1} s_{j_2} \dots s_{j_k}$ of a cyclic element $\beta \in [\text{Br}^+_{\mathfrak{g}}]$, there is the moduli space of G-orbits of flags $(\text{B}_1, \dots, \text{B}_k)$ such that $w(\text{B}_{i-1}, \text{B}_i) = s_{j_i}$, where $i \in \mathbb{Z}/k\mathbb{Z}$. The moduli spaces assigned to different reduced decompositions of β are canonically isomorphic.

2. The Moduli Space $\mathcal{X}(\text{G}, \beta; S)$

Let $\mathcal{B}$ be the flag variety of G. A G-local system $\mathcal{L}$ on S gives rise to the associated local system of flags $\mathcal{L}_{\mathcal{B}} := \mathcal{L} \times_{\text{G}} \mathcal{B}$. Denote by S^1_p the circle of rays at $T_p S$. Given a reduced decomposition $s_{j_1} \dots s_{j_k}$ of the cyclic word $\beta_p \in [\text{Br}^+_{\mathfrak{g}}]$, take distinct points $x_1, \dots, x_k \in S^1_p$ going cyclically around the circle, and put the elements s_{j_i} at the points x_i (Fig. 69).

Definition 8.4.1 Let $\beta := \{\beta_p\}$ be a collection of elements of $[\text{Br}^+_{\mathfrak{g}}]$ assigned to the punctures $\{p\}$ of a surface S. A *β_p-framing at a puncture p* on a G-local system $\mathcal{L}$ on S is given by flat sections $\varphi = (\varphi_1, \dots, \varphi_k)$ of the local system of flags $\mathcal{L}_{\mathcal{B}}$ over intervals $x_1 x_2, \dots, x_k x_1$ on S^1_p so that $w(\varphi_i, \varphi_{i+1}) = s_{\alpha_i}$:

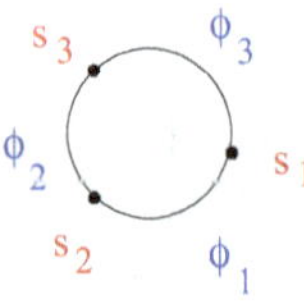

Fig. 69 Cyclic framing of type $[s_1 s_2 s_3]$

$$\varphi_1 \xrightarrow{s_{j_1}} \varphi_2 \xrightarrow{s_{j_2}} \dots \varphi_k \xrightarrow{s_{j_k}} \varphi_1.$$

A β-framing on $\mathcal{L}$ is a collection of local framings $\{\beta_p\}$ at the punctures of S.
The moduli space $\mathcal{X}(\mathrm{G}, \beta; S)$ parametrizes β-framed G-local systems $\mathcal{L}$ on S.

The stack $\mathcal{X}(\mathrm{G}, \beta; S)$ depends on the elements $\beta_p \in [\mathrm{Br}_{\mathfrak{g}}^+]$, rather than their reduced decompositions.

3. Examples of Stacks $\mathcal{X}(\mathrm{G}, \beta; S)$

1. If $\mathrm{G} = \mathrm{GL}_m$, a local *framed* Stokes data is just a β-framing at the puncture on an m-dimensional local system of vector spaces on the punctured disc for an element $\beta \in [\mathrm{Br}_{\mathfrak{gl}_m}^+]$.
2. If $\beta_p = e$ is the unit, then a local β_p-framing at the puncture p is just a monodromy invariant flag near the puncture.
3. If all $\beta_p = e$ are the unit elements, then $\mathcal{X}(\mathrm{G}, \beta; S)$ is the stack $\mathcal{X}(\mathrm{G}, S)$ [7] parametrizing G-local system $\mathcal{L}$ on S with a flat section of the local system of flags $\mathcal{L}_{\mathcal{B}}$ near each puncture of S.
4. Let $\mathbb{S}$ be a decorated surface with k boundary components, with $b_1, \dots, b_k$ special points on them.

   If $\beta_p = [\mu(w_0) \cdot \dots \cdot \mu(w_0)] \in [\mathrm{Br}_{\mathfrak{g}}^+]$ is the cyclic product of b_p elements $\mu(w_0)$ assigned to the longest element $w_0 \in W$, then $\mathcal{X}(\mathrm{G}, \beta; S)$ is the stack $\mathcal{X}_{\mathrm{G},\mathbb{S}}$ for from [7], as we show in Sect. 8.5.

   The cluster Poisson structure on moduli spaces (1) and (2) was defined for $\mathrm{G} = \mathrm{PGL}_m$ in [7], for classical groups, and G_2 by case-by case study in [27, 28] and in the general case in [22].
5. If $S = S^2 - \{0, \infty\}$ is a punctured disc, and $\beta_0 = e$, $\beta_\infty = \beta$, then the stack $\mathcal{X}(\mathrm{G}, \{e, \beta\}; \mathrm{D}^*)$ carries a cluster Poisson variety structure assigned to the cyclic braid semigroup element β in [10].
6. If $S = S^2 - \{0, \infty\}$, then the stack $\mathcal{X}(\mathrm{G}, \beta_0, \beta_p; \mathrm{D}^*)$ contains as a Zariski open part the double Bott-Samelson variety [32] assigned to a pair cyclic braid semigroup elements $\beta_0, \beta_\infty \in [\mathrm{Br}_{\mathfrak{g}}^+]$.

Theorem 8.4.2 *Any stack of framed* G*-Stokes data on a surface S is the stack $\mathcal{X}(\mathrm{G}, \beta; S)$ for some β.*

Proof This can be deduced from the formal classification in the G-case just as in Sect. 8.1. □

Theorem 8.4.3 *Let S be a punctured surface, and $S \neq S^2 - \{\infty\}$. Then the stack $\mathcal{X}(\mathrm{G}, \beta; S)$ has a cluster Poisson structure, equivariant under the wild mapping class group. Therefore, the stack $\mathcal{X}(\mathrm{G}, \beta; S)$ admits a cluster quantization, equivariant under the wild mapping class group.*

For each puncture p, the braid group $\mathrm{Br}_{\mathfrak{g}}$ *acts by automorphisms of the stack $\mathcal{X}(\mathrm{G}, \beta; S)$, preserving the cluster Poisson structure.*

Proof Assume first that $S \neq S^2 - \{0, \infty\}$. Then there is an ideal triangulation $\mathcal{T}$ of S.

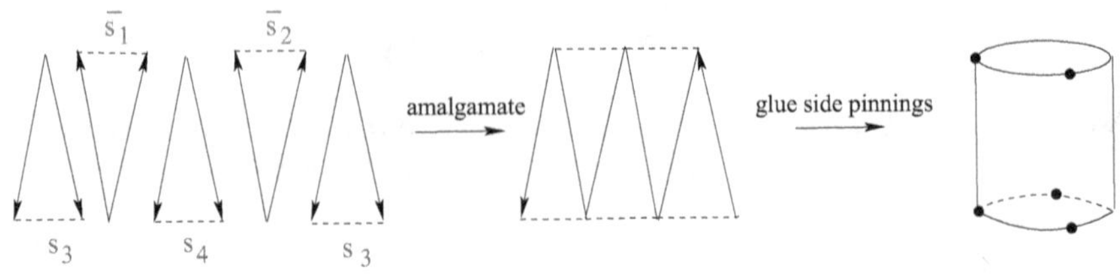

Fig. 70 Getting a cluster chart with $\beta_0 = s_1 s_2$, $\beta_\infty = s_3 s_4 s_3$ on a cylinder by the amalgamation

Take an ideal triangulation $\mathcal{T}$ of S, and inflate each of its edges E to a pair of thin bigons b_E, as we did in Sect. 8.3. For each puncture p, put all the points $x_1, \ldots, x_{b_p}$ inside of the arc on the circle S^1_p located inside of the left bigon at one of the edges sharing p. Then the moduli space $\mathcal{X}(\mathrm{G}, \beta; S)$ is birationally equivalent to the result of amalgamation of the moduli spaces $\mathcal{P}_{\mathrm{G},t}$ from [22] assigned to the triangles t of the triangulation and the moduli spaces $\mathcal{P}_{\mathrm{G},r_E}$ assigned to the rectangles corresponding to the bigons, that is, the edges of $\mathcal{T}$. The independence of the choice of $\mathcal{T}$ follows from [10] and [22].

Now let $S = S^2 - \{0, \infty\}$ be a cylinder. Recall the decorated flag variety $\mathcal{A} := \mathrm{G/U}$. Recall the elementary moduli space $\mathcal{P}_{\mathrm{G},s}$ paramterizing triples $(\mathrm{B}_1, \mathrm{A}_2, \mathrm{A}_3) \in \mathrm{G}\backslash(\mathcal{B} \times \mathcal{A} \times \mathcal{A})$, such that the pairs $(\mathrm{B}_1, \mathrm{A}_2)$ and $(\mathrm{B}_1, \mathrm{A}_2)$ are generic, and $w(\mathrm{B}_2, \mathrm{B}_3) = s$ for the pair of flags $(\mathrm{B}_2, \mathrm{B}_3)$ underlying the decorated flags $(\mathrm{A}_2, \mathrm{A}_3)$. We picture such a triple by a wedge with the flag B_1 at the vertex and two arrows pointing out to the decorated flags at the vertices on the short edge. The space $\mathcal{P}_{\mathrm{G},s}$ carries a cluster Poisson structure [22, Section 6]. The same space with the opposite Poisson structure is denoted by $\mathcal{P}_{\mathrm{G},\bar{s}}$.

Take reduced decompositions for the braid semigroup elements $\beta_0 = s_{i_1} \ldots s_{i_a}$ and $\beta_\infty = \bar{s}_{j_1} \ldots \bar{s}_{j_b}$. Amalgamate the corresponding moduli spaces $\mathcal{P}_{\mathrm{G},s_i}$ and $\mathcal{P}_{\mathrm{G},\bar{s}_j}$ so that the order of the spaces $\mathcal{P}_{\mathrm{G},s_i}$ (respectively $\mathcal{P}_{\mathrm{G},\bar{s}_j}$) follows a reduced decomposition β_0 (respectively β_∞); see Fig. 70. Then glue the left and the right pairs (A, B) for the resulting moduli spaces if they oriented the same way. Otherwise alter one of them by $(\mathrm{B}_1, \mathrm{A}_2) \longrightarrow \overline{w}_0(\mathrm{B}_1, \mathrm{A}_2)$ where $\overline{w}_0$ is the canonical lift of w_0 to G and then glue.

This way we get a Zariski open part of $\mathcal{X}(\mathrm{G}, \beta; S)$. Therefore, the result follows from [22].

The second claim follows from this by applying the cluster Poisson quantization machine [10].

Let us define the action a_p of the braid group $\mathrm{Br}_\mathfrak{g}$ on the stack $\mathcal{X}(\mathrm{G}, \beta; S)$. Pick a generator s_i of the braid group $\mathrm{Br}_\mathfrak{g}$. Take a cyclic word $\mathbf{b} = [s_{j_1} \cdot \ldots \cdot s_{j_n}]$ assigned to the cyclic collection of points $x_1, \ldots, x_n$ on the circle $S^1_p = T_pS - \{0\}/\mathbb{R}_{>0}$. We write this schematically as $[s_{j_1}|x_1, \cdot \ldots \cdot s_{j_n}|x_n]$. Then we find its representative of $[\mathbf{b}]$ starting from s_1 and shift the cyclic word by one step to the right:

$$[s_1|x_1, s_{j_2}|x_2, \cdot \ldots \cdot s_{j_n}|x_n] \longmapsto [s_1|x_2, s_{j_2}|x_3, \cdot \ldots \cdot s_{j_n}|x_1]$$

which starts from s_i. □

4. Cluster Quantization of $\mathcal{M}_{\rm DR}(\mathrm{G}, \beta; S)$
To quantize the De Rham stack $\mathcal{M}_{\rm DR}(\mathrm{G}, \beta; S)(\mathbb{C})$, we apply the G-analog of the universal Riemann-Hilbert equivalence (197) and then develop the $\Gamma_{\mathrm{G},\beta;S}$-equivariant quantization of the Betti stack $\mathcal{M}_{\rm B}(\mathrm{G}, \beta; S)$.

To do this, we quantize first the bigger stack $\mathcal{X}(\mathrm{G}, \beta; S)$ of framed Stokes data. There is a finite group $W_{\mathrm{G},S,\beta}$ which acts birationally on $\mathcal{X}(\mathrm{G}, \beta; S)$ by altering filtrations. Forgetting the filtrations we get a map, which is a finite Galois cover at the generic point:

$$p : \mathcal{X}(\mathrm{G}, \beta; S) \longrightarrow \mathcal{M}_{\rm B}([S, \mathcal{L}]).$$

The space $\mathcal{X}(\mathrm{G}, \beta; S)$ has a $\Gamma_{\mathrm{G},\beta;S} \times W_{\mathrm{G},\beta;S}$-equivariant Poisson structure. Therefore, [10] gives its $\Gamma_{\mathrm{G},\beta;S} \times W_{\mathrm{G},\beta;S}$-equivariant cluster Poisson quantization, given by:

1. A q-deformed algebra of functions $\mathcal{O}_q(\mathcal{X}(\mathrm{G}, \beta; S))$.
2. A unitary projective representation of the group $\Gamma_{\mathrm{G},\beta;S} \times W_{\mathrm{G},\beta;S}$ in a Hilbert space $\mathcal{H}$.
3. A representation of the $*$-algebra given by the modular double

$$\mathcal{O}_q(\mathcal{X}(\mathrm{G}, \beta; S)) \otimes \mathcal{O}_{q^\vee}(\mathcal{X}(\mathrm{G}, \beta; S)), \quad q = e^{i\pi\hbar}, \ q^\vee = e^{i\pi/\hbar}. \tag{208}$$

by unbounded operators in the Hilbert space $\mathcal{H}$, intertwining the action of the group $\Gamma_{\mathrm{G},\beta;S} \times W_{\mathrm{G},\beta;S}$ on the $*$-algebra (208) with its projective unitary action in $\mathcal{H}$ from 2). Here $\hbar \in \mathbb{R}_{>0}$ or $|\hbar| = 1$.

The group $W_{\mathrm{G},\beta;S}$ acts by cluster Poisson transformations (the regular case is done in [22]). Set

$$\mathcal{O}_q(\mathcal{M}_{\rm B}(\mathrm{G}, \beta; S)) := \mathcal{O}_q(\mathcal{X}(\mathrm{G}, \beta; S))^{W_{\mathrm{G},\beta;S}}.$$

We consider the representation of the modular double of this subalgebra in the Hilbert space $\mathcal{H}$. It commutes with the unitary action of the group $W_{\mathrm{G},\beta;S}$ in $\mathcal{H}$.

8.5 More Examples

1. Non-Commutative Stacks $\mathcal{X}_m(\mathbb{S})$ and $\mathcal{A}_m(\mathbb{S})$ as Stacks of Stokes Data
Below $\mathbb{S}$ is a decorated surface with boundary. We assume that the local systems on the strands of the webs, provided by the microlocal supports of admissible sheaves, are one-dimensional.

We assign to each boundary component π of $\mathbb{S}$, with b_π special points, an ideal Legendrian link $\mathcal{L}_{\pi,m}$ in a little cylinder $C_\pi \subset \mathbb{S}$ containing π, obtained by

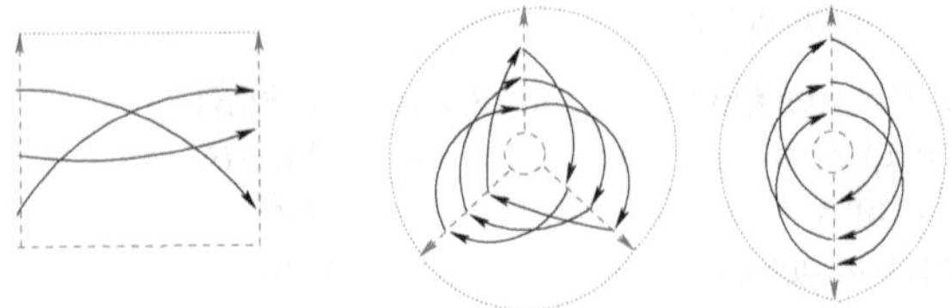

Fig. 71 A Legendrian realizing permutation $(1, 2, 3) \to (3, 2, 1)$ and examples of concatenation

concatenating b_π copies of the Legendrian with m strands on a rectangle realizing the longest permutation $w_0 : (1, 2, \dots, m) \longmapsto (m, \dots, 2, 1)$; see Fig. 71.

Let us assign to each puncture p on $\mathbb{S}$ a collection $\mathcal{L}_{p,m}$ of m small disjoint loops around the puncture.

Denote by $\mathcal{L}_{\mathbb{S},m}$ the union of these Legendrians:

$$\mathcal{L}_{\mathbb{S},m} := \bigcup_{\pi \in \pi_0(\partial\mathbb{S})} \mathcal{L}_{\pi,m} \cup \bigcup_{\text{punctures } p} \mathcal{L}_{p,m}. \tag{209}$$

Definition 8.5.1 The non-commutative stack $\mathcal{M}^{(1)}_{\mathcal{X}}(\mathbb{S}, \mathcal{L}_{\mathbb{S},m})$ parametrizes admissible sheaves on $\mathbb{S}$ from $C(\mathbb{S}, \mathcal{L}_{\mathbb{S},m})$, vanishing near marked points, whose microlocal support restricted to the Legendrian $\mathcal{L}_{\mathbb{S},m}$ is given by one-dimensional non-commutative local systems.

The stack $\mathcal{M}^{(1)}_{\mathcal{A}}(\mathbb{S}, \mathcal{L}_{\mathbb{S},m})$ parametrizes similar data with trivialized local systems on the Legendrian.

Theorem 8.5.2 *The stack $\mathcal{M}^{(1)}_{\mathcal{X}}(\mathbb{S}, \mathcal{L}_{\mathbb{S},m})$ is birationally isomorphic to the stack $\mathcal{X}_m(\mathbb{S})$. It has a non-commutative $\Gamma_{\mathbb{S},\mathcal{L}_{\mathbb{S},m}}$-equivariant cluster Poisson structure.*

The stack $\mathcal{M}^{(1)}_{\mathcal{A}}(\mathbb{S}, \mathcal{L}_{\mathbb{S},m})$ is birationally isomorphic to the stack $\mathcal{A}_m(\mathbb{S})$. It has a $\Gamma_{\mathbb{S},\mathcal{L}_{\mathbb{S},m}}$-equivariant non-commutative cluster $\mathcal{A}$-structure.

Proof The complement $\mathbb{S} - \mathcal{L}_{\mathbb{S},m}$ contains the unique internal component. It is homotopic to $\mathbb{S}$. Its boundary is a union of circles $\{S^1_p\}$ and $\{S^1_\pi\}$ parametrized by punctures p and boundary components π.

Restricting an admissible sheaf $\mathcal{S} \in \mathcal{M}^{(1)}(\mathbb{S}, \mathcal{L}_{\mathbb{S},m})$ to the internal component of $\mathbb{S} - \mathcal{L}_{\mathbb{S},m}$ we get an m-dimensional local system $\mathcal{V}$. For each puncture p, its restriction to the circle S^1_p has a complete filtration, provided by the Legendrian $\mathcal{L}_p$.

Given a boundary component π, consider b_π generic normal rays cutting the cylinder C_π into b_π rectangles (see Fig. 71), so that the link $\mathcal{L}_{\pi,m}$ is a concatenation of Legendrians in these rectangles realizing the longest permutation w_0. For each of the rays r, the admissible sheaf $\mathcal{S}$ provides a flag $\mathcal{F}_{\pi,r}$ in the fiber of $\mathcal{V}$ over $S^1_\pi \cap r$. The local system $\mathcal{V}$ with the flags on the circles $\{S^1_\pi\}$ and filtrations on the circles $\{S^1_p\}$ is exactly the data which determines a framed local system on $\mathbb{S}$ of dimension m.

Evidently, this construction can be inverted starting with a framed local system with the flags $\mathcal{F}_{\pi,r}$ in generic position. So we get a birational isomorphism. The $\mathcal{A}$-case is completely similar.

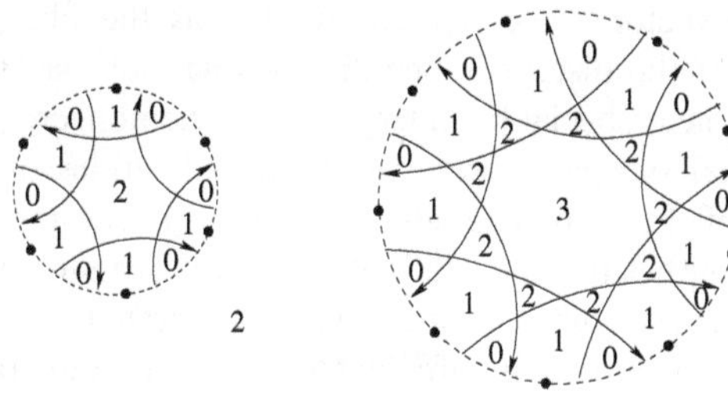

Fig. 72 The webs $\mathcal{W}_{2,5}$ and $\mathcal{W}_{3,9}$ on a disc describe configurations of 5 and 9 lines/vectors in non-commutative vector spaces of dimension 2 and 3, respectively, as stacks of admissible sheaves. The numbers tell dimensions of the fibers of admissible sheaves in each component of the complement to the web

The rest follows from the already established results about the spaces $\mathcal{X}_m(\mathbb{S})$ and $\mathcal{A}_m(\mathbb{S})$.

Alternatively, we can deform a Legendrian $\mathcal{L}_{\mathbb{S},m}$ to a web of zig-zags of a bipartite graph of type $\mathbb{S}$, say by using an ideal triangulation of $\mathbb{S}$ for a start. So we get a cluster torus of each of the two flavors. □

2. Non-Commutative Grassmannians as Stacks of Stokes Data

Consider a web $\mathcal{W}_{k,n}$ given by a collection of oriented arcs on a disc, cooriented away from the boundary, as shown on Fig. 72. We can describe it as follows. Take n special points on the boundary, and consider arcs which encircle $k-1$ subsequent points, so that for each point there is just one arc starting a bit to the right of a special points and ending a bit to the left of another special point.

Consider the stack $\mathrm{C}^{\circ}_{\mathcal{X}}(\mathcal{W}_{k,n})$ of admissible dg-sheaves with the following two properties:

(i) They vanish in the middle of each interval between the special points.
(ii) The local system on each arc describing the microlocal support is one-dimensional.

In fact these dg-sheaves are sheaves: they are concentrated in the degree zero.

The $\mathcal{A}$-flavor of this space is the stack $\mathrm{C}^{\circ}_{\mathcal{A}}(\mathcal{W}_{k,n})$ of admissible sheaves as above such that the one-dimensional local systems on each curve are trivialized.

Lemma 8.5.3

(i) The stack $\mathrm{C}^{\circ}_{\mathcal{X}}(\mathcal{W}_{k,n})$ describes cyclically ordered collections of n one-dimensional subspaces in a k-dimensional R-vector space such that any subsequent k of them generate the space. It has a non-commutative cluster Poisson variety structure.

(ii) The stack $\mathrm{C}^{\circ}_{\mathcal{A}}(\mathcal{W}_{k,n})$ describes cyclic collections of n vectors in a k-dimensional R-vector space such that any subsequent k of them generate the space. It has a cluster $\mathcal{A}$-variety structure.

Proof The ambient vector space is reconstructed as the fiber of an admissible sheaf in the center of the disc. The one-dimensional subspaces are the fibers of the admissible sheaf near special boundary points. The nonzero vectors generating these one-dimensional subspaces are given by trivializations of the local systems sitting on the arcs of the web. The admissibility at a crossing point is given by Lemma 8.1.6. This implies that the admissibility at the crossing points is equivalent to the condition that each k subsequent subspaces generate the space.

To get a cluster description, it is sufficient to find a bipartite graph whose zig-zag web is admissibly isotopic to a web $\mathcal{W}_{k,n}$. Then our general result implies that the torus $\mathrm{Loc}_1(\Gamma)$ sits inside of the stack we consider. This kind of zig-zag webs is the one considered by Postnikov [31]. □

9 Theorem 4.2.7 and the Steinberg Relation for Non-Commutative 2-Forms

In Sect. 9 we prove Theorem 4.2.7 and then observe that one of the key steps of the calculation can be viewed as an identity between non-commutative 2-forms generalizing the Steinberg relation. It would be very interesting to find similar identities for the non-commutative n-forms generalizing the Steinberg identity $d\log(z_1) \wedge \ldots \wedge d\log(1+x) \wedge \ldots \wedge d\log(x) \wedge \ldots \wedge d\log(z_{n-2}) = 0$.

9.1 The Proof of Theorem 4.2.7

Recall the transformation formulas, where $A_1 = a_1a_2a_3a_4$, etc.:

$$
\begin{aligned}
b_1 = (1+A_3^{-1})a_3 &= a_3(1+A_4^{-1}),\\
b_2 = (1+A_4)^{-1}a_4 &= a_4(1+A_1)^{-1},\\
b_3 = (1+A_1^{-1})a_1 &= a_1(1+A_2^{-1}),\\
b_4 = (1+A_2)^{-1}a_2 &= a_2(1+A_3)^{-1}.
\end{aligned}
\tag{210}
$$

Recall the identities:

$$
\begin{aligned}
&\{a,b\} = -\{b^{-1},a^{-1}\},\\
&\{b,c\} - \{ab,c\} + \{a,bc\} - \{a,b\} = 0,\\
&\{xy,y^{-1}\} = -\{x,y\}, \qquad \{x^{-1},xy\} = -\{x,y\}.
\end{aligned}
\tag{211}
$$

Lemma 9.1.1

$$\{b_2, b_3\} = \{a_2a_3a_4, (1 + A_1)^{-1}\} + \{a_2a_3, a_4\} + \{1 + A_1, a_4^{-1}\}. \tag{212}$$

Proof By the definition (145), we have

$$\{b_2, b_3\} = \{a_4(1 + A_1)^{-1}, (1 + A_1^{-1})a_1\}. \tag{213}$$

By the cocycle identity in (211) for the triple $(a_4, (1 + A_1)^{-1}, (1 + A_1^{-1})a_1)$,

$$\{b_2, b_3\} = \{(1+A_1)^{-1}, (1+A_1^{-1})a_1\} + \{a_4, A_1^{-1}a_1\} - \{a_4, (1+A_1)^{-1}\} \tag{214}$$

By the cocycle identity for the triple $((1 + A_1)^{-1}, 1 + A_1^{-1}, a_1)$ we can rewrite the first term and get

$$\begin{aligned}\{b_2, b_3\} =&\{A_1^{-1}, a_1\} - \{1 + A_1^{-1}, a_1\} + \{(1 + A_1)^{-1}, 1 + A_1^{-1}\}\\ &+ \{a_4, A_1^{-1}a_1\} - \{a_4, (1 + A_1)^{-1}\}.\end{aligned} \tag{215}$$

Writing the middle term as $-\{1 + A_1, A_1^{-1}\}$ by using the last identiy in (211), applying the cocycle identity for the triple $(1 + A_1, A_1^{-1}, a_1)$, and observing that $\{a_4, A_1^{-1}a_1\} = \{a_2a_3, a_4\}$ and $\{a_3, A_4^{-1}\} = -\{A_4, a_3^{-1}\} = \{a_4a_1a_2, a_3\}$ using the last line in (211), we get the claim. □

Lemma 9.1.2

$$-\{b_1, b_2\} =\{a_3A_4^{-1}, (1 + A_4)\} - \{a_4, a_1a_2\} - \{a_4^{-1}, 1 + A_4\}. \tag{216}$$

Proof By the definition (145), we have

$$\{b_2^{-1}, b_1^{-1}\} =\{a_4^{-1}(1 + A_4), (1 + A_4^{-1})^{-1}a_3^{-1}\}. \tag{217}$$

Using the cocycle identity for the triple $(a_4^{-1}(1 + A_4), (1 + A_4^{-1})^{-1}, a_3^{-1})$ we write this as

$$\begin{aligned}-\{b_1, b_2\} =& - \{(1 + A_4^{-1})^{-1}, a_3^{-1}\}+\{a_4^{-1}A_4, a_3^{-1}\}+\{a_4^{-1}(1 + A_4), (1 + A_4^{-1})^{-1}\}\\ =&\{a_3, 1 + A_4^{-1}\} - \{a_1a_2, a_3\} - \{1 + A_4^{-1}, (1 + A_4)^{-1}a_4\}.\end{aligned} \tag{218}$$

The cocycle identity in (211) allows to write the last term as

$$\{1 + A_4, (1 + A_4^{-1})^{-1}\} - \{a_4^{-1}, A_4\} - \{a_4^{-1}, 1 + A_4\}. \tag{219}$$

Since the first term here is equal to $\{A_4^{-1}, 1 + A_4\}$ and $\{a_3, A_4^{-1}a_4\} = \{a_1a_2, a_3\}$, we get

$$
\begin{aligned}
-\{b_1, b_2\} =& \{a_3, A_4^{-1}(1+A_4)\} - \{a_1a_2, a_3\} + \{A_4^{-1}, (1+A_4)\} \\
&- \{a_4, a_1a_2a_3\} - \{a_4^{-1}, 1+A_4\}.
\end{aligned} \tag{220}
$$

Using the cocycle relation for the triple $(a_3, A_4^{-1}, 1+A_4)$ and then the cocycle relation for (a_4, a_1a_2, a_3), we get the claim. □

Denote by Cycl_2 the operator on functions in $a_1, \ldots, a_4$ given by the identity + the cyclic shift by two:

$$
\mathrm{Cycl}_2\mathrm{F}(a_1, a_2, a_3, a_4) = \mathrm{F}(a_1, a_2, a_3, a_4) + \mathrm{F}(a_3, a_4, a_1, a_2).
$$

We split the sum to calculate as (A) + (B), where:

(A) The total contribution of the middle terms in the Lemmas is

$$
\begin{aligned}
&\mathrm{Cycle}_2\Big(-\{a_4, a_1a_2\} + \{a_2a_3, a_4\}\Big) \\
=& -\{a_4, a_1a_2\} + \{a_2a_3, a_4\} - \{a_2, a_3a_4\} + \{a_4a_1, a_2\} \\
=& \{a_1, a_2\} - \{a_2, a_3\} + \{a_3, a_4\} - \{a_4, a_1\}.
\end{aligned} \tag{221}
$$

The last equality uses cocycle identities for (a_4, a_1, a_2) and (a_2, a_3, a_4). We get exactly what we wanted.

(B) It is the rest:

$$
\begin{aligned}
\mathrm{Cycle}_2\Big(&\{1+A_1, a_4^{-1}\} + \{a_2a_3a_4, (1+A_1)^{-1}\} \\
&- \{a_4^{-1}, 1+A_4\} - \{(1+A_4)^{-1}, a_4a_1a_2\}\Big).
\end{aligned} \tag{222}
$$

This can be rewritten as

$$
\begin{aligned}
\mathrm{Cycle}_2\Big(&\Big(-dA_1a_4^{-1}da_4 + dA_1(a_2a_3a_4)^{-1}d(a_2a_3a_4)\Big)(1+A_1)^{-1} + \\
&\Big(da_4a_4^{-1}dA_4 - d(a_4a_1a_2)(a_4a_1a_2)^{-1}dA_4\Big)(1+A_4)^{-1}\Big).
\end{aligned} \tag{223}
$$

Expanding this, we get

$$
\begin{aligned}
&d(a_1a_2a_3a_4)(a_2a_3a_4)^{-1}d(a_2a_3)a_4(1+A_1)^{-1} \\
&+ d(a_3a_4a_1a_2)(a_4a_1a_2)^{-1}d(a_4a_1)a_2(1+A_3)^{-1} \\
&- a_4d(a_1a_2)(a_4a_1a_2)^{-1}dA_4(1+A_4)^{-1} \\
&- a_2d(a_3a_4)(a_2a_3a_4)^{-1}dA_2(1+A_2)^{-1}.
\end{aligned} \tag{224}
$$

Interchanging lines 2 and 3, writing $1 + A_4 = a_4(1 + A_1)a_4^{-1}$, and similarly for $1 + A_3$ and $1 + A_2$, we get

$$\begin{aligned} & d(a_1a_2a_3a_4)(a_2a_3a_4)^{-1}d(a_2a_3)a_4(1 + A_1)^{-1} \\ & - d(a_1a_2)(a_4a_1a_2)^{-1}d(a_4a_1a_2a_3)a_4(1 + A_1)^{-1} \\ & + (a_3a_4)^{-1}d(a_3a_4a_1a_2)(a_4a_1a_2)^{-1}d(a_4a_1)a_2a_3a_4(1 + A_1)^{-1} \\ & - (a_3a_4)^{-1}d(a_3a_4)(a_2a_3a_4)^{-1}d(a_2a_3a_4a_1)a_2a_3a_4(1 + A_1)^{-1}. \end{aligned} \tag{225}$$

Lemma 9.1.3 *Expression (225) is equal to zero.*

Proof Let us write the first two lines as

$$\begin{aligned} & \Big(d(a_1a_2)a_2^{-1}d(a_2a_3)a_4 + a_1a_2d(a_3a_4)(a_2a_3a_4)^{-1}d(a_2a_3)a_4 \\ & - d(a_1a_2)(a_4a_1a_2)^{-1}d(a_4a_1)a_2a_3a_4 - d(a_1a_2)a_2^{-1}d(a_2a_3)a_4\Big)(1 + A_1). \end{aligned} \tag{226}$$

The terms 1 and 4 cancel, so we get

$$\begin{aligned} (226) = \Big(& a_1a_2d(a_3a_4)(a_2a_3a_4)^{-1}d(a_2a_3)a_4 \\ & - d(a_1a_2)(a_4a_1a_2)^{-1}d(a_4a_1)a_2a_3a_4\Big)(1 + A_1). \end{aligned} \tag{227}$$

A similar calculation with the lines 3 and 4 in (225) gives

$$\begin{aligned} & \Big((a_3a_4)^{-1}d(a_3a_4)a_4^{-1}d(a_4a_1) + d(a_1a_2)(a_4a_1a_2)^{-1}d(a_4a_1) \\ & - (a_3a_4)^{-1}d(a_3a_4)(a_2a_3a_4)^{-1}d(a_2a_3)a_4a_1 \\ & (a_3a_4)^{-1}d(a_3a_4)a_4^{-1}d(a_4a_1)\Big)a_2a_3a_4(1 + A_1). \end{aligned} \tag{228}$$

The terms 1 and 4 cancel, so we get

$$\begin{aligned} (228) = \Big(& d(a_1a_2)(a_4a_1a_2)^{-1}d(a_4a_1) \\ & - (a_3a_4)^{-1}d(a_3a_4)(a_2a_3a_4)^{-1}d(a_2a_3)a_4a_1\Big)a_2a_3a_4(1 + A_1). \end{aligned} \tag{229}$$

Therefore, (226) + (228) is equal to

$$\Big(a_1a_2d(a_3a_4)(a_2a_3a_4)^{-1}d(a_2a_3)a_4$$
$$-\,d(a_1a_2)(a_4a_1a_2)^{-1}d(a_4a_1)a_2a_3a_4$$
$$+\,d(a_1a_2)(a_4a_1a_2)^{-1}d(a_4a_1)a_2a_3a_4 \tag{230}$$
$$-\,(a_3a_4)^{-1}d(a_3a_4)(a_2a_3a_4)^{-1}d(a_2a_3)a_4A_1\Big)(1+A_1).$$

Terms 2 and 3 evidently cancel. Terms 1 and 4 cancel if we permute A_1 and $1+A_1$ in term 4. □

Therefore, sum (B) is equal to zero. Theorem 4.2.7 is proved.

9.2 An Analog of the Steinberg Relation for Non-Commutative 2-Forms

Note that

$$\{x, y^{-1}\} = -x^{-1}dxy^{-1}dy, \quad \{x^{-1}, y\} = -dxx^{-1}dyy^{-1}.$$

Theorem 9.2.1 *Let us introduce the notation:*

$$\langle a|b\rangle := \{1+ab, b^{-1}\} - (b^{-1}, 1+ba\}. \tag{231}$$

Then one has

$$\text{Cycle}_2\Big(\langle a_1a_2a_3|a_4\rangle - \langle a_1|a_2a_3a_4\rangle\Big) = 0. \tag{232}$$

Proof We can write (222) as

$$\text{Cycle}_2\Big(\{1+A_1, a_4^{-1}\} - \{1+A_1, (a_2a_3a_4)^{-1}\}$$
$$-\{a_4^{-1}, 1+A_4\} + \{(a_4a_1a_2)^{-1}, 1+A_4\}\Big). \tag{233}$$

This coincides with (232). But we already proved that the sum (222) is equal to zero. □

Remark If variables a_i commute, then $\langle a|b\rangle = -2\cdot d\log(1+ab)\wedge d\log b$, and identity (232) reads as

$$d\log(1+a_1a_2a_3a_4)\wedge d\log(a_1a_2a_3a_4) = 0.$$

However, for a general matrix X we have $d\log(1+X) \wedge d\log X \neq 0$. The identity (232) is the analog of the Steinberg relation $d\log(1+z) \wedge d\log(z) = 0$ for the non-commutative 2-forms. It is interesting that it involves four non-commutative variables a_1, a_2, a_3, a_4 rather than one.

It would be interesting to find a similar non-commutative generalization of the de Rham realization of the defining relation in the Milnor K-group: $d\log(1+z) \wedge d\log(z) \wedge d\log(z_1) \wedge \ldots \wedge d\log(z_n) = 0$.

Acknowledgements We are grateful to A. Berenstein and V. Retakh for useful discussions during the Summer 2011 in IHES, when the project was started, and all results of Sect. 1.2 were obtained. The hospitality of IHES was crucial for developing the project. We are very grateful to the referee, who read the chapter very carefully and made many comments, incorporated in the text, which improved and clarified the exposition.

A.G. was supported by the NSF grants DMS-1059129, DMS-1564385, DMS-1900743, and DMS-2153059. He was a Fellow of the Simons Foundation in 2019 and 2023. He is grateful to the NSF, Simons Foundation, and IHES for hospitality and support.

References

1. A. Berenstein, V. Retakh, Non-commutative marked surfaces. arXiv:1510.02628
2. P. Boalch, Topology of the Stokes phenomenon. arXive:1903.12612
3. P. Boalch, Geometry and braiding of Stokes data; fission and wild character varieties. Ann. Math. arXive:1111.6228
4. T. Bridgeland, I. Smith, Quadratic differentials as stability conditions. arXiv:1302.7030.
5. P. Deligne, *Equations différentielles à points singuliers réguliers* (Springer, Berlin, 1970). Lecture Notes in Mathematics, vol. 163
6. P. Deligne, B. Malgrange, J-P. Ramis, *Singularités irréguliére: correspondance et documents*. Mathematical documents, vol. 5 (SMF 2007)
7. V. Fock, A. Goncharov, Moduli spaces of local systems and higher Teichmüller theory. Publ. Math. IHES **103**, 1–212 (2006). arXive:0311149
8. V. Fock, A. Goncharov, Cluster ensembles, quantization and the dilogarithm. Ann. Sci. L'Ecole Norm. Sup. (2009). arXive:0311245
9. V. Fock, A. Goncharov, Cluster X-varieties, amalgamation and Poisson-Lie groups. arXive:0508408
10. V.V. Fock, A.B. Goncharov, The quantum dilogarithm and representations of quantum cluster varieties. Invent. Math. **175**, 223–286 (2009). arXiv:math/0702397
11. V. Fock, A. Marshakov, Loop groups, Clusters, Dimers and Integrable systems. arXive:1401.1606
12. H. Gao, L. Shen, D. Weng, Augmentations, fillings and clusters. arXiv:2008.10793
13. S. Fomin, A. Zelevinsky, Cluster algebras. I. JAMS **15**(2), 497–529 (2002)
14. D. Gaiotto, G. Moore, A. Neitzke, Spectral networks. arXive:1204.4824
15. D. Gaiotto, G. Moore, A. Neitzke, Spectral networks and snakes. arXiv:1209.0866
16. S. Ganatra, J. Pardon, V. Shende, Microlocal Morse theory of wrapped Fukaya categories. arXiv:1809.08807
17. M. Gekhtman, M. Shapiro, A. Vainshtein, Cluster algebras and Weil-Petersson forms. arXiv:0309138
18. I. Gelfand, V. Retakh, Quasideterminants, I. Selecta Math. **3**, 517–546 (1997)
19. A.B. Goncharov, *Ideal webs moduli spaces of local systems, and 3d Calabi-Yau categories*, in *Algebra, Geometry, and Physics in the 21st Century*. Kontsevich Festshrift. Prog. in Math., vol. 324, 2017, pp. 31–99. arXive:1607.05228

20. A.B. Goncharov, R. Kenyon, Dimers and cluster integrable systems. Ann. Sci. L'Ecole Norm. Sup. **46**(5), 747–813 (2013). arXive:107.5588
21. A.B. Goncharov, L. Shen, Donaldson-Thomas trasnsformations of moduli spaces of G-local systems. arXiv:1602.06479
22. A.B. Goncharov, L. Shen, Quantum geometry of moduli spaces of local systems and representation theory. arXiv:1904.10491
23. S. Guillermou, Quantization of conic Lagrangian submanifolds of cotangent bundles. arXiv:1212.5818
24. S. Guillermou, M. Kashiwara, P. Schapira, Sheaf quantization of Hamiltonian Isotopies and applications to nondisplaceability problems. Duke Math J. **161**, 201–245 (2012). arXiv:1005.1517
25. M. Kashiwara, P. Schapira, *Sheaves on Manifolds*, Grundlehren der Mathematischen Wissenschafte, vol. 292 (Springer, 1994)
26. M. Kontsevich, Non-commutative identities. Arbeitstagung talk. arXiv:1109.2469
27. I. Le, Cluster structures on higher Teichmüller spaces for classical groups. arXiv:1603.03523
28. I. Le, An Approach to Cluster Structures on Moduli of Local Systems for General Groups. arXiv:1606.00961
29. B. Malgrange, *La classification des connexions irrégulières à une variable*, in *Mathematics and Physics* (Paris, 1979/1982), Prog. Math., vol. 37, pp. 381–399 (Birkhäuser Boston, Boston, MA, 1983)
30. B. Malgrange, *Equations différentielles a coefficients polynomiaux*, vol. 96 of *Progress in Mathematics* (Birkhäuser Boston Inc., Boston, MA, 1991)
31. A. Postnikov, Total positivity, Grassmannians, and networks. arXiv:0609764
32. L. Shen, D. Weng, Cluster structures on double Bott-Samelson cells. Forum Math. Sigma (2021). arXiv:1904.07992
33. V. Shende, D. Treumann, H. Williams, Legendrian knots and constructible sheaves. arXive:1402.0490
34. V. Shende, D. Treumann, H. Williams, E. Zaslow, Cluster varieties from Legendrian knots. arXive:1512.08942
35. V. Shende, D. Treumann, H. Williams, *On the combinatorics of exact Lagrangian surfaces*. arXive:1603.07449
36. I. Smith, Quiver algebras as Fukaya categories. arXiv:1309.0452
37. D. Thurston, From domino to hexagons. arXive:0405482

Semi-Stable Reduction of Foliations

Michael McQuillan

Contents

1 Introduction

In a historical quirk (cf. [16, Intro.]), the study of the canonical bundle of higher-dimensional varieties initiated by Mori [23], and, as such, often called Mori theory, has long ceased the original focus on rational curves in favour of a co-homological approach which would be better described as Kawamata theory. It is, therefore, not without irony that the study of rational curves on varieties foliated by curves is, arguably, both Mori theory as Mori intended and as Manin would have liked and leads to a complete minimal model programme.

Everything takes place in characteristic zero, so, say a projective variety $X/\mathbb{C}$ and a foliation by curves, $\mathcal{F}$, is just a (usually saturated) rank 1 sub-sheaf of the tangent sheaf, (22). Locally, where both X and $\mathcal{F}$ are smooth, this corresponds, by the classical Frobenius theorem, to a smooth fibration in the analytic topology. We therefore adopt the notation (and it's only notation) $X \to [X/\mathcal{F}]$ for foliations in order to reflect better the underlying geometry/real definition of a quotient of X by the holonomy groupoid; cf. Fact/Definition 3.1.1 and Fact/Definition 3.1.3. Irrespectively, there is, under mild hypothesis, e.g. X smooth, a well-defined bundle, $K_{\mathcal{F}}$, of forms along the leaves and corresponding notions, Definition 2.2.1, of foliated Gorenstein, resp. $\mathbb{Q}$-Gorenstein, singularities. Similarly, there are, functori-

M. McQuillan (✉)
Università degli studi di Roma 'Tor Vergata' & H.S.E. University, Moscow, Russia
e-mail: mcquilla@mat.uniroma2.it

Y. Tschinkel (ed.), *Arithmetic and Algebraic Geometry*, Simons Symposia,
https://doi.org/10.1007/978-3-031-74134-0_6

ally with respect to the ideas, notions of foliated terminal, log-terminal, canonical and log-canonical singularities, Definition 2.2.3. Unlike their classical counterparts, however, these definitions always admit a simple description in terms of local algebra. For example, terminal (Gorenstein) is equivalent, Revision 2.2.13, to smooth along the foliation, or equivalently given everywhere locally by a non-vanishing vector field, ∂, while a Gorenstein log-canonical singularity is a point, p, where, although ∂ vanishes, the implied linearisation

$$\partial : \frac{\mathfrak{m}(p)}{\mathfrak{m}^2(p)} \to \frac{\mathfrak{m}(p)}{\mathfrak{m}^2(p)} \tag{1}$$

is non-nilpotent, Revision 2.2.5.

Already this local global translation is highly indicative of why Mori theory of foliations by curves is that much more tractable than that of varieties. Nevertheless, there is no free lunch, i.e. it transpires that from ambient dimension 3 on that there are foliations by curves which never have log-canonical singularities on any smooth bi-rational model of the ambient space. The phenomenon is quite general, [24, §.III.iii], and, in se, straightforward enough, i.e. there are certain finite group actions on vector fields whose fixed points cannot be separated from the singularities while preserving smoothness of the ambient space. In practice, however, it means that if one wants a model of a foliation $X \to [X/\mathcal{F}]$ with (foliated) log-canonical singularities and X smooth, then one is obliged to pass from the category of varieties to the 2-category of Deligne-Mumford champs. In this context, the main theorem of [24] is the existence of log-canonical resolutions in ambient dimension 3, and the reader should be aware that for the moment the existence of log-canonical resolutions in higher dimension is open.

Irrespectively, we are obviously obliged to take as our starting point a smooth foliated champ $\mathcal{X} \to [\mathcal{X}/\mathcal{F}]$ with log-canonical singularities- from the existence of the Gorenstein covering champ, Fact/Definition 2.2.7 and [4]: if there is a model with log-canonical singularities, then there is one in which the ambient champ is smooth. This begins, however, to show signs of a rather pleasing loop since the natural context of the classification, [20], of foliated algebraic surfaces is exactly foliated smooth bi-dimensional champs, while the universal algebraic foliation in (hyperbolic) curves $\mathcal{M}_{g,1} \to \mathcal{M}_g$ is again, naturally, a smooth Deligne-Mumford champ.

To say that this begs the question of whether the minimal model programme for foliations by curves could be run wholly inside the 2-category in which the ambient champ is smooth may, to experts in the Mori theory of varieties, seem rather absurd. It transpires, however, to be the case in a way highly reminiscent of the structure of $\mathcal{M}_{g,1} \to \mathcal{M}_g$. The precise theorem is:

Theorem (5.5.6, 5.5.7, 5.5.8) *Let $\mathcal{X} \to [\mathcal{X}/\mathcal{F}]$ be a foliated champ which enjoys the following further properties:*

smooth; projective moduli; log-canonical, resp. canonical, foliation singularities, (2) *then there is a sequence of contractions and flips*

$$\begin{array}{ccccc} \mathcal{X} = \mathcal{X}_0 & & \mathcal{X}_1 & \cdots\cdots & \mathcal{X}_n = \mathcal{X}_{\min} \\ \downarrow & \dashrightarrow & \downarrow & \dashrightarrow & \dashrightarrow \downarrow \\ [\mathcal{X}/\mathcal{F}] = [\mathcal{X}_0/\mathcal{F}_0] & & [\mathcal{X}_1/\mathcal{F}_1] & & [\mathcal{X}_n/\mathcal{F}_n] = [\mathcal{X}_{\min}/\mathcal{F}_{\min}] \end{array} \tag{3}$$

such that each $\mathcal{X}_i \to [\mathcal{X}_i/\mathcal{F}_i]$ *enjoys all the (respective) properties* (2), *and exactly one of the following occurs:*

(a) $K_{\mathcal{F}_{\min}}$ *is nef.*
(b) $\mathcal{X}_{\min} \to [\mathcal{X}_{\min}/\mathcal{F}_{\min}]$ *is a Mori fibre space, i.e. the locus of a single extremal ray is all of* $\mathcal{X}_{\min}$, *and the foliation is a bundle of foliated varieties where the universal cover of a fibre is the radial (supposed saturated in dimension 1) foliation on a weighted projective champ, Definition 2.4.2, whose dimension is* 1 *iff the foliation singularities are canonical.*

Here a radial foliation is just the champ/weighted projective space variant of a pencil of lines through a point of projective space, and in a further irony, the harder part of the theorem is (b) in which the use of the word flip is slightly loose since it may, when the singularities are canonical, involve "very exceptional flips", Fact/Definition 5.5.5, i.e. a little invariant blowing up in the final stage, to preserve projectivity. The content of the theorem, however, should be clear: i.e. either we get a minimal model or a bundle of Fano objects, and the Fano objects are particularly simple, in fact, to all intents and purposes, rational curves if the singularities are canonical.

This said, let us give a brief breakdown both of the paper and the proof:

I. The first chapter is preliminary in nature. It contains generalities, Sect. 2.1, on Deligne-Mumford champs; a revision of foliation singularities, Sect. 2.2; the theory of weighted projective champs, Sect. 2.3, and their radial foliations, Sect. 2.4; a non-embedded variant of completion, Sect. 2.5; and some remarks on the analytic topology, Sect. 2.6. Technically, it's worth flagging the last 2 sections since the fact that many things fail to be an embedding for (separated) champs which are trivially so in the world of varieties, e.g. graphs of maps, is an issue, albeit sometimes this happens for the trivial reason that the étale topology is non-classical, while in the analytic topology one can still embed.
II. The second chapter is the critical one. It first proves the cone theorem, Fact 3.4.1, in maximal generality. This was already done in [5] for foliated Gorenstein varieties, and its extension to foliated Gorenstein champ, Sects. 3.1–3.4, may, largely, be considered technical in nature. In any case, it reveals that the $K_{\mathcal{F}}$-negative extremal rays are invariant parabolic (i.e. dominated by a rational curve) champs, $\mathcal{L}$, not factoring through the singular locus. Their particularly simple intersection with the singular locus, which occurs at a unique point $p : \mathrm{pt} \to \mathcal{L}$, of the foliation is described in Sect. 3.5, their normal bundle (should they have only nodes) by Sect. 3.6 and their formal

neighbourhoods (again for singularities no worse than nodes) in Sect. 3.7. The key point here, Proposition/Summary 3.7.3, is not only that the normal bundle determines the formal neighbourhood, but that everything is determined by the linearisation, (1), at the singularity p whose eigenvalues are, up to scaling, the slopes of the Harder-Narasimhan filtration of the normal bundle. The section concludes with an examination of the functoriality of the relationship between (1) and the Harder-Narasimhan filtration, Sect. 3.8, i.e. the said scaling is ambiguous in a non-trivial way up to ± 1, and this has a global manifestation, along with the necessary preliminaries, Sect. 3.9, for studying extremal rays with cusps.

III. The third chapter globalises the infinitesimal information of the second to describe the sub-champs swept out by extremal rays beginning with the general discussion Sect. 4.1 which leads to a definition in the specific, Fact 4.1.4, of extremal champs. As such Sects. 4.2–4.4 are devoted to describing their structure, which, as one might imagine from item (b) of the main theorem, is, Large Fact 4.4.7, basically that of a bundle of radially projective champs. The base of this bundle is essentially a smooth component of the singular locus, but the aforesaid issue of ± 1 in the scaling of (1) means that even when it has sense for it to be a bundle in the Zariski topology, it may not be.

IV. Finally, we construct contractions and flips, or, better, flaps, since everything is just a question of blowing up and down. Indeed, as one might imagine, contractions, Sects. 5.1–5.2, are easy. A critical fact, however, emerges, Proposition/Summary 5.1.4, that although a contraction renders the ambient champ less space like, i.e. can increase the local monodromy, it renders the foliation completely smooth about the contracted locus. As such, when one brings the full weight of the infinitesimal knowledge of Sect. 3 to bear in order to describe the formal neighbourhoods of extremal champ in a similar manner, Sect. 5.3, to that of a single ray in order to flip, Sect. 5.4, by the simple expedient of weighted blowing up and down, one concludes that flipping must terminate because it destroys a component of the singular locus at each stage. This leaves only loose ends, Sect. 5.5, to tie up related to scaling by ± 1 of (1), all of which can only occur when the generic leaf of the foliation is dominated by a rational curve. Consequently, we conclude the demonstration of the main Theorem in Sect. 5.5 and provide a log-variant in Sect. 5.6.

I am indebted to Bogomolov for pointing out that the language of algebraic champs was the correct setting for the main theorem, to Brunella for explaining to me the role of holonomy, to McKernan for furnishing an example that the issue of (1) with integer eigenvalues being only well defined up to ± 1 is genuine, to Marie Claude for the figures and to Cécile for the original typesetting, with any subsequent flaws being the result of my own clumsy modification. Ultimately, however, I am particularly grateful to Yuri Tschinkel for asking me whether I would like to contribute to this volume in memory of Yu. Manin. Indeed it was 33 years ago, while seated a few rows behind Yuri, that I first learned about bend and break in his course at M.I.T., and I can still remember his expression of pleasure at having learned such a "beautiful subject" in order to explain it to us. Requiescat in pace.

2 Preliminaries

2.1 Normal-folds

A normal fold is a particularly simple kind of champ, to wit:

Definition 2.1.1 A normal-fold is a not necessarily tame (although this will always be our context) excellent normal separated Noetherian Deligne-Mumford champ every generic point of which is scheme like.

A particularly important class of examples is given by

Fact/Definition 2.1.2 ([28, 2.8]) Following standard usage a smooth (over an implicit base S) normal-fold will be referred to as an orbifold. In particular, a (separated) algebraic space, X, of finite type over a field k has strict (or even non-strict if the action is tame) quotient singularities iff there is an almost étale map, $\mu : \mathcal{X} \to X$, from a smooth (over k) orbifold. In this case X is the moduli, [14, 1.3], of $\mathcal{X}$, and conversely, $\mathcal{X}$ is unique up to equivalence. As such $\mathcal{X}$ will be referred to as the *Vistoli covering champ* of X.

The following is a tiny variation on [28, 2.8]'s treatment of the Vistoli covering champ.

Lemma 2.1.3 *Let $\mu : \mathcal{X} \to X$ be the moduli of a normal-fold, with $U \to \mathcal{X}$ an étale atlas; then*

$$R := \big(\textit{normalisation of } U \times_X U \big) \rightrightarrows U \tag{4}$$

defines a groupoid and $\mathcal{X}$ is equivalent to the classifier $[U/R]$.

Proof Let $V \hookrightarrow \mathcal{X}$ be the everywhere scheme like embedded dense Zariski open guaranteed by Definition 2.1.1, and $U' := U \times_{\mathcal{X}} V$, then V is embedded in X, so $U' \times_V U'$ is a Zariski dense open of R. It is, however, also a Zariski dense open of $R_1 := U \times_{\mathcal{X}} U$, and we have a fibre square:

$$\begin{array}{ccc} U \times_X U & \longleftarrow & R_1 \\ \downarrow & & \downarrow \\ \mathcal{X} \times_X \mathcal{X} & \xleftarrow{\Delta_{\mathcal{X}/X}} & \mathcal{X} \end{array} \tag{5}$$

where by hypothesis the lower horizontal is finite. Consequently, $R_1 \to R$ is a finite bi-rational map of excellent normal schemes so they're equal. □

Irrespective of normality we have the further simplification

Lemma 2.1.4 *Let $\mu : \mathcal{X} \to X$ be the moduli of a separated excellent Deligne-Mumford champ, $X' \hookrightarrow X$ the (open, possibly empty) locus where μ is an*

isomorphism, and $f : \mathcal{Y} \to X$ a map such that $f^{-1}(X')$ meets every generic point; then f lifts to a composition $\mathcal{Y} \to \mathcal{X} \xrightarrow{\mu} X$ iff it lifts everywhere locally, i.e. for every étale neighbourhood $U \to \mathcal{X}$ of the image $f(y)$ of a geometric point y, there is an étale neighbourhood V_y of y and a lifting $V_y \to U$ of f.

Proof Necessity is obvious. By [14, 1.3] and [28, 2.8], there is, independently of any normal-fold hypothesis, an étale atlas $U = \coprod_\alpha U_\alpha$ of $\mathcal{X}$ and finite groups G_α acting on U_α such that $V := \coprod_\alpha V_\alpha := U_\alpha/G_\alpha$ is an étale atlas of X with $[U_\alpha/G_\alpha] = \mathcal{X} \times_X V_\alpha$. Now for anything with a well-defined map to X denote with a $'$ the fibre over X', so we have open embeddings:

$$\mathcal{Y}' \hookrightarrow \mathcal{Y}, \quad \mathcal{Y}'_\alpha \hookrightarrow \mathcal{Y}_\alpha := \mathcal{Y} \times_X V_\alpha. \tag{6}$$

Consequently, by hypothesis and refining U_α if necessary, there is an étale atlas $Y_\alpha \to \mathcal{Y}_\alpha$ and maps $f_\alpha : Y_\alpha \to U_\alpha$ such that

$$\begin{array}{ccc} Y_\alpha & \xrightarrow[f_\alpha]{} & U_\alpha \\ \downarrow & & \downarrow \\ \mathcal{Y}_\alpha & \longrightarrow & V_\alpha \end{array} \tag{7}$$

commutes. In particular, the G_α torsor $Y_\alpha \times_{V_\alpha} U'_\alpha$ is trivial, and we consider

$$\begin{array}{ccc} Y_0 := \coprod Y_\alpha \times G_\alpha & \xrightarrow[\mapsto \sigma . f_\alpha(y)]{y \times \sigma} & U_\alpha \\ \text{left vertical}\big\downarrow\text{in (7)} & & \\ \mathcal{Y} & & \end{array} \tag{8}$$

which leads (it's here [cf. Remark 2.1.5]; we use generically scheme like) to a commutative square:

$$\begin{array}{ccc} Y'_0 & \xrightarrow[\text{in (8)}]{\text{horizontal}} & U' \\ \text{via vertical}\big\downarrow\text{in (8)} & & \big\downarrow \\ X' & \longrightarrow & X \end{array} \tag{9}$$

As such, if we form the groupoids $R := U \times_{\mathcal{X}} U \rightrightarrows U$, and $Y_1 := Y_0 \times_{\mathcal{Y}} Y_0 \rightrightarrows Y_0$, then (9) ensures that $\mathcal{Y}' \to X' \hookrightarrow \mathcal{X}$ is equivalent to the composition of functors:

$$Y'_1 \to R' = U' \times_X U' \hookrightarrow R, \tag{10}$$

while by hypothesis Y_1' is dense in Y_1 and $\mathcal{X}$ is separated, so the simple expedient of taking the closure in (10) defines a functor $Y_1 \to R$. □

This is sufficiently close to optimal as to merit

Remark 2.1.5 One cannot replace X' by a Zariski open sub-champ $\mathcal{X}' \hookrightarrow \mathcal{X}$ in Lemma 2.1.4. Indeed take $\mathcal{X}$ to be the weighted projective champ $\mathbb{P}(n, n)$, Definition 2.3.1, $n > 1$. Its moduli is $\mathbb{P}^1$, so the fibre, $\mathcal{X}'$, over a standard $\mathbb{A}^1$ is an embedded Zariski open. Moreover, it's isomorphic to $\mathbb{A}^1 \times \mathrm{B}_{\mu_n}$, so in particular admits a section, and we could try to take $\mathcal{Y} = \mathbb{P}^1$. The gerbe $\mathbb{P}(n, n) \to \mathbb{P}^1$ is, however, non-trivial so the map $\mathcal{Y}' \to \mathcal{X}'$ cannot be extended to $\mathcal{Y} \to \mathcal{X}$ even though it is locally trivial, whence a fortiori without local obstruction. The problem is that if one replaces the moduli X, resp. X', by $\mathcal{X}$, resp. $\mathcal{X}'$, in (9), then the diagram needn't 2-commute in a slightly unusual way. Specifically, it's 2-commutative on geometric points, p say, by way of a natural transformation η_p between either possible composition, which, in the specific example, if say U_0, U_∞ are points in the standard affines around 0 and infinity, is

$$\eta_p = \begin{cases} 1_p & \text{if } p \in U_0, \\ p^{-1} \xrightarrow{p^{-1/n}} p & \text{if } p \in U_\infty \backslash 0, \end{cases} \tag{11}$$

where the latter arrow is to be understood in the presentation (35). Plainly, however, $p \to \eta_p$ isn't even continuous for p in $U_\infty \backslash 0$, and (9) fails to be 2-commutative.

This can often be combined with

Fact 2.1.6 Let $\mathcal{X}$ be a (connected) normal (or slightly more general uni-branch) excellent Deligne-Mumford champ; then there is a unique normal-fold $\mathcal{X}^0$ (slightly more generally uni-branch-fold with the obvious definition of that notion) such that $\mathcal{X} \to \mathcal{X}^0$ is a locally constant gerbe under some finite group B_G.

Proof Since $\mathcal{X}$ is excellent and uni-branch, one can insist, [10, 7.6.3], that the atlas $U = \coprod_\alpha U_\alpha$ encountered at the beginning of the proof of Lemma 2.1.4 consists solely of irreducible (affine) schemes U_α. Now for G_α of op. cit. define G_α' as the kernel of the representation $G_\alpha \to \mathrm{Aut}(U_\alpha)$ with G_α'' the image, then since $\mathcal{X}$ is uni-branch $\coprod_\alpha U_\alpha \times G_\alpha'$ is a normal (groupoid sense [14, 7.1]) U-group scheme of the stabiliser, so for $R := U \times_{\mathcal{X}} U \rightrightarrows U$, there is, op. cit. 7.4, a well-defined quotient $R \to R''$ where the latter is locally of the form $[U_\alpha / G_\alpha'']$. As such define $\mathcal{X}^0$ to be $[U/R'']$, and observe that all the G_α' are isomorphic. □

Finally, another important application of normality. Specifically let U be the spectrum of a Noetherian local ring, A, with closed point x and $j : U' \to U$ a Zariski open whose complement is defined by a regular sequence of length at least 2. As such, for $n \in \mathbb{N}$ the Kummer sequence,

$$0 \to \mu_n \to \mathbb{G}_m \xrightarrow{n} \mathbb{G}_m \to 0, \tag{12}$$

applied to U and U' combine to afford a short exact sequence:

$$0 \to \mathrm{H}^1(U, \boldsymbol{\mu}_n) \to \mathrm{H}^1(U', \boldsymbol{\mu}_n) \to \mathrm{Pic}(U')[n] \to 0. \tag{13}$$

In particular therefore, if A is strictly Henselian and $n^{-1} \in A$,

$$\mathrm{H}^1(U', \boldsymbol{\mu}_n) \xrightarrow{\sim} \mathrm{Pic}(U')[n]. \tag{14}$$

Now in the particular case that A is normal excellent we can take U' to be the regular locus and identify (primitive) generators of the right-hand side of (14) with $\mathbb{Q}$-Cartier divisors, L, on U of index $n = n(x)$, i.e. a Weil divisor, L, on U such that nL, but no smaller multiple, mL, $1 \leq m < n$, is a line bundle, while the elements of order n on the left are just $\boldsymbol{\mu}_n$-torsors $V' \to U'$ of order exactly n, and we assert:

Fact/Definition 2.1.7 For a $\mathbb{Q}$-Cartier divisor, L, of index n on a normal strictly Henselian U over which n is invertible, the associated *index 1-cover*, $V \to U$, is the integral closure of U in the corresponding $\boldsymbol{\mu}_n$-torsor $V' \to U'$. By construction $L \mid V$ is the trivial bundle and, in a sense, universally so, i.e. if $W \to U$ is any finite map from a normal scheme W every component of which is dominant such that $L \mid W$ is trivial, then it factors uniquely as $W \to V \to U$. In particular if $\Delta \to U$ is the strict Henselisation of some (scheme) point u of U of index $m|n$, then the normalisation, N, of $V \times_U \Delta$ is the trivial $\boldsymbol{\mu}_{\frac{n}{m}}$-torsor over the index 1-cover, M, of Δ.

Proof It remains to address the universal property, wherein, without loss of generality, W is connected. As such all of U, V, W are the spectra of normal Henselian local rings, so they are all domains, while the function field of V over that of U is Galois by construction, so the factorisation is unique if it exists. Now let W' be the fibre over U', then by (14) the $\boldsymbol{\mu}_n$-torsor $W' \times_{U'} V'$ has a section, which gives the factorisation $W' \to V' \to U'$, and since everything is S_2 the simple expedient of taking global functions on these opens gives $W \to V \to U$. Applying this to the in particular: there is a map from N to M, while $V' \times_{U'} \Delta$ is a Zariski dense open of the former which is the trivial $\boldsymbol{\mu}_{\frac{n}{m}}$ torsor over the fibre of U' in the latter. □

In the category of spaces it is rare that index 1-covers can be glued, whereas:

Fact 2.1.8 Let L be a $\mathbb{Q}$-Cartier divisor on an excellent normal Deligne-Mumford champ $\mathcal{X}$; then there is a finite map, $f : \mathcal{Y} \to \mathcal{X}$, from a normal Deligne-Mumford champ such that f^*L is Cartier enjoying the following universal property: if $g : \mathcal{Z} \to \mathcal{X}$ is a finite map from a normal champ every component of which is dominant and such that g^*L is Cartier, then there is a 2-commutative factorisation:

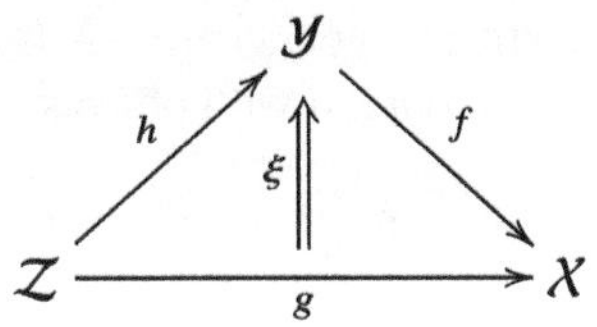

(15)

such that for any other factorisation, $\bar{\xi} : g \Rightarrow f\bar{h}$ there is a unique $\theta : h \Rightarrow \bar{h}$ for which $(g_*\theta)\xi = \bar{\xi}$.

Proof For every closed point x of $\mathcal{X}$ let $n(x)$ be the index of L at x and $U_x \to \mathcal{X}$ a sufficiently small étale neighbourhood such that the index 1-cover $V_x \to U_x$ of Fact/Definition 2.1.7 is well defined, with U'_x, V'_x as per op. cit.. Now, for $U = \coprod_x U_x$, we can without loss of generality suppose that $\mathcal{X}$ is the classifying champ of the étale groupoid $R_0 := U \times_{\mathcal{X}} U \rightrightarrows U$ and that $U' := \coprod_x U'_x$ is the locus where U is regular. As such, the restriction, $R'_0 \rightrightarrows U'$, is a dense Zariski open of R_0 equivalent to the restrictions $R' \rightrightarrows V' := \coprod V'_x$, where $R' \to R'_0$ is both étale and finite, and we define $R \rightrightarrows V$ to be the integral closure of R_0 in R'. Consequently, from the commutative diagram of fibre squares

$$\begin{array}{ccccc} R_0 & \longleftarrow & R'_0 & \longleftarrow & R' \\ \downarrow & & \downarrow & & \downarrow \\ U\times U & \longleftarrow & U'\times U' & \longleftarrow & V'\times V' \end{array} \tag{16}$$

and $V \times V \to U \times U$ finite, $R \rightrightarrows V$ defines a groupoid which, by the in particular in Fact/Definition 2.1.7, has étale source and sink.

Now let $g : \mathcal{Z} \to \mathcal{X}$ be given, then, up to equivalence, we can identify this with a functor of groupoids, $g : W_1 \to R_0$, where $W_1 = W \times_{\mathcal{Z}} W \rightrightarrows W$ for some étale cover $W \to \mathcal{Z}$ finer than the pre-image of U. By Fact/Definition 2.1.7, $W \to U$ factors (uniquely) through V affording a (unique) map:

$$h_1 : W_1 \to R_0 \times_{U\times U} V \times V \tag{17}$$

and R is the normalisation of the latter, while every local ring of W_1 is finite over $U \times U$ so this actually factors as a functor (because everything is unique) $h : W_1 \to R$. As such we get a unique strictly commutative factorisation $g = hf$ given $W \to U$. This supposes, however, that all of $\mathcal{X}$, $\mathcal{Y}$, $\mathcal{Z}$ were the classifying champ of the said groupoids, whereas they may be no better than equivalent to such, and whence the uniqueness statement (15). □

In the same vein one has

Fact/Definition 2.1.9 Let $\mathcal{D} \hookrightarrow \mathcal{X}$ be an effective Cartier divisor on a normal champ $\mathcal{X}$. As such for a sufficiently fine atlas $U \to \mathcal{X}$ we may identify $\mathcal{X}$ with the classifier of a groupoid $(s,t) : R_0 \rightrightarrows U$ and suppose that $\mathcal{D}|_U$ is defined by

$z = 0$ where $s^*z = gt^*z$ for some co-cycle $g : R_0 \to \mathbb{G}_m$. Now for $n \in \mathbb{N}$ invertible in every local ring of $\mathcal{X}$ define a groupoid with objects

$$\text{normalisation of } (T^n = z) \hookrightarrow U \times \mathbb{A}^1. \tag{18}$$

and arrows the normalisation, R', of the base change groupoid $R'_0 \rightrightarrows V$, i.e. the fibre

$$\begin{array}{ccc} R'_0 & \longrightarrow & V \times V \\ \downarrow & & \downarrow \\ R_0 & \xrightarrow{s \times t} & U \times U \end{array} \tag{19}$$

so that $R' \rightrightarrows V$ is a groupoid because $R'_0 \rightrightarrows V$ is, and everything is, normal. Equally, R' admits the explicit description:

$$\text{normalisation of } (T_1^n = s^*z,\ T_2^n = t^*z) \hookrightarrow R_0 \times \mathbb{A}^2, \tag{20}$$

which is the same thing as taking normalised nth roots of s^*z and the (invertible) transition function g. By hypothesis, however, n is everywhere invertible, so $R' \rightrightarrows V$ has étale source and sink, and we define $\mathcal{X}^{1/n} = [V/R'] \to \mathcal{X}$ to be the (extraction of a) nth root of $\mathcal{D}$. Observe, moreover, that a section of $s : R' \to V$ is a choice of nth root of g, so from the Cěch boundary in (12), the class of the fibration $\mathcal{D}' = \mathcal{D} \times_{\mathcal{X}} \mathcal{X}' \to \mathcal{D}$ in B_{μ_n}'s is exactly

$$\mathrm{c}_1(\mathcal{D}) \in \mathrm{H}^2(\mathcal{D}, \mu_n) \tag{21}$$

2.2 *Foliation Singularities*

This section is largely a summary, for the convenience of the reader of the relevant parts of [24]. The one exception to this rule is the concluding digression from Definition 2.2.12 to Construction 2.2.15, on how to avoid the study of boundaries altogether. Our interest is exclusively in *foliations by curves*, i.e. if $\mathcal{X}$ is a Deligne-Mumford champ of finite type over a field k (so $\Omega^1_{\mathcal{X}/k}$ is well defined), a torsion-free quotient:

$$\Omega^1_{\mathcal{X}/k} \to \mathcal{Q} \to 0 \tag{22}$$

which is locally free of rank 1 at every generic point. Certainly this is not the right definition in positive or mixed characteristic, [21], since in such situations (22) is not likely to be locally integrable in any meaningful sense. Fortunately, we never have to worry about this, so we proceed directly from (22) to

Definition 2.2.1 If $\mathcal{X}$ is normal and the double dual $Q^{\vee\vee}$ is a bundle, resp. a $\mathbb{Q}$-Cartier divisor, then we say that the foliation, $\mathcal{F}$, is Gorenstein, resp. $\mathbb{Q}$-Gorenstein, or possibly foliated Gorenstein, resp. foliated $\mathbb{Q}$-Gorenstein, if there is any danger (which there won't be) of confusion. In either case and indeed even if $\mathcal{X}$ were only normal, we write $K_{\mathcal{F}}$ instead of $Q^{\vee\vee}$ for the canonical divisor, so that in the Gorenstein case there is an ideal I_Z supported in the co-dimension 2 (schematic) singular locus of the foliation Z such that

$$Q = K_{\mathcal{F}} \cdot I_Z \hookrightarrow Q^{\vee\vee} = K_{\mathcal{F}}. \tag{23}$$

As such, even in the analytic topology, the classifying champ, $[\mathcal{X}/\mathcal{F}]$, may have no sense, albeit analytically (and with probability zero in any algebraic topology) $[\mathcal{X}\backslash Z/\mathcal{F}]$ has sense. Nevertheless, to better convey the idea we write

$$\mathcal{X} \to [\mathcal{X}/\mathcal{F}] \tag{24}$$

as a short hand for (22) and $\Omega^1_{\mathcal{X}/\mathcal{F}}$ for the kernel in op. cit..

Unfortunately, it is not technically correct to view a quasi-projective variety as a proper champ with infinite monodromy on the boundary, so we make

Remark 2.2.2 All of this is equally valid for champs with boundary, i.e. a couple $(\mathcal{X}, \mathcal{D})$, for $\mathcal{D} \hookrightarrow \mathcal{X}$ a reduced Weil divisor. Usually there'll be some further regularity, e.g. $\mathcal{X}$ and $\mathcal{D}$ smooth over k, but all that's a priori required is that we can give a sense to the sheaf $\Omega^1_{\mathcal{X}}(\log \mathcal{D})$, so $\mathcal{X}$ normal is sufficient. In any case, it therefore follows that the canonical bundle of the foliation $\mathcal{F}$ may have competing definitions according as to whether a boundary is involved, $K_{\mathcal{F}}$, or not, $K_{\mathcal{F}}^{\text{nolog}}$. These are related by:

$$K_{\mathcal{F}} = K_{\mathcal{F}}^{\text{nolog}} + \sum_i \epsilon(\mathcal{D}_i)\mathcal{D}_i \tag{25}$$

where $\mathcal{D}_i$ are the irreducible components of $\mathcal{D}$ and for W a Weil divisor

$$\epsilon(W) = \begin{cases} 0 & \text{if } W \text{ is } \mathcal{F} \text{ (in the sense of Definition 2.2.1) invariant,} \\ 1 & \text{otherwise.} \end{cases} \tag{26}$$

Similarly, there may also be competing definitions of invariant according as to whether this is understood for a saturated sub-sheaf of $\mathcal{T}_{\mathcal{X}}$ or $\mathcal{T}_{\mathcal{X}}(-\log \mathcal{D})$ so that should there be any risk of confusion the former, equivalently, Definition 2.2.1, will, following [24, I.i.2], be referred to as *strictly invariant*. Regardless, almost always our boundary will be empty, but when it isn't: $K_{\mathcal{F}}$ will, as suggested by (25), be reserved for the canonical with log-poles since this is more natural and the resulting formulae are cleaner.

A case in point is the following cut and paste of [24, I.ii.1]:

Definition 2.2.3 Let $(U, D, \mathcal{F})$ be an irreducible local germ of a $\mathbb{Q}$-Gorenstein foliated logarithmic geometrically normal k-variety, i.e. the germ about the generic point of a sub-variety Y of a geometrically normal variety such that the log-canonical bundle $K_{\mathcal{F}}$ is a $\mathbb{Q}$-divisor, then for v a divisorial valuation of $k(U)$ centred on Y the *log discrepancy*, $a_{\mathcal{F}}(v)$ is defined as follows:
By hypothesis there is a normal modification $\pi : \tilde{U} \to U$ of finite type, together with a divisor E on $\tilde{U}$ such that $\mathcal{O}_{\tilde{U},E}$ is the valuation ring of v. In particular, bearing in mind (25), there is an induced foliation $\tilde{\mathcal{F}}$ with log-canonical bundle $K_{\tilde{\mathcal{F}}}$, i.e. whose dual is saturated in $T_{\tilde{U}}(-\log E)$. Thus, there is a unique rational number $a_{\mathcal{F}}(v)$ such that

$$K_{\tilde{\mathcal{F}}} = \pi^* K_{\mathcal{F}} + a_{\mathcal{F}}(v)E \tag{27}$$

and for ϵ as in (26) we say that the local germ $(U, D, \mathcal{F})$ is:

$$\begin{aligned} &(1)\ \text{Terminal if } a_{\mathcal{F}}(v) > \epsilon(v) \\ &(2)\ \text{Canonical if } a_{\mathcal{F}}(v) \geq \epsilon(v) \\ &(3)\ \text{Log-terminal if } a_{\mathcal{F}}(v) > 0 \\ &(4)\ \text{Log-canonical if } a_{\mathcal{F}}(v) \geq 0, \end{aligned} \tag{28}$$

where the slightly unsettling shift of the definitions by $\epsilon(v)$ occurs as a result of the convention adopted in Remark 2.2.2 together with their correct functorial interpretation.

In contrast to this functorial framework, there is a "competing" local notion of what ought to be a good class of foliation singularities, viz:

Set Up 2.2.4 Let ∂ be a singular derivation of a local ring, $\mathcal{O}$, with residue field k. Thus, by definition, if $\mathfrak{m}$ is the maximal ideal of $\mathcal{O}$, $\partial : \mathcal{O} \to \mathfrak{m}$ and

$$\bar{\partial} : \frac{\mathfrak{m}}{\mathfrak{m}^2} \to \frac{\mathfrak{m}}{\mathfrak{m}^2} : x \mapsto \partial(x) \tag{29}$$

is k-linear by Leibniz's rule.

The relation between the linearisation (29) and (28) is as good as possible

Revision 2.2.5 [24, I.ii.3]. A Gorenstein foliation over the complex numbers is log-canonical iff every point is either smooth or, its linearisation, (29), is non-nilpotent.

Better still, one can always reduce to the Gorenstein case thanks to the specifics of one-dimensional leaves, i.e.

Revision 2.2.6 Let $(V, \tilde{D}, \tilde{\mathcal{F}}) \to (U, D, \mathcal{F})$ be the index 1-cover of the germ in Definition 2.2.3 associated to the log-canonical bundle $K_{\mathcal{F}}$ in the sense of Fact/Definition 2.1.7, or, more generally, an almost étale map, then for any (n) in (28), $1 \leq n \leq 4$, $(U, D, \mathcal{F})$ is (n) iff $(V, \tilde{D}, \tilde{\mathcal{F}})$ is.

Proof The easy ones are $n = 4$, [24, I.ii.5] and the if direction for $1 \leq n \leq 3$, [24, III.i.5], which also covers the subtler converse. □

Manifestly, therefore,

Fact/Definition 2.2.7 Let $\mathcal{X}\backslash\mathcal{D} \to [\mathcal{X}\backslash\mathcal{D}/\mathcal{F}]$ be a $\mathbb{Q}$-foliated Gorenstein logarithmic champ; then the index 1-cover, $\pi : \tilde{\mathcal{X}} \to \mathcal{X}$, defined by the log-canonical divisor $K_{\mathcal{F}}$, Fact 2.1.8, will be referred to as the Gorenstein covering champ. The map π is étale in co-dimension 2; there is an identity $K_{\tilde{\mathcal{F}}} = \pi^* K_{\mathcal{F}}$ of log-canonical divisors; $\tilde{\mathcal{X}}\backslash\tilde{\mathcal{D}} \to [\tilde{\mathcal{X}}\backslash\tilde{\mathcal{D}}/\tilde{\mathcal{F}}]$ is Gorenstein; and the cover enjoys (n), $1 \leq n \leq 4$, of (28) iff $\mathcal{X}\backslash\mathcal{D} \to [\mathcal{X}\backslash\mathcal{D}/\mathcal{F}]$ does.

As such, we work almost exclusively with Gorenstein foliations. Similarly, the already small difference between log-canonical and canonical becomes close to irrelevant for minimal model theory, i.e.

Definition 2.2.8 Let $(U, D, \mathcal{F})$ be a germ of a normal foliated Gorenstein log-variety about a point p such that a generator (in the sense of Definition 2.2.1) vanishes along a sub-variety Y; then a singularity is called *radial* iff after completion in the maximal ideal we can find a generator of the foliation of the form

$$\partial = n_1 x_1 \frac{\partial}{\partial x_1} + \ldots + n_r x_r \frac{\partial}{\partial x_r} + \delta \tag{30}$$

where $x_i = 0$ defining Y are linearly independent modulo $\mathfrak{m}^2_{U,p}$, $n_i \in \mathbb{N}$, and $\delta \in \mathrm{Der}(K, I_Y)$ for some quasi-coefficient field K. In particular for U smooth, D is strictly invariant, Remark 2.2.2, iff $\mathrm{codim}(Y) = r \geq 2$.

By way of clarification let us make

Remark 2.2.9 This isn't quite a cut and paste from [24], since op. cit. III.i.2 insists that Y of Definition 2.2.8 has co-dimension at least 2, which, although entirely a question of convention, isn't right for doing minimal model theory. In particular, therefore, when Y has co-dimension 1, e.g. Revision 2.2.10.(c), $D = Y$.

Irrespectively, the above definition of a radial singularity should not be confused with the closely related notion of a radial foliation encountered in Definition 2.4.2, and in any case the important point is:

Revision 2.2.10 [24, III.i.3]. For $(U, D, \mathcal{F})$ a germ of a normal foliated Gorenstein variety over a field k of characteristic 0, the following are equivalent:

(a) The singularity is radial.
(b) The singularity is log-canonical but not canonical.

(c) Y is the centre of a divisorial valuation of $k(U)$ of (log)-discrepancy zero and exceptional divisor (cf. Remark 2.2.9) not strictly invariant.

It thus follows that the passage from log-canonical to canonical is exactly

Revision 2.2.11 [24, III.ii.2]. If $\mathcal{X}\backslash\mathcal{D} \to [\mathcal{X}\backslash\mathcal{D}/\mathcal{F}]$ is a foliated smooth champ over a field of characteristic zero which has log-canonical but not canonical singularities; then every component of sing($\mathcal{F}$) where this occurs is smooth, and there is a smoothed weighted blow-up, [24, I.iv.3], in each of which such that the induced log-foliation on the resulting bi-rational modification $\tilde{\mathcal{X}} \to \mathcal{X}$ has everywhere log-canonical logarithmic foliation singularities, which amounts to the rather strong: at every point of the exceptional divisor, $\mathcal{E}$, the induced foliation is smooth and everywhere transverse to $\mathcal{E}$.

Such attention to the details of the logarithmic case notwithstanding our ultimate intention is to work almost exclusively with an empty boundary. In order to do this we introduce

Definition 2.2.12 A foliated space with orbifold boundary is a triple $(U, \Delta, \mathcal{F})$, where $U \to [U/\mathcal{F}]$ is a foliation in the sense of Definition 2.2.1 and Δ is a formal linear combination, $\sum_i a_i \Delta_i$, twice of effective Weil divisors, where $a_i = 1 - n_i^{-1}$ for some positive integers $n_i < \infty$; and we say (slightly contrary to standard usage) that $(U, \Delta, \mathcal{F})$ is $\mathbb{Q}$-Gorenstein if $U \to [U/\mathcal{F}]$ is and each Δ_i is $\mathbb{Q}$-Cartier. Moreover, if D is the Weil divisor $\sum_i \Delta_i$, then the discrepancy, $a_{\mathcal{F}}^{\Delta}(v)$, of $(U, \Delta, \mathcal{F})$ along a divisorial valuation v is defined to be

$$a_{\mathcal{F}}^{\Delta}(v) := a_{\mathcal{F}}(v) - \sum_i \epsilon(\Delta_i) m_i (1 - a_i) \tag{31}$$

where $a_{\mathcal{F}}(v)$ are the logarithmic discrepancies, (27), of the foliated log-variety $(U, D, \mathcal{F})$; ϵ is as (26); and m_i are the multiplicities of the Δ_i along the exceptional divisor E encountered in Definition 2.2.3. As such, we then say that $(U, \Delta, \mathcal{F})$ satisfies the corresponding properties (28) if the respective inequalities hold for $a_{\mathcal{F}}^{\Delta}(v)$ rather than $\alpha_{\mathcal{F}}(v)$.

The introduction of such orbifold boundaries is very much temporary since

Revision 2.2.13 [24, III.i.1]. Let $(U, D, \mathcal{F})$ be a foliated germ of a smooth log-variety supported at Z, then the following are equivalent:

(1) $(U, D, \mathcal{F})$ is terminal.
(2) $(U, D, \mathcal{F})$ is log-terminal.
(3) D is strictly (i.e. in the sense of Definition 2.2.1) invariant and $\mathcal{F}$ is smooth transverse to the generic point of Z.

which in turn affords

Corollary 2.2.14 *Let $(U, \Delta, \mathcal{F})$ be a germ of a log-canonical foliation singularity with $\mathcal{F}$-Gorenstein and non-empty orbifold boundary every component, Δ_i, of which is Cartier, then in fact it is canonical, and exactly one of the following holds:*

1. *Not only $(U, \mathcal{F})$ but also $(U, \Delta, \mathcal{F})$ is terminal while the non-invariant part of Δ has multiplicity 1 and is everywhere transverse to $\mathcal{F}$.*
2. *$(U, \mathcal{F})$, but not $(U, \Delta, \mathcal{F})$, is terminal, the weight of every non-invariant component of Δ (of which there are at most 2) is $1/2$, and the non-invariant part of D is defined by a single equation f of multiplicity 2 such that for a local generator, ∂, of the foliation $\partial^2(f)$ is a unit.*
3. *As per item (2) except that f has multiplicity 1 and enjoys a simple tangency with $\mathcal{F}$, i.e. $\partial^2(f)$ is again a unit.*

Proof From (27) and (31), the singularity $(U, \mathcal{F})$ without boundary is log-terminal, while it is Gorenstein by hypothesis. Thus, by Revision 2.2.13 it is defined by a no-where vanishing vector field ∂, and, [24, III.i.1], every valuation, v, centred on the singularity has $\epsilon(v) = 0$. In particular, therefore, $(U, \Delta, \mathcal{F})$ is always canonical, and it is terminal iff it is log-terminal.

Now, supposing, without loss of generality, that no component, Δ_i, is invariant consider the effect of blowing up in the maximal ideal of the germ. The discrepancy of $(U, \mathcal{F})$ is 1, so the only way for the multiplicity of D to be more than 1 is if it is 2 and all the weights $a_i = 1/2$. In this latter case the initial modification of $(U, \Delta, \mathcal{F})$ is, therefore, crepant, so the proper transform must itself be log-canonical, and whence the proper transform of D must only cut the exceptional divisor in smooth points of the induced foliation, i.e. $\partial^2(f)$ is a unit for f of multiplicity 2 defining D. To see that such a singularity is indeed canonical observe (proof of [24, III.i.1]) that in the local ring, R, of a divisorial valuation v, we can write

$$\partial = \pi^{-m}\tilde{\partial},\ f = \pi^n \tilde{f},\ \tilde{\partial}(\pi) = 0,\ v(\pi) = 1,\ m, n \in \mathbb{N} \tag{32}$$

for $\tilde{\partial}$ a derivation of R. As such,

$$\epsilon(v) = 0 = v(\partial^2 f) = (n - 2m) + v(\tilde{\partial}^2(\tilde{f})) \geq n - 2m \tag{33}$$

which is exactly the canonical condition.

Alternatively, therefore, the multiplicity of D is exactly 1, and if it is not everywhere transverse to the induced foliation, then the proper transform of D must cut the exceptional divisor in the singular locus of the transformed foliation, and a blow-up in this (singular) locus affords a valuation of negative discrepancy unless the weight is $1/2$. As such, we are in case (1) of Corollary 2.2.14 or most of case (3), i.e. it remains to prove that the tangency is simple. Observe, however, that D cuts the exceptional divisor in a smooth invariant sub-space, and blowing up in this not only yields a second exceptional divisor along which the discrepancy is zero, but separates the proper transform of D from the proper transform of the initial exceptional divisor. Consequently, if the tangency were not simple, the doubly transformed D would contain an invariant subspace of the induced foliation in the second exceptional divisor, and a blow-up in this would afford a valuation of negative discrepancy. Conversely a simple tangency with weight $1/2$ is canonical for the same reason as (32)–(33), while an everywhere transverse divisor of any weight is log-terminal because the "weight 1 case", i.e. $r = 1$ in (30) is, Revision 2.2.10, log-canonical. □

This can be applied to reduce to an empty boundary in the obvious way, to wit:

Construction 2.2.15 Suppose $(U, \Delta, \mathcal{F})$ is a $\mathbb{Q}$-Gorenstein log-canonical foliated germ with orbifold boundary, with no boundary component invariant. Then composing the index 1-covers associated to $\mathcal{F}$ and the boundary components Δ_i, we find a foliated germ with orbifold boundary $(U', \Delta', \mathcal{F}')$ satisfying the hypothesis of Corollary 2.2.14 such that $U' \to U$ is almost étale. By op. cit. and [24, III.i.1], the proof of [24, III.i.5] goes through verbatim, and the obvious variant of Revision 2.2.6 holds, i.e. for any (n) in (28), $1 \leq n \leq 4$, $(U, \Delta, \mathcal{F})$ is (n) iff $(U', \Delta', \mathcal{F}')$ is. Ignoring, for the sake of argument, cases (2) and (3) of Corollary 2.2.14, the latter boundary is, in the presence of log-canonical singularities, an everywhere transverse Cartier divisor of multiplicity 1 together with a weight $1 - n^{-1}$. As such if $f = 0$ is a local equation for Δ', then we could extract an nth root $\pi : V \to U'$ to obtain a Gorenstein foliation $V \to [V/\tilde{\mathcal{F}}]$ such that

$$K_{\tilde{\mathcal{F}}} = \pi^*(K_{\mathcal{F}} + \Delta) \tag{34}$$

and again the obvious variant of Revision 2.2.6 holds – for any (n) in (28), $1 \leq n \leq 4$, $(V, \tilde{\mathcal{F}})$ is (n) iff $(U', \Delta', \mathcal{F}')$ is – for exactly the same reason as above. Plainly all such local constructions will glue as champs by much the same argument as Fact/Definition 2.2.7, so all this is just the obvious fact that minimal model theory for foliations with orbifold boundary can be deduced from the minimal model theory of champs without boundary. The slightly subtler point, however, is that if one were to begin with a foliated champ $\mathcal{X}\backslash\mathcal{D} \to [\mathcal{X}\backslash\mathcal{D}/\mathcal{F}]$ with (integral) boundary, then extracting an $n(> 2)$th root, $\mathcal{X}^{1/n} \to \mathcal{X}$ of $\mathcal{D}$ yields a foliation $\mathcal{X}^{1/n} \to [\mathcal{X}^{1/n}/\mathcal{F}^{1/n}]$ which has log-canonical singularities iff $\mathcal{X}\backslash\mathcal{D} \to [\mathcal{X}\backslash\mathcal{D}/\mathcal{F}]$ does, so that not only the minimal model theory for foliations with orbifold boundary, but also with integral boundary, Sect. 5.6, can be deduced from the champ theorem without boundary.

2.3 Weighted Projective Champs

All of this section works in arbitrary generality, so over a base, say Spec(k), where k is a ring, with the object of interest being

Definition 2.3.1 For $\underline{a} = (a_0, \dots, a_n) \in \mathbb{Z}^{n+1}_{>0}$, $n > 0$, let $A_k := \mathbb{A}^{n+1}_k\backslash 0$, then by the weighted projective champ $\mathbb{P}_k(a_0, \dots, a_n)$, or just $\mathbb{P}_k(\underline{a})$, is to be understood the classifying champ $[A_k/\mathbb{G}_{m,k}]$ of the action:

$$\begin{aligned} R_k := \mathbb{G}_{m,k} \times A_k \rightrightarrows A_k : \\ (x_0, \dots, x_n) \leftarrow \lambda \times (x_0, \dots, x_n) \mapsto x^\lambda := (\lambda^{a_0} x_0, \dots, \lambda^{a_n} x_n). \end{aligned} \tag{35}$$

Just like any quotient space under a group there is a tautological torsor, i.e. $A_k \times \mathbb{G}_m$ with $\mathbb{G}_m$ action:

$$\mathbb{G}_m \times \left(A_k \times \mathbb{G}_m\right) : \lambda \times (x \times z) \mapsto x^\lambda \times (\lambda z) \tag{36}$$

which one extends to a line bundle in the usual way, to wit:

Fact/Definition 2.3.2 Choose an embedding $\mathbb{G}_m \hookrightarrow \mathbb{G}_a : z \mapsto z$, then by the tautological line bundle, $\mathcal{O}(1)$, on $\mathbb{P}_k(\underline{a})$ is to be understood the line bundle $\mathbb{G}_a \times A_k$ with $\mathbb{G}_m$ action given by (36) and by multiplication on the first factor via our aforesaid choice of embedding. In particular, therefore, we have defined $\mathbb{V}(\mathcal{O}(1)) \mid_{A_k}$ – EGA notation – whence as an equivariant $\mathcal{O}_{A_k}$-module $\mathcal{O}(1)$ has generator T where

$$T^\lambda = \lambda^{-1} T \tag{37}$$

so that the bundle $\omega_{A_k/k}$ of volume forms on A_k descends to the bundle $\omega := \mathcal{O}(-a_0 - \dots - a_n)$ on $\mathbb{P}_k(\underline{a})$.

Unsurprisingly, Serre's explicit calculation generalises to:

Fact 2.3.3 The bundle $\mathcal{O}(1)$ freely generates the Picard group of $\mathbb{P}_k(\underline{a})$; there are, for $p \geq 0$, canonical (dual) isomorphisms of free k-modules:

$$\begin{aligned} \mathrm{H}^0(\mathbb{P}_k(\underline{a}), \mathcal{O}(p)) = S_p := \coprod k \cdot x_0^{p_0 a_0} \cdots x_n^{p_n a_n} \\ \mathrm{H}^n(\mathbb{P}_k(\underline{a}), \omega(-p)) = S'_p := \prod k \cdot \frac{dx_0 \cdots dx_n}{x_0 \cdots x_n} \cdot x_0^{-p_0 a_0} \cdots x_n^{-p_n a_n} \end{aligned} \tag{38}$$

where the sum, resp. product, is taken over integers such that $p_0 a_0 + \cdots + p_n a_n = p$, and any other co-homology of any other line bundle in any other degree vanishes.

Proof The Picard group of A_k is trivial, so a line bundle on $\mathbb{P}_k(\underline{a})$ is the same thing as a map $\phi : R_k \to \mathbb{G}_m$ from the groupoid (35) satisfying the co-cycle condition $\phi(gf) = \phi(g)\phi(f)$. There are, however, no (algebraic) maps from A_k to $\mathbb{G}_m$, so all such co-cycles are integer multiples of the tautological one. As to the second part: if $\pi : A_k \to \mathbb{P}_k(\underline{a})$ is the projection, then for any sheaf $\mathcal{F}$ on A_k the Leray spectral sequence reads

$$\mathrm{H}^i(\mathbb{P}_k(\underline{a}), R^j \pi_* \mathcal{F}) \Rightarrow \mathrm{H}^{i+j}(A_k, \mathcal{F}). \tag{39}$$

Now the co-homology of the right-hand side of (39) is known, i.e. there are canonical dual, [12, Exposé IV.5.5], isomorphisms:

$$\mathrm{H}^0(A_k, \mathcal{O}_{A_k}) = \coprod_p S_p, \quad \mathrm{H}^n(A_k, \omega_{A_k/k}) = \coprod_p S'_p \tag{40}$$

while on the left-hand side there are canonical isomorphisms:

$$\pi_* \mathcal{O}_{A_k} \xrightarrow{\sim} \coprod_{q \in \mathbb{Z}} \mathcal{O}(q), \quad \pi_* \omega_{A_k/k} \xrightarrow{\sim} \prod_{q \in \mathbb{Z}} \omega(q) \tag{41}$$

and all higher direct images in (39) vanish, whence (38) by identifying the weight of the action of $\mathbb{G}_m$ in the equivariant isomorphism between (40) and (41) afforded by (39). □

In addition the bundle ω is the bundle of volume forms on $\mathbb{P}_k(\underline{a})$ when this has sense, i.e.

Claim 2.3.4 The moduli of any $\mathbb{P}_k(\underline{a})$ is projective; in fact better there is a finite flat map:

$$\mathbb{P}_k^n \to \mathbb{P}_k(\underline{a}) : [x_0, \ldots, x_n] \mapsto [x_0^{a_0}, \ldots, x_n^{a_n}] \tag{42}$$

and $\mathbb{P}_k(\underline{a})$ is Deligne-Mumford iff all the a_i are invertible in k. In addition the coordinate functions, $\partial_i = \frac{\partial}{\partial x_i}$, afford a $\mathbb{G}_m$ equivariant isomorphism:

$$\textstyle\coprod_i \mathcal{O}(a_i) \xrightarrow[\partial_0 + \cdots + \partial_n]{\sim} T_{A_k/k} \tag{43}$$

leading to the Euler sequence of $\mathbb{G}_m$-modules on A_k, equivalently bundles on $\mathbb{P}_k(\underline{a})$:

$$0 \to T_{\mathbb{G}_m} \xrightarrow{\sim} \mathcal{O} \xrightarrow{a_i x_i \partial_i} \coprod_i \mathcal{O}(a_i) \to \pi^* T_{\mathbb{P}_k(\underline{a})/k} \to 0, \tag{44}$$

whenever $\mathbb{P}_k(\underline{a})$ is Deligne-Mumford, so in particular

$$\Lambda^n \Omega_{\mathbb{P}_k(\underline{a})/k} \xrightarrow{\sim} \omega. \tag{45}$$

Proof The functor $\lambda \times x_i \mapsto \lambda \times x_i^{a_i}$ of the corresponding groupoids in (35) yields (42), while the stabiliser of the point with all but the ith coordinate 0 is $\boldsymbol{\mu}_{a_i}$ so the Deligne-Mumford criteria is plainly necessary, and similarly, it is sufficient since slicing (35) along $x_i = 1$ covers $\mathbb{P}_k(\underline{a})$ by affines with $\boldsymbol{\mu}_{a_i}$-action. The rest just amounts to λ acting on ∂_i by λ^{-a_i}. □

The triviality of (42) notwithstanding we have

Corollary 2.3.5 *If k is simply connected, then every $\mathbb{P}_k(\underline{a})$ is simply connected, i.e. irrespectively of any Deligne-Mumford criteria, there are no non-trivial Γ-torsors over $\mathbb{P}_k(\underline{a})$ for every finite group Γ.*

Proof By hypothesis $\mathbb{P}_k^n$ is simply connected, so it is sufficient by (42) to prove vanishing of a suitable Cěch group, i.e. that the groupoid

$$R := \mathbb{P}_k^n \times_{\mathbb{P}_k(\underline{a})} \mathbb{P}_k^n \rightrightarrows \mathbb{P}_k^n \tag{46}$$

does not admit any non-trivial functors to Γ. The space R may, by (35), be expressed as the classifier of the $\mathbb{G}_m$ action $(x_i, y_i) \mapsto (\lambda x_i, \lambda y_i)$ on the product of affine curves:

$$x_i^{a_i} = y_i^{a_i} \subset \mathbb{A}_k^2 \tag{47}$$

complemented in 0×0. Now the curves in (47) are geometrically connected, so their product is connected. It is also l.c.i. of dimension at least 2, so its homotopy depth is at least 2, whence the complement in 0 of the product is connected, and we're done a fortiori – the fact that the projections in (47) are the source and sink in (46) isn't even needed. □

Of which we will require the following variant:

Corollary 2.3.6 *If k is simply connected, and $\pi : \mathcal{P} \to \mathbb{P}_k(\underline{a})$ is a fibration in locally constant gerbes B_G for some finite group G such that $\mathcal{P}$ is simply connected, then G is a cyclic group of order a (invertible in k) and $\mathcal{P} \xrightarrow{\sim} \mathbb{P}_k(a\underline{a})$ in such a way that π is just $\lambda \mapsto \lambda^a$ in Definition 2.3.1.*

Proof The right way to prove this is the long exact sequence of homotopy groups of a fibration, which may be done wholly algebraically [22, III.g]. However, for convenience here is an ad hoc argument.

From Corollary 2.3.5, $\mathbb{P}_k(\underline{a})$ is simply connected, so by [8, IV.3.4] the locally constant gerbes up to isomorphism in B_G's over $\mathbb{P}_k(\underline{a})$ are canonically isomorphic to

$$\mathrm{H}^2(\mathbb{P}_k(\underline{a}), Z) \tag{48}$$

where Z is the centre of G. In particular if $\mathcal{P} \to \mathcal{P}'$ is $\mathcal{P}$ modulo the centre (cf. Fact 2.1.6), then $\mathcal{P}' \xrightarrow{\sim} \mathbb{P}_k(\underline{a}) \times \mathrm{B}_{G/Z}$, which isn't simply connected. As such, without loss of generality $G = Z$ is abelian, and the Leray spectral for π affords an isomorphism:

$$\mathrm{E}_2^{0,1} = \mathrm{End}(Z) \xrightarrow[d_2^{0,1}]{\sim} \mathrm{H}^2(\mathbb{P}_k(\underline{a}), Z) = \mathrm{E}_2^{2,0} \tag{49}$$

If, however, p is the characteristic of k, then from inductive application of the Artin-Schrier sequence

$$0 \to \mathbb{F}_p \to \mathbb{G}_a \to \mathbb{G}_a \to 0, \tag{50}$$

the latter group in (49) is the prime to p part of Z, so our initial G is cyclic of some order a prime to p. We have, however, a fibration:

$$\mathbb{P}_k(a\underline{a}) \to \mathbb{P}_k(\underline{a}) \tag{51}$$

in B_{μ_a}'s by the simple expedient of sending λ to λ^a in Definition 2.3.1, which is the generator of (48). □

Another very important fact which generalises is

Fact 2.3.7 Let $n = 1$ and E a vector bundle on $\mathbb{P}_k(\underline{a})$, then there are unique integers b_j such that (non-canonically)

$$E \xrightarrow{\sim} \coprod_j \mathscr{O}(b_j). \tag{52}$$

Proof We have done the rank 1 case in Fact 2.3.3, and we go by induction on the rank, $r > 1$. The push-forward of E to the moduli of $\mathbb{P}_k(\underline{a})$ is coherent, so there are plenty of meromorphic sections. As such, choose one of maximal degree to get a short exact sequence of bundles:

$$0 \to \mathscr{O}(b_r) \to E \to E'' \to 0. \tag{53}$$

Now by the induction hypothesis and Fact 2.3.3 this is split unless there is some $b_j > b_r$, $j < r$, such that

$$\mathrm{H}^0(\mathbb{P}_k(\underline{a}), \mathscr{O}(b_j - b_r - a_0 - a_1)) \neq 0. \tag{54}$$

Consequently, if we twist (53) by $\mathscr{O}(-b_r - a_0)$, then the kernel has no co-homology by Fact 2.3.3, while the co-kernel has a direct summand $\mathscr{O}(b_j - b_r - a_0)$ which has a non-trivial section given by tensoring anything in (54) with $X_1^{a_1}$, and we contradict the maximality of b_r. □

We've passed over the unicity since

Remark 2.3.8 The uniqueness of the integers b_j in (52) is just an easy version of the uniqueness of the Harder-Narasimhan filtration which, for β_j a complete repetition free list of the b_j ordered by $\beta_1 < \beta_2 < \cdots < \beta_m$, takes the form

$$E = E^0 \supset E^1 = \coprod_{b_j > \beta_1} \mathscr{O}(b_j) \supset \cdots \supset E^{m-1} = \coprod_{b_j > \beta_{m-1}} \mathscr{O}(b_j) \supset E^m = 0.$$

2.4 Radial Foliations

In this section we work over $\mathbb{C}$, and unfortunately, we will need

Notation 2.4.1 The vector $\underline{a} \in \mathbb{Z}_{>0}^{n+1}$ will be written (at least for this section) as the n-tuple of positive integers $(a_0, aa_1, \ldots, aa_n)$, $n \geq 1$, where $a_1, \ldots, a_n$ are relatively prime, and $a \in \mathbb{N}$.

The lack of symmetry in the notation is in the nature of

Definition 2.4.2 The radial foliation, $\mathcal{R}$, on $\mathbb{P}_{\mathbb{C}}(\underline{a})$ is equivalently:

(a) The foliation defined by the 0th coordinate $\mathcal{O}(a_0) \to T_{\mathbb{P}_{\mathbb{C}}(\underline{a})}$ in the Euler sequence (44).
(b) The foliation defined by the (rational) projection $\mathbb{P}_{\mathbb{C}}(\underline{a}) \dashrightarrow \mathbb{P}_{\mathbb{C}}(aa_1, \ldots, aa_n)$.

In the particular case that $n = 1$ there is a certain ambiguity in the definition according as to whether one saturates the map in (a) at the centre of the projection in (b), albeit, fortunately this tends to be clear according to context.

The following bunch of properties will aid in radial foliation recognition:

Facts 2.4.3 Given a radial foliation $\mathbb{P}_{\mathbb{C}}(\underline{a}) \to [\mathbb{P}_{\mathbb{C}}(\underline{a})/\mathcal{R}]$:

(a) Its canonical bundle, $K_{\mathcal{R}}$ (understood logarithmically if $n = 1$) is $\mathcal{O}(a_0)$.
(b) On the étale neighbourhood of the (unique) singular point given by $x_0 = 1$, $x_i = 0, i \geq 1$ in (35), $\mathcal{R}$ is generated by the vector field $a_1 x_1 \frac{\partial}{\partial x_1} + \cdots a_n x_n \frac{\partial}{\partial x_n}$.
(c) The ith coordinate axis in (b) is a smooth embedded $\mathcal{R}$-invariant $\mathbb{P}_{\mathbb{C}}(a_0, aa_i)$ with $K_{\mathcal{R}}$ degree $-1/aa_i$, while the degree of the generic invariant champ is $-1/a$.
(d) The smoothed weighted blow-up, [24, I.iv.3], $\mathcal{P} \to \mathbb{P}_{\mathbb{C}}(\underline{a})$ in the singularity with weights $a_1, \ldots, a_n$ resolves the rational map of Definition 2.4.2.(b). Indeed (cf. Revision 2.2.11), the induced foliation $\mathcal{P} \to [\mathcal{P}/\tilde{\mathcal{R}}]$ is a bundle of $\mathbb{P}_{\mathbb{C}}(a_0, a)$s over a $\mathbb{P}_{\mathbb{C}}(aa_1, \ldots, aa_n)$, and $K_{\tilde{\mathcal{R}}}(+\mathcal{E}) = K_{\mathcal{R}}$ for $\mathcal{E}$ the exceptional divisor.

Proof Of these only (d) merits comment. Specifically smoothed weighted blow-ups in [24, I.iv.3] are understood to have weights without a common divisor, so in the first place by the formulae of [24, pg. 89] and Lemma 2.1.4, we have a resolution:

$$\begin{array}{ccc} \mathcal{P}_0 & \xrightarrow[\rho_0]{} & \mathbb{P}_{\mathbb{C}}(a_1, \ldots, a_n) \\ \pi_0 \Big\downarrow \text{ weighted blow up with weights } a_i & & \\ \mathbb{P}_{\mathbb{C}}(a_0, aa_1, \ldots, aa_n) & & \end{array} \tag{55}$$

in which the exceptional divisor $\mathcal{E}_0$ is isomorphic to $\mathrm{B}_{\mu_{a_0}} \times \mathbb{P}_{\mathbb{C}}(a_1, \ldots, a_n)$, and the various bundles are related by

$$\rho_0^* \mathcal{O}_{\mathbb{P}_{\mathbb{C}}(a_1, \ldots, a_n)}(1) = \pi_0^* \mathcal{O}(a) - \mathcal{E}_0. \tag{56}$$

All of which becomes much cleaner if, the common divisor notwithstanding, one permits the weights $aa_1, \ldots, aa_n$. This is equivalent to taking an ath root of $\mathcal{E}_0$, so we get a diagram in which the square (whose horizontals are rightmost) is fibred:

$$\begin{array}{ccccc} \mathbb{P}_{\mathbb{C}}(a_0, aa_1, \ldots, aa_n) & \xleftarrow[\text{with weights } aa_i]{\text{weighted blow up}} & \mathcal{P} & \xrightarrow[\text{of } \mathcal{E}_0]{\text{extract } a\text{th root}} & \mathcal{P}_0 \\ \rho\downarrow & & & & \downarrow\rho_0 \\ \mathbb{P}_{\mathbb{C}}(aa_1, \ldots, aa_n) & & & \xrightarrow[\text{of order } a]{\text{non-trivial gerbe}} & \mathbb{P}_{\mathbb{C}}(a_1, \ldots, a_n) \end{array} \tag{57}$$

by (56), i.e. the gerbe of the bottom horizontal is the class of $\mathcal{O}(1)$ in

$$\mathrm{H}^2(\mathbb{P}_{\mathbb{C}}(a_1, \ldots, a_n), \boldsymbol{\mu}_a). \tag{58}$$

In particular, therefore, if $\mathcal{E}$ is the new exceptional divisor, then (56) becomes

$$\rho^* \mathcal{O}_{\mathbb{P}_{\mathbb{C}}(aa_1, \ldots, aa_n)}(1) = \pi^* \mathcal{O}(1) - \mathcal{E}, \tag{59}$$

while the fibres of ρ are identically those of ρ_0. The latter, however, are simply connected since ρ_0 has a section, so, [6, 1.1], a local calculation of their non-scheme-like points implies that they're all $\mathbb{P}_{\mathbb{C}}(a_0, a)$s. □

By way of disambiguation let us present the next proposition in the form

Fact/Definition 2.4.4 Every deformation of a radial foliation is locally trivial, i.e. if for a (geometrically) pointed scheme $\mathrm{pt} \xrightarrow{s} S$ we have a map $\mathcal{X} \to [\mathcal{X}/\mathcal{F}] \to S$ (equivalently of foliations indexed by the points of S) for which the special fibre $\mathcal{X}_s \to [\mathcal{X}_s/\mathcal{F}_s]$ is a radial foliation, then there is an étale neighbourhood $U \to S$ such that

$$\begin{array}{ccc} \mathcal{X} \times_S U & \xrightarrow{\sim} & \mathcal{X}_s \times U \\ \downarrow & & \downarrow \\ [\mathcal{X} \times_S U/\mathcal{F}] & \xrightarrow{\sim} & [\mathcal{X}_s/\mathcal{F}_s] \times U \end{array} \tag{60}$$

commutes, with the horizontal arrows' isomorphisms.

Proof By [2] it will suffice to replace S, resp. $\mathcal{X}$, by its completion in s, resp. the fibre, and to prove (60) in the formal category – so, keeping the same notation, $U \xrightarrow{\sim} S$. Consequently, if $\mathfrak{m}$ is the ideal of s and $S_n = \mathrm{Spec}(\mathcal{O}_S/\mathfrak{m}^n)$, it will even suffice to prove (60) with $U = S_n$, where, by way of notation, $\mathcal{X}_n := \mathcal{X} \times_S S_n$. Proceeding by induction on $n \geq 1$, the case $n = 1$ is given, while [11, Exposé III.5] applies as written to show that the obstruction to extending an isomorphism from $\mathcal{X}_n$ to $\mathcal{X}_0 \times S_n$ to the $n + 1$th thickening lies in

$$\mathrm{H}^1(\mathcal{X}_0, T_{\mathcal{X}_0} \otimes \mathfrak{m}^n/\mathfrak{m}^{n+1}). \tag{61}$$

By the Euler sequence, (44), and Serre's explicit calculation, Fact 2.3.3, this is zero. As such, we can certainly find an isomorphism $f : \mathcal{X}_n \xrightarrow{\sim} \mathcal{X}_0 \times S_n$, but it may not be foliated, i.e. the composition

$$f^*\Omega_{\mathcal{X}_0 \times S_{n+1}/\mathcal{F}} \to \Omega_{\mathcal{X}_{n+1}} \to K_{\mathcal{F}} \otimes \mathcal{O}_{\mathcal{X}_{n+1}} \tag{62}$$

may be non-trivial. We have, however, a foliated isomorphism at the nth level, and $\mathcal{X}_0$ is S_2 so (62) is, equivalently, a non-trivial map:

$$T_{\mathcal{F}}|_{\mathcal{X}_0} \xrightarrow{\sim} \mathcal{O}(a_0) \to \mathcal{T}_{\mathcal{X}_0/\mathcal{F}} \tag{63}$$

where the normal sheaf to the radial foliation is by (44) described by the commutative diagram with exact rows and columns:

$$\begin{array}{ccccccccc}
 & & & & 0 & & 0 & & \\
 & & & & \downarrow & & \downarrow & & \\
 & & & & \mathcal{O} & = & \mathcal{O} & & \\
 & & & & \downarrow & & \downarrow & & \\
0 & \longrightarrow & \mathcal{O}(a_0) & \longrightarrow & \coprod_j \mathcal{O}(aa_j) & \longrightarrow & \coprod_{j>0} \mathcal{O}(aa_j) & \longrightarrow & 0 \\
 & & \| & & \downarrow & & \downarrow & & \\
0 & \longrightarrow & \mathcal{O}(a_0) & \longrightarrow & T_{\mathcal{X}_0} & \longrightarrow & T_{\mathcal{X}_0/\mathcal{F}} & \longrightarrow & 0 \\
 & & & & \downarrow & & \downarrow & & \\
 & & & & 0 & & 0 & &
\end{array} \tag{64}$$

(64)

Twisting by $\mathcal{O}(-a_0)$ an arrow (63) is, therefore, a quotient of the space of global sections in the middle of the rightmost column of (64), i.e. the $\mathbb{C}$-vector space of vector fields with, in the notation of (35), basis:

$$x_0^{i_0} x_1^{i_1} \cdots x_n^{i_n} \cdot \frac{\partial}{\partial x_j}, \quad a_0 i_0 + aa_1 + \cdots aa_n = a_j - a_0 \;\; j > 0,\; i_k \geq 0 \tag{65}$$

On the other hand – [11, Exposé III.5] again – the possibilities for changing the isomorphism f are a principal homogeneous space under

$$\mathrm{H}^0(\mathcal{X}_0, T_{\mathcal{X}_0} \otimes \mathfrak{m}^n/\mathfrak{m}^{n+1}) \tag{66}$$

whose effect on (63) is given by the Lie bracket:

$$T_{\mathcal{X}_0} \to \mathrm{Hom}(T_{\mathcal{F}}, T_{\mathcal{X}_0/\mathcal{F}}) : D \mapsto [-, D] \tag{67}$$

which at the level of global sections has, by explicit calculation, image exactly (65), so a suitable twist of f under (66) is a foliated isomorphism. □

We will equally need a slight generalisation, to wit:

Remark 2.4.5 The same statement is equally true under the hypothesis that the universal cover of $\mathcal{X}_s$ is a radial foliation. Indeed for π_1 a finite group, all modules in which the cardinality of π_1 is invertible are acyclic, and we're in characteristic zero, so the obstruction (61) still vanishes and (67) is still surjective on global sections.

2.5 *Net Completion*

The entire contents of this section should be standard, but it isn't in the EGAs, so we give the details. We begin with the easiest case, viz, a local embedding $f : Y \to X$ of (not necessarily separated) schemes. Thus, by definition, [9, 4.2.1 & 4.5.1], for every $y \in Y$ there are (Zariski) open neighbourhoods $Y \supseteq U \ni y$, resp. $X \supseteq V \ni f(y)$, such that

$$f : U \hookrightarrow V \tag{68}$$

is a closed embedding. In particular, therefore, we have a short exact sequence:

$$0 \to \mathcal{I} \to f^{-1}\mathcal{O}_X \to \mathcal{O}_Y \to 0 \tag{69}$$

of sheaves, for some ideal $\mathcal{I}$, and we observe

Fact/Definition 2.5.1 For every $n \in \mathbb{Z}_{>0}$, define $\mathcal{O}_{f_n} := f^{-1}\mathcal{O}_X/\mathcal{I}^n$, then the ringed space $Y_n := (Y, \mathcal{O}_{f_n})$ is a scheme.

Proof The question is local on Y, so modulo notation we can, (68), suppose $f : Y \hookrightarrow X$ is a closed embedding of affines. In particular, therefore, it is defined by a quasi-coherent sheaf of ideals $\mathcal{J}$. As such $\mathcal{O}_{f_n}$ is the sheaf (on Y) associated to the pre-sheaf:

$$U \mapsto \varinjlim_{V \cap Y = U} \Gamma(V, \mathcal{O}_X)/\Gamma(V, \mathcal{J})^n. \tag{70}$$

This is, however, already not only a sheaf, but the structure sheaf, $\mathcal{O}_X/\mathcal{J}^n$, of the nth thickening of Y in X, so Y_n is a scheme. □

For the avoidance of possibly competing definitions when (without relevance to our current considerations) things fail to be Noetherian or excellent or whatever, let us make

Fact/Definition 2.5.2 A morphism $f : \mathcal{Y} \to \mathcal{X}$ of Deligne-Mumford champs is net if it is étale locally a closed embedding, i.e. for every geometric point y of Y there are étale neighbourhoods $U \to Y$ of y, resp. $V \to X$ of $x = f(y)$, together with a closed embedding $U \hookrightarrow V$ such that

$$\begin{array}{ccc} U & \longrightarrow & V \\ \downarrow & & \downarrow \\ Y & \xrightarrow{f} & X \end{array} \tag{71}$$

commutes. Consequently, if everything is Noetherian, then f is net iff the strict Henselisation $\mathcal{O}^h_{\mathcal{Y},y}$ is a quotient of $\mathcal{O}^h_{\mathcal{X},x}$ in every point; cf. [11, Exposé I,3.7].

Now suppose $f : Y \to X$ is a net map of algebraic spaces. Replacing X by a suitable (embedded) Zariski open, we may by (71) find étale covers $U \to Y$, resp. $V \to X$, affording (a not necessarily fibred) square of the form (71) in which $U \hookrightarrow V$ is a closed embedding. As such $R_0 := U \times_Y U \rightrightarrows U$, resp. $R := V \times V \rightrightarrows V$ are (not necessarily closed unless Y, resp. X is separated) embedded in $U \times U$, resp. $V \times V$ so that the induced functor $R_0 \to R$ is a not necessarily closed embedding, and we make

Fact/Definition 2.5.3 For every $n \in \mathbb{N}$, $R_n \hookrightarrow R$, resp. $U_n \hookrightarrow V$ is the nth thickening of $R_0 \hookrightarrow R$, resp. $U \hookrightarrow V$, in the sense of Fact/Definition 2.5.1. In particular $R_n \rightrightarrows U_n$ is an étale equivalence relation, and we define the nth thickening, Y_n, of Y along f to be the quotient U_n/R_n. Consequently, if Y is a scheme, then Y_n is too.

Proof Consider the diagram:

$$\begin{array}{ccccc} R_0 & \longrightarrow & R_U & \longrightarrow & R \\ {\scriptstyle s}\downarrow & & \downarrow & & \downarrow{\scriptstyle s} \\ U & = & U & \longrightarrow & V \end{array} \tag{72}$$

where the rightmost square is fibred. Thus, all the verticals are étale, the rightmost horizontals are closed embeddings, while the composition of the top row is an embedding, so $R_0 \hookrightarrow R$ is an open embedding, and whence the source and sink of $R_n \rightrightarrows U_n$ are étale. Finally, for any scheme T, the sets $R(T) \rightrightarrows V(T)$ form an equivalence relation, and we can identify the T-points of R_n with those of R such that the nth power of the ideal of the fibre over R_0 is 0, which since everything is étale implies that $R_n(T) \rightrightarrows U_n(T)$ is an equivalence relation. □

This brings us to a net map, $f : \mathcal{Y} \to \mathcal{X}$, of champs, then proceeding exactly as above, ($U \to \mathcal{X}$, $V \to \mathcal{Y}$ étale covers, etc.), we find that f is equivalent to a functor $R_0 \xrightarrow{F} R$ between groupoids, which as a map is itself net, and whence

Fact/Definition 2.5.4 The nth thickening of $\mathcal{Y}_n$ along f is the classifying champ $[U_n/R_n]$ of the étale groupoid $R_n \rightrightarrows U_n$ where R_n is the nth thickening of R_0 along the functor F, so inter alia there is a natural extension $f_n : \mathcal{Y}_n \to \mathcal{X}$ of f.

Proof The fact that $R_n \rightrightarrows U_n$ is an étale groupoid is mutatis mutandis the proof of Fact/Definition 2.5.3, and the description of the T-points therein also suffices to conclude that f_n exists. Finally, refining the covers U, V as necessary, the definition of $\mathcal{Y}_n$ is, up to equivalence, independent of the given presentation. □

It therefore only remains to make

Fact/Definition 2.5.5 The completion, $\mathfrak{Y}$, along a net map $f : Y \to X$ of schemes is the direct limit, $\varinjlim_n Y_n$, in the category of formal schemes of the nth thickening $f_n : Y_n \to X$ of Fact/Definition 2.5.3. Similarly, the completion, $\hat{\mathcal{Y}}$, along a net map $f : \mathcal{Y} \to \mathcal{X}$ of champ is the classifier of the étale groupoid which is the completion, $\mathfrak{R} \rightrightarrows \mathfrak{U}$, along the net functor $F : R_0 \to R$ of Fact/Definition 2.5.4. Consequently, by construction, f factors as

$$\mathcal{Y} \hookrightarrow \hat{\mathcal{Y}} \xrightarrow{\hat{f}} \mathcal{X} \tag{73}$$

where the former map is an embedding, and the latter is net.

2.6 *Trivial Remarks on the Analytic Topology*

As we've observed in the proof of Lemma 2.1.4 every separated Deligne-Mumford champ is étale locally the classifier, $[U/G]$, of a (not necessarily faithful) finite group action $G \times U \rightrightarrows U$. An étale neighbourhood is, however, rarely embedded, so this isn't quite as convenient as the corresponding analytic statement, i.e.

Fact 2.6.1 If $\mathcal{X}/\mathbb{C}$ is a separated Deligne-Mumford champ of finite type, then for every geometric point, x, there is an étale neighbourhood $x \in \Delta \to \mathcal{X}$ in the analytic topology together with a finite group action $G_x \times \Delta \rightrightarrows \Delta$ of the stabiliser such that $[\Delta/G_x] \hookrightarrow \mathcal{X}$ is an open embedding.

Proof From the algebraic statement: the coarse moduli U/G_x is an étale neighbourhood of the moduli $\mu : \mathcal{X} \to X$ such that we have a fibre square:

$$\begin{array}{ccc} \mathcal{X} & \longleftarrow & [U/G_x] \\ {\scriptstyle\mu}\big\downarrow & & \big\downarrow \\ X & \longleftarrow & U/G_x \end{array} \tag{74}$$

There is, however, an open embedding $\Delta' \hookrightarrow U/G_x$ whose composition with the lower horizontal in (74) is an embedding, so $\mu^{-1}(\Delta')$ is embedded in both $\mathcal{X}$ and $[U/G_x]$, while its pre-image, Δ, in U is both embedded and G_x equivariant. □

We will only ever have to consider smooth champs in the analytic topology, but as it happens, everything works in maximal generality. We require:

Lemma 2.6.2 *If X is a reduced complex space, then the sheaf, $\mathcal{R}_X$, of real analytic functions on X is coherent (as a module over itself).*

Proof The discussion is local, so we can suppose that X is a closed analytic subset of $U \subset \mathbb{C}^n$ with finitely many irreducible components $X_1, \cdots, X_r$. Each X_i has a conjugate $\bar{X}_i$ and by [25, V, Prop. 8] for any $x \in X_i$ the complexification of X_i at x in the real manifold $\mathbb{R}^n \times \mathbb{R}(1)^n$ is $X_i \times \bar{X}_i$. Consequently, op. cit. V, Prop. 1, $\cup_i X_i \times \bar{X}_i$ contains the complexification of X at any $x \in X$; and each X_i is everywhere locally Zariski dense in $X_i \times \bar{X}_i$, so X is everywhere locally Zariski dense in $\cup_i X_i \times \bar{X}_i$. Consequently, by op. cit., $\cup_i X_i \times \bar{X}_i$ is everywhere the complexification of X, so by op. cit. V, Prop. 5, $\mathcal{R}_X$ is coherent. □

This combines with Malgrange's preparation theorem to afford:

Fact/Definition 2.6.3 If $C_\bullet$ is the sheaf of continuous functions on a topological space, and $X/\mathbb{C}$ is a reduced complex space, then, functorially in X, there is a well-defined sub-sheaf, $\mathcal{A}_X \hookrightarrow C_X$, of smooth functions. In the particular case that $\mu : \mathcal{X} \to X$ is the moduli of a separated Deligne-Mumford champ,

$$\mu_* \mathcal{A}_{\mathcal{X}} \subseteq \mathcal{A}_X \subseteq \mu_* C_{\mathcal{X}} = C_X. \tag{75}$$

Proof First pass to the real analytic functions $\mathcal{R}_X$ and for a local embedding $i : X \hookrightarrow M$ in a smooth about $x \in X$, with ideal I_X in $\mathcal{R}_M$, we have, by Lemma 2.6.2 and [17, VI.3.10], an exact sequence:

$$0 \leftarrow \mathcal{R}_X \otimes_{\mathcal{R}_M} \mathcal{A}_M \leftarrow \mathcal{A}_M \leftarrow \mathcal{A}_M \otimes_{\mathcal{R}_M} I_X \leftarrow 0 \tag{76}$$

wherein $\mathcal{A}_M \otimes_{\mathcal{R}_M} I_X$ is equally the ideal of smooth functions, $\mathcal{A}_M$, vanishing on X. In particular, therefore, we have an embedding:

$$\mathcal{A}_X^M := \mathcal{R}_X \otimes_{\mathcal{R}_M} \mathcal{A}_M \hookrightarrow C_X. \tag{77}$$

Now observe (by way of the obvious diagram chase implied by (76)) that if M has the embedding dimension of X at x, then for any other smooth embedding $X \hookrightarrow N$ at x, there is a unique isomorphism which fills the right-hand side of

$$\begin{array}{ccc} \mathcal{R}_X & \longrightarrow & \mathcal{A}_X^N \\ \| & & \\ \mathcal{R}_X & \longrightarrow & \mathcal{A}_X^M \end{array} \tag{78}$$

in such a way that the diagram commutes. As such $X \mapsto \mathcal{A}_X$ is well defined and functorial, while (75) is immediate from Fact 2.6.1 and (77). □

In order to apply this we need another

Lemma 2.6.4 *Let $\mu : \mathcal{X} \to X$ be the moduli of a Deligne-Mumford champ and $\coprod_{\alpha\in A} W_\alpha \to X$ an open cover (in the classical sense), then up to passing to a locally finite refinement, there are functions*

$$\rho_\alpha \in \Gamma(X, \mu_*\mathcal{A}_{\mathcal{X}}) \text{ with support in } W_\alpha \text{ such that } \sum_\alpha \rho_\alpha = 1. \tag{79}$$

In particular for $\mathcal{M}$ any sheaf of $\mathcal{A}_{\mathcal{X}}$-modules,

$$\mathrm{H}^q(\mathcal{X}, \mathcal{M}) = 0, \quad \forall q > 0. \tag{80}$$

Proof Refining as necessary we can suppose, without loss of generality, that we have covers $\coprod_{\alpha\in A} U_\alpha$, $\coprod_{\alpha\in A} V_\alpha$ with $\bar{U}_\alpha \subset V_\alpha$; $\bar{V}_\alpha \subset W_\alpha$ and each of U_α, V_α, W_α satisfies Fact 2.6.1, i.e. there are étale covers $\coprod_{\alpha\in A} U'_\alpha \to \mathcal{X}$, etc.; finite group actions $G_\alpha \rightrightarrows U'_\alpha$ etc.; G_α equivariant inclusions $\bar{U}'_\alpha \subset V'_\alpha$ etc.; and compatible identifications of U_α with U'_α/G_α etc. As such if $f_\alpha : W'_\alpha \to [0, 1]$ is a smooth (in the sense of Fact/Definition 2.6.3) function which is identically 1 on U'_α, resp. identically 0 off V'_α, then its trace, g_α, is a global section of $\mu_*\mathcal{A}_{\mathcal{X}}$ supported in W_α which is identically 1 on U_α, resp. identically 0 off V_α, and

$$\rho_\alpha(x) := \frac{g_\alpha(x)}{\sum_\beta g_\beta(x)} \tag{81}$$

does the job. Consequently, any sheaf of $\mu_*\mathcal{A}_X$ modules is flasque, while μ_* is acyclic on $\mathbb{Q}$-vector spaces and whence (80). □

We come therefore to the point of the discussion, by way of

Fact 2.6.5 If $\mathcal{Y} \hookrightarrow \mathcal{X}$ is an embedding of smooth complex Deligne-Mumford champ with $\mathcal{Y}$ proper, then there are a family of open embeddings $\mathcal{Y} \hookrightarrow \mathcal{U}_t \hookrightarrow \mathcal{X}$ with $\cap_t \mathcal{U}_t = \mathcal{Y}$ and each $\mathcal{Y} \xrightarrow{i_t} \mathcal{U}_t \xrightarrow{r_t} \mathcal{Y}$ a deformation retract with $i_t r_t$ homotopic to the identity.

Proof The expedient of taking the trace under G_x in Fact 2.6.1 affords locally equivariant metrics which by (79) can be patched to a smooth metric, ω, on $\mathcal{X}$. As such at every geometric point x there is a G_x equivariant neighbourhood $V_x \hookrightarrow T_{\mathcal{X},x}$ of 0 such that the exponential afforded by ω yields an embedding:

$$\exp : [V_x/G_x] \to \mathcal{X} \tag{82}$$

On the other hand, by (80) the exact sequence

$$0 \to T_{\mathcal{Y}} \to T_{\mathcal{X}} \to N_{\mathcal{Y}/\mathcal{X}} \to 0 \tag{83}$$

has a smooth splitting, $n : N_{\mathcal{Y}/\mathcal{X}} \to T_{\mathcal{X}}$, so $\exp(n)$ restricted to appropriate neighbourhoods of the zero section in $N_{\mathcal{Y}/\mathcal{X}}$ gives what we want. □

This is, of course, just the usual proof of the corresponding fact for smooth manifolds so it's worth making

Remark 2.6.6 Slightly, but not much, more subtly if X is Kähler, then so is $\mathcal{X}$.

Finally, we require a baby GAGA:

Fact 2.6.7 Let $\mathcal{X}/\mathbb{C}$ be a normal complex analytic champ, i.e. the classifier of an étale groupoid $R \rightrightarrows U$ in the analytic topology, whose moduli $\mu : \mathcal{X} \to X$ is a finite map to an algebraic space with algebraic ramification in co-dimension 1, then $\mathcal{X}$ is an algebraic Deligne-Mumford champ. Similarly, if $\mathcal{Y}'_t \to \mathcal{U}_t$ is a smooth champ finite over the neighbourhoods of Fact 2.6.5, then there is an algebraic champ $\mathcal{Y}' \to \mathcal{Y}$ such that (in the notation of op. cit.) $\mathcal{Y}'_t$ is equivalent to $r_t^* \mathcal{Y}'$.

Proof Without loss of generality $\mathcal{X}$ is connected, so exactly as in Fact 2.1.6, there is a map $\mathcal{X} \to \mathcal{X}^0$ expressing $\mathcal{X}$ as a locally constant gerbe in B_Γ's for some finite group Γ wherein the stabiliser of the generic point of $\mathcal{X}^0$ is trivial, and by [1, 5.1] $\mathcal{X}_0$ is algebraic. As to $\mathcal{X} \to \mathcal{X}^0$, we must first consider the link in the sense of Giraud, [8, IV.1.1.7.3], i.e. the representation of $\pi_1(\mathcal{X}^0)$ in the outer automorphisms of Γ, but these are the same in the algebraic and analytic categories, so the next port of call is the obstruction to the existence of a champ with a given link. This is, [8, VI.2.3], a class in $\mathrm{H}^3(\mathcal{X}^0, \mathrm{Z})$, where Z is the centre of the link, i.e. the locally constant sheaf in the centre of Γ with induced $\pi_1(\mathcal{X}^0)$ action. By [13, Exposé XVI.4.1], étale and analytic cohomology coincide, while the obstruction vanishes analytically, so there is at least one algebraic champ, $\mathcal{X}' \to \mathcal{X}^0$, which is a locally constant gerbe in B_Γ's for the same link. Equally, $\mathcal{X}'$ is an analytic champ, so, in either case the equivalence class of all possible champs with this link is, [8, IV.3.4], the orbit of $\mathcal{X}'$ under $\mathrm{H}^2(\mathcal{X}^0, \mathrm{Z})$, and whence $\mathcal{X} \to \mathcal{X}^0$ is algebraic by another application of [13, Exposé XVI.4.1]. The second part about the $\mathcal{U}_t$'s proceeds mutatis mutandis given Fact 2.6.5. □

3 $K_{\mathcal{F}}$-negative curves

3.1 Foliations as Groupoids

As we've already remarked prior to Definition 2.2.1 the point of view of a foliation as an integrable quotient of the cotangent sheaf is misleading. Rather a foliation should, [21], be considered as an infinitesimal equivalence relation outside of its singularities and the equivalence of this definition to that involving linear 1st-order data as a non-trivial theorem (notwithstanding the triviality of the proof) specific to characteristic zero. In any case we work over $\mathbb{C}$ and begin by reviewing the aforesaid equivalence; whence let U be a complex affine scheme and $\mathcal{F}$ a smooth foliation on U. Notice that U may be singular, or even non-reduced, so $\mathcal{F}$ smooth means that (22) is everywhere a locally free rank 1 quotient, or, equivalently, in the notation of op. cit. for some (thus any) embedding of U in a smooth M the composition,

$$T_{\mathcal{F}} = \mathcal{H}om(Q, \mathcal{O}_U) \to \mathcal{T}_U \to T_M \otimes \mathcal{O}_U, \tag{84}$$

is, since Q is supposed torsion free, an injection of bundles, which, by hypothesis (immediately post (22)), is the case at every generic point. Now observe:

Fact/Definition 3.1.1 Let $U/\mathbb{C}$ be a foliated affine scheme such that $T_{\mathcal{F}}$ in (84) is a line bundle, but not that the composition therein is an injection of bundles, i.e. if U were normal and $\mathcal{F}$ saturated, a Gorenstein foliation by curves, with $\mathcal{D}_{\mathcal{F}}^{-\infty}$ the $\mathcal{O}_U$ sub-algebra of $\mathcal{D}iff_U^{-\infty}$ generated by $T_{\mathcal{F}}$, then for ∂ a local generator the dual of the co-product,

$$\mathcal{D}_{\mathcal{F}}^{-\infty} \to \mathcal{D}_{\mathcal{F}}^{-\infty} \otimes_{\mathcal{O}_U} \mathcal{D}_{\mathcal{F}}^{-\infty} : \frac{\partial^n}{n!} \mapsto \sum_{i+j=n} \frac{\partial^i}{i!} \otimes \frac{\partial^j}{j!}, \quad n \in \mathbb{Z}_{\geq 0}, \tag{85}$$

respectively, the product,

$$\mathcal{D}_{\mathcal{F}}^{-\infty} \otimes_{\mathcal{O}_U} \mathcal{D}_{\mathcal{F}}^{-\infty} \to \mathcal{D}_{\mathcal{F}}^{-\infty} : \partial^i \otimes \partial^j \mapsto \partial^{i+j} \quad i, j \in \mathbb{Z}_{\geq 0}, \tag{86}$$

define a sheaf of adic algebras, $\mathcal{O}_{\mathfrak{F}} = \varprojlim_n \mathcal{H}om(\mathcal{D}_{\mathcal{F}}^{-n}, \mathcal{O}_U)$, whence a relatively smooth formal scheme,

$$p : \mathfrak{F} := \mathrm{Spf}\mathcal{O}_{\mathfrak{F}} \to U, \tag{87}$$

respectively, an infinitesimal groupoid structure,

$$(p_1, p_2) : \mathfrak{F} \rightrightarrows U, \tag{88}$$

factoring through the infinitesimal equivalence relation (all points are equivalent) defined by Grothendieck's jets, i.e.

$$\begin{gathered}(p_1, p_2) : \mathfrak{P}_U = \mathrm{Spf}\, \mathcal{P}_U^{\infty} \rightrightarrows U, \quad \mathcal{P}_U^{\infty} = \varprojlim_n \mathcal{P}_U^{(n)} \\ \mathfrak{F} \xrightarrow{p_1 \times p_2} \mathfrak{P}_U \xrightarrow{p_1 \times p_2} U \times U.\end{gathered} \tag{89}$$

As such (88) defines an infinitesimal equivalence relation around $x \in U$ iff $\mathfrak{F}$ embeds via $p_1 \times p_2$ in $\mathfrak{P}_U$, which in turn is iff (84) is an embedding, i.e. $\mathcal{F}$ is smooth at x.

Proof The infinitesimal version of the relation all points are equivalent, i.e.

$$(p_1, p_2) : U \times U \rightrightarrows U \tag{90}$$

is the completion of the same in the diagonal, $\Delta \hookrightarrow U \times U$, to wit, in the notation of [10][16.3.1], the formal spectrum of the I_Δ-adic algebra in the first line in (89). In particular, therefore, the co-product of differential operators is just the dual of the product in $\mathscr{P}_U^\infty$. Furthermore, $\mathfrak{P}_U$ of (89) satisfies the axioms for a relation in the category of formal schemes, and, op. cit. 16.8.9.1, the product of differential operators is dual to the pull-back along composition of arrows:

$$c^* : \mathscr{P}_U^\infty \to \mathscr{P}_{U\, p_2^*}^\infty \otimes_{\mathscr{O}_X\, p_1^*} \mathscr{P}_U^\infty \tag{91}$$

wherein, as indicated, one should be careful about the bi-module structure on $\mathscr{P}_U^\infty$. Specifically (since working with sheaf valued operators slightly obscures op. cit. §16.7–8) given differential operators A, B, we have a map:

$$\begin{aligned}&\mathscr{P}_{U\, p_2^*}^\infty \otimes_{\mathscr{O}_U\, p_1^*} \mathscr{P}_U^\infty \xrightarrow{A\cdot B} \mathscr{O}_U : j\otimes k \mapsto A(jB(k));\ fA\cdot B(j\otimes k) = \\ &A\cdot B(p_1^* fj\otimes k),\ A\cdot B(jp_2^* f\otimes k) = A\cdot B(j\otimes p_1^* fk),\ f\in\mathscr{O},\end{aligned} \tag{92}$$

and the operator composition, AB, is the composite of (92) with (91). In any case, dualising just puts everything back the way it was modulo the fact that things aren't reflexive at singularities. More precisely, by (85),

$$\mathscr{O}_{\mathfrak{F}} = \mathscr{O}_U[[\partial^\vee]],\ \partial^\vee(\partial) = 1,\ p_1^* f = f,\ p_2^*(f) = \sum_{n=01}^{\infty} \frac{\partial^n(f)\partial^{-n}}{n!} \tag{93}$$

so not only is (87) formally smooth (even if U were nowhere reduced), the graded algebra in the adic filtration afforded by its trace is canonically $\mathrm{Sym}K_{\mathcal{F}}$. On the other hand, the graded algebra of the I_Δ-adic structure on the jet groupoid is a quotient:

$$\mathrm{Sym}\Omega_U \twoheadrightarrow \mathrm{gr}\mathscr{P}_U^\infty \tag{94}$$

so $\mathfrak{F}$ embeds in $\mathfrak{P}_U$ iff $K_{\mathcal{F}}$ is a quotient of Ω_U. □

Rather more picturesquely, what we have done at the smooth points is to add an infinitesimal germ of the leaf in the $p_2^* T_{\mathcal{F}}$ direction for each point in the diagonal, which we have completed as a groupoid at the singularities. However, as noted at the end of Fact/Definition 3.1.1 (cf. Further Claim 3.3.2), this is never a relation at the singular points, so the image of the smooth groupoid (88) in the jet groupoid, (89), always degenerates at such, to wit (Fig. 1):

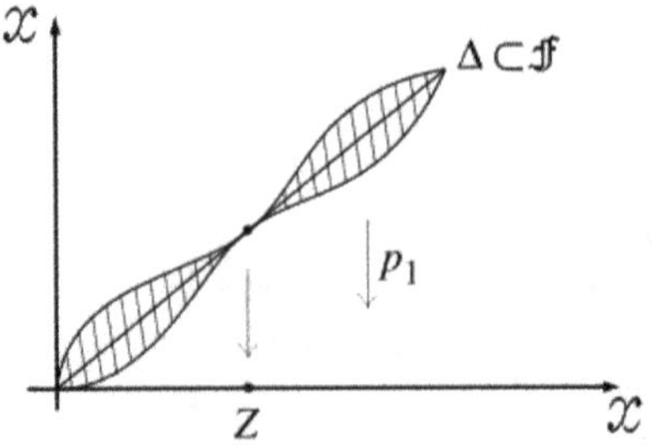

Fig. 1 Relation afforded by a groupoid with an essential singularity

In any case the infinitesimal nature of Fact/Definition 3.1.1 renders its extension to a Deligne-Mumford champ, $\mathcal{X}$, automatic since by definition this is equivalent to a groupoid with étale source and sink so there are well-defined sheaves of nilpotent $\mathcal{O}_{\mathcal{X}}$-algebras, $\mathscr{P}^{(n)}_{\mathcal{X}}$, of n-jets and, of course idem, modulo replacing nilpotent by topologically so, for the inverse limit $\mathscr{P}^{\infty}_{\mathcal{X}}$. Equally, the formation of the formal spectrum is a local construction, while both the projectors and the diagonal embedding patch, i.e. the étale local nature of (89), affords

Definition 3.1.2 The jet groupoid of a Deligne-Mumford champ $\mathcal{X}$ is the formal champ:

$$\mathfrak{P}_{\mathcal{X}} = \operatorname{Spf} \mathscr{P}^{\infty}_{\mathcal{X}} \rightrightarrows \mathcal{X} \tag{95}$$

with (formally representable) source map p_1, sink p_2 and identity the diagonal.

We have, however, already met a better description of this in Fact/Definition 2.5.4 to wit:

Fact/Definition 3.1.3 For every $n \in \mathbb{Z}_{\geq 0}$,

$$\operatorname{Spec}(\mathscr{P}^{(n)}_{\mathcal{X}}) \rightrightarrows \mathcal{X} \tag{96}$$

is the nth thickening, in the sense of Fact/Definition 2.5.4, of $\mathcal{X}$ along the diagonal $\Delta : \mathcal{X} \to \mathcal{X} \times \mathcal{X}$, so the jet groupoid of Definition 3.1.2 along with the projections p_1, p_2 is the net completion of $\mathcal{X}$ along the diagonal. Consequently, (95) is an honest groupoid, i.e. the source p_1, sink p_2, identity, inverse and composition are representable maps identically satisfying the axioms of a groupoid- and not just up to equivalence. In particular the composition is strictly associative. Similarly, if $\mathcal{X} \to [\mathcal{X}/\mathcal{F}]$ is a foliation in curves given by a line bundle:

$$T_{\mathcal{F}} \to \mathcal{T}_{\mathcal{X}}, \tag{97}$$

then (85)–(88) are valid as stated, so there is an *infinitesimal groupoid*:

$$(p_1, p_2) : \mathfrak{F} \rightrightarrows \mathcal{X} \tag{98}$$

where, again, source, sink, identity, inverse and composition are representable maps satisfying the groupoid axioms identically. In particular the identity/diagonal, $\Delta : \mathcal{X} \to \mathfrak{F}$, is a section of either projection and there is an isomorphism:

$$N_{\Delta/\mathfrak{F}} \xrightarrow{\sim} \mathcal{O}_\Delta(T_{\mathcal{F}}). \tag{99}$$

Proof Let $(s, t) : R \rightrightarrows U$ be a presentation of $\mathcal{X}$, then the sheaf of algebras $\mathscr{P}^{(n)}_{\mathcal{X}}$, $n \in \mathbb{Z}_{\geq 0}$, defined, immediately prior to Definition 3.1.2, by étale local patching may be identified with the sheaf of functions on the classifier of the groupoid:

$$\begin{array}{ccc} \mathrm{Spec}(\mathscr{P}^{(n)}_U) & \underset{\mathrm{Spec}(s^*)}{\overset{\mathrm{Spec}(t^*)}{\longleftleftarrows}} & \mathrm{Spec}(\mathscr{P}^{(n)}_R) \\ \downarrow & & \downarrow \\ U & \underset{s}{\overset{t}{\longleftleftarrows}} & R \end{array} \tag{100}$$

and in the limit $n \to \infty$ we can replace Spec in (100) by Spf to obtain a presentation in formal schemes of (95). At the same time, by Fact/Definition 2.5.4, a presentation of the nth thickening of $\mathcal{X}$ along its diagonal is given by the nth thickening of the diagonal in:

$$\begin{array}{ccc} R & \xrightarrow[\Delta]{} & R \times R \\ {\scriptstyle s}\downarrow{\scriptstyle t} & & {\scriptstyle s\times s}\downarrow{\scriptstyle t\times t} \\ U & \xrightarrow{\Delta} & U \times U \end{array} \tag{101}$$

so that such a thickening is exactly (100). As to the groupoid axioms, the composition of the diagonal in the top row of (101) with either projection is, rather than just equivalent to, the identity functor, so a fortiori, the diagonal affords the identity after net completion. Similarly, the transposition,

$$T : R \times R \to R \times R : (f_1, f_2) \mapsto (f_2, f_1), \tag{102}$$

respects the nth thickening and satisfies $T^2 = \mathrm{id}$, while the composition given by:

$$(R \times R)_{p_2} \times_{p_1} (R \times R) \to (R \times R) : (f_1, f_2) \times (g_1, g_2) \mapsto (f_1, g_2) \tag{103}$$

respects net completion and is strictly associative. Similarly, the bundle $T_{\mathcal{F}}$ of (97) generates sheaves of modules:

$$\begin{aligned} &\mathscr{D}_{\mathcal{F}}^{-\infty} = \varinjlim_{n} \mathscr{D}_{\mathcal{F}}^{-(n)} \to \varinjlim_{n} \mathscr{D}iff_{\mathcal{X}}^{-(n)} = \mathscr{D}iff_{\mathcal{X}}^{-\infty}, \text{ whose duals,} \\ &\mathscr{O}_{\mathfrak{F}} = \varprojlim_{n} \mathscr{O}_{\mathfrak{F}_n} = \mathscr{H}om(\mathscr{D}_{\mathcal{F}}^{-(n)}, \mathscr{O}_{\mathcal{X}}), \quad \mathfrak{F} := \mathrm{Spf}(\mathscr{O}_{\mathfrak{F}}), \end{aligned} \tag{104}$$

are sheaves of nilpotent $\mathscr{O}_{\mathcal{X}}$-algebras, supported on the diagonal, with limit an adic algebra with trace $\mathcal{X}$. In particular, (99) is immediate from (104), while the fact that we have a groupoid in representable maps satisfying the groupoid axioms identically is just the independence of p_2^* from the choice of ∂ in (93). Alternatively, if the foliation is smooth at every generic point, then the identity, inverse and composition satisfy the groupoid axioms identically because the jet groupoid does. □

Notice in particular that the diagonal is actually embedded in the jet groupoid, so it's worth making:

Remark 3.1.4 At a geometric diagonal point $x \times x$ in $\mathcal{X} \times \mathcal{X}$, its automorphism group is simply $\mathrm{Aut}(x) \times \mathrm{Aut}(x)$. Inside this group we have a copy of $\mathrm{Aut}(x)$ sitting diagonally. Now any attempt to define diagonal-type subgroups of automorphisms for off-diagonal points, and whence define an actual étale "neighbourhood" in which $\mathcal{X}$ embeds in some sort of diagonal way, is doomed to failure. At the infinitesimal level, however, this can be, and is, achieved by net completion in the diagonal, or equivalently the étale local description of Definition 3.1.2.

3.2 *Chow's Lemma*

We will confine ourselves to that which is strictly necessary for applications. Our interest centres on smooth formal champs $\mathfrak{F}$ whose trace C is a smooth proper champ of dimension 1 over $\mathbb{C}$. From our utilitarian point of view we'll confine ourselves to the case where $\dim \mathfrak{F} = 2$. Irrespectively, there is a well-defined normal bundle $N_{C/\mathfrak{F}}$, and we make:

Definition 3.2.1 $\mathfrak{F}$ is a concave formal neighbourhood of C if $\deg(N_{C/\mathfrak{F}}) > 0$.

Unsurprisingly, the classical Chow lemma continues to hold, i.e.

Lemma 3.2.2 (Chow, Grauert et al.) *Let L be a line bundle on $\mathfrak{F}$, then there is a quadratic polynomial P_L, depending on L, such that for all $n \in \mathbb{N}$,*

$$\mathrm{h}^0(\mathfrak{F}, L^{\otimes n}) \leq P_L(n)\,. \tag{105}$$

Proof Let $\mathfrak{F}_m$ be the mth-thickening of C, then we have an exact sequence:

$$0 \to \mathrm{Sym}^m N_{C/\mathfrak{F}}^{\vee} \to \mathscr{O}_{\mathfrak{F}_{m+1}} \to \mathscr{O}_{\mathfrak{F}_m} \to 0\,. \tag{106}$$

On the other hand, if $\mathrm{h}^0(C, L^n \otimes \mathrm{Sym}^m N^\vee_{C/\mathfrak{F}}) \neq 0$, then,

$$m \deg(N_{C/\mathfrak{F}}) \leq n \deg_C(L)\,. \tag{107}$$

Consequently, for any $n \in \mathbb{N}$,

$$\mathrm{H}^0(\mathcal{O}_{\mathfrak{F}_{m+1}} \otimes L^n) \hookrightarrow \mathrm{H}^0(\mathcal{O}_{\mathfrak{F}_m} \otimes L^n) \tag{108}$$

is injective, provided $m > M := n \deg_C(L) \deg(N_{C/\mathfrak{F}})^{-1}$ and

$$\begin{aligned} \mathrm{H}^0(\mathfrak{F}, L^{\otimes n}) &= \varprojlim_m \mathrm{H}^0(\mathfrak{F}_m, L^{\otimes n}), \text{ so} \\ \mathrm{h}^0(\mathfrak{F}, L^{\otimes n}) &\leq \sum_{k=0}^{M} \mathrm{h}^0(C, L^n \otimes N^{-k}_{C/\mathfrak{F}}). \end{aligned} \tag{109}$$

Equally, by [6, 1.1] there is a map, $\rho : C \to C$ from an honest curve, while for any bundle E, $\mathrm{h}^0(C, E) \leq \mathrm{h}^0(C, \rho^* E)$, so we conclude by Riemann-Roch. □

3.3 *Bend and Break*

We are now in a position to extend the results of [5], so to this end let $\mathcal{X} \to [\mathcal{X}/\mathcal{F}]$ be a foliated Gorenstein normal champ with projective moduli space $\mu : \mathcal{X} \to X$ and H an ample bundle on the latter. As ever the basic object of study is $K_{\mathcal{F}}$ negative curves on $\mathcal{X}$, i.e., profiting once more from [6, 1.1], maps $f : C \to \mathcal{X}$ from a smooth curve such that $K_{\mathcal{F}} \cdot_f C < 0$. Consequently, if we consider the infinitesimal groupoid of (98) as fibred over $\mathcal{X}$ via $p := p_1$, then base changing along f affords:

$$\begin{array}{ccccc} \mathfrak{F} & \longleftarrow & \mathfrak{F}_C & \xrightarrow[p \times p_2]{} & C \times \mathcal{X} \\ {\scriptstyle p:=p_1}\big\downarrow & & \big\downarrow{\scriptstyle p} & & \\ \mathcal{X} & \xleftarrow[\tilde{f}]{} & C & & \end{array} \tag{110}$$

Furthermore, the identity map of the groupoid gives a section s of p of the left, thus a fortiori of the right, vertical arrow in (110), so by (99), $\mathfrak{F}_C$ is a concave neighbourhood of $s(C)$ in the sense of Definition 3.2.1, and we assert:

Claim 3.3.1 The Zariski closure of the image of $\mathfrak{F}_C$ in $C \times X$ is irreducible of dimension at most 2 and exactly 2 if f does not factor through the singular locus $\mathcal{Z} = \mathrm{sing}(\mathcal{F})$.

Proof Let Y be the Zariski closure, which is irreducible since $\mathfrak{F}_C$ is. Moreover, X is projective by hypothesis, thus $C \times X$ admits an ample line bundle L, and

$$\mathrm{H}^0(Y, L) \to \mathrm{H}^0(\mathfrak{F}_C, L) \tag{111}$$

is injective by the definition of Y, so the Chow lemma, Lemma 3.2.2, does the first part. As to the second, by hypothesis there is an étale neighbourhood $U \to \mathcal{X}$ of the generic point of $f(C)$ on which the foliation is smooth. As such, by (89) et seq., we have an embedding:

$$\mathfrak{F}_U \xrightarrow{p_1 \times p_2} U \times U; \tag{112}$$

thus for $B = f^{-1}(U) \to C$, we have an embedding of $\mathfrak{F}_B := \mathfrak{F}_C \times_C B$ in the étale neighbourhood $Y \times_{C \times X} (B \times U)$ of Y so the dimension of the latter is at least two. □

Now let $\mathcal{Y}_0 := Y \times_X \mathcal{X}$, then, bearing in mind that C is smooth, we additionally assert:

Further Claim 3.3.2 Let $\phi_0 : \mathfrak{F}_C \to \mathcal{Y}_0$ be the natural map afforded by the definition of fibre products and $s : C \to \mathfrak{F}_C$ the trace, then around $\phi_0 s(C)$, the singularities of $\mathcal{Y}_0$ can be resolved by a (possibly zero length) chain:

$$\mathcal{Y}_0 \leftarrow \mathcal{Y}_1 \leftarrow \cdots \leftarrow \mathcal{Y}_n \tag{113}$$

of blow-ups in closed points of $\phi_0 s(C)$ and proper transforms thereof. Moreover, if $\mathcal{Y} := \mathcal{Y}_n$ is the shortest such resolution, then for all $0 \le i \le n$, there are liftings $\phi_i : \mathfrak{F}_C \to \mathcal{Y}_i$, and $\phi := \phi_n$ is étale.

Proof If the claim is true locally, then it is true since a posteriori the singularities of $\mathcal{Y}$ around $\phi_0 s(C)$ and its proper transforms are isolated so (113) is just the minimal resolution obtained by blowing up in the same, while any ϕ_i agrees with ϕ_0 on a dense set, so it lifts globally iff it lifts locally. As such we can replace $\mathcal{X}$ by an étale neighbourhood $U \to \mathcal{X}$ of the image of a closed point c of B, for some étale $B \to C$, $\mathfrak{F}$ by a formal scheme, F, with trace B and identify $\mathcal{Y}$ with a closed subvariety $V \hookrightarrow B \times U$. Now, let ∂ be a local generator of the Gorenstein foliation $\mathcal{F}$ on U and ∂_2 its pull-back to $B \times U$ via the 2nd projection, then, (93), V is ∂_2 invariant, which in turn lifts to F, and either implied foliation is just the projection to B. Consequently, from:

$$\begin{array}{ccccccccc}
0 & \longrightarrow & \Omega_B & \longrightarrow & \Omega_F\,|_{s(B)} & \longrightarrow & K_{\mathcal{F}}\,|_{s(B)} & \longrightarrow & 0 \\
 & & \| & & \uparrow \phi_0^* & & \| & & \\
0 & \longrightarrow & \Omega_B & \longrightarrow & \Omega_V\,|_{\phi_0 s(B)} & \longrightarrow & K_{\mathcal{F}}\,|_{\phi_0 s(B)} & &
\end{array} \tag{114}$$

wherever ∂_2 is non-singular, the bottom arrow in (114) is surjective, so $\mathcal{Y}$ is smooth and ϕ_0 is étale. Equally, if it is singular around a maximal ideal $\mathfrak{m}$ of $\mathcal{O}_V$, then, by definition, $\mathfrak{m}$ is ∂_2 invariant, and whence $\phi_0^{-1}\mathfrak{m}$ is too. However, ∂_2 is everywhere non-singular on F, so the latter ideal is the pull-back along $F \to B$ of an ideal on B. In particular, therefore, it is Cartier and ϕ_0 lifts to a map ϕ_1 to the blow-up of V in any singular point of ∂_2 contained in the image of $s(B)$. Now, plainly, none of the above changes if we replace ϕ_0 by ϕ_1, so it suffices to observe, [4, 1.3.3], that, after a finite sequence of such blow-ups, ∂_2 is everywhere non-singular around the proper transform of $\phi_0 s(B)$. □

As such, on replacing $\mathcal{Y}_0$ by $\mathcal{Y} = \mathcal{Y}_n$ and ϕ_0 by $\phi = \phi_n$, of our Further Claim 3.3.2, we arrive to:

Final Claim 3.3.3 For $\mathcal{Y}/C$ viewed as a C-champ via $\mathcal{Y} \to Y \to C \times X \to C$ there is an étale C-map $\phi : \mathfrak{F}_C \to \mathcal{Y}$. In particular, therefore, if for ease of notation we confuse s and ϕs:

(a) $\mathcal{Y}$ is smooth in a neighbourhood of $s(C)$.
(b) $K_{\mathcal{Y}/C} \cdot_s C = K_{\mathcal{F}} \cdot_f C$.
(c) The 2nd projection yields a map of foliated champs $(\mathcal{Y}/C) \to (\mathcal{X} \to [\mathcal{X}/\mathcal{F}])$.

Proof We've done everything in (3.3.2) except possibly (b) which is (99). □

Before proceeding to produce rational curves in Fact 3.3.5 from our Final Claim 3.3.3, it is opportune to make

Remark 3.3.4 The resolution procedure in Further Claim 3.3.2 is really a resolution procedure for the foliation $p_2^*\mathcal{F}$ in a neighbourhood of the diagonal $\Delta \to \mathcal{X}\times\mathcal{X}$. As with all resolution procedures, it is more demanding to achieve it with smooth centres, but the key point here is to make $p_2^*\mathcal{F}$ smooth around the proper transform of the diagonal, and this can be achieved with a single blow-up in the diagonal of the singularities. Specifically, the question is local, so say, $U \hookrightarrow M$ an embedding of an affine scheme into a smooth, with coordinates $x_1, \ldots, x_n$ on M restricting to functions on U, then since $\mathcal{F}|_U$ is Gorenstein, we may, on shrinking U as necessary, suppose that the foliation is defined by a vector field ∂ on U, which we write using the summation convention as

$$\partial = a_i \frac{\partial}{\partial x_i}. \tag{115}$$

As such the ideal I_Z of the singular locus Z is (a_i). Now introduce x_i, y_i as coordinates on $U \times U$ via the first, respectively second, projection, and put $z_i = x_i - y_i$, then in z_i, x_i coordinates,

$$p_2^*\partial = p_2^* a_i \frac{\partial}{\partial y_i} = -p_2^* a_i \frac{\partial}{\partial z_i}. \tag{116}$$

Now the diagonal embedding of Z has ideal $(p_1^* a_i, z_i)$ so on the proper transform $\tilde{\Delta}$ of Δ not only can we locate each point in some $p_1^* a_i \neq 0$ patch for an appropriate i, but even,

$$\partial \left(\frac{z_i}{p_1^* a_i} \right) = \frac{-p_2^* a_i}{p_1^* a_i} = 1 + \frac{(p_1^* a_i - p_2^* a_i)}{p_1^* a_i}, \tag{117}$$

so $p_2^* \partial$ is everywhere smooth around $\tilde{\Delta}$.

In any case, the following, together with our Final Claim 3.3.3, affords invariant rational curves through a generic point of the image of C.

Fact 3.3.5 Suppose in addition to items (a) and (b) of Final Claim 3.3.3 a family $p : \mathcal{Y} \to C$ of uni-dimensional champs with a section s satisfies $K_{\mathcal{Y}/C} \cdot_s C < 0$; then there is a finite extension $\mathbb{C}(C) \to K$ such that $\mathcal{Y}_K$ is dominated by $\mathbb{P}^1_K$.

Proof We may, without loss of generality, suppose that $\mathcal{Y}$, and indeed any base change, thereof, is normal. In particular, therefore, Fact 2.1.6, there is a fibration $\mathcal{Y} \to \mathcal{Y}^0$ expressing the former as a locally constant gerbe over a normal-fold, so that by [6, 1.1] we may further suppose that $\mathcal{Y} = \mathcal{Y}^0$. As such if the generic fibre of p is not dominated by a rational curve, then, op. cit., there is a finite extension $\mathbb{C}(C) \to K$ such that $\mathcal{Y} \times_C K$ is an orbifold of the form $[S_K/G]$ for some non-rational K-curve S_K and finite group G acting generically freely. Denoting by Y the moduli of $\mathcal{Y}$ and identifying K with the function field of a smooth curve B, we can suppose that S_K is the generic fibre over B of the integral closure S of Y in the function field of S_K. The normalisation $\mathcal{S}$ of the fibre $\mathcal{Y} \times_Y S$ is, therefore, a gerbe over S with generic fibre S_K. Consequently, by purity, $q : \mathcal{S} \to \mathcal{Y}$ is ramified only in components of fibres of $\mathcal{Y} \to C$. In addition, q is étale locally Galois since $S \to Y$ is and $\mathcal{Y} \to C$ is smooth in a neighbourhood of the section $s(C)$, so by [11, Exp. XIII, Cor. 5.3], q is étale locally around $s(C)$ the extraction of roots of fibres. As such, by the simple expedient of taking $\mathbb{C}(C) \to K$ sufficiently large, we can suppose – around $s(C)$ and it's pre-image – that q is scheme like and S is smooth. Better still since q is only ramified in fibres:

$$K_{S/B} = q^* K_{\mathcal{Y}/C}, \text{ and thus, } K_{S/B} \cdot_{\tilde{s}} B < 0, \tag{118}$$

for any lifting $\tilde{s}$ of s. Consequently, we may from either [5] or the classical theorem of Arakelov, [27], conclude to the absurdity that the generic fibre of $S \to B$ is a rational curve. □

The fibres of p in Fact 3.3.5 may not themselves be rational curves, and it is convenient to give them a name, to wit:

Fact/Definition 3.3.6 A smooth connected 1-dimensional Deligne-Mumford champ, $\mathcal{L}$, over a field k is said to be parabolic if its geometric fibre is dominated by a rational curve. Rather conveniently this occurs, [6, 1.1], iff the topological Euler-characteristic, $\chi(\mathcal{L})$, is strictly positive.

From which we can proceed to our conclusion:

Proposition 3.3.7 *Let $\mathcal{X} \to [\mathcal{X}/\mathcal{F}]$ be a foliated normal gerbe over a projective variety X, which is foliated Gorenstein along some $K_{\mathcal{F}}$ negative curve $C_0 \subset X$ around the generic point of which $\mathcal{F}$ is a non-singular foliation of $\mathcal{X}$; then for a generic $c \in C_0$ there is an invariant parabolic champ, $g_c : \mathcal{L}_c \to \mathcal{X}$ such that for M any nef. $\mathbb{R}$-divisor on $\mathcal{X}$ and $||$ the moduli,*

$$M_{\cdot_{|g_c|}} |\mathcal{L}_c| \leq 2 \frac{M_{\cdot_f} C}{-K_{\mathcal{F}_{\cdot_f}} C}. \tag{119}$$

Proof We apply the Final Claim 3.3.3 with C a curve mapping to the normalisation of the gerbe over C_0 in $\mathcal{X}$. By Fact 3.3.5 the generic fibre of $\mathcal{Y} \to C$ is an invariant parabolic champ, so it only remains to produce the degree bound. To this end identify the image of the section s with a curve C such that $C^2 = -K_{\mathcal{F}} \cdot C$ in the normal surface which is the moduli. Whence if L is a generic fibre of the same, M is notationally confused with the restriction of the given nef. $\mathbb{R}$-divisor, and $x \in \mathbb{R}_{>0}$, then by the Hodge index theorem,

$$2x \cdot (L \cdot M)\, C^2 \leq (L + xM)^2\, C^2 \leq \{C \cdot (L + xM)\}^2, \tag{120}$$

so taking $x = (M \cdot C)^{-1}$ we conclude. □

The same proof works, under the weaker hypothesis that only a neighbourhood of C_0 in the moduli is projective. More interestingly, the presence of even the most mild non-scheme-like structure on $\mathcal{X}$ can necessitate the precision of Proposition 3.3.7 that the existence of a parabolic invariant champ $\mathcal{L} \ni c$ is only guaranteed for generic c. Indeed:

Remark 3.3.8 Take a section C with positive square of a Hirzebruch surface $P \to C$. In the fibre through some $c \in C$, choose some set Q of points off C, and for $q \in Q$ let $n_q \in \mathbb{N}_{>1}$ be given. Choose a germ of a smooth curve, γ transverse to the fibre P_c at q. Blowing up in q, we get the proper transform γ_1 of γ, then blow up in the point where this crosses the exceptional divisor and repeat this process n_q times before blowing down the first $n_q - 1$ curves. The resulting surface S then has isolated cyclic quotient singularities with monodromy $\mathbb{Z}/n_q$ at each q in the proper transform of P_c, which itself meets at each q a rational curve in the fibre, but the said proper transform is the only component of the fibre meeting the section. Passing to the Vistoli covering champ, we see the necessity for taking $c \in C$ generic in Proposition 3.3.7, since the gerbe over the proper transform fails to be parabolic as soon as $\sum_q (1 - 1/n_q) > 2$.

3.4 The Cone of Curves

We may now apply the basic estimate of Proposition 3.3.7 to the cone of curves of a foliated Gorenstein normal champ $\mathcal{X} \to [X/\mathcal{F}]$ over $\mathbb{C}$. Indeed, more precisely, we have:

Fact 3.4.1 Let $\mathcal{X} \to [X/\mathcal{F}]$ be a foliated Gorenstein normal champ with log-canonical singularities in dimension 1 and projective moduli, then there are countably many $\mathcal{F}$-invariant parabolic, champ $\mathcal{L}_i$, with $0 < -K_{\mathcal{F}}. \mathcal{L}_i \leq 2$ such that

$$\overline{\mathrm{NE}}\,(\mathcal{X})_{\mathbb{R}} = \overline{\mathrm{NE}}\,(\mathcal{X})_{K_{\mathcal{F}} \geq 0} + \sum_i \mathbb{R}_+\, \mathcal{L}_i \tag{121}$$

where $\overline{\mathrm{NE}}\,(\mathcal{X})_{K_{\mathcal{F}} \geq 0}$ is the sub-cone of the closed cone of curves on which $K_{\mathcal{F}}$ is non-negative. Better still the $\mathbb{R}_+\, \mathcal{L}_i$ are locally discrete, and if $R \subset \overline{\mathrm{NE}}\,(\mathcal{X})_{\mathbb{R}}$ is an extremal ray in the half-space $\mathrm{NE}_{K_{\mathcal{F}} < 0}$, then it is of the form $\mathbb{R}_+\, \mathcal{L}_i$.

This is a wholly formal consequence, as per [16, III.1.2], of the following variant of Proposition 3.3.7.

Variant 3.4.2 Let $\mathcal{X} \to [X/\mathcal{F}]$ be as above and $C_0 \subset X$ a $K_{\mathcal{F}}$-negative curve in the moduli, then for generic $c \in C_0$ there is a $\mathcal{F}$-invariant parabolic champ $\mathcal{L}_c \ni f(c)$ in $\mathcal{X}$ with $0 < -K_{\mathcal{F}} \cdot \mathcal{L}_c \leq 2$ such that for all nef. $\mathbb{R}$-divisors M on X and $||$ the moduli,

$$M.\, |\mathcal{L}_c| \leq 2\, \frac{(M.\, C_0)}{-K_{\mathcal{F}}.\, C_0}\,. \tag{122}$$

The variant requires a couple of facts of independent interest to wit:

Fact 3.4.3 If $\mathcal{Z}$ is the singular locus of a foliated Gorenstein-champ $\mathcal{X} \to [X/\mathcal{F}]$ with log-canonical singularities in dimension 1, then $\mathcal{O}_{\mathcal{Z}}(K_{\mathcal{F}})$ is semi-ample.

Proof Consider the linearisation map, i.e. the composition of

$$D : I_{\mathcal{Z}}/I_{\mathcal{Z}}^2 \xrightarrow{d} \Omega_{\mathcal{X}} \otimes_{\mathcal{O}_X} \mathcal{O}_{\mathcal{Z}} \longrightarrow K_{\mathcal{F}} \otimes I_{\mathcal{Z}}/I_{\mathcal{Z}}^2. \tag{123}$$

By the Leibniz rule, this map is $\mathcal{O}_{\mathcal{Z}}$ linear, and since the singularities are log-canonical in dimension 1, for $z \in \mathcal{Z}$ outside a finite set, some symmetric function of D defines a section over $\mathcal{Z}$, non-vanishing at z, of some power $K_{\mathcal{F}}^{\otimes n}$, [24, I.ii.4], and we conclude by the Zariski-Fujita theorem. □

Fact/Definition 3.4.4 Let $\mathcal{X} \to [X/\mathcal{F}]$ be a foliated Gorenstein champ, $f : \mathcal{L} \to \mathcal{X}$ the normalisation of an invariant uni-dimensional champ not factoring through the singular locus $\mathcal{Z}$, $\chi(\mathcal{L})$ its topological Euler-characteristic and $\mathrm{s}_{\mathcal{Z}}(f)$ the Segre class of f along $\mathcal{Z}$, i.e. the multiplicity (counted with stabilisers) of the pre-image

$f^{-1}I_{\mathcal{Z}}$ of the ideal of singularities, then

$$\begin{aligned} K_{\mathcal{F}} \cdot_f \mathcal{L} &= -\chi(\mathcal{L}) - \mathrm{Ram}_f + \mathrm{s}_{\mathcal{Z}}(f) \\ &\geq -\chi(\mathcal{L}) + \sum_{l \in f^{-1}(\mathcal{Z})} |\mathrm{Aut}_{\mathcal{L}}(l)|^{-1}. \end{aligned} \tag{124}$$

Proof The image of $f^*\Omega^1_{\mathcal{X}}$ in $\Omega^1_{\mathcal{L}}$ is always $\Omega^1_{\mathcal{L}}(-\mathrm{Ram}_f)$, while in the particular, (23), it is also $f^*K_{\mathcal{F}}.f^{-1}I_{\mathcal{Z}}$, which proves the 1st line in (124). To get the second, observe that in characteristic 0 f can only ramify where it meets $\mathcal{Z}$, while if $f : \Delta \to U$ is a local branch of f meeting a singularity in $f(0)$, and

$$\partial = a_1 \frac{\partial}{\partial x_1} + \cdots + a_n \frac{\partial}{\partial x_n} \tag{125}$$

is a local generator of $\mathcal{F}$ with x_i coordinates on a smooth embedding of U, then the local contribution to $-\mathrm{Ram}_f + \mathrm{s}_Z(f)$ is

$$\begin{aligned} &-\min_i\{\mathrm{ord}(\dot{x}_i(t))\} + \min_i\{\mathrm{ord}(f^*(a_i))\}, \quad f : t \mapsto x_i(t) \\ &= 1 + (\min_i\{\mathrm{ord}(f^*(a_i))\} - \min_i\{\mathrm{ord}(f^*(x_i))\}) \geq 1 \end{aligned} \tag{126}$$

whence the 2nd line on correcting for the order of the stabiliser. □

At which point we can return to

Proof of Variant 3.4.2 By Fact 3.4.3 we need only prove the variant under the additional condition present in Proposition 3.3.7 that the foliated champ is non-singular over a generic point of C_0. As such retaking the notation of the proof of op. cit., we have a bi-dimensional champ $p : \mathcal{Y} \to C$ whose fibres map invariantly by g to $\mathcal{X}$, which is the normalisation of its image. The said image, $\mathcal{A}$, say, admits a possibly non-saturated, injection $T_{\mathcal{F}} \to \mathcal{T}_{\mathcal{A}}$. Every component of the singular locus is invariant by every vector field, while by [4], normalisation (in characteristic 0) can be realised in co-dimension 1 by a sequence of blow-ups in $\mathcal{F}$-invariant centres. Thus, $g^*T_{\mathcal{F}}$ maps to $\mathcal{T}_{\mathcal{Y}/C}$ in co-dimension 1, whence, everywhere since $\mathcal{Y}$, and therefore $\mathcal{T}_{\mathcal{Y}/C}$, is S_2. Consequently, for generic $c \in C$,

$$-K_{\mathcal{F}} \cdot_{g_c} C \leq \mathcal{T}_{\mathcal{Y}/C} \cdot C, \tag{127}$$

while by the adjunction formula of Fact/Definition 3.4.4 and smoothness of $\mathcal{Y}$ in co-dimension 2, we have

$$\mathcal{T}_{\mathcal{Y}/C} \cdot \mathcal{Y}_c = -\chi(\mathcal{Y}_c) \leq 2, \tag{128}$$

so indeed $-K_{\mathcal{F}} \cdot_{g_c} C \leq 2$ for generic c as required. □

In particular, under the hypothesis of log-canonical singularities in dimension 1, $K_{\mathcal{F}}$-negative curves are never contained in the singular locus of the foliation, and we proceed to examine the possibilities for $K_{\mathcal{F}}$ negative invariant parabolic champs outside the same. Whence let $f : \mathcal{L} \to \mathcal{X}$ be the normalisation of such, which we express as a locally constant gerbe, $\pi : \mathcal{L} \to \mathcal{L}_0$, over a champ without generic stabiliser, then by (124),

$$0 > \frac{K_{\mathcal{F}} \cdot_f \mathcal{L}}{(\mathcal{L} : \mathcal{L}_0)} \geq -2 + \sharp f^{-1}(\mathcal{Z}) + \sum_{q \notin \pi f^{-1}(\mathcal{Z})} (1 - \frac{1}{d_q}) \tag{129}$$

where in the sum, d_q, is the order of the local monodromy, and $\sharp$ means integer-valued cardinality of a set. As such,

Fact/Definition 3.4.5 For a Gorenstein foliation $\mathcal{X} \to [\mathcal{X}/\mathcal{F}]$ in the presence of log-canonical singularities in dimension 1, an irreducible $K_{\mathcal{F}}$-negative invariant champ (or just an irreducible $K_{\mathcal{F}}$-negative invariant champ whose generic point meets the smooth locus of the foliation if there are no hypothesis on the singularities of $\mathcal{X} \to [\mathcal{X}/\mathcal{F}]$) has a normalisation, $f : \mathcal{L} \to \mathcal{X}$, with $\mathcal{L}$ parabolic, and furthermore:

(a) The pre-image under f of the singular locus $\mathcal{Z}$ is supported in at most 1 point.
(b) If this pre-image is $\neq \emptyset$, then $\mathcal{L}_0$ has at most one non-scheme-like point outside it.
(c) If there is no such singular point, $\mathcal{X} \to [\mathcal{X}/\mathcal{F}]$ is generically a fibration in parabolic champs.

Proof Items (a) and (b) are clear from (129) which leaves (c). In this case f is an embedding whose normal bundle is flat via the representation afforded by the linear holonomy, while $\pi_1(\mathcal{L})$ is finite, so for $\tilde{f} : \tilde{\mathcal{L}} \to \mathcal{X}$ the composition with the universal cover, the deformations of $\tilde{f}$ are (locally) a smooth space of dimension $\dim(\mathcal{X}) - 1$, and every deformation of $\tilde{f}$ is invariant. □

3.5 *Singular Structure of $K_{\mathcal{F}}$—Negative Curves*

Throughout this section $f : \mathcal{L} \to \mathcal{X}$ is a map from a smooth invariant $K_{\mathcal{F}}$-negative curve with the further specifications of Fact/Definition 3.4.5. In particular f is an embedding everywhere except possibly at a point $p \in f^{-1}(\mathcal{Z})$. At p, however, not only may the monodromy exceed that of the generic point of $\mathcal{L}$, but f may fail to be an embedding because it has a cusp and/or because the image is not uni-branched. Nevertheless, there is a certain limit to the complication, whose description is the goal of this section, i.e.

Fact/Definition 3.5.1 Let everything be as in Fact/Definition 3.4.5 albeit we insist that $\mathcal{X} \to [\mathcal{X}/\mathcal{F}]$ has log-canonical singularities, and suppose moreover that

$f^{-1}(\mathcal{Z}) \neq \emptyset$ with $p : \mathrm{pt} \to \mathcal{L}$ the resulting geometric point; then the étale local contribution, (126), to $-\mathrm{Ram}_f + s_{\mathcal{Z}}(f)$ at p is exactly 1. As such by (124) and (129)

$$K_{\mathcal{F}}.\mathcal{L} = -1/d \tag{130}$$

where d is the maximum value of a stabiliser of $\mathcal{L}$ outside p, which is either attained at a unique point or is the same everywhere in the complement of p, and we refer to such curves as $-\frac{1}{d}\,\mathbb{F}$ *curves*.

We proceed by a series of lemmas beginning with

Claim 3.5.2 The foliation $\mathcal{F}$, by way of restriction over the generic point, affords a singular derivation of $\mathcal{L}$.

Proof We retake the notation of (125)–(126) in the proof of Fact/Definition 3.4.4. It therefore follows exactly as in the proof of Variant 3.4.2 that ∂ defines a derivation of $\mathcal{O}_\Delta$, and it remains to prove that it is actually singular at p. To see this observe that if $b : \tilde{\mathcal{X}} \to \mathcal{X}$ were the blow-up in p, then the induced foliation (understood without saturation if the singularities are not canonical, i.e. locally defined by $b^*\partial$) cannot (by the Frobenius theorem) be smooth where the proper transform of f crosses the exceptional divisor. On the other hand, a sequence of blow-ups in singular points resolves any singularity of any branch of f, so for $b : \tilde{\mathcal{X}} \to \mathcal{X}$ now a chain of such, we can suppose that the proper transform $\tilde{f} : \mathcal{L} \to \tilde{\mathcal{X}}$ is an embedding crossing the exceptional divisor in a singular point, $\tilde{f}(p)$, of the regular derivation $b^*\partial$, i.e. ∂ affords a singular derivation of $\mathcal{O}_\Delta$. □

Applying Claim 3.5.2, we can, in the said notation, write the restriction to $\mathcal{L}$ of a generator étale locally as

$$\partial = y^{r+1}u(y)\frac{\partial}{\partial y}, \quad u(y) \in \mathcal{O}_\Delta^\times,\ r \in \mathbb{Z}_{\geq 0}, \tag{131}$$

and the content of Fact/Definition 3.5.1 is that $r = 0$, all of which is a useful, if non-essential, point of reference in establishing our next

Claim 3.5.3 Understanding $\mathcal{X} \to [\mathcal{X}/\mathcal{F}]$ in the log-sense of Remark 2.2.2, if necessary (cf. Revision 2.2.10), without loss of generality $\mathcal{X}$ in Fact/Definition 3.5.1, is a smooth champ.

Proof By [4] there is a $\mathcal{F}$-equivariant resolution of singularities:

$$b : \tilde{\mathcal{X}} \to \mathcal{X}, \tag{132}$$

so that understanding $\tilde{\mathcal{X}} \to [\tilde{\mathcal{X}}/\tilde{\mathcal{F}}]$ in the log-sense if necessary the canonical bundle is unchanged. As such if b is an isomorphism at the generic point of f, there is a unique lifting $\tilde{f} : \mathcal{L} \to \tilde{\mathcal{X}}$ satisfying the hypothesis of Fact/Definition 3.5.1, and there is nothing to do. It may, however, happen that b is a modification around the image of f. Nevertheless, every component of the fibre over the said image is invariant, amongst which we choose one over the generic point of f and normalise it to get a not necessarily fibred square:

$$\begin{array}{ccc} \tilde{\mathcal{X}} & \xleftarrow[F]{} & \mathcal{Y} \\ {\scriptstyle b}\downarrow & & \downarrow{\scriptstyle B} \\ \mathcal{X} & \xleftarrow{\tilde{f}} & \mathcal{L} \end{array} \tag{133}$$

wherein any vector field along $\mathcal{F}$ on the bottom left-hand corner lifts naturally everywhere else. In particular, therefore, there is a possibly very far (even logarithmically) from saturated (cf. Claim 3.5.2) bundle of derivations:

$$F^* b^* T_{\mathcal{F}} \to \mathcal{T}_{\mathcal{Y}} \tag{134}$$

whose singular locus is contained in $B^{-1}(p)$, so that the restriction

$$F^* b^* T_{\mathcal{F}} |_{\mathrm{sing}(F^* b^* T_{\mathcal{F}})} \tag{135}$$

is trivial. On the other hand, b, and whence B, is relatively projective, so $\mathcal{Y}$ has projective moduli, and since (135) provides an appropriate variant of Fact 3.4.3, we may, since it makes no other use of saturation, apply Fact 3.4.1 to conclude that there are $F^* b^* K_{\mathcal{F}} = B^* f^* K_{\mathcal{F}}$-negative invariant curves:

$$\tilde{f} : \tilde{\mathcal{L}} \to \mathcal{Y} \to \tilde{\mathcal{X}} \tag{136}$$

lifting f. Of course, plausibly, $\tilde{\mathcal{L}} \to \mathcal{L}$ is ramified over p, but this would only cause a non-zero value of r in (131) to go up. □

Now, as we've said, and item (a) of Fact/Definition 3.4.5 notwithstanding, the image of f in $\mathcal{X}$ can even fail to be uni-branch. However,

Claim 3.5.4 Hypothesis as in Claim 3.5.3, then without loss of generality f is an embedding.

Proof In an easier variant of the proof of Claim 3.5.3: given $f : \mathcal{L} \to \mathcal{X}$ with $\mathcal{X}$ smooth, we can find a composition, $b : \tilde{\mathcal{X}} \to \mathcal{X}$, of blow-ups in singular points of the foliation such that the unique lifting $\tilde{f} : \mathcal{L} \to \tilde{\mathcal{X}}$ is an embedding. □

At which juncture we have a well-defined normal bundle $N_{\mathcal{L}/\mathcal{X}}$ and a specialised foliation to the same. Indeed somewhat more generally

Fact/Definition 3.5.5 Let $f : \mathcal{Y} \to \mathcal{X}$ be net, Fact/Definition 2.5.2, albeit that much more, Fact/Definition 2.5.5, is true, descent yields a well-defined normal cone $C_{\mathcal{Y}/\mathcal{X}}$. Specifically if $V \to \mathcal{X}$ is étale, then there is a sufficiently small étale neighbourhood U of any geometric point of $\mathcal{Y} \times_{\mathcal{X}} V$ such that $U \hookrightarrow V$, and the pull-back to U of the associated cone is

$$\mathrm{Spec}\, S := \bigoplus_{n=0}^{\infty} \frac{I_{U,V}^{n}}{I_{U,V}^{n+1}}\,. \tag{137}$$

In particular if the image of f is invariant, then the foliation leaves $I_{U,V}$ invariant, so a local generator ∂ of $T_{\mathcal{F}}$ passes to a graded derivation of S by way of applying it to any lifting of an element in the nth-graded piece, then reducing modulo $I_{U,V}^{n+1}$. This process may not immediately lead to a foliation, but only a pre-foliation, i.e. the specialisation may not be saturated. Nevertheless, for ease of notation (cf. Claim 3.5.3), we continue to ignore such a distinction, which, in any case, we'll clear up in Claim 3.6.1. Irrespectively, if $\mathcal{Y}$ is a smooth invariant curve not factoring through the singular locus, $\mathcal{Z}$, for y a coordinate along U around a point of $f^{-1}(\mathcal{Z})$ and x_i normal coordinates, the specialisation of ∂ takes, by (131), the form

$$\partial : y \mapsto b(y) = y^{r+1}u(y)\partial y\,,\; x_i \mapsto a_{ij}(y)\,x_j = \partial x_i \text{ both mod } I_{U,V} \tag{138}$$

where the summation convention is employed, so equivalently, the specialisation may be viewed as a connection on $N_{\mathcal{Y}/\mathcal{X}}$ with singularities.

Via Claim 3.5.3 and Claim 3.5.4 this may be applied to the case in point, i.e.

Fact 3.5.6 Let $\mathcal{X} \to [\mathcal{X}/\mathcal{F}]$ be a foliated smooth champ and $f : \mathcal{L} \to \mathcal{X}$ an invariant net map from a (smooth) parabolic champ not factoring through the singularities such that $K_{\mathcal{F}} \cdot_f \mathcal{L} < 0$; then either $r = 0$ or the linearisation, (29), $\bar{\partial}$ of a generator at the singular point is nilpotent.

Proof Without loss of generality, $\mathcal{L}$ is simply connected so $\mathcal{L} \xrightarrow{\sim} \mathbb{P}(d, e)$ for some $d, e \in \mathbb{N}$, [6, 1.1]. We thus have an explicit description of $\mathcal{L}$, to wit:

$$\begin{array}{ccccccccc}
\mathbb{G}_m & \xrightarrow[\beta]{t \mapsto t^{-e}} & U' & \xrightarrow{\sim} & \mathbb{A}^1 & \to & [\mathbb{A}^1/\mu_d] & \hookrightarrow & \mathcal{L} \\
{\scriptstyle t\mapsto t^d}\big\downarrow{\scriptstyle \alpha} & & & & & & & & \\
U & & \xrightarrow{\sim} & & \mathbb{A}^1 & \longrightarrow & [\mathbb{A}^1/\mu_e] & \hookrightarrow & \mathcal{L}
\end{array} \tag{139}$$

Furthermore, by Fact/Definition 3.4.5, we may suppose that the pre-image of the singular locus is a point p which we identify with 0 (the origin in U) while ∞ will denote the origin in U'. Consequently, by Fact 2.3.7 there is a longitudinal coordinate y, resp. η, and normal coordinates x_i, resp. ξ_i in neighbourhoods of 0, resp. ∞, such that

$$\beta^*\eta = t^{-e}, \quad \alpha^* y = t^d, \; \beta^*\xi_i = t^{-a_i}\alpha^* x_i \tag{140}$$

where the integers a_i are afforded by the Harder-Narasimhan filtration:

$$N_{\mathcal{L}/\mathcal{X}} \xrightarrow{\sim} \coprod_i \mathcal{O}_{\mathcal{L}}(a_i) \tag{141}$$

so the basis x_i, resp ξ_i, may even be supposed $\boldsymbol{\mu}_e$, resp $\boldsymbol{\mu}_d$, invariant, i.e.

$$(\epsilon, x_i) \mapsto \epsilon^{-a_i} x_i\,, \quad (\delta, \xi_i) \mapsto \delta^{-a_i} \xi_i\,, \quad \epsilon \in \boldsymbol{\mu}_e,\ \delta \in \boldsymbol{\mu}_d. \tag{142}$$

Irrespectively, $T_{\mathcal{F}} \xrightarrow{\sim} \mathcal{O}_{\mathcal{L}}(e - dr)$, where by hypothesis $e > dr$, and we normalise generators around 0 and ∞ according to

$$\partial_0(y) = dy^{r+1}, \quad \partial_\infty(\eta) = -e \tag{143}$$

so that for a specialised foliation described (cf. (138)), by matrices A, resp. B, over U, resp. U',

$$A(t^d) - t^{dr} \Delta = \frac{1}{t^{e-dr}} DB(t^{-e}) D^{-1} \tag{144}$$

where Δ, resp. D, is the diagonal matrix with entries a_i, resp t^{a_i}. Consequently, if we order the a_i to be decreasing in i, then every $i \leq j$th entry of DBD^{-1} on the right of (144) is a polynomial in t^{-1}, so from $e > dr$, $A(t^d)$ is an upper semi-triangular matrix with diagonal $a_i t^{dr}$, and whence the said linearisation is nilpotent if $r > 0$. □

By Revision 2.2.5 this completes the proof of Fact/Definition 3.5.1 and merits

Remark 3.5.7 The difficulty in Fact/Definition 3.5.1 comes from the fact that if $\mathfrak{X}$ were the completion in the singularity p,

$$\mathrm{H}^0(\mathfrak{X}, T_{\mathcal{F}}) \tag{145}$$

may not contain a generator, ∂. Indeed supposing f an embedding (just to fix ideas since it is of no importance) so that the monodromy, G, at p acts on the coordinate y of (131) by a character, γ, then

$$\partial^\sigma = \partial \Rightarrow \gamma(\sigma)^r = 1 \quad \sigma \in G. \tag{146}$$

On the other hand, from the adjunction formula of Fact/Definition 3.4.4, in the notation of (129),

$$\frac{K_{\mathcal{F}} \cdot \mathcal{L}}{(\mathcal{L} : \mathcal{L}_0)} = \frac{r}{\mathrm{ord}(\gamma)} - \frac{1}{d_q} \tag{147}$$

which from (146) is non-negative as soon as $r > 0$. There is, however, not only no way to guarantee that (145) contains a generator, but this may well be impossible on every bi-rational model with log-canonical singularities since this is the root cause, [24, III.iii.3.bis], of why log-canonical resolutions need not exist in the category of smooth varieties.

3.6 Linear Holonomy of, at Worst, Nodal $-\frac{1}{d}\,\mathbb{F}$ Curves

Throughout $\mathcal{X} \to [\mathcal{X}/\mathcal{F}]$ is a (saturated) foliation of a smooth complex Deligne-Mumford champ; $f : \mathcal{L} \to \mathcal{X}$ is a $-\frac{1}{d}\,\mathbb{F}$ curve, with f net and $\mathcal{L}$ smooth. As such we have the specialised foliation of Fact/Definition 3.5.5 to the normal bundle, $N_{\mathcal{L}/\mathcal{X}}$, and

Claim 3.6.1 The specialised foliation is in fact saturated.

Proof Suppose $\mathcal{L}$ is simply connected (which one can always reduce to by [11, Exposé I, 8.3], or [22, IV.a.2] in a slightly more appropriate generality and Fact/Definition 2.5.5), then by the definition of a $-\frac{1}{d}\,\mathbb{F}$ curve we have $r = 0$ in (144), while lack of saturation is equivalent to the matrix A of op. cit. being divisible by t which can only happen if the matrix Δ therein is 0, i.e. the normal bundle is trivial. Thus, exactly as in the proof of Fact/Definition 3.4.5, f moves in a covering family, f_t, of disjoint invariant parabolic champ each of which must meet sing($\mathcal{F}$) for numerical reasons – i.e. $K_{\mathcal{F}} \cdot_f \mathcal{L} = K_{\mathcal{F}} \cdot_{f_t} \mathcal{L}$, (124), and generic f_t a generic embedding – so the singular locus of $\mathcal{F}$ must be a divisor. □

Thus, we can cease to worry about whether the foliation is a saturated or not, and

Set Up 3.6.2 We further suppose that $\mathcal{L}$ is simply connected, i.e. it is the weighted projective champ $\mathbb{P}^1(d, e)$ of Definition 2.3.1.

In particular therefore we can describe the total space, $N := N_{\mathcal{L}/\mathcal{X}} \to \mathcal{L}$, of the normal bundle as the classifying champ $[E/\mathbb{G}_m]$ of the action

$$(y_0, y_1) \times (x_1, \cdots, x_n) \times \lambda \mapsto (\lambda^d y_0, \lambda^e y_1) \times (\lambda^{a_i} x_i),\ \lambda \in \mathbb{G}_m \tag{148}$$

on $E := (\mathbb{A}^2\backslash 0) \times \mathbb{A}^n)$, where as in Fact 3.5.6 the weights

$$a_1 \geq a_2 \cdots \geq a_n \tag{149}$$

are those of the Harder-Narasimhan filtration of Fact 2.3.7 of the normal bundle. Consequently, if $\pi : E \to N$ is the projection, then the tangent space to the normal bundle is described by an Euler sequence of $\mathbb{G}_m$-equivariant (cf. Fact/Definition 2.3.2) bundles:

$$0 \to \mathcal{O} \xrightarrow{1 \mapsto \rho} T_E = \mathcal{O}(d) \amalg \mathcal{O}(e) \amalg_i \mathcal{O}(a_i) \to \pi^* T_N \to 0 \tag{150}$$

where ρ is the radial (cf. Definition 2.4.2) vector field:

$$\rho := dy_0\frac{\partial}{\partial y_0} + ey_1\frac{\partial}{\partial y_1} + a_1x_1\frac{\partial}{\partial x_1} + \cdots + a_nx_n\frac{\partial}{\partial x_n}. \tag{151}$$

Now by Fact/Definition 3.5.1 the canonical bundle of the specialised foliation is $\mathscr{O}(-e)$, while for any $\mathbb{G}_m$-equivariant coherent sheaf $\mathcal{E}$ we have, in the notation of Definition 2.3.1, a Höschild-Serre spectral sequence:

$$\mathrm{H}^p(\mathrm{B}_{\mathbb{G}_m}, \mathrm{H}^q(A_k, \mathcal{E})) \Rightarrow \mathrm{H}^{p+q}(\mathbb{P}^1(d,e), \mathcal{E}) \tag{152}$$

and whence by Fact 2.3.3

$$\mathrm{H}^1(\mathrm{B}_{\mathbb{G}_m}, \pi^* K_{\mathcal{F}}) = 0. \tag{153}$$

Combining this with (150) implies that the specialised foliation on the normal bundle is defined by a vector field ∂ on the total space E such that

$$\partial^{\lambda} = \lambda^{-e}\partial, \quad \lambda \in \mathbb{G}_m. \tag{154}$$

At the same time, by construction, (138), there are functions F_p, A_{ij} in $\mathbb{C}[\mathbb{A}^2]$ such that

$$\partial = F_0\frac{\partial}{\partial y_0} + F_1\frac{\partial}{\partial y_1} + A_{ij}x_j\frac{\partial}{\partial x_i}, \quad 1 \le i, j \le n \tag{155}$$

where as per op. cit. we employ the summation convention. As such from (149), (154), and our normalisation, (139), there is a singularity at (0, 1):

$$F_0 = 0, \; F_1 \in \mathbb{C}^{\times}, \; A_{ij}, \text{ is } a_i - a_j - e \text{ weighted homogeneous.} \tag{156}$$

In particular therefore, by (149), A_{ij} is an upper semi-triangular matrix with 0 diagonal. We can, however, do better, to wit:

Fact 3.6.3 For a possibly different splitting of the Harder-Narasimhan filtration, Remark 2.3.8, of $N_{\mathcal{L}/\mathcal{X}}$ and after a trivial renormalisation by a constant,

$$\partial = -e\frac{\partial}{\partial y_1}. \tag{157}$$

Proof Consistent with the notation of (149) the Harder-Narasimhan filtration may be written as

$$[0] = N_0 \subsetneq N_1 \subsetneq N_2 \subsetneq \cdots \subsetneq N_k = N_{\mathcal{L}/\mathcal{X}} \tag{158}$$

where the normal bundle of N_i in N_{i+1} restricted to the zero section is a trivial bundle twisted by some $\mathcal{O}_{\mathcal{L}}(\alpha_i)$ for α_i a complete repetition free list of the a_is, thus strictly decreasing as one proceeds up the chain. By (156) this is equally a filtration by $\mathcal{F}$-invariant sub-bundles, so understanding the induced foliation on a sub-bundle logarithmically, Remark 2.2.2, if necessary (i.e. if $a_1 = 0$) we prove (157) by induction on the length of the chain (158). The case $k = 1$ is immediate by (156), so by induction the matrix A_{ij} is an idempotent of the form

$$\begin{bmatrix} 0 & A \\ 0 & 0 \end{bmatrix}, \quad A \in \mathrm{Hom}_{\mathbb{G}_m}(N_k/N_{k-1}, N_{k-1})(-e). \tag{159}$$

Plainly, we aim for (156) via a change of coordinates of the form

$$\begin{bmatrix} \tilde{x}_{a_i<\alpha_k} \\ \tilde{x}_{a_i=\alpha_k} \end{bmatrix} = \begin{bmatrix} 1 & B \\ 0 & 1 \end{bmatrix} \begin{bmatrix} x_{a_i<\alpha_k} \\ \tilde{x}_{a_i=\alpha_k} \end{bmatrix}, \quad B \in \mathrm{Hom}_{\mathbb{G}_m}(N_k/N_{k-1}, N_{k-1}) \tag{160}$$

so that what we have to solve (in matrices of function in $\mathbb{C}[\mathbb{A}^2]$) is

$$e\frac{\partial B}{\partial y_1} = A \tag{161}$$

while respecting the $\mathbb{G}_m$-equivariance of (159)–(160), which, (156), is clear. □

To reinterpret this in terms of the standard affine patches U, U' of (139)–(140), one simply splits (150) along the inclusion of the respective (quasi) transversals $y_p = 1$, for $p = 0$ or 1, i.e.

Summary 3.6.4 Suppose the (embedded) $-\frac{1}{d}\,\mathbb{F}$ curve has at worst nodes, equivalently that its normalisation is net over $\mathcal{X}$ and that the universal cover, $\mathcal{L}$, of the same is a $\mathbb{P}^1(d, e)$, then $\mathcal{L} \to \mathcal{X}$ is net with a well-defined normal bundle $N_{\mathcal{L}/\mathcal{X}}$ such that after pulling back to the universal cover of Set Up 3.6.2 we have in the étale description, (139)–(140), of the normal bundle:

(1) On $U \xrightarrow{\sim} \mathbb{A}^1$ an étale neighbourhood of the singularity a μ_e invariant generator of the specialised foliation and $\zeta \in \mu_e$-action given by

$$\partial = dy\frac{\partial}{\partial y} + a_i x_i \frac{\partial}{\partial x_i}, \quad y \mapsto \zeta^d y \quad x_i \mapsto \zeta^{a_i} x_i,\ a_i \in \mathbb{N} \tag{162}$$

(2) On $U' \xrightarrow{\sim} \mathbb{A}^1$ a complementary neighbourhood of the singularity, a basis ξ_i of functions invariant by the specialised foliation, on which $\zeta \in \mu_d$ acts via $\xi_i \mapsto \zeta^{a_i}\xi_i$
(3) A patching $t^{a_i}\beta^*\xi_i = \alpha^* x_i$ in the notation of (139) and whence an isomorphism

$$N_{\mathcal{L}/\mathcal{X}} \xrightarrow{\sim} \coprod_i \mathcal{O}_{\mathcal{L}}(a_i) \tag{163}$$

In particular the canonical or Harder-Narasimhan filtration of $N_{\mathcal{L}/\mathcal{X}}$, (158), is a filtration by $\mathcal{F}$-invariant sub-bundles whose slopes and rank may be read directly from the generator (162) at the singularity.

3.7 Formal Holonomy

We wish to extend the previous discussion of linear holonomy of smooth $-\frac{1}{d}\mathbb{F}$ champs to the rather more delicate case of formal holonomy. Plainly when the curve, $\mathcal{L}$, is smooth and simply connected, the calculations are easier, and we denote by $\mathfrak{X} \to [\mathfrak{X}/\mathcal{F}]$ a foliated smooth formal champ whose trace $\mathcal{L}$ is a $-\frac{1}{d}\mathbb{F}$ champ isomorphic to $\mathbb{P}^1(d, e)$. In practice $\mathfrak{X}$ will, by [11, Exposé I, 8.3] or [22, IV.a.2], be the universal cover of the net completion of Fact/Definition 2.5.5 of a smooth foliated algebraic champ $\mathcal{X} \to [\mathcal{X}/\mathcal{F}]$ along the net map afforded by the normalisation of an at worst nodal $-\frac{1}{d}\mathbb{F}$ curve. For the moment, however, this is logically irrelevant. Supposing no risk of confusion with the notation of (137)–(138), we replace U' by V in (139) and take $\mathfrak{U} \to \mathfrak{X}$, resp. $\mathfrak{V} \to \mathfrak{X}$, to be formal étale neighbourhoods in the analytic topology with trace the $\mathbb{A}^1$'s U, resp. V of (139). In particular, therefore, $\mathfrak{U}$, resp. $\mathfrak{V}$, has a $\boldsymbol{\mu}_e$, resp. $\boldsymbol{\mu}_d$, action and there are open analytic embeddings $[\mathfrak{U}/\boldsymbol{\mu}_e] \hookrightarrow \mathfrak{X}$, resp. $[\mathfrak{V}/\boldsymbol{\mu}_d] \hookrightarrow \mathfrak{X}$, extending $[U/\boldsymbol{\mu}_e] \hookrightarrow \mathcal{L}$, resp. $[V/\boldsymbol{\mu}_d] \hookrightarrow \mathcal{L}$. Whence $\mathfrak{V}$ is simply connected, and, in the analytic topology, there is a certain strengthening of Summary 3.6.4, viz, the foliation may be supposed trivial over $\mathfrak{V}$, i.e. we have analytic coordinate functions ξ_i, η normal and parallel to our $\mathbb{A}^1$ respectively such that in $\mathfrak{V}$ the foliation is just the formal fibration $\xi_1 \times \cdots \times \xi_n : \mathfrak{V} \to \hat{\Delta}^n$, where the latter space is a n-polydisc completed in the origin. The algebra $\mathbb{C}[[\xi_1, \ldots, \xi_n]]$ comes equipped with a $\boldsymbol{\mu}_d$ action – the formal holonomy representation – which, modulo the square of the maximal ideal, is nothing other than that of the linear holonomy, (142). The said algebra is, however, an inverse limit of finite-dimensional vector spaces over a field in which d is invertible, so the action may be written $\xi_i \to \zeta^{-a_i}\xi_i$ without prejudice to items (1)–(3) of Summary 3.6.4.

Now, we can choose ∂ on $\mathfrak{U}$ to be $\boldsymbol{\mu}_e$ invariant, and inductively, we further suppose: for $m \in \mathbb{N}$ given and a possibly different $\boldsymbol{\mu}_e$-invariant generator, ∂, on $\mathfrak{U}$ that there is a coordinate function y restricting to that of item (1) of Summary 3.6.4 such that

$$\partial y \equiv dy(I^m_{\mathcal{L}}),\ (\zeta, y) \mapsto \zeta^d y\,(I^m_{\mathcal{L}}),\ (\zeta, \partial) \mapsto \partial \in \mathrm{Der}(\mathcal{O}_U), \quad \zeta \in \boldsymbol{\mu}_e. \tag{164}$$

The space $\mathfrak{U}$, unlike its trace U, has non-trivial units, so a priori this isn't equivalent to the weaker

$$\partial y = duy(I^m_{\mathcal{L}}),\ (\zeta, y) \mapsto \zeta^d y\,(I^m_{\mathcal{L}}),\ (\zeta, \partial) \mapsto \partial \in \mathrm{Der}(\mathcal{O}_U), \quad \zeta \in \boldsymbol{\mu}_e \tag{165}$$

for u invertible modulo $I^m_{\mathcal{L}}$. Nevertheless, we're in characteristic zero, so in fact

Claim 3.7.1 The conditions (164) and (165) are equivalent.

Proof Supposing (165), we have

$$\partial(y) - duy = f,\ y^{\zeta} - \zeta^d y = g \quad f, g \in \Gamma(U, I^m_{\mathcal{L}}) \tag{166}$$

from which the invariance of ∂ affords:

$$d\zeta^d y(u^{\zeta} - u) = \zeta^d f - f^{\zeta} + \partial(g) - du^{\zeta} g \in \Gamma(U, I^m_{\mathcal{L}}) \tag{167}$$

and we conclude that $u^{\zeta} - u \in \mathrm{H}^1(\boldsymbol{\mu}_e, I^m_{\mathcal{L}})$. Since everything is tame, however, such a co-homology group vanishes, so we can find a $\boldsymbol{\mu}_e$-invariant unit v equal to u modulo $I^m_{\mathcal{L}}$, and replacing ∂ by $v^{-1}\partial$ we deduce (164) from (166). □

Denoting by $\mathfrak{X}_m$, U_m, V_m, etc. the reduction of whatever modulo $I^m_{\mathcal{L}}$, observe that by item (3) of Summary 3.6.4 for y as in (164), there is a function t_0 on $U_m \times_{\mathfrak{X}} V_m$ such that yt_0^{-d} is congruent to 1 modulo nilpotents. Now, we're in characteristic 0, so from the power series of the logarithm, yt_0^{-d} has a dth root. Thus,

$$\exists t \in \Gamma(U_m \times_{\mathfrak{X}} V_m) \ni y\,|_{U_m \times_{\mathfrak{X}} V_m} = t^d, \tag{168}$$

and we further assert:

Claim 3.7.2 Suppose that (164) holds; then for a possibly different $\boldsymbol{\mu}_d$ linear basis ξ_i of the algebra $\mathbb{C}[[\xi_1, \ldots, \xi_n]]$ compatible with any previous choice of the same modulo $I^{m'}_{\mathcal{L}}$ for $m' < m$, then:

(1) There are coordinates x_i normal to $\mathcal{L}$ on U such that in $I_{\mathcal{L}}/I^{m+1}_{\mathcal{L}}$,

$$\partial(x_l) = a_l x_l \quad x_i \mapsto \zeta^{a_i} x_i,\ \zeta \in \boldsymbol{\mu}_e,\ a_i \in \mathbb{N},\ 1 \le i \le n \tag{169}$$

(2) The x_i glue to the ξ_i via $\alpha^* x_i = t^{a_i} \beta^* \xi_i$ as global sections of the $\mathcal{O}_{\mathfrak{X}_m}$ module

$$I_{\mathcal{L}}/I^{m+1}_{\mathcal{L}}(a_i) \tag{170}$$

where $\mathcal{O}_{\mathfrak{X}_m}(1)$ is the bundle with transition function t on $U_m \times_{\mathfrak{X}} V_m$.

Proof We proceed inductively on m, the case $m = 1$ being Summary 3.6.4, so by the first item of the induction hypothesis for $m - 1$, $m \ge 2$, we can find coordinate functions x_i normal to $\mathcal{L}$ whose reduction modulo $I^2_{\mathcal{L}}$ are a basis of the normal bundle over U such that

$$\partial(x_i) - a_i x_i = a_{iJ}(y)\, x^J \in \Gamma(U_m, I^m_{\mathcal{L}}/I^{m+1}_{\mathcal{L}}), \tag{171}$$

where x^J is the monomial $x_1^{j_1} \dots x_n^{j_n}$, $j_1 + \cdots + j_n = m$, the summation convention is employed, and $a_{iJ}(y)$ is an entire function. Similarly, by the second part of the inductive hypothesis,

$$t^{a_i} \xi_i - x_i \,|_{U_m \times_X V_m} = b_{iJ}(t)\, x^J \in \Gamma(U_m \times_X V_m,\, I_{\mathcal{L}}^m / I_{\mathcal{L}}^{m+1}) \tag{172}$$

with the same conventions, but where, now, $b_{iJ}(t)$ are only holomorphic for $t \in \mathbb{G}_m$. Combining (171) and (172), we obtain

$$t\, \dot{b}_{iJ} + b_{iJ}(a_J - a_i) = -a_{iJ}(t^d) \in \mathcal{O}_{\mathbb{G}_m} \tag{173}$$

where $a_J = \sum_i j_i\, a_i$, and no summation is implied. Again we can integrate this, by way of

$$\frac{d}{dt}\,(t^{(a_J - a_i)}\, b_{iJ}) = -\, a_{iJ}(t^d)\, t^{a_J - a_i - 1}\,. \tag{174}$$

A priori, however, the b_{iJ} are holomorphic for $t \in \mathbb{G}_m$, so the b_{iJ} are, in fact, meromorphic, and no $a_{iJ}\, t^{a_J - a_i - 1}$ has a residue, whence

$$b_{iJ} = h_{iJ}(t^d) + \frac{\lambda_{iJ}}{t^{a_J - a_i}} \tag{175}$$

where h_{iJ} is entire, and λ_{iJ} is a constant. In particular,

$$\tilde{x}_i := x_i + h_{iJ}(t^d) x^J \text{ satisfies } \partial(\tilde{x}_i) = a_i \tilde{x}_i \pmod{I_{\mathcal{L}}^{m+1}} \tag{176}$$

and defines n normal coordinate functions on $\mathfrak{U}$, such that

$$\tilde{x}_i = t^{a_i} \tilde{\xi}_i, \text{ where, } \tilde{\xi}_i := \xi_i - \lambda_{iJ} \xi^J. \tag{177}$$

The far left-hand side of (176) is entire in t^d, so $\tilde{\xi}_i$ is still a $\boldsymbol{\mu}_d$-linear basis of $I_{\mathcal{L}} / I_{\mathcal{L}}^{n+1} \,|_{V_m}$ (compatible with our previous choices), and $a_J \equiv a_i(d)$ if $\lambda_{iJ} \neq 0$ by the coincidence of the formal holonomy with the linear holonomy (142). It therefore only remains to guarantee the $\boldsymbol{\mu}_e$ linearity, (142). To this end, supposing the change of basis in (176) and (177) already made so as to momentarily drop the ˜ from the notation, we have for $\zeta \in \boldsymbol{\mu}_e$ a generator:

$$x_i^{\zeta} - \zeta^{a_i} x_i = g_{iJ}(y) x^J \in \Gamma(U_m,\, I_{\mathcal{L}}^m / I_{\mathcal{L}}^{m+1}). \tag{178}$$

Applying the invariance of ∂ in (164), the right-hand side of (178) must belong to the eigenspace of a_i for ∂ viewed as a $\mathbb{C}$-linear map. As such,

$$g_{iJ}(y) = \sum_{nd+a_J=a_i} g_{iJn} y^n \tag{179}$$

and we can suppose x_i has been rendered $\boldsymbol{\mu}_e$-linear by a coordinate change:

$$\tilde{x}_i = x_i + \sum_{nd+a_J=a_i} f_{iJn} y^n x^J \tag{180}$$

which yields new functions over ∞:

$$\tilde{\xi}_i := t^{-a_i}\tilde{x}_i = \xi_i + \sum_{nd+a_J=a_i} f_{iJn}\xi^J, \tag{181}$$

and since J now has cardinality ≥ 2, this is also a $\boldsymbol{\mu}_d$ linear coordinate change. □

Let us now observe how to bootstrap in the presence of Claim 3.7.2, by finding some y satisfying (164) modulo $\mathcal{I}_{\mathcal{L}}^{m+1}$, $m \geq 1$. Over U we have, in the notation/spirit of the proof of Claim 3.7.2,

$$\partial y = dy + c_J\, x^J\, (\mathcal{I}_{\mathcal{L}}^{m+1}), \quad \partial x_i = a_i\, x_i + c_{iK}\, x^K\, (\mathcal{I}_{\mathcal{L}}^{m+2}) \tag{182}$$

where the summation convention is back in force, with respect to multi-indices J and K of degrees m, $m+1$, respectively, all the c_*s are regular functions of y, and by tameness of the monodromy $(\zeta, y) \mapsto \zeta^d y$, $\zeta \in \boldsymbol{\mu}_e$ in all of $\mathcal{O}_U$. We know that the holonomy of the system (182) is a quotient of $\boldsymbol{\mu}_d$, so we again take t as in (168), and at a presumably negligible risk of notational confusion, let

$$\xi_i - t^{-a_i}\, x_i + b_{iK}(t)\, x^K \pmod{\mathcal{I}_{\mathcal{L}}^{m+2}} \tag{183}$$

be a basis of invariant functions on an analytic étale neighbourhood of $\mathbb{G}_m$, with summation over the multi-index K of degree $m+1$ being implied. Combining these, yields for any i,

$$\begin{aligned} &c_{iK}\, t^{(a_K - a_i - 1)} + \frac{d}{dt}(t^{a_K}\, b_{iK}) \\ &= \begin{cases} \frac{a_i}{d}\cdot c_J\, t^{a_J-(d+1)} & \text{if } x^K = x^J x_i \text{ for some } J, \\ 0 & \text{otherwise.} \end{cases} \end{aligned} \tag{184}$$

By Claim 3.6.1, we know there is some i with $a_i \neq 0$, while b_{iK} must be holomorphic in $\mathbb{G}_m$, so

$$\text{if } a_J = d \text{ then } c_J(0) = 0, \tag{185}$$

since in such an eventuality the exponent of the leftmost term, $a_J - 1$, is non-negative. Similarly, if much more straightforwardly, the $\boldsymbol{\mu}_e$ invariance of ∂, and our insistence that $y \mapsto \zeta^d y$ implies,

$$c_J^\zeta = \zeta^{d-b_J} c_J, \quad b_J = \sum_i b_i j_i, \text{ for, } J = (j_1, \dots, j_n) \tag{186}$$

with b_i as per item (1) of Claim 3.7.2, and whence,

$$\text{if } c_J(0) \neq 0, \text{ then } b_J \equiv d(e). \tag{187}$$

On the other hand, consider the obstruction to finding a coordinate $\tilde{y}$ over U restricting to the same on $\mathcal{L}$ such that

$$\partial \tilde{y} = d(1+\lambda)\,\tilde{y}\,(\mathcal{I}_{\mathcal{L}}^{m+1}), \quad \lambda = \lambda_J x^J \in \mathcal{I}_{\mathcal{L}}^m, \quad \lambda_J \in \mathcal{O}_{U \cap \mathcal{L}}. \tag{188}$$

If we look for such a $\tilde{y}$ in the form $y + \Lambda_J x^J$, with Λ_J constants, then we require to solve

$$(a_J - d)\Lambda_J - d\lambda_J\, y = -c_J \tag{189}$$

for all J. However, if $a_J \neq d$, then $\Lambda_J = -c_J(0)(a_J - d)^{-1}$, and λ_J whatever will do, while if $a_J = d$, then by (185) we can take $\Lambda_J = 0$, and $\lambda_J = c_J\, y^{-1}$. Whether trivially in the latter case, or by (186)–(187) in the former case, $\tilde{y}^\zeta \equiv \zeta^d \tilde{y}\,(\mathcal{I}_{\mathcal{L}}^{m+1})$, so we obtain (165) and whence (164) by Claim 3.7.1.

As per Claim 3.7.2, the coordinate $\tilde{y}$ also restricts to the previous choice modulo $\mathcal{I}_{\mathcal{L}}^m$, so we obtain in the limit an extension of the canonical/Harder-Narasimhan filtration to the whole neighbourhood, i.e.

Proposition/Summary 3.7.3 Let $\mathfrak{X} \to [\mathfrak{X}/\mathcal{F}]$ be a foliated smooth formal champ whose trace is a smooth simply connected $-\frac{1}{d}\,\mathbb{F}$ curve, $\mathcal{L} \xrightarrow{\sim} \mathbb{P}^1(d, e)$, then:

(1) There is a bundle $\mathcal{O}_{\mathfrak{X}}(1)$ lifting $\mathcal{O}_{\mathcal{L}}(1)$ and a smooth formal invariant divisor D, with $\mathcal{O}_{\mathfrak{X}}(D) = \mathcal{O}_{\mathfrak{X}}(d)$ transverse to $\mathcal{L}$ which restricts to the unique point z of $\mathcal{L} \cap \operatorname{sing}(\mathcal{F})$.

(2) There is a filtration of formal invariant sub-champs:

$$\mathcal{L} = \mathfrak{X}_0 \subsetneqq \mathfrak{X}_1 \subsetneqq \cdots \subsetneqq \mathfrak{X}_k = \mathfrak{X} \tag{190}$$

such that if $\alpha_1 > \cdots > \alpha_k$ are the distinct eigenvalues of ∂ considered linearised in $\operatorname{End}(N_{\mathcal{L}/\mathfrak{X}} \otimes \mathbb{C}(z))$ and normalised via item (1) of Summary 3.6.4 with $n_1, \dots, n_k$ the dimensions of the corresponding eigenspaces, then $\mathfrak{X}_i$ is defined by $\mathcal{F}$-invariant global sections γ_j of $\mathcal{O}_{\mathfrak{X}}(\alpha_j)$, $j > i$, and n_j-sections for each j. In particular,

$$N_{\mathcal{L}/\mathfrak{X}_i} \xrightarrow{\sim} \coprod_{j \leq i} \mathcal{O}_{\mathcal{L}}(\alpha_j)^{n_j}. \tag{191}$$

(3) All of this is encoded in a particular $\boldsymbol{\mu}_e$ linear coordinate system, $y, x_i, y \mapsto \zeta^d y$, $x_i \mapsto \zeta^{a_i} x_i$ of an étale neighbourhood $\mathfrak{U} \to \mathfrak{X}$ with trace $\mathbb{A}^1$ containing the singular point over which we have a $\boldsymbol{\mu}_e$-invariant generator:

$$\partial = dy \frac{\partial}{\partial y} + a_i\, x_i \frac{\partial}{\partial x_i} \tag{192}$$

where the summation convention is in force, so that the α_j are a complete repetition free list of the a_i.

3.8 Jordan Decomposition

We briefly interrupt our discussion of $K_{\mathcal{F}}$-negative invariant champs to recall some salient facts on Jordan decomposition which will be relevant both to our study of cusps and the local uniqueness of the Harder-Narasimhan filtration. The situation is entirely local and, initially, scheme-like, i.e. $\mathcal{O}$ is the ring of formal power series $\mathbb{C}[[x_1, \ldots, x_n]]$, $\mathfrak{m}$ its maximal ideal and ∂ a $\mathbb{C}$-derivation of $\mathcal{O}$ with a singularity at the origin. Recall that since $\mathcal{O}$ is an inverse limit of finite-dimensional vector spaces, ∂ admits a Jordan decomposition, i.e. $\partial = \partial_S + \partial_N$, where the semi-simple part ∂_S acts as a semi-simple matrix on each $\mathcal{O}/\mathfrak{m}^n$, $n \in \mathbb{N}$, ∂_N is nilpotent, and of course $[\partial_S, \partial_N] = 0$. In particular if $\partial_S = \lambda_i\, x_i \frac{\partial}{\partial x_i}$, summation convention, then a conventional choice of basis for the nilpotent fields commuting with ∂_S is

Revision 3.8.1 (cf. [18]) Notations as above then$\partial_N = \sum_{i=1}^{n} \sum_{Q_i} a_{Q_i} x^{Q_i} x_i \frac{\partial}{\partial x_i}$, $a_{Q_i} \in \mathbb{C}$, where for $\Lambda = (\lambda_1, \ldots, \lambda_n)$ the inner product $\Lambda \cdot Q_i = 0$ and either:

(i) $Q_i = (q_1, \ldots, q_n)$, $q_j \in \mathbb{N} \cup \{0\}$, $x^{Q_i} = x_1^{q_1} \ldots x_n^{q_n}$, or
(ii) $Q_i = (q_1, \ldots, q_n)$, $q_i = -1$, $q_j \in \mathbb{N} \cup \{0\}$, $j \neq i$, $x^{Q_i} = x_1^{q_1} \ldots x_n^{q_n}$.

Now the Jordan decomposition of a vector field is certainly unique, and whence the property of semi-simplicity of a vector field is wholly unambiguous. For a foliation however, the situation is rather more delicate since there is a question of rescaling by units. Whence suppose our field ∂ is semi-simple, and consider a field $\tilde{\partial} = u\partial$, where $u \equiv 1(\mathfrak{m})$ to avoid stupidity. Furthermore, let's say, without loss of generality, that $\partial = \partial_S = \lambda_i\, x_i \frac{\partial}{\partial x_i}$, then we assert:

Claim 3.8.2 Notations as above, there is a change of coordinates of the form $\xi_i = u_i\, x_i$, $u_i \equiv 1(\mathfrak{m})$, and $\varepsilon \equiv 0(\mathfrak{m})$ with $\partial\varepsilon = 0$ such that the Jordan decomposition of $\tilde{\partial}$ is

$$\tilde{\partial} = \lambda_i\, \xi_i \, \frac{\partial}{\partial \xi_i} + \varepsilon\, \lambda_i\, \xi_i\, \frac{\partial}{\partial \xi_i}, \tag{193}$$

i.e. $\tilde{\partial}$ may not be semi-simple, but the extent to which it is not is very particular.

Proof Consider the following inductive proposition for $k \in \mathbb{N}$:
there are coordinates $x_{ik} = u_{ik}\, x_i$, $u_{ik} \equiv 1(\mathfrak{m})$, $\tilde{\partial} = u_k\, \partial_k$, $\partial_k = \lambda_i\, x_{ik}\, \frac{\partial}{\partial x_{ik}}$, $u_k \equiv 1(\mathfrak{m})$ such that $u_k^{-1} = 1 + \varepsilon_k + \delta_k$, where ε_k, δ_k are defined by way of the Jordan decomposition of $\mathfrak{m}$ as $\operatorname{Ker} \partial_k \oplus \operatorname{Im} \partial_k$, and $\delta_k \in \mathfrak{m}^k$.
The case $k = 1$ is simply our given data. Otherwise consider trying to improve the situation by putting, $x_{ik+1} = v_{ik}\, x_{ik}$, $v_{ik} \equiv 1(\mathfrak{m})$ to be chosen. If such a change were to actually render $\tilde{\partial}$ semi-simple, then we would have to solve

$$\partial_k \log v_{ik} = \lambda_i \Big(\frac{1}{u_k} - 1 \Big) = \lambda_i (\varepsilon_k + \delta_k) \tag{194}$$

which may not be possible if $\lambda_i \neq 0$, and $\varepsilon_k \neq 0$. However, we can solve $\partial_k \log v_{ik} = \lambda_i\, \delta_k$, so in particular, $v_{ik} \equiv 1(\mathfrak{m}^k)$, while in the new coordinates,

$$\tilde{\partial} = \frac{1 + \delta_k}{1 + \varepsilon_k + \delta_k}\, \lambda_i\, x_{ik+1}\, \frac{\partial}{\partial x_{ik+1}} \tag{195}$$

which is indeed what we are looking for, since putting $u_{k+1} = (1 + \delta_k)\, u_k$, then

$$u_{k+1}^{-1} = 1 + \varepsilon_k (1 + \delta_k)^{-1} = 1 + \varepsilon_k + \sum_{n=1}^{\infty} (-1)^n\, \varepsilon_k\, \delta_k^n \tag{196}$$

so that $\delta_{k+1} \in \mathfrak{m}^{k+1}$. Certainly therefore $\delta_k \to 0$, but the proof also shows that for each i the infinite product, $\prod_k v_{ik}$, converges to some u_i, so putting $\xi_i = u_i\, x_i$ we are certainly done on observing that $\partial \varepsilon = 0$ obliges:

$$\Big[\lambda_i\, \xi_i\, \frac{\partial}{\partial \xi_i}\,,\ \varepsilon\, \lambda_i\, \xi_i\, \frac{\partial}{\partial \xi_i} \Big] = 0\,. \tag{197}$$

□

The consequence of the fact that not only can Jordan decomposition of a rescaling of semi-simple only fail in a very controlled way, but also that Jordan decompositions of rescalings are related in such a straightforward way suggests that we introduce

Definition 3.8.3 On a formal disc $\hat{\Delta}^n$, i.e. $\mathrm{Spf}(\mathbb{C}[[t_1, \cdots, t_n]])$, a foliation $(\hat{\Delta}^n, \mathcal{F})$, with a not necessarily isolated singularity at the origin, is said to be semi-simple, if $T_{\mathcal{F}} = \mathcal{O}_{\hat{\Delta}^n} \partial$ for some semi-simple vector field ∂.

As an important example/application consider the situation of blowing up in the origin, i.e. $\rho : (X, \tilde{\mathcal{F}}) \to (\hat{\Delta}^n, \mathcal{F})$ is the said modification with induced foliation and X is the completion in the exceptional divisor of the blow-up of Spec $\mathcal{O}$. Denoting by $\partial = \partial_S + \partial_N$ a Jordan decomposition of any generator of $T_{\mathcal{F}}$, we have

Fact 3.8.4 Suppose $\partial_S \neq 0$ and $(X, \tilde{\mathcal{F}})$ is not everywhere smooth (which in any case could only happen if in suitable coordinates $\partial = x_i \frac{\partial}{\partial x_i}$); then the following are equivalent:

(1) $(\hat{\Delta}^n, \mathcal{F})$ is semi-simple.
(2) $(X, \tilde{\mathcal{F}})$ is semi-simple at all of its singular points.
(3) $(X, \tilde{\mathcal{F}})$ is semi-simple at one of its singular points, and $(\hat{\Delta}^n, \mathcal{F})$ is semi-simple modulo $\mathfrak{m}^2$.
(4) $(X, \tilde{\mathcal{F}})$ is semi-simple at one of its singular points.

Before proceeding, we will require a lemma, to wit:

Lemma 3.8.5 *Notations as above, then at every point of its singular locus, $\rho^* \partial_N$ is nilpotent.*

Proof Without loss of generality we can suppose the projective coordinates of some singular point, p, in the exceptional divisor to be $[1, 0, \ldots, 0]$. Thus, if $\partial_N = a_{ij} x_j \frac{\partial}{\partial x_i}$, summation convention in force, then $a_{i1}(0) = 0$ for every $i \geq 2$. This is equivalent, however, to the column vector defined by p being an eigenvector, so $a_{11}(0) = 0$ too. Now, observe that a square matrix, $[c_{ij}]_{i,j \geq 1}$ with a zero first column is nilpotent iff the matrix $[c_{ij}]_{i,j \geq 2}$ is nilpotent, while the linearisation of $\rho^* \partial_N$ in p is

$$\begin{bmatrix} a_{11}(0) & 0 & \ldots & 0 \\ \frac{\partial a_{21}}{\partial x_1}(0) & & \ldots\ldots & \\ \ldots & & \ldots a_{ij}(0) \ldots & \\ \frac{\partial a_{n1}}{\partial x_1}(0) & & \ldots\ldots & \end{bmatrix} \tag{198}$$

which has a zero first row, so it's also nilpotent. □

Proof of Fact 3.8.4 Since $(X, \tilde{\mathcal{F}})$ is not everywhere smooth, the induced foliation is given everywhere by $\rho^* \partial$ (cf. Revision 2.2.10) so trivially (1) implies everything else, while both (2) and (3) trivially imply (4). Consider therefore (4) $\Rightarrow$ (1). As in the above proof of Lemma 3.8.5, a singular point of the singular locus of $\rho^* \partial$ is an eigenvector of its linearisation, whence an eigenvector of the linearisations of $\rho^* \partial_S$ and $\rho^* \partial_N$ and thus a singularity of both $\rho^* \partial_S$ and $\rho^* \partial_N$. We know, however, that every singularity of the former is semi-simple, so by Lemma 3.8.5, $\rho^* \partial = \rho^* \partial_S + \rho^* \partial_N$ remains a Jordan decomposition at every point of the singular locus of $\tilde{\mathcal{F}}$. By hypothesis, at such a point p, there is some semi-simple generator $\tilde{\partial}$, so an application of Claim 3.8.2 yields $\varepsilon \in \hat{\mathcal{O}}_{X,p}$ such that $\rho^* \partial(\varepsilon) = 0$, and

$$\varepsilon\,\rho^*\,\partial_S = \rho^*\,\partial_N\,. \tag{199}$$

As such, if x_1 is an eigenvector of ∂_S, with eigenvalue $\lambda_1 \neq 0$, then for $f = \partial_N(x_1)$, $\varepsilon = \rho^*(f/\lambda_1 x_1)$, while

$$0 = x_1 \partial(f/x_1) = \partial f - \frac{f}{x_1}\cdot(\lambda_1\,x_1 + f) \tag{200}$$

so $x_1 \mid f$, and ε is actually a function on $\hat{\Delta}^n$, from which we conclude. □

A further question which we may reasonably address here is the uniqueness, or lack thereof, of the Jordan decomposition. Even without rescaling the particular choice of coordinates in which we may write a semi-simple field as $\lambda_i\,x_i\,\partial/\partial x_i$ may be catastrophically non-unique. Plainly the worst possible case is when all the λ_i are rational, or equivalently up to a harmless rescaling integers. Even this is of course not unique but it's not too bad since of course any rational point in some $\mathbb{P}^N(\mathbb{Q})$ is up to multiplication by ± 1 uniquely represented by a tuple of relatively prime integers; consequently, let's establish some notation:

Notation 3.8.6 Let ∂ be a semi-simple derivation of $\mathscr{O}$ with integer eigenvalues $a_1, \ldots, a_r, -b_1, \ldots, -b_t, a_i, b_j \in \mathbb{Z}_{>0}$, s zeroes, $r \geq 1$, although possibly $t = 0$, i.e. no negatives, and $(a_1, \ldots, a_r, b_1, \ldots, b_t) = 1$, then we will suppose these ordered by decreasing size, i.e.

$$a_1 \geq a_2 \geq \cdots \geq a_r > 0 > -b_t \geq \cdots \geq -b_1, \tag{201}$$

and by $\alpha_1, \ldots, \alpha_k$, $k \leq r$, $\beta_1, \ldots, \beta_l$, $l \leq t$ a complete repetition free list of the same, so that

$$\begin{aligned} a_1 = \alpha_1 > \alpha_2 > \cdots > \alpha_k > 0 \\ 0 > -\beta_l > \cdots - \beta_2 > \cdots > -\beta_1 = -b_1\,. \end{aligned} \tag{202}$$

Now for a given choice of basis of a semi-simple derivation ∂ with the said eigenvalues i.e. a particular way of writing it as $a_i\,y_i\,\frac{\partial}{\partial y_i} - b_j\,x_j\,\frac{\partial}{\partial x_j}$, with say z_k the null vectors, we can introduce

Definition 3.8.7 The Harder-Narasimhan pair of $(\hat{\Delta}^n, \mathcal{F})$ with respect to the data $(\partial,\ y_i,\ x_j)$ is the invariant formal sub-schemes, $\hat{X}_{>0}$, $\hat{X}_{<0}$ whose ideals are generated by the non-positive, respectively non-negative, eigenvectors of ∂. If instead we take strictly negative, respectively strictly positive, eigenvectors then the resulting sub-schemes, $\hat{X}_{\geq 0}$, $\hat{X}_{\leq 0}$, will be called the non-strict Harder-Narasimhan pair.

Manifestly, apart from abbreviating Harder-Narasimhan to H-N, what's important is that the H-N pairs are well defined up to ± 1, i.e.

Fact 3.8.8 Fix a choice of semi-simple ∂ with integer eigenvalues normalised as per Notation 3.8.6; then the following are equivalent:

(1) $\{\hat{X}_{>0},\ \hat{X}_{<0}\}$, respectively $\{\hat{X}_{\geq 0},\ \hat{X}_{\leq 0}\}$, is the H-N, resp. non-strict H-N, pair with respect to ∂ in the basis $\{x_i,\ y_j\}$.
(2) $\{\hat{X}_{>0},\ \hat{X}_{<0}\}$, resp. $\{\hat{X}_{\geq 0},\ \hat{X}_{\leq 0}\}$, is the H-N, resp. non-strict H-N, pair with respect to ∂ in any semi-simple basis.
(3) $\{\hat{X}_{>0},\ \hat{X}_{<0}\}$, resp. $\{\hat{X}_{\geq 0},\ \hat{X}_{\leq 0}\}$, is the H-N, resp. non-strict H-N, pair of any semi-simple $\tilde{\partial} = u\partial$ in any semi-simple basis for the same, where $u \equiv 1(\mathfrak{m})$.

Proof (3) $\Rightarrow$ (2) $\Rightarrow$ (1) are all trivial, so consider (1) $\Rightarrow$ (3). By Claim 3.8.2, we know that we can find units $u_i,\ v_j \equiv 1(\mathfrak{m})$ such that if $\eta_i = u_i\, y_i,\ \xi_j = v_j\, x_j$, then $\tilde{\partial} = a_i\, \eta_i \frac{\partial}{\partial \eta_i} - b_j\, \xi_j \frac{\partial}{\partial \xi_j}$. As such $\{\hat{X}_{>0},\ \hat{X}_{<0}\}$, resp. $\{\hat{X}_{\geq 0},\ \hat{X}_{\leq 0}\}$, is the H-N, resp. strict, pair of $\tilde{\partial}$ in the basis $\{\xi_i,\ \eta_j\}$. Now suppose $\tilde{\partial} = a_i\, f_i \frac{\partial}{\partial f_i} - b_j\, g_j \frac{\partial}{\partial g_j}$ in some other basis $f_i,\ g_j$. At the mod $\mathfrak{m}^2$ level this is just a question of the uniqueness of diagonalisation/the commutator of a diagonal matrix, so without loss of generality let's say $f_i \equiv \xi_i$, and $g_j \equiv \eta_j(\mathfrak{m}^2)$. On the other hand, for higher-order terms, consider the Taylor expansion:

$$f_i = \xi_i + \sum_{\#J+\#K\geq 2} c_{iJKL}\, \xi^J\, \eta^K\, \zeta^L\,, \tag{203}$$

where, as ever, ξ^J, etc. is the monomial $\xi_1^{j_1} \ldots \xi_r^{j_r}$, etc., and $\zeta_1, \ldots, \zeta_s$ are the null vectors. Now $\tilde{\partial}\, f_i = a_i\, f_i$, so

$$c_{iJKL} \neq 0 \Rightarrow \sum_{\alpha} a_\alpha\, j_\alpha - \sum_{b} b_\beta\, k_\beta = a_i\,. \tag{204}$$

Consequently, if $f_i \notin (\xi_1, \ldots, \xi_r)$, then we have a manifest absurdity and so conclude by symmetry. □

The dependence on ± 1 is, however, unavoidable. Indeed, let $\hat{\Delta}^n \to [\hat{\Delta}^n/\mathcal{T}]$ be a singular foliation invariant by a finite group G, or, equivalently for ∂ a generator,

$$\partial^\sigma = \sigma\partial\sigma^{-1} = u(\sigma)\partial, \quad u : G \to \mathcal{O}^\times_{\hat{\Delta}^n} \tag{205}$$

where u is a group co-cycle, so better, by the acyclicity of B_G on torsion free abelian groups, a character χ on replacing ∂ by $v\partial$ for a suitable unit. At which point, however, if $\partial = \partial_S + \partial_N$ is a Jordan decomposition of ∂, then $\partial^\sigma = \partial^\sigma_S + \partial^\sigma_N$ is a Jordan decomposition of ∂^σ, so by unicity of the same,

$$\partial^\sigma_S = \chi(\sigma)\partial_S, \text{ and, } \partial^\sigma_N = \chi(\sigma)\partial_N. \tag{206}$$

As such, if in addition $\hat{\Delta}^n \to [\hat{\Delta}^n/\mathcal{F}]$ is semi-simple, then, by Claim 3.8.2, ∂ and ∂_S generate the same foliation, so:

Fact 3.8.9 If $\hat{\Delta}^n \to [\hat{\Delta}^n/\mathcal{F}]$ is a germ of a singular semi-simple foliation invariant by a finite group G, then there is a character $\chi : G \to \mathbb{Q}(1)/\mathbb{Z}(1)$ of G and a semi-simple generator ∂ of the foliation such that $\partial^\sigma = \chi(\sigma)\partial$, for all $\sigma \in G$. In particular, if the eigenvalues of a linearisation in $\mathfrak{m}/\mathfrak{m}^2$ are in $\mathbb{P}^{n-1}(\mathbb{Q})$, then χ takes values in $\{\pm 1\}$, and:

(a) If χ is trivial, all of $\hat{X}_{>0}$, $\hat{X}_{<0}$, $\hat{X}_{\geq 0}$, $\hat{X}_{\leq 0}$ are G invariant, and there is a H-N pair, respectively non-strict H-N pair, of embedded $\mathcal{F}$-invariant formal sub-champs, $\{[\hat{X}_{>0}/G], [\hat{X}_{<0}/G]\}$, respectively $\{[\hat{X}_{\geq 0}/G], [\hat{X}_{\leq 0}/G]\}$, in $[\hat{\Delta}^n/G]$.

(b) Otherwise, in the notation of Notation 3.8.6, $a_i = b_i$, $r = t$, etc., and $[\hat{X}_{>0}/\mathrm{Ker}_\chi]$, respectively $[\hat{X}_{\geq 0}/\mathrm{Ker}_\chi]$, is isomorphic to $[\hat{X}_{<0}/\mathrm{Ker}_\chi]$, respectively $[\hat{X}_{\leq 0}/\mathrm{Ker}_\chi]$, but is only net in $[\hat{X}_{>0} \cup \hat{X}_{<0}/G]$, respectively $[\hat{X}_{\geq 0} \cup \hat{X}_{\leq 0}/G]$, which in turn are embedded in $[\hat{\Delta}^n/G]$, being defined by the G-invariant ideal $(x_i y_i, z_k)$, respectively $(x_i y_i)$.

Proof If x is an eigenvector of ∂ with eigenvalue λ, then for any $\sigma \in G$, x^σ is an eigenvector of ∂ with eigenvalue $\lambda\chi(\sigma)^{-1}$, so when the eigenvalues are rational, χ must take values in rational roots of unity. □

Consequently, even in a purely scheme-like situation, we have two canonical pairs rather than two pairs of canonical sub-schemes, and we make:

Remark/Definition 3.8.10 Let $\hat{\Delta}^n \to [\hat{\Delta}^n/\mathcal{F}]$ be a germ of a singular semi-simple foliation such that the eigenvalues of a linearisation in $\mathfrak{m}/\mathfrak{m}^2$ are in $\mathbb{P}^{n-1}(\mathbb{Q})$ then there are two canonical pairs of invariant formal sub-schemes, the H-N pair, $\{\hat{X}_{>0}, \hat{X}_{<0}\}$, and the non-strict H-N pair $\{\hat{X}_{\geq 0}, \hat{X}_{\leq 0}\}$, where the former intersect in the origin, the latter in the whole singular locus. If no-confusion is likely, the suffices may be dropped.

In the particular case of Proposition/Summary 3.7.3, the trace of the formal neighbourhood $\mathfrak{X}$ affords a distinguished eigenvector, so the character appearing in Fact 3.8.9 around the singularity, p, is trivial. As such, by op. cit., the H-N pair, respectively non-strict H-N pair, extends from a formal neighbourhood of p to a pair of embedded invariant formal sub-champs $\{\mathfrak{X}_{>0}, \mathfrak{X}_{<0}\}$, respectively, $\{\mathfrak{X}_{\geq 0}, \mathfrak{X}_{\leq 0}\}$ of $\mathfrak{X}$. An important further task will be to extend this to cusps.

3.9 Cusps

We consider the consequences of the previous discussion for cuspidal $-\frac{1}{d}\,\mathbb{F}$ curves, $f : \mathcal{L} \to \mathcal{X}$, where, as ever, $\mathcal{X} \to [\mathcal{X}/\mathcal{F}]$ is a foliated smooth champ. In the first instance the discussion is purely local, so say, $f : \hat{\Delta}^1 \to \hat{X}$, the map between completions in the singularity $0 \in f^{-1}(\mathcal{Z})$, for $\mathcal{Z} = \mathrm{sing}(\mathcal{F})$. By, for example, [4], the cusp may (cf. Claim 3.5.4) be resolved by the étale local operation of blowing up in the sequence of closed points:

$$\tilde{\mathcal{X}} = \mathcal{X}_N \to \ldots \to \mathcal{X}_1 \to \mathcal{X}_0 = \mathcal{X} \tag{207}$$

of which the first is $z := f(0)$ and subsequently where the proper transform of f meets the exceptional divisor until such time that f becomes an embedding, $\tilde{f}$, say, meeting the proper transform in $\tilde{z}$. Necessarily, each blow-up in (207) is in a point where the foliation is singular, so $K_{\tilde{\mathcal{F}}} \le K_{\mathcal{F}}|_{\tilde{X}}$, and $\tilde{f}$ can only fail to be a $-\frac{1}{d}\mathbb{F}$ curve if $\tilde{\mathcal{F}}$ is smooth everywhere around $\tilde{f}$. Now although such an occurrence is highly simplifying, e.g. $\mathcal{F}$ is algebraic in conics by item (c) of Fact/Definition 3.4.5; the foliation has a first integral in a (finite) étale neighbourhood of $\mathcal{L}$, etc.; it's preferable to avoid a separation of cases by viewing such a final situation as a $-\frac{1}{d}\mathbb{F}$ curve for $K_{\tilde{\mathcal{F}}} + E$, equivalently working logarithmically, in the sense of Remark 2.2.2, around the final exceptional divisor, E, in (207). In this way, item (3) of Proposition/Summary 3.7.3 and Fact 3.8.4 are always valid, from which

Lemma 3.9.1 *Let $f : \mathcal{L} \to \mathcal{X}$ be a $-\frac{1}{d}\mathbb{F}$ curve meeting the singular locus in z, then around z the foliation is semi-simple.*

Consequently, let's say $\partial = \lambda_i x_i \frac{\partial}{\partial x_i}$, a semi-simple generator of the foliation in the complete local ring $\mathcal{O}_{\hat{X},z}$, with $f : t \mapsto x_i(t) = t^{v_i} u_i(t)$ an expression for the cusp in terms of some local parameter t, with $v_i \in \mathbb{Z}_{>0}$ and u_i units whenever $f^* x_i$ is not identically zero. As such, for any pair of indices i, j for which $f^*(x_i x_j)$ is not identically zero,

$$\frac{\dot{x}_i(t)}{\lambda_i x_i} = \frac{\dot{x}_j(t)}{\lambda_j x_j}. \tag{208}$$

Whence, if we re-label the coordinate system as y_i for those non-zero on the curve, x_j for those identically zero and $y_1(t) = t^{v_1}$, then:

$$\frac{\dot{y}_i(t)}{y_i} = \frac{v_i}{t} + \text{holomorphic} - \frac{\lambda_i v_1}{\lambda_1 t}, \tag{209}$$

so $y_i(t) = \eta_i t^{v_i}$, for some constant η_i; thus, without loss of generality $\eta_i = 1$ and $\lambda_i - v_i$. Proceeding thus, there may be some mild redundancy. Indeed, the cusp has an embedding dimension k and re-labelling so that v_1 is minimal; then if $v_1 | v_i$ one can replace any x_i by something in the same eigenspace (of ∂ qua operator on $\mathcal{O}$) which vanishes identically, viz, $y_i - y_1^{v_i/v_1}$, and in general, one can achieve:

$$v_1 < v_2 < \cdots < v_k, \text{ with } v_{i+1} \notin \mathbb{Z}_{\ge 0}\, v_1 + \cdots + \mathbb{Z}_{\ge 0}\, v_i \tag{210}$$

for each $1 \le i \le k$, so we get exactly k y_is, the v_i have gcd 1 since f is bi-rational, and every other coordinate is a x_j vanishing identically.

Now, by hypothesis the local monodromy group, G, preserves the foliation on the formal completion, $\hat{X}$, of $\mathcal{X}$ at the singular point. Appealing to (206), we may suppose that it acts on the above ∂ by a character χ, and we denote by H the stabiliser of the image C of (the irreducible branch) $f : L \to \hat{X}$ obtained by completing the local ring of $\mathcal{L}$ at p. Consequently, there is a factorisation:

$$f : [L/H] \xrightarrow{\nu} [C/H] \xrightarrow{\phi} [\hat{X}/G], \tag{211}$$

and since everything is convergent in the étale topology, this can be glued to a global factorisation:

$$f : \mathcal{L} \xrightarrow{\nu} C \xrightarrow{\phi} \mathcal{X} \tag{212}$$

where the first map is the normalisation of C, ϕ is net, and C is uni-branch. As such, outwith the unique singular point p, ν is an isomorphism and ϕ a closed embedding. Equally, the wholly general Fact/Definition 2.5.5 applies, so there is a formal champ $\mathfrak{X}$ with trace C such that the map $\mathfrak{X} \to \hat{\mathcal{X}}$ onto the completion of $\mathcal{X}$ in the image of f is étale representable, and

Fact 3.9.2 Let $f : \mathcal{L} \to C \hookrightarrow \mathfrak{X} \to \mathcal{X}$ be the above factorisation of the normalisation, $f : \mathcal{L} \to \mathcal{X}$ of a $-1/d\,\mathbb{F}$ cusp, with v_i as (210), and y_i, x_j, as above, suitable formal coordinates (on $\hat{X}$) about the singular point; then there are $a_j \in \mathbb{Z}$ such that the foliation is generated by

$$\partial = dv_i\, y_i \frac{\partial}{\partial y_i} + a_j x_j \frac{\partial}{\partial x_j}. \tag{213}$$

Proof If there is a divisorial valuation of negative discrepancy passing through the closed singular point, then the proposition follows from Revision 2.2.10, (210) and the fact that the v_i have g.c.d. 1, so we may suppose that the singularity is canonical rather than just log-canonical.

Now we require a certain reappraisal of (208)–(210) in the presence of the action of H in (211). To this end let I be the ideal of the image, C, of the cusp in the completion $\hat{X}$ in the singular point p, whose maximal ideal we denote by $\mathfrak{m}$, then we have an H-equivariant exact sequence:

$$0 \to I/I \cap \mathfrak{m}^2 \to \Omega_{\hat{X}} \otimes \mathbb{C}(p) \to \Omega_C \otimes \mathbb{C}(p) \to 0 \tag{214}$$

which is equally equivariant under a semi-simple generator ∂ of the foliation. In particular, therefore, the induced endomorphism

$$\partial : \Omega_C \otimes \mathbb{C}(p) \to \Omega_C \otimes \mathbb{C}(p) \tag{215}$$

may be supposed to have eigenvalues the (distinct) v_i of (210) with multiplicity (both geometric and algebraic) equal to 1. As such, although H acts on ∂ a priori by a character, (206), such an action must (cf. Fact 3.8.9) be trivial. Consequently, the $\mathbb{C}$-linear decomposition of $\mathfrak{m}$ into eigenspaces of ∂ is also H-equivariant. On the other hand, all exact sequences of $\mathbb{C}[H]$-modules are split exact, so from

$$0 \to (I(\lambda) \cap \mathfrak{m}^2)/(I(\lambda) \cap \mathfrak{m}^k) \to I(\lambda)/(I(\lambda) \cap \mathfrak{m}^k) \to I(\lambda)/(I(\lambda) \cap \mathfrak{m}^2) \to 0,$$

$k \geq 2$, etc. for any eigenvalue λ of ∂, we can write the H-action as blocks of $\mathbb{C}$-linear actions:

$$H \ni \sigma : x_{i,\lambda} \mapsto A_{ij}(\sigma)x_{j,\lambda}, \quad y_i \mapsto \chi_i(\sigma)y_i \tag{216}$$

for a coordinate system $\{x_{i,\lambda}, y_i\}$ in which $x_{i,\lambda} \in I(\lambda)$, the y_is afford eigenvectors of (215) with eigenvalues v_i, and the χ_i are characters. In particular there is a filtration which is both H and $\mathcal{F}$ equivariant:

$$F^p = \Big(\prod_i y_i^{b_i} \prod_{j,\lambda} x_{j,\lambda}^{c_{j,\lambda}} : \sum_i (b_i v_i) + \sum_{j,\lambda} c_{j,\lambda} \geq p\Big) \tag{217}$$

of the complete local ring. Plainly, however, the filtration (217) is actually the completion of a bi-equivariant filtration of the Henselian local ring of $\mathfrak{X}$ (in fact even that of $\mathcal{X}$, albeit here, (211), the invariance under the possibly larger local monodromy may fail) so it affords, [24, I.iv.3], a smoothed $\mathcal{F}$-invariant weighted blow-up:

$$\rho : \tilde{\mathfrak{X}} \to \mathfrak{X} \tag{218}$$

which is an isomorphism off p. In particular, therefore, the unique lift $\tilde{f} : \mathcal{L} \to \tilde{\mathfrak{X}}$ of f of (212) is a $-\frac{1}{d}\mathbb{F}$ curve with smoothly embedded image, and Proposition/Summary 3.7.3 holds. By direct calculation, (220) (cf. [24, pg. 180]), however, the eigenvalues (in an étale patch) of ∂ and $\rho^*\partial$ along the proper transforms of the $x_{i,\lambda}$s differ by 1, so (213) follows from (192) applied to $\rho^*\partial$. □

Of course, we also proved that not just the linear holonomy, but actually all of the holonomy is cyclic of order dividing d, so although Fact 3.9.2 is sufficient for applications, we can actually do better thanks to

Fact 3.9.3 Let $\mathfrak{X} \to [\mathfrak{X}/\mathcal{F}]$ be a foliated smooth formal champ whose trace has an étale neighbourhood the invariant affine cusp, C, i.e. image of $t \mapsto (t^{v_1}, \ldots, t^{v_k})$, for v_i as per (210), $t \in \mathbb{A}^1$, with the origin the unique point where C meets the foliation singularities; then the formal holonomy is cyclic of order at most d iff we can find formal holomorphic functions, $y_1, \ldots, y_k, x_1, \ldots, x_\ell$, restricting to a coordinate system on an analytic neighbourhood in $\mathfrak{X}$ of the singular point with the y_is embedding coordinates respecting (210), x_j vanishing on the cusp and a generator ∂ for the foliation all of which are holomorphic on an étale neighbourhood (in the analytic topology) of $\mathfrak{X}$ with trace C such that for some $a_j \in \mathbb{Z}$,

$$\partial = dv_i\, y_i \frac{\partial}{\partial y_i} + a_j x_j \frac{\partial}{\partial x_j} \tag{219}$$

holds on an analytic étale neighbourhood of the singularity where the y_i, x_j form a system of coordinates.

Proof The if direction is trivial, and for smooth curves, this is the conclusion of Proposition/Summary 3.7.3, or more accurately, a slight re-phrasing thereof. In any case, the affine cusp has no (holomorphic) Picard group, so a global holomorphic generator, ∂, of the foliation on $\mathfrak{X}$ exists, and we proceed to combine Proposition/Summary 3.7.3 with Fact 3.9.2 to achieve the said form. In particular, by the latter and Claim 3.8.2, there are coordinates y_i, x_j and an invariant function ε, all in the completion in the singularity, 0, which render $\tilde{\partial} := (1+\varepsilon)^{-1}\partial$ in the given form. A local coordinate system for the weighted blow-up (218) is given by:

$$y_1 = \tilde{y}_1^{v_1}, \ y_i = \tilde{y}_i \tilde{y}_1^{v_i}, \ x_j = \tilde{x}_j \tilde{y}_1, \tag{220}$$

in which $\tilde{y}_i = 1, i > 1$ where our cusp crosses the exceptional divisor, p, say, so the v_1th roots of unity act without fixed points in a neighbourhood of p, and $\tilde{y}_1$, $\tilde{z}_i := \tilde{y}_i - 1$, $\tilde{x}_j$ furnish coordinates in which $\rho^*\tilde{\partial}$ is semi-simple at p.

Now we appeal to the conclusion of Proposition/Summary 3.7.3, to find a possibly different generator, $D = v\rho^*\partial$, for v a (holomorphic) unit on an étale neighbourhood of $\tilde{\mathfrak{X}}$ with trace the resolution $\tilde{C}$ of C, such that

$$D = d\eta \frac{\partial}{\partial \eta} + (a_j - d)\xi_j \frac{\partial}{\partial \xi_j} \tag{221}$$

along with some other coordinate ζ_i, such that η, ζ_i, ξ_j agree with $\tilde{y}_1, \tilde{z}_i, \tilde{x}_j$ modulo $\mathfrak{m}(p)^2$, but the former are defined on all of $\tilde{\mathfrak{X}}$. Both coordinate systems are semi-simple, so Claim 3.8.2 applies to yield units u, $u_i, i > 1$, w_j in the completion at p such that $\tilde{\eta} = u\tilde{y}$, $\tilde{\zeta}_i = u_i\tilde{z}_i$, $\tilde{\xi}_j = w_j\tilde{x}_j$ are a semi-simple coordinate system for D in the complete local ring at p. Effecting an appropriate linear change, this latter coordinate system is related to that in the η, ζ_i, ξ_j by

$$\tilde{\eta} = \eta + \sum_{m \geq 2} \sum_{|I|+|J|=m} c_{IJ}(\eta)\zeta^I \xi^J \tag{222}$$

and, similarly, employing the notation of (171) et seq., for the $\tilde{\zeta}_i$ and the $\tilde{\xi}_j$. Both the left- and right-hand sides in (222) have the same eigenvalue, viz, d, so for all I, J we must have

$$d\eta c'_{IJ}(\eta) + (a_J - d|J|)c_{IJ}(\eta) = 0 \tag{223}$$

and a_J takes only finitely many values for $|I| + |J|$ bounded. Consequently, for every m such that $|I| + |J| = m$, c_{IJ} is a polynomial in η, from which $\tilde{\eta}$ converges not just in the completion at p, but in the full étale neighbourhood with trace $\tilde{C}$.

Arguing similarly for the $\tilde{\zeta}_i$, $\tilde{\xi}_j$s and ε, we may, without loss of generality, suppose $\tilde{\eta} = \eta$, $\zeta_i = \tilde{\zeta}_i$, $\xi_j = \tilde{\xi}_j$, and that ε is defined in a neighbourhood with trace C. Thus, we may suppose that $\varepsilon = 0$, and whence

$$\frac{Du}{u} = d(1-v), \quad \frac{Du_i}{u_i} = 0, \quad \frac{Dw_j}{w_j} = (a_j - d)(1-v) \tag{224}$$

where, without loss of generality, all of u, u_i, w_j are congruent to 1 modulo $I_{\tilde{C}}$. Thus, for example, we can write

$$u = \exp\Big(\sum_{m\geq 2}\sum_{|I|+|J|=m} u_{IJ}(\eta)\zeta^I\xi^J\Big) \tag{225}$$

so if $\tilde{v}_{IJ}(\eta)$ are the coefficients of a similar Taylor expansion for $v(1+\epsilon)$, then from (224)–(225),

$$d\eta u'_{IJ}(\eta) + a_J u_{IJ}(\eta) = -d\tilde{v}_{IJ}(\eta) \tag{226}$$

where the right-hand side is holomorphic in η, while, a priori, the left-hand side is formal, whence, a posteriori, holomorphic. Consequently, u is well defined on our étale neighbourhood with trace $\tilde{C}$, so idem for $\tilde{y}_1$ and by an identical argument, all of the $\tilde{z}_i, \tilde{x}_j$ are equally so defined on the said neighbourhood. The relation of these to the original coordinates y_i, x_j defined on completing $\mathfrak{X}$ in the singular point is given by (220), so not just the normalising factor $(1+\varepsilon)$, but also the y_i, x_j, are defined on an étale neighbourhood of $\mathfrak{X}$ with trace C. By construction, however, y_i, x_j are already a formal coordinate system at the singularity, so they are in fact coordinates on at worst an analytic neighbourhood of the same, while on any such (219) holds by construction. □

The role of the analytic topology in Fact 3.9.3 and its proof merits a clarifying,

Remark 3.9.4 Profiting from the Euclidean algorithm to solve $c_1v_1 + \cdots + c_kv_k = 1$, for some integers c_i, one would like to make a more strict analogue of Proposition/Summary 3.7.3. Indeed in the above notation, $\phi = y_1^{c_1}\ldots y_k^{c_k}$ is a meromorphic function on the affine cusp which, close to the singularity, restricts to a coordinate function on the normalisation, and one might hope to form an explicit patch with an étale affine neighbourhood of the non-scheme-like point at infinity according to the relation $\phi^e = s^{-d}$, for s a coordinate on the $\mathbb{A}^1 \ni \infty$ (cf. (139) and (220)). In principle, however, ϕ so constructed has an essential singularity at infinity. Plainly the problem is the intervention of v in (224), which unlike the smooth case cannot be avoided. Specifically, as in the smooth case, for t the unique (up to scaling by a constant) coordinate on the normalisation of the affine cusp, one would like to normalise (cf. (143)) a generator ∂ of the foliation restricted to the cusp according to

$$\partial(t) = t \tag{227}$$

so that a posteriori $\phi = t$ and everything is meromorphic over ∞. It can, however, happen under the hypothesis of Fact 3.9.3 that (227) doesn't admit a solution. If one follows the proof of Fact 3.9.3 and takes ∂ to be holomorphic, then this is equivalent to asking that the unit v which appears restricts to a unit defined on the affine cusp rather than just its normalisation. Similarly, if one works algebraically this is equivalent to $K_{\mathcal{F}}$ restricting to an algebraically (rather than just holomorphically) trivial bundle on the affine cusp. Consequently, a counterexample where (227) cannot be solved is

$$\partial = 2(x+y)\frac{\partial}{\partial x} + 3(y+x^2)\frac{\partial}{\partial y}, \quad y^2 = x^3 \subset \mathbb{A}^2. \tag{228}$$

Since for $t = \sqrt{x}$, $\partial(t) = t(1+t)$ and $T_{\mathcal{F}}$ defined by gluing this to the unique (up to scaling by a constant) nowhere vanishing field, ∂_∞, on $V \xrightarrow{\sim} \mathbb{A}^1 \ni \infty$ along the open set $V\backslash\{-1, \infty\}$ by way of

$$\partial = \frac{(1+t)}{t} \cdot \partial_\infty \tag{229}$$

defines a bundle whose restriction to the affine cusp is algebraically non-trivial. As such:

Warning 3.9.5 Formal neighbourhoods of cusps, even though the problem is wholly at the level of the bundle of derivations defined by restricting the foliation to the reduced cuspidal curve, do not admit a description comparable to Proposition/Summary 3.7.3. Ultimately, therefore, our treatment of cusps, Sect. 4.3, requires global hypothesis, Claim 4.3.1, rather than the local hypothesis of Fact 3.9.3

On the bright side, however:

Remark 3.9.6 We've complemented Proposition/Summary 3.7.3 in the course of the proof even in the smooth case, since, in principle, even if a generator ∂ on the étale neighbourhood U of item (3) of Proposition/Summary 3.7.3 were semi-simple at the singular point, there might have been an obstruction to expressing ∂ in terms of semi-simple coordinates on an analytic neighbourhood of $0 \in U$, as found in op. cit., due to a possible re-scaling by a unit implicit in (188). We see, however, from the proof of Fact 3.9.3 that there is no such obstruction.

4 Extremal Subvarieties

4.1 Generalities

Throughout this chapter $\mathcal{X} \to [\mathcal{X}/\mathcal{F}]$, unless specified otherwise, will be a foliated non-singular champ, with log-canonical foliation singularities. We switch our attention from $K_{\mathcal{F}}$ negative curves to $K_{\mathcal{F}}$ negative extremal rays R. The moduli X is of course supposed projective so if H_R is a nef. Cartier divisor supporting the ray, i.e. $H_R.\,\alpha = 0$, and α in the closed cone of curves iff $\alpha \in R$, then for sufficiently large $m \in \mathbb{N}$, $A_R := m\,H_R - K_{\mathcal{F}}$ is ample. In any case following Kollàr, Mori, et al. (cf. [16] III.1), we introduce our main object of study, by way of

Definition 4.1.1 The locus of R, Loc (R) is the set of closed points $x \in X(\mathbb{C})$ such that there is a curve $x \in C \subset X$ with $[C] \in R \subset \mathrm{NS}_1(X)$.

Observe that a priori Loc (R) is not a subvariety of X. Indeed for $m \in \mathbb{N}$, we can filter Loc (R) by sub-schemes $\mathrm{Loc}_m\,(R)$ on demanding that $x \in \mathrm{Loc}_m\,(R)$ if we can take the curve C of the definition to have $A_R.\,C \leq m$. That $\mathrm{Loc}_m\,(R)$ is a sub-scheme is immediate from the existence of the Hilbert scheme. To remedy this let us introduce

Definition 4.1.2 An R-pre-extremal subvariety is an irreducible subvariety $Y \subset$ Loc (R) maximal amongst the set of irreducible varieties contained in the locus.

Trivially, the dimension in chains of proper inclusions of irreducible varieties must increase so R-pre-extremal subvarieties exist; any $x \in$ Loc (R) is contained in one; and Loc (R) is the a priori countable union of all of them. Now if Y is R-pre-extremal, and $y \in Y$, then there is a C_y with $[C_y] \subset R$ containing y. However, applying Variant 3.4.2, we know, for y generic, there is an invariant parabolic champ $f_y : \mathcal{L}_y \to \mathcal{X}$ through y with moduli L_y such that

$$H_R.\,L_y \leq 2\,\frac{H_R.\,C_y}{-K_{\mathcal{F}}.\,C_y} = 0\,. \tag{230}$$

So in fact $L_y \in R$, and $A_R.\,L_y \leq 2$. Additionally, L_y cannot be contained in sing $(\mathcal{F})$ since it has $K_{\mathcal{F}}$-negative degree, so we can make a $\mathcal{F}$-invariant subvariety W by adding to generic points of Y an appropriate L_y. On the other hand, Y is by hypothesis R-pre-extremal, so $W = Y$, i.e. Y is $\mathcal{F}$ invariant, with the induced foliated variety $Y \to [Y/\mathcal{F}]$ being a pencil of rational curves of A_R degree at most 2. Hilbert schemes, however, exist and being invariant is a closed condition so in fact there are at most finitely many R-pre-extremal subvarieties for a given R. Better still the Hilbert scheme yields for any R-pre-extremal subvariety Y a flat family, $L \to T$, for some irreducible sub-scheme T of the Hilbert scheme such that the projection of L to X factors as a generically finite map over Y. An awkward case occurs when X is itself a R-pre-extremal subvariety, i.e. $\mathcal{X} \to [\mathcal{X}/\mathcal{F}]$ is a pencil in parabolic champs. As a result we introduce

Definition/More Terminology 4.1.3 An R-extremal subvariety Y is a subvariety of an R-pre-extremal subvariety Y' which is maximal amongst the subvarieties of Y' which are covered by invariant curves passing through at least one point of the image in X of the singular locus of $\mathcal{X} \to [\mathcal{X}/\mathcal{F}]$.

So indeed unless $\mathcal{X} \to [\mathcal{X}/\mathcal{F}]$ is a pencil in parabolic champs, then extremal and pre-extremal coincide, while in the awkward case an extremal variety will be specified by taking the invariant curves passing through an appropriate component of the singular locus. Now pulling everything back by the moduli map, $\mu : \mathcal{X} \to X$, define a *R-extremal champ* as the fibre over an extremal sub-variety, idem whether for pre-extremal or the locus, denoted $\mathcal{L}\mathrm{oc}\,(R)$, and observe

Fact 4.1.4 The locus $\mathcal{L}\mathrm{oc}\,(R)$ of an extremal ray is a finite union of R-pre-extremal champs. Denote by $\mathcal{L}\mathrm{oc}'\,(R)$ the sub-champ which is the union of R-extremal champs; then any $\mathcal{Y} \subset \mathcal{L}\mathrm{oc}'\,(R)$ making up this union is covered by $-1/d\,\mathbb{F}$ curves, where d may vary from curve to curve. There is however a family $\mathcal{L} \to T$ of champs, possibly non-flat at the non-scheme like points, such that $(\mathcal{L} \to T) \to (\mathcal{Y} \to [Y/\mathcal{F}])$ is a generically finite map of foliated champs.

In a similar, albeit more refined, vein we will also employ:

Fact 4.1.5 Suppose $\mathcal{X}$ is a smooth separated champ (over a field for ease of exposition) and $f : Y \to \mathcal{X}$ a map from a proper algebraic space; then there is a separated (Deligne-Mumford) champ $\mathcal{T}$ and a deformation $F : Y \times \mathcal{T} \to \mathcal{X}$ of f such that if $G : Y \times M \to \mathcal{X}$ is any deformation of f parametrised by an algebraic space M, then there is a map $g : M \to \mathcal{T}$ and a natural transformation $\gamma : G \Rightarrow F(\mathrm{id}_Y \times g)$ such that if $h : M \to \mathcal{T}$ is any other map for which there is a natural transformation $\theta : G \Rightarrow F(\mathrm{id}_Y \times h)$, then there is a unique natural transformation $\alpha : g \Rightarrow h$ for which $\theta = F_*(\mathrm{id} \times \alpha)\gamma$. In addition the dimension of $\mathcal{T}$ at the point afforded by the trivial deformation and the above universal property is at least

$$\mathrm{h}^0(f^*\,T_{\mathcal{X}}) - \mathrm{h}^1(f^*\,T_{\mathcal{X}}). \tag{231}$$

Proof The existence of $\mathcal{T}$ is a special (if key) case of the main theorem of [26]. As such the dimension computation is infinitesimal and wholly space like in nature (cf. Definition 3.1.2), i.e. deformations of the trace of the formal space:

$$\mathfrak{P} := \mathrm{Spf}(f^*\mathscr{P}_{\mathcal{X}}) \tag{232}$$

so we can replace $\mathcal{X}$ in (231) by $\mathfrak{P}$ and appeal to [16, I.2.16] – we only need the case Y projective. □

4.2 Finding Weighted Projective Spaces

As ever let $\mathcal{X} \to [\mathcal{X}/\mathcal{F}]$ be a foliated smooth champ with log-canonical foliation singularities, albeit with projective moduli, and $f : \mathcal{L} \to \mathcal{X}$ the normalisation of a $-\frac{1}{d}\,\mathbb{F}$ curve with at worst nodes and, in the notation of Proposition/Summary 3.7.3, eigenvalues $a_1 \geq a_2 \geq \cdots \geq a_n$ of a generator ∂, in the normal directions, at the unique point p where f meets the singular locus. If $a_1 \leq 0$, then we simply have nothing to say for the moment. Otherwise, consider the net completion, $q : \mathfrak{X} \to \hat{\mathcal{X}}$, of Fact/Definition 2.5.5 of $\mathcal{X}$ along the composite of f with the the universal cover, $q : \tilde{\mathcal{L}} \to \mathcal{L}$. By Proposition/Summary 3.7.3 (cf. Remark/Definition 3.8.10), there is a unique invariant closed formal sub-champ, $\mathfrak{X}_{>0} \hookrightarrow \mathfrak{X}$, such that

$$N_{\tilde{\mathcal{L}}/\mathfrak{X}_{>0}} \xrightarrow{\sim} \coprod_{a_i>0} \mathcal{O}_{\tilde{\mathcal{L}}}(a_i). \tag{233}$$

By the Chow lemma, Lemma 3.2.2, there is an irreducible sub-variety $X_{>0}$ of the moduli X of $\mathcal{X}$ of the same dimension as $\mathfrak{X}_{>0}$ obtained by taking the Zariski closure of the image of this in X. We therefore have maps:

$$\begin{array}{ccc} \mathfrak{X}_{>0} & \longrightarrow & X_{>0} \\ \downarrow & & \downarrow \\ \mathcal{X} & \longrightarrow & X \end{array} \tag{234}$$

so the leftmost vertical factors through the gerbe $\mathcal{X}_{>0} := \mathcal{X} \times_X X_{>0} \to X_{>0}$, and even through the normalisation, $\tilde{\mathcal{X}}_{>0} \to \mathcal{X}_{>0}$, since $\mathfrak{X}_{>0}$ is smooth. The said vertical arrow is, however, net so $\mathfrak{X}_{>0} \to \tilde{\mathcal{X}}_{>0}$ is étale. Indeed the assertion is local, and everything is excellent, so it suffices to work with the corresponding complete local rings in geometric points, but then $\mathfrak{X}_{>0}$ can be identified with an irreducible component of $\mathcal{X}_{>0}$, from which it's isomorphic to its image in the normalisation, and we assert:

Claim 4.2.1 There is a smoothed weighted, [24, I.iv.3], blow-up $\beta : \mathcal{X}_b \to \tilde{\mathcal{X}}_{>0}$ supported in the point p such that the induced (after saturation) foliation $\mathcal{X}_b \to [\mathcal{X}_b/\mathcal{F}_b]$ is smooth and everywhere transverse to the exceptional divisor.

Proof Since p is isolated and, as above, $\mathfrak{X}_{>0}$ and $\tilde{\mathcal{X}}_{>0}$ have isomorphic complete local rings, it will suffice to prove that there is a smoothed weighted blow-up of the complete local ring, $\hat{\mathcal{O}}$, of $\mathfrak{X}_{>0}$ completed in p which is independent of any automorphism, σ, of $\hat{\mathcal{O}}$ preserving the foliation.

Now by Proposition/Summary 3.7.3.(3) there are coordinates $y_0, y_1, \cdots, y_r$ in $\hat{\mathcal{O}}$; positive integers $a_i > 0$, $0 \le i \le r$; and a generator ∂ of the foliation such that

$$\partial = a_0 y_0 \frac{\partial}{\partial y_0} + a_i\, y_i \frac{\partial}{\partial y_i} \tag{235}$$

wherein $y_i = 0$, $i > 0$ define $\mathcal{L}$, so that for $i > 0$, a_i are as in (233), while $a_0 = d$ in the notation of (192). As such if σ is an automorphism of $\hat{\mathcal{O}}$ preserving the foliation, then there is a unit u_σ such that

$$\partial^\sigma = \sigma \partial \sigma^{-1} = u_\sigma \partial \tag{236}$$

and y_i^σ is an eigenvector of the linearisation of ∂ with eigenvalue $u_\sigma(0)^{-1} a_i$, for all $0 \le i \le r$, so $u_\sigma(0) = 1$. Consequently, by Claim 3.8.2 and $a_i > 0$, ∂^σ is not only semi-simple but

$$\partial^{\sigma} = a_0 \eta_0 \frac{\partial}{\partial \eta_0} + a_i\, \eta_i \frac{\partial}{\partial \eta_i} \tag{237}$$

for a coordinate system of the form $\eta_i = u_i y_i$, u_i a unit, $0 \leq i \leq r$. If, therefore, we define a filtration of $\hat{\mathcal{O}}$ by the ideals

$$I_n = (y_0^{t_0} \cdots y_r^{t_r} \mid a_0 t_0 + \cdots + a_r t_r \geq n), \tag{238}$$

then this is independent of the choice of y_i in (235) since a basis of the eigenvectors of ∂ with eigenvalue a_i are monomials $y_0^{t_0} \cdots y_r^{t_r}$ with $a_0 t_0 + \cdots + a_r t_r = a_i$, and it is independent whether of σ, resp. the choice of ∂, by (237), resp. mutatis mutandis. The filtration, (238), defines a weighted blow-up exactly as in (247) with smoothing as per (248). □

Now let us apply this to a qualitative description of $\mathcal{X}_{>0}$, i.e.

Corollary 4.2.2 *If $\mathcal{L}$ corresponds to an extremal ray R in Néron-Severi, with supporting function H_R and ample bundle $A_R = m H_R - K_{\mathcal{F}}$, then for all $x \in \mathcal{X}_{>0}$, there is a $-1/d(x)\mathbb{F}$ – so a fortiori from Fact/Definition 3.4.5 – an invariant parabolic champ $\mathcal{L}_x \ni x$ in $\mathcal{X}$ which, in addition, meets the singular locus in the same singular point p as $\mathcal{L}$; and every invariant curve is not only of this form, but is parallel to R in Néron-Severi. In particular the singular locus of the induced foliation in $\mathcal{X}_{>0}$ is the isolated point p.*

Proof The in particular follows from the antecedents. Otherwise, without loss of generality, we can replace $\mathcal{X}_{>0}$ by $\tilde{\mathcal{X}}_{>0}$; form the weighted blow-up $\beta : \mathcal{X}_b \to \tilde{\mathcal{X}}_{>0}$ of Claim 4.2.1; lift f to $\tilde{f} : \tilde{\mathcal{L}} \to \mathcal{X}_b$, for a possibly different but still parabolic $\tilde{\mathcal{L}}$ by item (b) of Fact/Definition 3.4.5; and argue as in item (c) of op. cit. to find a deformation $\mathcal{M}/T$, T proper, of $\tilde{f}$ composed with the universal cover of $\tilde{\mathcal{L}}$ which covers $\tilde{\mathcal{X}}_b$, so equivalently, the push-forward of which covers $\tilde{\mathcal{X}}_{>0}$.

If, however, $\sum a_i C_i$ is some effective invariant 1-cycle numerically equivalent to a rational multiple of $\mu_*[\mathcal{L}]$, then every C_i generates R, so the gerbe C_i over any such C_i is a $K_{\mathcal{F}}$-negative invariant curve. Consequently, we require, in the first instance, to show that every $K_{\mathcal{F}}$ negative invariant curve, with $f : C \to \tilde{\mathcal{X}}_{>0}$ its normalisation is a $-1/d\,\mathbb{F}$ curve for some d, so equivalently, avoiding the possibilities:

(a) $f(C) \subseteq \mathrm{sing}(\mathcal{F}) \cap \mathcal{X}_{>0} \subseteq \mathrm{sing}(\mathcal{F})$, which is impossible by the definition of log-canonical singularities as encountered in the proof of Variant 3.4.2.

(b) $f(C) \cap \mathrm{sing}(\mathcal{F}) = \emptyset$. Should this occur, then f is an embedding, and for $\tilde{C} \to C$ the universal cover, another application of item (c) of Fact/Definition 3.4.5, affords a finite étale neighbourhood $\tilde{V} \to V$ of the completion in C with trace $\tilde{C}$, such that the induced foliation in $\tilde{V}$ is a smooth fibration. From which, the generic invariant curve misses p, which is absurd.

Now, a fortiori, the singularities of the induced foliation in $\mathcal{X}_{>0}$ are contained in $\mathrm{sing}(\mathcal{F}) \cap \mathcal{X}_{>0}$, and by construction this has at least the isolated point p. The leaves of $\mathcal{F}$ in $\mathcal{X}_{>0}$ afford, however, a family of connected curves $C \to T$ in X over an irreducible base T, the gerbes over each component of each fibre of which have been seen to be $-1/d\,\mathbb{F}$ curve for some d. As such, suppose there is another singular point q, then there is a $-1/d\,\mathbb{F}$ curve through it, and this must be the gerbe over some component C_i of some fibre C_t. By definition, however, a $-1/d\,\mathbb{F}$ curve cannot meet $\mathrm{sing}(\mathcal{F})$ in any other point, while meeting p is a closed condition, so there is a different curve C_j in the fibre C_t through p. The fibre is, however, connected, so there must be a third curve C_k meeting the singular locus twice, which is nonsense. □

From which we deduce a series of corollaries:

Corollary 4.2.3 *The champ $\tilde{\mathcal{X}}_{>0}$, but (cf. the pre-amble to Sect. 3.5), maybe not $\mathcal{X}_{>0}$, is smooth.*

Proof The singular locus, $\mathcal{B}$, of $\mathcal{X}_{>0}$ is invariant by every vector field, so a fortiori by $\mathcal{F}$, while every leaf meets p, so $\mathcal{B}$ must meet it, yet, by construction the complete local rings at p of $\tilde{\mathcal{X}}_{>0}$ and $\mathfrak{X}_{>0}$ coincide, while the latter is smooth. □

Corollary 4.2.4 *The moduli $Y_{>0}$ of any representable étale cover $\mathcal{Y}_{>0} \to \tilde{\mathcal{X}}_{>0}$ has exactly one point over p, so in particular if $C \hookrightarrow \tilde{\mathcal{X}}_{>0}$ is any embedded $-1/d\,\mathbb{F}$ curve, then the natural map, $\pi_1(C) \twoheadrightarrow \pi_1(\mathcal{X}_{>0})$, be it of analytic or algebraic fundamental groups, is surjective.*

Proof Any étale cover $\mathcal{Y}_{>0} \to \tilde{\mathcal{X}}_{>0}$ still has étale neighbourhoods around a cover of $\mathcal{L}$ satisfying Proposition/Summary 3.7.3 with $\tilde{\mathcal{X}}_{>0}$ instead of $\mathcal{X}$ in op. cit. As such the proof of Corollary 4.2.2 certainly applies to deduce that $\mathcal{Y}_{>0}$, or more correctly $Y_{>0}$, has foliation singularities supported in an isolated point whenever $Y_{>0}$ is algebraic. It applies, however, even if $Y_{>0}$ were a priori analytic since the deformations of smooth parabolic invariant champs in the weighted blow-up guaranteed by Claim 4.2.1 are certainly open, but they're also closed by the simple expedient of taking the limit algebraically and lifting to the universal cover.

As to the in particular, otherwise, $C \times_{\tilde{\mathcal{X}}_{>0}} \mathcal{Y}_{>0}$ is disconnected, and $C \to \tilde{\mathcal{X}}_{>0}$ is supposed an embedding, so there would be at least two singular points in $\mathcal{Y}_{>0}$. □

Corollary 4.2.5 *For each eigendirection $\frac{\partial}{\partial x_i}$ of the linearisation of a foliation generator in* End $(N_{\mathcal{L}/\tilde{\mathcal{X}}_{>0}} \otimes \mathbb{C}(p))$, *there is an at worst nodal $-1/d_i\,\mathbb{F}$ invariant champ $f_i : \mathcal{L} \to \mathcal{X}$ through p with a branch parallel to $\frac{\partial}{\partial x_i}$ and a rational multiple of R in Néron-Severi.*

Proof There is a formal invariant curve in the said direction in the formal étale neighbourhood $\mathfrak{X}_{>0}$, but every leaf is a $-1/d\,\mathbb{F}$ curve for some d, and all branches of the embedded image are isomorphic. □

Additionally, points in $\mathbb{P}^t(\mathbb{Q})$, $t \in \mathbb{N}$, are, up to ± 1, uniquely represented by $t+1$ tuples of integers with gcd $= 1$, so if we change to a more homogeneous notation, viz:

New Notation 4.2.6 Linearise a local generator ∂ of $T_{\mathcal{F}}$ in the completion of $\hat{\mathcal{O}}_{\mathcal{X},p}$ of $\mathcal{O}_{\mathcal{X},p}$ in $\mathfrak{m}_{\mathcal{X}}(p)$ by way of $\partial = a_1\, y_1 \frac{\partial}{\partial y_1} + \cdots + a_r\, y_r \frac{\partial}{\partial y_r} - b_i\, x_i \frac{\partial}{\partial x_i}$, $a_i \in \mathbb{N}$, $b_i \in \mathbb{N} \cup \{0\}$, $(a_1, \ldots, a_r, b_1, \ldots, b_t) = 1$, with $x_i = 0$ local equations for $\tilde{\mathcal{X}}_{>0}$, the summation convention in the obvious way and t the codimension of $\mathcal{X}_{>0}$. As such in the above situation of Corollary 4.2.5, $a_i \mid d_i$.

By Corollary 4.2.4 we can (since otherwise Corollary 2.3.5 will do) conclude that $\tilde{\mathcal{X}}_{>0}$ has finite analytic, and whence finite algebraic, fundamental group on establishing

Claim 4.2.7 Let $\mathcal{C} \to \mathbb{P}^1$ be a gerbe with at most 2 points whose monodromy exceeds that of the generic point and which has a unique singular point, p, every branch of which is smooth; then the topological fundamental group $\pi_1(\mathcal{C})$ is finite.

Proof The local model, C, of $\mathcal{C}$ is b-smooth branches through p on which a finite group, G, acts transitively on the branches while fixing p. In particular, the monodromy of the generic point is isomorphic to the stabiliser of any point other than p, which, in turn, is a proper sub-group of G since its image in the permutation representation on branches fixes at least one such. Consequently, p is a point of $\mathcal{C}$ with non-generic monodromy, and we denote by q the other such, should it exist, or some point distinct from p otherwise. In either case, let $U \ni p$ be the complement of q in $\mathbb{P}^1$.

Now, observe that if $\mathcal{L} \to \mathcal{C}$ is the normalisation, and $B \hookrightarrow C$ a branch whose stabiliser in the permutation representation is H, then $[B/H]$ is a local model for $\mathcal{L}$, and $\mathcal{L}_U := \mathcal{L} \times_{\mathbb{P}^1} U$ has fundamental group H and universal cover isomorphic to U, with H acting linearly. In particular, if we identify a branch with a disc, Δ, in U, embed C in V where the latter is b copies of U through the point p, and for good measure observe that all of this is necessarily compatible with a linearisation of G in appropriate coordinates, we find a commutative diagram of fibre squares with vertical embeddings:

$$\begin{array}{ccccc} C & \longrightarrow & [C/G] & \longrightarrow & \Delta \\ \downarrow & & \downarrow & & \downarrow \\ V & \longrightarrow & \mathcal{C} \times_{\mathbb{P}^1} U & \longrightarrow & U \end{array} \tag{239}$$

The upper left horizontal arrow is, however, the universal cover, and all the verticals are homotopy equivalences since the rightmost is, so the lower left is a universal covering. As in (139), the mapping $U \to \mathcal{L}_U$ may not extend over q as a map from $\mathbb{P}^1$ to $\mathcal{L}$, but this holds over some cyclic Galois cover $\tilde{U} \to U$ ramified exactly in p which respects the commutativity of

$$\begin{array}{ccc} \tilde{U} & \longrightarrow & \mathbb{P}^1 \\ \downarrow & & \downarrow \\ U & \longrightarrow & \mathcal{L} \end{array} \tag{240}$$

Better still, taking b copies $\tilde{V}$ of $\tilde{U}$, the resulting composition $\tilde{V} \to C \times_{\mathbb{P}^1} U$ with the lower left map in (239) now admits an extension, $\bar{V} \to C$, over b copies of $\mathbb{P}^1$ meeting in a single point since the upper horizontal in (240) is an embedding. By construction, $\bar{V} \to C$ is open in the origin, and everywhere else it's flat, so it's open everywhere. As such if $\mathcal{M} \to \mathcal{L}$ is any (not necessarily finite) representable connected étale covering with group Γ, then $\mathcal{M} \times_{\mathcal{L}} \bar{V} \xrightarrow{\sim} \bar{V} \times \Gamma$, and the image of any $\bar{V} \times \gamma \to \mathcal{M}$, $\gamma \in \Gamma$ is open and closed, so it's all of $\mathcal{M}$. □

Now let $\mathcal{Y} \to \tilde{\mathcal{X}}_{>0}$ be the finite universal cover assured by Corollary 4.2.4 and Claim 4.2.7; then we further assert:

Claim 4.2.8 $\mathrm{Pic}(\mathcal{Y}) \xrightarrow{\sim} \mathbb{Z}$.

Proof By construction $\mu : \mathcal{Y} \to Y$ is a gerbe over a projective variety, and the proof of [7] that the Hodge-De Rham spectral sequence degenerates at E_1 is valid mutatis mutandis since it only requires local smoothness and the co-homological criteria for ampleness both of which hold on $\mathcal{Y}$. As such, since $\mathcal{Y}$ is simply connected and μ_* is acyclic,

$$\mathrm{H}^1(\mathcal{Y}, \mathcal{O}_{\mathcal{Y}}) = \mathrm{H}^1(Y, \mu_* \mathcal{O}_{\mathcal{Y}}) = \mathrm{H}^1(Y, \mathcal{O}_Y) = 0. \tag{241}$$

Now quite generally we have that $\mathrm{Pic}\,(\mathcal{Y})_{\mathbb{Q}} = \mathrm{Pic}\,(Y)_{\mathbb{Q}}$, and by (241), these are equally their respective Néron-Severi groups with $\mathbb{Q}$-coefficients. The Néron-Severi group, $\mathrm{NS}_1(Y)_{\mathbb{Q}}$, of Y is, however, known, e.g. [16] II.4.21, to be of rank 1, so $\mathrm{Pic}\,(\mathcal{Y})_{\mathbb{Q}} \xrightarrow{\sim} \mathbb{Q}$, which is equally the image of the Picard group under $(\mathrm{c}_1)_{\mathbb{Q}}$ in $\mathrm{H}^2(\mathcal{Y}, \mathbb{Q}(1))$ as deduced from the exponential sequence:

$$\mathrm{H}^1(\mathcal{Y}, \mathcal{O}_{\mathcal{Y}}) = \mathrm{H}^1(Y, \mathcal{O}_Y) \to \mathrm{Pic}(\mathcal{Y}) \xrightarrow{\mathrm{c}_1} \mathrm{H}^2(\mathcal{Y}, \mathbb{Z}(1)) \tag{242}$$

while the remaining possibility of torsion is excluded by $\mathcal{Y}$ simply connected, and the exact sequence,

$$\mathrm{H}^1(\mathcal{Y}, \mathbb{Q}(1)/\mathbb{Z}(1)) \to \mathrm{H}^2(\mathcal{Y}, \mathbb{Z}(1)) \to \mathrm{H}^2(\mathcal{Y}, \mathbb{Q}(1)).$$ □

We will need some auxiliary constructions, so initially, $\mathcal{Y}^0$, i.e. $\mathcal{Y}$ modulo its generic stabiliser as defined in Fact 2.1.6 and their common moduli, Y. At the singular point p identified with the origin as found in the New Notation 4.2.6, we have, therefore, its stabiliser G in $\mathcal{Y}$, of which the stabiliser G^0 in $\mathcal{Y}^0$ is a quotient acting faithfully on the local ring. Furthermore, in a minor variant of the New Notation 4.2.6, we have, étale locally at p, a foliation generator ∂ with co-prime positive integer eigenvalues a_i which is invariant by the action of G. Under this action, however, eigenvectors must go to eigenvectors, so the linear representation, ρ of G, which is equally its local action, splits as a direct sum of ρ_αs, where $\alpha \in A$ is a complete repetition free list of the a_is, and ρ_α permutes the eigenvectors of ∂ with eigenvalue α. In particular, therefore, following the New Notation 4.2.6, the action of G commutes with the action of $\mathbb{G}_m$ defined by

$$\lambda \times (y_1, \ldots, y_r) \mapsto \underline{y}^\lambda = (\lambda^{a_1} y_1, \ldots \lambda^{a_r} y_r), \tag{243}$$

while the leaves may be identified with the images of

$$\phi_{\underline{c}} : t \mapsto (c_1 t^{a_1}, \ldots, c_r t^{a_r}), \text{ where, } \underline{c} \in \mathbb{A}^r \backslash 0, \tag{244}$$

with two such functions $\phi_{\underline{c}}, \phi_{\underline{c}'}$ defining the same leaf in Y iff

$$\underline{c}' = (\rho(g)\underline{c})^\lambda, \quad g \in G, \lambda \in \mathbb{G}_m, \tag{245}$$

with $\mathbb{G}_m$ action as per (243), which, as we have said, commutes with G, so if H is the image of the representation

$$G \to \text{Aut}(P(a_1, \ldots, a_r)) \tag{246}$$

in automorphisms of the moduli of the weighted projective champ $\mathbb{P}(a_1, \ldots, a_r)$, then the leaf space is $P(a_1, \ldots, a_r)/H$.

Similarly, if we consider the weighted blow up,

$$\mathcal{Y}^1 := \text{Proj}(\coprod_n I_n) \to \mathcal{Y}^0, \quad I_n = (y_1^{t_1} \ldots y_r^{t_r} : a_1 t_1 + \ldots a_r t_r \geq n), \tag{247}$$

then the moduli, E, of the exceptional divisor is equally the said leaf space, so we have a map $\mathcal{Y}^1 \to E$. In addition, $\mathcal{Y}^1$ has only quotient singularities, so we can form the smoothed weighted blow-up $\mathcal{Y}^2 \to \mathcal{Y}^1$, [24, I.iv.3], or if one prefers not to cross reference, replace $\mathcal{Y}^1$ by what is locally its Vistoli covering champ of Fact/Definition 2.1.2. In particular $\mathcal{Y}^2$ is smooth, with smooth connected exceptional divisor $\mathcal{E}^2$. Certainly, the moduli of $\mathcal{E}^2$ is E, but it's usually false that $\mathcal{Y}^2$ maps to $\mathcal{E}^2$ because the latter is highly non-scheme like. Indeed since $\rho|_{G^0}$ is faithful, the stabiliser of a generic point is the kernel, K of $G^0 \to H$, which by (245) and (243) is isomorphic under the restriction of ρ to some finite group of roots of unity $\boldsymbol{\mu}_{a_0}$ acting according to (243), albeit for $\lambda \in \boldsymbol{\mu}_{a_0}$. Alternatively; in the stabiliser of every geometric point of $\mathcal{E}^2$, K may be identified with the normal sub-group of pseudo-reflections in $\mathcal{E}^2$, and killing such reflections affords a map $\mathcal{Y}^2 \to \tilde{\mathcal{Y}}$, where $\tilde{\mathcal{Y}}$ is smooth, still a gerbe over the moduli of $\mathcal{Y}^1$, and $\mathcal{Y}^2 \to \tilde{\mathcal{Y}}$ is the extraction of an a_0th root of a smooth divisor $\mathcal{E} \xrightarrow{\sim} [\mathbb{P}(a_1, \ldots, a_r)/H]$ – this latter notation being absolutely unambiguous since H acts on $\mathbb{P}(a_1, \ldots, a_r)$ because of the commutativity of G with (243). Consequently, we have a diagram:

$$\begin{array}{ccc} (\mathcal{Y}^1, \mathcal{E}^1) & \xleftarrow{\text{Vistoli covering}} & (\mathcal{Y}^2, \mathcal{E}^2 = \frac{1}{a_0}\cdot \tilde{\mathcal{E}}) \\ \text{Weighted blowup}\downarrow & & \downarrow a_0\text{th root} \\ \mathcal{Y}^0 \ni p & \underset{\text{if } a_0 > 1.}{\xleftarrow{\text{not defined at } p}} & (\tilde{\mathcal{Y}}, \tilde{\mathcal{E}} \xrightarrow{\sim} [\mathbb{P}(a_1, \ldots, a_r)/H]) \end{array} \quad (248)$$

where, to be precise, the final arrow is an isomorphism off $\tilde{\mathcal{E}}$ and is defined over p iff $a_0 = 1$. This final pair is the good one for extending the map $\mathcal{Y}^1 \to E$, to wit:

Claim 4.2.9 The map $\mathcal{Y}^1 \to E$ to the moduli lifts to a map $\pi : \tilde{\mathcal{Y}} \to \tilde{\mathcal{E}}$, and better still, not only is this the quotient $\tilde{\mathcal{Y}} \to [\tilde{\mathcal{Y}}/\tilde{\mathcal{F}}]$ but there is an $a \in \mathbb{N}$ such that this expresses that foliation as a fibration in $\mathbb{P}(1, a)$s in the étale site of $\tilde{\mathcal{E}}$, while the identity

$$K_{\mathcal{F}}|_{\mathcal{Y}^2} = K_{\mathcal{F}^2} + \mathcal{E}^2 = K_{\tilde{\mathcal{F}}} + \tilde{\mathcal{E}}|_{\mathcal{Y}^2} \quad (249)$$

with implied pull-backs those in (248) not only gives sense to $K_{\mathcal{F}}$ on $\tilde{\mathcal{Y}}$, but is a well-defined tautological bundle, i.e. of degree $1/a$ on geometric fibres.

Proof We will give a proof in the analytic topology, since by [8, IV.3.4] and [13, XVI.4.1] (cf. [22, IV.a.3]), it is equivalent, and trying to avoid this just leads to repeating variations on the steps in op. cit.

The smoothed weighted blow-up operation – left vertical followed by top horizontal in (248) – smooth thc foliation, and dropping to $\tilde{\mathcal{Y}}$ it remains smooth since $\mathcal{E}$ is everywhere transverse. Now let q be a geometric point of $\tilde{\mathcal{E}}$, with S_q its stabiliser in $\tilde{\mathcal{Y}}$; then we can find a polydisc Δ^r centred on q with coordinates y_i, $y_1 = 0$ an equation for $\tilde{\mathcal{E}}$, $\tilde{\partial} = \frac{\partial}{\partial y_1}$ generating the foliation and S_q acting linearly via

$$y_1 \times \sigma \mapsto \chi(\sigma) y_1, \quad y_i \times \sigma \mapsto \theta_{ij}(\sigma) y_j. \quad (250)$$

From which, we can naturally identify $\theta : S_q \to \mathrm{GL}(r-1, \mathbb{C})$ with the full (not just linear) holonomy of the piece – $[\Delta/S_q]$ – of the leaf $\mathcal{L}_q \ni q$ through q in $\tilde{\mathcal{Y}}$, and θ is faithful because there are no pseudo-reflections in $\tilde{\mathcal{E}}$.

The foliation is smooth with proper leaves, so their universal cover is constant, and since the leaves are $-\frac{1}{d}$ $\mathbb{F}$-curves in $\mathcal{Y}^0$ without generic monodromy, and the generic point of $\tilde{\mathcal{E}}$ has no-monodromy, this is $\mathbb{P}(1, a)$ for some $a \in \mathbb{N}$, and the monodromy representation extends to

$$S_q = \pi_1([\Delta/S_q]) \to \pi_1(\mathcal{L}_q) \to \mathrm{GL}(r-1) \quad (251)$$

so the first arrow in (251) is an injection. By either the long exact sequence of a fibration or, more algebraically [22, III.c.3], $\pi_1(\mathcal{L}_q)$ is an extension of the fundamental group of the orbifold over which it is a locally constant gerbe by a quotient of the generic monodromy by a central element, so S_q is also surjective by item (b) of Fact/Definition 3.4.5. As such, the holonomy covering of $\mathcal{L}_q$ is its universal covering, so that for S_q acting diagonally, we have an embedding:

$$[\mathbb{P}(1, a) \times \Delta^{r-1}/S_q] \hookrightarrow \tilde{\mathcal{Y}} \tag{252}$$

for some possibly smaller transversal polydisc, and the natural projection,

$$[\mathbb{P}(1, a) \times \Delta^{r-1}/S_q] \to [\Delta^{r-1}/S_q] \hookrightarrow \tilde{\mathcal{E}}, \tag{253}$$

is the unique analytic continuation of our initial projection $[\Delta^r/S_q] \to \tilde{\mathcal{E}}$. This latter exists everywhere in a neighbourhood of $\tilde{\mathcal{E}}$ – in fact everywhere in a formal neighbourhood would be enough which follows from the normal form of the New Notation 4.2.6 – so the projections (253) glue by Lemma 2.1.4 to a projection on all of $\tilde{\mathcal{Y}}$. Finally, (249) is an easy local calculation at the singularity. □

The fibration in Claim 4.2.9 has connected and simply connected fibres, so

$$\pi_1(\tilde{\mathcal{Y}}) \xrightarrow{\sim} \pi_1(\tilde{\mathcal{E}}) \xrightarrow{\sim} \pi_1([\mathbb{P}(a_1, \ldots, a_r)/H]), \tag{254}$$

and by Corollary 2.3.5, a weighted projective space is simply connected, so this latter group is H, which in turn affords a connected H-covering of $\tilde{\mathcal{Y}}\backslash\tilde{\mathcal{E}}$ since this is embedded as a representable Zariski open of $\tilde{\mathcal{Y}}$. Further, the diagram (248) can be formed locally with $\mathcal{Y}^0$ either $[\Delta^r/G^0]$ or $[\Delta^r/K]$, yielding a pair of diagrams with the obvious commutativity between them. Consequently, the above H-covering of $\tilde{\mathcal{Y}}\backslash\tilde{\mathcal{E}}$ implied by (254) glues to the H-covering $[\Delta^r/K] \to [\Delta^r/G^0]$, while $\mathcal{Y}$ is simply connected, so we must have $H = 1$, and we further assert:

Claim 4.2.10 The foliation $\mathcal{Y}^0 \to [\mathcal{Y}^0/\mathcal{F}]$ is isomorphic to the radial foliation, $\mathcal{R}$, on the weighted projective champ $\mathbb{P}(a_0, aa_1, \ldots, aa_n)$, Definition 2.4.2. In particular, since $\mathcal{Y}^0$ is generically scheme like, a, and a_0 are relatively prime.

Proof The start of the Leray spectral sequence applied to the fibration π of Claim 4.2.9 yields an exact sequence:

$$0 \to \mathrm{H}^1(\tilde{\mathcal{E}}, \mathbb{G}_m) \xrightarrow{\pi^*} \mathrm{Pic}(\tilde{\mathcal{Y}}) \to \mathrm{H}^0(\tilde{\mathcal{E}}, R^1\pi_*\mathbb{G}_m) \xrightarrow{d_2^{0,1}} \cdots \tag{255}$$

and by (249) this latter group is generated by the image of $K_{\mathcal{F}}$, so $d_2^{0,1} = 0$, and for a as per Claim 4.2.9 we can write

$$\mathcal{O}_{\tilde{\mathcal{Y}}}(\tilde{\mathcal{E}}) = T_{\mathcal{F}}^a \otimes \mathcal{O}_{\tilde{\mathcal{E}}}(-m) \tag{256}$$

for some $m \in \mathbb{N}$, with the latter bundle the tautological bundle of Fact/Definition 2.3.2 on our weighted projective space. Forming, the exact sequence,

$$0 \to \mathcal{O}_{\tilde{\mathcal{Y}}}(aT_{\mathcal{F}} - \mathcal{E}) \to \mathcal{O}_{\tilde{\mathcal{Y}}}(aT_{\mathcal{F}}) \to \mathcal{O}_{\tilde{\mathcal{E}}} \to 0, \tag{257}$$

and pushing forward by π, affords

$$0 \to \mathcal{O}_{\tilde{\mathcal{E}}}(m) \to \pi_* \mathcal{O}_{\tilde{\mathcal{Y}}}(aT_{\mathcal{F}}) \to \mathcal{O}_{\tilde{\mathcal{E}}} \to 0 \tag{258}$$

which by Fact 2.3.3 is a split rank 2 vector bundle, V, with the splitting even being canonical if $a > 1$. Indeed, we already know by Claim 4.2.9 that if there were extra monodromy at ∞, then it forms a smooth divisor on $\tilde{\mathcal{Y}}$ admitting a group of reflections of order a, so equivalently, if we killed these pseudo reflections, then all of the above is equally valid for some $\mathcal{Y}^0_a$, $\tilde{\mathcal{Y}}_a$, etc., and $\tilde{\mathcal{Y}} \to \tilde{\mathcal{Y}}_a$ is an extraction of an ath root of a section, ∞, of the $\mathbb{P}^1$ bundle, $\mathbb{P}(V) = \tilde{\mathcal{Y}}_a$.

Now, by (254) et. seq. $G^0 = K$, and the important thing to observe is that because of the commutativity of the action of G^0 with the $\mathbb{G}_m$-action (243), the locally constant gerbe $\mathcal{E}^2 \to \tilde{\mathcal{E}}$ of (248) in B_Ks is in fact trivial, so $a_0|m$ by (21). If, however, a_0 and a were to have a non-trivial gcd, $\alpha > 1$, then the leafwise universal cover,

$$\mathbb{P}(\frac{a_0}{\alpha}, \frac{a_1}{\alpha}) \to \mathbb{P}(a_0, a), \tag{259}$$

of the fibres of $\mathcal{Y}^2 \to \tilde{\mathcal{E}}$ is globally well defined, i.e. by (258): raising to the power α on the $\mathbb{G}_m$ torsor $\mathcal{O}_{\tilde{\mathcal{E}}}(\frac{m}{\alpha})$ and extending over 0 and ∞. The resulting covering $\tilde{\mathcal{Y}}^2 \to \mathcal{Y}^2$ is étale representable and, locally about the singularity, patches to the μ_α covering $[\Delta^r/\mu_{\frac{a_0}{\alpha}}] \to [\Delta^r/\mu_{a_0}]$ and whence the absurdity that $\mathcal{Y}$ isn't simply connected.

Having thus established the in particular, everything else follows quickly. The fact that a_0 and a are relatively prime imply that in an embedded neighbourhood (formal will do) of the singularity p, $\mathcal{Y} \to [\mathcal{Y}/\mathcal{F}]$ is isomorphic to the radial foliation, $\mathcal{R}$, of Definition 2.4.2, on the said weighted projective champ, $\mathcal{P}$. All of the above, and specifically (248), apply (cf. Facts 2.4.3), if our starting point is $\mathcal{P} \to [\mathcal{P}/\mathcal{R}]$. The fact that we have an isomorphism at p, and the same monodromy at infinity, obliges us to have the same $\mathbb{P}(1, a)$ bundle, so $\tilde{\mathcal{Y}} \to [\tilde{\mathcal{Y}}/\tilde{\mathcal{F}}]$ and $\tilde{\mathcal{P}} \to [\tilde{\mathcal{P}}/\tilde{\mathcal{R}}]$ are isomorphic in a way compatible with the initial isomorphism at p, and whence, item (d) of Facts 2.4.3, $\mathcal{Y}^0 \to [\mathcal{Y}^0/\mathcal{F}]$ is isomorphic to $\mathcal{P} \to [\mathcal{P}/\mathcal{R}]$.

□

Our initial $\mathcal{Y}$ is simply connected and a locally constant gerbe over $\mathcal{Y}^0$, so by Corollary 2.3.6, it is again a weighted projective champ and only the notation changes, to wit:

Fact 4.2.11 The foliation $\mathcal{Y} \to [\mathcal{Y}/\mathcal{F}]$ is isomorphic to the radial foliation on the weighted projective space $\mathbb{P}(a_0, aa_1, \ldots, aa_n)$, where a_0 is the order of the stabiliser of the singularity p, and the generic leaf is a $-\frac{1}{a}$-curve. In particular the generic stabiliser is cyclic of order the gcd of a_0 and a.

4.3 Ignoring Cusps

So far we haven't discussed what may happen if our extremal ray R is represented by an invariant champ $f : \mathcal{L} \to \mathcal{X}$ which has a cusp at the unique singular point z where f meets sing $(\mathcal{F})$. This is, however, easily reduced to the previous case by way of

Claim 4.3.1 Let z be a geometric point of the singular locus of a foliated smooth champ, $\mathcal{X} \to [\mathcal{X}/\mathcal{F}]$, with log-canonical foliation singularities and projective moduli; then if there is a $K_{\mathcal{F}}$-negative extremal ray, R, represented by a $-1/d\,\mathbb{F}$ curve through z, there exists a $-1/d'\,\mathbb{F}$ curve through z with at worst nodes.

Proof Let $\pi : \tilde{\mathcal{X}} \to \mathcal{X}$ be the blow-up in z; then the exceptional divisor, $\mathcal{E}$, is invariant and $\pi^* K_{\mathcal{F}}$ is again the foliated canonical bundle unless perhaps all the eigenvalues in the New Notation 4.2.6 are equal, but then there are no $-1/d\,\mathbb{F}$ cusps through z by Fact 3.9.2, and we're done. As such, by the cone theorem, Fact 3.4.1, there is a $-1/d'\,\mathbb{F}$ curve whose class, $\tilde{R}$, in $\mathrm{NE}_1(\tilde{\mathcal{X}})$ is extremal and $\pi^* R \geq \tilde{R}$. Consequently, without loss of generality, we may suppose that there is a $-1/d\,\mathbb{F}$ curve $f : \mathcal{L} \to \mathcal{X}$ which has a cusp at z and whose class, resp. that of its proper transform $\tilde{f}$, is extremal in $\mathrm{NE}_1(\mathcal{X})$, resp. $\mathrm{NE}_1(\tilde{\mathcal{X}})$. The local structure of a branch of a cusp is described by (210) and Fact 3.9.2, and in the notation of op. cit., $\tilde{f}$ meets the exceptional divisor with a (local) multiplicity v_1 in any scheme-like chart. Now consider $\pi' : \mathcal{X}' \to \mathcal{X}$ where $\mathcal{X}'$ is the extraction of a v_1th root of $\mathcal{E}$; then the induced map $f' : \mathcal{L} \to \mathcal{X}'$ has at worst nodes. On the other hand, $\mathcal{E}$ is invariant so the canonical class is the same and f' is still extremal, and, somewhat superfluously, the singularities $\mathcal{X}' \to [\mathcal{X}'/\mathcal{F}']$ are still log-canonical since $\mathcal{E}$ is smooth. In any case, at the point z' where f' crosses the exceptional divisor, we can apply Corollary 4.2.5 to find $-1/d_i\,\mathbb{F}$ curves with smooth branches parallel to every axis afforded by the embedding dimensions of the original cusp, any of which represent the extremal ray. In particular if one takes the $-1/d'\,\mathbb{F}$ curve in an eigendirection normal to the exceptional divisor in the local coordinates at z' implied by those of Fact 3.9.2 at z, then the projection of this curve to $\mathcal{X}$ has at worse nodes. □

4.4 Structure of Extremal Champs

We begin with exactly the same preliminaries as Sect. 4.2 prior to (233) except that in the notation of op. cit. our interest is the unique formal champ $\mathfrak{X}_{\geq 0} \hookrightarrow \mathfrak{X}$ with normal bundle:

$$N_{\tilde{\mathcal{L}}/\mathfrak{X}_{\geq 0}} \xrightarrow{\sim} \coprod_{a_i \geq 0} \mathcal{O}_{\tilde{\mathcal{L}}}(a_i). \tag{260}$$

Now $\mathfrak{X} \to \mathcal{X}$ is net so the tangent space to the deformation space (wherein we insist that the deformation meets sing($\mathcal{F}$)) whether of $\tilde{f} : \tilde{\mathcal{L}} \to \mathcal{X}$ or any composition with $\mathbb{P}^1 \to \tilde{\mathcal{L}}$ is the tangent space to the deformation space whether of $\tilde{\mathcal{L}} \hookrightarrow \mathfrak{X}_{\geq 0}$, or such a composition. The latter are however un-obstructed, (231), so by way of $\mathfrak{X}_{\geq 0} \hookrightarrow \mathfrak{X} \to \mathcal{X}$ the former are too. Consequently, there is a Zariski closed sub-variety, $X_{\geq 0}$, of the moduli – the variety swept out by the deformations of $\tilde{f}$ or compositions thereof with $\mathbb{P}^1 \to \tilde{\mathcal{L}}$ – of the same dimension as $\mathfrak{X}_{\geq 0}$ and containing its image. Exactly as in (234) we therefore get maps:

$$
\begin{array}{ccccccc}
\mathfrak{X}_{\geq 0} & \longrightarrow & \tilde{\mathcal{X}}_{\geq 0} & \longrightarrow & \mathcal{X}_{\geq 0} & \longrightarrow & X_{\geq 0} \\
 & & & & \downarrow & & \downarrow \\
 & & & & \mathcal{X} & \longrightarrow & X
\end{array}
\tag{261}
$$

wherein the square is fibred, $\tilde{\mathcal{X}}_{\geq 0} \to \mathcal{X}_{\geq 0}$ is normalisation, and the top leftmost arrow is an étale cover over its image. As ever we normalise a local generator ∂ of the foliation in the complete local ring $\hat{\mathcal{O}}_{\mathcal{X},p}$, for $p = f^{-1}(\operatorname{sing}\mathcal{F})$, according to the New Notation 4.2.6 with $d_1 = a_1 d$, $d \in \mathbb{N}$ albeit with the further refinement:

New Notation 4.4.1 Linearise a local generator ∂ of $T_{\mathcal{F}}$ in the completion $\hat{\mathcal{O}}_{\mathcal{X},p}$ of $\mathcal{O}_{\mathcal{X},p}$ in $\mathfrak{m}_{\mathcal{X}}(p)$ by way of, $\partial = a_1\, y_1 \frac{\partial}{\partial y_1} + \cdots + a_r\, y_r \frac{\partial}{\partial y_r} - b_i\, x_i \frac{\partial}{\partial x_i}$, $a_i \in \mathbb{N}$, $b_i \in \mathbb{N}$, $(a_1, \ldots, a_r, b_1, \ldots, b_t) = 1$, with $x_i = 0$ local equations for $\tilde{\mathcal{X}}_{\geq 0}$, and $z_1, \ldots, z_s$ the additional (formally invariant) functions which cut out $\tilde{\mathcal{X}}_{>0}$, (234), so that t is now the codimension of $\mathcal{X}_{\geq 0}$ and $s + t$ the co-dimension of $\mathcal{X}_{>0}$.

Now let us suppose that the $-1/d\,\mathbb{F}$ curve $f : \mathcal{L} \to \mathcal{X}$ affording (260) is an extremal ray, R, then we have constructed an integral invariant sub-champ $\mathcal{X}_{\geq 0}$ of $\mathcal{X}$ through every point of which there is a $-1/e\,\mathbb{F}$ champ, for varying e, parallel to R in Néron-Severi, and we assert:

Claim 4.4.2 Let $\mathcal{Z}$ be the intersection of $\mathcal{X}_{\geq 0}$ with the singular locus of $\mathcal{F}$; then $\mathcal{Z}$ is smooth and connected.

Proof Firstly, suppose $\mathcal{Z}$ is a disjoint union of components $\mathcal{Z}_1$, $\mathcal{Z}_2$; then we may consider the sub-champs $\mathcal{Y}_1, \mathcal{Y}_2$ whose moduli is covered by $K_{\mathcal{F}}$-negative extremal 1-dimensional champs parallel to R through $\mathcal{Z}_1$ and $\mathcal{Z}_2$, respectively. Consequently, if $y \in \mathcal{Y}_1 \cap \mathcal{Y}_2$, it is a singular point of some extremal 1-dimensional invariant champs $\mathcal{L}_1, \mathcal{L}_2$, so in $\mathcal{Z}_1 \cap \mathcal{Z}_2$ by Fact/Definition 3.4.5.(a), which is nonsense, and $\mathcal{Z}$ is connected. Better still at the singularity, p, of the initial curve f, we know from Fact 3.9.2 that $\mathcal{Z}$ is irreducible and smooth of dim $= s$, so there is some irreducible component $\mathcal{Z}_0$ of sing($\mathcal{F}$) of dimension s contained wholly in $\mathcal{Z}$. However, $\forall \zeta \in \mathcal{Z}$, there is a $-1/e(\zeta)\,\mathbb{F}$ champ $\mathcal{L}_\zeta \ni \zeta$ contained in $\mathcal{X}_{\geq 0}$, so sing($\mathcal{F}$) is smooth at ζ by another application of Fact 3.9.2. Consequently, $\zeta \mapsto \dim_\zeta \operatorname{sing}(\mathcal{F})$ is not just upper semi-continuous but continuous, i.e. the constant s in the New Notation 4.4.1, and since $\mathcal{Z}$ is connected: $\mathcal{Z}_0 = \mathcal{Z}$ is smooth irreducible of dimension s. □

Now consider the ideal $I_{\mathcal{Z}}$ of $\mathcal{Z}$ in $\mathcal{X}$; then the composition

$$I_{\mathcal{Z}} \underset{d}{\longrightarrow} \Omega_X \longrightarrow K_{\mathcal{F}}.I_{\mathcal{Z}} \tag{262}$$

affords an $\mathcal{O}_{\mathcal{Z}}$-linear map

$$I_{\mathcal{Z}} \underset{d}{\longrightarrow} \Omega_X \longrightarrow K_{\mathcal{F}}.I_{\mathcal{Z}}\,.$$

affords an $\mathcal{O}_{\mathcal{Z}}$-linear map

$$I_{\mathcal{Z}}/I_{\mathcal{Z}}^2 \xrightarrow[D_{\mathcal{Z}}]{} I_{\mathcal{Z}}/I_{\mathcal{Z}}^2 \otimes K_{\mathcal{F}} \tag{263}$$

of which the trace gives a global section of $\mathcal{O}_{\mathcal{Z}}(K_{\mathcal{F}})$. Plausibly this is zero, but by Fact 3.9.2 it may, on normalising in the direction of some smooth branch guaranteed by Claim 4.3.1 and (260), be identified locally with a $\mathbb{Q}$-valued function, so by Claim 4.4.2, it's non-zero iff the trace of a generator normalised according to the New Notation 4.4.1 is non-zero at some $p \in \mathcal{Z}$. Similarly, the 2nd symmetric function is a global section of $\mathcal{O}_{\mathcal{Z}}(2K_{\mathcal{F}})$ which may locally be identified with a $\mathbb{Q}$-valued function, whose expression in the notation of op. cit. is

$$\frac{1}{2}\Big(\sum_i a_i - \sum_j b_j\Big)^2 - \frac{1}{2}\Big(\sum_i a_i^2 + \sum_j b_j^2\Big) \tag{264}$$

so if the trace doesn't define a nowhere vanishing section of $\mathcal{O}_{\mathcal{Z}}(K_{\mathcal{F}})$, there is at worst an étale double cover $\mathcal{Z}^{+-} \to \mathcal{Z}$ such that $K_{\mathcal{F}}|_{\mathcal{Z}^{+-}}$ is trivial. As a result the eigenvalues of $D_{\mathcal{Z}}$ are well-defined constant functions up to a choice of generator of $\mathcal{O}_{\mathcal{Z}}(K_{\mathcal{F}})$ when this is possible, and otherwise they're well defined on $\mathcal{Z}^{+-}$. Thus, if necessary we choose a lifting p^+ of the singularity p of the curve of (260) to the double cover and subsequently choose our local generator in such a way to have compatibility with our formal linearisation at p (identified locally with p^+ if necessary), i.e. the eigenvalues of $D_{\mathcal{Z}}$ are everywhere $a_1, \dots, a_r, -b_1, \dots, -b_t$, with $a_i, b_j \in \mathbb{N}$, and $\gcd(a_1, \dots, a_r, b_1, \dots, b_t) = 1$. In any case, for every $\zeta \in \mathcal{Z}$, there is a well-defined pair of eigenspaces, $\{T_{+(\zeta)}, T_{-(\zeta)}\}$ of $T_{\mathcal{X}} \otimes \mathbb{C}(\zeta)$, and every $K_{\mathcal{F}}$-negative 1-dimensional invariant champ has tangent space at ζ contained in precisely one of these. To fully profit from this we will have to extend from the normal bundle to a formal neighbourhood of $\mathcal{Z}$, which probably shows that being lazy about convergence wasn't perhaps an optimal use of time. The discussion is local over affine neighbourhoods U covering $\mathcal{Z}$ over which the normal bundle and $K_{\mathcal{F}}$ trivialise and which we consider centred on a point ζ of $\mathcal{Z}$. To momentarily simplify the notations let λ_i denote the necessarily non-zero eigenvalues of the normal bundle, and consider the following inductive proposition,

Claim 4.4.3 Let $\hat{\mathcal{O}}_U$ be the completion of $\mathcal{O}_U$ in $\mathfrak{m}_{\mathcal{X}}(\zeta)$, then for $k \in \mathbb{N}$, we have coordinates x_i normal to Z (evidently giving a basis for $N^{\vee}_{Z/\mathcal{X}}$) and a generator ∂ of $\mathcal{F}$ over U such that:

(1) $\partial x_i = \lambda_i\, x_i \,(\text{mod} I_Z^k)$
(2) There is a semi-simple generator $\hat{\partial}$ of $T_{\mathcal{F}} \otimes \hat{\mathcal{O}}_{U,\zeta}$ of the form $\lambda_i\, \xi_i\, \frac{\partial}{\partial \xi_i}$, for $\xi_i \in \hat{\mathcal{O}}_{U,\zeta}$ and $\xi_i = x_i \,(\text{mod} I_Z^k)$.

Proof The case $k = 2$ trivially follows from the previous discussion, so consider going from k to $k+1$, which evidently we wish to be compatible with restriction so that things converge. In any case, in terms of our usual notations about monomials and summation conventions, we have, mod I_Z^{k+1},

$$\partial x_i = \lambda_i\, x_i + a_{iJ}\, x^J\ ,\ a_{iJ} \in \mathcal{O}_U\ ,\ \xi_i = x_i + b_{iJ}\, x^J\ ,\ b_{iJ} \in \hat{\mathcal{O}}_{U,\zeta}\ . \tag{265}$$

Furthermore, $\hat{\partial} = u\partial$, $u \in \hat{\mathcal{O}}_{U,\zeta}$, and, $u = 1 + u_{iK}\, x^K$, $u_{iK} \in \hat{\mathcal{O}}_{U,\zeta}$, with $\# J = k$, $\# K = k-1$, so if we put these equations together, then we obtain

$$a_{iJ} = \begin{cases} (\lambda_i - \lambda_J)\, b_{iJ} - u_{iK}\, \lambda_i & \text{if } x^K\, x_i = x^J, \\ (\lambda_i - \lambda_J)\, b_{iJ} & \text{otherwise} \end{cases} \tag{266}$$

without any summations. The latter is rather good since if $\lambda_i \neq \lambda_J := j_p\, \lambda_p$, we conclude that the b_{iJ} are algebraic, so replacing, without loss of generality, x_i, by

$$x_i \mapsto x_i + \sum_{\substack{\lambda_i \neq \lambda_J \\ x_i \nmid x^J}} b_{iJ}\, x^J, \tag{267}$$

then in fact we conclude that $a_{iJ} = 0$ if $x_i \nmid x^J$. As for the 1st case we do what we can. Specifically, again without loss of generality we can replace x_i by

$$x_i \mapsto x_i + \sum_{\lambda_i \neq \lambda_J} \frac{a_{iJ}}{\lambda_i - \lambda_J}\, x^J, \tag{268}$$

so that $a_{iJ} = 0$ if $\lambda_i \neq \lambda_J$, while if $\lambda_i = \lambda_J$ we conclude that u_{iK} is algebraic. Thus, if we replace ∂ by

$$\partial \mapsto \Big(1 + \sum_{\lambda_K = 0} u_{iK}\, x^K\Big)\partial, \tag{269}$$

then $u_{iK} = 0$ if $\lambda_K = 0$, so in fact we can suppose $a_{iJ} = 0$ for all J. Consequently, $\hat{\partial}$ has the form

$$\Big(1+\sum_{\lambda_K\neq 0} u_{iK}\, x^K\Big)\partial\,. \tag{270}$$

However, if we replace $\hat{\partial}$ by

$$\tilde{\partial}=\Big(1+\sum_{\lambda_K\neq 0}\tilde{u}_{iK}\,\xi^K\Big)^{-1}\hat{\partial} \tag{271}$$

for $\tilde{u}_{iK}$ appropriate functions of coordinates z in $\hat{\mathcal{O}}_{Z,\zeta}$ which restrict from coordinates in $\hat{\mathcal{O}}_{U,\zeta}$ annihilated by ∂ and of course $\tilde{u}_{iK} = u_{iK} \pmod{I_Z}$, then by Claim 3.8.2 $\tilde{\partial}$ is still semi-simple, with respect to a possibly different basis $\tilde{\xi}_i$ of the form $v_i\,\xi_i$, $v_i \equiv 1\,(I_Z^{k-1})$. To complete the induction, therefore, it suffices to observe, on supposing without loss of generality that $\xi_i = \tilde{\xi}_i$, that

$$\xi_i \mapsto \xi_i - \sum_{\substack{\lambda_K=0\\ x^J=x_i\,x^K}} \tilde{b}_{iJ}(z)\,\xi^J \tag{272}$$

for $\tilde{b}_{iJ}$ satisfying much the same prescriptions as the $\tilde{u}_{iK}$ is still a trivialising basis for $\hat{\partial}$. □

Consequently, over an appropriately small affine U containing ζ and bearing in mind that for any $\zeta' \in Z$ we know we can find appropriate coordinates in $\hat{O}_{\mathcal{X},\zeta'}$ annihilated by ∂, we obtain formal sub-schemes $U_{>0}$, $U_{<0}$ of the completion $\hat{U}$ of U in Z, whose subsequent completion at any $\zeta' \in Z\cap U$ is the non-strict Harder-Narasimhan pair of Remark/Definition 3.8.10. The monodromy of the pair $\{U_{>0}, U_{<0}\}$ is precisely the monodromy of the pair $\{T_{\hat{X}_{>0}}, T_{\hat{X}_{<0}}\}$, so either these patch to formal sub-champs, $\mathfrak{X}_{\geq 0}$, $\mathfrak{X}_{\leq 0}$ of the completion of $\mathcal{X}$ in $\mathcal{Z}$, which completed at any point is the non-strict H-N pair, and of course we normalise so that $\forall\,\zeta \in Z$, $T_{>0}(\zeta) = T_{\hat{X}_{>0}} \otimes \mathbb{C}(\zeta)$, $T_{<0}(\zeta) = T_{\hat{X}_{<0}} \otimes \mathbb{C}(\zeta)$, or we get the same conclusion on a double covering of the completion.

With this out of the way we can quickly proceed to a conclusion. To begin with complete $\mathcal{X}_{\geq 0}$ in $\mathcal{Z}$, call it $\mathfrak{Y}$. By Claim 4.3.1 there is, for every $\zeta \in \mathcal{Z}$, a $-1/d\,\mathbb{F}$ curve through ζ with at worst nodes and parallel to the given extremal ray. By the unicity of the New Notation 4.4.1 up to $\pm$ such a curve must factor through $\mathfrak{X}_{\geq 0} \cup \mathfrak{X}_{\leq 0}$, which is always well defined even if $\mathfrak{X}_{\geq 0}$, $\mathfrak{X}_{\leq 0}$ are only well defined on a cover. In addition, exactly as post (260), the deformation space of the universal cover of the normalisation of such a curve is un-obstructed, so locally, it covers whichever of $\mathfrak{X}_{\geq 0}$, $\mathfrak{X}_{\leq 0}$ it factors through, and we've normalised so that our initial curve factors through $\mathfrak{X}_{\geq 0}$, so $\mathfrak{Y}$ is either $\mathfrak{X}_{\geq 0}$ or $\mathfrak{X}_{\geq 0}\cup\mathfrak{X}_{\leq 0}$. In particular, there is a smooth Zariski open $\mathcal{U} \hookrightarrow \mathcal{X}_{\geq 0}\backslash\mathcal{Z}$, which close to $\mathcal{Z}$ is just the complement of the same, so all leaves in $\mathcal{X}_{\geq 0}$ meet $\mathcal{U}$. On the other hand, the singular locus of $\mathcal{X}_{\geq 0}$ is invariant by the induced foliation, so it's at worst contained in $\mathcal{Z}$, and indeed it is either empty or all of $\mathcal{Z}$ according to whether its completion $\mathfrak{Y}$ is smooth or not, i.e. iff the H-N pair is without monodromy or not. In the latter case, $\mathfrak{Y} = \mathfrak{X}_{\geq 0} \cup \mathfrak{X}_{\leq 0}$ so the normalisation $\tilde{\mathcal{X}}_{\geq 0}$ is smooth, and indeed $\tilde{\mathcal{X}}_{\geq 0} \to \mathcal{X}_{\geq 0}$ is everywhere an isomorphism except over $\mathcal{Z}$ where it's the double cover $\mathcal{Z}^{+-} \to \mathcal{Z}$, and for the unity of notation we put $\tilde{\mathcal{Z}} \hookrightarrow \tilde{\mathcal{X}}_{\geq 0}$ to be $\mathcal{Z}^{+-}$ or $\mathcal{Z}$ as appropriate.

We next wish to consider the operation of "projecting to $\mathcal{Z}$", by sending an invariant 1-dimensional champ to its unique singular point. To this end, we introduce the moduli, $X_{\geq 0}$, and the associated orbifold $\tilde{\mathcal{X}}^0_{\geq 0}$ of Fact 2.1.6, of the normalisation $\tilde{\mathcal{X}}_{\geq 0}$. Again the issue is that we have to be careful about the gerbe structure on $\mathcal{Z}$, so say

$$\begin{aligned}&\mathcal{Z}'' \hookrightarrow \tilde{\mathcal{X}}^0_{\geq 0} \text{ the fibre over the moduli } Z \hookrightarrow X_{\geq 0} \text{ of } \mathcal{Z}, \text{ and } \mathcal{Z}^0 \text{ the}\\ &\text{associated orbifold, so that } \mathcal{Z}'' \to \mathcal{Z}^0 \text{ is a locally constant gerbe.}\end{aligned} \tag{273}$$

We now proceed as in Claim 4.2.9. In the first instance (247) again affords a (well defined by Claim 4.4.3) weighted blow-up $\tilde{\mathcal{X}}^1_{\geq 0} \to \mathcal{X}^0_{\geq 0}$, whose exceptional divisor, $\mathcal{E}$, is the projectivisation of the graded $\mathcal{O}_{\mathcal{Z}''}$-algebra:

$$A := \coprod A_n := I_n \otimes \mathcal{O}_{\mathcal{Z}''}, \quad I_n = (y_1^{t_1} \dots y_r^{t_r} : a_1 t_1 + \dots a_r t_r \geq n). \tag{274}$$

In particular therefore the automorphism group of any geometric point of $\mathcal{Z}''$ has a projective representation in the automorphisms of $\mathrm{Proj}(A)$, and better still

Claim 4.4.4 The kernel, $\mathcal{S}'$, of the representation of the stabiliser, $\mathcal{S} \to \mathcal{Z}''$, in automorphisms of $\mathrm{Proj}(A)$ is locally constant, and the operation of quotienting by this kernel (cf. Fact 2.1.6), affords a factorisation $\mathcal{Z}'' \to \mathcal{Z}' \to \mathcal{Z}^0$ of locally constant gerbes.

Proof Let $U \to \tilde{\mathcal{X}}^0_{\geq 0}$ be a small étale neighbourhood of $\zeta \in \mathcal{Z}''$, with G the local monodromy; then by definition any $\sigma \in G$ which acts trivially on $\mathrm{Proj}(A)$ acts trivially on the pre-image $Z \hookrightarrow U$ of $\mathcal{Z}''$, i.e. σ is a well-defined element of every stabiliser $G_z \hookrightarrow G$ of every $z \in Z$, which stabilises $\mathrm{Proj}(A)$ around z by the uniform definition of the y_is in (274), i.e. Claim 4.4.3. □

Now, modulo notation, the diagram (248) and the proof of Claim 4.2.9 (which doesn't employ the simple connectedness of $\mathcal{Y}$ in (248)) are valid as stated, so "projection along a leaf" certainly yields, in the notation of op. cit.,

$$\tilde{\mathcal{Y}} \to \tilde{\mathcal{E}}. \tag{275}$$

On the other hand, $\tilde{\mathcal{E}}$ maps (cf. (243) et seq.), to $\mathrm{Proj}(A)$ understood as a cone over $\mathcal{Z}'$, and whence (275) affords a composition:

$$\tilde{\mathcal{X}}^0_{\geq 0} \backslash \mathcal{Z}'' \to \mathcal{Z}' \to \mathcal{Z}^0 \tag{276}$$

which may, plainly, be extended everywhere locally around $\mathcal{Z}''$ while $\mathcal{Z}^0$ itself is an orbifold so by Lemma 2.1.4 we finally get a projection:

$$\pi^0 : \tilde{\mathcal{X}}^0_{\geq 0} \to \mathcal{Z}^0 \text{ and a composition } \pi : \tilde{\mathcal{X}}_{\geq 0} \to \tilde{\mathcal{X}}^0_{\geq 0} \xrightarrow{\pi^0} \mathcal{Z}^0. \tag{277}$$

Before proceeding, let us emphasise the need for caution by way of

Warning 4.4.5 In general (277) needn't extend to a map to $\mathcal{Z}''$ or even $\mathcal{Z}'$. As such the extent to which one can profit from Fact 4.2.11 is limited according to whether we can glue together the universal covers of the fibres of π in (277), or some variant thereof for a different champ structure over the base, which *de facto* requires that π, or the said variant, has a section.

Consequently, we confine our description of π to

Claim 4.4.6 Let $\tilde{\mathcal{X}}_{\geq 0} \to [\tilde{\mathcal{X}}_{\geq 0}/\tilde{\mathcal{F}}]$ be the induced foliation; then π is a smooth $\tilde{\mathcal{F}}$-equivariant (foliated) fibre bundle (in the étale topology) with fibre a foliated champ whose (finite) universal cover is described by Fact 4.2.11, i.e. a weighted projective champ in its radial foliation.

Proof By construction, (276), functions on $\mathcal{Z}^0$ are invariant, i.e. π is certainly a $\tilde{\mathcal{F}}$-equivariant morphism of smooth champs. As such the map

$$d\pi : \Omega^1_{\mathcal{Z}^0} \to \Omega^1_{\tilde{\mathcal{X}}_{\geq 0}} \tag{278}$$

is given, locally, by a $s \times (r+s)$ matrix, P, say such that for ∂ a local generator of the foliation there is a $(r+s) \times (r+s)$ matrix B for which $\partial P = PB$, so the locus where $d\pi$ fails to have full rank is $\tilde{\mathcal{F}}$-invariant. By definition, however, every leaf of $\tilde{\mathcal{F}}$ meets $\tilde{\mathcal{Z}}$, and, Claim 4.4.3, π is smooth in a formal neighbourhood of $\tilde{\mathcal{Z}}$, whence the co-kernel of (278) is a vector bundle of rank s everywhere, and a surjective map of smooth varieties is flat as soon as the fibres are equidimensional, so π is smooth. As such the condition, (61), for π to be a bundle of champs is true by Fact 4.2.11, Fact 2.3.3 and (because we're in characteristic zero) the Höschild-Serre spectral sequence. Consequently, by Fact 4.2.11, for $V \to \mathcal{Z}^0$ a sufficiently small étale neighbourhood, $\pi^{-1}(V)$ is of the form $[V \times_{\mathbb{C}} \mathbb{P}_{\mathbb{C}}(\underline{a})/G]$ for G a finite group of automorphisms of a weighted projective champ $\mathbb{P}_{\mathbb{C}}(\underline{a})$. Again, however, by the Höschild-Serre spectral sequence, the representation of G cannot be deformed, so the only obstruction to having a bundle of foliated champs is that radial foliations on weighted projective champs might deform. This is, however, excluded by Fact/Definition 2.4.4. □

We have, therefore, established

Large Fact 4.4.7 Given a $-1/d\,\mathbb{F}$ curve $f : \mathcal{L} \to \mathcal{X}$ parallel to an extremal ray R in Néron-Severi meeting sing($\mathcal{F}$) with p the unique geometric point of their intersection, then after multiplication by a suitable constant, a linearisation in $\mathrm{End}(\Omega_{\mathcal{X}} \otimes \mathbb{C}(p))$ of a generator ∂ of the foliation is a diagonal matrix $\mathrm{diag}\{a_1, \ldots, a_r, 0, -b_1, \ldots, -b_t\}$, $a_i, b_j \in \mathbb{N}$ without common divisor and s zeroes. Better still, normalising so that the tangent space to $f(\mathcal{L})$ lies in the positive eigenspace, there is an R-extremal champ $\mathcal{X}_{\geq 0} \hookrightarrow \mathcal{X}$ containing f such that:

(a) $\mathcal{X}_{\geq 0}$ contains a unique, smooth s-dimensional component $\mathcal{Z}$ of the singular locus of $\mathcal{F}$.

(b) The normalisation $\tilde{\mathcal{X}}_{\geq 0}$ retracts onto $\mathcal{Z}^0$ where the pre-image $\tilde{\mathcal{Z}} \hookrightarrow \tilde{\mathcal{X}}_{\geq 0}$ of the singular locus is a locally constant gerbe over $\mathcal{Z}^0$, (273), via π of (277), and we have exactly one of:

(i) $K_{\mathcal{F}}|_{\mathcal{Z}}$ is trivial, and $\tilde{\mathcal{X}}_{\geq 0} \xrightarrow{\sim} \mathcal{X}_{\geq 0}$.

(ii) $K_{\mathcal{F}}^{\otimes 2}|_{\mathcal{Z}}$ is trivial, but $K_{\mathcal{F}} \mid \mathcal{Z}$ is not; then $\tilde{\mathcal{Z}} \to \mathcal{Z}$ is an étale μ_2 covering which is exactly where $\tilde{\mathcal{X}}_{\geq 0} \to \mathcal{X}_{\geq 0}$ fails to be an isomorphism.

(c) The fibration π is actually an étale bundle of foliated varieties where the transition functions are automorphisms of a foliated variety $\mathcal{Y} \to [\mathcal{Y}/\mathcal{F}]$ whose (finite) universal cover is the radial foliation on some $\mathbb{P}(a_0, aa_1, \ldots, aa_n)$ for a_0, a as per Fact 4.2.11.

(d) Every extremal champ meeting sing($\mathcal{F}$) is of this form.

There are a few loose ends here which we'll tidy up via

Remark 4.4.8 All of the above includes the case that sing($\mathcal{F}$) has dimension zero at z but non-trivial monodromy (cf. Fact 3.8.9). Indeed, by item (c) of Fact/Definition 3.4.5, the only way that an extremal ray can fail to meet sing($\mathcal{F}$) is if the foliation is generically a fibration in parabolic champs. This is also the only way that not just (b).(ii) (so inter alia an isolated singularity with monodromy switching the H-N pair) can occur, but that (possibly different) extremal rays can factor through both the positive and negative parts of the H-N pair. This is, however, more subtle, so its proof is postponed. It is, therefore, not unreasonable to paraphrase Large Fact 4.4.7 as "every" extremal sub-champ is a smoothly embedded bundle of radially foliated weighted projective spaces.

Irrespectively, however, of clarifying when Large Fact 4.4.7.b.(ii) does occur, we have

Corollary 4.4.9 *The number of extremal rays in the half-space,* $\mathrm{NE}_{K_{\mathcal{F}}<0}$, *is finite.*

Proof An extremal ray which meets a singularity is described by Large Fact 4.4.7, and by Fact 3.8.9 it must factor through either $\hat{\mathcal{X}}_{>0}$ or $\hat{\mathcal{X}}_{<0}$ of the H-N pair, so every connected component of sing($\mathcal{F}$) meets at most two such sub-champs which themselves are maximal amongst those covered by extremal rays meeting sing($\mathcal{F}$). By item (c) of Fact/Definition 3.4.5 we're therefore done unless $\mathcal{X} \to [\mathcal{X}/\mathcal{F}]$ is generically a fibration in rational curves, but in this case (*cf. op. cit.*), the component of the deformation space of an invariant curve which doesn't meet sing($\mathcal{F}$) cover $\mathcal{X}$ with leaves, so all such rays are equivalent. □

5 Flip, Flap, Flop

5.1 Contractions

We will profit from a number of simplifications afforded by the analytic topology. As such we spell out our

Set Up 5.1.1 Let $\mathcal{X} \to [\mathcal{X}/\mathcal{F}]$ be a foliated champ with projective moduli and $\mathcal{Y} \hookrightarrow \mathcal{X}$ an embedded invariant sub-champ equal to $\tilde{\mathcal{X}}_{\geq 0}$ of Large Fact 4.4.7 in case b.(i) for some extremal ray. Fix a (not necessarily scheme-like) point $z \hookrightarrow \mathcal{Z}^0$, i.e. the pre-image of a point in the moduli, for $\mathcal{Z}^0$ as in op. cit., and let $\mathcal{X}_! \hookrightarrow \mathcal{X}$ be an embedded analytic open neighbourhood of $\mathcal{Y}_z$ whose intersection $\mathcal{Y}_! \hookrightarrow \mathcal{Y}$ with $\mathcal{Y}$ is (as a foliated variety) of the form $\pi^{-1}(\mathcal{Z}_!)$, with π as per item (c) of Large Fact 4.4.7, and $\mathcal{Z}_! \hookrightarrow \mathcal{Z}^0$ a small embedded analytic neighbourhood of z. Such data admits, therefore, arbitrarily small shrinking around $\mathcal{Y}_z$ which, in order to employ Fact 2.6.5, we will make without warning. This is equally the good set up for constructing flips, so in our immediate context we add the precision that uniquely for this section, Sect. 5.1, $\mathcal{Y}$ is a divisor.

Profiting from Fact 4.2.11 and shrinking as necessary, we have, from item (c) of Large Fact 4.4.7 and Fact 2.6.5, that for some polydisc V there is a fibred square:

$$\begin{array}{ccc} \mathcal{Y}^! := \mathbb{P}(a_0, aa_1, \ldots, aa_r) \times V & \longrightarrow & \mathcal{X}^! \\ \downarrow & & \downarrow \\ \pi^{-1}(\mathcal{Z}_!) & \longrightarrow & \mathcal{X}_! \end{array} \tag{279}$$

where the horizontal arrows are embeddings, the vertical arrows étale Galois coverings under $\pi_1(\mathcal{Y}_z)$, a_i as in the New Notation 4.4.1 and a, a_0 as in Fact 4.2.11. In particular, therefore, for $\mathcal{O}(1)$ the tautological bundle on the weighted projective space in the left-hand corner of (279), Proposition/Summary 3.7.3, implies

$$N_{\mathcal{Y}^!/\mathcal{X}^!} \xrightarrow{\sim} \mathcal{O}(-ab) \tag{280}$$

for $b = b_1$ of the New Notation 4.4.1. Now consider the operation of extracting a dth root of the Cartier divisor $\mathcal{Y}^!$, as defined in Fact/Definition 2.1.9:

$$\begin{array}{ccc} \mathcal{Y}_{!!} & \longrightarrow & \mathcal{X}_{!!} \\ \downarrow & & \downarrow \\ \mathcal{Y}^! & \longrightarrow & \mathcal{X}^! \end{array} \tag{281}$$

then, for any d the left-hand vertical is a locally constant gerbe under B_{μ_d} and if, moreover $d = ab$ this gerbe is trivial, so by Fact 2.6.5 again, after appropriate shrinking there is a fibre square:

$$\begin{array}{ccc} \mathcal{Y}^! & \longrightarrow & \mathcal{X}^{!!} \\ \downarrow & & \downarrow \\ \mathcal{Y}_{!!} & \longrightarrow & \mathcal{X}_{!!} \end{array} \tag{282}$$

where, once more, the horizontals are embeddings and the verticals étale coverings, but now under μ_{ab}. This construction has a number of convenient properties, to wit:

Claim 5.1.2 The complement $X^* := \mathcal{X}^{!!} \backslash \mathcal{Y}^!$ is everywhere space like and an étale Galois covering of $\mathcal{X}_! \backslash \mathcal{Y}_!$ with group an extension of the form

$$1 \to \mu_{ab} \to E_z \to \pi_1(\mathcal{Y}_z) \to 1, \quad i.e.\ \mathcal{X}_! \backslash \mathcal{Y}_! \xrightarrow{\sim} [X^*/E_z]. \tag{283}$$

Proof That we have a covering with the said group is immediate from (279), (281) and (282), while by (280) $N_{\mathcal{Y}^!/\mathcal{X}^{!!}}$ is isomorphic to $\mathcal{O}(-1)$. As such the local monodromy acts faithfully on the complement of the zero section $\mathcal{Y}^! \hookrightarrow N_{\mathcal{Y}^!/\mathcal{X}^{!!}}$, so a fortiori X^* is space like. □

Before profiting from this let us make:

Remark 5.1.3 One could certainly take an abth root globally of $\mathcal{Y} \hookrightarrow \mathcal{X}$, say

$$\mathcal{X}^{1/ab} \to \mathcal{X}, \tag{284}$$

with $\mathcal{X}_!^{1/ab} \hookrightarrow \mathcal{X}_! \times_{\mathcal{X}} \mathcal{X}^{1/ab}$ the resulting neighbourhood. This does not imply, however, that (283) is split since there may be torsion effects in $\mathrm{Pic}(\mathcal{Y})$ – cf. (21). Similarly, if one is prepared to assume that $\mathrm{Pic}(\mathcal{X}^!) \xrightarrow{\sim} \mathrm{Pic}(\mathcal{Y}^!)$, then one can do steps (281)–(282) in a single move, viz: extract the abth root of the section of $\mathcal{O}(-ab)$ defined by $\mathcal{Y}^!$. This is easy if one completes in $\mathcal{Y}^!$, i.e. the exponential sequence for the nth thickening:

$$0 \to I^n/I^{n+1} \xrightarrow{x \mapsto 1+x} \mathcal{O}_{n+1}^\times \to \mathcal{O}_n^\times \to 0 \tag{285}$$

and Fact 2.3.3, but otherwise would require a little analysis that can reasonably be avoided, via Fact 2.6.5, by confining ourselves to purely topological statements.

Irrespectively, having arrived to this juncture, we can complete $\mathcal{X}^{!!}$ in $\mathcal{Y}^{!}$ to a formal champ, $\hat{\mathcal{X}}^{!!}$, with trace $\mathcal{Y}^{!}$ and argue as in (285) to deduce

$$\mathrm{Pic}(\hat{\mathcal{X}}^{!!}) \xrightarrow{\sim} \mathrm{Pic}(\mathcal{Y}^{!}) = \mathbb{Z}\mathscr{O}(1). \tag{286}$$

As such the $\mathbb{G}_m$-torsor $\mathfrak{X} \to \hat{\mathcal{X}}^{!!}$ defined by $\mathscr{O}(1)$ has trace a product with V of the $\mathbb{G}_m$-torsor, (36), in the definition of a weighted projective space. The latter is space like, and the formal space, $\mathfrak{X}$, may be described wholly explicitly, i.e.

$$\mathfrak{X} \xrightarrow{\sim} (\mathbb{A}^{r+1}\backslash\{0\}) \times \hat{\Delta} \times V, \quad \hat{\Delta} := \mathrm{Spf}\mathbb{C}[[x]] \tag{287}$$

on which $\lambda \in \mathbb{G}_m$ acts according to

$$\begin{aligned}(\mathbb{A}^{r+1}\backslash\{0\}) \times \hat{\Delta} \times V \ni &(y_0, y_1, \dots, y_r) \times x \times z \mapsto \\ &(\lambda^{a_0} y_0, \lambda^{aa_1} y_1, \dots, \lambda^{aa_n} y_r) \times \lambda^{-1} x \times z.\end{aligned} \tag{288}$$

Now observe that the ring, A, of $\mathbb{G}_m$ invariant functions affords maps

$$\mathfrak{X} \to \hat{\mathcal{X}}^{!!} \xrightarrow{\text{moduli}} \hat{X}^{!!} \to V \times \mathrm{Spf}A\,(\xrightarrow{\sim} \mathbb{C}[[x^{a_0} y_0, x^{aa_1} y_1, \dots, x^{aa_n} y_r]]) \tag{289}$$

where, by definition, A is equally the ring of formal functions on $\hat{X}^{!!}$. Consequently, the final map in (289) is a formal contraction in the sense of [3] and whence by op. cit. is the completion in V of the contraction of analytic spaces:

$$\begin{array}{ccccc} \mathcal{Y}^{!} & \longrightarrow & \mathcal{X}^{!!} & \xrightarrow{\text{moduli}} & X^{!!} \\ {\scriptstyle \text{projection } \pi \text{ of (277)}}\downarrow & & \rho\downarrow & & \rho_0\downarrow{\scriptstyle \text{contraction}} \\ V & \longrightarrow & X_z & = & X_z \end{array} \tag{290}$$

In particular, by (289), X_z is smooth, and we're well advanced in proving:

Proposition/Summary 5.1.4 There is a Galois covering $\mathcal{X}^{!!} \to \mathcal{X}_{!}$, under the group E_z of (283), ramified uniquely over $\mathcal{Y}^{!} \to \mathcal{Y}_{!}$, and there (in the New Notation of 4.4.1 and that of Fact 4.2.11) to order exactly ab such that:

(a) The contraction, X_z, of (290) is smooth.
(b) It is E_z equivariant, and although $\mathcal{X}_{!} \to [X_z/E_z]$ may not be defined at $\mathcal{Y}$, $\mathcal{X}_{!}^{1/ab}$, as encountered in (284) et seq., to $[X_z/E_z]$ is everywhere defined.
(c) The contraction is bi-rational, i.e. $\mathcal{X}_{!}\backslash\mathcal{Y} \xrightarrow{\sim} [X_z\backslash V/E_z] = [X^{*}/E_z]$.

Better still, all of this globalises, i.e. there is a foliated smooth champ $\mathcal{X}_0 \to [\mathcal{X}_0/\mathcal{F}_0]$ fitting into a diagram (the contraction of $\mathcal{Y}$) – described locally by the above items (a)–(c) – and an isomorphism off $\mathcal{Y}$, to wit:

$$\begin{array}{ccc} \mathcal{X}^{1/ab} & \xrightarrow[\rho]{} & \mathcal{X}_0 \\ {\scriptstyle K_{\mathcal{F}}\text{ unramified}}\big\downarrow{\scriptstyle (284)} & & \\ \mathcal{X} & & \end{array} \tag{291}$$

Proof We've done (a) and (c), and as per Remark 5.1.3 we have from the construction, (18), of extracting roots a map $\mathcal{X}_{!!} \to \mathcal{X}^{1/ab}$. If, however, $\mathcal{Y}_!^{1/ab} \hookrightarrow \mathcal{X}_!^{1/ab}$ is the reduced fibre over $\mathcal{Y}$, then

$$\begin{array}{ccc} \mathcal{Y}_{!!} & \longrightarrow & \mathcal{Y}^! \\ \big\downarrow & & \big\downarrow \\ \mathcal{Y}_!^{1/ab} & \longrightarrow & \mathcal{Y}_! \end{array} \tag{292}$$

is not just commutative but the top horizontal is the pull-back of the locally constant gerbe defined by the bottom horizontal. As such, the square is fibred so the left vertical is a representable étale cover, and whence, by (282), $\mathcal{Y}^!$ is the universal cover of $\mathcal{Y}_!^{1/ab}$, so that shrinking as necessary, $\mathcal{X}^{!!} \to \mathcal{X}_!^{1/ab}$ is equally the universal cover. In particular, therefore, we have a diagram:

$$\begin{array}{ccc} \mathcal{X}^{!!} & \longrightarrow & X_z \\ \big\downarrow & & \big\downarrow \\ \mathcal{X}_!^{1/ab} & & [X_z/E_z] \end{array} \tag{293}$$

wherein the left-hand vertical is an E_z-torsor, and the pull-back of $\mathcal{O}_{\mathcal{X}_!^{1/ab}}(\mathcal{Y}_!^{1/ab})$ to $\mathcal{Y}^! \hookrightarrow \mathcal{X}^{!!}$ is $\mathcal{O}(-1)$, so there is an E_z action on the torsor $\mathfrak{X}$ commuting with the $\mathbb{G}_m$-action (288). Consequently, the top horizontal is E_z equivariant, so by the definition of the bottom right-hand corner, the square can be completed along the bottom horizontal, i.e. (b) holds.

Turning to globalisation, the unicity of contractions ensures that the contraction of the subspace of the moduli, X, of $\mathcal{X}$ defined by the moduli of $\mathcal{Y}$ to that of $\mathcal{Z}^0$ is an algebraic space X_0. Now denote by $*$ the complement of $\mathcal{Y}$, or the contracted locus as appropriate, then for ζ another point of $\mathcal{Z}^0$ the normalisation of $X_z^* \times_{X_0} X_\zeta^*$ is equally that of $X_z^* \times_X X_\zeta^*$ so by Lemma 2.1.3, either projection of

$$R := \big(\text{ normalisation of } U \times_{X_0} U\big) \rightrightarrows U, \quad U = \coprod_z X_z, \tag{294}$$

is unramified in co-dimension 1. Consequently, by purity, they're unramified everywhere, and since $R^* \rightrightarrows U^*$ is both a groupoid and dense in R, (294) defines an étale groupoid, or, equivalently, an orbifold $\mathcal{M}_0 \to X_0$ with atlas U. At the same time, we can express $\mathcal{X}^{1/ab}$ as a locally constant gerbe in B_Γ's over an orbifold $\mathcal{M}$ for some finite group Γ. Thus, $\mathcal{M}$ and $\mathcal{M}_0$ agree on an open dense set, so by Lemma 2.1.4 and (290), there is a map $\rho : \mathcal{M} \to \mathcal{M}_0$. Next, observe that the contracted locus is an embedded smooth sub-champ of real co-dimension at least 4, whence the homotopy depth about the same, [12, Exposé XIII.6], is also at least 4, so the locally constant gerbe $\mathcal{X}^* \to \mathcal{M}^*$ extends uniquely to a locally constant gerbe $\mathcal{X}_0 \to \mathcal{M}_0$. Equally, locally the universal cover is generically scheme like by Claim 5.1.2, so from the long exact sequence of a fibration we must have

$$1 \to \Gamma \to E_z = \pi_1(\mathcal{X}_!^{1/ab}) \to \pi_1(\mathcal{M}_!) \to 1 \tag{295}$$

for $\mathcal{M}_!$ a small neighbourhood of $\mathcal{Y}_!^{1/ab}$. On the other hand in the diagram

$$\begin{array}{ccc} X_z & \longleftarrow & \mathcal{X}^{!!} \\ \downarrow & & \downarrow \\ \mathcal{M}_0 & \longleftarrow & \mathcal{M}_! \end{array} \tag{296}$$

the left hand is the universal cover of its image under the group E_z/Γ, so by (295), the diagram (296) is a pull-back of a covering along the bottom horizontal. In particular, therefore, (296) is fibred so for a locally constant sheaf, Λ, $R^1\rho_*\Lambda = 0$, and the Leray spectral sequence yields an exact sequence:

$$0 \to \mathrm{H}^2(\mathcal{M}_0, \Lambda) \to \mathrm{H}^2(\mathcal{M}, \Lambda) \to \mathrm{H}^0(\mathcal{M}_0, R^2\rho_*\Lambda). \tag{297}$$

In addition $\pi_1(\mathcal{M}) \xrightarrow{\sim} \pi_1(\mathcal{M}_0)$, so $\mathcal{X}$ and $\rho^*\mathcal{X}_0$ are locally constant gerbes for the same link in the sense of [8, IV.1.1.7.3], and their difference, op. cit. IV.3.4, defines a class in $\mathrm{H}^2(\mathcal{M}, \Lambda)$ for Λ the centre of the link – so locally the centre of the aforesaid Γ. Finally, $\mathcal{X}_0 \to \mathcal{M}_0$ is locally trivial by definition, so the image of this class in the far right of (297) is zero by (296) and (290), while the resulting class in the leftmost group is trivial because this is the same as $\mathrm{H}^2(\mathcal{M}^*, \Lambda)$. □

It follows that we've actually proved a little more, to wit:

Remark/Definition 5.1.5 From (293), the fibre of the horizontal arrow in (291) has fibre $\mathcal{X}^{!!}$ over X_z, which is also the smoothed weighted blow-up (composition of left vertical and top horizontal in (248)) with weights $a_0, aa_1, \dots, aa_n$ in the obvious coordinates suggested by (289) while by purity the left vertical in (291) is exactly the same as the rightmost vertical in (248), i.e. killing a group (here μ_{ab}) of pseudo reflections. Moreover, since $\rho : \mathcal{X} \to \mathcal{X}_0$ needn't be everywhere defined, it's more technically correct to call the bi-rational map ρ a flip, which, in turn, has the very specific structure of (291), which might reasonably be described as a *flap*.

The resulting foliation on $\mathcal{X}_0$ is described by

Corollary 5.1.6 *The canonical bundles of the various foliations are related by*

$$K_{\mathcal{F}^{1/ab}} = K_{\mathcal{F}}|_{\mathcal{X}^{1/ab}} = \rho^* K_{\mathcal{F}_0} + a_0 \mathcal{Y}^{1/ab} \tag{298}$$

so in particular, $\mathcal{F}_0$ is smooth and everywhere transverse to the contracted locus.

Proof The first identity in (298) is just that the left vertical in (291) is unramified along the foliation because $\mathcal{Y}$ is invariant, while the 2nd identity follows, for purely numerical reasons, from (280) and Fact 4.2.11. Now say D is a local generator of $\mathcal{F}_0$ on X_z, and s_0 is the coordinate function of weight a_0 in (289), then, by (298), $\rho^*(s_0 D)$ is an everywhere regular derivation which coincides with a local generator of $\mathcal{F}^{1/ab}$ at every point where $\rho^*(s_0)$ only vanishes along the exceptional divisor. In particular, therefore, it coincides by Fact 4.2.11 with a local generator close to $\mathrm{sing}(\mathcal{F}^{1/ab})$, where by op. cit. a local equation $x = 0$ for the exceptional divisor may be supposed of the form $x^{a_0} = \rho^* s_0$. Now the exceptional divisor is invariant and by (3.7.3) defines a non-zero eigenspace at the singularity, so $\rho^*(D(s_0))$ is non-zero everywhere, whence idem $D(s_0)$, □

5.2 Projectivity of the Contraction

We start with a general projectivity criteria:

Lemma 5.2.1 *Let X be a proper algebraic space over a field k; then X is projective iff both of the following conditions hold:*

(a) For every irreducible subspace $Y \hookrightarrow X$

$$\left(\overline{\mathrm{NE}}_1(Y) \ni \alpha \mapsto 0 \in \overline{\mathrm{NE}}_1(X)\right) \Rightarrow \alpha = 0.$$

(b) The cone $\mathrm{NE}_1(X) \subseteq \mathrm{NS}_1(X)_{\mathbb{R}}$ does not contain a line.

Proof The conditions are clearly necessary. The second condition is equivalent to the existence of a Cartier divisor H non-negative on $\mathrm{NE}_1(X)$ such that

$$\left(\overline{\mathrm{NE}}_1(X) \ni \alpha \mapsto H.\alpha = 0\right) \Rightarrow \alpha = 0. \tag{299}$$

Thus, if (a) and (b) hold for X, they hold for every sub-variety, so by induction we can suppose $H^{\dim(Y)}.Y > 0$ for every non-trivial sub-variety of dimension smaller than that of X. Consequently, by the Nakai-Moishezon criteria, [15, 3.11], we require to prove for every irreducible component of X of maximal dimension the top power of H is positive. As such, say, without loss of generality, X irreducible of dimension $d+1$ and $p : X' \to X$ a projective modification, then p^*H is nef. Better still some Zariski open of X is a scheme, whence it contains sub-varieties of all possible dimensions; thus, $H^d = p_*(p^*H^d)$ is a non-zero class in $\overline{\mathrm{NE}}_1(X)$, and so by (b) $H^{d+1} > 0$. □

A less general, but more relevant variation of the same is

Corollary 5.2.2 *Let $p : X' \to X$ be proper, an isomorphism off $Z \hookrightarrow X$, with X' projective and X a $\mathbb{Q}$-factorial algebraic space over a field k; then X is projective iff both of the following conditions hold:*

(a) For every irreducible subspace $Y \hookrightarrow Z$

$$\big(\overline{\mathrm{NE}}_1(Y) \ni \alpha \mapsto 0 \in \overline{\mathrm{NE}}_1(X)\big) \Rightarrow \alpha = 0.$$

(b) The cone $\mathrm{NE}_1(X) \subseteq \mathrm{NS}_1(X)_{\mathbb{R}}$ *does not contain a line.*

Proof Again necessity is obvious and (b) affords a Cartier divisor H non-negative on $\mathrm{NE}_1(X)$ satisfying (299) which we prove satisfies op. cit. (and whence Claim Lemma 5.2.1.(a)) for all sub-varieties $Y \hookrightarrow X$ by induction on their dimension. In dimension 1, there are two cases a curve, Y, factors through Z so $H \cdot Y > 0$ by Corollary 5.2.2.(a), or it doesn't. In the latter case, however, $Y \backslash Z$ is a non-empty curve in the quasi-projective variety $X \backslash Z$, so it certainly intersects non-trivially some divisor $D \hookrightarrow X \backslash Z$ without being contained in it. By hypothesis, however, the closure $\bar{D} \hookrightarrow X$ of D is $\mathbb{Q}$-Cartier so $\bar{D} \cdot Y \neq 0$ and item (a) of Lemma 5.2.1 holds. Similarly, for Y of dimension $d+1 \leq \dim(X)$ we again distinguish 2-cases. If Y factors through Z, we're done by hypothesis; otherwise, we prove $H\mid_Y$ is ample. In the latter case, by Nakai-Moishezon and our induction hypothesis it is sufficient to prove $H^{d+1} \cdot Y > 0$. As before, however, there is a Cartier divisor, $\bar{D}$, on X intersecting Y non-trivially, so $H^d \cdot \bar{D} \cdot Y > 0$, while for all $\epsilon > 0$ sufficiently small, $H - \epsilon \bar{D}$ satisfies (299), so $p^*(H - \epsilon \bar{D})$ is nef., and $(H - \epsilon \bar{D})^{d+1} \cdot Y \geq 0$, whence $H^{d+1} \cdot Y > 0$. □

Of which a corollary to the corollary is

Corollary 5.2.3 *Let everything be as in Corollary 5.2.2; then we can replace condition (a) by*

$$Z \text{ is projective and} \big(\overline{\mathrm{NE}}_1(Z) \ni \alpha \mapsto 0 \in \overline{\mathrm{NE}}_1(X)\big) \Rightarrow \alpha = 0, \tag{300}$$

which may be applied to the case in point, i.e.

Fact 5.2.4 The contraction, Proposition/Summary 5.1.4, has projective moduli.

Proof Observe that under the hypothesis of Set Up 5.1.1 the locus of the extremal ray R must be the connected smooth divisor $\mathcal{Y}$ because $\mathcal{Y} \cdot R < 0$. Now, let $\rho : X \to X_0$ be the moduli of the contraction (291), with Z the moduli of the singular locus in X meeting the extremal ray; then since X_0 is $\mathbb{Q}$-factorial, $\rho^* : \mathrm{NS}^1(X_0) \to \mathrm{NS}^1(X)$ is injective with image classes in the latter annihilated by R. Consequently, by duality there is an exact sequence:

$$0 \to R \to \mathrm{NS}_1(X) \xrightarrow{\rho_*} \mathrm{NS}_1(X_0) \to 0 \tag{301}$$

while $\overline{\mathrm{NE}}_1(X) \twoheadrightarrow \overline{\mathrm{NE}}_1(X_0)$, so Corollary 5.2.2.(b) holds because R is extremal. Now although there may be ambiguity, Warning 4.4.5, about the champ structure on the singular locus and the base of the contraction, there is no such ambiguity at the level of the moduli, i.e. Z is a section of the locus where ρ fails to be an isomorphism, so by (300) and (301) we need only check that a non-zero class in $\overline{\mathrm{NE}}_1(Z)$ cannot belong to R, which is clear, e.g. $K_{\mathcal{F}}|_Z$ is nef. □

5.3 *The H-N Filtration Again*

We will require knowledge of the normal bundle of the extremal smooth sub-champ $\mathcal{Y} \hookrightarrow \mathcal{X}$ of Set Up 5.1.1 akin to that of Proposition/Summary 3.7.3 so, without loss of generality, $\dim(\mathcal{Y}) > 1$. Our primary interest is the local variation of $N_{\mathcal{Y}/\mathcal{X}}$ over a small embedded analytic open, $\mathcal{Z}_!$, of the base/singular locus, so to begin with, and essentially without loss of generality, we'll restrict attention to the case $s = 0$ of Large Fact 4.4.7. As ever we first carry out our analysis at the level of the universal cover of $\mathcal{Y}$, i.e. a radially foliated weighted projective champ by Fact 4.2.11, and so abuse notation slightly, i.e. replace $\mathcal{Y}$ by its universal cover, $\mathcal{X}$ by a small neighbourhood of the former, etc. There are two tautological bundles of relevance, namely, $\mathcal{O}(1)$, on the weighted projective champ $\mathcal{Y}$, which, op. cit., is related to the radial foliation, $\mathcal{R}$, by

$$K_{\mathcal{R}} \xrightarrow{\sim} \mathcal{O}(-a_0) \tag{302}$$

and the relative tautological bundle of $\rho : \mathcal{P} := \mathbb{P}(N^{\vee}_{\mathcal{Y}/\mathcal{X}}) = P(N_{\mathcal{Y}/\mathcal{X}}) \to \mathcal{Y}$ which we'll denote H, while P, Y, etc. will be the corresponding moduli. Now say $\mathcal{F}$ is the specialisation (cf. Fact/Definition 3.5.5 and Claim 3.6.1) of our original foliation to the projective normal cone; then $K_{\mathcal{F}} = \rho^* K_{\mathcal{R}}$ so a $K_{\mathcal{F}}$-negative extremal ray, R, of $\mathcal{P}$, has, by Fact 3.4.1, to be an invariant curve of $\mathcal{F}$ lying over an invariant curve of $\mathcal{R}$. By Claim 4.3.1 we may suppose that the former has at worst nodes and whence also the latter from our explicit knowledge, Fact 3.9.2, of the singularity. The moduli of such a champ is the moduli of its normalisation, so without loss of generality, R is an extremal ray of $P(f^* N_{\mathcal{Y}/\mathcal{X}})$ for $f : \mathcal{L} \to \mathcal{Y}$ some coordinate axis of the radial foliation $\mathcal{R}$ – all of which are smooth and embedded on a weighted projective space. By Proposition/Summary 3.7.3 we know exactly what these are, and in terms of Fact 3.9.2 and the New Notation 4.4.1 we may describe them as follows: the local monodromy at the singularity, p, of the radial foliation is μ_{a_0}, and by hypothesis, Large Fact 4.4.7.(b).(i), there is a local generator,

∂, of the ambient foliation on $\mathcal{X}$ which is $\boldsymbol{\mu}_{a0}$-invariant so that the eigenvectors of Jordan decomposition of ∂ at p afford a $\boldsymbol{\mu}_{a_0}$ equivariant decomposition:

$$N_{\mathcal{Y}/\mathcal{X}} \otimes \mathbb{C}(p) = \coprod_{1 \le i \le l} V_i, \tag{303}$$

for V_i the subspace generated by the eigenvectors of weight $-\beta_i$ for β_i a complete repetition free list of the b_i, amongst which, in the Notation 3.8.6, $-\beta_l$ is largest. The decomposition (303) then describes the singular locus of the specialised foliation exactly, i.e. it is a disjoint union:

$$\operatorname{sing}(\mathcal{F}) = \coprod_{1 \le i \le l} P(V_i) \times \mathrm{B}_{\boldsymbol{\mu}_{a_0}} \tag{304}$$

and the extremal ray in question is any invariant section over $\mathcal{L}$ which cuts $P(V_l)$, or, to be more precise, cuts $P(V_l) \times \mathrm{B}_{\boldsymbol{\mu}_{a_0}} \hookrightarrow \mathcal{P}$ which is the embedded component of the singular locus. We can, therefore, apply Large Fact 4.4.7 to conclude that the extremal rays define a sub-champ $\mathcal{Y}_l \hookrightarrow \mathcal{P}$ together with a projection:

$$\mathcal{Y}_l \to P(V_l) \tag{305}$$

whose fibres have universal cover a weighted projective champ $\mathbb{P}(c_0, c_1, \dots, c_r)$ for some weights c_i to be determined, radially foliated by $\mathcal{R}'$, say. Now, by (304), (305) has a section so $\mathbb{P}(\underline{c}) \times P(V_1)$ is the universal cover of $\mathcal{Y}_l$. We have, however, by Proposition/Summary 3.7.3, $\mathcal{F}$-invariant embeddings $\mathcal{L}_i \to \mathcal{Y}_l$ lifting any coordinate axis $f_i : \mathcal{L}_i \hookrightarrow \mathcal{Y}$, and each $\mathcal{L}_i$ is simply connected, so there are $\mathcal{R}'$-invariant embeddings $f_i' : \mathcal{L}_i \hookrightarrow \mathbb{P}(\underline{c})$ of every $\mathcal{L}_i \xrightarrow{\sim} \mathbb{P}(a_0, aa_i)$, and whence $\mathbb{P}(\underline{c}) \xrightarrow{\sim} \mathcal{Y}$. Better still,

$$\begin{aligned} K_{\mathcal{R}'} &= K_{\mathcal{F}} |_{\mathbb{P}(\underline{c}) \times P(V_1)} = K_{\mathcal{R}} |_{\mathbb{P}(\underline{c}) \times P(V_1)} \quad \text{and} \\ K_{\mathcal{R}'} \cdot_{f_i'} \mathcal{L}_i &= K_{\mathcal{R}} \cdot_{f_i} \mathcal{L}_i \text{ by Fact/Definition 3.4.5} \end{aligned} \tag{306}$$

so $\mathbb{P}(\underline{c}) \to [\mathbb{P}(\underline{c})/\mathcal{R}']$ is, unsurprisingly, the radial foliation $\mathcal{Y} \to [\mathcal{Y}/\mathcal{R}]$ that we started with. Consequently, the map

$$\mathcal{Y}_l \to \mathcal{Y} \times P(V_l) \tag{307}$$

afforded by the structural projection ρ and (305) is an étale cover. As such (307) exhibits the former as a locally constant gerbe over the latter. By explicit local calculation, however (cf. (304)) (307) is an isomorphism in a neighbourhood of the fibre over the singularity p, so it's an isomorphism everywhere. We have, therefore, proved most of

Fact 5.3.1 Suppose, as above, that $\mathcal{Y}_!$ of Set Up 5.1.1 is simply connected (whence isomorphic to the product of a polydisc with a radially foliated weighted projective space), then there is a filtration of $N_{\mathcal{Y}/\mathcal{X}}\,|_{\mathcal{Y}_!}$ by invariant sub-bundles for the induced foliation:

$$\mathcal{O} = N_l \subsetneq N_{l-1} \subsetneq N_{l-2} \subsetneq \cdots \subsetneq N_0 = N_{\mathcal{Y}/\mathcal{X}},$$

such that if $\beta_1 > \cdots > \beta_l$ is a complete repetition free list of the $b_1, \ldots, b_t$ of the New Notation 4.4.1 and q_j, $1 \leq j \leq k$ the corresponding multiplicities, then for a as per Fact 4.2.11, locally over $\mathcal{Z}$

$$N_{j-1}/N_j \xrightarrow{\sim} \mathcal{O}_{\mathcal{Y}}(-a\beta_j)^{\oplus q_j}\,.$$

Proof $\mathcal{Y}_l$ of (305) is, by our Large Fact 4.4.7, the image of a deformation space of extremal rays, which is constant on taking products with a small polydisc, whence this addition changes nothing, and for notational convenience we'll continue to ignore it. In any case, the embedding $\mathcal{Y}_l \hookrightarrow \mathcal{P}$ affords a sub-bundle:

$$\left(\rho_* H\,|_{\mathcal{Y}_l}\right)^{\vee} \hookrightarrow N_{\mathcal{Y}/\mathcal{X}} \tag{308}$$

which is the N_{l-1}th term in the above filtration. Moreover, there is a canonical isomorphism:

$$N_{\mathcal{Y}_l/\mathcal{P}} \xrightarrow{\sim} \rho^*\left(N_{\mathcal{Y}/\mathcal{X}}/N_{l-1}\right) \otimes H \tag{309}$$

and so we conclude by induction. □

Unsurprisingly, we continue to refer to this as the H-N filtration and observe

Corollary 5.3.2 *Let $\mathcal{Y}_! \hookrightarrow \mathcal{X}_!$ be simply connected; then there is a non-canonical splitting:*

$$N_{\mathcal{Y}/\mathcal{X}}\,|_{\mathcal{Y}_!} \xrightarrow{\sim} \coprod_j \mathcal{O}_{\mathcal{Y}}(-a\beta_j)^{\oplus q_j} \tag{310}$$

and, better still, any section over $\mathcal{Y}_!$ of $I_{\mathcal{Y},\mathcal{X}}/I^2_{\mathcal{Y}/\mathcal{X}} \otimes \mathcal{O}_{\mathcal{Y}_!}(-a\beta_j)$ can be lifted to a (formal) section of $I_{\mathcal{Y},\mathcal{X}} \hat{\otimes} \mathcal{O}_{\hat{\mathcal{X}}_!}(-a\beta_j)$ over the completion $\hat{\mathcal{X}}_!$ of $\mathcal{X}_!$ in $\mathcal{Y}_!$.

Proof The non-trivial case, given Fact 5.3.1, is when the fibres of $\mathcal{Y} \to \mathcal{Z}_!$ have dimension 1. This is, however, Proposition/Summary 3.7.3, and otherwise it's immediate by Fact 5.3.1 and Fact 2.3.3. □

The apparent arbitrarity of such sections notwithstanding, choose some, say

$$\underline{\xi} := \xi_j : \mathcal{O}_{\hat{\mathcal{X}}_!}(a\beta_j) \to \hat{I} := I_{\mathcal{Y},\mathcal{X}} \hat{\otimes} \mathcal{O}_{\hat{\mathcal{X}}_!},\ 1 \leq j \leq t \tag{311}$$

and define (cf. (247)) a filtration on $\hat{I}$ by way of

$$F^p_{\underline{\xi}}\hat{I} := \left(\xi_1^{j_1}\cdots\xi_t^{j_t} \mid b_1 j_1 + \cdots + b_t j_t \geq p\right)\; b_j := a\beta_j, \tag{312}$$

i.e. the ideal generated by the images of the $\mathcal{O}_{\hat{\mathcal{X}}_!}(j_1 b_1 + \cdots j_t b_t)$ under (311), and observe

Claim 5.3.3 The filtration (312) is algebraic, i.e. shrinking as necessary, there is a filtration $F^p I_{\mathcal{Y},\mathcal{X}}|_{\mathcal{X}_!}$ whose completion is (312). Better still this is independent of the choice (311) and $\mathcal{F}$ invariant.

Proof Plainly F^p contains some power, say q, of $\hat{I}$, so the first part just amounts to the coherence of $F^p/\hat{I}^q$ on the qth thickening of $\mathcal{Y}_!$. As to the first part of the better still: say η_j is another choice, then either this is the same as ξ_j, or there is a smallest $p > 0$ such that

$$0 \neq \xi_j - \eta_j \in \left(F^p_{\underline{\xi}}/F^{p+1}_{\underline{\xi}}\right) \otimes \mathcal{O}_{\mathcal{Y}_!}(-b_j) \xrightarrow{\sim} \coprod_{b_1 j_1 + \cdots + b_t j_t = p} \mathcal{O}_{\mathcal{Y}_!}(p - b_j) \tag{313}$$

so $p \geq b_j$, whence $\eta_j : \mathcal{O}_{\hat{\mathcal{X}}_!}(b_j) \to F^{b_j}_{\underline{\xi}}$, and we're done by symmetry. Similarly, suppose the composition

$$\mathcal{O}_{\hat{\mathcal{X}}_!}(b_j) \xrightarrow{\xi_j} F^{b_j}\hat{I} \to \mathcal{O}_{\hat{\mathcal{X}}_!} \to K_{\mathcal{F}} \tag{314}$$

doesn't factor through $K_{\mathcal{F}} \otimes F^{b_j}\hat{I}$, then there is a smallest $b_j > p \geq 0$ through which it does factor, so (314) affords a non-zero $\mathcal{O}_{\mathcal{Y}_!}$-linear map:

$$\mathcal{O}_{\hat{\mathcal{Y}}_!}(b_j) \to K_{\mathcal{F}} \otimes \mathcal{O}_{\mathcal{Y}_!}(p)^{\oplus N} \xrightarrow{\sim} \mathcal{O}_{\mathcal{Y}_!}(p - a_0)^{\oplus N} \tag{315}$$

where N is the number of integers $j_1, \ldots, j_t$ such that $b_1 j_1 + \cdots + b_t j_t = p$, which is nonsense. □

Putting this all together we have therefore

Fact/Definition 5.3.4 Let $\mathcal{Y} \hookrightarrow \mathcal{X}$ be as in Set Up 5.1.1; then there is a $\mathcal{F}$-invariant filtration:

$$\cdots \subset F^p \cdots \subset F^{>0} = I_{\mathcal{Y},\mathcal{X}} \subset \mathcal{O}_{\mathcal{X}} \tag{316}$$

such that:

(a) The restriction of (316) to a small embedded analytic neighbourhood $\mathcal{X}_!$ as defined in the Set Up 5.1.1 pulls back to (312) on the universal cover of $\mathcal{X}_!$.

(b) For $f : \tilde{\mathcal{L}} \hookrightarrow \mathfrak{X} \to \mathcal{X}$ the normalisation of an extremal ray with at worst nodes embedded in its net completion, the pull-back of (316) is the filtration defined by the invariant divisors in item (2) of Proposition/Summary 3.7.3 combined in the (obvious) way suggested by (312).

Proof The filtration has already been defined on the universal cover, say $\mathcal{X}^! \to \mathcal{X}_!$ with Galois group π_1. As such, it descends to $\mathcal{X}_!$ provided (312) admits a π_1 action, which is clear from the proof of Claim 5.3.3 because by Fact 5.3.1 the H-N filtration is π_1-equivariant. Similarly, to compare the filtrations on 2-small analytic open embeddings $\mathcal{X}_{!\alpha} \hookrightarrow \mathcal{X}$, $\mathcal{X}_{!\beta} \hookrightarrow \mathcal{X}$, we only need to compare them on any (faithfully flat) étale covering of $\mathcal{X}_{!\alpha} \cap \mathcal{X}_{!\beta}$, so again this is just Claim 5.3.3 and the definition (312) as in (b). □

5.4 Existence of Flips

Let $\mathcal{Y} \hookrightarrow \mathcal{X}$ be as in Set Up 5.1.1 then by (316) there is a $K_{\mathcal{F}}$-invariant smoothed weighted blow-up, [24, I.iv.3], defined as in (248), to wit:

$$\begin{array}{ccc} \mathcal{X}^1 := \mathrm{Proj}\left(\coprod_p F^p\right) & \xleftarrow{\text{Vistoli covering}} & \mathcal{X}^2,\ K_{\mathcal{F}^2} = K_{\mathcal{F}}\,|_{\mathcal{X}^2} \\ {\scriptstyle \text{weighted blowup}}\Big\downarrow{\scriptstyle \text{Everything } \mathcal{F} \text{ invariant}} & & \\ \mathcal{X} & & \end{array} \tag{317}$$

Before progressing let us make a clarifying:

Remark 5.4.1 The implied weights in (317) are not the $a\beta_j$ of (312) but $b_i' := b_i/b$ where b_i are as per the New Notation 4.4.1 and b is their gcd. Following (291), however, we will be taking the covering

$$\mathcal{X}^2 \xleftarrow{ab\text{th root of } \mathcal{E}^2} \mathcal{X}^{1/ab}, \quad \mathcal{Y}^{1/ab} := \frac{1}{ab} \cdot \mathcal{E}^2 \tag{318}$$

and the totality, i.e. the horizontal in (317) composed with (318), is, functorially with respect to the ideas, *the smoothed weighted blow-up with weights ab_i* where a is given by Fact 4.2.11 and the b_i by the New Notation 4.4.1. Consequently, there's a certain convenience in doing both steps at once, or, at least, referring, as we will, to their totality in terms of the unifying idea.

This said the exceptional divisor $\mathcal{E}^2$ on $\mathcal{X}^2$ is described by:

Claim 5.4.2 The weighted projective bundle $\mathcal{E}^2 \to \mathcal{Y}$ enjoys the following triviality property: for $\mathcal{Y}_! \hookrightarrow \mathcal{Y}$ as per Set Up 5.1.1 (with $\mathcal{Z}_! \hookrightarrow \mathcal{Z}^0$ understood sufficiently small) and $\mathcal{Y}^! \to \mathcal{Y}_!$ its finite universal cover:

$$\mathcal{E}^2 \mid \mathcal{Y}^! \xrightarrow{\sim} \mathcal{Y}^! \times \mathbb{P}(b_1', \dots, b_t'). \tag{319}$$

Moreover, the induced foliation (understood either logarithmically, Remark 2.2.2, or, equivalently, without saturation if the fibres $\mathcal{Y} \to \mathcal{Z}^0$ have dimension 1) has canonical bundle the restriction of $K_{\mathcal{F}}$ and singular locus the fibre over the unique connected component $\mathcal{Z}$ of sing($\mathcal{F}$) contained in $\mathcal{Y}$.

Proof The pth factor of the graded algebra associated to (312) is $\mathcal{O}_{\mathcal{Y}^!}(p)$ tensored with the pth factor of the trivial graded algebra freely generated by generators of weights b_i, $1 \leq i \leq t$ (cf. (313) and (315)), which has the same Proj as that which is freely generated after cancelling the common factors, Remark 5.4.1, whence (319). As to the moreover: the exceptional divisor of a (weighted) blow-up in an invariant centre is always smooth in the foliation direction, so we only have to compute what happens over the singular locus which we can do explicitly using Fact/Definition 5.3.4 by way of its relation, Proposition/Summary 3.7.3.(3), with the Jordan decomposition, and appropriate local coordinates (cf. (220)). □

Now, irrespectively of whether $\mathcal{E}^2$ is extremal in $\mathcal{X}^2$, the cone theorem applies to $\mathcal{E}^2$ in its induced foliation, while extremal rays in $\mathcal{Y}$ with at worst nodes lift (cf. the preamble to the proof of Fact 5.3.1) to the same in $\mathcal{E}^2$ by Proposition/Summary 3.7.3. As such Large Fact 4.4.7 applies to $\mathcal{E}^2$ in se (i.e. as the locus of its own extremal ray) to imply

Fact/Definition 5.4.3 The champ $\mathcal{E}^2$ is a bundle of foliated varieties (whose fibres have universal covers radial foliations on a $\mathbb{P}(a_0, aa_1, \dots, aa_n)$) over an orbifold $\mathcal{Z}^0$ which by Claim 5.4.2 is itself a bundle of $\mathbb{P}(b_1', \dots, b_t')$'s over the orbifold structure on the singular locus of $\mathcal{Y}$. Consequently, for $\mathcal{X}^{1/ab}$ as in (318) there is a contraction $\rho_+ : \mathcal{X}^{1/ab} \to \mathcal{X}_+$ of $\mathcal{Y}^{1/ab}$ to a locally constant gerbe over $\mathcal{Z}^0$ such that the induced foliation $\mathcal{X}_+ \to [\mathcal{X}_+/\mathcal{F}_+]$ is smooth and everywhere transverse to the locus where ρ_+ is not an isomorphism. The bi-rational map $\rho_+ : \mathcal{X} \to \mathcal{X}_+$ will, irrespectively of whether the moduli of $\mathcal{X}_+$ is projective, be referred to as a *flip* and the more precise data:

$$\begin{array}{ccc} \mathcal{X}^{1/ab} & \xrightarrow[\text{(291)}]{\text{Blow down, } \rho_+, \text{ with weights } aa_i} & \mathcal{X}_+ \\ {\scriptstyle \text{Blow up, } \rho_-} \big\downarrow {\scriptstyle \text{with weights } ab_j, \text{ (317) \&(318)}} & & \\ \mathcal{X}_- := \mathcal{X} & & \end{array} \tag{320}$$

of a weighted blow-up followed by a weighted blow down as a *flap*.

Proof As observed the structure of $\mathcal{E}^2$ is implied by Large Fact 4.4.7 given the structure, Claim 5.4.2, of the singular locus. This is, however, the sum total of what we need to deduce the existence of the contraction ρ from Proposition/Summary 5.1.4, i.e. the condition that $\mathcal{Y}^{1/ab}$ is covered by extremal rays of the ambient space is necessary for the projectivity of the moduli of the contraction, but not for its existence as an algebraic space. □

Before proceeding let us avoid any ambiguity by way of

Remark 5.4.4 A priori the sense of flip in 5.4.3 is a bi-rational transformation in codimension 2, which is weaker and, according to Wikipedia, a less common usage, than a contraction followed by a small modification. Indeed if following (285) et seq. we denote by $\hat{\mathcal{X}}^!$ the completion of the universal cover $\mathcal{X}^!$ of $\mathcal{X}_!$ in $\mathcal{Y}^!$, then (286) is valid as stated as we again have the $\mathbb{G}_m$ torsor $\mathfrak{X} \to \hat{\mathcal{X}}^!$ defined by $\mathcal{O}(1)$. As such by Proposition/Summary 3.7.3 in the difficult case, i.e. $r = 1$, and (311) et seq. otherwise, (287) becomes

$$\mathfrak{X} \xrightarrow{\sim} (\mathbb{A}^{r+1}\backslash\{0\}) \times \hat{\Delta} \times V, \quad \hat{\Delta} := \mathrm{Spf}\mathbb{C}[[x_1, \ldots, x_t]], \tag{321}$$

where upon $\lambda \in \mathbb{G}_m$ now acts by

$$\begin{aligned} &(\mathbb{A}^{r+1}\backslash\{0\}) \times \hat{\Delta} \times V \ni (y_0, y_1, \ldots, y_r) \times (x_1, \ldots, x_t) \times z \mapsto \\ &(\lambda^{a_0} y_0, \lambda^{a_1} y_1, \ldots, \lambda^{a_n} y_r) \times (\lambda^{-b_1} x_1, \ldots, \lambda^{-b_t} x_t) \times z. \end{aligned} \tag{322}$$

Consequently, following (289), the ring, A, of $\mathbb{G}_m$ invariant functions now affords maps:

$$\mathfrak{X} \to \hat{\mathcal{X}}^! \xrightarrow{\text{moduli}} \hat{X}^! \to V \times \mathrm{Spf}A\,(\xrightarrow{\sim} \mathbb{C}[[X^I Y^J \mid \underline{a} \cdot I + \underline{b} \cdot J = 0]]), \tag{323}$$

where in the final arrow is again a formal contraction in the sense of [3], and whence, exactly as in (290), it is, equally, the completion in V of the contraction in analytic spaces of the moduli of $\mathcal{Y}^!$ in $X^!$ to V. Plainly, this is equally a contraction of the moduli of the $\mathbb{P}(b_1, \ldots, b_t)$ bundle to which $\mathcal{Y}^!$ is flapped in (320), so that gluing our contractions we obtain an algebraic space X_0 such that the moduli X_- and X_+ of op. cit. belong to the more familiar diagram:

$$\begin{array}{ccc} X_- & & \\ \Big\downarrow {\scriptstyle \text{contraction of a weighted projective bundle with weights } \underline{a} \text{ to its base } Z} & & \\ X_0 & \xleftarrow[\text{with weights } \underline{b} \text{ to the same base } Z]{\text{contraction of a weighted projective bundle}} & X_+ \end{array} \tag{324}$$

To examine the projectivity of the moduli of the flipped champ of Fact/Definition 5.4.3, let us suppose in addition to the Set Up 5.1.1:

Set Up 5.4.5 Fix an extremal ray R and suppose that every $-1/d\,\mathbb{F}$ curve equivalent to R belongs to a connected smooth embedded sub-champ, $\mathcal{Y}_p \hookrightarrow \mathcal{X}$, of the form encountered in Set Up 5.1.1, and that all such sub-champs are disjoint, equivalently none of the following occur:

(a) For some smooth connected component $\mathcal{Z} \to \text{sing}(\mathcal{F})$ there are 2 such sub-champs (for the same R) meeting in $\mathcal{Z}$, Fact 3.8.9.
(b) For some smooth connected component $\mathcal{Z} \to \text{sing}(\mathcal{F})$ and, again, the same R, item (b).(ii) of the Large Fact 4.4.7, occurs.
(c) There is a representative of R avoiding the singular locus.

Observe that the criteria for the projectivity of the flip is particularly simple, i.e.

Claim 5.4.6 In the context of (320), the following are equivalent:

(a) The moduli of the flipped champ $\mathcal{X}_+$ is projective.
(b) The cone $\overline{\text{NE}}_1(\mathcal{X}_+)$ does not contain a line.
(c) The $-1/d\,\mathbb{F}$ curve contracted by ρ_+ is extremal.

Proof Plainly (a) implies (b), and (300) always holds – same argument as the end of the proof of Fact 5.2.4 – whence, conversely, Corollary 5.2.3, (b) implies (a), while (b) iff (c) is the general duality considerations of (301). □

The same applies, a little more generally, if one flips several sub-champs in $\mathcal{X}$ at the same time, provided, as is our context, Set Up 5.4.5, the champs being flipped are all disjoint, which we'll employ without further comment in

Claim 5.4.7 The flip, (320), of any of the $\mathcal{Y}_p$ has projective moduli.

Proof Since the horizontal arrows in (320) are (étale locally) weighted blow downs, it will suffice to do everything at once, which is all we need anyway. As such, consider the totality, at the level of the moduli, of the flaps (320) performed in all of the $\mathcal{Y}_p$, i.e.

$$\begin{array}{ccc} X(R) & \xrightarrow[\rho_+]{} & X_+ \\ {\scriptstyle \rho_-}\downarrow & & \\ X_- & & \end{array} \tag{325}$$

with E^p the exceptional divisors, C_-^p curves in the same contracted by ρ_- and $C_+^p \hookrightarrow E^p$ a $K_{\mathcal{F}}$-negative invariant curve contracted by ρ_+. Fix p; then by the cone theorem, Fact 3.4.1, there are finitely many extremal rays represented by (multiples of) $K_{\mathcal{F}}$-negative invariant curves, R_i, and a (pseudo-) effective class Z_p on which $K_{\mathcal{F}}$ is non-negative such that on $X(R)$

$$C_+^p = \sum_i R_i + Z_p. \tag{326}$$

Now, by construction, (290), $(\rho_-)_*(C_+^p)$ is parallel to R, so $(\rho_-)_*(Z_p)$ is too. However, ρ_- is unramified in the foliation direction, so $(\rho_-)_*(Z_p) = 0$. Consequently, by the projectivity of X_-, Z_p is a sum:

$$\sum_q c_-^q C_-^q, \quad c_-^q \geq 0. \tag{327}$$

On the other hand, all the R_i lie over R, so by the hypothesis of our Set Up 5.4.5 and (290) every R_i is parallel to some C_+^q for some q. Thus, we equally have

$$\sum R_i = \sum_q c_+^q C_+^q, \quad c_+^q \geq 0. \tag{328}$$

Combining all of (326)–(328) we have therefore

$$C_+^p = \sum_q \left(c_+^q C_+^q + c_-^q C_-^q\right), \text{ every} C_+^q \text{ extremal by (328),} \tag{329}$$

while all the divisors E_q are disjoint and strictly negative on both C_+^q, C_-^q, so the only index that can occur on the right of (329) is p. Consequently, C_+^p is extremal and we're done by Claim 5.4.6. □

5.5 Exceptional Flips and Termination

The first case to be considered is:

Set Up 5.5.1 $\mathcal{Y} \hookrightarrow \mathcal{X}$ is an extremal sub-champ satisfying item (b).(ii) of the Large Fact 4.4.7 with $\mathcal{Z} \hookrightarrow \mathcal{Y}$ the unique (smooth) connected component of sing($\mathcal{F}$) contained in it.

Now observe that by the unicity and local uniformity of Jordan decomposition in Claim 4.4.3, there is a well-defined (smoothed) weighted blow-up supported in $\mathcal{Z}$ whose weights in the New Notation 4.4.1 are

$$y_i, \text{ resp. } x_i, \text{ has weight } a_i, \text{ resp. } b_i, \text{ where } a_i = b_i \text{ and } r = t \tag{330}$$

and whose effect is described by:

Claim 5.5.2 Let $\mathcal{X}^1 \to \mathcal{X}$ be the smoothed weighted blow-up defined by (330) with $\mathcal{E}^1$ its exceptional divisor and $\mathcal{Y}^1$ the proper transform of $\mathcal{Y}$; then:

(a) The singular locus of $\mathcal{F}^1$ over $\mathcal{Z}$ is the intersection of $\mathcal{E}^1$ and $\mathcal{Y}^1$. It is smooth connected, and, for good measure, a $\mathbb{P}(a_1, \ldots, a_r)$-bundle over the μ_2 covering of $\mathcal{Z}$ defined in item (b).(ii) of Large Fact 4.4.7.
(b) The embedded sub-champ $\mathcal{Y}^1 \hookrightarrow \mathcal{X}^1$ is the locus of (rather than just a connected component of) an extremal ray R_1 satisfying item (b).(i) of Large Fact 4.4.7.
(c) The exceptional divisor $\mathcal{E}^1$ is covered by $K_{\mathcal{F}}$-nil invariant parabolic champs.

Proof To calculate the singular locus we use the Jordan coordinates of the New Notation 4.4.1, so [24, I.iv.3], on, say the $y_1 \neq 0$ chart, we have local coordinates η_i, ξ_j defined by

$$y_1 = \eta_1^{a_1},\ y_2 = \eta_2\eta_1^{a_2}, \cdots, y_r = \eta_r\eta_1^{a_r},\ x_1 = \xi_1\eta_1^{b_1}, \cdots, x_r = \xi_r\eta_1^{b_r} \tag{331}$$

which gives that étale locally there are 2 smooth components of the singular locus in the fibre of $\mathcal{E}^1$ over $\mathcal{Z}$, which in turn are the intersection of $\mathcal{Y}^1$ and $\mathcal{E}^1$. Plainly (paragraph prior to (274)) the local system defined by these components is the same as the μ_2 cover $\mathcal{Z}^{+-} \to \mathcal{Z}$, so the singular locus is connected, and the good measure part is clear. As to (b) this is just an easy variation on (326)–(327). Specifically, suppose the proper transform, L_1, of an invariant curve isn't extremal, then op. cit. and $\mathcal{E}^1 \cdot L_1 > 0$ imply the absurd. Finally, (c) follows from the explicit coordinates (331) and the fact that the canonical, $K_{\mathcal{F}^1}$, is just $K_{\mathcal{F}}|_{\mathcal{X}^1}$. □

We can, therefore, combine this with (320) to make

Fact/Definition 5.5.3 By an *exceptional flip* (or, better, exceptional flap) is to be understood, for $\mathcal{Y} \hookrightarrow \mathcal{X}$ as in Set Up 5.5.1, the diagram

$$\begin{array}{ccc} \mathcal{X}^{1/ab} & \xrightarrow[(320)]{\rho_+} & \mathcal{X}_+ \\ \text{flip of } \mathcal{Y}^1 \text{ in Claim 5.5.2} \Big\downarrow \text{by way of } \rho_- \text{ of (320)} & & \\ \mathcal{X}_- = \mathcal{X}^1 \hookleftarrow \mathcal{E}^1 & & \\ \text{Weighted blow up} \Big\downarrow \text{of Claim 5.5.2} & & \\ \mathcal{X} & & \end{array} \tag{332}$$

Better still:

(a) The moduli of $\mathcal{X}_+$ is projective.
(b) The image $\mathcal{E}_+$ of $\mathcal{E}^1$ is covered by invariant parabolic champs (it's a bundle of such over a $\mathbb{P}(a_i) \times \mathbb{P}(b_j)$-bundle over $\mathcal{Z}^{+-}$) none of which meet the singular locus, so the generic fibre of $\mathcal{X}_+ \to [\mathcal{X}_+/\mathcal{F}_+]$ is a smooth parabolic champ.

Proof Part (a) follows from Claim 5.4.6 and item (b) of Claim 5.5.2, while $\mathcal{E}_+$ is contained in the smooth locus of $\mathcal{F}_+$ by Fact/Definition 5.4.3 whence (b) by Claim 5.5.2.(c) and Fact/Definition 3.4.5.(c). □

Similarly, should item (a) of Set Up 5.4.5 occur, or slightly more generally:

Claim 5.5.4 If there are 2 extremal (not necessarily for the same ray) champs meeting in the same component of sing($\mathcal{F}$), then the generic fibre of $\mathcal{X} \to [\mathcal{X}/\mathcal{F}]$ is a parabolic champ. Moreover, if both varieties arise from the same extremal ray, i.e. Set Up 5.4.5.(a), then the flip of (320) does not have projective moduli, and there are invariant parabolic champ in (the original $\mathcal{X}$) which do not meet sing($\mathcal{F}$) and are parallel to the given extremal rays, i.e. Set Up 5.4.5.(c) also holds.

Proof Choose one, say $\mathcal{Y}'$, of the extremal varieties, and flip it, then whether or not the moduli is projective, Fact/Definition 5.4.3 and Fact/Definition 3.4.5.(c) still apply. Furthermore, if both rays are extremal, then as in the proof of Claim 5.5.2 the proper transform, R_1, of an invariant curve in the other, say, $\mathcal{Y}''$, is an extremal ray. Plainly, however, the invariant curves, L, in the fibre over $\mathcal{Y}'$ have the form $R_1 + C_-$ where C_- is contracted by ρ_-, while the exceptional divisor, $\mathcal{E}_-$, of ρ_- is negative on L and positive on R_1, whence it's negative on C_-, so C_- is effective; L isn't extremal, and the moduli of $\mathcal{X}_+$ isn't projective. On the other hand, $(\rho_-)_* R_1$ is an invariant parabolic champ missing sing($\mathcal{F}_+$), so it can be moved off the flipped locus to some R_+. As such the proper transform $\tilde{R}_+$ (in $\mathcal{X}^{1/ab}$ of (320)) is a linear combination of L and R, so $(\rho_-)_*(\tilde{R}_+)$ is parallel to the original extremal ray. □

Given the well-defined way in which it occurs, the loss of projectivity in Claim 5.5.4 is very far from deadly. However, it is better avoided, so we make:

Fact/Definition 5.5.5 By a *very exceptional flip* (or, better, very exceptional flap) is to be understood, for $\mathcal{Y}' \hookrightarrow \mathcal{X}$ and $\mathcal{Y}'' \hookrightarrow \mathcal{X}$ a pair of extremal varieties meeting in the same component of the singular locus of $\mathcal{F}$ and parallel to the same extremal ray as per Claim 5.5.4, the diagram (332) with the further proviso

$$\begin{gathered}\text{the arrow } \rho_-, \text{ resp. } \rho_+, \text{ is the weighted blow-up,}\\ \text{resp. down, in both } \mathcal{Y}' \text{ and } \mathcal{Y}''.\end{gathered} \tag{333}$$

The moduli of the resulting champ $\mathcal{X}_+$ is projective, while the resulting foliation $\mathcal{F}_+$ is smooth and everywhere transverse to the locus where ρ_+ is not an isomorphism for exactly the same reasons that the corresponding statements hold for the exceptional flips of Fact/Definition 5.5.3.

Now flipping, exceptional or otherwise, terminates for the simple reason that the number of connected components of the singular locus decreases by at least 1 with the flip of any extremal ray, and so in increasing order of difficulty we have

Proposition/Summary 5.5.6 Let $\mathcal{X} \to [\mathcal{X}/\mathcal{F}]$ be a foliated champ which is not a foliation in parabolic champs and which enjoys the following further properties:

$$\begin{gathered}\text{smooth; projective moduli; log-canonical,}\\ \text{resp. canonical, foliation singularities;}\end{gathered} \tag{334}$$

then there is a sequence of contractions and flips in the sense of Proposition/Summary 5.1.4 and Fact/Definition 5.4.3 (or alternatively just flaps (291) and (320)):

$$\begin{array}{ccccc} \mathcal{X} = \mathcal{X}_0 & & \mathcal{X}_1 & \cdots\cdots & \mathcal{X}_n = \mathcal{X}_{\min} \\ \downarrow & \dashrightarrow & \downarrow & \dashrightarrow & \dashrightarrow \downarrow \\ [\mathcal{X}/\mathcal{F}] = [\mathcal{X}_0/\mathcal{F}_0] & & [\mathcal{X}_1/\mathcal{F}_1] & & [\mathcal{X}_n/\mathcal{F}_n] = [\mathcal{X}_{\min}/\mathcal{F}_{\min}] \end{array} \tag{335}$$

such that each $\mathcal{X}_i \to [\mathcal{X}_i/\mathcal{F}_i]$ enjoys the properties (334), and $K_{\mathcal{F}_{\min}}$ is nef.

Proof The hypothesis that the foliation isn't in parabolic champs implies by item (b) of Fact/Definition 5.5.3 and Claim 5.5.4 that we must, at every stage, be in the situation of (5.4.5), i.e. Large Fact 4.4.7.(b).(i). Consequently, we eventually run out of components of the singular locus through which a $-1/d$ $\mathbb{F}$-curve can pass, and we terminate with $K_{\mathcal{F}}$ nef. by the cone theorem of Fact 3.4.1. □

The alternative to which is

Proposition/Summary 5.5.7 Let everything be as per the hypothesis of Proposition/Summary 5.5.6 with the exception of the hypothesis "not a foliation in parabolic champs" which we replace by "no model has nef. (foliated) canonical bundle"; then after a sequence of contractions and flips in the sense of Proposition/Summary 5.1.4, resp. Fact/Definition 5.4.3, as described in (335) all of (334), continues to hold (i.e. we're still excluding the exceptional cases Fact/Definition 5.5.3 and Fact/Definition 5.5.5) and exactly one of the following happens:

(a) $\mathcal{X}_n \to [\mathcal{X}_n/\mathcal{F}_n]$ is a Mori fibre space, i.e. the locus of a single extremal ray is all of $\mathcal{X}_n$ and the foliation is a bundle of foliated varieties where the universal cover of a fibre is the radial (supposed saturated in dimension 1) foliation on a weighted projective space whose dimension is 1 iff the foliation singularities are canonical.
(b) At least one of items (a) or (b) of Set Up 5.4.5 occurs at every connected component of the singular locus. In particular, therefore, all of the foliation singularities are canonical.

Proof If we exclude (b), then the only other thing that can happen is that the locus of an extremal ray is everything with the champ itself described by item (b).(i) of Large Fact 4.4.7, i.e. (a), while the various facts about canonical vs. log-canonical singularities are just the definitions. □

This leaves us to elaborate the final case:

Proposition/Summary 5.5.8 Should case (b) of Set Up 5.4.5 occur then, without loss of generality, there are no further occurrences of either the contractions of Proposition/Summary 5.1.4 or the flips of Fact/Definition 5.4.3, and should there be any exceptional flips, we continue by

$$\left(\mathcal{X}_n \to [\mathcal{X}_n/\mathcal{F}_n]\right) \dashrightarrow \left(\mathcal{X}_{n+1} \to [\mathcal{X}_{n+1}/\mathcal{F}_{n+1}]\right) \tag{336}$$

wherein all possible exceptional flips of Fact/Definition 5.5.3 are performed at once with all of (334) being preserved. If we are still not done, i.e. $\mathcal{F}_{n+1}$ is not smooth, then item (a) of Set Up 5.4.5 occurs, and we have the following choices for $(\mathcal{X}_{n+1} \to [\mathcal{X}_{n+1}/\mathcal{F}_{n+1}]) \dashrightarrow (\mathcal{X}_{n+2} \to [\mathcal{X}_{n+1}/\mathcal{F}_{n+2}])$,

(a) For each component of the singular locus of $\mathcal{F}_{n+1}$ choose an extremal sub-champs and flip it according to Fact/Definition 5.4.3. This will necessarily result in the loss of projectivity, Claim 5.5.4, but otherwise list (334) is conserved.
(b) Perform at the same time all possible very exceptional flips of Fact/Definition 5.5.5 and thus preserve list (334) in its entirety.

In either case $\mathcal{X}_{n+2} \to [\mathcal{X}_{n+2}/\mathcal{F}_{n+2}]$ is a bundle of 1-dimensional parabolic champs which is identically its own Mori fibre space.

Proof All exceptional or very exceptional flips can only occur at smooth connected components of the singular locus so the extremal sub-champs that they determine cannot intersect (except, of course, in a very exceptional flip wherein $\mathcal{Y}' \cup \mathcal{Y}''$ of Fact/Definition 5.5.5 should be thought of as a single entity) so without loss of generality, all these operations can be combined into one. Better still both the extremal champ, $\mathcal{Y}$, of an exceptional flip of Fact/Definition 5.5.3, or $\mathcal{Y}' \cup \mathcal{Y}''$ in the case of the very exceptional flip of Fact/Definition 5.5.5, are the only invariant sub-champs meeting their respective components of the singular locus, whence the two exceptional cases commute with the contractions of Proposition/Summary 5.1.4 and the (non-exceptional) flips of Fact/Definition 5.4.3, so there's no loss of generality in a priori exhausting all such operations. □

5.6 Logarithmic Remarks

In order to reference it we spell out our:

Set Up 5.6.1 By hypothesis $\mathcal{D} \hookrightarrow \mathcal{X}$ will be a divisor, no generic point of which is invariant, in a connected smooth proper champ and $\mathcal{X}\backslash\mathcal{D} \to [\mathcal{X}\backslash\mathcal{D}/\mathcal{F}]$ a foliation with log-canonical singularities.

As such, by Revision 2.2.10, $\mathcal{D}$ is smooth and everywhere transverse to $\mathcal{F}$. In particular, therefore, for every $e \in \mathbb{Z}_{>1}$, the extraction $\epsilon : \mathcal{X}^{1/e} \to \mathcal{X}$ of an eth root, Fact/Definition 2.1.9, of $\mathcal{D}$ is smooth, and the induced foliation $\mathcal{X}^{1/e} \to [\mathcal{X}^{1/e}/\mathcal{F}^{1/e}]$ has, by Construction 2.2.15, log-canonical singularities which, Revision 2.2.13, are terminal around the pre-image of $\mathcal{D}$. Furthermore, we assert:

Claim 5.6.2 Let everything be as above with $f : C \to \mathcal{X}$ a map from a (smooth irreducible) curve such that $(K_{\mathcal{F}} + \mathcal{D}) \cdot_f C < 0$; then f does not factor through $\mathcal{D}$. In particular, there is a lifting $f^{1/e} : C^{1/e} \to \mathcal{X}^{1/e}$ and $K_{\mathcal{F}^{1/e}} \cdot_{f^{1/e}} C^{1/e} < 0$.

Proof The tangency between $\mathcal{D}$ and $\mathcal{F}$ always yields a section of $\mathcal{O}_{\mathcal{D}}(K_{\mathcal{F}} + \mathcal{D})$, which by hypothesis is trivial, i.e. in a highly degenerate case of Fact 3.4.3 the trace is a constant section over $\mathcal{D}$, so f certainly cannot factor through it. As such there is certainly a lifting $f^{1/e} : C^{1/e} \to \mathcal{X}^{1/e}$, while

$$K_{\mathcal{F}^{1/e}} \cdot_{f^{1/e}} C^{1/e} \leq (K_{\mathcal{F}^{1/e}} + \mathcal{D}^{1/e}) \cdot_{f^{1/e}} C^{1/e} = (K_{\mathcal{F}} + \mathcal{D}) \cdot_f C < 0 \quad (337)$$

where $\epsilon^* \mathcal{D} = e\mathcal{D}^{1/e}$, and $\mathcal{D}^{1/e}$ is smooth. □

It certainly therefore follows that if $K_{\mathcal{F}^{1/e}}$ is nef., then $K_{\mathcal{F}} + \mathcal{D}$ is nef., but plausibly in running the minimal model programme for $\mathcal{X}^{1/e} \to [\mathcal{X}^{1/e}/\mathcal{F}^{1/e}]$, we could lose the hypothesis of Set Up 5.6.1. Observe, however, that the operations of flipping and extracting roots commute, i.e.:

Fact 5.6.3 For any contraction, resp. flip,

$$\left(\mathcal{X}^{1/e} \to [\mathcal{X}^{1/e}/\mathcal{F}^{1/e}]\right) \dashrightarrow \left(\mathcal{X}_+^{1/e} \to [\mathcal{X}_+^{1/e}/\mathcal{F}_+^{1/e}]\right), \quad (338)$$

in the sense of Proposition/Summary 5.1.4, resp. Fact/Definition 5.4.3, there is a contraction, resp. flip:

$$\left(\mathcal{X} \to [\mathcal{X}/\mathcal{F}]\right) \dashrightarrow \left(\mathcal{X}_+ \to [\mathcal{X}_+/\mathcal{F}_+]\right) \quad (339)$$

such that the proper transform, $\mathcal{D}_+ \hookrightarrow \mathcal{X}_+$, satisfies Set Up 5.6.1, and $\mathcal{X}_+^{1/e} \to \mathcal{X}_+$ is the extraction of an eth root of $\mathcal{D}_+$.

Proof That a contraction, resp. flip, of $\mathcal{X}^{1/e} \to [\mathcal{X}^{1/e}/\mathcal{F}^{1/e}]$ determines the same of $\mathcal{X} \to [\mathcal{X}/\mathcal{F}]$ is immediate from Claim 5.6.2 and the definitions if $\mathcal{X}$ has projective moduli. However, even without this, it still follows since projectivity is only used (cf. Claim 4.4.2) to ensure that the contracted, resp. flipped, sub-champ $\mathcal{Y}$ meets a unique component of sing($\mathcal{F}$) through which each of the $-1/d\,\mathbb{F}$ curves which cover $\mathcal{Y}$ must pass. Irrespectively, what we need to do in the first instance is to prove that there is a map:

$$\mathcal{X}_+^{1/e} \to \mathcal{X}_+ \quad (340)$$

To this end observe, exactly as in the final steps of the proof, (297) et seq., of Proposition/Summary 5.1.4, the expression of either side of (340) as a locally constant gerbe over an orbifold, Fact 2.1.6, is determined in co-dimension 2, so without loss of generality, there is no generic stabiliser. Furthermore, flips are actually flaps, so by the unicity of contraction both sides of (340) have the same moduli X_+ and whence they equally factor through the same Vistoli covering champ $\mathcal{X}_+^v \to X_+$ of Fact/Definition 2.1.2. Now, the Vistoli covering of the moduli of a smooth champ is obtained, [28, 2.8], by killing pseudo reflections, while a pseudo reflection of a foliated champ stabilises exactly one of an invariant divisor or a generically transverse divisor, so we have further factorisations such as

$$\mathcal{X}_+^{1/e} \xrightarrow[\text{reflections}]{\text{kill transverse}} \mathcal{X}_+^i \xrightarrow[\text{reflections}]{\text{kill invariant}} \mathcal{X}_+^v \tag{341}$$

and similarly for $\mathcal{X}_+ \to \mathcal{X}_+^v$. Now let x be a geometric point of the proper transform $\mathcal{D}_+^{1/e} \hookrightarrow \mathcal{X}_+^{1/e}$ of $\mathcal{D}$, G_x its stabiliser, and $U \to \mathcal{X}_+^{1/e}$ an étale neighbourhood, then there is a non-trivial normal sub-group, S_x, generated by pseudo-reflections fixing smooth branches of $\mathcal{D}_+^{1/e}$, while $\mathcal{X}_+^{1/e} \to [\mathcal{X}_+^{1/e}/\mathcal{F}_+^{1/e}]$ is smooth at x by the hypothesis of Set Up 5.6.1 and Proposition/Summary 5.1.4. As such by Revision 2.2.13 and (the non-subtle) part of Revision 2.2.6, the induced foliation, $\mathcal{G}$, on $V := U/S_x$ is also smooth. Equally, $U \to V$ is ramified uniquely in the image, Δ, of $\mathcal{D}$, to order e, so there is a factorisation:

$$U \to V^{1/e} \to V \tag{342}$$

through an eth root of Δ in which the first map is almost étale, so by op. cit. $V^{1/e}$ in the induced foliation $\mathcal{G}^{1/e}$ is log-terminal. Consequently, $V \to [V/\mathcal{F}]$ with the orbifold boundary $(1 - 1/e)\Delta$ is also log-terminal, whence by Corollary 2.2.14, Δ is smooth and everywhere transverse to $\mathcal{G}$, and so $\mathcal{D}_+^{1/e}$ is too. This is, however, equivalent to: S_x is a cyclic normal sub-group of G_x and the restriction of the character, $\chi_x : G_x \to \mathbb{G}_m$ afforded by $\mathcal{D}_+^{1/e}$ to S_x is an isomorphism, so every sub-group of S_x is normal. The monodromy of every generic point of $\mathcal{D}_+^{1/e}$ is, moreover, of the form

$$0 \to \mu_e \to \mu_{ee'} \to \mu_{e'} \to 0 \tag{343}$$

where e' is the order of the corresponding stabiliser in the original $\mathcal{X}$. Consequently, the μ_e in (343) afford a well-defined normal sub-group scheme of the stabiliser $\mathcal{S} \to \mathcal{D}_+^{1/e}$ which just as in (341) can be killed to yield a factorisation:

$$\mathcal{X}_+^{1/e} \xrightarrow[\text{in } \mu_e]{\text{kill reflections}} \mathcal{X}_+^j \xrightarrow[\text{reflections}]{\text{kill all further}} \mathcal{X}_+^v \tag{344}$$

in which the image in $\mathcal{X}_+^j$ of $\mathcal{D}_+^{1/e}$ is smooth everywhere transverse to the foliation, and the first map in (344) is just the extraction of an eth root. By definition, however, $\mathcal{X}_+^j$ and $\mathcal{X}_+$ coincide in co-dimension 1, and since they're both smooth, they're equal by purity and Lemma 2.1.4. □

Next, observe that we equally have a log cone theorem, i.e.:

Fact 5.6.4 Let $\mathcal{X}\backslash\mathcal{D} \to [\mathcal{X}\backslash\mathcal{D}/\mathcal{F}]$ be a logarithmic foliated normal champ with both $K_{\mathcal{F}}$ and $\mathcal{D}$ Cartier; log-canonical singularities in dimension 1 and projective moduli, then there are countably many $\mathcal{F}$-invariant parabolic, champ $\mathcal{L}_i$, with, $0 < -(K_{\mathcal{F}} + \mathcal{D}) \cdot \mathcal{L}_i \leq 2$ such that

$$\overline{\mathrm{NE}}\,(\mathcal{X})_{\mathbb{R}} = \overline{\mathrm{NE}}\,(\mathcal{X})_{K_{\mathcal{F}}+\mathcal{D}\geq 0} + \sum_i \mathbb{R}_+\,\mathcal{L}_i \tag{345}$$

where $\overline{\mathrm{NE}}\,(\mathcal{X})_{K_{\mathcal{F}}+\mathcal{D}\geq 0}$ is the sub-cone of the closed cone on which $K_{\mathcal{F}} + \mathcal{D}$ is non-negative. Better still the $\mathbb{R}_+\,\mathcal{L}_i$ are locally discrete, and if $R \subset \overline{\mathrm{NE}}\,(\mathcal{X})_{\mathbb{R}}$ is an extremal ray in the half-space $\overline{\mathrm{NE}}_{K_{\mathcal{F}}+\mathcal{D}<0}$, then it is of the form $\mathbb{R}_+\,\mathcal{L}_i$.

Proof By Corollary 2.2.14, Claim 5.6.2 is independent of any smoothness hypothesis, so by Fact 3.4.1, we have a cone theorem for $K_{\mathcal{F}^{1/e}}$. On the other hand, if $R \subset \overline{\mathrm{NE}}\,(\mathcal{X})_{\mathbb{R}}$ is an extremal ray in the half-space $\mathrm{NE}_{K_{\mathcal{F}}+\mathcal{D}<0}$, then it's an extremal ray in the half-space $\mathrm{NE}_{K_{\mathcal{F}^{1/e}}<0}$ for all $e \gg 0$, whence by Fact 3.4.1 there is an invariant parabolic champ $f : \mathcal{L} \to \mathcal{X}$ with $K_{\mathcal{F}^{1/e}} \cdot_f \mathcal{L} \geq -2$ parallel to it. In particular, therefore, the extremal rays in this half-space are locally discrete. Similarly, if ρ is the dimension of Néron-Severi, with $\alpha \in \overline{\mathrm{NE}}_1(\mathcal{X})$, then there are a sequence of classes $\alpha_e \in \overline{\mathrm{NE}}\,(\mathcal{X})_{K_{\mathcal{F}^{1/e}}\geq 0}$ and generators R_{ei} of extremal rays, $0 \leq i \leq n_e \leq \rho$, in the half-space $\mathrm{NE}_{K_{\mathcal{F}^{1/e}}<0}$ such that

$$\alpha = \alpha_e + \sum_{i=1}^{n_e} R_{ei}. \tag{346}$$

Subsequencing in e as necessary, we may suppose $n = n_e$ is independent of n, and all of α_e, R_{ei} converge. Plainly, however, the α_e converge to a class in the half-space $K_{\mathcal{F}^{1/e}} \geq 0$, which, equally, is either true of a given R_{ei} or it belongs to a half-space $K_{\mathcal{F}} + \mathcal{D} + \epsilon H < 0$- H is ample, $\epsilon > 0$- in which, as noted, extremal rays are discrete, so $\mathbb{R}_+ R_{ei}$ is independent of e. □

This can be combined with Fact 5.6.3 to yield:

Proposition/Summary 5.6.5 Let $\mathcal{X}\backslash\mathcal{D} \to [\mathcal{X}\backslash\mathcal{D}/\mathcal{F}]$ be as in Set Up 5.6.1 with projective moduli and non-empty boundary $\mathcal{D}$; $\mathcal{X}^{1/2} \to [\mathcal{X}^{1/2}/\mathcal{F}^{1/2}]$ the square root of $\mathcal{D}$; $\mathcal{X}^{1/2}_{\mathrm{final}} \to [\mathcal{X}^{1/2}_{\mathrm{final}}/\mathcal{F}^{1/2}_{\mathrm{final}}]$ the result of a maximal sequence of contractions and flips in the sense of Proposition/Summary 5.1.4, resp. Fact/Definition 5.4.3, as described in (335) (i.e. we exclude the exceptional cases Fact/Definition 5.5.3 and Fact/Definition 5.5.5); then there is a foliated logarithmic champ $\mathcal{X}_{\mathrm{final}}\backslash\mathcal{D}_{\mathrm{final}} \to [\mathcal{X}_{\mathrm{final}}\backslash\mathcal{D}_{\mathrm{final}}/\mathcal{F}_{\mathrm{final}}]$ satisfying Set Up 5.6.1 with projective moduli and non-empty boundary of which $\mathcal{X}^{1/2}_{\mathrm{final}} \to \mathcal{X}_{\mathrm{final}}$ is the square root of $\mathcal{D}_{\mathrm{final}}$, and exactly one of the following happens:

(a) $K_{\mathcal{F}^{1/2}_{\mathrm{final}}}$, so by Claim 5.6.2, a fortiori $K_{\mathcal{F}_{\mathrm{final}}} + \mathcal{D}_{\mathrm{final}}$ is nef.
(b) The foliation $\mathcal{X}^{1/2}_{\mathrm{final}} \to [\mathcal{X}^{1/2}_{\mathrm{final}}/\mathcal{F}^{1/2}_{\mathrm{final}}]$ is a bundle of foliated varieties where the universal cover of a fibre is the radial foliation on a weighted projective space of dimension at least 2, so idem for $\mathcal{X}_{\mathrm{final}}\backslash\mathcal{D}_{\mathrm{final}} \to [\mathcal{X}_{\mathrm{final}}\backslash\mathcal{D}_{\mathrm{final}}/\mathcal{F}_{\mathrm{final}}]$; $\mathcal{D}$ the

hyperplane at infinity, i.e. after a weighted projective coordinate change: $x_0 = 0$ on the universal cover in the notation of Definition 2.4.2; and $K_{\mathcal{F}_{\text{final}}} + \mathcal{D}_{\text{final}}$ is torsion.

(c) As per item (b) but with fibres of the bundle weighted projective space of dimension one, and the implied Mori fibre space is exactly the foliation $\mathcal{X}_{\text{final}} \backslash \mathcal{D}_{\text{final}} \to [\mathcal{X}_{\text{final}} \backslash \mathcal{D}_{\text{final}} / \mathcal{F}_{\text{final}}]$, i.e. on each parabolic fibre $K_{\mathcal{F}_{\text{final}}} + \mathcal{D}_{\text{final}}$ is negative.

Proof By Fact/Definition 3.4.5 the structure of a $K_{\mathcal{F}} + \mathcal{D}$ negative invariant champ $f : \mathcal{L} \to \mathcal{X}$ is particular, i.e. either it misses $\mathcal{D}$ completely or it misses the singular locus completely and cuts $\mathcal{D}$ in one point. If, however, Proposition/Summary 5.5.7.(b) were to occur for $\mathcal{X}^{1/2} \to [\mathcal{X}^{1/2}/\mathcal{F}^{1/2}]$, then the foliation is in parabolic champ; the generic champ must meet $\mathcal{D}$, but none of the smooth invariant champ in the exceptional flipped locus – $\mathcal{E}_+$ in Fact/Definition 5.5.3.(b) – can meet $\mathcal{D}$ because an extremal subvariety satisfying Set Up 5.5.1 must meet the singularities. Consequently, by Claim 5.6.2, Fact 5.6.3, and Proposition/Summary 5.5.6 it remains to show that Proposition/Summary 5.5.7.(a) implies items (b) or (c) above, but this is clear since by Definition 2.4.2.(a) and Fact 2.3.3 the only divisors everywhere transverse to the radial foliation are, in the notation of op. cit., defined by a weighted homogeneous function, F, of weight a_0 such that $\frac{\partial F}{\partial x_0} \neq 0$. □

Finally, let us conclude with:

Remark 5.6.6 While it's true, (25), that the only part of a divisor which is relevant to minimal model theory are the components whose generic points are transverse to the foliation, it may well be the case that one starts with a divisor $\mathcal{D} = \mathcal{D}' + \mathcal{D}''$ where, say, $\mathcal{D}'$ satisfies Set Up 5.6.1, $\mathcal{D}''$ is invariant, and whether $\mathcal{D}$, or just $\mathcal{D}''$, is simple normal crossing, and, for whatever reason, one wants to have a similar situation on $\mathcal{X}_{\text{final}}$ after running the minimal model programme of Proposition/Summary 5.6.5. Now, certainly, hypotheses such as $\mathcal{D}''$ simple normal crossing are nothing to do with the definitions of log-canonical singularities, so there's no reason for them to be conserved by Proposition/Summary 5.6.5. On the other hand, simple normal crossings whether of $\mathcal{D}$ or $\mathcal{D}''$ can, by [4] and the definition of log-canonical singularities, be restored by invariant blowing up without prejudice to the $K_{\mathcal{F}}+\mathcal{D}$ nefness conclusion, resp. the smooth fibration in parabolic champ statement of item (a), resp. (b), of Proposition/Summary 5.6.5.

References

1. M. Artin, Etale coverings of schemes over Hensel rings. Am. J. Math. **88**, 915–934 (1966)
2. M. Artin, The implicit function theorem in algebraic geometry, in *Algebraic Geometry (Internat. Colloq., Tata Inst. Fund. Res., Bombay, 1968)* (Oxford University Press, London, 1969), pp. 13–34
3. M. Artin, Algebraization of formal moduli. II. Existence of modifications. Ann. Math. (2) **91**, 88–135 (1970)

4. E. Bierstone, P.D. Milman, Canonical desingularization in characteristic zero by blowing up the maximum strata of a local invariant. Invent. Math. **128**(2), 207–302 (1997)
5. F. Bogomolov, M. McQuillan, Rational curves on foliated varieties, in *Foliation Theory in Algebraic Geometry*, volume 2 of *Simons Symposia* (Springer, Berlin, 2016), pp. 21–51
6. K. Behrend, B. Noohi, Uniformization of Deligne-Mumford curves. J. Reine Angew. Math. **599**, 111–153 (2006)
7. P. Deligne, L. Illusie, Relèvements modulo p^2 et décomposition du complexe de de Rham. Invent. Math. **89**(2), 247–270 (1987)
8. J. Giraud, Cohomologie non abélienne (Springer, Berlin-New York 1971). Die Grundlehren der mathematischen Wissenschaften, Band 179
9. A. Grothendieck, with the collaboration of Jean Dieudonné. Éléments de géométrie algébrique. I. Le langage des schémas. Inst. Hautes Études Sci. Publ. Math. (4), 228 (1960)
10. A. Grothendieck, with the collaboration of Jean Dieudonné. Éléments de géométrie algébrique. IV. Étude locale des schémas et des morphismes de schémas. Inst. Hautes Études Sci. Publ. Math., (20 (§0–1), 24 (§2–7), 28 (§8–15), 32 (§16–21)), 1101, 1964–1967. Rédigés avec la collaboration de Jean Dieudonné
11. A. Grothendieck, Revêtements étales et groupe fondamental (SGA 1). Documents Mathématiques (Paris) [Mathematical Documents (Paris)], 3. Société Mathématique de France, Paris, 2003. Séminaire de géométrie algébrique du Bois Marie 1960–61, Augmenté de deux exposés de M. Raynaud, Updated and annotated reprint of the 1971 original [Lecture Notes in Mathematics, 224, Springer, Berlin; MR0354651 (50 #7129)]
12. A. Grothendieck, Cohomologie locale des faisceaux cohérents et théorèmes de Lefschetz locaux et globaux (SGA 2). Documents Mathématiques (Paris) [Mathematical Documents (Paris)], 4. Société Mathématique de France, Paris, 2005. Séminaire de Géométrie Algébrique du Bois Marie, 1962, Augmenté d'un exposé de Michèle Raynaud. With a preface and edited by Yves Laszlo, Revised reprint of the 1968 French original
13. A. Grothendieck et al., Théorie des topos et cohomologie étale des schémas. Tome 3. Lecture Notes in Mathematics, vol. 305 (Springer, Berlin-New York, 1973). Séminaire de Géométrie Algébrique du Bois-Marie 1963–1964 (SGA 4), Dirigé par M. Artin, A. Grothendieck et J. L. Verdier. Avec la collaboration de P. Deligne et B. Saint-Donat
14. S. Keel, S. Mori, Quotients by groupoids. Ann. Math. (2) **145**(1), 193–213 (1997)
15. J. Kollár, Projectivity of complete moduli. J. Differ. Geom. **32**(1), 235–268 (1990)
16. J. Kollár, *Rational curves on algebraic varieties*, volume 32 of *Ergebnisse der Mathematik und ihrer Grenzgebiete. 3. Folge. A Series of Modern Surveys in Mathematics [Results in Mathematics and Related Areas. 3rd Series. A Series of Modern Surveys in Mathematics]* (Springer, Berlin, 1996)
17. B. Malgrange, On nonlinear differential Galois theory. Chin. Ann. Math. Ser. B **23**(2), 219–226 (2002). Dedicated to the memory of Jacques-Louis Lions
18. J. Martinet, Normalisation des champs de vecteurs holomorphes (d'après A.-D. Brjuno), in *Bourbaki Seminar, Vol. 1980/81*, volume 901 of *Lecture Notes in Mathematics* (Springer, Berlin-New York, 1981), pp. 55–70
19. M. McQuillan, Foliated mori theory & hyperbolicity of algebraic surfaces. http://www.mat.uniroma2.it/~mcquilla
20. M. McQuillan, Canonical models of foliations. Pure Appl. Math. Q. **4**(3, part 2), 877–1012 (2008)
21. M. McQuillan, Formal groupoids. Eur. J. of Math. **8**(Suppl. 2), 629–673 (2022)
22. M. McQuillan, Elementary topology of champs (2015). arXiv:1507.00797
23. S. Mori, Threefolds whose canonical bundles are not numerically effective. Ann. of Math. (2) **116**(1), 133–176 (1982)
24. M. McQuillan, D. Panazzolo, Almost étale resolution of foliations. J. Differ. Geom. **95**(2), 279–319 (2013)
25. R. Narasimhan, *Introduction to the Theory of Analytic Spaces*. Lecture Notes in Mathematics, No. 25 (Springer, Berlin-New York, 1966)

26. M.C. Olsson, Hom-stacks and restriction of scalars. Duke Math. J. **134**(1), 139–164 (2006)
27. L. Szpiro, *Séminaire sur les Pinceaux de Courbes de Genre au Moins Deux*, volume 86 of *Astérisque* (Société Mathématique de France, Paris, 1981)
28. A. Vistoli, Intersection theory on algebraic stacks and on their moduli spaces. Invent. Math. **97**(3), 613–670 (1989)

Notation & Definitions

Sheaves of Differentials, and Jets

$K_{\mathcal{F}}$ canonical divisor of a foliation, page 249

$K_{\mathcal{F}}^{\text{nolog}}$ canonical divisor of a foliation without boundary, page 249

$K_{\mathcal{F}}$ in the presence of a boundary, page 249

$\Omega^1_{\mathcal{X}/\mathcal{F}}$ bundle of invariant differentials, page 249

$\mathscr{D}_{\mathcal{F}}^{-\infty}$ algebra of operators along the foliation, page 268

$T_{\mathcal{F}}$ tangent sheaf along the foliation, page 268

$\mathscr{P}_U^{(n)}$ sheaf of Grothendieck n-jets, page 268

$\mathscr{P}_U^{\infty}$ sheaf of adic algebras of Grothendieck jets, page 268

Weighted Projective Spaces

$\mathbb{P}_k(\underline{a})$ shorthand for weighted projective champ over a ring k, page 254

$\mathbb{P}_k(a_0, \ldots, a_n)$ weighted projective champ over a ring k, page 254

$\mathbb{P}_{\mathbb{C}}(\underline{a}) \to [\mathbb{P}_{\mathbb{C}}(\underline{a})/\mathcal{R}]$ notation for a radial foliation, page 259

$\mathscr{O}(1)$ tautological line bundle on a weighted projective champ, page 255

$A_k := \mathbb{A}_k^{n+1} \backslash 0$ punctured affine space over a ring k, page 254

Flips and Flaps

$\mathcal{X}^{!!}$, $\mathcal{Y}^{!!}$ etc. universal cover of $\mathcal{X}_{!!}$, $\mathcal{Y}_{!!}$ etc., see equation (282), page 325

$\mathcal{X}^{!}$, $\mathcal{Y}^{!}$ etc. universal cover of $\mathcal{X}_{!}$, $\mathcal{Y}_{!}$ etc., see equation (279), page 324

$\mathcal{X}_+$ result of a flip, or, equivalently a flap, see equation (320), page 336

$\mathcal{X}_-$ flipped, or, equivalently flapped champ, see equation (320), page 336

$\mathcal{X}_{!!}$, $\mathcal{Y}_{!!}$ etc. result of extracting a root of $\mathcal{X}^{!}$, $\mathcal{Y}^{!}$ etc., see equation (281), page 325

$\mathcal{X}_{!}$, $\mathcal{Y}_{!}$ etc. arbitrarily small embedded analytic opens in $\mathcal{X}$, $\mathcal{Y}$ etc., page 324

ρ_+ weighted blowing down occurring in a flap, see equation (320), page 336

ρ_- weighted blowing up occurring in a flap, see equation (320), page 336

Harder-Narismhan Pairs

$\mathcal{X}_{>0}$ gerbe over the moduli, $X_{>0}$, of the Zariski closure of the strictly positive part, $\mathfrak{X}_{>0}$, of the HN pair about a $-\frac{1}{d}$ $\mathbb{F}$ curve, page 307

Miscellaneous Notation

Index of Definitions

The Hasse Principle for 9-Nodal Cubic Threefolds

Nicholas I. Shepherd-Barron

Contents

1 Introduction

It has been known since the nineteenth century that if X is a cubic threefold over a field K of characteristic zero and $X \otimes \overline{K}$ has just d isolated singularities, then $d \leq 10$. Coray [6] has shown that if d is prime to 3, then $X(K)$ is non-empty, while Colliot–Thélène and Salberger [2] have shown that if $d = 3$ and K is a number field then X satisfies the Hasse principle. More recently Coray et al. [7] have proved the Hasse principle if $d = 6$. This paper fills the gap, as far as singular cubic threefolds are concerned, by showing that if $d = 9$, then (Theorem 4.5) the only obstruction to the Hasse principle is the Brauer–Manin obstruction described in [4]. We do this by descent. More precisely, we show first that, provided that X has points everywhere locally, universal torsors over the smooth locus X^0 of X exist, that they are K-birational to cones over certain singular cubic sevenfolds, and that all of them satisfy the Hasse principle.

We also prove (Theorem 7.5) that on 10-nodal cubic threefolds the only obstruction to weak approximation is the Brauer–Manin one, and prove a partial such result (Proposition 5.2) in the 9-nodal case.

N. I. Shepherd-Barron (✉)
Department of Mathematics, King's College, London, UK
e-mail: Nicholas.Shepherd-Barron@kcl.ac.uk

Y. Tschinkel (ed.), *Arithmetic and Algebraic Geometry*, Simons Symposia,
https://doi.org/10.1007/978-3-031-74134-0_7

As mentioned, the proofs depend on the consideration of various torsors, under tori. What often makes various such torsors computable is that the base variety is given by particularly simple (e.g. linear) equations inside some torus embedding, or rather an equivariant compactification of some torsor under a torus. Nine-nodal cubic threefolds follow this pattern: they turn out to be hyperplane sections of some Galois twist of the *Perazzo cubic fourfold* P, defined in $\mathbb{P}^5$ by $x_1x_2x_3 = y_1y_2y_3$. However, 10-nodal cubic threefolds X are different. They are not, apparently, usefully embeddable in a toric variety, and yet universal torsors over X^0 are simple; they are birational to cones over the Grassmannian $G(2, 6)$.

2 Universal Torsors over the Smooth Locus of a Perazzo Cubic

The object that renders 9-nodal cubic threefolds tractable is the Perazzo cubic fourfold P [10]. This is given by the equation $x_1x_2x_3 = y_1y_2y_3$. It contains nine 3-planes L_{ij}, given by $x_i = y_j = 0$. The complement $P^0 = P - \cup L_{ij}$ is a torsor under a trivial 4-dimensional torus T, so P is, by abuse of language, a torus embedding. This makes it easy to compute universal torsors over it and those of its twists that are also torus embeddings (i.e. those that have K-points); we shall do this explicitly in the next section.

Note that the nine 3-planes L_{ij} are conjugate under the wreath product $\Gamma = S_3 \wr S_2$; this is the subgroup of the symmetric group S_6 generated by $S_3 \times S_3$ and the involution ι of $S_3 \times S_3$ that switches the two factors. The first factor S_3 permutes the x_i, and the second permutes the y_i and ι switches x_i with y_i. The singular locus $\mathrm{Sing}(P)$ consists of nine lines l_{pq}, where l_{pq} is given by $x_i = y_j = 0$ for $i \neq p, j \neq q$. The six points $w_1 = (1, 0, \dots, 0), \dots, w_6 = (0, \dots, 0, 1)$ are Γ-conjugate; note that under the subgroup $S_3 \times S_3$ of Γ of index two they fall into two orbits. Because the complement T of the 3-planes is a 4-dimensional torus and the embedding $T \hookrightarrow P$ is T-invariant, the class group $\mathrm{Pic}\, P^0 = \mathrm{Cl}(P \otimes \overline{K})$ is torsion-free of rank $9 - \dim S = 5$; it is generated by the classes L_{ij}, subject to the relations of the form

$$0 = (x_1/y_2) = L_{11} + L_{13} - L_{22} - L_{32}.$$

Lemma 2.1 The automorphism group scheme $G = \mathrm{Aut}(P, \mathcal{O}(1))$ over $\mathbb{Q}$ is a split extension of Γ by T.

Proof The six points w_i are distinguished as those points where two or more of the l_{pq} meet. So the connected component $\mathrm{Aut}^0(P, \mathcal{O}(1))$ preserves each of them, and so preserves the 3-planes L_{ij}. Now the result is obvious, with T being the complement in P of $\cup L_{ij}$. □

More generally, we define a *Perazzo cubic fourfold* to be a cubic fourfold Y that is a Gal_K-twist of P. Its smooth locus will be denoted by Y^0.

Proposition 2.2 Every Perazzo cubic Y satisfies the Hasse principle.

Proof Quadratic base extensions are harmless, so that we can assume that the six distinguished points fall into two Galois orbits of three points each. Since it is known [2] that cubics with three conjugate nodes satisfy the Hasse principle, we are done. □

Definition 2.3 A *double-three* is a configuration of six 2-planes L_i, M_j in $\mathbb{P}^8$, where $i, j = 1, 2, 3$, such that $\cup L_i$ and $\cup M_j$ each span $\mathbb{P}^8$ and each intersection $L_i \cap (\cup M_j)$ and $M_j \cap (\cup L_i)$ consists of three non-collinear points.

Note that any double-three has a unique decomposition into the union of two threes, where the 2-planes in each three are mutually disjoint.

Proposition 2.4 Suppose that Y is a Perazzo cubic with a K-point.

(1) There exist universal torsors over Y^0.
(2) Every such universal torsor is K-birational to $\mathbb{A}^9_K$.
(3) The corresponding rational map $\mathbb{A}^9_K - \to Y$ factors through the standard projection $\mathbb{A}^9 - \to \mathbb{P}^8_K$.
(4) The rational map $\mathbb{P}^8_K - \to Y \hookrightarrow \mathbb{P}^5$ is given by a linear system of cubics passing doubly through a double-three.

Proof We know that Y^0 is a torus embedding $S_1 \hookrightarrow Y^0$, where S_1 is the complement of nine 3-planes in Y and a torsor under the torus $S = \mathrm{Aut}^0(Y)$, and that $\overline{K}[Y^0]^* = \overline{K}^*$. Hence universal torsors over Y^0 exist and can be constructed according to the procedure described in [3]. That is, there is an exact sequence

$$1 \to S_0 \to M \to S \to 1$$

of tori, where $\hat{S}_0 \cong \mathrm{Pic}(Y^0)$ as Gal_K-modules and $\hat{M}$ is the free module spanned by the classes of the 3-planes. The coboundary map $S(K) \to H^1(K, S_0)$ is surjective and the universal torsors over Y^0 form a torsor under this H^1, so any universal torsor $\mathcal{T} \to Y^0$ has the property that $\mathcal{T}|_S \to S$ is the pull-back of $\alpha : M \to S$ via the translation ϕ_x by $-x$, for some $x \in S(K)$.

It therefore remains to describe the map $M \to S$. We first do this in the untwisted case, where $Y = P$ and $S = T$.

The group $\mathrm{Pic}(P^0)$ is generated by the planes L_{ij} subject to the relations $\sum_q L_{iq} - \sum_p L_{pj} = (x_i/y_j) \sim 0$. Introduce nine new variables z_{ij}; then there is a morphism $\pi : \mathbb{A}^9_K = \mathrm{Spec}\, K[\{z_{ij}\}] \to P$ defined by $x_i = \prod_q z_{iq}$ and $y_j = \prod_p z_{pj}$. Identify M with the open subset of $\mathbb{A}^9$ given by $\prod z_{ij} \neq 0$; then π restricts to α.

Consider the variables z_{ij} as the entries of a 3×3 matrix. Then there are six triples of variables z_{ab}, z_{cd}, z_{ef} such that no two of any triple lie in the same row or the same column. For each such triple, let $L_{ab,cd,ef}$ be the 2-plane defined by the vanishing of the other variables. Then every cubic in the linear system defining $\mathbb{P}^8 - \to Y$ is double along each $L_{ab,cd,ef}$; it is clear that these 2-planes form a double-three, with $L_{11,22,33}$, $L_{23,12,31}$, and $L_{21,32,13}$ forming one three.

For the general case, note that S is the twist of T by some cocycle (in fact, homomorphism) $\mathrm{Gal}_K \to \Gamma$. The action of Γ on T lifts to a π-equivariant linear action on $\mathbb{A}^9$; taking the twists gives the results. □

3 Geometry

In this section we investigate the basic geometry of 9-nodal cubic threefolds. Some of this material can also be found in [1].

We let K denote a perfect field of characteristic zero, $\overline{K}$ an algebraic closure of K, and X a cubic threefold over K whose singular locus $\mathrm{Sing}\, X$ consists of exactly nine $\overline{K}$-points. We also assume that every Galois-conjugate subset of $\mathrm{Sing}\, X$ has at least *two* members, or else X is K-rational and there is nothing more to be said. The smooth locus of X will be denoted by X^0.

Lemma 3.1

(1) Every singularity of X is an ordinary node.
(2) X contains just nine 2-planes.
(3) The 2-planes in X and the points of $\mathrm{Sing}\, X$ form a $(9, 4)$-configuration G which is the 1-skeleton of the cell decomposition of the 2-torus formed as the product of two triangles.
(4) The symmetry group of G is Γ.

Proof For (1), we can assume that $K = \mathbb{C}$.

Since cubic surfaces have at most *four* isolated singularities, X is not a cone and so, by I, Theorem 1.7 of [14], the dual variety $X^\vee$ of X is a hypersurface. Moreover,

$$\deg X^\vee = 3.2^3 - \sum_{v \in \mathrm{Sing}(X)} m(v),$$

where the class $m(v)$ of the singularity (X, v) has the property that $m(v) \geq 2$ and $m(v) = 2$ if and only if v is an ordinary node. Precisely, $m(v) = \mu(v) + \mu'(v)$, where μ is the Milnor number of an isolated hypersurface singularity (X, v) and μ' is the Milnor number of a general hyperplane section [13]. If (X, v) is defined locally analytically (or formally) in $\mathbb{C}^n$ by $f(x_1, \dots, x_n) = 0$, then

$$\mu(v) = \dim_{\mathbb{C}} O_{\mathbb{C}^n}/(\partial f/\partial x_1, \dots, \partial f/\partial x_n).$$

If $\mu(v') \geq 2$ for some v then $m(v) \geq 4$ for at least two points v, so that

$$\deg X^\vee \leq 24 - 7 \times 2 - 2 \times 4 = 2.$$

But $\deg X^\vee \geq 3$, by the biduality theorem, so that $\mu(v') = 1$ for all v; this means that every v' is simple, of type A_1, and v is simple of type A.

Say that X has n_r points of type A_r, so that $\sum n_r = 9$. Note that $m(A_r) = r + 1$, so that

$$3 \leq 24 - \sum_{r \geq 1} n_r(r+1) = 6 - \sum_{r \geq 2} (r-1)n_r.$$

This gives $n_r = 0$ for all $r \geq 5$ and $n_2 + 2n_3 + 3n_4 \leq 3$, so that $n_1 \geq 6$.

Choose an A_1-point P. Projecting from P identifies $\mathrm{Bl}_P X$ with $\mathrm{Bl}_C \mathbb{P}^3$, where C is a reduced curve of bidegree $(3, 3)$ on a smooth quadric with $n_1 - 1$ points of type A_1 and n_r of type n_r for all $r \geq 2$. Suppose that $\widetilde{C} = \sqcup \widetilde{C}_j \to C$ is the normalization; then

$$\sum \chi(O_{\widetilde{C}_j}) = \chi(O_C) + (n_1 - 1).1 + n_2.1 + n_3.2 + n_4.2 = 5 + n_3 + n_4.$$

So, if $n_3 + n_4 > 0$, then C is the transverse union of six lines, and then $n_3 = n_4 = 0$, contradiction. Therefore $n_3 = n_4 = 0$, so that $C = \sum_1^5 C_j$ where C_5 is a conic and the other components are lines. Then also $n_2 = 0$ and (1) is proved.

Remark 3.2 If there is a tenth singular point, then it is known classically that X is a form of the Segre cubic.

For (2), note that the description of the curve C in the proof of (1) shows that there are exactly four 2-planes on X through each node. Each plane in X contains four nodes of X, so that there are just nine planes in X.

(3) From the description of C it also follows that through each node of X there are four lines in X that pass through a further node and every such line is the intersection of a unique pair of 2-planes in X. Each of the four nodes in a 2-plane L lies on two lines of the form $L \cap M$, since each line in C meets two others. This proves (3).

(4) it is clear that the symmetry group of G is isomorphic to $S_3 \sim S_2$. □

Remark 3.3 We can also view G as a graph whose vertices are the planes in X and whose edges are the pairs of planes that meet in a point.

Proposition 3.4 Any 9-nodal cubic threefold X over K is K-isomorphic to a hyperplane section of some Perazzo fourfold.

Proof Assume first $K = \overline{K}$. By Lemma 3.1 there are nine 2-planes in X; denote them by L_{ij}, where $i, j \in \mathbb{Z}/3$ and L_{ij} meets $L_{i\pm1,j}$ and $L_{i,j\pm1}$, each in a line. So for all i and j, $\sum_i L_{ij}$ and $\sum_j L_{ij}$ are hyperplane sections. Say $\sum_j L_{ij} = (x_i = 0)$ and $\sum_i L_{ij} = (y_j = 0)$. Then $\sum_{ij} L_{ij}$ is the complete intersection $\prod x_i = \prod y_j = 0$, so that X is a member of the pencil $|3H - \sum_{ij} L_{ij}|$. Hence the equation of X is $\alpha \prod x_i = \beta \prod y_j$, as required.

Next, we check (still assuming $K = \overline{K}$) that if X, X' are isomorphic sections of P, then there is an automorphism of P taking X to X'.

There exists $\sigma \in PGL_6$ such that $\sigma(X) = X'$. Put $P' = \sigma(P)$; then there is a hyperplane H such that $P.H = P'.H$, and it is enough to find $\tau \in PGL_6$ such that

$\tau(P) = P'$ and $\tau|_H = 1$. Put $G = \{\tau \in PGL_6 | \tau|_H = 1\}$; then G acts transitively on $\mathbb{P}^5 - H$.

Both P and P' contain nine 3-planes M_{ij}, M'_{ij}, respectively; we can take them to be ordered so that $M_{ij}.H = M'_{ij}.H$ for all i, j. Suppose that P, resp. P', is given by $F = 0$, resp. $F' = 0$, where $F = x_1x_2x_3 - y_1y_2y_3$; then we can take M_{ij} to be given by $x_i = y_j = 0$. Put $v = (1, 0, 0, 0, 0, 0)$, so that $v = \cap_{i \neq 1} L_{ij}$. So, after applying a suitable $\tau \in G$, we have $\cap_{i\neq 1} L_{ij} = \cap_{i\neq 1} L'_{ij}$. Since $L_{ij}.H = L'_{ij}.H$, we now have $L_{ij} = L'_{ij}$ for all $i \neq 1$. So $F' \in \cap_{i\neq 1}(x_i, y_j) = (x_2x_3, y_1y_2y_3)$; the equality of these two ideals is verified by noting that both define subschemes of codimension 2 and degree 6 and that obviously $(x_2x_3, y_1y_2y_3) \subseteq \cap_{i\neq 1}(x_i, y_j)$. Hence $F' = lx_2x_3 + \alpha y_1y_2y_3$, where $\alpha \in K$ and l is linear. Say that H is defined by $m = 0$; it is then easy to see, since X is irreducible, that, after rescaling, $m = l - x_1$ and $F' = (x_1 + m)x_2x_3 - y_1y_2y_3$. Then $\tau \in GL_6$ given by

$$x_1 \mapsto x_1 + m, x_i \mapsto x_i \quad \text{for} \quad i \neq 1, y_j \mapsto y_j$$

has the required effect.

Now drop the assumption that $K = \overline{K}$. There is an embedding $\phi : X_{\overline{K}} \hookrightarrow P_{\overline{K}}$ with the property that for all $\sigma \in \mathrm{Gal}_K$, there exists $\psi_\sigma \in \mathrm{Aut}\, P(\overline{K})$ such that $\phi^\sigma = \psi_\sigma \circ \phi$. Then $\psi_{\sigma\tau}\phi = (\psi_\sigma)^\tau \psi_\tau \phi$ for $\sigma, \tau \in \mathrm{Gal}_K$.

Write $\omega = \psi_{\sigma\tau}^{-1}(\psi_\sigma)^\tau\psi_\tau$. So $\omega\phi = \phi$. That is, ω is an automorphism of $P_{\overline{K}}$ that acts trivially on the hyperplane H cutting out $X_{\overline{K}}$.

We want to show that any such ω is the identity. For this, we can assume once more that $K = \overline{K}$ and that $\omega \neq 1$. After moving H by an element of S we can suppose also that $(1, \ldots, 1) \in H$. Then $\omega \in \Gamma_0$, where $\Gamma_0 \subseteq \mathrm{Aut}\, P$ is a copy of Γ that splits the surjection $\mathrm{Aut}\, P \to \Gamma$. Since ω is trivial on H, it is conjugate to $(s, 1) \in S_3 \times S_3 \subseteq \Gamma_0$, where s is a transposition; we can take $s = (12)$. Then H is given by $x_1 = x_2$, which contradicts the fact that $P.H$ has isolated singularities. So $\omega = 1$ and $(\psi_\sigma) \in Z^1(\mathrm{Gal}_K, \mathrm{Aut}\, P(\overline{K}))$. It is now easy to see that X embeds into the twist of P by (ψ_σ). □

Now suppose that X is a hyperplane section of Y and that $\mathrm{Sing}\, X = X \cap \mathrm{Sing}\, Y$. So X is a 9-nodal cubic threefold. Let $\widetilde{X} \to X$ be the blow-up of the nodes. It is then easy to see that the Betti numbers of $\widetilde{X}$ are determined by $e(\widetilde{X}) = -6 + 9.4 = 30$ and $b_3 = 0$, so that $\mathrm{rank}\, \mathrm{Cl}(X \otimes \overline{K}) = 5$.

Lemma 3.5

(1) The natural map $\mathrm{Cl}(Y_{\overline{K}}) \to \mathrm{Cl}(X_{\overline{K}})$ is a Gal_K-isomorphism.
(2) $\mathrm{Cl}(X_{\overline{K}})$ is generated by the classes L_{ij} subject to the relations $R_i - C_j = 0$, where $R_i = \sum_i L_{ij}$ and $C_j = \sum_j L_{ij}$.

Proof We check first that the 2-planes on X generate $\mathrm{Cl}(X)$.

Fix a node P on X; then, as before, $\mathrm{Bl}_P\, X$ is identified with $\mathrm{Bl}_C\, \mathbb{P}^3$ and we see that $\mathrm{Cl}(\mathrm{Bl}_P\, X)$ is generated by $L_1, \ldots, L_4, Q, H$, where L_i is the strict transform of a plane that projects to a line C_i in C, Q the strict transform of a quadric cone projecting to the conic C_5 in C, and H is the pull-back of the hyperplane class on $\mathbb{P}^3$.

Let H_1 be the hyperplane class on X. Then $H_1 - E \sim H$ in $\mathrm{Cl}(\mathrm{Bl}_P X)$ and there are 2-planes L', L'' on X such that $H_1 \sim L_1 + L_3 + L'$ and $H_1 \sim Q + L''$ in $\mathrm{Cl}(X)$. Since $\mathrm{Cl}(X) \cong \mathrm{Cl}(\mathrm{Bl}_P X)/\mathbb{Z}.[E]$, we get relations $H \sim H_1 \sim L_1 + L_3 + L'$ and $Q \sim H_1 - L''$ in $\mathrm{Cl}(X)$, and the planes on X do generate $\mathrm{Cl}(X)$. The relations $R_i = C_j$ follow from the observation that $R_i \sim H_1 \sim C_j$.

In Y, the 3-planes form the boundary $Y - U$, as described above, and so generate $\mathrm{Cl}(Y)$. Hence $\mathrm{Cl}(Y) \to \mathrm{Cl}(X)$ is surjective. Since both groups have rank 5, it is enough to prove that $\mathrm{Cl}(X)$ is torsion-free. Since X has isolated hypersurface singularities and is 3-dimensional, $\pi_1(X^0) \to \pi_1(X)$ is an isomorphism. Since $\pi_1(X) = 1$, we are done. □

Lemma 3.6 The image W of Gal_K on $\mathrm{Cl}(X_{\overline{K}})$ equals its image in Γ.

Proof This is a consequence of the fact, which has already been remarked, that the nine 2-planes in X generate $\mathrm{Cl}(X_{\overline{K}})$. □

4 The Hasse Principle

Now suppose that K is a number field and that X is a 9-nodal cubic threefold over K. We know that X is a hyperplane section of some Perazzo cubic fourfold Y over K.

Lemma 4.1 Suppose that $\mathcal{T} \to Y^0$ is a universal torsor. Then so is $\mathcal{T} \times_{Y^0} X^0 \to X^0$. Moreover, every universal torsor on X^0 arises in this way.

Proof Let S_0 denote the torus whose character group $\hat{S}_0$ is $\mathrm{Pic}(Y^0)$. There is a commutative diagram (cf. [4])

$$\begin{array}{ccccccc}
0 & \longrightarrow & H^1(K, S_0) & \longrightarrow & H^1(Y^0, S_0) & \xrightarrow{\chi} & \mathrm{Hom}_{\mathrm{Gal}_K}(\hat{S}_0, \mathrm{Pic}\, Y^0) \\
 & & \Big\downarrow{=} & & \Big\downarrow & & \Big\downarrow{\cong} \\
0 & \longrightarrow & H^1(K, S_0) & \longrightarrow & H^1(X^0, S_0) & \xrightarrow{\chi} & \mathrm{Hom}_{\mathrm{Gal}_K}(\hat{S}_0, \mathrm{Pic}\, X^0).
\end{array}$$

By definition, a torsor under S_0 is universal if the image of its class under χ is the identity, and now the result is immediate. □

Now assume that X has K_v–points for all places v of K. That is, $X(\mathbb{A}_K)$ is not empty, where $\mathbb{A}_K$ is the ring of adèles of K.

Proposition 4.2 There is a universal torsor over X^0.

Proof We know that X is a section of Y; since Y satisfies the Hasse principle it has a K-point, and so there is a universal torsor over Y^0. Now use Lemma 4.1. □

Proposition 4.3 Every universal torsor over X^0 is K-birational to the cone over a cubic sevenfold that is singular along a double-three.

Proof Immediate from Proposition 2.4. □

Proposition 4.4 The universal torsors $\mathcal{T}$ over X^0 satisfy the Hasse principle.

Proof Say that $\mathcal{T}$ is K-birational to the cone over the cubic sevenfold Z. Since quadratic extensions are harmless, we can suppose that the given double-three along which Z is singular splits into two threes. Then Z has three conjugate singular points, and so [2] satisfies the Hasse principle. □

For any K-variety V, there is a pairing

$$\mathrm{Br}(V) \times \prod V(\mathbb{A}_K) \to \mathbb{Q}/\mathbb{Z}$$

given by

$$(\mathcal{A}, (P_v)) \mapsto \sum_v \mathrm{inv}_v(\mathcal{A}(P_v)).$$

We denote by $V(\mathbb{A}_K)^{\mathrm{Br}}$ the subset of $V(\mathbb{A}_K)$ that is the kernel of this pairing. The set $V(K)$ lies naturally in $V(\mathbb{A}_K)^{\mathrm{Br}}$. If the non-emptiness of $V(\mathbb{A}_K)^{\mathrm{Br}}$ implies that of $V(K)$ then we say that "the only obstruction to the Hasse principle on V is the Brauer–Manin obstruction".

Theorem 4.5 The Brauer–Manin obstruction to the Hasse principle on X is the only one.

Proof By definition [4], the Brauer–Manin obstruction to the existence of a K-point on X (equivalently, since X is a cubic, on X^0) is the existence, for all $(x_v) \in X^0(\mathbb{A}_K)$, of an element A of $\mathrm{Br}(X^0)$ such that $\sum_v \mathrm{inv}_v(A) \neq 0$. Assume that this obstruction vanishes; then the proof of Théorème 3.8.1 of *loc. cit.* shows that there is a universal torsor over X^0 with a point everywhere locally. By Proposition 4.4, we are done. □

Colliot–Thélène points out that the following variant of Th. 3.8.1 of *loc. cit.* is valid, where $\widetilde{X}$ denotes a smooth compactification of X^0 (e.g. the blow-up of the nodes of X).

Lemma 4.6 Suppose that Z is a projective variety over K with only nodes and that $\dim Z \geq 2$. Denote by Z^0 its smooth locus and $\widetilde{Z} \to Z$ the blow-up of the nodes. Then the natural map $\mathrm{Br}(\widetilde{Z}) \to \mathrm{Br}(Z^0)$ is an isomorphism.

Proof We first prove this when $K = \overline{K}$.

Brauer groups are torsion, so it is enough to prove this for the n-torsion subgroups. The Kummer sequence shows that then it is enough to prove the surjectivity of $H^2(\widetilde{Z}, \mu_n) \to H^2(Z^0, \mu_n)$. Let $E = \sum E_i$ be the exceptional locus in $\widetilde{Z}$ and $j : Z^0 \to \widetilde{Z}$ the inclusion. Then the purity theorem shows that $j_*\mu_n = (\mu_n)_{\widetilde{Z}}$, $R^1 j_*\mu_n$ is locally isomorphic to $(\mu_n)_E$ and that $R^q j_*\mu_n = 0$ for $q \geq 2$. Since each E_i is simply connected, it follows that $H^1(\widetilde{Z}, R^1 j_*\mu_n) = 0$. Now the Leray spectral sequence $E_2^{pq} = H^p(\widetilde{Z}, R^q j_*\mu_n) \Rightarrow H^{p+q}(Z^0, \mu_n)$ gives the result.

The general case then follows from the fact that $\mathrm{Br}(V)/\mathrm{Br}(K)$ is naturally isomorphic to $H^1(\mathrm{Gal}_K, \mathrm{Pic}(V_{\overline{K}}))$ and that the homomorphism $\mathrm{Br}(\widetilde{Z}_{\overline{K}}) \to \mathrm{Br}(Z^0_{\overline{K}})$ is Gal_K-equivariant. □

Proposition 4.7 (Colliot–Thélène.) If there is no Brauer–Manin obstruction using $\mathrm{Br}(\widetilde{X})$ to the existence of a K-point on X, then there is none using $\mathrm{Br}(X^0)$.

Proof Choose a finite set $\mathcal{A}_1, \ldots, \mathcal{A}_n$ of representatives of the elements of the finite group $\mathrm{Br}(X^0)/\mathrm{Br}(K)$. Over an open subscheme $\mathrm{Spec}(O)$ of the spectrum of the ring of integers of K, the variety X^0/K has a model $\mathcal{X}^0/O$ such that $\mathcal{X}^0(O_v) \neq \emptyset$ for each $v \in \mathrm{Spec}(O)$, and such that each $\mathcal{A}_i$ extends to an element of $\mathrm{Br}(\mathcal{X}^0)$. Using Harari's "formal lemma" (Lemma 2.6.1 of [9], but see also p. 225 of [4]) and the hypothesis that there is no Brauer–Manin obstruction on $\widetilde{X}$, we find a finite set S of places, which we may assume contains all places not in $\mathrm{Spec}(O)$, and local points $M_v \in X^0(K_v)$ for $v \in S$, such that

$$\sum_{v \in S} \mathrm{inv}_v(\mathcal{A}_i(M_v)) = 0$$

for each $i = 1, \ldots, n$.

Now pick any set of integral points $M_v \in \mathcal{X}^0(O_v)$ for $v \notin S$. Then for each $i \in \{1, \cdots, n\}$, the sum $\sum_v \mathrm{inv}_v(\mathcal{A}_i(M_v))$ vanishes (the sum is over all places of K), thus completing the proof. □

5 Weak Approximation

Recall that a variety V over K satisfies *weak approximation* (WA) if for every finite set S of places of K, $V(K)$ is dense in $\prod_{v \in S} V(K_v)$. If V is complete, this is equivalent to $V(K)$ being dense in $V(\mathbb{A}_K)$. Moreover, we say that the Brauer–Manin obstruction to WA on V is the only one if $V(K)$ is dense in $V(\mathbb{A}_K)^{\mathrm{Br}}$. This is equivalent to the density of $V(K)$ in the image of $V(\mathbb{A}_K)^{\mathrm{Br}}$ under every projection $V(\mathbb{A}_K) \to \prod_{v \in S} V(K_v)$, for every finite set S.

Now assume that X and K are as in Sect. 4.

Lemma 5.1 $X^0(\mathbb{A}_K)^{\mathrm{Br}}$ is dense in $X(\mathbb{A}_K)^{\mathrm{Br}}$.

Proof This is Corollary 1.2 of [5]. □

Proposition 5.2 Assume that W lies in the index 2 subgroup $S_3 \times S_3$ of Γ. Then the only obstruction to weak approximation on X is the Brauer–Manin one.

Proof Take a point $(P_v) \in X(\mathbb{A}_K)^{\mathrm{Br}}$. By the lemma, we can approximate (P_v) by $(M_v) \in X^0(\mathbb{A}_K)^{\mathrm{Br}}$. By the version of descent given in Proposition 1.3 of [5], there is a universal torsor $\pi : T \to X^0$ such that (M_v) lifts to a point $(\widetilde{M}_v) \in T(\mathbb{A}_K)$. We know that T is K-birational to the cone over a cubic sevenfold Z and that Z is singular along a double-three. The hypothesis on the Galois group means that the double-three splits into two threes, so that Z has three conjugate singularities. Since this gives WA for Z, by [2], T then satisfies WA, so that there exists $t \in T(K)$ close to $(\widetilde{M}_v)$. Then $x := \pi(t) \in X^0(K)$ is close to (M_v), and then x is close to (P_v). □

6 Computing the Brauer Group and Obstructions

Suppose that X is a 9-nodal cubic threefold over K and $\widetilde{X}$ its blow-up at the nodes. Put $\mathrm{Pic}(X^0 \otimes \overline{K}) = P$. Then, by Lemma 4.6 and the Hochschild–Serre spectral sequence, $\mathrm{Br}(X^0)/\mathrm{Br}(K)$ is isomorphic to $H^1(\mathrm{Gal}_K, P) = H^1(W, P)$.

Proposition 6.1 $\mathrm{Br}(X^0)/\mathrm{Br}(K)$ is either trivial or of order 3.

Proof Pick a general projection $\pi : X \to \mathbb{P}^1_K$, so that the generic fibre X_η is a smooth cubic surface and all geometric fibres are irreducible. Put $K(\mathbb{P}^1) = L$. The main result of [11] is that $\mathrm{Br}(X_\eta)/\mathrm{Br}(L)$ is a subgroup of either $(\mathbb{Z}/2\mathbb{Z})^2$ or $(\mathbb{Z}/3\mathbb{Z})^2$.

We show first that the natural homomorphism

$$\phi : \mathrm{Br}(X^0)/\mathrm{Br}(K) \to \mathrm{Br}(X_\eta)/\mathrm{Br}(L)$$

is injective. For this, suppose that $\alpha \in \mathrm{Br}(X^0)$ restricts to $\pi^*\beta$ in $\mathrm{Br}(X_\eta)$. For any closed point M of $\mathbb{P}^1_K$, the image of α under the residue map $\partial_{\pi^{-1}(M)} : \mathrm{Br}(K(X)) \to H^1(K(\pi^{-1}(M)), \mathbb{Q}/\mathbb{Z})$ is zero. Since $K(M)$ is algebraically closed in $K(\pi^{-1}(M))$, it follows that $\partial_M(\beta) = 0$, so that $\beta \in \mathrm{Br}(\mathbb{P}^1_K) = \mathrm{Br}(K)$. So ϕ is injective.

Suppose next that there is 2-torsion in $\mathrm{Br}(X^0)/\mathrm{Br}(K)$. Then the same is true after any base extension of odd degree, so that we can assume that X has a K-rational node. Then X is K-rational, so that, by Lemma 4.6, $\mathrm{Br}(X^0)/\mathrm{Br}(K) = 0$, contradiction. So $\mathrm{Br}(X^0)/\mathrm{Br}(K)$ has odd order.

Since $\Gamma \cong S_3 \wr S_2$, we can assume that W is non-trivial and lies in the unique Sylow 3-subgroup of Γ, which is $A_3 \times A_3$. So, up to Γ-conjugacy, there are three possibilities for W: $A_3 \times 1$; the diagonal copy Δ of A_3 in $A_3 \times A_3$; and $A_3 \times A_3$.

Recall that P is generated by the L_{ij} subject to the relations $R_i - C_j = 0$. We consider the three possibilities separately.

1. $W = A_3 \times 1$. Put $P_1 = \oplus_{j \neq 3}\mathbb{Z}.L_{ij}$. This is a permutation W-module and there is a short exact sequence

$$0 \to \mathbb{Z}.(C_1 - C_2) \to P_1 \to P \to 0.$$

Since $H^2(W, \mathbb{Z}) \cong \mathbb{Z}/3\mathbb{Z}$ and $H^i(W, P_1) = 0$ for $i = 1, 2$, it follows that $H^1(W, P)$ is isomorphic to $\mathbb{Z}/3\mathbb{Z}$.

2. $W = \Delta$. Put $T = s - 1$, where $s = ((123), (123))$ is a generator of Δ, and $N = \sum s^i$. So, for any W-module B, $H^1(W, B) \cong \ker N_B / \operatorname{im} T_B \cong \operatorname{Tors}(\operatorname{coker} T_B)$. Let F denote the free $\mathbb{Z}$-module on the L_{ij}, so that there is a short exact sequence

$$0 \to Q \to F \to P \to 0$$

of W-modules, which defines the submodule Q of F. Inspection shows that for $i \neq j$ $L_{ii} - L_{jj} = \pm T(L_{ii})$ or $\pm T(L_{jj})$ and $L_{ik} - L_{kj} = \pm T(L_{ik})$ or $\pm T(L_{kj})$. Hence $Q \subseteq \operatorname{im}(T_F)$, so that $\operatorname{coker} T_P \cong F/(Q + \operatorname{im} T_F) \cong \operatorname{coker} T_F$. Since $H^1(W, F)$ vanishes, so does $H^1(W, P)$.

3. $W = A_3 \times A_3$. By Lemma 5 of [11] $\operatorname{Br}(X_\eta)/\operatorname{Br}(L)$ is then of order at most 3, so the same is true of $\operatorname{Br}(X^0)/\operatorname{Br}(K)$. □

Proposition 6.2 Suppose that $W \subseteq A_3 \times A_3$. Then $\operatorname{Br}(X^0)/\operatorname{Br}(K) \cong \mathbb{Z}/3\mathbb{Z}$ if W is Γ-conjugate to either $A_3 \times A_3$ or $A_3 \times 1$ and is trivial otherwise. Moreover, when $\operatorname{Br}(X^0)/\operatorname{Br}(K)$ is non-trivial and K contains a cube root of unity, there is an explicit description (given in the course of the proof) of a non-trivial element.

Proof We use the results and notation of [12], especially Lemma 2 of *loc. cit.*

Assume that $\operatorname{Br}(X^0)/\operatorname{Br}(K)$ is non-trivial. There is a cubic extension K_1/K over which the divisors L_{11}, L_{22}, L_{33} are defined. The image W' of Gal_{K_1} in Γ is of order 3 and is generated by $\sigma = ((123), 1)$. Choose a hyperplane section H defined over K and put $D = \sum L_{ii} - H$. Then $\sum_j \sigma^j(D)$ is principal; say $\sum \sigma^j(D) = (f)$, with $f \in K(X)$. Then, according to *loc. cit.*, there is a non-trivial element $\mathcal{A}$ of $\operatorname{Br}(X^0)$ such that, for every adelic point (P_v) on X^0, with no P_v in the support of (f), $\sum_v \operatorname{inv}_v(\mathcal{A}(P_v)) = \sum_v (f(P_v), K_1/K)_v$, where the summands on the right are the norm residue symbols. □

7 An Aside: Weak Approximation for 10 Nodes

Coray has shown [6] that a 10-nodal cubic threefold X has K-points. We show here that also weak approximation holds.

We start by recalling a construction from [8]. There the ground field is $\mathbb{C}$, but this part of *loc. cit.* is valid over any field K, or even over $\mathbb{Z}$.

Regard $T = \mathbb{G}_{m,K}^6$ as the group of 6×6 diagonal matrices acting in the obvious way on the 6-dimensional K-vector space V. Let U be the standard 2-dimensional representation of SL_2 over K. Then $G = (SL_2 \times T)/\mu_2$ acts on the space $M = U \otimes V \cong \mathbb{A}^{12}$ of 2×6 matrices, where μ_2 is embedded diagonally. Regard SL_2 as acting on the left and T on the right. Let M^0 be the locus of matrices of rank 2; then SL_2 acts freely on M^0 and there is a geometric quotient $SL_2 \backslash M^0$ isomorphic to the punctured cone $\widetilde{Gr}$ over the Grassmannian $Gr(2, V)$ in its Plücker embedding. If M^{00} is the locus of matrices where no column is zero and no three columns are proportional, then G acts freely on M^{00}. There is a geometric quotient M^{00}/G, which is isomorphic to the smooth locus Σ^0 of the Segre cubic threefold Σ given by $\sum_1^6 x_i^3 = \sum_1^6 x_i = 0$. We can identify $\Sigma^0 = M^{00}/G$ with $(SL_2\backslash M^{00})/S$, where $S = T/\mu_2$ with μ_2 embedded in the diagonal copy of $\mathbb{G}_m$ in T. It follows that the S-torsor $\widetilde{Gr} \to \Sigma^0$ is a universal torsor over Σ^0.

The basic geometry of Σ is described in [10], p.169. It is the image of $\mathbb{P}^3$ under the rational map defined by the linear system of quadrics that pass through five given points in general position. It has 10 nodes and 15 planes, all defined over $\mathbb{Q}$, and over $\overline{\mathbb{Q}}$ it is the unique 10-nodal cubic threefold. That is, every 10-nodal cubic threefold X over K is a twist of Σ.

Lemma 7.1

(1) $\mathrm{Aut}(\Sigma) = \mathrm{Aut}(\Sigma \otimes \bar{\mathbb{Q}}) \cong S_6$.
(2) Every 1-cocycle $\mathrm{Gal}_Q \to \mathrm{Aut}(\Sigma)$ is a homomorphism.
(3) Every 10-nodal cubic threefold X over K is the twist of Σ by a homomorphism $\mathrm{Gal}_K \to S_6$.
(4) Given such an X over K, there is a separable sextic K-algebra L such that X is defined in $\mathbb{P}(L)$ by the equation $\mathrm{Tr}(z) = \mathrm{Tr}(z^3) = 0$.

Proof

(1) The projective dual of $\Sigma \otimes \overline{\mathbb{Q}}$ is a quartic threefold T that is the Satake compactification of the moduli space of principally polarized Abelian surfaces with level 2 structure. The Satake boundary is $\mathrm{Sing}\, T$, which is a $(15_3, 15_3)$ configuration of points and lines. The automorphism group of this is clearly $Sp_4(\mathbb{F}_2) \cong S_6$, so that there is a homomorphism $f : \mathrm{Aut}(\Sigma \otimes \overline{\mathbb{Q}}) \to S_6$. Restrict f to the copy of S_6 in $\mathrm{Aut}(\Sigma) \subseteq \mathrm{Aut}(\Sigma \otimes \overline{\mathbb{Q}})$; this is clearly an isomorphism, so that f is split. Since the lines in $\mathrm{Sing}\, T$ are the images of the planes in Σ, any $s \in \ker f$ preserves every plane in Σ. Since each node of Σ is (in many ways) the intersection of two of these planes, s fixes every node. It is then clear that $s = 1$.
(2) This follows at once from Lemma 2.1 and the definition of a 1-cocycle.
(3), (4) Now immediate. □

Now suppose that X/K is a 10-nodal cubic threefold with smooth locus X^0. We have just seen that X is K-isomorphic to a twist Σ_ρ of Σ by a homomorphism $\rho : \mathrm{Gal}_K \to S_6$. Then the construction above can be twisted to show that X^0 is the geometric quotient M_ρ^{00}/G_ρ.

Proposition 7.2 There is a universal torsor T_0 over X^0 whose total space is isomorphic to $\widetilde{Gr}$.

Proof The morphism $\widetilde{Gr} \to \Sigma^0$ can be twisted by ρ. The twist of Gr by ρ is $Gr(2, V_\rho)$, which is isomorphic to Gr, by Hilbert's Theorem 90. It is also possible to twist the action of the torus S on $\widetilde{Gr}$ by ρ, and we are done, since $\widetilde{Gr}$ has trivial Picard group. □

Corollary 7.3 Every universal torsor T over X^0 is K-birational to a line bundle over a twist of Gr.

Proof Given T_0, the other universal torsors over X^0 are classified by $H^1(K, S)$. Since S acts by right multiplication on Gr and the Plücker line bundle $O(1)$ over Gr is S-linearized, the result follows. □

Corollary 7.4 Both the Hasse principle and weak approximation hold for every universal torsor T over X^0.

Proof Every twist V of Gr is homogeneous under its automorphism group, and the point stabilizers are connected. Hence the Hasse principle and weak approximation hold for V, and then for line bundles over it. □

Theorem 7.5 The only obstruction to weak approximation on X is the Brauer–Manin one.

Proof Exactly as for the class of 9-nodal cubics that we handled earlier, in 5.2. □

References

1. I. Cheltsov, Y. Tschinkel, Z. Zhang, Equivariant geometry of singular cubic threefolds. To appear
2. J.-L. Colliot-Thélène, P. Salberger, Arithmetic on some singular cubic hypersurfaces. Proc. LMS **58**, 519–549 (1989)
3. J.-L. Colliot-Thélène, J.-J. Sansuc, *La descente sur les variétés rationelles I* (Géométrie d'Angers 1979, 1980)
4. J.-L. Colliot-Thélène, J.-J. Sansuc, La descente sur les variétés rationelles II. Duke Math. J. **54**, 375–492 (1987)
5. J.-L. Colliot-Thélène, A.N. Skorobogatov, Descent of fibrations over $\mathbb{P}^1_k$ revisited. Math. Proc. Camb. Philos. Soc. **128**, 383–393 (2000)
6. D. Coray, Cubic hypersurfaces and a result of Hermite. ibidem
7. D. Coray, D. Lewis, N.I. Shepherd-Barron, H.P.F. Swinnerton-Dyer, Cubic threefolds with six double points. Number Theory in Progress (1999)
8. I. Dolgachev, D. Ortland, Point sets in projective space and theta functions. Astérisque **165** (1988)
9. D. Harari, Méthode des fibrations et obstruction de Manin. Duke Math. J. **75**, 221–260 (1994)
10. J.G. Semple, L. Roth, *Introduction to Algebraic Geometry* (Oxford, 1949)
11. H.P.F. Swinnerton-Dyer, The Brauer group of cubic surfaces. Math. Proc. Camb. Philos. Soc. **113**, 449–460 (1993)
12. H.P.F. Swinnerton-Dyer, Brauer–Manin obstructions on some del Pezzo surfaces. Math. Proc. Camb. Philos. Soc. **125**, 193–198 (1999)
13. B. Teissier, Cycle evanéscents, sections planes et conditions de Whitney. Astérisque **7,8**, 285–362 (1973)
14. F.L. Zak, *Tangents and Secants of Algebraic Varieties*. Translations of Mathematical Monographs, vol. 27 (AMS, 1993)

Manin's Work in Birational Geometry

Vyacheslav V. Shokurov

Contents

V. V. Shokurov (✉)
Department of Mathematics, the Johns Hopkins University, Baltimore, MD, USA
Steklov Mathematical Institute RAS, Moscow, Russia

Fundamental Mathematics Center MPTI, Dolgoprudny, Russia
e-mail: shokurov@math.jhu.edu

Y. Tschinkel (ed.), *Arithmetic and Algebraic Geometry*, Simons Symposia,
https://doi.org/10.1007/978-3-031-74134-0_8

1 Rational Surfaces

One of Manin's main contributions to birational geometry of algebraic surfaces was given in two subsequent papers in the late 1960s, concerning rational surfaces over a perfect field [16, 17].

In both papers, Manin uses very classical methods but with a new cohomological style and also adds some fresh ideas, inherited by modern birational geometry. (Both publications were in Russian. Unfortunately, in the first, there are many misprints—it was published in France and lacked a Russian editor. We hope that its English translation is better.) We try to present the most important results from the modern perspective and with slightly updated terminology.

Both papers are about algebraic surfaces F over a perfect field k which are *rational* over an algebraic closure $\overline{k}$ of k. Such surfaces are called *rational*. This terminology is still used in modern birational geometry. In Grothendieck style, it would be better to say that F is *geometrically rational*, that is, $F_{\overline{k}} = F \times_{\operatorname{Spec} k} \operatorname{Spec} \overline{k}$ is rational over $\overline{k}$. Equivalent cohomological characterization of those smooth complete surfaces is due to Castelnuovo: F is rational if and only if $q(F) = P_2(F) = 0$, where $q(F) = \dim_k H^1(F, \mathcal{O}_F)$ is the irregularity, $P_2(F) = \dim_k H^0(F, \omega_F^{\otimes 2})$ is the 2-nd genus, and ω_F is the canonical invertible sheaf of F [2] [11, §2]. The main problems discussed in both papers are concerned with the birational classification of rational surfaces F and their birational invariants. Recall that two surfaces F, F' over k are *birationally equivalent* if there exists a birational isomorphism $F \dashrightarrow F'$ over k; equivalently, the fields of rational functions $k(F)$ and $k(F')$ are isomorphic over k. Thus every surface F' in the birational equivalence class is a *model* of F or of its field of rational functions $k(F)$; and the birational invariants are invariants of the field $k(F)$. For the surface F, the field $k(F)$ is a finitely generated field of transcendence degree 2 over k. Actually, Manin considers also the quasi-birational equivalence which adds purely inseparable dominant morphisms, *radical dominant* in terminology of [16, 1.1], e.g., Frobenius morphisms, to usual birational isomorphisms. The quasi-birational equivalence preserves rationality of varieties, in particular, of surfaces. For the field k of characteristic 0 both equivalences coincide. Notice that in algebraic geometry in positive characteristic Frobenius morphisms play a very important role. However, in our review, we consider only the usual birational equivalence.

A good model F' of F should be *smooth* and *complete*, so projective. Manin assumes this, and to construct such a model, especially, in the positive characteristic, he uses the Abhyankar result [1].

2 Minimal Models

Another important property of a good model is to be *minimal*. In modern terminology, a surface F is a *minimal model* if it is smooth complete and its canonical divisor K_F is nef (numerically eventually free). Since F is smooth, every canonical divisor K_F is Cartier and its intersection number $(K_F.D)$ with any other divisor D is well-defined. The nef property means that $(K_F.D) \geq 0$ for any effective divisor D or, equivalently, $(K_F.C) \geq 0$ for any curve C on F. Notice that the canonical divisor K_F is defined up to a linear equivalence on F and $(K_F.D)$ is invariant under the equivalence. Manin uses the canonical invertible sheaf $\omega_F = \mathcal{O}_F(K_F)$ instead of K_F. Surfaces F under consideration in [16, 17] and in other papers are not such minimal models because they are rational over $\overline{k}$ and their Kodaira dimension is negative. However, it is well-known as a slightly different concept: a surface F over k is *minimal* or k-minimal, or, we can say, a *minimal model in the Italian sense* if it is smooth and complete and if every birational contraction $F \to F'$ over k to another smooth surface F' over k is an isomorphism. The minimal property over algebraically closed field k means the absence of *exceptional curves of the first kind* C on F, that is, nonsingular curves with $C^2 = (K_F.C) = -1$. The first condition $C^2 = -1$ is more traditional but as Mori explained the second condition $(K_F.C) = -1$ is really crucial. After him the negativity of intersection property with K_F is one of the cornerstones of modern birational geometry [19]. For a general field the minimal property has a similar interpretation [16, Lemma in 0.4]: there are no exceptional curves of the first kind X on F over $\overline{k}$ such that for every $s \in G$, either the conjugated curve $s(X) = X$ or $s(X)$ does not intersect X, where G denotes the Galois group $G(\overline{k}/k)$.

The minimal property ditto makes sense for every proper relative surface F/Z, that is, with respect to a proper morphism $F \to Z$.

Notice that the minimal model surfaces are also minimal in the Italian sense. The converse does not hold and the difference between these two classes is exactly *Mori fibrations for surfaces*. They have two types:

(1) A contraction $f\colon F \to C$, where C is a complete geometrically irreducible and reduced smooth curve over k, $-K_F$ is ample over C, and F over C is a relative minimal surface.
(2) F is a del Pezzo surface with the trivial contraction $F \to \text{pt.} = \operatorname{Spec} k$ and the Picard number 1 over k.

The minimal surface property and its interpretation [16, Lemma in 0.4] do not have direct generalizations in higher dimensions. Except for the Hirzebruch surface $\mathbb{F}_1$ with its natural contraction $\mathbb{F}_1 \to \mathbb{P}^1$ in (1), this property for rational surfaces F is equivalent to the *extremal* property of contractions $F \to C$ in (1) and of $F \to \text{pt.}$ in (2): the relative Picard number of F over C in (1) and of F/k in (2) is 1. This is the Mori property. It is easy to generalize it in any dimension and it is another cornerstone of modern birational geometry. Italian algebraic geometers and Manin already well understood its importance.

Notice that the generic fiber F_η in type (1) is a complete, geometrically irreducible and reduced curve. For every closed point $t \in F$, the fiber F_t is irreducible over k and geometrically reduced with $H^1(F_t, \mathcal{O}_t) = 0$. Every smooth geometric fiber F_t is isomorphic to $\mathbb{P}^1_{\overline{k}}$ and nonsmooth geometric fiber is isomorphic to the pair of lines intersecting in one point $\mathbb{P}^1_{\overline{k}} \vee \mathbb{P}^1_{\overline{k}}$. In other words, in type (1) the surface F over C is geometrically a conic bundle. (In modern terminology $F, F_{\overline{k}}$ over $C, C_{\overline{k}}$ are *central models* over $k, \overline{k}$ of ranks $1, \rho(F_{\overline{k}}/C_{\overline{k}})$, respectively, where $\rho(F_{\overline{k}}/C_{\overline{k}})$ denotes the relative Picard number of $F_{\overline{k}}$ over $C_{\overline{k}}$.)

Recall that, by definition, $-K_F$ is ample on a del Pezzo surface F. Notice that the extremal property is very restrictive. For example, if k is algebraically closed then $\mathbb{P}^2$, up to isomorphism, is the only del Pezzo surface with Picard number 1 over k.

3 Minimal Models of Rational Surfaces

In [16] Manin shows that every rational surface F has a model F' and a contraction $f: F' \to C$ such that C is a proper, geometrically irreducible and reduced, smooth curve over k of the genus 0, F' is a minimal surface over C, and the reduction of generic fiber of f has the arithmetic genus 0 or 1. In the case of genus 0 fibers, f is a Mori fibration of type (1) with geometrically reduced fibers. In the case genus 1 fibers, the generic fiber of f is not necessarily reduced that is why Manin uses quasi-birational equivalence. Under this equivalence he can assume that the generic fiber of f is also reduced in the case of genus 1 fibers. The rational surfaces with such a fibration or pencil in more traditional terminology have the Picard number ≤ 10 and form a bounded family over k. The boundedness of this class of surfaces with a genus 1 fibration and the rational curve base has interesting generalization in higher dimensions which plays an important role in modern mathematical physics. But still it is not a final product as a Mori fibration. And really on every such surface there exists another fibration which is a Mori fibration of type (1) or (2) possibly after a contraction of exceptional curves as in [16, Lemma in 0.4]. But this fact was established later by Manin's former student Iskovskihk [11, Theorem 1] and even for any field k. That was done 2 years before the Mori theory [19]. Conversely, every del Pezzo surface has a pencil of genus 1 curves which can be converted into a genus 1 fibration up to quasi-birational equivalence after a blowup of the fixed points of the pencil.

However, as we understand now, it is better to replace fibrations of genus 1 curves by del Pezzo surfaces with the Picard number 1 as in (2). Actually, Manin himself uses a reduction to del Pezzo surfaces in the proof of [16, Theorem 1.7] about fibrations of genus 1 curves. The class of Mori fibrations for surfaces gives the complete description of minimal rational surfaces.

4 Cohomological Approach to Enriques Results

To prove the existence of genus 0 and 1 fibrations over a genus 0 curve, Manin develops an updated version of Enriques results based on modern cohomological technique. Today this approach is known as Enriques-Manin method. It considers adjoint linear systems $|nK_F + H|$, where H is a very ample divisor and n is a non-negative integral threshold such that $\dim |nK_F + H| \geq 1$, while $\dim |(n+1)K_F + H| \leq 0$. A linear subsystem of dimension 1, a *pencil*, in the complete linear system $|nK_F + H|$ gives a required contraction to a curve C. This method is very classical and in general also known as an "adjunction method" and respectively linear system $|nK_F + H|$ is called *adjoint*; we adjoin here a canonical divisor K_F to a divisor H. Using this method Manin constructed a contraction $f: F' \to C$ where F' is a projective resolution of F and C is a rational curve [16, Theorem 1.2]. After that, he makes the fibration minimal over C. If F' is minimal over C and the generic fiber f has genus 0, then f gives a Mori fibration of type (1). Its properties are described in [16, Theorem 1.6]: the fibers over $\overline{k}$ are isomorphic to $\mathbb{P}^1_{\overline{k}}$ or to the pair of lines $\mathbb{P}^1_{\overline{k}} \vee \mathbb{P}^1_{\overline{k}}$ intersecting in one point. In other words f is a conic bundle over $\overline{k}$.

Twenty-five years later, Mori proposed a numerical version of the Enriques-Manin method [19], where n is not necessarily integral.

5 Minimal Genus 1 Fibrations

In the case of minimal f with the generic fiber of genus 1, after reduction in positive characteristics, we can suppose that the generic fiber is geometrically smooth, in particular, reduced up to quasi-birational equivalence. Under the last assumption, Manin proved that F' has Picard number 10 over $\overline{k}$ [16, Theorem 1.7]. Notice that in this case f over $\overline{k}$ is a usual minimal fibration with the generic genus 1 curve. Moreover, for a positive integer $a = (X.F'_x)$, a linear equivalence

$$aK_{F'} \sim -F'_x \tag{1}$$

holds, where X is an exceptional curve of the first kind and F'_x is a geometric fiber of $F'_{\overline{k}}$ over a closed point $x \in C_{\overline{k}}$. Actually relation (1) shows that a is independent of X and of $x \in C_{\overline{k}}$ because the Picard group of F' and of $F_{\overline{k}}$ do not have torsions.

6 Forms of $\mathbb{P}^2$ and of $\mathbb{F}_n$

The simplest examples of non-trivial rational surfaces are Severi-Brauer surfaces, forms of $\mathbb{P}^2$ over k, that is, such surfaces F that $F_{\overline{k}}$ is isomorphic to $\mathbb{P}_{\overline{k}}$ over $\overline{k}$. Similarly, we can consider forms over k of other rational surfaces over $\overline{k}$. For instance, the forms of $\mathbb{F}_0 = \mathbb{P}^1 \times \mathbb{P}^1$ over k are quadrics of dimension 2. Manin classified the forms of Hirzebruch surfaces $\mathbb{F}_n$ for $n \geq 1$ in [16, Theorem 1.10] (cf. with [15, Theorem 2]):

(a) If $n \equiv 1 \mod 2$, then any form of $\mathbb{F}_n$ over k is trivial, that is, isomorphic to $\mathbb{F}_n$ over k.
(b) If $n \equiv 0 \mod 2$, then the forms F of $\mathbb{F}_n$ over k are in 1-to-1 correspondence with the forms of $\mathbb{P}^1$ over k; the correspondence is given by mapping F to a curve $C \subset F$ with $C^2 = -n$.

Recall that the forms of $\mathbb{P}^1$ over k are in 1-to-1 correspondence with the quaternion algebras over k.

7 Del Pezzo Surfaces of Degrees 9, 8, and 7

Manin also classifies del Pezzo surfaces of degree $n = K_F^2 \geq 3$. In this case, $-K_F$ is *very* ample and F has the anti-canonical embedding $F \hookrightarrow \mathbb{P}^3$. This gives the 1-to-1 correspondence between the lines on F and the exceptional curves of the first kind [16, Corollary 3.3]. To bound the del Pezzo surfaces Manin uses the Noether formula:

$$K_F^2 + n(F_{\overline{k}}) = 10,$$

where $n(F_{\overline{k}})$ is the rank of Neron-Severi group $N(F_{\overline{k}})$ for $F_{\overline{k}}$ or the Picard number of $F_{\overline{k}}$ because $N(F_{\overline{k}}) = \mathrm{Pic}(F_{\overline{k}})$ for rational surfaces F. Thus, $9 \geq n \geq 3$ [16, Theorem 3.4] and:

(a) For $n = 9$, $F_{\overline{k}}$ is isomorphic to $\mathbb{P}^2$.
(b) For $n = 8$, $F_{\overline{k}}$ is isomorphic to $\mathbb{P}^1 \times \mathbb{P}^1$ or to a blowup of $\mathbb{P}^2$ in a closed point.
(c) For $3 \leq n \leq 7$, $F_{\overline{k}}$ is isomorphic to a blowup of $\mathbb{P}^2$ in $9 - n$ closed points, for which none of three points lie on a line and none of six points lie on a conic.
(d) Every exceptional curve of the first kind on $F_{\overline{k}}$ is a blowup of a closed point in $\mathbb{P}^2$, or a proper birational transformation of a line through two of such points, or of a conic through five of such points.

This classification implies that, for $n \leq 7$, the exceptional curves of the first kind of a del Pezzo surface $F_{\overline{k}}$ generate the Neron-Severi group $N(F_{\overline{k}})$ of $F_{\overline{k}}$ (cf. [16, следствие стр. 73]). This also implies that every del Pezzo surface over k is rational (over $\overline{k}$) and is a form of del Pezzo surfaces in the above classification for degrees $n \geq 6$. The large part of [16] is devoted to the description of forms of simplest del Pezzo surfaces of degree $9 \geq n \geq 4$ and of line configurations on them.

For $9 \geq n \geq 7$, Manin obtained the following [16, Theorem 3.7]: let F be a del Pezzo surface over k of degree n then

(a) For $n = 9$, F is isomorphic to a Severi-Brauer surface.
(b) For $n = 8$, F is a form of $\mathbb{P}^1 \times \mathbb{P}^1$, or is a blowup of $\mathbb{P}^2$ in a closed k-point, i.e., isomorphic to $\mathbb{F}_1$.

(c) For $n = 7$, F is a blowup of $\mathbb{P}^2$ of two closed k-points, or of one closed point $x \in \mathbb{P}^2$ of degree 2, that is, $[k(x) : k] = 2$.

This implies that every del Pezzo surface of degree 7 has a closed k-point [16, следствие 1, стр. 75]; if k is a number field of finite degree or a functional field of transcendence degree 1 over a finite field, then the Hasse principle holds for del Pezzo surfaces over k of degrees 8 and 9 [16, следствие 2, стр. 75]; a del Pezzo surface of degree 7, 8, or 9 is rational over k if and only if it has a k-point [16, следствие 3, стр. 75]. Actually, every del Pezzo surface of degree 7 has a closed k-point and is rational over k.

8 Del Pezzo Surfaces of Degree 6

The del Pezzo surfaces of degree 6 are related to toric geometry. Indeed, any such surface F is a form of a blowup of $\mathbb{P}^2$ in three closed points in general position which is a toric variety with the open torus $(\overline{k}^\times)^2$ of dimension 2, the complement to three lines in general position in $\mathbb{P}^2_{\overline{k}}$. Respectively, for surface F, the open subset $U = F \setminus X$ has a natural structure of a principal homogeneous space for $(\overline{k}^\times)^2$, where X denotes the divisor $\sum_{i=1}^6 X_i$ on F which is a sum of the exceptional divisors X_i on $F_{\overline{k}}$, a hexagon [16, Theorem 3.10].

This implies that a del Pezzo surface of degree 6 is rational over k if and only if it has a closed k-point [16, следствие 1, стр. 77].

9 Del Pezzo Surfaces of Degree 5

For del Pezzo surfaces of degree 5, Manin generalizes the Enriques theorem: every del Pezzo surface of degree 5 over an infinite field k is rational over k [5], and partially revises its proof. For any field k, Manin shows that if a del Pezzo surface F of degree 5 has a closed k point then it is rational over k [16, теорема 3.15]. He uses for this the double projection from a closed k-point of F. On the other hand, for any finite field k, it was already known that every rational surface has a closed k-point due to A. Weil formula for the number of such points [27]. In the case of an infinite field k, Manin gave a sufficient general condition for the existence of such a k-point on a del Pezzo surface F in terms of the group $N(F_{\overline{k}})$ with the action of Galois group G of $\overline{k}$ over k [16, теорема 3.12, a)]. However, it was insufficient for the existence of a closed k-point on a del Pezzo surface of degree 5 [16, замечание на стр. 83]. A few years later the existence of such a point was established by Swinnerton-Dyer [26]. For a short proof of the Enriques theorem, see also in [24].

10 Cohomological Invariants

To investigate birational geometry of del Pezzo surfaces of degree 4 and of other algebraic surfaces, Manin develops certain birational invariants of algebraic surfaces (and actually it works in higher dimensions too). Let F be a smooth complete algebraic surface over a perfect field k. Recall that $N(F_{\overline{k}})$ denotes the Neron-Severi group of $F_{\overline{k}}$; equivalently, $N(F_{\overline{k}})$ is the group of classes of divisors on $F_{\overline{k}}$ modulo numerical equivalence. The group has the following additional structures:

(a) A continuous G-module structure, where G is the Galois group of $\overline{k}$ over k. As a $\mathbb{Z}$-module $N(F_{\overline{k}})$ is free of a finite rank.
(b) A symmetric G-invariant pairing $N(F_{\overline{k}}) \times N(F_{\overline{k}}) \to \mathbb{Z}$ given by intersection of divisors on $F_{\overline{k}}$.
(c) The numerical equivalence class ω_F of a canonical divisor of $F_{\overline{k}}$. For the rational surfaces the notation is correct because in this case the numerical equivalence is the linear one, and the classes can be identified with invertible sheaves up to isomorphism.

Denote by $\mathfrak{C}(k)$ the category of $\mathbb{Z}$-free continuous G-modules with a G-invariant pairing. A module in $\mathfrak{C}(k)$ is called *trivial* if it is a finite direct sum of modules $\mathbb{Z}[G]_{\mathbb{Z}[H]} \otimes \mathbb{Z}$, where H are open subgroups in G.

Manin established the following criterion [16, теорема 2.2]: if smooth complete surfaces F, F' over k are birationally equivalent over k then there exist trivial G-modules M, M' such that

$$N(F_{\overline{k}}) + M' \simeq N(F'_{\overline{k}}) + M.$$

This is an easy corollary of factorization for a birational isomorphism between two smooth complete surfaces into elementary modification: blowups of closed points and inverse modifications to them.

Changes of the canonical class ω_F do not have such a simple description in general. However, Italian geometers and Manin already understood that the difference for an elementary modification $f\colon F' \to F$, blowup in $x \in F$, is easy computable and related to the discrepancy in the exceptional divisor in modern terminology:

$$\omega_{F'} = f^*\omega_F + f^{-1}x.$$

Ono described some invariants of G-modules in $\mathfrak{C}(k)$ which are independent of adding trivial G-modules [20]. For instance, the Galois cohomology group $H^1(G/H, N)$ is independent of an open subgroup H and

$$H^1(G/H, N + M) = H^1(G/H, N),$$

if M is trivial [20, §3]. This group is denoted by $H^1(K, N)$. As it was suggested to Manin by Shafarevich, the above criterion implies the canonical isomorphism

$$H^1(K, N(F_{\overline{k}})) = H^1(K, N(F'_{\overline{k}}))$$

for birationally equivalent smooth complete surfaces F, F', where K is any finite extension of k.

The group $H^1(k, N(F_{\overline{k}}))$ can be explicitly computed in many interesting cases and gives a powerful tool to distinguish birational equivalence classes. Manin [16] also discussed some other invariants of Ono's paper [20]. See also his survey with Tsfasman [18].

The necessary cohomological condition does not imply non-trivial results for del Pezzo surfaces of degree ≥ 5. For del Pezzo surfaces of degree $n \leq 4$ the existence of a closed k-point is not equivalent anymore to its rationality over k. In particular, B. Segre discovered [22] that minimal del Pezzo surfaces of degree 3, *cubic* surfaces, are not rational over k, while it is easy to construct examples of such surfaces with a closed k-point. We discuss this result later in connection with the Noether-Fano-Iskovskihk-Manin method [17, 14]. Actually, this method works also for del Pezzo surfaces of degree 4.

11 Del Pezzo Surfaces of Degree 4

To compute $H^1(k, N(F_{\overline{k}})$ for a minimal del Pezzo surface F of degree 4, Manin uses decomposition of lines on $F_{\overline{k}}$ into orbits under the Galois action (cf. [16, лемма 2.9]). By a result similar to Segre's, rigidity does not hold for those del Pezzo surfaces. As we know today, all such del Pezzo surfaces are not rational; this is an easy corollary (a graduate student-level exercise) of the classification of Sarkisov links for algebraic surfaces [12, теорема 2.6]. The paper of Iskovskikh was dedicated to Manin's 60th birthday, and a better title for it would have been "Factorization of birational morphisms of rational algebraic surfaces from the Sarkisov point of view." For example, every del Pezzo surface F of degree 4 with the Picard number 1 over k and a closed k-point $x \in F$ has a link $f : F \dashrightarrow F'$ of type I into a relative minimal conic bundle $F' \to C$ on a cubic surface F' with a unique line on F', where f is the blowup of x. The surface F can be treated as a Mori fibration of type (2), and the conic bundle $F' \to C$ can be treated as a Mori fibration of type (1) over a rational curve C. That is, the link transforms one Mori fibration into another one. Conversely, every cubic surface F' with a unique line has a relative minimal conic bundle structure and a link f^{-1} of type III, the blowdown of the line into the k-point x [12, ibid]. This implies the positive answer to a question of Segre [22]: whether a cubic surface with a unique line over k is non-rational. Contracting the line we get a del Pezzo surface of degree 4 with the Picard number 1 over k. Thus, the non-rationality question for cubics is equivalent to the same question for such a del Pezzo surface.

Manin classifies the configurations of lines and their orbits on minimal del Pezzo surfaces F of degree 4 and computes respectively cohomology $H^1(k, N(F_{\overline{k}}))$ [16, таблица 3.28]. He obtained 19 types and for 12 of them the cohomology is non-trivial. (Corrections to the classification are in [18, 7.3].) Thus, the corresponding surfaces are non-rational over k. This gives a partial answer to the Segre's question. This also shows that, for this type of problems, the cohomological (formal) method is good, but not as effective as the (geometric) Noether-Fano-Iskovskikh-Manin method.

12 Central Models

A *central model* is a non-birational contraction $Y \to T$ of projective algebraic varieties over a field k such that Y has only terminal singularities and an anti-canonical divisor $-K_Y$ of Y is ample over T. As usual, Y means Y/pt. The rank of the model Y/T is the relative *class number*, that is, the rank of the group of Weil divisors of Y modulo numerical or $\mathbb{Q}$-linear equivalence over T. If $\dim X = 2$, then X is $\mathbb{Q}$-factorial and the relative class number is the relative Picard number $\rho(Y/T)$. The central models of rank 1 are exactly Mori fibrations. Notice that they are $\mathbb{Q}$-factorial. Central models of rank $r \geq 2$ are not necessarily $\mathbb{Q}$-factorial in dimension ≥ 3. The central models of rank 2 are not Sarkisov links but every Sarkisov can be constructed from a central model Y/T. Indeed, let $Y' \to Y$ be a projective $\mathbb{Q}$-factoralization of Y. Then the Picard number of Y'/T is also 2 and we can apply the two-ray game to Y'/T. This gives us two Mori fibrations $Y_1/T_1, Y_2/T_2$ over T in the following commutative diagram:

$$\begin{array}{ccccc}
Y_1 & \dashleftarrow & Y' & \dashrightarrow & Y_2 \\
\downarrow & & \downarrow & & \downarrow \\
T_1 & & Y & & T_2 \\
& \searrow & \downarrow & \swarrow & \\
& & T & &
\end{array}$$

with birational transformations $Y' \dashrightarrow Y_1$, $Y' \dashrightarrow Y_2$. Their composition $Y_1 \dashrightarrow Y_2$ (actually with inverse $Y_1 \dashrightarrow Y'$) is a birational transformation of the Mori fibration Y_1/T_1 into the Mori fibration Y_2/T_2 *over* T which is called a *Sarkisov link over* T (in modern terminology). Notice that the link is never an isomorphism, i.e., it is not regular. Actually, each central model of rank 2 gives two Sarkisov links: $Y_1 \dashrightarrow Y_2$ and its inverse $Y_1 \dashleftarrow Y_2$. Conversely, any Sarkisov link over T has a *unique* central model over T of rank 2 which gives the link [25, Central model p. 525]. A type of Sarkisov link is determined by the dichotomy: contraction $T_i \to T, i = 1, 2$, is an isomorphism or not. Thus, we have four types: I–IV. For example, a type I link has a canonical isomorphism $T_1 = T$ and a non-isomorphism $T_2 \to T$.

Central models of rank 3 give *elementary relations* of Sarkisov links and *elementary syzygies* for higher ranks [9].

It is easy to classify central models for algebraic surfaces over an algebraically closed field k (cf. [17, теорема 3.2]):

(a) Conic bundles $X \to C$ over a curve C with the rank equal to 1 plus the number of degenerate fibers $\mathbb{P}^1_{\bar{k}} \vee \mathbb{P}^1_{\bar{k}}$
(b) Del Pezzo surfaces $X \to$ pt. with the rank equal to the Picard number of X

In this situation, the only Mori fibrations are $\mathbb{P}^1$-bundles over a curve and $\mathbb{P}^2$ over a point pt. Respectively, the only rank 2 central models are conic bundles $X \to C$ over a curve C with a single degenerate fiber and two del Pezzo surfaces $\mathbb{P}^1 \times \mathbb{P}^1, \mathbb{F}_1$ over pt. It easy to find corresponding links. For example, $\mathbb{F}_1$ gives two well-known links over pt.:

$$\begin{array}{ccccc} \mathbb{P}^2 & \leftarrow \mathbb{F}_1 & & \mathbb{F}^1 & \rightarrow \mathbb{P}^2 \\ \downarrow & \downarrow & \text{and inverse} & \downarrow & \downarrow \\ \text{pt.} & \leftarrow \mathbb{P}^1 & & \mathbb{P}^1 & \rightarrow \text{pt.} \end{array}$$

of types I and III, respectively. In [17, теорема 3.2] Manin stumbles at $\mathbb{F}_1$ because it is minimal over $\mathbb{P}^1$ but not over pt. Today we clearly understand that $\mathbb{F}_1$ over pt. or itself is a central model over pt. related to the above link which actually appears in [17, теорема 3.2].

If k is not algebraically closed, the complete classification of Sarkisov links is given in [12, теорема 2.6]: about 40 classes and about 20 rank 2 central model classes. Unfortunately, in dimension 3 we have hundreds already known classes of Mori fibrations and of Sarkisov links, and it is expected that there are tens of thousands of them. Of course, the most interesting in this case are the classes of Mori fibrations and of Sarkisov links of rational threefolds over an algebraically closed fields k, say, $\mathbb{C}$. It is a similar but much more challenging task as Manin's one to classify all rational minimal models for surfaces over a perfect or general field k. This was essentially done in [16, 17].

Let X be an algebraic variety and Y/T be a Mori fibration or, more generally, a central model with a birational isomorphism $Y \dashrightarrow X$. Then we say that Y/T is a *Mori fibration* model or, respectively, a *central model of* X. The central models of X of rank 2 give the Sarkisov links of X. If $Y_1/T_1, Y_2/T_2$ are two models of X with birational isomorphisms $Y_1 \dashrightarrow X, Y_2 \dashrightarrow X$ then a *canonical* birational isomorphism $Y_1 \dashrightarrow Y_2$ is defined. Models $Y_1/T_1, Y_2/T_2$ are *equal* and we write $Y_1/T_1 = Y_2/T_2$ if the canonical birational isomorphism is an isomorphism, that is, regular, and gives the commutative diagram

$$\begin{array}{ccc} Y_1 & = & Y_2 \\ \downarrow & & \downarrow \\ T_1 & = & T_2 \end{array},$$

where $Y_1 = Y_2, T_1 = T_2$ denote canonical isomorphisms. Notice that the isomorphism $T_1 = T_2$ is unique if it exists. For any Sarkisov link $Y_1/T_1 \dashrightarrow Y_2/T_2$, the birational isomorphism $Y_1 \dashrightarrow Y_2$ is canonical and models $Y_1/T_1, Y_2/T_2$ are not equal. Even if $Y_1 = Y_2$, one stll has $Y_1/T_1 \neq Y_2/T_2$. For example, $\pi_1 \dashrightarrow \pi_2$ is the link of type IV over pt. from the projection π_1 on the first factor of $\mathbb{P}^1 \times \mathbb{P}^1$ into the projection π_2 on the second one:

$$\begin{array}{ccc} \mathbb{P}^1 & \times & \mathbb{P}^1 \\ \pi_1 \downarrow & & \downarrow \pi_2 \\ \mathbb{P}^1 & & \mathbb{P}^1 \end{array}$$

with $\pi_1 \neq \pi_2$.

The same applies to central models. Notice that if $Y_1/T_1 = Y_2/T_2$ and Y_1/T_1 is a Mori fibration, then Y_2/T_2 is also a Mori fibration.

13 Sarkisov Theory

The main result of a weak version of the Sarkisov theory states that for any algebraic variety X the canonical birational isomorphism $Y/T \dashrightarrow Y'/T'$ of two Mori fibrations $Y/T, Y'/T'$ of X can be factorized into a sequence of Sarkisov links of X: there exist Sarkisov links $Y_i/T_i \dashrightarrow Y_{i+1}/T_{i+1}, i = 1, \dots, n-1$, of X such that $Y \dashrightarrow Y'$ is the following composition:

$$Y = Y_1 \dashrightarrow Y_2 \dashrightarrow \cdots \dashrightarrow Y_{n-1} \dashrightarrow Y_n = Y'.$$

It was established over any field k of characteristic 0 and even in the relative case [25, Theorem 7.2] [8]. According to the geography of log models [25] and the LMMP, it holds in dimension ≤ 3 in positive characteristic too. In particular, this holds for algebraic surfaces at least over any perfect field k.

The Sarkisov theory or Sarkisov Program states a little bit stricter result that the above sequence also decreases the Sarkisov degree which is given by an appropriate very ample linear system on Y'. This theory is actually a direct generalization of the Noether-Fano-Iskovskikh-Manin method to untwist the birational isomorphism $Y \dashrightarrow Y'$. We will discuss the method below. The Sarkisov theory is still established in dimensions ≤ 4 in characteristic 0 [10] and in dimensions ≤ 3 in positive characteristic. Thus, the weak Sarkisov theory shows that any two Mori fibrations of a variety X can be connected by elementary birational transformations, called Sarkisov links, while in the Sarkisov theory the choice of links should be stricter. For most of applications and, in particular, for Manin and Manin-Iskovskikh, and even for Sarkisov results on rigidity [21], it is enough for the weak Sarkisov theory. However, the Noether-Fano-Iskovskikh-Manin inequality from the method and its generalizations play a very important role in detecting rigid and superrigid varieties.

14 Birational Rigidity

An algebraic variety X is called *birationally rigid* if it has only one Mori fibration, up to a possibly non-canonical isomorphism. Moreover, it is *superrigid* if it has only one Mori fibration, i.e., its any two Mori fibrations are canonically isomorphic. It is easy to see that a birationally rigid algebraic variety X does not have Mori fibrations Y/T with $\dim T \geq 2$. There are a lot of rigid varieties with a Mori fibration over a point: $T = \text{pt}$. We do not know whether there exist any such examples over a curve. At least they do not exist for surfaces. Thus, in the following, for a birationally rigid variety X we usually suppose that $X = Y/\text{pt}$. is itself a Mori-Fano variety: terminal $\mathbb{Q}$-factorial Fano variety with the Picard number 1 over k.

If X is a rigid variety then its group $\text{Bir}(X)$ of birational automorphisms is generated by birational automorphisms given by isomorphisms of Mori models. Moreover, it suffices to consider isomorphisms of Mori models $\gamma\colon Y_1 \simeq Y_2$ for Sarkisov links $\beta\colon Y_1/T_1 \dashrightarrow Y_2/T_2$. Such an isomorphism γ exists by the rigidity. The corresponding birational automorphism of X is

$$X \overset{\alpha^{-1}}{\dashrightarrow} Y_1 \overset{\beta}{\dashrightarrow} Y_2 \overset{\gamma^{-1}}{\simeq} Y_1 \overset{\alpha}{\dashrightarrow} X,$$

where $\alpha\colon Y_1 \dashrightarrow X$ is the canonical (structure) birational isomorphism for the model Y_1/T_1. The only non-canonical isomorphism in the above chain is γ. The constructed isomorphism $X \dashrightarrow X$ depends on its choice. The isomorphism γ is unique up to automorphism of Y_1 and of Y_2; it is enough to take automorphisms on one side. If $\delta\colon Y/T \dashrightarrow X$ is a Mori fibration model of X and $\varepsilon\colon X \dashrightarrow X$ a birational isomorphism of X then

$$\begin{array}{c} Y' = Y \overset{\delta}{\dashrightarrow} X \overset{\varepsilon}{\dashrightarrow} X \\ \downarrow \\ T \end{array}$$

is also a Mori fibration model $\varepsilon \circ \delta\colon Y' = Y/T \dashrightarrow X$ of X. This allows to convert ε into a canonical birational isomorphism

$$Y' = Y \overset{\delta}{\dashrightarrow} X \overset{\varepsilon}{\dashrightarrow} X \overset{\delta^{-1}}{\dashrightarrow} Y$$

between models $Y/T, Y' = Y/T$ and conversely. Under this correspondence, the Sarkisov links go into above generators in $\text{Bir}(X)$ and composition of links into product in $\text{Bir}(X)$. Thus, the factorization theorem into Sarkisov links implies the generation of $\text{Bir}(X)$. Notice that if $X = Y$ for a Mori fibration Y/T of X then for each Sarkisov link we can choose one generator but to these generators we need to add the automorphisms of $X = Y$.

15 Superrigidity

For a superrigid variety X every birational automorphism of X is regular when $X = Y/\text{pt.}$ is a Mori-Fano variety (model of itself): $\text{Bir}(X) = \text{Aut}(X)$. Indeed, in this case the set of generators coming from Sarkisov links is empty. Thus, we have only generators from $\text{Aut}(X)$. Moreover, if $-K_X$ is very ample then $\text{Aut}(X)$ is a subgroup of $\text{Aut}(\mathbb{P}^n)$ under the monomorphism corresponding to the anti-canonical imbedding $X \hookrightarrow \mathbb{P}^n$.

16 G-Varieties

The same applies to G-varieties and, in particular, to G-surfaces, where G is a finite group of automorphisms of X. In positive characteristic, G can be a group scheme of finite type over k of dimension 0. In [17, §§1,2,4] Manin develops birational geometry of G-surfaces. A G-variety according to Manin is an algebraic variety V over a field k and group G acting on $V_{\overline{k}} = V \times_k \text{Spec}(\overline{k})$ such that one of the following holds:

(a) *Algebraic case.* The field k is perfect, and G is the Galois group of $\overline{k}$ over k and acts on $V_{\overline{k}}$.
(b) *Geometric case.* The field k is algebraically closed, and G is a finite group and acts on V.

A *G-morphism* $V_1 \to V_2$ of G-varieties V_1, V_2 in the algebraic case is a morphism $V_1 \to V_2$ of algebraic varieties V_1, V_2 over k. Respectively, a *G-morphism* $V_1 \to V_2$ of G-varieties V_1, V_2 in the geometric case is a G-equivariant morphism $V_1 \to V_2$ of algebraic varieties V_1, V_2 over k. Similarly, we can define *rational G-morphisms*, *birational G-isomorphisms*, *G-contractions*, *G-invariant curves of the first kind* (not necessarily irreducible) on G-surfaces, *rational* G-varieties, *minimal* G-surfaces, *Mori G-fibrations*, *central G-models*, etc. The algebraic surfaces over a perfect field k are now included into the more general notion of a G-surface.

Manin shows that the adjunction Enriques-Manin method works for rational G-surfaces and the classification of minimal rational G-surfaces is similar to the geometric case with $G = \{1\}$ and the case of rational surfaces over a field k. For this Manin introduces *standard* G-surfaces [17, 4.1]:

(a) *G-surfaces with a rational pencil.* G-surface F is smooth and has a G-contraction $F \to C$ on a rational G-curve with ample $-K_F$ over C and F is relative minimal G-surface over C. The generic fiber of contraction is rational and equivalently is geometrically reduced irreducible of genus 0.
(b) *Non-degenerate del Pezzo G-surfaces.* The G-surface F is smooth, its anti-canonical divisor $-K_F$ is ample, and the G-invariant Picard number of F is 1.

They are exactly Mori G-fibrations for rational G-surfaces and prototypes of Mori G-fibrations in higher dimensions.

Manin also added to standard G-surfaces degenerate del Pezzo G-surfaces [17, стр. 175]. They are weak del Pezzo G-surfaces in modern terminology and not considered as good (minimal) G-models. Manin actually struggled with the question of what is the right class of del Pezzo surfaces (see [17, замечание на стр. 175]).

Manin proves that every rational G-surface is birationally G-isomorphic to a standard G-surface up to unessential drawbacks [17, теорема 4.2]. Namely, Manin supposes here that G is abelian in the geometric case, and to standard G-surfaces he adds also degenerate del Pezzo G-surfaces. However, he obtains a more precise and geometrical result (cf. [17, основная лемма 2.1]). The result on an isomorphism with a standard G-surface as it was stated above was established later by Iskovskikh [11, теорема 1G]. Notice also that the groups G acting effectively on geometric G-surfaces satisfy the Jordan property and so are close to (finite) abelian groups. Thus, the groups should have an explicit description. For this, in [17, §4] Manin investigates the representation of G in the group $N(F_{\overline{k}})$ for standard G-surfaces. He gave the complete description of such representations in terms of root systems [17, теорема 4.5]. For del Pezzo G-surfaces F of degree ≤ 5 the representation is faithful if G acts effectively on F. On the other hand, degenerate del Pezzo G-surfaces can be G-contracted to non-degenerate ones. The standard G-surfaces with a rational pencil have the G-invariant Picard number 2 and are minimal G-surfaces if $K_F^2 < 0$ (actually ≤ 2) [17, теорема 4.3] [12, теорема 2.6].

17 First Examples of Rigid and Superrigid Varieties

After this lengthy intermission we are ready to continue our discussion of Manin's contributions to birational rigidity. Manin constructed first examples of rigid varieties—minimal del Pezzo surfaces of degree $\leq$ 3, and first examples of superrigid varieties—minimal del Pezzo surfaces of degree 1 [17], the part II of [16], and (with Iskovskikh) quartic threefolds [14]. Of course, such examples of minimal del Pezzo surfaces exist only for algebraically non-closed fields. In dimension 3 there are a lot of rigid and superrigid varieties.

The main results of [17] are about cubic surfaces—del Pezzo surfaces of degree 3. Under the anti-canonical imbedding such a surface is a cubic hypersurface in $\mathbb{P}^3$. Manin proves that minimal cubic surfaces over k are not rational over k [17, теорема A, a)]. Moreover, two minimal cubic surfaces are birationally isomorphic if and only if they are isomorphic under a projective transformation of $\mathbb{P}^3$ [17, теорема A, б)]. The first statement belongs to Segre [22] and is an immediate corollary of the second one in a slightly more general form (cf. [17, теоремы 5.7, 5.8]). This more general form means that every minimal Mori fibration of a minimal cubic surface F is a minimal cubic surface F' too; in particular, there are no Mori fibration models over curves. Thus, degenerate del Pezzo surfaces in [17, следствие стр. 185] are impossible. This follows from the table of Sarkisov links for algebraic surfaces [12, теорема 2.6] or the Noether-Fano-Iskovskikh-Manin method. But the second statement is much stronger: every cubic model F'

is projectively isomorphic to F; equivalently, F' is isomorphic to F. This means that F is rigid. Actually, F is superrigid if there exist no closed points of degrees 1 and 2 on F. Indeed, if F is not superrigid then there exists a Sarkisov link into another model. Thus, there exists a central model F'' over F which is a blowup of a closed point $x \in F$, and $K_{F''}^2 = 3 - d$. Since F'' is also a del Pezzo surface, we get $3 - d \geq 1$ and $d \leq 2$. Notice now that if x has degree 1, i.e., is a k-point, then there exists a birational automorphism $t_x : F \dashrightarrow F$—Geiser involution: x does not lie on a line on F, and for any line l through x in $\mathbb{P}^3$, the divisor $F_{|l} - x$ is invariant under the action of t_x. Actually, this involution corresponds to the link given by the central model F''. Similarly, every closed point $x \in F$ of degree 2 does not lie on a line on F and determines a Bertini involution $s_x : F \dashrightarrow F \in \mathrm{Bir}(F)$ which corresponds to F''. Thus, we have only links of these two types. In particular, this shows the stated rigidity if we know the weak form of Sarkisov theory: every canonical birational isomorphisms between Mori fibrations models can be factorized into Sarkisov links. This explains also the next Manin result [17, теорема Б]: $\mathrm{Bir}(F)$ is generated by involutions t_x, s_x and by the finite group of projective automorphisms of F. Actually, every automorphism of F is projective and the group $\mathrm{Aut}(F)$ is finite, e.g., because the set of exceptional curves of the first kind on $F_{\overline{k}}$ is finite. The generation implies that F has a closed k-point if and only if $\mathrm{Bir}(F)$ is infinite. Indeed, even when $\mathrm{Bir}(F)$ is finitely generated (e.g., over a finite field [17, 5.16]) subgroup $< t_x, s_x >$ generated by involutions t_x, s_x is infinite. The subgroup $< s_x, t_x >$ is normal in $\mathrm{Bir}(F)$ and the group $\mathrm{Bir}(F)$ is a semidirect product of $< t_x, s_x >$ with the finite group $\mathrm{Aut}(F)$ permuting separately Geiser and Bertini involutions. In introduction [17, стр. 162], Manin states also that $< s_x, t_x >$ is close to a free product of cyclic groups $\mathbb{Z}/2\mathbb{Z}$ of order 2. Every Bertini involution s_x does not have relations with the other involutions and relations are only possible for Geiser involutions. Indeed, the only central models of rank 3 for minimal cubics are blowups of two closed k-points $x, x' \in F$. A more conclusive statement will be given below for del Pezzo surfaces of degree 2.

The rigidity holds for the minimal G-cubics and, moreover, for the minimal Del Pezzo G-surfaces F of degree ≤ 3 [17, теорема 5.9]. The group $\mathrm{Bir}(F)$ is generated by $\mathrm{Aut}(F)$ and by Geiser and Bertini involutions t_x and s_x for F of degree 3, where x is a G-invariant closed point of $F_{\overline{k}}$ for t_x and x is a G-invariant pair of closed points of $F_{\overline{k}}$; the point x should be *non-special*, i.e., not on a line on $F_{\overline{k}}$ and the pair x should not be on a line nor on a conic on $F_{\overline{k}}$. Respectively, for degree 2, the group $\mathrm{Bir}(F)$ is generated by $\mathrm{Aut}(F)$ and by Bertini involutions s_x where x are the non-special G-invariant closed points of $F_{\overline{k}}$. Additionally, for degree 2, the subgroup $< s_x >$ generated by Bertini involutions is normal in $\mathrm{Bir}(F)$, a free product of order 2 subgroups corresponding to each involution and $\mathrm{Bir}(F)$ is a semidirect product of $< s_x >$ with the finite group $\mathrm{Aut}(F)$ permuting Bertini involutions [17, замечание 5.15]. Finally, the minimal del Pezzo G-surfaces of degree 1 are superrigid and $\mathrm{Bir}(F) = \mathrm{Aut}(F)$ is finite. Thus, for every minimal del Pezzo surface of degree ≤ 3 over a finite field k, the group $\mathrm{Bir}(F)$ is finitely generated.

In [17, теорема 6.1], Manin shows unirationality of del Pezzo surfaces F over a field k of degree $n = 2, 3$, and 4. More precisely, if F has a non-special closed k-point, then there exists a rational dominant morphism $\mathbb{P}^2 \dashrightarrow F$ of degree 2 for $n = 4$ and of degree 6 for $n = 3$. Thus if k is infinite and such an F has a non-special k-point then the set of closed k-points is dense in the Zariski topology. For $n = 3$ the same result was established by Segre without the non-special assumption [22]. Manin also proved that established degrees of unirationality are optimal [17, теорема 6.3].

18 Quartic Threefolds

The year 1971 was a turning point for higher-dimensional birational algebraic geometry. Independently, there were constructed three types of counterexamples to Lüroth problem: whether or not a unirational algebraic variety is rational. Lüroth proved that this is true in dimension 1: for any field k, any finitely generated subfield of the field $k(x)$ of rational functions in one variable x is isomorphic to $k(x)$ over k; equivalently, for any rational dominant morphism $\mathbb{P}^1 \dashrightarrow C$ over k, C is rational over k or $C =$ pt.. In dimension 2, Lüroth problem has a positive answer if k is algebraically closed. This follows from the Castelnuovo cohomological characterization of smooth rational projective surfaces F over an algebraically closed field k. But this is not true for surfaces over any algebraically non-closed field k, e.g., for the minimal del Pezzo surfaces of degrees 3 and 2. Each of the counterexamples [3], [4], [14] constitutes a new direction in birational algebraic geometry. The direction due to Iskovskikh-Manin is related to the Noether-Fano-Iskovskikh-Manin method and to the Minimal Model Program.

The main theorem of [14]: if $\chi : V \dashrightarrow V'$ is a birational isomorphism of two smooth hypersurfaces of degree 4 (quartics) in $\mathbb{P}^4$ over any field, then χ is a projective isomorphism, i.e., is regular. Thus, the group of birational automorphisms of any smooth quartic V is finite and V is not rational [14, следствие стр. 141]. On the other hand, Segre constructed examples of smooth unirational quartics at least for fields k with characteristic $\neq 2, 3$ [23]. This gives examples of unirational but not rational varieties in dimension 3.

It was noticed in 1988 that every smooth quartic threefold V is superrigid; this follows from the proof in [14]. Since the Picard number of V is 1, V is a Mori fibration over pt. For example, it is not birationally isomorphic to a Mori fibration of conic bundle type because the latter has infinitely many birational isomorphisms. After this example, there appeared many other examples of rigid and superrigid varieties. Actually, Iskovskihk and Manin noticed that quartics do not have non-trivial birational automorphisms and this means extremal "rigidity" of these unirational and possibly non-unirational varieties. Still, we do not know whether or not every smooth quartic threefold is unirational.

The statement of [14, основная теорема] and many fundamental ideas in its proof belongs to Fano [6, 7]. The Noether-Fano-Iskovskikh method uses the singularities of linear systems of good models, e.g., of Mori fibrations. It is based on the uniqueness of canonical model [13, теорема 2.3]. We explain this method for a Mori-Fano variety X. Suppose that $\varphi'\colon X' \to T'$ is its Mori fibration model with a canonical birational isomorphism $\chi\colon X \dashrightarrow X'$. To prove a superrigidity of X we need to verify that χ is an isomorphism, that is, is regular. Let

$$\mathcal{H}' = \left|-\mu' K_{X'} + \varphi'^* A'\right|$$

be a very ample complete linear system on X', where μ' is a positive integer ≥ 2 and A' is a very ample divisor on T'; such a liner system exists for every Mori fibration. Let $\mathcal{H} = \chi_*^{-1}\mathcal{H}'$ be its proper birational transform on X. Then $\mathcal{H}$ is a mobile linear subsystem in $|-\mu K^s|$ with a positive rational number μ and K^s is a semi-canonical divisor, that is, a canonical divisor up to $\mathbb{Q}$-linear or numerical equivalence; μ is integral if X is smooth and $K^s = K_X$ generates the Picard group of X up to linear equivalence. For sufficiently general effective divisor $H' \in \mathcal{H}'$, the pair $(X', \frac{1}{\mu'}H')$ is a minimal model of $(X, \frac{1}{\mu'}H)$ where H is the proper birational transform of H', i.e., the pair $(X', \frac{1}{\mu'}H')$ has only log canonical, terminal in codimension ≥ 2 singularities, and $K_{X'} + \frac{1}{\mu'}H'$ is nef. If the $(X, \frac{1}{\mu})$ is weak log canonical, i.e., has only log canonical, canonical in codimension ≥ 2 singularities, and $K_X + \frac{1}{\mu}H$ is nef, then χ is an isomorphism by [13, теорема 2.3]. Thus, to prove the superrigidity it is sufficient to show the weak log canonical property of $(X, \frac{1}{\mu}H)$. Otherwise X is not superrigid and we can construct a Sarkisov link of $X/\text{pt.}$ to another Mori fibration with a smaller invariant μ, the first component in the Sarkisov degree. After finitely many such steps conjecturally we get X'/T'. This holds in dimensions ≤ 3 but is unknown in higher dimensions—the Sarkisov theory.

For a rigid variety X we should check that each Sarkisov link gives an isomorphic variety. This type of constructions Manin develops in [17] for minimal del Pezzo surfaces of degree ≤ 3.

In the superrigid case, since $K^s + \frac{1}{\mu}H \equiv K_X + \frac{1}{\mu}H \equiv 0$ and nef, we need to verify that $(X, \frac{1}{\mu}H)$ is log canonical and canonical in codimension ≥ 2, where $\equiv$ denotes the numerical equivalence. This means to investigate singularities of H. The tool used by Iskovskikh and Manin goes back to Noether and Fano and after its sharpening becomes the Noether-Fano-Iskovskikh-Manin inequality. It depends on a situation. Thus, we consider now a smooth quartic threefold $X = V$. In this case $K^s = K_V$ and generates the Picard group of V up to the linear equivalence. Thus, μ is a positive integer and $H \in \mathcal{H} \subseteq |-\mu K_V|$. In particular, every component of the effective divisor H has multiplicity $\leq \mu$ and $(V, \frac{1}{\mu}H)$ has only log canonical singularities in codimension 1. The remaining possible non-canonical singularities are along curves or in points of V. On the other hand, V is smooth and has only terminal singularities. This implies the existence of the *canonical threshold*

$$c(V, H) = \max\{t \in \mathbb{Q} \mid K_X + tH \text{ has only canonical singularities in codim} \geq 2\}.$$

This is the second component of the Sarkisov degree. Instead of $c((V, H)$, Iskovskikh and Manin, as Noether and Fano, use the *maximal multiplicity* $\lambda = 1/c(V, H)$ *of* H or *of the linear system* $\mathcal{H}$ [14, §5]. The threshold is attained on an exceptional divisor E, and it, or its center on V, is called the *maximal singularity*. By our construction, $c(V, H) < 1/\mu$ and we need to verify absence of maximal singularities.

19 Noether Inequality

If the maximal singularity is a curve C on V then in the case $c(V, H) < 1/\mu$ we get the Noether inequality $\operatorname{ord}_C H > \mu$. The same holds for any smooth Fano variety X, an effective divisor $H \in |-\mu K_X|$ and a maximal point P of codimension ≥ 2 (closed or not): if $c(X, H; P) < 1/\mu$ then

$$\operatorname{ord}_P H > \mu \quad \text{(Noether)},$$

where $c(X, D; P)$ is the canonical threshold at P. The modern proof is very simple: it follows from the fact that the discrepancy in blown-up divisor in a point P of codimension ≥ 2 is ≥ 1, and for any effective divisor D on X

$$c = c(X, D; P) \geq \frac{1}{\operatorname{ord}_P D} \quad \text{or } \lambda \leq \operatorname{ord}_P D,$$

where $\lambda = 1/c$ is the *maximal multiplicity at* P. For P of codimension 2, the equality holds

$$c = c(X, D; P) = \frac{1}{\operatorname{ord}_P D} \quad \text{or } \lambda = \operatorname{ord}_P D.$$

This explains the terminology. Now for a quartic threefold $X = V$ and a line l on V but not on H, passing through a point P on a maximal singularity curve C, $(l, -K_V) = 1$ and $\mu = (l, H) \geq \operatorname{ord}_C H > \mu$ hold, a contradiction by the Noether inequality where $(-, -)$ denotes the intersection form for curves with divisors.

20 Fano Inequality

Thus, every maximal singularity for H on V is a closed point P. In this case we have stronger inequality: $\operatorname{ord}_P H > 2\mu$. As above, more generally, for a maximal points P of codimension ≥ 3 if $c(X, H) < 1/\mu$ then

$$\operatorname{ord}_P H > 2\mu \quad \text{(Fano)}.$$

In this situation the discrepancy in blown-up divisor in P is ≥ 2 and for any effective divisor D on X

$$c = c(X, D; P) \geq \frac{2}{\operatorname{ord}_P D} \text{ or } \lambda \leq \frac{\operatorname{ord}_P D}{2}.$$

Unfortunately, the Fano inequality is not optimal in codimensions ≥ 3 vs. codimension 2 and leads to the absence of maximal singularities only in general cases, e.g., when the maximal singularity is on the first blowup of a point.

21 Iskovskikh-Manin Inequality

Iskovskikh and Manin discovered a quadratic version of the Fano inequality that works for all maximal singularities. Let $\mathcal{H}$ be a mobile linear system on a smooth threefold X and a closed point $P \in X$ be a maximal singularity of $(X, \frac{1}{\mu}H)$ for sufficiently general $H \in \mathcal{H}$. Let H_1, H_2 be two members of $\mathcal{H}$ without common irreducible components and S be an effective surface through P not containing any irreducible component of $H_1 \cap H_2$. Then the local intersection number $(H_1, H_2, S)_P$ at P is well defined and

$$(H_1, H_2, S)_P > 4m^2 \quad \text{(Iskovskikh-Manin)}.$$

For the quartic V take sufficiently general divisors $H_1, H_2 \in \mathcal{H}$ and a sufficiently general hyperplane section S through P. Then

$$4m^2 = (H_1, H_2, S) \geq (H_1, H_2, S)_P > 4m^2,$$

a contradiction, where $(-, -, -)$ is the intersection form for divisors on V.

22 Conclusion

The impact of Manin's work in birational geometry is enormous and produced many further generalizations and developments. The flow of mathematical research is like a river with many contributions as influxes which make the flow full-flowing. In the case of birational rigidity: Noether, Fano, Manin, Iskovskikh, Sarkisov, Pukhlikov, etc. And this flow goes on and on to the sea of perfection.

References

1. S. Abhyankar, *Ramification Theoretic Methods in Algebraic Geometry* (Princeton, 1959)
2. Алгебраические поверхности (под ред. И.Р.Шафаревича). – 1965. – (Труды МИАН СССР; Т. 75)
3. M. Artin, D. Mumford, Some elementary examples of unirational varieties which are not rational. Proc. Lond. Math. Soc. **25**, 75–95 (1972)
4. H. Clemens, Ph. Griffiths, The intermediate Jacobian of the cubic threefold. Ann. Math. **95**, 281–356 (1972)
5. F. Enriques, Sulle irrazionalita da cui puo farsi dipendere la resoluzione d'un' equazione algebrica $f(x, y, z) = 0$ con funzioni razionali di due parametri. Math. Annal. **49**, 1–23 (1897)
6. G. Fano, Sopra aleune varietà algebriche a tre dimensioni aventi tutti i generi nulli. Atti Acc. Torino **43**, 973–977 (1907–1908)
7. G. Fano, Osservazioni sopra aleune vatetà non razionaliaventi tutti i genera nulli. Atti Acc. Torino **50**, 1067–1071 (1915)
8. Ch. Hacon, J. McKernan, Sarkisov program. J. Alg. Geom. **22**, 389–405 (2013)
9. Y. He, On syzygies of Mori fibre spaces. arXuve:2311.08609 78pp
10. Y. He, On strong Sarkisov program. arXuve:2311.08750 24pp
11. В.А. Исковских, Минимальные модели рациональных поверхностей над произвольными полями. Изв. АН СССР. Сер. мат. **43**, 19-43 (1979)
12. В.А. Исковских, Факторизация бирациональных отображений рациональных поверхностей с точки зрения теории Мори. УМН **51**, 3–72 (1996)
13. В.А. Исковских, В.В. Шокуров, Бирациональные модели и перестройки. УМН **60**, 29–98 (2005)
14. В.А. Исковских, Ю.И. Манин, Трехмерные квартики и контрпримеры к проблеме Люрота. Матем. сб. **128**, 140–166 (1971)
15. Ю.И. Манин, Об арифметике рациональных поверхностей. Доклады АН СССР **152**, 46–49 (1963)
16. Ю.И. Манин, Рациональные поверхности над совершенными полями. Publ. Math. IHES **30**, 55–113 (1966)
17. Ю.И. Манин, Рациональные поверхности над совершенными полями II. Матем. сб. **114**, 161–192 (1967)
18. Ю.И. Манин, М.А. Цфасман, Рациональные многообразия: алгебра, геометрия, арифметика. УМН **41**, 43–94 (1986)
19. Sh. Mori, Threefolds whose canonical bundles are not numerically effective. Ann. Math. **116**, 133–176 (1982)
20. T. Ono, On the Tamagawa number of algebraic tori. Ann. Math. **78**, 47–73 (1963)
21. В.Г. Саркисов, Бирациональные автоморфизмы расслоений на коники. Изв. АН СССР. Сер. мат. 44, 918–945 (1980)
22. B. Segre, The rational solutions of homogeneous cubic equations in four variables. Math. Notae, Universidad Rosario, anno II, fasc. 1–2, 1–68
23. B. Segre, Variazione continua ad omotopia in geometria algebrica. Ann. mat. pura ed appl., Ser. IV **L**, 149–186 (1960)
24. N.I. Shepherd-Barron, The rationality of quintic del Pezzo surfaces – a short proof. Bull. Lond. Math. Soc. **24**, 249–250 (1992)
25. V. Shokurov, S.R. Choi, Geography of log models: theory and applications. Cent. Eur. J. Math. **9**, 489–534 (2011)
26. H.P.F. Swinnerton-Dyer, Rational Points on del Pezzo Surfaces of Degree, vol. 5. Algebraic Geometry, 287–290. (Oslo, 1970)
27. A. Weil, Abstract versus classical algebraic geometry. Proc. Int. Congr. Math. **3**, 550–558 (1954)

Endomorphism Algebras and Automorphism Groups of Certain Complex Tori

Yuri G. Zarhin

Contents

1 Introduction

As usual, we write $\mathbb{Q}, \mathbb{R}, \mathbb{C}$ for the fields of rational, real, complex numbers and $\mathbb{Z}$ for the ring of integers. We write $\bar{\mathbb{Q}}$ for the subfield of all algebraic numbers in $\mathbb{C}$, which is an algebraic closure of $\mathbb{Q}$. If p is a prime then $\mathbb{F}_p, \mathbb{Z}_p, \mathbb{Q}_p$ stand for the p-element (finite) field, the ring of p-adic integers, and the field of p-adic numbers, respectively. If E is a number field of degree $n = [E : \mathbb{Q}]$ then r_E and s_E are nonnegative integers such that the $\mathbb{R}$-algebra $E_{\mathbb{R}} = E \otimes_{\mathbb{Q}} \mathbb{R}$ is isomorphic to a product $\mathbb{R}^{r_E} \times \mathbb{C}^{s_E}$. (In other words, r_E is the number of "real" field embeddings $E \hookrightarrow \mathbb{R}$ and $2s_E$ is the number of "imaginary" field embeddings $E \hookrightarrow \mathbb{C}$, whose images do *not* lie in $\mathbb{R}$.) In particular,

$$[E : \mathbb{Q}] = r_E + 2s_E. \tag{1}$$

Y. G. Zarhin (✉)
Department of Mathematics, Pennsylvania State University, University Park, PA, USA
e-mail: zarhin@math.psu.edu

Y. Tschinkel (ed.), *Arithmetic and Algebraic Geometry*, Simons Symposia,
https://doi.org/10.1007/978-3-031-74134-0_9

Let X be a connected compact complex Kähler manifold and $\mathrm{H}^2(X, \mathbb{Q})$ its second rational cohomology group equipped with the canonical rational Hodge structure, i.e., there is the Hodge decomposition

$$\mathrm{H}^2(X, \mathbb{Q}) \otimes_{\mathbb{Q}} \mathbb{C} = \mathrm{H}^2(X, \mathbb{C}) = \mathrm{H}^{2,0}(X) \oplus \mathrm{H}^{1,1}(X) \oplus \mathrm{H}^{2,0}(X)$$

where $\mathrm{H}^{2,0}(X) = \Omega^2(X)$ is the space of holomorphic 2-forms on X, $\mathrm{H}^{0,2}(X)$ is the "complex-conjugate" of $\mathrm{H}^{2,0}(X)$, and $\mathrm{H}^{1,1}(X)$ coincides with its own "complex-conjugate" (see [14, Sections 2.1–2.2], [29, Chs. VI–VII])). The following property of X was introduced and studied by F. Campana [10, Definition 3.3]. (Recently, it was used in the study of coisotropic and lagrangian submanifolds of symplectic manifolds [1].)

Definition 1.1 A manifold X is *irreducible in weight 2* (irréductible en poids 2) if it enjoys the following property.

Let H be a rational Hodge substructure of $\mathrm{H}^2(X, \mathbb{Q})$ such that

$$H_{\mathbb{C}} \cap \mathrm{H}^{2,0}(X) \neq \{0\}$$

where $H_{\mathbb{C}} := H \otimes_{\mathbb{Q}} \mathbb{C}$.

Then $H_{\mathbb{C}}$ contains the whole $\mathrm{H}^{2,0}(X)$.

Our aim is to study complex tori T that enjoy this property. Namely, we discuss their endomorphism algebras, automorphism groups, and Hodge groups.

Let $T = V/\Lambda$ be a complex torus of positive dimension g where V is a g-dimensional complex vector space and Λ is a discrete lattice of rank $2g$ in V. One may naturally identify Λ with the first integral homology group $\mathrm{H}_1(T, \mathbb{Z})$ of T and

$$\Lambda_{\mathbb{Q}} = \Lambda \otimes \mathbb{Q} = \{v \in V \mid \exists n \in \mathbb{Z} \setminus \{0\} \text{ such that } nv \in \Lambda\}$$

with the first rational homology group $\mathrm{H}_1(T, \mathbb{Q})$ of T. There are also natural isomorphisms of real vector spaces

$$\Lambda \otimes \mathbb{R} = \Lambda_{\mathbb{Q}} \otimes_{\mathbb{Q}} \mathbb{R} \to V, \ \lambda \otimes r \mapsto r\lambda$$

that may be viewed as isomorphisms related to the first real cohomology group $\mathrm{H}_1(T, \mathbb{R})$ of T:

$$\mathrm{H}_1(T, \mathbb{R}) = \mathrm{H}_1(T, \mathbb{Z}) \otimes \mathbb{R} = \mathrm{H}_1(T, \mathbb{Q}) \otimes_{\mathbb{Q}} \mathbb{R} \to V.$$

In particular, there is a canonical isomorphism of real vector spaces

$$\mathrm{H}_1(T, \mathbb{R}) = V, \tag{2}$$

and a canonical isomorphism of complex vector spaces

$$\mathrm{H}_1(T,\mathbb{C}) = \mathrm{H}_1(T,\mathbb{Q}) \otimes_{\mathbb{Q}} \mathbb{C} = \mathrm{H}_1(T,\mathbb{R}) \otimes_{\mathbb{R}} \mathbb{C} = V \otimes_{\mathbb{R}} \mathbb{C} =: V_{\mathbb{C}} \tag{3}$$

where $\mathrm{H}_1(T,\mathbb{C})$ is the first complex homology group of T.

There are natural isomorphisms of $\mathbb{R}$-algebras

$$\mathrm{End}_{\mathbb{Z}}(\Lambda) \otimes \mathbb{R} \cong \mathrm{End}_{\mathbb{R}}(V),\ u \otimes r \mapsto ru,$$

$$\mathrm{End}_{\mathbb{Q}}(\Lambda_{\mathbb{Q}}) \otimes \mathbb{R} \cong \mathrm{End}_{\mathbb{R}}(V),\ u \otimes r \mapsto ru,$$

which give rise to the natural ring embeddings

$$\mathrm{End}_{\mathbb{Z}}(\Lambda) \subset \mathrm{End}_{\mathbb{Q}}(\Lambda_{\mathbb{Q}}) \subset \mathrm{End}_{\mathbb{R}}(V) \subset \mathrm{End}_{\mathbb{R}}(V) \otimes_{\mathbb{R}} \mathbb{C} = \mathrm{End}_{\mathbb{C}}(V_{\mathbb{C}}). \tag{4}$$

Here the structure of a $2g$-dimensional *complex* vector space on $V_{\mathbb{C}}$ is defined by

$$z(v \otimes s) = v \otimes zs \quad \forall v \otimes s \in V \otimes_{\mathbb{R}} \mathbb{C} = V_{\mathbb{C}},\ z \in \mathbb{C}.$$

If $u \in \mathrm{End}_{\mathbb{R}}(V)$ then we write $u_{\mathbb{C}}$ for the corresponding $\mathbb{C}$-linear operator in $V_{\mathbb{C}}$, i.e.,

$$u_{\mathbb{C}}(v \otimes z) = u(v) \otimes z \quad \forall u \in V, z \in \mathbb{C}, v \otimes z \in V_{\mathbb{C}}. \tag{5}$$

Remark 1.2 Sometimes, we will identify $\mathrm{End}_{\mathbb{R}}(V)$ with its image $\mathrm{End}_{\mathbb{R}}(V) \otimes 1 \subset \mathrm{End}_{\mathbb{C}}(V_{\mathbb{C}})$ and write u instead of $u_{\mathbb{C}}$, slightly abusing notation.

As usual, one may naturally extend the complex conjugation $z \mapsto \bar{z}$ on $\mathbb{C}$ to the $\mathbb{C}$-antilinear involution

$$V_{\mathbb{C}} \to V_{\mathbb{C}},\ w \mapsto \bar{w},\ v \otimes z \mapsto \overline{v \otimes z} = v \otimes \bar{z},$$

which is usually called the complex conjugation on $V_{\mathbb{C}}$. Clearly,

$$u_{\mathbb{C}}(\bar{w}) = \overline{u(w)} \quad \forall u \in \mathrm{End}_{\mathbb{R}}(V), w \in V_{\mathbb{C}}. \tag{6}$$

This implies easily that the set of fixed points of the involution is

$$V = V \otimes 1 \subset V_{\mathbb{C}}.$$

Let $\mathrm{End}(T)$ be the endomorphism ring of the complex commutative Lie group T and $\mathrm{End}^0(T) = \mathrm{End}(T) \otimes \mathbb{Q}$ the corresponding endomorphism algebra, which is a finite-dimensional algebra over $\mathbb{Q}$; see [21, 6, 3]. There are well-known canonical isomorphisms

$$\mathrm{End}(T) = \mathrm{End}_{\mathbb{Z}}(\Lambda) \cap \mathrm{End}_{\mathbb{C}}(V), \quad \mathrm{End}^0(T) = \mathrm{End}_{\mathbb{Q}}(\Lambda_{\mathbb{Q}}) \cap \mathrm{End}_{\mathbb{C}}(V).$$

Let $g \geq 2$ and

$$\mathrm{H}^2(T, \mathbb{Q}) = \wedge^2_{\mathbb{Q}}(\Lambda_{\mathbb{Q}}, \mathbb{Q})$$

be the *second rational cohomology group* of T, which carries a natural rational Hodge structure of weight two:

$$\mathrm{H}^2(T, \mathbb{C}) = \mathrm{H}^2(T, \mathbb{Q}) \otimes_{\mathbb{Q}} \mathbb{C} = \mathrm{H}^{2,0}(T) \oplus \mathrm{H}^{1,1}(T) \oplus \mathrm{H}^{0,2}(T)$$

where $\mathrm{H}^{2,0}(T) = \Omega^2(T)$ is the $g(g-1)/2$-dimensional space of holomorphic 2-forms on T.

Definition 1.3 Let $g = \dim(T) \geq 2$. We say that T is 2-*simple* if it is *irreducible in weight 2*, i.e., enjoys the following property.

Let H be a rational Hodge substructure of $\mathrm{H}^2(T, \mathbb{Q})$ such that

$$H_{\mathbb{C}} \cap \mathrm{H}^{2,0}(T) \neq \{0\}$$

where $H_{\mathbb{C}} := H \otimes_{\mathbb{Q}} \mathbb{C}$.

Then $H_{\mathbb{C}}$ contains the whole $\mathrm{H}^{2,0}(T)$.

Remark 1.4 We call such complex tori 2-simple, because they are simple in the usual meaning of this word if $g > 2$; see Theorem 1.7(i).

Example 1.5 (See [10, Example 3.4(2)].) If $g = 2$ then $\dim_{\mathbb{C}}(\mathrm{H}^{2,0}(T)) = 1$. This implies that (in the notation of Definition 1.3) if $H_{\mathbb{C}} \cap \mathrm{H}^{2,0}(T) \neq \{0\}$ then $H_{\mathbb{C}}$ contains the whole $\mathrm{H}^{2,0}(T)$. Hence, every 2-dimensional complex torus is 2-simple.

In what follows we write $\mathrm{Aut}(T) = \mathrm{End}(T)^*$ for the automorphism group of the complex Lie group T. We will need the following well-known definition.

Definition 1.6 A number field E is called *primitive* if either $E = \mathbb{Q}$ or the only proper subfield of E is $\mathbb{Q}$.

Our main result is the following assertion.

Theorem 1.7 *Let T be a complex torus of dimension $g \geq 3$. Suppose that T is 2-simple.*

Then T enjoys the following properties:

(i) *T is simple.*
(ii) *If E is any subfield of* $\mathrm{End}^0(T)$ *then it is a number field, whose degree* $[E : \mathbb{Q}]$ *over* $\mathbb{Q}$ *is either* 1 *or* g *or* $2g$.
(iii) $\mathrm{End}^0(T)$ *is a number field, whose degree* $[\mathrm{End}^0(T) : \mathbb{Q}]$ *over* $\mathbb{Q}$ *is either* 1 *or* g *or* $2g$.
(iv) *If* $[\mathrm{End}^0(T) : \mathbb{Q}] = 1$ *then*

$$\mathrm{End}^0(T) = \mathbb{Q}, \mathrm{End}(T) = \mathbb{Z}, \ \mathrm{Aut}(T) = \{\pm 1\}.$$

(v) *If* $E = \mathrm{End}^0(T)$ *and* $[E : \mathbb{Q}] = 2g$ *then* E *is a purely imaginary number field and* $\mathrm{Aut}(T) \cong \{\pm 1\} \times \mathbb{Z}^{g-1}$. *In addition, if* E *is not primitive then it contains precisely one proper subfield except* $\mathbb{Q}$ *and this subfield has degree* g.

(vi) *If* $E = \mathrm{End}^0(T)$ *and* $[E : \mathbb{Q}] = g$ *then* E *is a primitive number field and* $\mathrm{Aut}(T) \cong \mathbb{Z}^d \times \{\pm 1\}$ *where the positive integer* d *equals* $r_E + s_E - 1$. *In particular,*

$$\frac{1}{2} \leq \frac{g}{2} - 1 \leq d \leq g - 1.$$

In addition, if T *is a complex abelian variety then* E *is a primitive totally real number field and* $d = g - 1$.

Remark 1.8

(i) It is well known (and can be easily checked) that T is *simple* if and only if the rational Hodge structure on $\Lambda_{\mathbb{Q}} = \mathrm{H}_1(T, \mathbb{Q})$ is *irreducible*.[1]

(ii) We may view $\mathrm{H}^2(T, \mathbb{Q})$ as the $\mathbb{Q}$-vector subspace $\mathrm{H}^2(T, \mathbb{Q}) \otimes 1$ of

$$\mathrm{H}^2(T, \mathbb{Q}) \otimes_{\mathbb{Q}} \mathbb{C} = \mathrm{H}^2(T, \mathbb{C}).$$

Let us consider the $\mathbb{Q}$-vector (sub)space

$$\mathrm{H}^{1,1}(T, \mathbb{Q}) := \mathrm{H}^2(T, \mathbb{Q}) \cap \mathrm{H}^{1,1}(T)$$

of 2-dimensional *Hodge cycles* on T. Below the *irreducibility* of the rational Hodge structure on $\Lambda_{\mathbb{Q}}$ implies the *complete reducibility*[2] of the rational Hodge structure on $\mathrm{H}^2(T, \mathbb{Q}) = \mathrm{Hom}_{\mathbb{Q}}\left(\wedge^2_{\mathbb{Q}} \Lambda_{\mathbb{Q}}, \mathbb{Q}\right)$. (It follows from the reductiveness of the Mumford-Tate group of a simple torus [11, Sect. 2.2].) In light of (i) and Theorem 1.7(i), a complex torus T of dimension > 2 is 2-simple if and only if it is simple and $\mathrm{H}^2(T, \mathbb{Q})$ splits into a direct sum of $\mathrm{H}^{1,1}(T, \mathbb{Q})$ and an *irreducible* rational Hodge substructure.

Theorem 1.9 *Let* $g \geq 3$ *be an integer. Let* $\mathbf{r}, \mathbf{s}$ *be nonnegative integers such that*

$$\mathbf{r} + 2\mathbf{s} = g.$$

Then there exists a 2-simple torus T *of degree* g *that enjoys the following properties.*

[1] A rational Hodge structure H is called *irreducible* or *simple* if its only rational Hodge substructures are H itself and $\{0\}$ [11, Sect. 2.2].

[2] A rational Hodge structure is called completely reducible if it splits into a direct sum of simple rational Hodge structures.

The endomorphism algebra $\mathrm{End}^0(T)$ *is a number field* E *such that*

$$[E:\mathbb{Q}] = g, \ r_E = \mathbf{r}, \ s_E = \mathbf{s}.$$

In particular, if d *is an integer such that*

$$\frac{g}{2} - 1 \leq d \leq g - 1$$

then there exists a g*-dimensional* 2*-simple complex torus* T *such that*

$$\mathrm{Aut}(T) \cong \mathbb{Z}^d \times \{\pm 1\}.$$

The paper is organized as follows. We prove Theorem 1.7 in Sect. 3, using explicit constructions related to the Hodge structure on $\Lambda_{\mathbb{Q}}$ that will be discussed in Sect. 2. Section 4 deals with (mostly well known) results about number fields that will be used in the computations of Hodge groups of complex tori. In Sect. 5 we discuss general properties of Hodge groups of 2-simple tori. In Sect. 6 we concentrate on the case when the endomorphism algebra is a number field of degree g. Section 7 contains more or less standard results from "Semilinear Algebra" that we use in Sect. 8, in order to study representations of semisimple Lie algebras over arbitrary fields of characteristic zero. Using results of Sect. 8, we study in Sect. 9 Hodge groups of 2-simple tori without nontrivial endomorphisms.

This paper may be viewed as a follow-up of [21] and [3].

2 Hodge Structures

2.1 It is well known that $\Lambda_{\mathbb{Q}} = \mathrm{H}_1(T, \mathbb{Q})$ carries the natural structure of a rational Hodge structure of weight -1. Let us recall the construction. Let $J : V \to V$ be the multiplication by $\mathbf{i} = \sqrt{-1}$, which is viewed as an element of $\mathrm{End}_{\mathbb{R}}(V)$ such that

$$J^2 = -1.$$

Hence, $J_{\mathbb{C}}^2 = -1$ in $\mathrm{End}_{\mathbb{C}}(V_{\mathbb{C}})$ and we define two mutually complex-conjugate $\mathbb{C}$-vector subspaces (of the same dimension) $\mathrm{H}_{-1,0}(T)$ and $\mathrm{H}_{0,-1}(T)$ of $V_{\mathbb{C}}$ as the eigenspaces $V_{\mathbb{C}}(\mathbf{i})$ and $V_{\mathbb{C}}(-\mathbf{i})$ of $J_{\mathbb{C}}$ attached to eigenvalues $\mathbf{i}$ and $-\mathbf{i}$, respectively. Clearly,

$$V_{\mathbb{C}} = V_{\mathbb{C}}(\mathbf{i}) \oplus V_{\mathbb{C}}(-\mathbf{i}) = \mathrm{H}_{-1,0}(T) \oplus \mathrm{H}_{0,-1}(T),$$

which defines the rational Hodge structure of weight -1 on $\Lambda_{\mathbb{Q}}$, in light of $V_{\mathbb{C}} = \Lambda_{\mathbb{Q}} \otimes_{\mathbb{Q}} \mathbb{C}$. It also follows that both $\mathrm{H}_{-1,0}(T)$ and $\mathrm{H}_{0,-1}(T)$ have the same complex dimension $2g/2 = g$.

Recall that V is a complex vector space. I claim that the map

$$\Psi : V \to V_{\mathbb{C}}(\mathbf{i}) = \mathrm{H}_{-1,0}(T),\ v \mapsto Jv \otimes 1 + v \otimes \mathbf{i} \tag{7}$$

is an isomorphism of complex vector spaces. Indeed, first, Ψ defines a homomorphism of real vector spaces $V \to V_{\mathbb{C}}$. Second, if $v \in V$ then

$$J_{\mathbb{C}}(Jv \otimes 1 + v \otimes \mathbf{i}) = J^2 v \otimes 1 + Jv \otimes \mathbf{i} = -v \otimes 1 + Jv \otimes \mathbf{i} = \mathbf{i}(Jv \otimes 1 + v \otimes \mathbf{i}),$$

i.e., $Jv \otimes 1 + v \otimes \mathbf{i} \in V_{\mathbb{C}}(\mathbf{i}) = \mathrm{H}_{-1,0}(T)$, and therefore the map (7) is defined correctly. Third, taking into account that J is an automorphism of V and $V_{\mathbb{C}} = V \otimes 1 \oplus V \otimes \mathbf{i}$, we conclude that Ψ is an injective homomorphism of real vector spaces and a dimension argument implies that it is actually an isomorphism. It remains to check that Ψ is $\mathbb{C}$-linear, i.e.,

$$\Psi(Jv) = \mathbf{i}\Psi(v).$$

Let us do it. We have

$$\Psi(Jv) = J(Jv) \otimes 1 + Jv \otimes \mathbf{i} = -v \otimes 1 + Jv \otimes \mathbf{i} = \mathbf{i}(Jv \otimes 1 + v \otimes \mathbf{i}) = \mathbf{i}\Psi(v).$$

Hence, Ψ is a $\mathbb{C}$-linear isomorphism and we are done.

Now suppose that $u \in \mathrm{End}_{\mathbb{R}}(V)$ commutes with J, i.e., $u \in \mathrm{End}_{\mathbb{C}}(V)$. Then

$$\Psi \circ u = u_{\mathbb{C}} \circ \Psi. \tag{8}$$

In particular, $\mathrm{H}_{-1,0}(T)$ is $u_{\mathbb{C}}$-invariant. Indeed, if $v \in V$ then

$$\begin{aligned}\Psi \circ u(v) &= Ju(v) \otimes 1 + u(v) \otimes \mathbf{i} = uJ(v) \otimes 1 + u_{\mathbb{C}}(v \otimes \mathbf{i})\\ &= u_{\mathbb{C}}(J(v) \otimes 1) + u_{\mathbb{C}}(v \otimes \mathbf{i}) = u_{\mathbb{C}} \circ \Psi(v),\end{aligned}$$

which proves our claim.

Similarly, there is an antilinear isomorphism of complex vector spaces

$$V \to V_{\mathbb{C}}(-\mathbf{i}) = \mathrm{H}_{0,-1}(T),\ v \mapsto Jv \otimes 1 - v \otimes \mathbf{i}.$$

It is also well known that there is a canonical isomorphism of rational Hodge structures of weight 2

$$\mathrm{H}^2(T, \mathbb{Q}) = \mathrm{Hom}_{\mathbb{Q}}(\wedge^2_{\mathbb{Q}} \mathrm{H}_1(T, \mathbb{Q}), \mathbb{Q})$$

where the Hodge components $\mathrm{H}^{p,q}(T)$ $(p, q \geq 0, p + q = 2)$ are as follows.

$$\mathrm{H}^{2,0}(T) = \mathrm{Hom}_{\mathbb{C}}(\wedge^2_{\mathbb{C}} \mathrm{H}_{-1,0}(T), \mathbb{C}),$$

$$\mathrm{H}^{0,2}(T) = \mathrm{Hom}_{\mathbb{C}}(\wedge^2_{\mathbb{C}}\mathrm{H}_{0,-1}(T), \mathbb{C}), \tag{9}$$

$$\mathrm{H}^{1,1}(T) = \mathrm{Hom}_{\mathbb{C}}(\mathrm{H}_{-1,0}(T), \mathbb{C}) \wedge \mathrm{Hom}_{\mathbb{C}}(\mathrm{H}_{0,-1}(T), \mathbb{C}) \cong$$

$$\mathrm{Hom}_{\mathbb{C}}(\mathrm{H}_{-1,0}(T), \mathbb{C}) \otimes_{\mathbb{C}} \mathrm{Hom}_{\mathbb{C}}(\mathrm{H}_{0,-1}(T), \mathbb{C}).$$

Clearly,

$$\dim_{\mathbb{C}}(\mathrm{H}^{2,0}(T)) = \frac{g(g-1)}{2}.$$

3 Endomorphism Fields and Automorphism Groups

Proof of Theorem 1.7 Let T be a 2-simple complex torus and assume that

$$g = \dim(T) \geq 3.$$

(i) Suppose that T is *not* simple. This means that there is a proper complex subtorus $S = W/\Gamma$ where W is a complex vector subspace of V with

$$0 < d = \dim_{\mathbb{C}}(W) < \dim_{\mathbb{C}}(V) = g$$

such that

$$\Gamma = W \cap \Lambda$$

is a discrete lattice of rank $2d$ in W. Then the quotient T/S is a complex torus of positive dimension $g - d$.

Let $H \subset \mathrm{H}^2(T, \mathbb{Q})$ be the image of the canonical *injective* homomorphism of rational Hodge structures $\mathrm{H}^2(T/S, \mathbb{Q}) \hookrightarrow \mathrm{H}^2(T, \mathbb{Q})$ induced by the quotient map $T \to T/S$ of complex tori. Clearly, H is a rational Hodge substructure of $\mathrm{H}^2(T, \mathbb{Q})$ and its (2, 0)-component

$$H^{2,0} \subset H_{\mathbb{C}}$$

has $\mathbb{C}$-dimension

$$\dim_{\mathbb{C}}(H^{2,0}) = \dim_{\mathbb{C}}(\mathrm{H}^{2,0}(T/S))$$
$$= \frac{(g-d)(g-d-1)}{2} < \frac{g(g-1)}{2} = \dim_{\mathbb{C}}(\mathrm{H}^{2,0}(T)).$$

In light of the 2-simplicity of T,

$$\dim_{\mathbb{C}}(\mathrm{H}^{2,0}) = 0,$$

which implies that

$$g - d = 1.$$

On the other hand, let $\tilde{H}$ be the kernel of the canonical *surjective* homomorphism of rational Hodge structures $\mathrm{H}^2(T, \mathbb{Q}) \twoheadrightarrow \mathrm{H}^2(S, \mathbb{Q})$ induced by the inclusion map $S \subset T$ of complex tori. Clearly, $\tilde{H}$ is a rational Hodge substructure of $\mathrm{H}^2(T, \mathbb{Q})$. Notice that the induced homomorphism of (2, 0)-components $\mathrm{H}^{2,0}(T) \to \mathrm{H}^{2,0}(S)$ is also surjective, because every holomorphic 2-form on S obviously extends to a holomorphic 2-form on T. This implies that the (2, 0)-component

$$\tilde{H}^{2,0} \subset \tilde{H}_{\mathbb{C}}$$

of $\tilde{H}$ has $\mathbb{C}$-dimension

$$\dim_{\mathbb{C}}(\tilde{H}^{2,0}) = \dim_{\mathbb{C}}(\mathrm{H}^{2,0}(T)) - \dim_{\mathbb{C}}(\mathrm{H}^{2,0}(S)) = \frac{g(g-1)}{2} - \frac{d(d-1)}{2} > 0.$$

In light of the 2-simplicity of T,

$$\dim_{\mathbb{C}}(\tilde{H}^{2,0}) = \dim_{\mathbb{C}}(\mathrm{H}^{2,0}(T)) = \frac{g(g-1)}{2},$$

which implies that $\frac{d(d-1)}{2} = 0$, i.e., $d = 1$. Taking into account that $g - d = 1$, we get $g = 1 + 1 = 2$, which is not true. The obtained contradiction proves that T is simple and (i) is proven. In particular, $\mathrm{End}^0(T)$ is a division algebra over $\mathbb{Q}$.

In order to handle (ii), let us assume that E is a subfield of $\mathrm{End}^0(T)$. The simplicity of T implies that $1 \in E$ is the identity automorphism of T. Then $\Lambda_{\mathbb{Q}}$ becomes a faithful E-module. This implies that E is a number field and $\Lambda_{\mathbb{Q}}$ is an E-vector space of finite positive dimension

$$d_E = \frac{2g}{[E : \mathbb{Q}]}.$$

This implies that $V_{\mathbb{C}} = \Lambda_{\mathbb{Q}} \otimes_{\mathbb{Q}} \mathbb{C}$ is a free $E \otimes_{\mathbb{Q}} \mathbb{C}$-module of rank d_E. Clearly, both $\mathrm{H}_{-1,0}(T)$ and $\mathrm{H}_{0,-1}(T)$ are $E \otimes_{\mathbb{Q}} \mathbb{C}$-submodules of its direct sum $V_{\mathbb{C}}$. Let

$$\mathrm{tr}_{E/\mathbb{Q}} : E \to \mathbb{Q}$$

be the trace map attached to the field extension $E/\mathbb{Q}$ of finite degree. Let

$$\mathrm{Hom}_E\left(\wedge_E^2 \Lambda_{\mathbb{Q}}, E\right)$$

be the $\frac{d_E(d_E-1)}{2}$-dimensional E-vector space of alternating E-bilinear forms on $\Lambda_{\mathbb{Q}}$; it carries the natural structure of a rational Hodge structure of $\mathbb{Q}$-dimension $[E:\mathbb{Q}]\cdot\frac{d_E(d_E-1)}{2}$. There is the natural embedding of rational Hodge structures

$$\mathrm{Hom}_E\left(\wedge_E^2\Lambda_{\mathbb{Q}}, E\right) \hookrightarrow \mathrm{Hom}_{\mathbb{Q}}\left(\wedge_{\mathbb{Q}}^2\Lambda_{\mathbb{Q}}, \mathbb{Q}\right) = \mathrm{H}^2(T,\mathbb{Q}),\ \phi_E \mapsto \phi := \mathrm{tr}_{E/\mathbb{Q}} \circ \phi_E, \quad (10)$$
$$\forall \phi_E \in \mathrm{Hom}_E\left(\wedge_E^2\Lambda_{\mathbb{Q}}, E\right),$$

i.e.,

$$\phi(\lambda_1,\lambda_2) = \mathrm{tr}_{E/\mathbb{Q}}\big(\phi_E(\lambda_1,\lambda_2)\big)\ \forall \lambda_1,\lambda_2 \in \Lambda_{\mathbb{Q}}. \quad (11)$$

The image of $\mathrm{Hom}_E\left(\wedge_E^2\Lambda_{\mathbb{Q}}, E\right)$ in $\mathrm{Hom}_{\mathbb{Q}}\left(\wedge_{\mathbb{Q}}^2\Lambda_{\mathbb{Q}}, \mathbb{Q}\right) = \mathrm{H}^2(T,\mathbb{Q})$ coincides with the $\mathbb{Q}$-vector subspace

$$H_E := \{\phi \in \mathrm{Hom}_{\mathbb{Q}}\left(\wedge_{\mathbb{Q}}^2\Lambda_{\mathbb{Q}}, \mathbb{Q}\right) \mid \phi(u\lambda_1,\lambda_2)=\phi(\lambda_1,u\lambda_2)\ \forall u \in E,\ \lambda_1,\lambda_2 \in \Lambda_{\mathbb{Q}}\}. \quad (12)$$

Indeed, it is obvious that the image lies in H_E. In order to check that the image coincides with the whole subspace H_E, let us construct the inverse map

$$H_E \to \mathrm{Hom}_E\left(\wedge_E^2\Lambda_{\mathbb{Q}}, E\right),\ \phi \mapsto \phi_E$$

to (10) as follows. If $\lambda_1,\lambda_2 \in \Lambda_{\mathbb{Q}}$ then there is a $\mathbb{Q}$-linear map

$$\Phi : E \mapsto \mathbb{Q},\ u \mapsto \phi(u\lambda_1,\lambda_2) = \phi(\lambda_1,u\lambda_2) = -\phi(u\lambda_2,\lambda_1) = -\phi(\lambda_2,u\lambda_1). \quad (13)$$

The properties of the trace map imply that there exists precisely one $\beta \in E$ such that

$$\Phi(u) = \mathrm{tr}_{E/\mathbb{Q}}(u\beta)\ \forall u \in E.$$

Let us put

$$\phi_E(\lambda_1,\lambda_2) := \beta.$$

It follows from (13) that $\phi_E \in \mathrm{Hom}_E\left(\wedge_E^2\Lambda_{\mathbb{Q}}, E\right)$. In addition,

$$\mathrm{tr}_{E/\mathbb{Q}}(\phi_E(\lambda_1,\lambda_2)) = \mathrm{tr}_{E/\mathbb{Q}}(\beta) = \mathrm{tr}_{E/\mathbb{Q}}(1\cdot\beta) = \Phi(1) = \phi(\lambda_1,\lambda_2),$$

which proves that $\phi \mapsto \phi_E$ is indeed the inverse map, in light of (11).

Clearly, H_E is a rational Hodge substructure of $\mathrm{H}^2(T,\mathbb{Q})$.

By 2-simplicity of T, the $\mathbb{C}$-dimension of the $(2, 0)$-component $H_E^{(2,0)}$ of H_E is either 0 or $g(g-1)/2$. Let us express this dimension explicitly in terms of g and $[E : \mathbb{Q}]$.

In order to do that, let us consider the set Σ_E of all field embeddings $\sigma : E \hookrightarrow \mathbb{C}$, which consists of $[E : \mathbb{Q}]$-elements. We have

$$E_{\mathbb{C}} := E \otimes_{\mathbb{Q}} \mathbb{C} = \bigoplus_{\sigma\in\Sigma_E} \mathbb{C}_\sigma \ \text{ where } \mathbb{C}_\sigma = E \otimes_{E,\sigma} \mathbb{C} = \mathbb{C}, \tag{14}$$

which gives us a splitting of $E_{\mathbb{C}}$-modules

$$V_{\mathbb{C}} = \oplus_{\sigma\in\Sigma_E} V_\sigma = \bigoplus_{\sigma\in\Sigma_E} \left(\mathrm{H}_{-1,0}(T)_\sigma \oplus \mathrm{H}_{0,-1}(T)_\sigma\right) \tag{15}$$

where for all $\sigma \in \Sigma_E$ we define

$$\mathrm{H}_{-1,0}(T)_\sigma := \mathbb{C}_\sigma \mathrm{H}_{-1,0}(T) = \{x \in \mathrm{H}_{-1,0}(T) \mid u_{\mathbb{C}}x = \sigma(u)x \ \forall u \in E\} \subset \mathrm{H}_{-1,0}(T);$$

$$n_\sigma := \dim_{\mathbb{C}}(\mathrm{H}_{-1,0}(T)_\sigma);$$

$$\mathrm{H}_{0,-1}(T)_\sigma := \mathbb{C}_\sigma \mathrm{H}_{0,-1}(T) = \{x \in \mathrm{H}_{0,-1}(T) \mid u_{\mathbb{C}}x = \sigma(u)x \ \forall u \in E\} \subset \mathrm{H}_{0,-1}(T);$$

$$m_\sigma := \dim_{\mathbb{C}}(\mathrm{H}_{0,-1}(T)_\sigma);$$

$$V_\sigma = \mathbb{C}_\sigma = \mathbb{C}_\sigma V_{\mathbb{C}} = \{x \in V_{\mathbb{C}} \mid u_{\mathbb{C}}x = \sigma(u)x \ \forall u \in E\} = \mathrm{H}_{-1,0}(T)_\sigma \oplus \mathrm{H}_{0,-1}(T)_\sigma.$$

Since $\mathrm{H}_{-1,0}(T) \oplus \mathrm{H}_{0,-1}(T) = V_{\mathbb{C}}$ is a free $E_{\mathbb{C}}$-module of rank d_E, its direct summand V_σ is a vector space of dimension d_E over $\mathbb{C}_\sigma = \mathbb{C}$ and therefore

$$n_\sigma + m_\sigma = d_E \tag{16}$$

for all σ. Since $\mathrm{H}_{-1,0}(T)$ and $\mathrm{H}_{0,-1}(T)$ are mutually complex-conjugate subspaces of $V_{\mathbb{C}}$, it follows from (6) that

$$m_\sigma = n_{\bar\sigma} \ \text{ where } \bar\sigma : E \hookrightarrow \mathbb{C}, \ u \mapsto \overline{\sigma(u)}$$

is the *complex-conjugate* of σ. Therefore, in light of (16),

$$n_\sigma + n_{\bar\sigma} = d_E \ \forall \sigma. \tag{17}$$

We have

$$\sum_{\sigma\in\Sigma_E} n_\sigma = \sum_{\sigma\in\Sigma_E} \dim_{\mathbb{C}}(\mathrm{H}_{-1,0}(T)_\sigma) = \dim_{\mathbb{C}}(\mathrm{H}_{-1,0}(T)) = g. \tag{18}$$

Let us consider the complexification of H_E

$$H_{E,\mathbb{C}} := H_E \otimes_{\mathbb{Q}} \mathbb{C} \subset \mathrm{Hom}_{\mathbb{Q}}\left(\wedge^2 \Lambda_{\mathbb{Q}}, \mathbb{Q}\right) \otimes_{\mathbb{Q}} \mathbb{C} =$$

$$\mathrm{Hom}_{\mathbb{C}}\left(\wedge^2_{\mathbb{C}}(\Lambda_{\mathbb{Q}} \otimes_{\mathbb{Q}} \mathbb{C}), \mathbb{C}\right) = \mathrm{Hom}_{\mathbb{C}}\left(\wedge^2 V_{\mathbb{C}}, \mathbb{C}\right).$$

In light of (12),

$$H_{E,\mathbb{C}} = \{\phi \in \mathrm{Hom}_{\mathbb{C}}\left(\wedge^2_{\mathbb{C}} V_{\mathbb{C}}, \mathbb{C}\right) \mid \phi(u_{\mathbb{C}}x, y) = \phi(x, u_{\mathbb{C}}y)\ \forall u \in E, ; x, y \in V_{\mathbb{C}}\} \tag{19}$$
$$= \{\phi \in \mathrm{Hom}_{\mathbb{C}}\left(\wedge^2_{\mathbb{C}} V_{\mathbb{C}}, \mathbb{C}\right) \mid \phi(u_{\mathbb{C}}x, y) = \phi(x, u_{\mathbb{C}}y)\ \forall u \in E_{\mathbb{C}}; x, y \in V_{\mathbb{C}}\}.$$

In particular, if $\sigma, \tau \in \Sigma_E$ are *distinct* field embeddings then for all $\phi \in H_{E,\mathbb{C}}$

$$\phi(V_\sigma, V_\tau) = \phi(V_\tau, V_\sigma) = \{0\}.$$

This implies that

$$H_{E,\mathbb{C}} = \oplus_{\sigma\in\Sigma_E} \mathrm{Hom}_{\mathbb{C}}\left(\wedge^2_{\mathbb{C}} V_\sigma, \mathbb{C}\right)$$
$$= \bigoplus_{\sigma\in\Sigma_E} \mathrm{Hom}_{\mathbb{C}}(\wedge^2_{\mathbb{C}}\left(\mathrm{H}_{-1,0}(T)_\sigma \bigoplus \mathrm{H}_{0,-1}(T)_\sigma\right), \mathbb{C}). \tag{20}$$

In light of (9), the (2, 0)-Hodge component of $H_{E,\mathbb{C}}$ is

$$H_E^{(2,0)} = \oplus_{\sigma\in\Sigma_E} \mathrm{Hom}_{\mathbb{C}}\left(\wedge^2_{\mathbb{C}} \mathrm{H}_{-1,0}(T)_\sigma, \mathbb{C}\right) \text{ and } \dim_{\mathbb{C}}(H_E^{(2,0)})$$
$$= \sum_{\sigma\in\Sigma_E} \frac{n_\sigma(n_\sigma - 1)}{2}. \tag{21}$$

This implies that $\dim_{\mathbb{C}}(H_E^{(2,0)}) = 0$ if and only if all n_σ are in $\{0, 1\}$. If this is the case then, in light of (17), $d_E \in \{1, 2\}$, i.e., $[E : \mathbb{Q}] = 2g$ or g.

On the other hand, it follows from (18) combined with the second formula in (21) that $\dim_{\mathbb{C}}(H_E^{(2,0)}) = g(g-1)/2$ if and only if there is precisely one σ with $n_\sigma = g$ (and all the other multiplicities n_τ are 0). This implies that either $d_E = 2g$ and $E = \mathbb{Q}$, or $d_E = g$ and E is an imaginary quadratic field with the pair of the field embeddings

$$\sigma, \bar{\sigma} : E \hookrightarrow \mathbb{C}$$

such that

$$n_\sigma = g,\ n_{\bar{\sigma}} = 0.$$

It is therefore enough to rule out the case $d_E = g$. By way of contradiction, assume that $d_E = g$. Then E is an imaginary quadratic field; in addition, if

$$u \in E \subset \mathrm{End}_{\mathbb{Q}}(\Lambda_{\mathbb{Q}}) \subset \mathrm{End}_{\mathbb{R}}(V)$$

then $u_{\mathbb{C}}$ acts on $\mathrm{H}_{-1,0}(T)$ as multiplication by $\sigma(u) \in \mathbb{C}$. In light of (6), $u_{\mathbb{C}}$ acts on the complex-conjugate subspace $\mathrm{H}_{0,-1}(T)$ as multiplication by $\overline{\sigma(u)} = \bar{\sigma}(u) \in \mathbb{C}$. Since E is an imaginary quadratic field, there are a positive integer D and $\alpha \in E$ such that $\alpha^2 = -D$ and $E = \mathbb{Q}(\alpha)$. It follows that $\sigma(\alpha) = \pm\mathbf{i}\sqrt{D}$. Replacing if necessary α by $-\alpha$, we may and will assume that

$$\sigma(\alpha) = \mathbf{i}\sqrt{D}$$

and therefore $\alpha_{\mathbb{C}}$ acts on $\mathrm{H}_{-1,0}(T)$ as multiplication by $\mathbf{i}\sqrt{D}$. Hence, $\alpha_{\mathbb{C}}$ acts on $\mathrm{H}_{0,-1}(T)$ as multiplication by $\overline{\mathbf{i}\sqrt{D}} = -\mathbf{i}\sqrt{D}$. Since

$$V_{\mathbb{C}} = \mathrm{H}_{-1,0}(T) \oplus \mathrm{H}_{0,-1}(T),$$

we get $\alpha_{\mathbb{C}} = \sqrt{D}J_C$ and therefore

$$\alpha = \sqrt{D}J.$$

This implies that the centralizer $\mathrm{End}^0(T)$ of J in $\mathrm{End}_{\mathbb{Q}}(\Lambda_{\mathbb{Q}})$ coincides with the centralizer of α in $\mathrm{End}_{\mathbb{Q}}(\Lambda_{\mathbb{Q}})$, which, in turn, coincides with the centralizer $\mathrm{End}_E(\Lambda_{\mathbb{Q}})$ of E in $\mathrm{End}_{\mathbb{Q}}(\Lambda_{\mathbb{Q}})$, i.e.,

$$\mathrm{End}^0(T) = \mathrm{End}_E(\Lambda_{\mathbb{Q}}) \cong \mathrm{Mat}_{d_E}(E).$$

This is the algebra of matrices, which is not a division algebra, because $d_E = g > 1$. This contradicts the simplicity of T. The obtained contradiction rules out the case $d_E = g$. This ends the proof of (ii).

In order to prove (iii), recall that $\mathrm{End}^0(T)$ is a division algebra over $\mathbb{Q}$, thanks to the simplicity of T [21]. Hence $\Lambda_{\mathbb{Q}}$ is a free $\mathrm{End}^0(T)$-module of finite positive rank and therefore

$$\dim_{\mathbb{Q}}(\mathrm{End}^0(T)) \mid 2g, \tag{22}$$

because $2g = \dim_{\mathbb{Q}}(\Lambda_{\mathbb{Q}})$. We will apply several times the already proven assertion (ii) to various subfields of $\mathrm{End}^0(T)$.

Suppose that $\mathrm{End}^0(T)$ is *not* a field and let $\mathcal{Z}$ be its center. Then $\mathcal{Z}$ is a number field and there is an integer $d > 1$ such that $\dim_{\mathcal{Z}}(\mathrm{End}^0(T)) = d^2$ and therefore

$$\dim_{\mathbb{Q}}(\mathrm{End}^0(T)) = d^2 \cdot [\mathcal{Z} : \mathbb{Q}]$$

divides $2g$, thanks to (22). Since $\mathcal{Z}$ is a subfield of $\mathrm{End}^0(T)$, the degree $[\mathcal{Z}:\mathbb{Q}]$ is either 1 or g or $2g$. If $[\mathcal{Z}:\mathbb{Q}]>1$ then $2g$ is divisible by

$$d^2\cdot[\mathcal{Z}:\mathbb{Q}]\ge 2^2g=4g,$$

which is nonsense. Hence, $[\mathcal{Z}:\mathbb{Q}]=1$, i.e., $\mathcal{Z}=\mathbb{Q}$ and $\mathrm{End}^0(T)$ is a central division $\mathbb{Q}$-algebra of dimension d^2 with $d^2|2g$. Then every maximal subfield E of the central division $\mathbb{Q}$-algebra $\mathrm{End}^0(T)$ has degree d over $\mathbb{Q}$ [22, Sect. 13.1, Cor. b]. By the already proven assertion (ii), $d\in\{1,g,2g\}$. Since $d>1$, we obtain that either $d=g$ and $g^2\mid 2g$ or $d=2g$ and $(2g)^2\mid 2g$. This implies that $d=g$ and $g=1$ or 2. Since $g\ge 3$, we get a contradiction, which implies that $\mathrm{End}^0(T)$ is a field.

It follows from the already proven assertion (ii) that the degree $\dim_{\mathbb{Q}}(\mathrm{End}^0(T))$ of the number field $\mathrm{End}^0(T)$ is either 1 or g or $2g$.

Assertion (iv) is obvious and was included just for the sake of completeness.

In order to handle the structure of $\mathrm{Aut}(T)$, let us check first that the only roots of unity in $\mathrm{End}^0(T)$ are 1 and -1. If this is not the case then the field $\mathrm{End}^0(T)$ contains either $\sqrt{-1}$ or a primitive pth root of unity ζ where p is a certain odd prime. In the former case $\mathrm{End}^0(T)$ contains the quadratic field $\mathbb{Q}(\sqrt{-1})$, which contradicts (ii). In the latter case $\mathrm{End}^0(T)$ contains either the quadratic field $\mathbb{Q}(\sqrt{-p})$ or the quadratic field $\mathbb{Q}(\sqrt{p})$: each of these outcomes contradicts (ii) as well.

Now recall that $\mathrm{End}(T)$ is an order in the number field $E=\mathrm{End}^0(T)$ and $\mathrm{Aut}(T)=\mathrm{End}(T)^*$ is its group of units. By the Theorem of Dirichlet about units [5, Ch. II, Sect. 4, Th. 5], the group of units is

$$\mathrm{Aut}(T)\cong\mathbb{Z}^d\times\{\pm1\}\ \text{ with }\ d=r_E+s_E-1 \tag{23}$$

where r_E is the number of real field embeddings $E\hookrightarrow\mathbb{R}$ and

$$r_E+2s_E=[E:\mathbb{Q}],\ \ \text{i.e.,}\ s_E=\frac{[E:\mathbb{Q}]-r_E}{2}. \tag{24}$$

Let us prove (v). Assume that the number field $E=\mathrm{End}^0(T)$ has degree $2g$. A dimension argument implies that $\Lambda_{\mathbb{Q}}$ is a 1-dimensional E-vector space and $V=\Lambda_{\mathbb{Q}}\otimes_{\mathbb{Q}}\mathbb{R}$ is a free $E_{\mathbb{R}}=E\otimes_{\mathbb{Q}}\mathbb{R}$-module of rank 1. Hence $E_{\mathbb{R}}$ coincides with its own centralizer $\mathrm{End}_{E_{\mathbb{R}}}(V)$ in $\mathrm{End}_{\mathbb{R}}(V)$. Since J commutes with $\mathrm{End}^0(T)=E$, it also commutes with $E_{\mathbb{R}}$ and therefore

$$J\in\mathrm{End}_{E_{\mathbb{R}}}(V)=E_{\mathbb{R}}.$$

Recall that the $\mathbb{R}$-algebra $E_{\mathbb{R}}$ is isomorphic to a product of copies of $\mathbb{R}$ and $\mathbb{C}$. Since $J^2=-1$, only copies of $\mathbb{C}$ appear in $E_{\mathbb{R}}$, i.e., E is purely imaginary, which means that $r_E=0$ and therefore $2g=[E:\mathbb{Q}]=2s_E$. This proves the first assertion of (v); the second one follows readily from (23) combined with (24).

In order to prove the last assertion, assume that E contains two distinct proper subfields E_1 and E_2, none of which coincides with $\mathbb{Q}$. Clearly,

$$[E_1 : \mathbb{Q}] = g = [E_2 : \mathbb{Q}],$$

which means that both field extensions E/E_1 and E/E_2 are quadratic. This implies that the (finite) automorphism group $G := \mathrm{Aut}(E/\mathbb{Q})$ of the field extension $E/\mathbb{Q}$ contains two distinct elements t_1 and t_2 of order 2 such that

$$E_1 = \{u \in E \mid t_1(u) = u\}, \quad E_2 = \{u \in E \mid t_2(u) = u\},$$

It follows that G is a group of order M where M is an even integer that is strictly greater than 2. Then the subfield $F := E^G$ of G-invariants is a proper subfield of E and its degree

$$[F : \mathbb{Q}] = \frac{[E : \mathbb{Q}]}{M} < \frac{2g}{2} = g.$$

It follows from (ii) that $F = \mathbb{Q}$ and therefore $M = [E : \mathbb{Q}] = 2g$.

If g is *not* a power of 2 then there is an odd prime p dividing g and therefore dividing M. It follows that G contains an element t of order p. Therefore the subfield E^t of t-invariants is a proper subfield of E and its degree $[E^t : \mathbb{Q}]$ is $2g/p < g$. By (ii), $E^t = \mathbb{Q}$ and therefore $2g = [E : \mathbb{Q}] = p$, which is wrong, since p is odd. Hence g is a power of 2 and therefore G is a finite 2-group. It follows that G has a normal subgroup H of index 2. Then the subfield $F := E^H$ of H-invariants is a proper subfield of E and its degree $[F : \mathbb{Q}]$ equals the index $[G : H] = 2$. This also contradicts (ii), which ends the proof of the last assertion of (v).

Let us prove (vi). Assume that $[E : \mathbb{Q}] = g$. Then the assertion about $\mathrm{Aut}(T)$ follows readily from (23) combined with (24). If $F \neq \mathbb{Q}$ is a proper subfield of E then

$$1 = [\mathbb{Q} : \mathbb{Q}] < [F : \mathbb{Q}] < [E : \mathbb{Q}] = g$$

and therefore $1 < [F : \mathbb{Q}] < g$, which contradicts (ii) applied to F instead of E. So, such an F does *not* exist, i.e., E is *primitive*.

Assume now that T is a complex abelian variety. By Albert's classification [19], $E = \mathrm{End}^0(T)$ is either a totally real number field or a CM field. If E is a CM field then it contains a subfield E_0 of degree $[E : \mathbb{Q}]/2 = g/2$. Since E_0 is a subfield of $\mathrm{End}^0(T)$ and $1 < g/2 < g$ (recall that $g \geq 3$), the existence of E_0 contradicts the already proven assertion (ii). This proves that E is a totally real number field, i.e., $s = 0, r = g$. Now the assertion about $\mathrm{Aut}(T)$ follows from (23). □

4 Number Fields and Transitive Permutation Groups

All the results of this section are standard and pretty well known except, may be, the notion of almost double transitivity.

Definition 4.1 Let $\mathcal{T}$ be a set that consists of at least three elements. We write $\mathrm{Perm}(\mathcal{T})$ for the group of all permutations of $\mathcal{T}$. Let G be a group that acts on $\mathcal{T}$, i.e., we are given a group homomorphism

$$G \to \mathrm{Perm}(\mathcal{T}),$$

whose image we denote by $\tilde{G}$, which is a subgroup of $\mathrm{Perm}(\mathcal{T})$. We say that a *transitive* action of G on $\mathcal{T}$ is *almost doubly transitive* if the action of G on the set of all two-element subsets of $\mathcal{T}$ is transitive.

Remarks 4.2

1. Every doubly transitive action is almost doubly transitive.
2. Every almost doubly transitive action of G on $\mathcal{T}$ is primitive, i.e., the stabilizer of a point is a maximal subgroup. Indeed, suppose the action is *not* primitive, i.e., that $\mathcal{T}$ partitions into a disjoint union of r sets $\mathcal{T}_1, \dots, \mathcal{T}_r$ such that $r \geq 2$, each $\mathcal{T}_i$ consists of $m \geq 2$ elements, and G permutes $\mathcal{T}_i$s. Let A be a 2-element subset of T_1. Pick $b_1 \in \mathcal{T}_1$ and $b_2 \in \mathcal{T}_2$, and consider a 2-element subset $B = \{b_1, b_2\}$ of $\mathcal{T}$. Clearly, no $s \in G$ sends A to B, i.e., the action is not almost doubly transitive.
3. If $\mathcal{T}$ consists of three elements then every transitive action on $\mathcal{T}$ of any group G is almost doubly transitive.
4. If $\mathcal{T}$ is a finite set then the group $\tilde{G}$ is a finite group of permutations of $\mathcal{T}$ that is primitive (resp. almost doubly transitive, resp. doubly transitive) if and only if G is primitive (resp. almost doubly transitive, resp. doubly transitive).
5. Suppose that $\mathcal{T}$ is a finite set that consists of $n \geq 3$ elements and G is a group that acts faithfully and almost doubly transitively on $\mathcal{T}$. Let N be the order of $\tilde{G}$.

   Then N is divisible by $n(n-1)/2$. If N is even then $\tilde{G}$ contains an element $\tilde{\sigma}$ of order 2 and therefore there are two distinct elements $s_1, s_2 \in \mathcal{T}$ such that

   $$\tilde{\sigma}(s_1) = s_2, \tilde{\sigma}(s_2) = s_1.$$

   It follows that the action of $\tilde{G}$ on $\mathcal{T}$ is doubly transitive and therefore the action of G on $\mathcal{T}$ is also doubly transitive. This implies that if either $4|n$ or $n \equiv 1 \mod 4$ then the action of G on $\mathcal{T}$ is doubly transitive, because in these cases $n(n-1)/2$ is even.
6. Let $n = q$ be a prime power that is congruent to 3 modulo 4. Let $\mathbb{F}_q$ be a q-element finite field and $\mathbb{F}_q^*$ the multiplicative group of nonzero elements of $\mathbb{F}_q$. Then $\mathbb{F}_q^*$ splits into a direct product $\mathbb{F}_q^* = H \times \{\pm 1\}$ where H is the cyclic subgroup in $\mathbb{F}_q^*$ of *odd* order $(q-1)/2$. Let us put $\mathcal{T} = \mathbb{F}_q$ and let G be the group of affine transformations of $\mathbb{F}_q$

$$x \mapsto ax + b, \ a \in H \subset \mathbb{F}_q^*, b \in \mathbb{F}_q.$$

Then the action of G on $\mathbb{F}_q$ is almost doubly transitive but *not* doubly transitive.

Let $\bar{\mathbb{Q}}$ be the algebraic closure of $\mathbb{Q}$ in $\mathbb{C}$ and

$$\mathrm{Gal}(\mathbb{Q}) = \mathrm{Gal}(\bar{\mathbb{Q}}/\mathbb{Q}) = \mathrm{Aut}(\bar{\mathbb{Q}}/\mathbb{Q})$$

the absolute Galois group of $\mathbb{Q}$. Let us consider the humongous group $\mathrm{Aut}(\mathbb{C})$ of all automorphisms of the field $\mathbb{C}$. Obviously, the subfield $\bar{\mathbb{Q}}$ is $\mathrm{Aut}(\mathbb{C})$-invariant, which gives rise to the (restriction) homomorphism of groups

$$\mathrm{Aut}(\mathbb{C}) \twoheadrightarrow \mathrm{Gal}(\mathbb{Q}), \quad s \mapsto \{\alpha \mapsto s(\alpha)\} \ \forall s \in \mathrm{Aut}(\mathbb{C}), \alpha \in \bar{\mathbb{Q}} \tag{25}$$

which is *surjective.*

Let E be a number field of degree $n = [E : \mathbb{Q}]$. We write Σ_E for the n-element set of all field embeddings $\sigma : E \hookrightarrow \mathbb{C}$. For each $\sigma \in \Sigma_E$ the image $\sigma(E)$ lies in $\bar{\mathbb{Q}}$. If t is an element of $\mathrm{Aut}(\mathbb{C})$ (or of $\mathrm{Gal}(\mathbb{Q})$) then the composition

$$t \circ \sigma : E \overset{\sigma}{\hookrightarrow} \bar{\mathbb{Q}} \overset{t}{\to} \bar{\mathbb{Q}} \subset \mathbb{C}$$

also lies in Σ_E. Then the map

$$\mathrm{Aut}(\mathbb{C}) \times \Sigma_E \to \Sigma_E, \quad (t, \sigma) \mapsto t \circ \sigma \tag{26}$$

is a *transitive group action* of $\mathrm{Aut}(\mathbb{C})$ on Σ_E, which factors through $\mathrm{Gal}(\mathbb{Q})$ via (25). Notice that the corresponding finite subgroup

$$\widetilde{\mathrm{Aut}(\mathbb{C})} = \widetilde{\mathrm{Gal}(\mathbb{Q})} \subset \mathrm{Perm}(\Sigma_E)$$

is isomorphic to the Galois group $\mathrm{Gal}(\tilde{E}/\mathbb{Q})$ where $\tilde{E}$ is a *normal closure* of E. This action of $\mathrm{Aut}(\mathbb{C})$ is *primitive* (i.e., the stabilizer of a point is a maximal subgroup) if and only if E is a primitive number field. Similarly, the map

$$\mathrm{Gal}(\mathbb{Q}) \times \Sigma_E \to \Sigma_E, \quad (t, \sigma) \mapsto t \circ \sigma \tag{27}$$

is a *transitive group action* of $\mathrm{Gal}(\mathbb{Q})$ on Σ_E, which is primitive if and only if E is a primitive number field.

Definition 4.3 We say that E is a *doubly transitive* (respectfully *almost doubly transitive*) number field if the action (26) (or equivalently the action (27)) is doubly transitive (respectfully almost doubly transitive).

Remark 4.4 Clearly, if E and F are isomorphic number fields then E is primitive (resp. doubly transitive) (resp. almost doubly transitive) if and only if F is primitive (resp. doubly transitive) (resp. almost doubly transitive).

Example 4.5

(i) Let $f(x) \in \mathbb{Q}[x]$ be an irreducible polynomial of degree $n \geq 2$ over $\mathbb{Q}$ and $E_f = \mathbb{Q}[x]/f(x)\mathbb{Q}[x]$ the corresponding number field of degree n. We write $\mathfrak{R}_f$ for the n-element set of roots of $f(x)$ in $\bar{\mathbb{Q}}$ and $\mathbb{Q}(\mathfrak{R}_f)$ for the subfield of $\bar{\mathbb{Q}}$ generated by $\mathfrak{R}_f$. By definition, $\mathbb{Q}(\mathfrak{R}_f)$ is a splitting field of $f(x)$ that is a finite Galois extension of $\mathbb{Q}$. We write $\mathrm{Gal}(f)$ for the Galois group $\mathrm{Gal}(\mathbb{Q}(\mathfrak{R}_f)/\mathbb{Q})$ of the field extension $\mathbb{Q}(\mathfrak{R}_f)/\mathbb{Q}$. It is well known that $\mathrm{Gal}(\mathbb{Q})$ acts transitively (through $\mathrm{Gal}(f)$) on $\mathfrak{R}_f$. There is a $\mathrm{Gal}(\mathbb{Q})$-equivariant bijection between Σ_{E_f} and $\mathfrak{R}_f$ that is defined as follows. To each $\alpha \in \mathfrak{R}_f$ corresponds the field embedding

$$\sigma_\alpha : E_f = \mathbb{Q}[x]/f(x)\mathbb{Q}[x] \hookrightarrow \bar{\mathbb{Q}} \subset \mathbb{C},\ h(x) + f(x)\mathbb{Q}[x] \mapsto h(\alpha)$$

(in particular, the coset of x goes to α). This implies that the field E_f is doubly transitive (respectfully almost doubly transitive) if and only if the action of $\mathrm{Gal}(f)$ on $\mathfrak{R}_f$ is doubly transitive (respectfully almost doubly transitive). The similar characterization of primitive number fields is well known:
the field E_f is primitive if and only if the action of $\mathrm{Gal}(f)$ *on $\mathfrak{R}_f$ is primitive.* On the other hand, obviously, r_{E_f} equals the number of *real roots* of $f(x)$ and $2s_{E_f} = n - r_{E_f}$.

(ii) Conversely, let F be a number field of degree n and $z \in F$ be a primitive element of F, i.e., the smallest subfield $\mathbb{Q}(z)$ of F that contains z coincides with F (such an element always exists). Let $f(x) \in \mathbb{Q}[x]$ be the minimal polynomial of z, i.e., $f(x)$ is irreducible over $\mathbb{Q}$ and $f(z) = 0$; in addition, $\deg(f) = n$. Then there is a field isomorphism $E_f \cong F$ such that the coset $x + f(x)\mathbb{Q}[x] \in E_f$ goes to $z \in F$. Therefore the number field F is (almost) doubly transitive if and only if $\mathrm{Gal}(f)$ acts (almost) doubly transitively on $\mathfrak{R}_f$.

Theorem 4.6 *Let $n \geq 3$ be an integer. Let $\mathbf{r}, \mathbf{s}$ be nonnegative integers such that*

$$\mathbf{r} + 2\mathbf{s} = n. \tag{28}$$

Then there exists a number field E of degree n that enjoys the following properties.

(i) $r_E = \mathbf{r}, \quad s_E = \mathbf{s}$.
(ii) *E is doubly transitive.*

Proof We will use *weak approximation* in $\mathbb{Q}$, approximating several polynomials of degree n in $\mathbb{Q}[x]$ with respect to several metrics in $\mathbb{Q}$.

First, fix a degree n monic polynomial $h_\infty(x) \in \mathbb{Z}[x]$ that has precisely $\mathbf{r}$ distinct real roots and $\mathbf{s}$ distinct pairs of non-real complex-conjugate roots. (E.g., one may take

$$h_\infty(x) = \left(\prod_{i=1}^{\mathbf{r}}(x - i)\right) \cdot \left(\prod_{j=1}^{\mathbf{s}}(x^2 + j^2)\right) \in \mathbb{Z}[x] \subset \mathbb{Q}[x].)$$

Second, take any prime p and choose a monic (p-adic Eisenstein) polynomial $h_p(x) \in \mathbb{Z}[x]$, all whose coefficients (except the leading one) are divisible by p while the constant term is *not* divisible by p^2. (E.g., one may take

$$h_p(x) = x^n - p \in \mathbb{Z}[x] \subset \mathbb{Q}[x].)$$

Third, take any prime $\ell \neq p$ and choose a monic *irreducible* polynomial

$$\tilde{u}_\ell(x) \in \mathbb{F}_\ell[x]$$

over $\mathbb{F}_\ell$ of degree $n-1$. (Such a polynomial always exists for any given prime ℓ and positive integer $n-1$.) Since $n-1 \geq 3-1 > 1$, the irreducibility of $\tilde{u}_\ell(x)$ implies that

$$\tilde{u}_\ell(0) \neq 0.$$

Let $u_\ell(x) \in \mathbb{Z}[x]$ be any monic degree $(n-1)$ polynomial with integer coefficients, whose reduction modulo ℓ coincides with $\tilde{u}_\ell(x)$. Let us put

$$h_\ell(x) := x \cdot u_\ell(x) \in \mathbb{Z}[x] \subset \mathbb{Q}[x].$$

It follows from a weak approximation theorem [2, Th. 1] that there is a monic degree n polynomial $f(x) \in \mathbb{Q}[x]$ that enjoys the following properties.

(a) $f(x)$ is so close to $h_\infty(x)$ in the archimedean topology that it also has precisely $\mathbf{r}$ distinct real roots and $\mathbf{s}$ distinct pairs of non-real complex-conjugate roots.

(b) $f(x) - h_p(x) \in p^2 \cdot \mathbb{Z}_p[x]$. By Eisenstein's criterion, $f(x)$ is irreducible over the field $\mathbb{Q}_p$ of p-adic numbers and therefore is irreducible over $\mathbb{Q}$.

(c) $f(x) - h_\ell(x) \in \ell \cdot \mathbb{Z}_\ell[x]$. This implies that

$$f(x) \in \mathbb{Z}_\ell[x], \quad f(x) \bmod \ell = x \cdot \tilde{u}_\ell(x) \subset \mathbb{F}_\ell[x].$$

By Hensel's Lemma, there are

$$\alpha \in \ell\mathbb{Z}_\ell \subset \mathbb{Z}_\ell$$

and a monic degree $(n-1)$ polynomial $v(x) \in \mathbb{Z}_\ell[x]$ such that

$$f(x) = (x - \alpha) \cdot v(x) \in \mathbb{Z}_\ell[x], \quad v(x) \bmod \ell = \tilde{u}_\ell(x) \in \mathbb{F}_\ell[x]. \tag{29}$$

By [31, Sect. 66], the irreducibility of $\tilde{u}_\ell(x)$ combined with (29) implies that $\mathrm{Gal}(f)$, viewed as the certain transitive subgroup of $\mathrm{Perm}(\mathfrak{R}_f)$, contains a permutation s that is a cycle of length $n-1$. In particular, if $\alpha \in \mathfrak{R}_f$ is the fixed point of s then the cyclic subgroup $< s >$ of $\mathrm{Gal}(f)$ generated by s acts transitively on

$\mathfrak{R}_f \setminus \{\alpha\}$. Now the transitivity of $\mathrm{Gal}(f)$ implies its double transitivity. It remains only to put $E := E_f$ and apply the results of Example 4.5(i). □

5 Hodge Groups

Recall that $\Lambda_{\mathbb{R}} = V$ carries the natural structure of a complex vector space. This gives rise to the injective homomorphism of real Lie groups

$$h : \mathbb{C}^* \hookrightarrow \mathrm{Aut}_{\mathbb{R}}(\Lambda_{\mathbb{R}})$$

where $h(z)$ is multiplication by a nonzero complex number z in $\Lambda_{\mathbb{R}} = V$. Let $\mathbb{S}^1 \subset \mathbb{C}^*$ be the subgroup of all complex numbers z with $|z| = 1$. Clearly, $h(\mathbb{S}^1)$ is a one-dimensional closed connected real Lie subgroup of $\mathrm{Aut}_{\mathbb{R}}(\Lambda_{\mathbb{R}})$; in addition, the Lie algebra of $h(\mathbb{S}^1)$ is

$$\mathbb{R} \cdot J \subset \mathrm{End}_{\mathbb{R}}(\Lambda_{\mathbb{R}}).$$

Actually, $h(\mathbb{S}^1)$ lies in the special linear group $\mathrm{SL}(\Lambda_{\mathbb{R}})$, while $\mathbb{R} \cdot J$ lies in the Lie algebra $\mathfrak{sl}(\Lambda_{\mathbb{R}})$ of traceless operators in $\Lambda_{\mathbb{R}}$.

By definition [18, 24] (see also [35]), the Hodge group $\mathrm{Hdg}(T)$ of the rational Hodge structure $\mathrm{H}_1(T, \mathbb{Q}) = \Lambda_{\mathbb{Q}}$ is the smallest algebraic $\mathbb{Q}$-subgroup G of $\mathrm{GL}(\Lambda_{\mathbb{Q}})$, whose group of real points

$$G(\mathbb{R}) \subset \mathrm{Aut}_{\mathbb{R}}(\Lambda_{\mathbb{R}})$$

contains $h(\mathbb{S}^1)$. One may easily check that $\mathrm{Hdg}(T)$ enjoys the following properties that we will freely use throughout the text.

(i) $\mathrm{Hdg}(T)$ is a *connected* algebraic $\mathbb{Q}$-group that is a subgroup of the special linear group $\mathrm{SL}(\Lambda_{\mathbb{Q}})$.
(ii) The centralizer of $\mathrm{Hdg}(T)$ in $\mathrm{End}_{\mathbb{Q}}(\Lambda_{\mathbb{Q}})$ coincides with $\mathrm{End}^0(T)$.
(iii) A $\mathbb{Q}$-vector subspace $H_{\mathbb{Q}}$ of $\Lambda_{\mathbb{Q}}$ is $\mathrm{Hdg}(T)$-invariant if and only if it is a rational Hodge substructure of $\Lambda_{\mathbb{Q}}$.
(iv) The subspace of $\mathrm{Hdg}(T)$-invariants

$$\mathrm{H}^2(T, \mathbb{Q})^{\mathrm{Hdg}(T)} \subset \mathrm{H}^2(T, \mathbb{Q}) = \mathrm{Hom}_{\mathbb{Q}}(\wedge^2_{\mathbb{Q}} \Lambda_{\mathbb{Q}}, \mathbb{Q})$$

coincides with the subspace $\mathrm{H}^2(T, \mathbb{Q}) \cap H^{1,1}(T)$ of two-dimensional Hodge classes on T.
(v) The group $\mathrm{Hdg}(T)(\mathbb{Q})$ of $\mathbb{Q}$-points is *Zariski dense* in $\mathrm{Hdg}(T)$, because Hdg is connected and the field $\mathbb{Q}$ is infinite (see [4, Cor. 18.3]).
(vi) If $T \neq \{0\}$ then $\mathrm{Hdg}(T)$ is a positive-dimensional connected algebraic group.

Let us consider the $\mathbb{Q}$-Lie algebra hdg_T of the linear *algebraic* $\mathbb{Q}$-group $\mathrm{Hdg}(T)$. By definition, hdg_T is a linear *algebraic* Lie subalgebra of $\mathrm{End}_{\mathbb{Q}}(\Lambda_{\mathbb{Q}})$.

Remark 5.1 Clearly, hdg_T is the *smallest algebraic* Lie $\mathbb{Q}$-subalgebra $\mathfrak{g}$ of $\mathrm{End}_{\mathbb{Q}}(\Lambda_{\mathbb{Q}})$ such that

$$J \in \mathfrak{g} \otimes_{\mathbb{Q}} \mathbb{R}. \tag{30}$$

Properties (i) and (ii) above imply that

$$\mathrm{hdg}_T \subset \mathfrak{sl}(\Lambda_{\mathbb{Q}}) \subset \mathrm{End}_{\mathbb{Q}}(\Lambda_{\mathbb{Q}}) \tag{31}$$

and the centralizer of hdg_T in $\mathrm{End}_{\mathbb{Q}}(\Lambda_{\mathbb{Q}})$ is described as follows.

$$\mathrm{End}_{\mathrm{hdg}_T}(\Lambda_{\mathbb{Q}}) = \mathrm{End}^0(T). \tag{32}$$

Clearly,

$$J \in \mathrm{hdg}_{T,\mathbb{R}} := \mathrm{hdg}_T \otimes_{\mathbb{Q}} \mathbb{R} \subset \mathrm{End}_{\mathbb{Q}}(\Lambda_{\mathbb{Q}}) \otimes \mathbb{R} = \mathrm{End}_{\mathbb{R}}(\Lambda_{\mathbb{R}}).$$

Let us consider the *complexification*

$$\mathrm{hdg}_{T,\mathbb{C}} := \mathrm{hdg}_T \otimes_{\mathbb{Q}} \mathbb{C} \subset \mathrm{End}_{\mathbb{Q}}(\Lambda_{\mathbb{Q}}) \otimes \mathbb{C} = \mathrm{End}_{\mathbb{C}}(\Lambda_{\mathbb{C}})$$

where

$$\Lambda_{\mathbb{C}} = \Lambda_{\mathbb{Q}} \otimes_{\mathbb{Q}} \mathbb{C} = \Lambda_{\mathbb{R}} \otimes_{\mathbb{R}} \mathbb{C}.$$

We have

$$J \in \mathrm{hdg}_{T,\mathbb{R}} = \mathrm{hdg}_{T,\mathbb{R}} \otimes 1 \subset \mathrm{hdg}_{T,\mathbb{R}} \otimes_{\mathbb{R}} \mathbb{C} = \mathrm{hdg}_{T,\mathbb{C}} \subset \mathrm{End}_{\mathbb{C}}(\Lambda_{\mathbb{C}}).$$

(See (5) and Remark 1.2.) In what follows, we will write J instead of $J_{\mathbb{C}}$, slightly abusing notation.

The group $\mathrm{Aut}(\mathbb{C})$ acts naturally, *semi-linearly* and compatibly on $\Lambda_{\mathbb{C}}$, $\mathrm{End}_{\mathbb{C}}(\Lambda_{\mathbb{C}})$ and $\mathrm{hdg}_{T,\mathbb{C}}$.

The *minimality property* of $\mathrm{Hdg}(T)$ allows us to give the following "explicit" description of the complexification $\mathrm{hdg}_{T,\mathbb{C}}$ (compare with [35, Lemma 6.3.1]).

Theorem 5.2 *The complex Lie algebra* $\mathrm{hdg}_{T,\mathbb{C}}$ *coincides with the Lie subalgebra* $\mathfrak{u}$ *of* $\mathrm{End}_{\mathbb{C}}(\Lambda_{\mathbb{C}})$ *generated by all* $s(J)$ *where* s *run over the group* $\mathrm{Aut}(\mathbb{C})$. *In particular,* hdg_T *coincides with the smallest* $\mathbb{Q}$*-Lie subalgebra* $\mathfrak{g} \subset \mathrm{End}_{\mathbb{Q}}(\Lambda_{\mathbb{Q}})$ *such that*

$$\mathfrak{g} \otimes_{\mathbb{Q}} \mathbb{R} \subset \mathrm{End}_{\mathbb{Q}}(\Lambda_{\mathbb{Q}}) \otimes_{\mathbb{Q}} \mathbb{R} = \mathrm{End}_{\mathbb{R}}(\Lambda_{\mathbb{R}})$$

contains J.

Proof Clearly, $\mathfrak{u} \subset \mathrm{hdg}_{T,\mathbb{C}}$. Let us prove that $\mathfrak{u}$ is an *algebraic* complex Lie subalgebra of $\mathrm{End}_{\mathbb{C}}(\Lambda_{\mathbb{C}})$.

Recall that

$$J \in \mathrm{End}_{\mathbb{R}}(\Lambda_{\mathbb{R}}) \subset \mathrm{End}_{\mathbb{C}}(\Lambda_{\mathbb{C}});\ J^2 = -1. \tag{33}$$

Clearly, $J : \Lambda_{\mathbb{C}} \to \Lambda_{\mathbb{C}}$ is a semisimple $\mathbb{C}$-linear operator, whose spectrum consists of eigenvalues, $\mathbf{i}$ and $-\mathbf{i}$, because $J^2 = -1$. Similarly, for all $s \in \mathrm{Aut}(\mathbb{C})$ the $\mathbb{C}$-linear operator $s(J) : \Lambda_{\mathbb{C}} \to \Lambda_{\mathbb{C}}$ is also semisimple and its spectrum is also $\{\mathbf{i}, -\mathbf{i}\}$, because (in light of (33))

$$s(J)^2 = s(J^2) = s(-1) = -1. \tag{34}$$

It follows that the $\mathbb{Q}$-vector subspace $\mathbb{Q}(s(J))$ of $\mathbb{C}$ generated by the *spectrum* of $s(J)$ coincides with $\mathbb{Q} \cdot \mathbf{i}$; in particular, the $\mathbb{Q}$-vector (sub)space $\mathbb{Q}(s(J))$ is one-dimensional. This implies that each $\mathbb{C} \cdot s(J)$ is an algebraic $\mathbb{C}$-Lie subalgebra of $\mathrm{End}_{\mathbb{C}}(\Lambda_{\mathbb{C}})$, because each *replica* of $s(J)$ is a scalar multiple of $s(J)$. Thus, the linear $\mathbb{C}$-Lie algebra $\mathfrak{u}$ is generated by the algebraic Lie subalgebras $\mathbb{C} \cdot \sigma(f)$ and therefore is algebraic itself, thanks to [12, volume 2, Ch. 2, Sect. 14, Th. 14]. Clearly, $\mathfrak{u}$ is $\mathrm{Aut}(\mathbb{C})$-invariant. By Lemma 7.4 (see below), $\mathfrak{u}$ is defined over $\mathbb{Q}$, i.e., there is a $\mathbb{Q}$-vector subspace

$$\mathfrak{u}_0 \subset \mathrm{End}_{\mathbb{Q}}(\Lambda_{\mathbb{Q}})$$

such that

$$\mathfrak{u} = \mathfrak{u}_0 \otimes_{\mathbb{Q}} \mathbb{C}, \quad \mathfrak{u}_0 = \mathfrak{u} \cap \mathrm{End}_{\mathbb{Q}}(\Lambda_{\mathbb{Q}}).$$

Since $\mathfrak{u}$ is a $\mathbb{C}$-Lie subalgebra, the latter equality implies that $\mathfrak{u}_0$ is a $\mathbb{Q}$-Lie subalgebra of $\mathrm{End}_{\mathbb{Q}}(\Lambda_{\mathbb{Q}})$, and this Lie algebra is algebraic, in light of [12, volume 2, Ch. 2, Sect. 14, Prop. 4]. Clearly,

$$\mathfrak{u} = \left(\mathfrak{u}_0 \otimes_{\mathbb{Q}} \mathbb{R}\right) \oplus \mathbf{i} \cdot \left(\mathfrak{u}_0 \otimes_{\mathbb{Q}} \mathbb{R}\right)$$

as a real vector space. This implies that

$$\mathfrak{u}_0 \otimes_{\mathbb{Q}} \mathbb{R} = \mathfrak{u} \cap \mathrm{End}_{\mathbb{R}}(\Lambda_{\mathbb{R}}). \tag{35}$$

Let $\mathcal{U}$ be the connected algebraic $\mathbb{Q}$-subgroup of $\mathrm{GL}(\Lambda_{\mathbb{Q}})$, whose Lie algebra coincides with $\mathfrak{u}_0$. We need to prove that

$$\mathfrak{u}_0 = \mathrm{hdg}_T.$$

Clearly, $\mathfrak{u}_0 \subset \mathrm{hdg}_T$, because the complexification of $\mathfrak{u}_0$ lies in the complexification of hdg_T. We know that

$$J \in \mathfrak{u}_0 \otimes_{\mathbb{Q}} \mathbb{C} = \mathfrak{u}.$$

Since $J \in \mathrm{End}_{\mathbb{R}}(\Lambda_{\mathbb{R}})$, it follows from (35) that

$$J \in \mathfrak{u}_0 \otimes_{\mathbb{Q}} \mathbb{R}.$$

In light of Remark 5.1, $\mathfrak{u}_0 \supset \mathrm{hdg}_T$. This implies that $\mathfrak{u}_0 = \mathrm{hdg}_T$, which ends the proof. □

Corollary 5.3 *Let us put*

$$f_T := \frac{1}{\mathbf{i}} J \in \mathrm{End}_{\mathbb{C}}(\Lambda_{\mathbb{C}}).$$

Then

$$s(f_T)^2 = 1 \ \forall s \in \mathrm{Aut}(\mathbb{C}) \tag{36}$$

and the complex Lie algebra $\mathrm{hdg}_{T,\mathbb{C}}$ *coincides with the Lie subalgebra* $\mathfrak{u}$ *of* $\mathrm{End}_{\mathbb{C}}(\Lambda_{\mathbb{C}})$ *generated by all* $s(f_T)$ *where s run over the group* $\mathrm{Aut}(\mathbb{C})$. *In particular,* hdg_T *coincides with the smallest* $\mathbb{Q}$*-Lie subalgebra* $\mathfrak{g} \subset \mathrm{End}_{\mathbb{Q}}(\Lambda_{\mathbb{Q}})$ *such that*

$$\mathfrak{g} \otimes_{\mathbb{Q}} \mathbb{C} \subset \mathrm{End}_{\mathbb{Q}}(\Lambda_{\mathbb{Q}}) \otimes_{\mathbb{Q}} \mathbb{C} = \mathrm{End}_{\mathbb{C}}(\Lambda_{\mathbb{C}})$$

contains f_T.

Proof Since $\mathbf{i} = \sqrt{-1}$,

$$s(\mathbf{i}) = \pm\mathbf{i}, s(f_T) = \pm\mathbf{i} \cdot s(J) \ \forall s \in \mathrm{Aut}(\mathbb{C}). \tag{37}$$

Therefore

$$s(f_T)^2 = (\pm\mathbf{i} \cdot s(J))^2 = -s(J)^2.$$

It follows from (34) that

$$s(f_T)^2 = -(-1) = 1,$$

which proves our first assertion. It follows from (37) that

$$\mathbb{C} \cdot s(f_T) = \mathbb{C} \cdot s(J) \ \forall s \in \mathrm{Aut}(\mathbb{C}).$$

Now our second assertion follows from Theorem 5.2. □

The following definition was actually introduced in [37, Definition 1.1].

Definition 5.4 Let C be a field, k a subfield of C, and $\mathfrak{g}$ be a k-Lie algebra. Let $\mathfrak{g}_C$ the corresponding C-Lie algebra $\mathfrak{g} \otimes_k C$.

If $f \in \mathfrak{g}_C$ then the *Hodge ideal* attached to f is the *smallest ideal* $\mathrm{hmt} = \mathrm{hmt}(f)$[3] of $\mathfrak{g}$ such that

$$\mathrm{hmt}_C := \mathrm{hmt} \otimes_k C$$

contains f. We call f a *Hodge element* of $\mathfrak{g}$ if $\mathrm{hmt}(f) = \mathfrak{g}$.

Remark 5.5 Let $\mathfrak{a}$ be the *smallest ideal* of hdg_T such that $\mathfrak{a}_{\mathbb{C}} := \mathfrak{a} \otimes_{\mathbb{Q}} \mathbb{C}$ contains f_T. Clearly, $\mathfrak{a}_{\mathbb{C}}$ contains $s(f_T)$ for all $s \in \mathrm{Aut}(\mathbb{C})$. It follows from Corollary 5.3 that $\mathfrak{a}_{\mathbb{C}} = \mathrm{hdg}_{T,\mathbb{C}}$. This implies that

$$\mathfrak{a} = \mathrm{hdg}_T.$$

The latter equality means that f_T is a *Hodge element* of the $\mathbb{Q}$-Lie algebra hdg_T where

$$k = \mathbb{Q}, \quad C = \mathbb{C}.$$

For the sake of simplicity, from now on let us assume that T is *simple.* This means that the natural faithful representation of $\mathrm{Hdg}(T)$ in

$$\Lambda_{\mathbb{Q}} = \mathrm{H}_1(T, \mathbb{Q})$$

is *irreducible* and therefore $\mathrm{End}^0(T)$ is a division algebra over $\mathbb{Q}$. This implies that the $\mathbb{Q}$-algebraic (sub)group $\mathrm{Hdg}(T)$ is *reductive.* In addition, the $\mathbb{Q}$-Lie (sub)algebra

$$\mathrm{hdg}_T \subset \mathfrak{sl}\,(\mathrm{H}_1(T, \mathbb{Q})) \subset \mathrm{End}_{\mathbb{Q}}\,(\mathrm{H}_1(T, \mathbb{Q}))$$

is *reductive algebraic*, the faithful hdg_T-module $\mathrm{H}_1(T, \mathbb{Q})$ is simple, and the centralizer of hdg_T in $\mathrm{End}_{\mathbb{Q}}\,(\mathrm{H}_1(T, \mathbb{Q}))$ is the division $\mathbb{Q}$-algebra $\mathrm{End}^0(T)$. Then the *center* $\mathcal{Z}(T)$ of $\mathrm{End}^0(T)$ is a number field.

Let us split the *reductive* $\mathbb{Q}$-Lie algebra hdg_T into a direct sum

$$\mathrm{hdg}_T = \mathrm{hdg}_T^{\mathrm{ss}} \oplus \mathfrak{c}_T$$

of the semisimple $\mathbb{Q}$-Lie algebra

$$\mathrm{hdg}_T^{\mathrm{ss}} = [\mathrm{hdg}_T, \mathrm{hdg}_T]$$

and the *center* $\mathfrak{c}_T$ of hdg_T with

$$\mathfrak{c}_T \subset \mathcal{Z}(T) \subset \mathrm{End}^0(T).$$

[3] Here hmt is short for Hodge-Mumford-Tate.

The following useful assertion is well known in the case of abelian varieties.

Lemma 5.6 *Suppose that T is simple and $\mathcal{Z}(T) = \mathbb{Q}$ (e.g.,* $\mathrm{End}^0(T) = \mathbb{Q}$*). Then $\mathfrak{c}_T = \{0\}$, i.e., the $\mathbb{Q}$-Lie algebra is semisimple and therefore* $\mathrm{Hdg}(T)$ *is a semisimple $\mathbb{Q}$-algebraic group.*

Proof The result follows readily from the combination of inclusions

$$\mathfrak{c}_T \subset \mathcal{Z}(T) = \mathbb{Q}, \ \mathfrak{c}_T \subset \mathrm{hdg}_T \subset \mathfrak{sl}\,(\mathrm{H}_1(T, \mathbb{Q}))\,.$$

□

The next example deals with the opposite case when the endomorphism algebra of a simple torus T is a number field of (largest possible) degree $2\dim(T)$.

Example 5.7 Suppose that a complex torus $T = V/\Lambda$ is simple of dimension g and $\mathrm{End}^0(T)$ is a number field E of degree $2g$. Then $\Lambda_{\mathbb{Q}}$ becomes a one-dimensional vector space over E. Therefore

$$E = \mathrm{End}_E(\Lambda_{\mathbb{Q}}), \ E^* = \mathrm{Aut}_E(\Lambda_{\mathbb{Q}}).$$

This implies that

$$\mathrm{Hdg}(T)(\mathbb{Q}) \subset E^{(1)} := \{e \in E^* \mid \mathrm{Norm}_{E/\mathbb{Q}}(e) = 1\} \subset E^* = \mathrm{Aut}_E(\Lambda_{\mathbb{Q}}). \quad (38)$$

Here

$$\mathrm{Norm}_{E/\mathbb{Q}} : E^* \to \mathbb{Q}^*, \ e \mapsto \prod_{\sigma \in \Sigma_E} \sigma(e)$$

is the *norm homomorphism* of the multiplicative groups of fields attached to the field extension $E/\mathbb{Q}$

Let $\mathcal{S}_E = \mathrm{Res}_{E/\mathbb{Q}}(\mathbb{G}_m)$ be the $2g$-dimensional *algebraic* torus over $\mathbb{Q}$ obtained from the multiplicative group $\mathbb{G}_m$ by the Weil's restriction of scalars from E to $\mathbb{Q}$. Then $\mathcal{S}_E(\mathbb{Q}) = E^*$ and for each $\sigma \in \Sigma_E$ there is a certain character

$$\delta_\sigma : \bar{\mathcal{S}}_E := \mathcal{S}_E \times_{\mathbb{Q}} \bar{\mathbb{Q}} \to \mathbb{G}_m \times_{\mathbb{Q}} \bar{\mathbb{Q}}$$

of the algebraic torus $\bar{\mathcal{S}}_E$ over $\bar{\mathbb{Q}}$ such that the restriction of δ_σ to

$$E^* = \mathcal{S}_E(\mathbb{Q}) \subset \mathcal{S}_E(\bar{\mathbb{Q}}) = \bar{\mathcal{S}}_E(\bar{\mathbb{Q}})$$

coincides with

$$\sigma : E^* \hookrightarrow \bar{\mathbb{Q}}^*.$$

In addition, the $2g$-element set $\{\delta_\sigma \mid \sigma \in \Sigma_K\}$ constitutes a *basis* of the free $\mathbb{Z}$-module $\mathrm{X}(\bar{\mathcal{S}}_E)$ of *characters of the algebraic torus* $\bar{\mathcal{S}}_E$ over $\bar{\mathbb{Q}}$. Since $\mathcal{S}_E$ is defined over $\mathbb{Q}$, the group $X(\bar{\mathcal{S}}_E)$ is provided with the natural structure of a $\mathrm{Gal}(\mathbb{Q})$-module in such a way that

$$s(\delta_\sigma) = \delta_{s(\sigma)} \ \forall \sigma \in \Sigma_E, s \in \mathrm{Gal}(\mathbb{Q}) \tag{39}$$

([26, Ch. II, Sect. 1], [30, Ch. III, Sects. 5 and 6]). Clearly, the character

$$\chi = \prod_{\sigma \in \Sigma_E} \delta_\sigma \in X(\bar{\mathcal{S}}_E)$$

is $\mathrm{Gal}(\mathbb{Q})$-invariant and may be viewed as the character of $\mathcal{S}_E$ such that

$$\chi(e) = \mathrm{Norm}_{E/\mathbb{Q}}(e) \in \mathbb{Q}^* \quad \forall e \in E^* = \mathcal{S}_E(\mathbb{Q}).$$

Let us put

$$\mathcal{S}_E^1 = \ker(\chi).$$

Since χ is obviously non-divisible in the group of characters, $\mathcal{S}_E^1$ is an algebraic $\mathbb{Q}$-subtorus of dimension $2g - 1$ in $\mathcal{S}_E$ such that

$$\mathcal{S}_E^1(\mathbb{Q}) = \ker(\mathrm{Norm}_{E/\mathbb{Q}}) = E^{(1)}. \tag{40}$$

Combining (38) and (40), and taking into account that $\mathrm{Hdg}(T)(\mathbb{Q})$ is Zariski dense in $\mathrm{Hdg}(T)$, we conclude that

$$\mathrm{Hdg}(T) \subset \mathcal{S}_E^1. \tag{41}$$

In particular, if $\mathcal{S}_E^1$ is a *simple* algebraic torus over $\mathbb{Q}$ then

$$\mathrm{Hdg}(T) = \mathcal{S}_E^1.$$

By definition of $\mathcal{S}_E^1$, the Galois module $X(\bar{\mathcal{S}}_E^1)$ of characters of the algebraic $\bar{\mathbb{Q}}$-torus

$$\bar{\mathcal{S}}_E^1 = \mathcal{S}_E^1 \times_{\mathbb{Q}} \bar{\mathbb{Q}}$$

is the quotient $X(\bar{\mathcal{S}}_E)/(\mathbb{Z} \cdot \chi)$. It follows from (39) that the $\mathrm{Gal}(\mathbb{Q})$-module $X(\bar{\mathcal{S}}_E^1)$ is isomorphic to the quotient $\mathbb{Z}^{\Sigma_E}/\mathbb{Z} \cdot \mathbf{1}$ where $\mathbb{Z}^{\Sigma_E}$ is the free $\mathbb{Z}$-module of functions $\phi : \Sigma_E \to \mathbb{Z}$ and $\mathbf{1}$ is the constant function 1. It follows easily that the Galois module $X(\bar{\mathcal{S}}_E^1) \otimes \mathbb{Q}$ is isomorphic to the $\mathbb{Q}$-vector space

$$\left(\mathbb{Q}^{\Sigma_E}\right)^0 := \{\phi : \Sigma_E \to \mathbb{Q} \mid \sum_{\sigma \in \Sigma_E} \phi(\sigma) = 0\}$$

of $\mathbb{Q}$-valued functions on Σ_E with zero "integral." Recall that the action of Gal(E) on Σ_E is transitive and this action induces the structure of the Galois module on $\left(\mathbb{Q}^{\Sigma_E}\right)^0$. Notice that if the action of Gal$(\mathbb{Q})$ on $\left(\mathbb{Q}^{\Sigma_E}\right)^0$ is *doubly transitive* (i.e., E is doubly transitive) then the representation of Gal$(\mathbb{Q})$ in $\left(\mathbb{Q}^{\Sigma_E}\right)^0$ is *irreducible*, i.e., the Galois module $\mathrm{X}(\bar{\mathcal{S}}^1_E) \otimes \mathbb{Q}$ is simple, which means that the algebraic $\mathbb{Q}$-torus $\mathcal{S}^1_E$ is simple and therefore $\mathrm{Hdg}(T) = \mathcal{S}^1_E$. So we have proven that

$$\mathrm{Hdg}(T) = \mathcal{S}^1_E \tag{42}$$

if E is *doubly transitive*. In particular, the algebraic $\mathbb{Q}$-torus is simple.

Theorem 5.8 *Let $T = V/\Lambda$ be a simple complex torus of dimension $g > 2$ such that its endomorphism algebra is a number field E of degree $2g$ that is doubly transitive.*

Then:

(i) *The Hodge group* $\mathrm{Hdg}(T)$ *of T coincides with $\mathcal{S}^1_E$. In addition,* $\mathrm{Hdg}(T)$ *is a simple algebraic $\mathbb{Q}$-torus of dimension $2g - 1$.*
(ii) *The* $\mathrm{Hdg}(T)$*-module* $\mathrm{H}^2(T, \mathbb{Q})$ *is simple. In particular, T is 2-simple.*

Proof We keep the notation of Example 5.7 where the assertion (i) and the simplicity of $\mathrm{Hdg}(T)$ are already proven.

In order to prove (ii), notice that $\mathrm{H}^2(T, \mathbb{Q}) = \mathrm{Hom}_{\mathbb{Q}}(\wedge^2_{\mathbb{Q}}\Lambda_{\mathbb{Q}}, \mathbb{Q})$, so, it suffices to check that the $\mathrm{Hdg}(T)$-module $\wedge^2_{\mathbb{Q}}\Lambda_{\mathbb{Q}}$ is simple.

If $\sigma \in \Sigma_E$ then let us consider the character $\delta^{(1)}_\sigma$ of $\bar{\mathcal{S}}^1_E$ that is the restriction of the character δ_σ to $\bar{\mathcal{S}}^1_E$. Clearly,

$$\prod_{\sigma \in \Sigma_E} \delta^{(1)}_\sigma = 1 \in \mathrm{X}(\bar{\mathcal{S}}^1_E)$$

and this is the only "nontrivial" multiplicative relation between $\delta^{(1)}_\sigma$. In particular, if A and B are two *distinct* 2-element subsets of Σ_E then

$$\delta^1_A := \prod_{\sigma \in A} \delta^{(1)}_\sigma \neq \prod_{\sigma \in B} \delta^{(1)}_\sigma =: \delta^1_B. \tag{43}$$

In other words, δ^1_A and δ^1_B are *distinct* characters of $\bar{\mathcal{S}}^1_E$.

Let us fix an order on the $2g$-element set Σ_E and consider the $\bar{\mathbb{Q}}$-vector space

$$\bar{\Lambda} = \Lambda_{\mathbb{Q}} \otimes_{\mathbb{Q}} \bar{\mathbb{Q}},$$

which is endowed with the natural faithful action of $\bar{\mathcal{S}}^1_E$, and splits into a direct sum

$$\bar{\Lambda} = \oplus_{\sigma \in \Sigma_E} \bar{\Lambda}_\sigma$$

of *one-dimensional* weight subspaces $\bar{\Lambda}_\sigma$ defined by the condition that $\bar{\mathcal{S}}_E^1$ acts on $\bar{\Lambda}_\sigma$ by the character $\delta_\sigma^{(1)}$.

We have

$$\wedge^2_{\mathbb{Q}}\Lambda_{\mathbb{Q}} \otimes_{\mathbb{Q}} \bar{\mathbb{Q}} = \wedge^2_{\bar{\mathbb{Q}}}\bar{\Lambda} = \oplus_A \bar{\Lambda}_A,$$

where

$$A = \{\sigma, \tau\} \subset \Sigma_E; \ \sigma < \tau$$

runs through the set of all 2-element subsets of Σ_E and

$$\bar{\Lambda}_A = \bar{\Lambda}_\sigma \wedge_{\bar{\mathbb{Q}}} \bar{\Lambda}_\tau \cong \bar{\Lambda}_\sigma \otimes_{\bar{\mathbb{Q}}} \bar{\Lambda}_\tau$$

is the corresponding one-dimensional $\bar{\mathcal{S}}_E^1$-invariant subspace; the action of $\bar{\mathcal{S}}_E^1$ on $\bar{\Lambda}_A$ is defined by the character δ_A^1.

It follows from (43) that if W is a nonzero $\mathcal{S}_E^1$-invariant $\mathbb{Q}$-vector subspace of $\Lambda_{\mathbb{Q}}$ then $\bar{W} = W \otimes_{\mathbb{Q}} \bar{\mathbb{Q}}$ is a direct sum of some of $\bar{\Lambda}_A$. The double transitivity condition implies that all $\bar{\Lambda}_A$'s are mutually Galois-conjugate over $\mathbb{Q}$. It follows that $\bar{W} = \wedge^2_{\bar{\mathbb{Q}}}\bar{\Lambda}$, i.e., $W = \wedge^2_{\mathbb{Q}}\Lambda_{\mathbb{Q}}$, and we are done. □

Remark 5.9 See [3] for explicit examples of complex tori that satisfy the conditions of Theorem 5.8.

In the case of arbitrary simple complex tori (or even abelian varieties) the Hodge group may be neither semisimple nor commutative (see [32, 33, 34] for explicit examples). This is not the case for 2-simple tori in dimensions > 2, in light of the following assertion.

Proposition 5.10 *Let T be a 2-simple torus of dimension $g > 2$. (In particular, T is simple.) Then* $\mathrm{Hdg}(T)$ *is either semisimple or commutative. The latter case occurs if and only if* $\mathrm{End}^0(T)$ *is a number field of degree $2g$.*

Proof We know (thanks to Theorem 1.7) that $E = \mathrm{End}^0(T)$ is a number field of degree

$$[E : \mathbb{Q}] \in \{1, g, 2g\}.$$

If $[E : \mathbb{Q}] = 1$ then $E = \mathbb{Q}$. In light of Lemma 5.6, $\mathrm{Hdg}(T)$ is semisimple.

If $[E : \mathbb{Q}] = 2g$ then $\Lambda_{\mathbb{Q}} = \mathrm{H}_1(T, \mathbb{Q})$ is a one-dimensional E-vector space, i.e.,

$$E = \mathrm{End}_E\left(\mathrm{H}_1(T, \mathbb{Q})\right).$$

This implies that

$$\mathrm{hdg}_T \subset \mathrm{End}_E\left(\mathrm{H}_1(T, \mathbb{Q})\right) \subset E$$

and therefore hdg_T is a commutative $\mathbb{Q}$-Lie algebra. It follows that $\mathrm{Hdg}(T)$ is commutative.

Assume that $[E:\mathbb{Q}]=g$, i.e., $\Lambda_{\mathbb{Q}}=\mathrm{H}_1(T,\mathbb{Q})$ is a two-dimensional E-vector space. Then

$$\mathrm{hdg}_T \subset \mathrm{End}_E\,(\mathrm{H}_1(T,\mathbb{Q})) \supset E \supset \mathfrak{c}_T.$$

Let

$$\mathrm{Tr}_E:\mathrm{End}_E\,(\mathrm{H}_1(T,\mathbb{Q}))\to E \tag{44}$$

be the (surjective) E-linear trace map, which is a homomorphism of $\mathbb{Q}$-Lie algebras (here we view E as a commutative $\mathbb{Q}$-Lie algebra); in addition, the restriction of Tr_E to E is multiplication by

$$\dim_E\,(\mathrm{H}_1(T,\mathbb{Q}))=2.$$

We write $\mathfrak{sl}\,(\mathrm{H}_1(T,\mathbb{Q})/E)$ for $\ker(\mathrm{Tr}_E)$, which is an absolutely simple E-Lie algebra of traceless E-linear operators in $\mathrm{H}_1(T,\mathbb{Q})$. (Viewed as the $\mathbb{Q}$-Lie algebra, $\mathfrak{sl}\,(\mathrm{H}_1(T,\mathbb{Q})/E)$ is simple but not absolutely simple.)

On the other hand, let

$$\det_E:\mathrm{Aut}_E\,(\mathrm{H}_1(T,\mathbb{Q}))\to E^*$$

be the multiplicative *determinant* map. Clearly,

$$\mathrm{Hdg}(T)(\mathbb{Q})\subset \mathrm{Aut}_E\,(\mathrm{H}_1(T,\mathbb{Q}))$$

and the group $\mathrm{Aut}_E\,(\mathrm{H}_1(T,\mathbb{Q}))$ acts naturally on the one-dimensional E-vector space $\mathrm{Hom}_E\left(\wedge_E^2\mathrm{H}_1(T,\mathbb{Q}),E\right)$ via the character $\det_E$. I claim that $\det_E$ kills $\mathrm{Hdg}(T)(\mathbb{Q})$. Indeed, if this is not the case, then the rational Hodge substructure $H_E\cong\mathrm{Hom}_E\left(\wedge_E^2\mathrm{H}_1(T,\mathbb{Q}),E\right)$ of $\mathrm{H}^2(T,\mathbb{Q})$ (defined in Sect. 3, see (12) and (10)) has *nonzero* $(2,0)$-component, whose $\mathbb{C}$-dimension

$$\le\frac{\dim_{\mathbb{Q}}(H_E)}{2}=\frac{[E:\mathbb{Q}]}{2}=\frac{g}{2}<g,$$

which contradicts the 2-simplicity of T. Hence, $\det_E$ kills $\mathrm{Hdg}(T)(\mathbb{Q})$. Taking into account that $\mathrm{Hdg}(T)(\mathbb{Q})$ is dense in $\mathrm{Hdg}(T)$ in Zariski topology and the minimality property in the definition of the Hodge group, we conclude that

$$\mathrm{Hdg}(T)\subset \mathrm{Res}_{E/\mathbb{Q}}\mathrm{SL}((\mathrm{H}_1(T,\mathbb{Q})/E)$$

where $\mathrm{SL}((\mathrm{H}_1(T,\mathbb{Q})/E)$ is the special linear group of the E-vector space $\mathrm{H}_1(T,\mathbb{Q})$, which is a *simple algebraic E-group*, and $\mathrm{Res}_{E/\mathbb{Q}}$ is the *Weil restriction* of scalars.

Taking into account that the $\mathbb{Q}$-Lie algebra $\mathfrak{sl}\,(\mathrm{H}_1(T,\mathbb{Q})/E)$ is the Lie algebra of the $\mathbb{Q}$-algebraic group $\mathrm{Res}_{E/\mathbb{Q}}\mathrm{SL}\,(\mathrm{H}_1(T,\mathbb{Q})/E)$, we conclude that

$$\mathrm{hdg}_T \subset \mathfrak{sl}\,(\mathrm{H}_1(T,\mathbb{Q})/E) \cong \mathfrak{sl}(2,E). \tag{45}$$

In particular, Tr_E kills $\mathfrak{c}_T$. Since $\mathfrak{c}_T \subset E$,

$$0 = \mathrm{Tr}_E(c) = 2c \ \forall c \in \mathfrak{c}_T \subset E.$$

This implies that $\mathfrak{c}_T = \{0\}$, i.e., hdg_T is semisimple, i.e., $\mathrm{Hdg}(T)$ is semisimple. This ends the proof. □

The following assertion may be viewed as a variant of a theorem of P. Deligne [15] about complex abelian varieties (see also [25, 35, 36]).

Theorem 5.11 *Let T be a simple complex torus. Let* $\mathrm{hdg}_T \subset \mathrm{End}_{\mathbb{Q}}(\Lambda_{\mathbb{Q}})$ *be the (reductive) $\mathbb{Q}$-Lie algebra of* $\mathrm{Hdg}(T)$, *whose natural representation in $\Lambda_{\mathbb{Q}}$ is irreducible.*

Let $\mathfrak{g}$ be a simple (non-abelian) factor of the complex reductive Lie algebra

$$\mathrm{hdg}_{T,\mathbb{C}} = \mathrm{hdg}_T \otimes_{\mathbb{Q}} \mathbb{C} \subset \mathrm{End}_{\mathbb{Q}}(\Lambda_{\mathbb{Q}}) \otimes_{\mathbb{Q}} \mathbb{C} = \mathrm{End}_{\mathbb{C}}(\Lambda_{\mathbb{C}}) = \mathrm{End}_{\mathbb{C}}\,(V_{\mathbb{C}})\,.$$

Then

(i) *The simple complex Lie algebra $\mathfrak{g}$ is of classical type* ($\mathsf{A}_l, \mathsf{B}_l, \mathsf{C}_l, \mathsf{D}_l$) *of a certain positive rank l.*
(ii) *Let W be a nontrivial simple $\mathfrak{g}$-submodule of $V_{\mathbb{C}}$. Then its highest weight is minuscule one.*

Proof The result follows readily from Corollary 5.3 combined with Proposition 2.4.1 of [35] applied to

$$k = \mathbb{C},\ k_0 = \mathbb{Q},\ W = \Lambda_{\mathbb{Q}},\ g = \mathrm{hdg}_T,\ f = f_T,\ A = \{1,-1\}$$

and

$$n = 1,\ a_0 = 1,\ a_1 = -1.$$

□

The following assertion may be viewed as a variant of a theorem of M.V. Borovoi about complex abelian varieties [7]; see also [37].

Theorem 5.12 *Suppose that T is a simple complex torus with* $\mathrm{End}^0(T) = \mathbb{Q}$.

Then its Hodge group $\mathrm{Hdg}(T)$ *is a $\mathbb{Q}$-simple linear algebraic group, i.e., its $\mathbb{Q}$-Lie algebra* hdg_T *is simple.*

Proof Clearly, hdg_T is a semisimple $\mathbb{Q}$-Lie algebra, whose natural faithful representation in $\Lambda_{\mathbb{Q}}$ is *absolutely irreducible*. By Remark 5.5, $f_T \in \mathrm{hdg}_{T,\mathbb{C}}$ is a Hodge element of hdg_T. The spectrum of the linear semisimple operator f_T in $\Lambda_{\mathbb{Q}}$ consists of precisely two eigenvalues, 1 and -1. Now it follows from Theorem 1.5 of [35] that hdg_T is simple. This means that $\mathrm{Hdg}(T)$ is a $\mathbb{Q}$-simple algebraic group. □

Corollary 5.13 *Suppose that T is a simple complex torus of dimension g with* $\mathrm{End}^0(T) = \mathbb{Q}$. *Assume also that $2g$ is not a power (e.g., g is odd).*

Then $\mathrm{Hdg}(T)$ *is an absolutely simple $\mathbb{Q}$-algebraic group that enjoys precisely one of the following two properties:*

(i) $\mathrm{Hdg}(T)$ *is of type* A_{2g-1} *or* C_g *or* D_g.

(ii) $\mathrm{Hdg}(T)$ *is of type* A_r *where r is a positive integer that enjoys the following properties.*
$1 < r < 2g-1$ *and there is an integer j such that* $1 < j < 2g-1$ *and* $2g = \binom{r+1}{j}$.

Proof By Theorem 5.12, hdg_T is a simple $\mathbb{Q}$-Lie algebra. Suppose that hdg_T is *not* absolutely simple, i.e., the complex Lie algebra $\mathrm{hdg}_{T,\mathbb{C}}$ splits into a direct sum

$$\mathrm{hdg}_{T,\mathbb{C}} = \oplus_{\ell=1}^{d} \mathfrak{g}_\ell$$

of d simple complex Lie algebras $\mathfrak{g}_\ell$ where $d > 1$. We are going to prove that $2g$ is a dth power, which gives us a desired contradiction.

The simplicity of hdg_T means that $\mathrm{Aut}(\mathbb{C})$ permutes the set $\{\mathfrak{g}_\ell\}_{\ell=1}^{d}$ *transitively*. Namely, each $s \in \mathrm{Aut}(\mathbb{C})$ gives rise to the semi-linear automorphism of the $\mathbb{C}$-vector space

$$\Lambda_{\mathbb{C}} \to \Lambda_{\mathbb{C}}, \quad x \otimes z \mapsto x \otimes s(z)\ \forall x \in \Lambda_{\mathbb{Q}}, z \in \mathbb{C}$$

and to the semi-linear automorphism of the $\mathbb{C}$-algebra

$$\mathrm{End}_{\mathbb{C}}(\Lambda_{\mathbb{C}}) \to \mathrm{End}_{\mathbb{C}}(\Lambda_{\mathbb{C}}), \quad u \otimes z \mapsto u \otimes s(z)\ \forall x \in \mathrm{End}_{\mathbb{Q}}(\Lambda_{\mathbb{Q}}), z \in \mathbb{C}.$$

We continue to denote those automorphisms by s.

The transitivity means that for each $\mathfrak{g}_\ell$ there is $s_\ell \in \mathrm{Aut}(\mathbb{C})$ such that

$$\mathfrak{g}_\ell = s_\ell\left(\mathfrak{g}_1\right).$$

We know that the $\mathrm{hdg}_{T,\mathbb{C}}$-module $\Lambda_{\mathbb{C}}$ is (absolutely) simple. Since each $\mathfrak{g}_\ell$ is a direct summand of $\mathrm{hdg}_{T,\mathbb{C}}$, the $\mathfrak{g}_\ell$-module $\Lambda_{\mathbb{C}}$ is isotypic, i.e., there is a simple $\mathfrak{g}_\ell$-submodule $W_\ell \subset \Lambda_{\mathbb{C}}$ such that all simple $\mathfrak{g}_\ell$-submodules of $\Lambda_{\mathbb{C}}$ are isomorphic to W_ℓ. In addition, the $\mathrm{hdg}_{T,\mathbb{C}}$ $\left(= \oplus_{\ell=1}^{d}\mathfrak{g}_\ell\right)$-module $\Lambda_{\mathbb{C}}$ splits into a tensor product $\otimes_{\ell=1}^{d} W_\ell$ of simple $\mathfrak{g}_\ell$-modules W_ℓ. Let us prove that $\dim_{\mathbb{C}}(W_\ell)$ does *not* depend on ℓ.

Indeed, $s_\ell(W_1)$ is a simple $\mathfrak{g}_\ell$-submodule of $\Lambda_{\mathbb{C}}$ and therefore is isomorphic to W_1. This implies that $\dim(W_1) = \dim(W_\ell)$ and therefore

$$2g = \dim_{\mathbb{C}}(\Lambda_{\mathbb{C}}) = \left(\dim_{\mathbb{C}}(W_1)\right)^d,$$

which gives us a desired contradiction. So, $\mathrm{hdg}_{T,\mathbb{C}}$ is a *simple* complex Lie algebra and $\Lambda_{\mathbb{C}}$ is a faithful simple $\mathrm{hdg}_{T,\mathbb{C}}$-module. By Theorem 5.11, $\mathrm{hdg}_{T,\mathbb{C}}$ is a classical Lie algebra (of type A_r, B_r, C_r or D_r), and the highest weight of $\Lambda_{\mathbb{C}}$ is *minuscule.* The remaining assertion follows readily from the inspection of dimensions of minuscule representations of classical Lie algebras [9, Tables]. □

Example 5.14 Suppose that T is a complex torus of dimension g such that one of the following conditions holds:

(i) $\mathrm{Hdg}(T) = \mathrm{SL}(\Lambda_{\mathbb{Q}})$.
(ii) There exists a nondegenerate quadratic form

$$\phi : \Lambda_{\mathbb{Q}} \to \mathbb{Q}$$

of even signature $(2p, 2q)$ with $p + q = g \geq 3$ such that $\mathrm{Hdg}(T)$ coincides with the corresponding special orthogonal group $\mathrm{SO}(\Lambda_{\mathbb{Q}})$.
(iii) There exists a nondegenerate alternating $\mathbb{Q}$-bilinear form

$$\Lambda_{\mathbb{Q}} \times \Lambda_{\mathbb{Q}} \to \mathbb{Q}$$

such that $\mathrm{Hdg}(T)$ coincides with the corresponding symplectic group $\mathrm{Sp}(\Lambda_{\mathbb{Q}})$.

Then T is 2-simple.

Indeed, in cases (i) and (ii), (in the obvious notation) the natural representation of $\mathrm{SL}(\Lambda_{\mathbb{C}})$ (resp. $\mathrm{SO}(\Lambda_{\mathbb{C}})$) in $\wedge^2_{\mathbb{C}}(\Lambda_{\mathbb{C}})$ is irreducible; see [9, Ch. 8, Sect. 13]. This implies that the natural representation of $\mathrm{SL}(\Lambda_{\mathbb{Q}})$ (resp. $\mathrm{SO}(\Lambda_{\mathbb{Q}}, \phi)$) in $\wedge^2_{\mathbb{Q}}(\Lambda_{\mathbb{Q}})$ is absolutely irreducible. By duality, the same is true for $\mathrm{Hom}_{\mathbb{Q}}(\wedge^2_{\mathbb{Q}}(\Lambda_{\mathbb{Q}}), \mathbb{Q}) = \mathrm{H}^2(T, \mathbb{Q})$, i.e., the $\mathrm{Hdg}(T)$-module $\mathrm{H}^2(T, \mathbb{Q})$ is simple. This implies that T is 2-simple. Notice that in these cases we deal with simple complex tori that are *not* abelian varieties (since they do not carry nonzero 2-dimensional Hodge classes) and whose endomorphism algebra is $\mathbb{Q}$.

In case (iii), the natural representation of $\mathrm{Sp}(\Lambda_{\mathbb{C}})$ in $\wedge^2_{\mathbb{C}}(\Lambda_{\mathbb{C}})$ is a direct sum of an irreducible representation and a trivial one-dimension representation [28, Tables]. This implies that the natural representation of $\mathrm{Sp}(\Lambda_{\mathbb{Q}})$ in $\wedge^2_{\mathbb{Q}}(\Lambda_{\mathbb{Q}})$ is a direct sum of an absolutely irreducible representation and a trivial one-dimension representation. By duality, the same is true for $\mathrm{Hom}_{\mathbb{Q}}(\wedge^2_{\mathbb{Q}}(\Lambda_{\mathbb{Q}}), \mathbb{Q}) = \mathrm{H}^2(T, \mathbb{Q})$, i.e., the $\mathrm{Hdg}(T)$-module $\mathrm{H}^2(T, \mathbb{Q})$ is a direct sum of an absolutely simple module and a trivial module of $\mathbb{Q}$-dimension 1. The latter consists of all $\mathrm{Hdg}(T)$-invariants in $\mathrm{H}^2(T, \mathbb{Q})$, i.e., coincides with $\mathrm{H}^{1,1}(T, \mathbb{Q})$. The former is an irreducible rational Hodge structure. It follows from Remark 1.8 (ii) that T is 2-simple.[4] See [39] for explicit examples (in all dimensions) of complex abelian varieties T with $\mathrm{Hdg}(T) = \mathrm{Sp}(\Lambda_{\mathbb{Q}})$.

[4] For abelian varieties T case (iii) was done in [1, Sect. 5.1].

Theorem 5.15 *Let $\Pi_{\mathbb{Q}}$ be a $\mathbb{Q}$-vector space of positive even dimension $2g$, and $\mathcal{G}$ a $\mathbb{Q}$-simple algebraic subgroup of $\mathrm{GL}(\Pi_{\mathbb{Q}})$, whose $\mathbb{Q}$-Lie algebra $\mathfrak{g}$ may be viewed as a simple $\mathbb{Q}$-Lie subalgebra of $\mathrm{End}_{\mathbb{Q}}(\Pi_{\mathbb{Q}})$. Let us consider the real Lie subalgebra*

$$\mathfrak{g}_{\mathbb{R}} = \mathfrak{g} \otimes_{\mathbb{Q}} \mathbb{R} \subset \mathrm{End}_{\mathbb{Q}}(\Pi_{\mathbb{Q}}) \otimes_{\mathbb{Q}} \mathbb{R} = \mathrm{End}_{\mathbb{R}}(\Pi_{\mathbb{R}})$$

where $\Pi_{\mathbb{R}} = \Pi_{\mathbb{Q}} \otimes_{\mathbb{Q}} \mathbb{R}$ is the corresponding $2g$-dimensional real vector space. Suppose that there exists an element

$$J_0 \in \mathfrak{g}_{\mathbb{R}} \subset \mathrm{End}_{\mathbb{R}}(\Pi_{\mathbb{R}})$$

such that $J_0^2 = -1$ in $\mathrm{End}_{\mathbb{R}}(\Pi_{\mathbb{R}})$. Then there exists $J \in \mathfrak{g}_{\mathbb{R}}$ that enjoys the following properties:

(i) $J^2 = -1$.

(ii) *Let us endow $\Pi_{\mathbb{R}}$ with the structure of a g-dimensional complex vector space by defining*

$$(a + b\mathbf{i})v := av + bJ(v) \quad \forall a + b\mathbf{i} \in \mathbb{C} \text{ with } a, b \in \mathbb{R}.$$

Then for every discrete subgroup Λ of rank $2g$ in $\Pi_{\mathbb{Q}}$ the corresponding complex torus $T = \Pi_{\mathbb{R}}/\Lambda$ has Hodge group $\mathcal{G}$.

Proof We will need the following auxiliary statement.

Lemma 5.16 *If $\mathfrak{v}$ is a proper nonzero $\mathbb{Q}$-vector subspace of $\mathfrak{g}$ then the real vector subspace*

$$\mathfrak{v}_{\mathbb{R}} = \mathfrak{v} \otimes_{\mathbb{Q}} \mathbb{R} \subset \mathfrak{g} \otimes_{\mathbb{Q}} \mathbb{R} = \mathfrak{g}_{\mathbb{R}}$$

does not contain a nonzero ideal of $\mathfrak{g}_{\mathbb{R}}$.

Notice that the set of proper nonzero $\mathbb{Q}$-Lie subalgebras L of $\mathfrak{g}$ is *countable*. By Lemma 5.16, every $L_{\mathbb{R}} = L \otimes_{\mathbb{Q}} \mathbb{R}$ does not contain a nonzero ideal of $\mathfrak{g}_{\mathbb{R}}$. By [38, Lemma 2 on p. 494], the closed subset

$$\mathcal{G}(L_{\mathbb{R}}, J_0) = \{u \in \mathcal{G}(\mathbb{R}) \mid J_0 \in u^{-1} L_{\mathbb{R}} u\}$$

is nowhere dense in $\mathcal{G}(\mathbb{R})$. It follows that there exists $u \in \mathcal{G}(\mathbb{R})$ such that J_0 does not lie in any of $u^{-1} L_{\mathbb{R}} u$. Let us put

$$J := u J_0 u^{-1} \in \mathfrak{g}_{\mathbb{R}} \subset \mathrm{End}_{\mathbb{R}}(\Pi_{\mathbb{R}}).$$

Then J does not lie in any of $L_{\mathbb{R}}$ and $J^2 = -1 \in \mathrm{End}_{\mathbb{R}}(\Pi_{\mathbb{R}})$. It follows that $\mathfrak{g}$ coincides with the smallest $\mathbb{Q}$-Lie subalgebra $\mathfrak{u}$ of $\mathfrak{g}$ such that $\mathfrak{u}_{\mathbb{R}}$ contains J. This implies that $\mathfrak{g}$ coincides with the smallest $\mathbb{Q}$-Lie subalgebra $\mathfrak{u}$ of $\mathrm{End}_{\mathbb{Q}}(\Pi_{\mathbb{Q}})$ such that $\mathfrak{u}_{\mathbb{R}}$ contains J. It follows readily that $\mathfrak{g}$ coincides with the Lie algebra of the

Hodge group of a complex torus $T = \Pi_{\mathbb{R}}/\Lambda$ where the complex structure on the real vector space $\Pi_{\mathbb{R}}$ is defined by J and Λ is any discrete subgroup Λ of rank $2g$ in $\Pi_{\mathbb{Q}}$. □

Proof of Lemma 5.16 Suppose that $L_{\mathbb{R}}$ contains a nonzero ideal $\mathfrak{a}$ of $\mathfrak{g}_{\mathbb{R}}$. Then the $\mathbb{C}$-vector subspace

$$L_{\mathbb{C}} = L_{\mathbb{R}} \otimes_{\mathbb{R}} \mathbb{C} = L \otimes_{\mathbb{Q}} \mathbb{C}$$

contains a nonzero ideal $\mathfrak{a}_{\mathbb{C}} = \mathfrak{a} \otimes_{\mathbb{R}} \mathbb{C}$ of the complex Lie algebra

$$\mathfrak{g}_{\mathbb{C}} = \mathfrak{g}_{\mathbb{R}} \otimes_{\mathbb{R}} \mathbb{C} = \mathfrak{g} \otimes_{\mathbb{Q}} \mathbb{C}.$$

Then

$$\tilde{\mathfrak{a}} = \sum_{s \in \mathrm{Aut}(\mathbb{C})} s(\mathfrak{a})$$

is a $\mathrm{Aut}(\mathbb{C})$-invariant ideal of $\mathfrak{g}_{\mathbb{C}}$ that lies in $L_{\mathbb{C}}$. Hence, there is a $\mathbb{Q}$-vector subspace $\mathfrak{a}_{\mathbb{Q}}$ such that

$$\tilde{\mathfrak{a}} = \mathfrak{a}_{\mathbb{Q}} \otimes_{\mathbb{Q}} \mathbb{C};$$

in addition, $\mathfrak{a}_{\mathbb{Q}}$ is an ideal of $\mathfrak{g}$, which contradicts the simplicity of the $\mathbb{Q}$-Lie algebra $\mathfrak{g}$. This ends the proof. □

Example 5.17 We keep the notation of Theorem 5.15. Let $g \geq 3$ be an integer and $\Pi_{\mathbb{Q}}$ a $2g$-dimensional vector space over $\mathbb{Q}$. Let $\mathcal{G}$ be a $\mathbb{Q}$-simple algebraic subgroup of $\mathrm{GL}(\Pi_{\mathbb{Q}})$ that enjoys one of the following properties:

(i) $\mathcal{G} = \mathrm{SL}(\Pi_{\mathbb{Q}})$.

(ii) There exists a nondegenerate quadratic form

$$\phi : \Pi_{\mathbb{Q}} \to \mathbb{Q}$$

of even signature $(2p, 2q)$ with $p + q = g$ such that $\mathcal{G}$ coincides with the corresponding special orthogonal group $\mathrm{SO}(\Pi_{\mathbb{Q}}, \phi)$.

In both cases there exists $J_0 \in \mathfrak{g}_{\mathbb{R}}$ with $J_0^2 = -1$. (Here $\mathfrak{g} \subset \mathrm{End}_{\mathbb{Q}}(\Pi_{\mathbb{Q}})$ is the Lie algebra of $\mathcal{G}$.) In light of Theorem 5.15, there exists a complex structure on the real vector space $\Pi_{\mathbb{R}}$ such that the Hodge group of the corresponding complex torus $T = \Pi_{\mathbb{R}}/\Lambda$ coincides with $\mathcal{G}$. In light of Example 5.14, T is 2-simple.

6 The Degree g Case

In this section we discuss g-dimensional 2-simple tori, whose endomorphism algebra is a number field of degree g.

Theorem 6.1 *Let T be a 2-simple torus of dimension $g > 2$. If $\mathrm{End}^0(T)$ is a number field E of degree g then*

$$\mathrm{Hdg}(T) = \mathrm{Res}_{E/\mathbb{Q}}\mathrm{SL}((\mathrm{H}_1(T, \mathbb{Q})/E).$$

Proof It suffices to check that

$$\mathrm{hdg}_T = \mathfrak{sl}\,(\mathrm{H}_1(T, \mathbb{Q})/E). \tag{46}$$

In light of (45), the desired equality (46) is an immediate corollary of the following observation applied to

$$k = \mathbb{Q},\ K = E, W = \mathrm{H}_1(T, \mathbb{Q}),\ \mathfrak{g} = \mathrm{hdg}_T.$$

□

Lemma 6.2 *Let g be a positive integer, W a $2g$-dimensional vector space over a field k of characteristic 0, and $\mathfrak{g} \subset \mathrm{End}_k(W)$ a linear semisimple k-subalgebra such that the centralizer*

$$K :- \mathrm{End}_{\mathfrak{g}} W \subset \mathrm{End}_k(W)$$

is an overfield of k with $[K : k] = g$. Then $\mathfrak{g}$ coincides with the Lie algebra $\mathfrak{sl}(W/K)$ of traceless K-linear operators in W.

Proof of Lemma 6.2 The semisimplicity of $\mathfrak{g}$ implies that

$$\mathfrak{g} \subset \mathfrak{sl}(W/K). \tag{47}$$

In what follows we mimic the arguments of [23, pp. 790–791, Proof of Th. 4.4.10] where ℓ-adic Lie algebras are treated.

Let $\bar{k}$ be an algebraic closure of k, and Σ_K the g-element set of field embeddings $\sigma : K \hookrightarrow \bar{k}$ that coincide with the identity map on k. Let us consider the $2g$-dimensional $\bar{k}$-vector space $\bar{W} = W \otimes_k \bar{k}$ and the $\bar{k}$-Lie algebra

$$\bar{\mathfrak{g}} = \mathfrak{g} \otimes_k \bar{k} \subset \mathrm{End}_k(W) \otimes_k \bar{k} = \mathrm{End}_{\bar{k}}(\bar{W}). \tag{48}$$

The semisimplicity of the k-Lie algebra $\mathfrak{g}$ implies the *semisimplicity* of the $\bar{k}$-Lie algebra $\bar{\mathfrak{g}}$.

Clearly, the *centralizer* $\mathrm{End}_{\bar{\mathfrak{g}}}(\bar{W})$ of $\bar{\mathfrak{g}}$ in $\mathrm{End}_{\bar{k}}(\bar{W})$ equals

$$\mathrm{End}_{\mathfrak{g}}(W) \otimes_k \bar{k} = K \otimes_k \bar{k} \tag{49}$$

and $\bar{W}$ is a free $K \otimes_k \bar{k}$-module of rank 2, because W is a vector space over K of dimension 2. We have

$$K \otimes_k \bar{k} = \oplus_{\sigma \in \Sigma_K} K \otimes_{K,\sigma} \bar{k} = \oplus_{\sigma \in \Sigma_K} \bar{k}_\sigma \tag{50}$$

where

$$\bar{k}_\sigma = K \otimes_{K,\sigma} \bar{k} = \bar{k}.$$

We have

$$\bar{W} = \oplus_{\sigma \in \Sigma_K} \bar{W}_\sigma \quad \text{where } \bar{W}_\sigma = \bar{k}_\sigma \bar{W} \subset \bar{W}.$$

The freeness of the $K \otimes_k \bar{k}$-module $\bar{W}$ with rank 2 implies that each $\bar{W}_\sigma$ is a $\bar{k}_\sigma$-vector space of dimension 2. Since

$$\bar{k}_\sigma \subset K \otimes_k \bar{k} = \mathrm{End}_{\bar{\mathfrak{g}}}(\bar{W}),$$

each $\bar{W}_\sigma = \bar{k}_\sigma \bar{W}$ is a $\bar{\mathfrak{g}}$-invariant subspace of $\bar{W}$ and the centralizer of $\bar{\mathfrak{g}}$

$$\mathrm{End}_{\bar{\mathfrak{g}}}(\bar{W}_\sigma) = \bar{k}_\sigma = \bar{k}. \tag{51}$$

Let

$$\bar{\mathfrak{g}}_\sigma \subset \mathrm{End}_{\bar{k}_\sigma}(\bar{W}_\sigma) = \mathrm{End}_{\bar{k}}(\bar{W}_\sigma)$$

be the image of the natural $\bar{k}$-Lie algebra homomorphism

$$\bar{\mathfrak{g}} \to \mathrm{End}_{\bar{k}}(\bar{W}_\sigma).$$

The semisimplicity of $\bar{\mathfrak{g}}$ implies the semisimplicity of the Lie algebra $\bar{\mathfrak{g}}_\sigma$, because the latter is isomorphic to a quotient of the former. This implies that

$$\bar{\mathfrak{g}}_\sigma \subset \mathfrak{sl}(\bar{W}_\sigma) \cong \mathfrak{sl}(2, \bar{k}_\sigma) = \mathfrak{sl}(2, \bar{k}).$$

Taking into account (51) and the semisimplicity of $\bar{\mathfrak{g}}_\sigma$, we conclude that

$$\bar{\mathfrak{g}}_\sigma = \mathfrak{sl}(\bar{W}_\sigma) \cong \mathfrak{sl}(2, \bar{k}). \tag{52}$$

This implies that

$$\bar{\mathfrak{g}} \subset \oplus_{\sigma \in \Sigma_K} \bar{\mathfrak{g}}_\sigma = \oplus_{\sigma \in \Sigma_K} \mathfrak{sl}(\bar{W}_\sigma) \subset \oplus_{\sigma \in \Sigma_K} \mathrm{End}_{\bar{k}}(\bar{W}_\sigma). \tag{53}$$

Let σ and τ be *distinct* elements of Σ_K. Clearly, $\bar{W}_\sigma \oplus \bar{W}_\tau$ is a $\bar{\mathfrak{g}}$-invariant subspace of $\bar{W}$. Let $\bar{\mathfrak{g}}_{\sigma,\tau}$ be the image of $\bar{\mathfrak{g}}$ in $\mathrm{End}_{\bar{k}}(\bar{W}_\sigma \oplus \bar{W}_\tau)$. Since $\bar{\mathfrak{g}}_{\sigma,\tau}$ is isomorphic to a quotient of $\bar{\mathfrak{g}}$, it is a semisimple $\bar{k}$-Lie algebra such that

$$\bar{\mathfrak{g}}_{\sigma,\tau} \subset \mathfrak{sl}(\bar{W}_\sigma) \oplus \mathfrak{sl}(\bar{W}_\tau) \subset \mathrm{End}_{\bar{k}}(\bar{W}_\sigma) \oplus \mathrm{End}_{\bar{k}}(\bar{W}_\tau) \subset \mathrm{End}_{\bar{k}}(\bar{W}_\sigma \oplus \bar{W}_\tau).$$

Notice that $\bar{\mathfrak{g}}_{\sigma,\tau}$ projects *surjectively* on both

$$\bar{\mathfrak{g}}_\sigma = \mathfrak{sl}(\bar{W}_\sigma) \text{ and } \bar{\mathfrak{g}}_\tau = \mathfrak{sl}(\bar{W}_\tau),$$

because $\bar{\mathfrak{g}}$ does. The simplicity of both pairwise isomorphic Lie algebras $\mathfrak{sl}(\bar{W}_\sigma)$ and $\mathfrak{sl}(\bar{W}_\tau)$ and the semisimplicity of $\bar{\mathfrak{g}}_{\sigma,\tau}$ implies that either

$$\bar{\mathfrak{g}}_{\sigma,\tau} = \mathfrak{sl}(\bar{W}_\sigma) \oplus \mathfrak{sl}(\bar{W}_\tau) \tag{54}$$

or

$$\bar{\mathfrak{g}}_{\sigma,\tau} \cong \mathfrak{sl}(\bar{W}_\sigma) \cong \mathfrak{sl}(\bar{W}_\tau) \cong \mathfrak{sl}(2, \bar{k}).$$

In the latter case the $\bar{\mathfrak{g}}_{\sigma,\tau}$-modules $\bar{W}_\sigma$ and $\bar{W}_\tau$ are isomorphic, because the Lie algebra $\mathfrak{sl}(2, \bar{k})$ has precisely one nontrivial 2-dimensional representation over $\bar{k}$, up to an isomorphism. This implies that the $\bar{\mathfrak{g}}$-modules $\bar{W}_\sigma$ and $\bar{W}_\tau$ are isomorphic as well and therefore the centralizer $\mathrm{End}_{\bar{\mathfrak{g}}}(\bar{W})$ is noncommutative, which is not the case. The obtained contradiction proves that equality (54) holds for any σ, τ. Now, it follows from lemma on p. 790–791 of [23] that

$$\bar{\mathfrak{g}} = \oplus_{\sigma\in\Sigma_K} \bar{\mathfrak{g}}_\sigma = \oplus_{\sigma\in\Sigma_K} \mathfrak{sl}(\bar{W}_\sigma).$$

This implies that

$$\dim_{\bar{k}}(\bar{\mathfrak{g}}) = 3g = \dim_k \mathfrak{sl}(W/K).$$

By (47), $\mathfrak{g} \subset \mathfrak{sl}(W/K)$. Taking into account that $\dim_k(\mathfrak{g}) = \dim_{\bar{k}}(\bar{\mathfrak{g}})$, we conclude that $\dim_k(\mathfrak{g}) = \dim_k \mathfrak{sl}(W/K)$. This implies that $\mathfrak{g} = \mathfrak{sl}(W/K)$, which ends the proof. □

Theorem 6.3 *Let T be a simple complex torus of dimension $g > 2$. Suppose that* $\mathrm{End}^0(T)$ *is a number field E of degree g and*

$$\mathrm{Hdg}(T) = \mathrm{Res}_{E/\mathbb{Q}}\mathrm{SL}((\mathrm{H}_1(T, \mathbb{Q})/E)\,.$$

Then the following conditions are equivalent.

(i) T *is 2-simple.*
(ii) E *is almost doubly transitive.*

It follows from Remark 1.8(ii) that Theorem 6.3 is an immediate corollary of the following observation applied to

$$k = \mathbb{Q},\ K = E,\ W = \mathrm{H}_1(T, \mathbb{Q}).$$

Lemma 6.4 *Let g be a positive integer ≥ 2, W a $2g$-dimensional vector space over a field k of characteristic 0, $\mathfrak{g} \subset \mathrm{End}_k(W)$ a linear semisimple k-subalgebra such that the centralizer of $\mathfrak{g}$*

$$K := \mathrm{End}_{\mathfrak{g}} W \subset \mathrm{End}_k(W)$$

is an overfield of k with $[K : k] = g$, and $\mathfrak{g} = \mathfrak{sl}(W/K)$ is the Lie algebra of traceless K-linear operators in W.

Then:

- *(a1) The subspace $\left(\wedge_k^2 W\right)^{\mathfrak{g}}$ of $\mathfrak{g}$-invariants does* not *coincide with the whole space $\wedge_k^2 W$.*
- *(a2) The subspace $\mathrm{Hom}(\wedge_k^2 W, k)^{\mathfrak{g}}$ of $\mathfrak{g}$-invariants does* not *coincide with the whole space $\mathrm{Hom}(\wedge_k^2 W, k)$.*
- *(b) The following conditions are equivalent:*
  - *(b1) The $\mathfrak{g}$-module $\wedge_k^2 W$ is a direct sum of its submodule $\left(\wedge_k^2 W\right)^{\mathfrak{g}}$ of $\mathfrak{g}$-invariants and a simple $\mathfrak{g}$-module.*
  - *(b2) The $\mathfrak{g}$-module $\mathrm{Hom}(\wedge_k^2 W, k)$ is a direct sum of its submodule $\mathrm{Hom}(\wedge_k^2 W, k)^{\mathfrak{g}}$ of $\mathfrak{g}$-invariants and a simple $\mathfrak{g}$-module.*
  - *(b3) Let $\mathrm{Gal}(k) = \mathrm{Aut}(\bar{k}/k)$ be the absolute Galois group of k. Let Σ_K be the set of k-linear field embeddings $K \hookrightarrow \bar{k}$. Then the natural action of $\mathrm{Gal}(k)$ on Σ_K is almost doubly transitive.*

Remark 6.5 The equivalence of properties (a1) and (a2) and the equivalence of conditions (b1) and (b2) follow readily from the semisimplicity of $\mathfrak{g}$.

Corollary 6.6 *Let T be a simple complex torus of dimension 3. Suppose that $\mathrm{End}^0(T)$ is a cubic number field E and*

$$\mathrm{Hdg}(T) = \mathrm{Res}_{E/\mathbb{Q}} \mathrm{SL}((\mathrm{H}_1(T, \mathbb{Q})/E).$$

Then T is 2-simple.

Proof of Corollary 6.6 The result follows readily from Theorem 6.3, since every transitive action on the 3-element set Σ_E is almost doubly transitive. □

Proof of Lemma 6.4 We use the notation of the Proof of Lemma 6.2. In particular,

$$\bar{\mathfrak{g}}_\sigma = \mathfrak{sl}(\bar{W}_\sigma),\ \bar{\mathfrak{g}}_{\sigma,\tau} = \mathfrak{sl}(\bar{W}_\sigma) \oplus \mathfrak{sl}(\bar{W}_\tau)\ \forall \sigma, \tau \in \Sigma_K, \sigma \neq \tau;$$

$$\bar{W} = \oplus_{\sigma \in \Sigma_K} \bar{W}_\sigma,\ \bar{\mathfrak{g}} = \oplus_{\sigma \in \Sigma_K} \bar{\mathfrak{g}}_\sigma\ .$$

Let us start with the $\mathfrak{g}$-module

$$W^{\otimes 2} := W \otimes_k W \to W.$$

There is an involution

$$\delta : W^{\otimes 2} \to W^{\otimes 2},\ u \otimes v \mapsto v \otimes u,$$

whose subspace of invariants is the symmetric square $\mathbf{S}_k^2 W$ of W and the subspace of anti-invariants is the exterior square $\wedge_k^2 W$. Clearly, δ commutes with the action of $\mathfrak{g}$; in particular, both $\mathbf{S}_k^2 W$ and $\wedge_k^2 W$ are $\mathfrak{g}$-invariant subspaces of the tensor square of W.

Let us consider the $\bar{\mathfrak{g}}$-module

$$\bar{W}^{\otimes 2} := \bar{W} \otimes_{\bar{k}} \bar{W}.$$

Extending by $\bar{k}$-linearity the involution δ, we get the involution

$$\bar{\delta} : \bar{W}^{\otimes 2} \to \bar{W}^{\otimes 2},\ u \otimes v \mapsto v \otimes u,$$

whose subspace of invariants is the symmetric square $\mathbf{S}_{\bar{k}}^2 \bar{W}$ of $\bar{W}$ and the subspace of anti-invariants is the exterior square $\wedge_{\bar{k}}^2 \bar{W}$. Clearly, $\bar{\delta}$ commutes with the action of $\bar{\mathfrak{g}}$; in particular, both $\mathbf{S}_{\bar{k}}^2 \bar{W}$ and $\wedge_{\bar{k}}^2 \bar{W}$ are $\bar{\mathfrak{g}}$-invariant subspaces of $\bar{W}^{\otimes 2}$.

Let us choose an *order* on Σ_K. Let $\Sigma_{K,2}$ be the set of all two-element subsets B of Σ_K with

$$B = \{\sigma, \tau\};\ \sigma, \tau \in \Sigma_K;\ \sigma < \tau. \tag{55}$$

Let us consider the following decomposition of the $\bar{\mathfrak{g}}$-module $\bar{W}^{\otimes 2}$ into a direct sum of $\bar{\delta}$-invariant $\bar{\mathfrak{g}}$-submodules

$$\bar{W}^{\otimes 2} = \left(\oplus_{\sigma \in \Sigma_K} \bar{W}_\sigma \otimes_{\bar{k}} \bar{W}_\sigma \right) \oplus \left(\oplus_{B = \{\sigma,\tau\} \in \Sigma_{K,2}} (\bar{W}_\sigma \otimes_{\bar{k}} \bar{W}_\tau) \oplus (\bar{W}_\tau \otimes_{\bar{k}} \bar{W}_\sigma) \right). \tag{56}$$

Clearly, the action of the Lie algebra $\bar{\mathfrak{g}}$ on the tensor product $\bar{W}_\sigma \otimes_{\bar{k}} \bar{W}_\sigma$ factors through

$$\bar{\mathfrak{g}}_\sigma = \mathfrak{sl}(\bar{W}_\sigma),$$

while the action of $\bar{\mathfrak{g}}$ on

$$\bar{W}^{(B)} := (\bar{W}_\sigma \otimes_{\bar{k}} \bar{W}_\tau) \oplus (\bar{W}_\tau \otimes_{\bar{k}} \bar{W}_\sigma)$$

factors through

$$\bar{\mathfrak{g}}_{(B)} := \bar{\mathfrak{g}}_{\sigma,\tau} = \mathfrak{sl}(\bar{W}_\sigma) \oplus \mathfrak{sl}(\bar{W}_\tau) \text{ with } B = \{\sigma, \tau\}.$$

We have

$$\bar{W}_\sigma \otimes_{\bar{k}} \bar{W}_\sigma = \mathbf{S}^2_{\bar{k}} \bar{W}_\sigma \oplus \wedge^2_{\bar{k}} \bar{W}_\sigma$$

where first summand is a simple $\bar{\mathfrak{g}}_\sigma$-module that lies in $\mathbf{S}^2_{\bar{k}} \bar{W}$ while the action of $\bar{\mathfrak{g}}_\sigma$ (and therefore of $\bar{\mathfrak{g}}$) on the second one is *trivial.*

Both $\bar{\mathfrak{g}}_{(B)} = \bar{\mathfrak{g}}_{\sigma,\tau}$-modules $\bar{W}_\sigma \otimes_{\bar{k}} \bar{W}_\tau$ and $\bar{W}_\tau \otimes_{\bar{k}} \bar{W}_\sigma$ are faithful simple; in addition, they are isomorphic. Let us split $\bar{W}^{(B)}$ into a direct sum

$$\bar{W}^{(B)} = \bar{W}^{(B)}_+ \oplus \bar{W}^{(B)}_-$$

of the subspaces $\bar{W}^{(B)}_+$ of $\bar{\delta}$-invariants and $\bar{W}^{(B)}_-$ of $\bar{\delta}$-anti-invariants. Clearly, both subspaces are nonzero $\bar{\mathfrak{g}}_{\sigma,\tau}$-invariant subspaces and therefore are *nontrivial simple* $\bar{\mathfrak{g}}_{\sigma,\tau}$-modules that are isomorphic to

$$\bar{W}_\sigma \otimes_{\bar{k}} \bar{W}_\tau \cong \bar{W}_\tau \otimes_{\bar{k}} \bar{W}_\sigma.$$

The last sentence remains true if we replace "$\bar{\mathfrak{g}}_{\sigma,\tau}$-modules" by "$\bar{\mathfrak{g}}$-modules." Obviously,

$$\bar{W}^{(B)}_+ \subset \mathbf{S}^2_{\bar{k}} \bar{W}, \; \bar{W}^{(B)}_- \subset \wedge^2_{\bar{k}} \bar{W}.$$

It follows from (56) that the $\bar{\mathfrak{g}}$-module $\wedge^2_{\bar{k}} \bar{W}$ splits into a direct sum of the trivial $\bar{\mathfrak{g}}$-module $\oplus_{\sigma \in \Sigma_K} \wedge^2_{\bar{k}} \bar{W}_\sigma$ and a direct sum $\oplus_{B \in \Sigma_{K,2}} \bar{W}^{(B)}_-$ of nontrivial mutually non-isomorphic simple $\bar{\mathfrak{g}}$-modules $\oplus_{B \in \Sigma_{K,2}} \bar{W}^{(B)}_-$. So,

$$\mathbf{S}^2_{\bar{k}} \bar{W} = \left(\oplus_{\sigma \in \Sigma_K} \mathbf{S}^2_{\bar{k}} \bar{W}_\sigma\right) \oplus \left(\oplus_{B \in \Sigma_{K,2}} \bar{W}^{(B)}_+\right); \tag{57}$$

$$\wedge^2_{\bar{k}} \bar{W} = \left(\oplus_{\sigma \in \Sigma_K} \wedge^2_{\bar{k}} \bar{W}_\sigma\right) \oplus \left(\oplus_{B \in \Sigma_{K,2}} \bar{W}^{(B)}_-\right). \tag{58}$$

Clearly,

$$\bar{W}^{\{0\}}_- := \oplus_{\sigma \in \Sigma_K} \wedge^2_{\bar{k}} \bar{W}_\sigma$$

coincides with the subspace of all $\bar{\mathfrak{g}}$-invariants in $\wedge^2_{\bar{k}} \bar{W}$.

Notice that $\mathrm{Gal}(k)$ acts naturally on both Σ_K and $\Sigma_{K,2}$ in such a way that for all $s \in \mathrm{Gal}(k)$

$$s\left(\wedge_{\bar{k}}^{2}\bar{W}_{\sigma}\right)=\wedge_{\bar{k}}^{2}\bar{W}_{s\sigma},\ s\left(\bar{W}_{-}^{(B)}\right)=\bar{W}_{-}^{(sB)}. \tag{59}$$

It follows that

$$s\bar{W}_{-}^{\{0\}}=\bar{W}_{-}^{\{0\}}.$$

It is also clear that if we put

$$\bar{U}:=\oplus_{B\in\Sigma_{K,2}}\bar{W}_{-}^{(B)}\subset\wedge_{\bar{k}}^{2}W. \tag{60}$$

then $s\bar{U}=\bar{U}$ for all $s\in\mathrm{Gal}(k)$. This implies that both $\bar{W}_{-}^{\{0\}}$ and $\bar{U}$ are *nonzero* $\bar{k}$-vector subspaces that are defined over k, i.e., there are *nonzero* vector k-subspaces $W_{-}^{\{0\}}$ and U of $\wedge_{k}^{2}W$ such that

$$\bar{W}_{-}^{\{0\}}=W_{-}^{\{0\}}\otimes_{k}\bar{k},\ \bar{U}=U\otimes_{k}\bar{k}.$$

It follows from (58) and (60) that

$$\wedge_{k}^{2}W=W_{-}^{\{0\}}\oplus U. \tag{61}$$

The $\bar{\mathfrak{g}}$-invariance of both $\bar{k}$-vector subspaces $\bar{W}_{-}^{\{0\}}$ and $\bar{U}$ implies that both k-vector subspaces $W_{-}^{\{0\}}$ and U are $\mathfrak{g}$-submodules of $\wedge_{k}^{2}W$. It is also clear that $W_{-}^{\{0\}}$ coincides with the subspace $\left(\wedge_{k}^{2}W\right)^{\mathfrak{g}}$ of all $\mathfrak{g}$-invariants in $\wedge_{k}^{2}W$. Since $U\neq\{0\}$, it follows from (61) that

$$\left(\wedge_{k}^{2}W\right)^{\mathfrak{g}}\neq\wedge_{k}^{2}W, \tag{62}$$

which proves (a1) that, in turn, implies (a2), in light of Remark 6.5.

Combining the equality $\left(\wedge_{k}^{2}W\right)^{\mathfrak{g}}=W_{-}^{\{0\}}$ with (61), we obtain that property (b1) of our lemma is equivalent to the simplicity of the $\mathfrak{g}$-module U.

Let O be a $\mathrm{Gal}(k)$-*orbit* in $\Sigma_{K,2}$. Let us consider the corresponding $\bar{\mathfrak{g}}$-*submodule* of $\bar{U}$ defined by

$$\bar{U}^{O}=\sum_{B\in O}\bar{W}_{-}^{(B)}. \tag{63}$$

Clearly, $s\bar{U}^{O}=\bar{U}^{O}$ for all $s\in\mathrm{Gal}(k)$. This means that $\bar{U}^{O}$ is defined over k, i.e., there is a $\mathfrak{g}$-submodule U^{O} of U such that

$$\bar{U}^{O}=U^{O}\otimes_{k}\bar{k}.$$

Since all the summands in the RHS of (63) are mutually non-isomorphic simple $\bar{\mathfrak{g}}$-modules that (in light of (59)) are permuted transitively by $\mathrm{Gal}(k)$, we conclude that U^O is a *simple* $\mathfrak{g}$-submodule of U. Clearly, $U^O = U$ if and only if $O = \Sigma_{K,2}$, i.e., if and only if the action of $\mathrm{Gal}(k)$ on $\Sigma_{K,2}$ is *transitive*. This implies that the $\mathfrak{g}$-module U is simple if and only if the action of $\mathrm{Gal}(k)$ on $\Sigma_{K,2}$ is *transitive*. It follows that conditions (b1) and (b3) of our lemma are equivalent. By Remark 6.5, conditions (b1) and (b2) are equivalent. This ends the proof. □

Theorem 6.7 *Let E be a number field of degree $g > 2$. Then there exists a simple g-dimensional complex torus $T = V/\Lambda$ such that*

$$\mathrm{End}^0(T) = E, \ \mathrm{Hdg}(T) = \mathrm{Res}_{E/\mathbb{Q}}\mathrm{SL}((\mathrm{H}_1(T, \mathbb{Q})/E).$$

In particular, T is 2-simple if and only if E is almost doubly transitive.

Proof Let us consider the matrix

$$J_0 = \begin{bmatrix} 0 & 1 \\ -1 & 0 \end{bmatrix} \in \mathrm{Mat}_2(\mathbb{Q}) \subset \mathrm{Mat}_2(E) \subset \mathrm{Mat}_2(E_{\mathbb{R}})$$

where $E_{\mathbb{R}} := E \otimes_{\mathbb{Q}} \mathbb{R}$ is the realification of E. By [21, Prop. 2.8 on p. 19], there is $u \in \mathrm{Mat}_2(E_{\mathbb{R}})$ such that

$$J = \exp(u) J_0 \exp(-u) = \exp(u) J_0 \, (\exp(u))^{-1} \in \mathrm{Mat}_2(E_{\mathbb{R}})$$

enjoys the following property. If D is a $\mathbb{Q}$-subalgebra of $\mathrm{Mat}_2(E)$ such that $D_{\mathbb{R}} = D \otimes_{\mathbb{Q}} \mathbb{R}$ contains J then $D = \mathrm{Mat}_2(E)$. Notice that

$$J_0^2 = -1, \quad J_0 \in \mathfrak{sl}_2(E_{\mathbb{R}}) \subset \mathrm{Mat}_2(E_{\mathbb{R}}).$$

It follows that

$$J^2 = -1, \quad J \in \mathfrak{sl}_2(E_{\mathbb{R}}) \subset \mathrm{Mat}_2(E_{\mathbb{R}}).$$

Let $\mathfrak{g}$ be the smallest $\mathbb{Q}$-Lie subalgebra of $\mathrm{Mat}_2(E)$ such that its realification

$$\mathfrak{g}_{\mathbb{R}} = \mathfrak{g} \otimes_{\mathbb{Q}} \mathbb{R} \subset \mathrm{Mat}_2(E) \otimes_{\mathbb{Q}} \mathbb{R} = \mathrm{Mat}_2(E_{\mathbb{R}})$$

contains J. Clearly,

$$\mathfrak{g} \subset \mathfrak{sl}_2(E) \tag{64}$$

and the $\mathbb{Q}$-subalgebra of $\mathrm{Mat}_2(E)$ generated by $\mathfrak{g}$ coincides with $\mathrm{Mat}_2(E)$. It makes the $2g$-dimensional $\mathbb{Q}$-vector space

$$E^2 = E \oplus E$$

a faithful simple $\mathfrak{g}$-module such that the centralizer of $\mathfrak{g}$ in $\mathrm{End}_{\mathbb{Q}}(E^2)$ coincides with E. This implies that $\mathfrak{g}$ is a reductive $\mathbb{Q}$-Lie algebra and its center lies in E. In light of (64), this center is $\{0\}$, i.e., $\mathfrak{g}$ is a semisimple $\mathbb{Q}$-Lie algebra. Applying Lemma 6.2 to

$$k = \mathbb{Q},\ K = E,\ W = E^2,$$

we conclude that

$$\mathfrak{g} = \mathfrak{sl}_2(E). \tag{65}$$

Now we are ready to construct the desired complex torus T. The operator J provides the structure of a complex vector space on

$$V := E^2 \otimes_{\mathbb{Q}} \mathbb{R} = E^2_{\mathbb{R}} = E_{\mathbb{R}} \oplus E_{\mathbb{R}}$$

such that $J \in \mathrm{End}_{\mathbb{R}}(V)$ defines multiplication by $\mathbf{i}$. Pick any $\mathbb{Z}$-lattice of rank $2g$ in E^2 and put $T := V/\Lambda$. One may naturally identify $\Lambda \otimes \mathbb{Q}$ with E^2. In light of Theorem 5.2, the $\mathbb{Q}$-Lie algebra hdg_T coincides with $\mathfrak{g}$, i.e., $\mathrm{hdg}_T = \mathfrak{sl}_2(E)$. It follows that $\mathrm{Hdg}(T) = \mathrm{Res}_{E/\mathbb{Q}}\mathrm{SL}((\mathrm{H}_1(T, \mathbb{Q})/E)$, which ends the proof of the first assertion of our theorem. Now the second one follows from Theorem 6.3. □

Proof of Theorem 1.9 The first assertion follows readily from Theorem 6.7 combined with Theorem 4.6 applied to $n = g$.

In order to prove the second assertion, one should take

$$\mathbf{s} = g - d - 1 \geq 0, \quad \mathbf{r} = g - 2\mathbf{s} = g - 2(g - d - 1) = 2(d + 1) - g \geq 0.$$

□

7 Semi-linear Algebra

This section contains auxiliary results that will be used for the study of Hodge groups of 2-simple tori without nontrivial endomorphisms. In what follows k stands for a field of characteristic 0 and K for an overfield of k such that the automorphism group $\mathrm{Aut}(K/k)$ of k-linear automorphisms of K enjoys the following property.

The subfield $K^{\mathrm{Aut}(K/k)}$ of $\mathrm{Aut}(K/k)$-invariants coincides with k. (This property holds if K is an algebraically closed field; see Proposition 8.2.)

Definition 7.1 Let $\mathcal{V}$ be a finite-dimensional vector space over K and $\sigma \in \mathrm{Aut}(K/k)$. Then the finite-dimensional vector space ${}^{\sigma}V$ over K is defined as follows. Viewed as an additive group, ${}^{\sigma}V$ coincides with V but multiplication by elements $a \in K$ is defined in ${}^{\sigma}V$ by the formula

$$a, v \mapsto \sigma^{-1}(a)v.$$

Clearly,

$$\dim_K(\mathcal{V}) = \dim_K({}^{\sigma}\mathcal{V}).$$

Remark 7.2

(1) If $x \in \mathrm{End}_K(\mathcal{V})$ is a K-linear operator in $\mathcal{V}$ then

$$x(\sigma^{-1}(a)v) = \sigma^{-1}(a)x(v) \quad \forall a \in K, v \in \mathcal{V}.$$

In other words, one may view x as the K-linear operator in ${}^{\sigma}\mathcal{V}$ that we denote by ${}^{\sigma}\mathrm{id}(x) \in \mathrm{End}_K({}^{\sigma}\mathcal{V})$.

(2) Let $m := \dim_K(\mathcal{V}) > 0$, and $\{e_1, \ldots, e_m\}$ be a basis of $\mathcal{V}$. Then one may view $\{e_1, \ldots, e_m\}$ as a basis of ${}^{\sigma}\mathcal{V}$.
If $A = (a_{ij})_{i,j=1}^m$ is the matrix of $x \in \mathrm{End}_K(\mathcal{V})$ with respect to $\{e_1, \ldots, e_m\}$ then obviously $\sigma(A) = (\sigma(a_{ij}))_{i,j=1}^m$ is the matrix of ${}^{\sigma}x \in \mathrm{End}_K({}^{\sigma}\mathcal{V})$ with respect to $\{e_1, \ldots, e_m\}$.

Lemma 7.3 *The formula*

$$\begin{gathered} {}^{\sigma}\mathrm{id} : \mathrm{End}_K(\mathcal{V}) \to \mathrm{End}_K({}^{\sigma}\mathcal{V}), x \mapsto {}^{\sigma}\mathrm{id}(x) = \{v \mapsto x(v)\} \\ \forall x \in \mathrm{End}_K(\mathcal{V}),\ v \in {}^{\sigma}\mathcal{V} = \mathcal{V} \end{gathered}$$

defines a ring isomorphism that enjoys the following properties:

(i) ${}^{\sigma}\mathrm{id}(ax) = \sigma(a) \cdot {}^{\sigma}\mathrm{id}(x) \quad \forall a \in K, x \in \mathrm{End}_K(\mathcal{V})$.

(ii) *Let*

$$\mathcal{P}_{x,\min}(t),\ \mathcal{P}_{x,\mathrm{char}}(t) \in K[t]$$

be the minimal and characteristic polynomials of x, respectively.
Then the minimal and characteristic polynomials of ${}^{\sigma}\mathrm{id}(x)$ coincide with $\sigma(\mathcal{P}_{x,\min}(t))$ and $\sigma(\mathcal{P}_{x,\mathrm{char}}(t))$, respectively.

(iii) *If $a \in K$ is the trace of $x \in \mathrm{End}_K(\mathcal{V})$ then $\sigma(a)$ is the trace of ${}^{\sigma}\mathrm{id}(x) \in \mathrm{End}_K({}^{\sigma}\mathcal{V})$.*

Proof (i) is obvious. Both assertions (ii) and (iii) follow from Remark 7.2. □

Let $\mathcal{V}_0$ be a finite-dimensional k-vector space and

$$\mathcal{V} := \mathbf{T}_{k,K}(\mathcal{V}) = \mathcal{V}_0 \otimes_k K$$

the corresponding K-vector space endowed by the following semi-linear action of $\mathrm{Aut}(K/k)$.

$$\sigma(v_0 \otimes a) = v_0 \otimes \sigma(a) \quad \forall a \in K, v_0 \in \mathcal{V}_0.$$

We will identify $\mathcal{V}_0$ with the k-vector subspace

$$\mathcal{V}_0 \otimes 1 = \{v_0 \otimes 1 \mid v_0 \in \mathcal{V}_0\} \subset \mathcal{V}_0 \otimes_k K = \mathcal{V}.$$

Clearly, the k-vector subspace $\mathcal{V}_0 = \mathcal{V}_0 \otimes 1$ coincides with the k-vector subspace $\mathcal{V}^{\mathrm{Aut}(K/k)}$ of $\mathrm{Aut}(K/k)$-invariants.

The next assertion is probably known but I was unable to find a reference.

Lemma 7.4 *Let $\mathcal{W}$ be a K-vector subspace of $\mathcal{V}$. Then the following conditions are equivalent:*

(i) $\mathcal{W}$ *is* $\mathrm{Aut}(K/k)$*-invariant.*
(ii) *There exists a k-vector subspace $\mathcal{W}_0$ of $\mathcal{V}_0$ such that*

$$\mathcal{W} = \mathbf{T}_{k,K}(\mathcal{W}_0) = \mathcal{W}_0 \otimes_k K \subset \mathcal{V}_0 \otimes_k K = \mathbf{T}_{k,K}(\mathcal{V}_0).$$

If this is the case then $\mathcal{W}_0 = \mathcal{W} \cap \mathcal{V}_0$.

Proof Let us put

$$m := \dim_k(\mathcal{V}_0) = \dim_K(\mathcal{V}); \quad n := \dim_K(\mathcal{W}) \le m.$$

If either $n = 0$ or $n = m$ then the desired result is obvious. So, we may and will assume that $0 < n < m$, i.e.,

$$1 \le n \le m-1; \quad m \ge 2.$$

Let us fix a k-basis $\{e_1, \dots, e_m\}$ of $\mathcal{V}_0$, which we will view as a K-basis of $\mathcal{V}$.

Step 1. Assume that $n = 1$. Take a nonzero vector $w \in \mathcal{W}$. Then at least one of its coefficients with respect to our basis is not 0, i.e.,

$$w = \sum_{i=1}^{n} a_i e_i, \ a_i \in K$$

and $\exists j \in \{1, \dots, n\}$ such that $a_j \ne 0$. Replacing w by $a_j^{-1} w \in \mathcal{W}$, we may and will assume that $a_j = 1$. Then

$$K \cdot w = \mathcal{W} \ni \sigma(w) = \sum_{i=1}^{n} \sigma(a_i) e_i \quad \forall \sigma \in \mathrm{Aut}(K/k).$$

We have

$$\sigma(a_j) = \sigma(1) = 1 = a_j$$

and

$$\sigma(w) \in \mathcal{W} = K \cdot w.$$

Since both w and $\sigma(w)$ have the same (nonzero) jth coordinate, we conclude that $\sigma(w) = w$ for all σ, i.e., all the coefficients $a_i \in k$ and therefore $w \in \mathcal{V}_0$ and $\mathcal{W} = \mathcal{W}_0 \otimes_k K$ with

$$\mathcal{W}_0 = k \cdot w \subset \mathcal{V}_0.$$

So, we have proven our assertion in the case of $n = 1$.

Step 2. Let us prove that the k-vector subspace of $\mathrm{Aut}(K)$-invariants

$$\tilde{\mathcal{W}}_0 := \mathcal{W}^{\mathrm{Aut}(K/k)} = \mathcal{W} \cap V_0 \tag{66}$$

is *not* $\{0\}$. Let us use induction by n and m. By Step 1, our assertion is true for $n = 1$. This implies its validity for $m = 2$. So, we may assume that

$$1 < n < m, \quad m > 2.$$

Let us consider the hyperplanes

$$\mathcal{H}_0 = \sum_{i=1}^{m-1} k \cdot e_i \subset V_0, \quad \mathcal{H} = \mathbf{T}_{k,K}(\mathcal{H}_0) = \mathcal{H}_0 \otimes_k K = \sum_{i=1}^{m-1} K \cdot e_i \subset \mathcal{V}.$$

Clearly, the intersection

$$\mathcal{W}_H = \mathcal{W} \cap \mathcal{H} \subset \mathcal{H}$$

is an $\mathrm{Aut}(K/k)$-invariant subspace of both $\mathcal{W}$ and $\mathcal{H}$. Clearly, either $\mathcal{W}_H = \mathcal{W}$ or $\dim_K(\mathcal{W}_H) = n - 1 > 0$. In the former case $\mathcal{W}_H$ is $\mathrm{Aut}(K/k)$-invariant subspace of the $(m-1)$-dimensional K-vector space $\mathcal{H} = \mathcal{H}_0 \otimes_k K$. Now the induction assumption for m (applied to $\mathcal{H}$ instead of $\mathcal{V}$) implies that

$$\tilde{\mathcal{W}}_0 = \mathcal{W}^{\mathrm{Aut}(K/k)} = (\mathcal{W}_H)^{\mathrm{Aut}(K/k)} \neq 0.$$

In the latter case, the induction assumption for n applied to $\mathcal{W}_H$ implies that $(\mathcal{W}_H)^{\mathrm{Aut}(K/k)} \neq 0$. Since $\mathcal{W} \supset \mathcal{W}_H$, we get $\mathcal{W}^{\mathrm{Aut}(K/k)} \neq 0$, which ends the proof.

Step 3. We have

$$\mathbf{T}_{k,K}(\tilde{\mathcal{W}}_0) = \tilde{\mathcal{W}}_0 \otimes_k K \subset \mathcal{W}.$$

This implies that

$$n_0 = \dim_k(\tilde{\mathcal{W}}_0) \leq \dim_K(W) = n.$$

The assertion of our lemma actually means that the equality holds. By Step 2, $n_0 > 0$. Suppose that $n_0 < n$ and choose in $\mathcal{V}_0$ a $(m - n_0)$-dimensional k-vector subspace $\mathcal{U}_0$ such that $\mathcal{U}_0 \cap \tilde{\mathcal{W}}_0 = \{0\}$ (i.e., $\mathcal{V}_0 = \mathcal{W}_0 \oplus \mathcal{U}_0$). Let us consider the $(m - n_0)$-dimensional K-vector subspace

$$\mathcal{U} = \mathbf{T}_{k,K}(\mathcal{U}_0) = \mathcal{U}_0 \otimes_k K \subset \mathcal{V}.$$

Clearly,

$$\mathcal{V} = \mathbf{T}_{k,K}(\tilde{\mathcal{W}}_0) \oplus \mathbf{T}_{k,K}(\tilde{\mathcal{U}}_0) = \mathbf{T}_{k,K}(\tilde{\mathcal{W}}_0) \oplus \mathcal{U},$$

and therefore

$$\mathcal{U} \cap \mathbf{T}_{k,K}(\tilde{\mathcal{W}}_0) = \{0\}.$$

Dimension arguments imply that $\mathcal{U}_1 := \mathcal{U} \cap \mathcal{W}$ is a nonzero $\mathrm{Aut}(K/k)$-invariant K-vector subspace of $\mathcal{V}$. By Step 2, the subspace $\tilde{\mathcal{U}}_1 := \mathcal{U}_1^{\mathrm{Aut}(K/k)} \neq \{0\}$; on the other hand, $\tilde{\mathcal{U}}_1$ obviously lies in $\mathcal{W}^{\mathrm{Aut}(K/k)}$ but meets the latter only at $\{0\}$. The obtained contradiction proves that $n_0 = n$, which ends the proof. □

Remark 7.5 Let us consider the dual vector spaces

$$\mathcal{V}_0^* = \mathrm{Hom}_k(\mathcal{V}_0, k), \quad \mathcal{V}^* = \mathrm{Hom}_K(\mathcal{V}, K).$$

Obviously, the restriction map

$$\mathrm{res}_{K,k} : \mathcal{V}^* = \mathrm{Hom}_K(\mathcal{V}, K) = \mathrm{Hom}_K(\mathcal{V}_0 \otimes_k K, K) \to \mathrm{Hom}_k(\mathcal{V}_0, K),$$
$$\phi \mapsto \{v_0 \mapsto \phi(v_0 \otimes 1)\}$$

is a $\mathrm{Aut}(K/k)$-equivariant isomorphism of K-vector spaces where the actions of $\mathrm{Aut}(K/k)$ are defined as follows:

$$\sigma : \phi \mapsto \sigma \circ \phi \circ \sigma^{-1} \quad \forall \phi \in \mathrm{Hom}_K(\mathcal{V}, K),$$

$$\sigma : \phi_0 \mapsto \{v_0 \mapsto \sigma(\phi_0(v_0)) \quad \forall \phi_0 \in \mathrm{Hom}_k(\mathcal{V}_0, K)$$

for all $\sigma \in \mathrm{Aut}(K/k)$. As usual, we have

$$\sigma(\phi)(\sigma(v)) = \sigma(\phi(v)) \quad \forall v \in V, \phi \in V^*, \sigma \in \mathrm{Aut}(K/k).$$

7.6 What is discussed in this section (and in Theorem 8.6) is pretty well known in the case of $k = \mathbb{R}$ and $K = \mathbb{C}$; see [20].

Let $\mathfrak{u}$ be a Lie k-algebra of finite dimension and

$$\bar{\mathfrak{u}} := \mathfrak{u} \otimes_k K$$

the corresponding finite-dimensional Lie K-algebra. Let

$$\rho : \mathfrak{u} \to \mathrm{End}_K(\mathcal{V})$$

be a homomorphism of Lie k-algebras. Extending ρ by K-linearity, we get the homomorphism of Lie K-algebras

$$\bar{\rho} : \bar{\mathfrak{u}} \to \mathrm{End}_K(\mathcal{V}),$$

which coincides with ρ on

$$\mathfrak{u} = \mathfrak{u} \otimes 1 \subset \mathfrak{u} \otimes_k K = \bar{\mathfrak{u}}.$$

Thus $\bar{\rho}$ endows $\mathcal{V}$ with the structure of a $\bar{\mathfrak{u}}$-module.

If $\sigma \in \mathrm{Aut}(K/k)$ then we may define the composition

$${}^{\sigma}\rho : \mathfrak{u} \stackrel{\rho}{\to} \mathrm{End}_K(\mathcal{V}) \stackrel{{}^{\sigma}\mathrm{id}}{\to} \mathrm{End}_K({}^{\sigma}\mathcal{V}),$$

which is a homomorphism of k-Lie algebras. Then the corresponding homomorphism of Lie K-algebras

$$\overline{{}^{\sigma}\rho} : \bar{\mathfrak{u}} \to \mathrm{End}_K({}^{\sigma}\mathcal{V}),$$

provides ${}^{\sigma}\mathcal{V}$ with the structure of a $\bar{\mathfrak{u}}$-module.

Remark 7.7 Let $\mathcal{W}$ be a K-vector subspace of $\mathcal{V}$. Clearly, $\mathcal{W}$ is $\mathfrak{u}$-invariant if and only if it is $\bar{\mathfrak{u}}$-invariant. It follows easily that $\mathcal{W}$ is a $\bar{\mathfrak{u}}$-submodule of $\mathcal{V}$ if and only if it is a $\bar{\mathfrak{u}}$-submodule of ${}^{\sigma}\mathcal{V}$. This implies that the $\bar{\mathfrak{u}}$-module $\mathcal{V}$ is *simple* if and only if the $\bar{\mathfrak{u}}$-module ${}^{\sigma}\mathcal{V}$ is *simple*.

7.8 Let $\mathcal{V}_0$ be a finite-dimensional k-vector space endowed with a homomorphism of k-Lie algebras

$$\rho_0 : \mathfrak{u} \to \mathrm{End}_k(\mathcal{V}_0)$$

that endowed $\mathcal{V}_0$ with the structure of a $\mathfrak{u}$-module. Let us consider the K-vector space $\mathcal{V} := \mathcal{V}_0 \otimes_k K$ and the obvious homomorphism of k-Lie algebras

$$\rho_0 \otimes 1 : \mathfrak{u} = \mathfrak{u} \otimes 1 \to \mathrm{End}_k(\mathcal{V}_0) \otimes_k K = \mathrm{End}_K(\mathcal{V}_0 \otimes K) = \mathrm{End}_K(\mathcal{V}).$$

obtained from ρ_0 by extension of scalars.

Let $\mathcal{W}$ be a $\mathfrak{u}$-invariant K-vector subspace of $\mathcal{V}$. If $\sigma \in \text{Aut}(K/k)$ then obviously $\sigma(\mathcal{W})$ is also a $\mathfrak{u}$-invariant K-vector subspace of $\mathcal{V}$. Clearly, both $\mathcal{W}$ and $\sigma(\mathcal{W})$ carry the natural structure of modules over the Lie K-algebra

$$\bar{\mathfrak{u}} = \mathfrak{u} \otimes_k K.$$

We will need the following assertion.

Proposition 7.9 *The* $\bar{\mathfrak{u}}$*-modules* $\sigma(\mathcal{W})$ *and* ${}^{\sigma}\mathcal{W}$ *are isomorphic.*

Proof It suffices to check that the $\mathfrak{u}$-modules $\sigma(\mathcal{W})$ and ${}^{\sigma}\mathcal{V}$ are isomorphic. Let us consider the k-linear isomorphism

$$\Pi : \sigma(\mathcal{W}) \to {}^{\sigma}\mathcal{W}, \quad \sigma(w) \mapsto w \;\forall w \in \mathcal{W}.$$

Actually, Π is K-linear, because for all $a \in K$, $w \in \mathcal{W}$ the vector

$$\sigma^{-1}(a\sigma(w)) = \sigma^{-1}(a)w \in \mathcal{W}$$

(recall that in ${}^{\sigma}\mathcal{W}$ multiplication by a is defined as multiplication by $\sigma^{-1}(a)$). Clearly, the actions of $\mathfrak{u}$ and $\text{Aut}(K/k)$ on $\mathcal{V}$ do commute. This implies that

$$\Pi \circ \sigma = \sigma \circ \Pi \quad \forall \sigma \in \text{Aut}(K/k).$$

It follows that Π is an isomorphism of $\mathfrak{u}$-modules, which ends the proof. □

8 Representations of Semisimple Lie Algebras and Highest Weights

Throughout this section, K is an *algebraically closed* field of characteristic 0 that contains a subfield k. We will need the following assertion.

Theorem 8.1 *Let* F *be a subfield of* K*. Then every automorphism of* F *can be extended to an automorphism of* K*.*

Proof This is a special case (in characteristic 0) of Theorem 3.19 in [13]. □

In order to be able to apply results of Sect. 7 to K/k, we need the following assertion.

Proposition 8.2 $K^{\text{Aut}(K/k)} = k$.

Proof Let $c \in K \setminus k$. Thanks to Theorem 8.1, in order to prove that c is *not* $\text{Aut}(K/k)$-invariant, it suffices to find an intermediate field F such that

$$k \subset F \subset K, \quad c \in F,$$

and an automorphism $\sigma \in \operatorname{Aut}(F/k)$ such that

$$\sigma(c) \neq c.$$

If c is an algebraic over k then it is a root of a certain monic irreducible polynomial $f(t) \in k[t]$. Let $F \subset K$ be the splitting field of $f(t)$ over k. Then $c \in F$ and F/k is a finite *Galois extension*. By Galois theory,

$$F^{\operatorname{Aut}(F/k)} = k.$$

Since $c \notin k$, there exists an automorphism $\sigma \in \operatorname{Aut}(F/k)$ such that $\sigma(c) \neq c$.

If c is transcendental over k then the subfield $k(c)$ of K generated by c over k is canonically isomorphic to the field $k(t)$ of rational functions in an independent variable t over the field of constants k. Namely, there is a k-linear field isomorphism

$$\mathrm{ev}_c : k(t) \to k(c), \ \frac{u(t)}{v(t)} \mapsto \frac{u(c)}{v(c)}$$

for any polynomials $u(t), v(t) \in k[t]$ with $v(t) \neq 0$. For example, ev_c sends the rational functions $t = \frac{t}{1}$ and $t+1 = \frac{t+1}{1}$ to c and $c+1$, respectively.

Now let us put $F = k(c)$ and define its k-linear automorphism σ by

$$\sigma : \frac{u(c)}{v(c)} \mapsto \frac{u(c+1)}{v(c+1)}.$$

Then $\sigma(c) = c + 1 \neq c$.

This ends the proof. □

Recall that K is *algebraically closed* (e.g., K is an algebraic closure of k). Let $\mathfrak{g}$ be a nonzero semisimple finite-dimensional Lie algebra over k of rank l and consider the corresponding semisimple finite-dimensional Lie algebra

$$\bar{\mathfrak{g}} := \mathfrak{g} \otimes_k K$$

over K. If $\mathfrak{h}$ is a Cartan subalgebra of $\mathfrak{g}$ then

$$\dim_k(\mathfrak{h}) = l.$$

We write

$$\bar{\mathfrak{h}} := \mathfrak{h} \otimes_k K \subset \mathfrak{g} \otimes_k K = \bar{\mathfrak{g}}$$

for the corresponding Cartan subalgebra of $\bar{\mathfrak{g}}$; we have

$$\dim_K(\bar{\mathfrak{h}}) = l.$$

As usual, let us consider the dual K-vector space

$$\bar{\mathfrak{h}}^* := \mathrm{Hom}_K(\bar{\mathfrak{h}}, K)$$

of K-dimension l endowed by the action of $\mathrm{Aut}(K/k)$ defined by the formula

$$\sigma \mapsto \{\phi :\mapsto \sigma \circ \phi \circ \sigma^{-1}\} \quad \forall \phi : \bar{\mathfrak{h}} \to K$$

and $\sigma \in \mathrm{Aut}(K/k)$. As above, the restriction map

$$\mathrm{res}_{K,k} : \bar{\mathfrak{h}}^* := \mathrm{Hom}_K(\bar{\mathfrak{h}}, K) \to \mathrm{Hom}_k(\mathfrak{h}, K)$$

is an isomorphism of K-vector spaces. (Here as above we identify $\mathfrak{h}$ with

$$\mathfrak{h} \otimes 1 \subset \mathfrak{h} \otimes_k K = \bar{\mathfrak{h}}.)$$

The inverse map

$$\mathrm{res}_{K,k}^{-1} : \mathrm{Hom}_k(\mathfrak{h}, K) \to \mathrm{Hom}_K(\bar{\mathfrak{h}}, K) = \mathrm{Hom}_K(\mathfrak{h} \otimes_k K, K),$$

is described explicitly by the formula

$$\mu \mapsto \{h \otimes a \mapsto a \cdot \mu(h)\} \quad \forall h \in \mathfrak{h}, a \in K.$$

Let $R \subset \bar{\mathfrak{h}}^*$ be the *root system* of $(\bar{\mathfrak{g}}, \bar{\mathfrak{h}})$ [9]. By definition, R consists of all *nonzero* $\alpha \in \bar{\mathfrak{h}}^*$ such that

$$\bar{\mathfrak{g}}_\alpha := \{x \in \bar{\mathfrak{g}} \mid [H, x] = \alpha(H)x \ \forall H \in \bar{\mathfrak{h}}\} \neq \{0\}.$$

Clearly,

$$\bar{\mathfrak{g}}_\alpha = \{x \in \bar{\mathfrak{g}} \mid [H, x] = \alpha(H)x \ \forall H \in \mathfrak{h}\}$$

and therefore

$$\sigma(\bar{\mathfrak{g}}_\alpha) = \bar{\mathfrak{g}}_{\sigma(\alpha)} \quad \forall \sigma \in \mathrm{Aut}(K/k).$$

It follows that the subset R of $\bar{\mathfrak{h}}^*$ is $\mathrm{Aut}(K/k)$-invariant. We write

$$\mathrm{W}(R) \subset \mathrm{Aut}_K(\bar{\mathfrak{h}}^*)$$

for the *Weyl group* of the root system R. Notice that $\mathrm{W}(R)$ permutes elements of R. Let us choose a *basis* (a *simple root system*) B of R. The l-element set B is a basis of the K-vector space $\bar{\mathfrak{h}}^*$. Every root $\alpha \in R$ is a linear combination of elements of B with integer coefficients; in addition, the nonzero coefficients are either all positive or all negative. (Actually, these properties characterize a *basis* of R.) This implies the equality of abelian subgroups

$$\mathbb{Z} \cdot R := \sum_{\alpha \in R} \mathbb{Z} \cdot \alpha = \sum_{\beta \in B} \mathbb{Z} \cdot \beta; \tag{67}$$

hence, $\mathbb{Z} \cdot R$ is a free abelian group of rank l that is a $\mathrm{W}(R)$-invariant subgroup of $\bar{\mathfrak{h}}^*$.

The set B does *not* have to be $\mathrm{Aut}(K/k)$-invariant. However, if $\sigma \in \mathrm{Aut}(K/k)$ then $\sigma(B)$ is a basis of R as well. Since the Weyl group $\mathrm{W}(R)$ acts transitively on the set of all simple root systems of R, there is $w_\sigma \in \mathrm{W}(R)$ such that

$$w_\sigma(\sigma(B)) = B$$

(compare with [27, p. 203]). In particular,

$$s_\sigma := w_\sigma \circ \sigma \in \mathrm{Aut}_K(\bar{\mathfrak{h}}^*)$$

permutes elements of B. It is also clear that s_σ permutes elements of R. Hence, $\mathbb{Z} \cdot R$ is s_σ-invariant.

8.3 Throughout this subsection we use the notation and constructions of Sect. 7.6 applied to

$$\mathfrak{u} = \mathfrak{g}, \quad \bar{\mathfrak{u}} = \bar{\mathfrak{g}}.$$

Let $\mathcal{V}$ be a nonzero finite-dimensional vector space over K endowed by the homomorphism of Lie K-algebras

$$\bar{\mathfrak{g}} \to \mathrm{End}_K(\mathcal{V}), \tag{68}$$

which may be viewed (in the notation of Sect. 7.6) as $\bar{\rho}$ where

$$\rho : \mathfrak{g} \to \mathrm{End}_K(\mathcal{V})$$

is the restriction of the homomorphism (68) to $\mathfrak{g} = \mathfrak{g} \otimes 1$. The homomorphism $\bar{\rho}$ that appeared in (68) provides $\mathcal{V}$ with the structure of a $\bar{\mathfrak{g}}$-module. Let us assume that this module is *simple*.

Let us consider the set

$$\mathrm{Supp}(\mathcal{V}) \subset \bar{\mathfrak{h}}^*$$

of weights of the $\bar{\mathfrak{g}}$-module $\mathcal{V}$, i.e., $\mu \in \bar{\mathfrak{h}}^*$ lies in $\mathrm{Supp}(\mathcal{V})$ if and only if the *weight subspace*

$$\mathcal{V}_\mu := \{v \in \mathcal{V} \mid \rho(H)(x) = \mu(H)v \quad \forall H \in \bar{\mathfrak{h}}\} \neq \{0\}.$$

Then (see [9])

$$\mathrm{Supp}(\mathcal{V}) \subset \mathbb{Q} \cdot R := \sum_{\alpha \in R} \mathbb{Q} \cdot \alpha =: \sum_{\beta \in B} \mathbb{Q} \cdot \beta \subset \bar{\mathfrak{h}}^*, \tag{69}$$

and there exists the *highest weight* λ of the $\bar{\mathfrak{g}}$-module $\mathcal{V}$ that enjoys the following properties:

(i) $\lambda \in \mathrm{Supp}(\mathcal{V})$.
(ii) If $\mu \in \mathrm{Supp}(\mathcal{V})$ then $\lambda - \mu$ is a linear combination of elements of B with nonnegative integer coefficients.

Remark 8.4 It is well known that:

(iii)

$$\mathrm{Supp}(\mathcal{V}) \subset \sum_{\beta \in B} \mathbb{Q} \cdot \beta = \mathbb{Q} \cdot R.$$

(iv) The subset $\mathrm{Supp}(\mathcal{V})$ is $\mathrm{W}(R)$-invariant.

Remark 8.5 It follows from the $\mathrm{W}(R)$-invariance of $\mathbb{Z} \cdot R$ (see (67)) that the l-dimensional $\mathbb{Q}$-vector (sub)space $\mathbb{Q} \cdot R$ is $\mathrm{W}(R)$-invariant.

Recall (Sect. 7.6) that one may attach to each $\sigma \in \mathrm{Aut}(K/k)$ the homomorphism of Lie K-algebras

$$\overline{{}^{\sigma}\rho} : \bar{\mathfrak{g}} \to \mathrm{End}_K({}^{\sigma}\mathcal{V}),$$

and the corresponding $\bar{\mathfrak{g}}$-module ${}^{\sigma}\mathcal{V}$ is simple.

Theorem 8.6 *Suppose that λ is the dominant weight of a simple $\bar{\mathfrak{g}}$-module $\mathcal{V}$ of finite dimension. If $\sigma \in \mathrm{Aut}(K/k)$ then $s_{\sigma^{-1}}(\lambda)$ is the dominant weight of the simple $\bar{\mathfrak{g}}$-module ${}^{\sigma}\mathcal{V}$.*

Proof First, notice that

$$\mathrm{Supp}({}^{\sigma}\mathcal{V}) = \sigma(\mathrm{Supp}(\mathcal{V})).$$

Indeed, for any

$$H \in \mathfrak{h} \subset \bar{\mathfrak{h}},$$

the spectrum of the diagonalizable operator $\rho(H)$ in $\mathcal{V}$ is the collection

$$\{\mu(H) \mid \mu \in \mathrm{Supp}(\mathcal{V})\} \text{ (with multiplicities).}$$

In light of Lemma 7.3, the spectrum of the diagonalizable operator ${}^{\sigma}\mathrm{id} \circ \rho(H)$ in ${}^{\sigma}\mathcal{V}$ is the collection

$$\{\sigma(\mu(H)) \mid \mu \in \mathrm{Supp}(\mathcal{V})\} \text{ (with multiplicities).}$$

More precisely, let $n = \dim(\mathcal{V})$ and $\{e_1, \ldots, e_n\}$ be a common (weight) eigenbasis of all elements of $\bar{\mathfrak{h}}$ in $\mathcal{V}$, i.e., for each index $i \in \{1, \ldots, n\}$ there is a weight

$$\mu_i \in \mathrm{Supp}(\mathcal{V}) \in \bar{\mathfrak{h}}^*$$

such that

$$\rho(H)(e_i) = \mu_i(H)e_i \ \forall H \in \bar{\mathfrak{h}}.$$

(Clearly, the collection $\{\mu_1, \ldots, \mu_n\}$ coincides with $\mathrm{Supp}(\mathcal{V})$.) In light of Remark 7.2, $\{e_1, \ldots, e_n\}$ is a basis of ${}^{\sigma}\mathcal{V}$, and if $H \in \mathfrak{h}$, then $H = \sigma^{-1}H$ and

$$^{\sigma}\mathrm{id} \circ \rho(H)(e_i) = \sigma(\mu_i(H))e_i = \sigma(\mu_i(\sigma^{-1}H))e_i = (\sigma(\mu_i))(H)e_i. \tag{70}$$

Since $\bar{\mathfrak{h}} = \mathfrak{h} \otimes_k K$, we conclude that

$$\overline{^{\sigma}\rho}(H)(e_i) = (\sigma(\mu_i))(H)e_i \ \forall H \in \bar{\mathfrak{h}}. \tag{71}$$

In other words,

$$\mathrm{Supp}({}^{\sigma}\mathcal{V}) = \{\sigma \circ \mu_i \mid i = 1, \ldots, n\} = \{\sigma \circ \mu \mid \mu \in \mathrm{Supp}(\mathcal{V})\}.$$

Second, the $\mathrm{W}(R)$-invariance of $\mathrm{Supp}({}^{\sigma}\mathcal{V})$ implies that

$$\mathrm{Supp}({}^{\sigma}\mathcal{V}) = w_\sigma \circ \sigma(\mathrm{Supp}(\mathcal{V})) = (w_\sigma \circ \sigma)(\mathrm{Supp}(\mathcal{V})) = s_\sigma(\mathrm{Supp}(\mathcal{V})).$$

It follows that $\mathrm{Supp}({}^{\sigma}\mathcal{V})$ contains $s_\sigma(\lambda)$, and all the other weights in $\mathrm{Supp}({}^{\sigma}\mathcal{V})$ are of the form $s_\sigma(\mu)$ where $\lambda - \mu$ is a linear combination of elements of B with nonnegative integer coefficients. Since s_σ permutes elements of B, the difference $s_\sigma(\lambda) - s_\sigma(\mu)$ is also a linear combination of elements of B with nonnegative integer coefficients. It follows that $s_\sigma(\lambda)$ is the dominant weight of the simple $\mathfrak{g}$-module ${}^{\sigma}\mathcal{V}$. □

9 2-Simple Complex Tori Without Nontrivial Endomorphisms

Theorem 9.1 *Suppose that T is a 2-simple complex torus of dimension $g \geq 3$ with $\mathrm{End}^0(T) = \mathbb{Q}$. Assume also that $g \neq 10$ and $2g$ is not a power (e.g., g is odd). Then $\mathrm{Hdg}(T)$ enjoys one of the following properties:*

(i) $\mathrm{Hdg}(T) = \mathrm{SL}(\Lambda_{\mathbb{Q}})$.

(ii) *There exists a nondegenerate symmetric* $\mathbb{Q}$*-bilinear form*

$$\Lambda_{\mathbb{Q}} \times \Lambda_{\mathbb{Q}} \to \mathbb{Q}$$

such that $\mathrm{Hdg}(T)$ *coincides with the corresponding special orthogonal group* $\mathrm{SO}(\Lambda_{\mathbb{Q}})$.

(iii) *There exists a nondegenerate alternating* $\mathbb{Q}$*-bilinear form*

$$\Lambda_{\mathbb{Q}} \times \Lambda_{\mathbb{Q}} \to \mathbb{Q}$$

such that $\mathrm{Hdg}(T)$ *coincides with the corresponding symplectic group* $\mathrm{Sp}(\Lambda_{\mathbb{Q}})$.

Proof It follows from Corollary 5.13 that $\mathrm{hdg}_{T,\mathbb{C}}$ is a complex *simple* classical Lie algebra, whose natural faithful representation in $\Lambda_{\mathbb{C}}$ has a minuscule weight as the highest weight, thanks to Theorem 5.11. Since $2g = \dim_{\mathbb{C}}(\Lambda_{\mathbb{C}})$ is not a power of 2, one should exclude the cases when either $\mathrm{hdg}_{T,\mathbb{C}}$ is of type B_l or $\mathrm{hdg}_{T,\mathbb{C}}$ is of type D_l and $\Lambda_{\mathbb{C}}$ is one of its two semi-spinorial representations. Let us list the remaining cases:

(i) $\mathrm{hdg}_{T,\mathbb{C}}$ is of type C_g or D_g, and there is a nondegenerate alternating or symmetric bilinear form on $\Lambda_{\mathbb{C}}$ such that $\mathrm{hdg}_{T,\mathbb{C}}$ coincides with the corresponding symplectic Lie algebra $\mathfrak{sp}(\Lambda_{\mathbb{C}})$ or the corresponding orthogonal Lie algebra $\mathfrak{so}(\Lambda_{\mathbb{C}})$.

(ii) $\mathrm{hdg}_{T,\mathbb{C}}$ is of type A_l, i.e., $\mathrm{hdg}_{T,\mathbb{C}}$ may be identified with the Lie algebra $\mathfrak{sl}(W)$ of a $(l+1)$-dimensional complex vector space W in such a way that the $\mathfrak{sl}(W)$-module $\Lambda_{\mathbb{C}}$ is isomorphic to the jth exterior power $\wedge^j_{\mathbb{C}}(W)$ of W for some integer j with $1 \le j \le l$. We may assume that $1 < j < l$.

Let us handle case (i). In this situation the $\mathrm{hdg}_{T,\mathbb{C}}$-module $\Lambda_{\mathbb{C}}$ is self-dual, which implies that there is a nonzero homomorphism between the $\mathrm{hdg}_{T,\mathbb{C}}$-module $\Lambda_{\mathbb{C}}$ and its dual. This, in turn, implies that there is a nonzero homomorphism between the $\mathrm{hdg}_{T,\mathbb{Q}}$-module $\Lambda_{\mathbb{Q}}$ and its dual. Now the simplicity of the hdg_T-module $\Lambda_{\mathbb{Q}}$ implies that $\Lambda_{\mathbb{Q}}$ and its dual are isomorphic, i.e., there is a nondegenerate hdg_T-invariant bilinear form

$$\Lambda_{\mathbb{Q}} \times \Lambda_{\mathbb{Q}} \to \mathbb{Q}.$$

The absolutely simplicity of the $\mathrm{hdg}_{T,\mathbb{Q}}$-module $\Lambda_{\mathbb{Q}}$ implies that this form is unique (up to multiplication by a nonzero rational number) and, therefore, is alternating if hdg_T is of type C_g or symmetric if hdg_T is of type D_g. Now the dimension arguments imply that $\mathrm{hdg}_T = \mathfrak{sp}(\Lambda_{\mathbb{C}})$ in the former case and $\mathrm{hdg}_T = \mathfrak{so}(\Lambda_{\mathbb{C}})$ in the latter case.

Let us handle the case (ii). We know that $T = V/\Lambda$ where the -module $V_{\mathbb{C}}$ over $\mathrm{hdg}_{T,\mathbb{C}} = \mathfrak{sl}(W)$ is isomorphic to $\wedge^j_{\mathbb{C}}(W)$.

If $l = 1$ then the inequality $1 < j < l = 1$ implies that this case does not occur.

If $l = 2$ then the inequality $1 < j < l = 2$ implies that $j = 1$ and $V_{\mathbb{C}}$ is isomorphic to W, which is a 3-dimensional complex vector space. Since 3 is an odd integer and the $\mathbb{C}$-dimension of $V_{\mathbb{C}}$ is even, this case also does not occur.

If $l = 3$ then the inequality $1 < j < l = 3$ implies that $j = 2$ and $V_{\mathbb{C}}$ is isomorphic to $\wedge^2_{\mathbb{C}} W$ where W is a 4-dimensional complex vector space and $\wedge^2_{\mathbb{C}} W$ is an irreducible 6-dimensional orthogonal representation of the Lie algebra $\mathfrak{sl}(W)$. This implies that the representation of $\mathrm{hdg}_{T,\mathbb{C}}$ in the 6-dimensional complex vector space $V_{\mathbb{C}}$ is orthogonal irreducible. It follows that

$$\dim_{\mathbb{Q}}(\Lambda_{\mathbb{Q}}) = \dim_{\mathbb{C}}(W) = 6$$

and hdg_T is a $\mathbb{Q}$-Lie subalgebra $\mathfrak{so}(\Lambda_{\mathbb{Q}})$ of the corresponding special orthogonal group $\mathrm{SO}(\Lambda_{\mathbb{Q}})$. It follows that the representation of hdg_T in the 6-dimensional $\mathbb{Q}$-vector space $\Lambda_{\mathbb{Q}}$ is orthogonal irreducible and therefore

$$\dim_{\mathbb{Q}}(\mathrm{hdg}_T) \le \dim_{\mathbb{Q}}(\mathfrak{so}(\Lambda_{\mathbb{Q}}) = 15.$$

However,

$$\dim_{\mathbb{Q}}(\mathrm{hdg}_T) = \dim_{\mathbb{C}}(\mathrm{hdg}_{T,\mathbb{C}}) = \dim_{\mathbb{C}}(\mathfrak{sl}(W)) = 15.$$

This implies that hdg_T coincides with $\mathfrak{so}(\Lambda_{\mathbb{Q}})$, i.e., Hdg_T coincides with $\mathrm{SO}(\Lambda_{\mathbb{Q}})$.

So, we may and will assume that $l > 3$. Then there is an element $u \in \mathfrak{sl}(W)$ that acts on $V_{\mathbb{C}}$ as J. Since J is a nonzero semisimple linear operator in $V_{\mathbb{C}}$, the element u is also a semisimple (i.e., diagonalizable) nonzero linear operator in W. Let $\{e_1, \ldots, e_{l+1}\}$ be an eigenbasis of W and $\{z_1, \ldots, z_{l+1}\} \subset \mathbb{C}$ be the corresponding eigenvalues of u, i.e.,

$$u(e_i) = z_i e_i \ i = 1, \ldots, l+1$$

and the trace of u (in W) is

$$\sum_{i=1}^{l+1} z_i = 0.$$

This implies that u has at least two distinct eigenvalues.

Then the collections of eigenvalues of J in $V_{\mathbb{C}} \cong \wedge^j_{\mathbb{C}}(W)$ listed with multiplicities coincide with

$$\{t_A := \sum_{i \in A} z_i\}_A$$

where A runs through all j-element subsets A of $\{1, \ldots, l+1\}$. On the other hand, we know that the spectrum of J in $V_{\mathbb{C}}$ consists of two eigenvalues $\mathbf{i}$ and $-\mathbf{i}$, whose multiplicities coincide. It follows almost immediately that u has precisely two (distinct) eigenvalues, say, **a** and **b**, and none of them is 0. Indeed, suppose that the spectrum of u contains (at least) three eigenvalues, say, **a**, **b**, **c**.

Reordering the eigenbasis if necessary, we may assume that

$$z_1 = \mathbf{a},\ z_2 = \mathbf{b},\ z_3 = \mathbf{c}.$$

Let B be any $(j-2)$-element subset of $\{4, \ldots, l+1\}$. Let us consider three distinct j-element subsets

$$A_1 = \{2,3\} \cup B,\ A_2 = \{1,3\} \cup B,\ A_3 = \{1,2\} \cup B$$

of $\{1, \ldots, l+1\}$. If we put

$$C := \{1,2,3\} \cup B \subset \{1, \ldots, l+1\}, \quad c := \sum_{l \in C} z_i \in \mathbb{C}$$

then we get three distinct eigenvalues

$$t_{A_1} = c - \mathbf{a},\ t_{A_2} = c - \mathbf{b},\ t_{A_3} = c - \mathbf{c}$$

of J, which do not exist. This proves that the spectrum of u consists of precisely two eigenvalues, say, $\mathbf{a}, \mathbf{b} \in \mathbb{C}$. Since the trace of *nonzero* u is 0, both **a** and **b** are *not* zero.

Let p be the multiplicity of the eigenvalue **a** and q the multiplicity of the eigenvalue **b**. Both p and q are positive integers, whose sum

$$p + q = l + 1 > 3 + 1 = 4.$$

Since u is traceless,

$$p\mathbf{a} + q\mathbf{b} = 0.$$

I claim that either $p = 1$ or $q = 1$. Indeed, suppose that

$$p \geq 2,\ q \geq 2.$$

Since $p + q > 4$, we may assume that $p \geq 3$. Notice also that all three complex numbers

$$2\mathbf{a},\ 2\mathbf{b},\ \mathbf{a} + \mathbf{b}$$

are distinct. Reordering the eigenbasis if necessary, we may assume that

$$z_1 = \mathbf{a},\ z_2 = \mathbf{a}, z_3 = \mathbf{a},\ z_l = \mathbf{b},\ z_{l+1} = \mathbf{b}$$

(recall that $l+1>4$).

Let $\mathbf{B}$ be a $(j-2)$-element subset of the $(l-3)$-element subset $\{3,4,\dots,l-1\}$ and $b:=\sum_{i\in \mathbf{B}} z_i$. Let us consider three distinct j-element subsets

$$A_1=\{1,2\}\cup B,\ A_2=\{l,l+1\}\cup B,\ A_3=\{1,l\}\cup B$$

of $\{1,\dots,l+1\}$. Then we get three *distinct* eigenvalues

$$t_{A_1}=b+2\mathbf{a},\ t_{A_2}=b+2\mathbf{b},\ t_{A_3}=b+(\mathbf{a}+\mathbf{b})$$

of J, which could not be the case. The obtained contradiction proves that either $p=1$ or $q=1$.

Without loss of generality we may assume that $p=1$. Reordering the eigenbasis if necessary, we may assume that $z_1=\mathbf{a}$ and all other $z_i=\mathbf{b}$ (for all $i>1$). It follows easily that the spectrum of J consists of two eigenvalues, namely, $j\mathbf{b}$ of multiplicity $\binom{l}{j}$ and $\mathbf{a}+(j-1)\mathbf{b}$ of multiplicity $\binom{l}{j-1}$. It follows that

$$\binom{l}{j}=\binom{l}{j-1},$$

i.e.,

$$\frac{l-j+1}{j}=1,\quad l-j+1=j,\quad l=2j-1.$$

It remains to put $m=j$ and we get that

$$l=2m-1,\ j=m,\ 2g=\binom{2m}{m}.$$

Since $2m-1=l>3$, we get

$$m\geq 3. \tag{72}$$

Now it is natural to look at the structure of the $\mathfrak{sl}(W)$-module $\wedge^2_{\mathbb{C}}\left(\wedge^m_{\mathbb{C}}(W)\right)$. We are going to apply results of Sect. 7 with

$$k=\mathbb{Q},\ K=\mathbb{C},\ \mathrm{Aut}(K/k)=\mathrm{Aut}(\mathbb{C}),$$

$$\mathfrak{g}=\mathrm{hdg}_T,\ \bar{\mathfrak{g}}=\mathrm{hdg}_{T,\mathbb{C}},\ \mathcal{V}=\wedge^2_{\mathbb{Q}}(\wedge^m_{\mathbb{Q}}\Lambda_{\mathbb{Q}}),\ \bar{\mathcal{V}}=\wedge^2_{\mathbb{C}}(\wedge^m_{\mathbb{C}}\Lambda_{\mathbb{C}}).$$

Let us fix a Cartan subalgebra $\mathfrak{h}$ of the simple Lie $\mathbb{Q}$-algebra hdg_T, which is an l-dimensional $\mathbb{Q}$-vector space. Then

$$\bar{\mathfrak{h}}=\mathfrak{h}\otimes_{\mathbb{Q}}\otimes\mathbb{C}$$

is a Cartan subalgebra of the complex simple Lie algebra $\mathrm{hdg}_{T,\mathbb{C}}$ that is an l-dimensional complex vector space endowed with the natural semi-linear Aut($\mathbb{C}$)-action; its subalgebra of invariants coincides with $\mathfrak{h} \otimes 1 = \mathfrak{h}$.

As in Sect. 7, let us consider the dual complex vector space

$$\bar{\mathfrak{h}}^* = \mathrm{Hom}_{\mathbb{C}}(\bar{\mathfrak{h}}, \mathbb{C}).$$

Let

$$R \subset \bar{\mathfrak{h}}^*$$

be the root system of $(\mathrm{hdg}_{T,\mathbb{C}}, \bar{\mathfrak{h}})$.

Let us choose a simple root system B of simple roots (basis) of R and let $P_{++}(R) \subset \bar{\mathfrak{h}}^*$ be the corresponding semigroup of dominant weights [8].

If $\mu \in P_{++}(R)$ then we write $\mathbf{V}(\mu)$ for the simple $\mathrm{hdg}_{T,\mathbb{C}}$-module with highest weight μ [9]. In particular, $\mathbf{V}(0)$ stands for the one-dimensional $\bar{\mathbb{Q}}$-vector space $\bar{\mathbb{Q}}$ with trivial (zero) action of $\mathrm{hdg}_{T,\mathbb{C}}$. Then the $\binom{2m}{m}$-dimensional $\bar{\mathbb{Q}}$-vector space

$$\bar{\Lambda} = \Lambda_{\mathbb{Q}} \otimes_{\mathbb{Q}} \bar{\mathbb{Q}}$$

becomes a simple $\mathrm{hdg}_{T,\mathbb{C}}$-module that is isomorphic to $\mathbf{V}(\bar{\omega}_m)$. Hereafter we use the notation of Bourbaki ([8, Tables], [9, Tables]). In particular,

$$B = \{\alpha_1, \ldots, \alpha_l\} = \{\alpha_1, \ldots, \alpha_{2m-1}\}$$

(see Root systems of type A_l in [8, Tables]), and $\bar{\omega}_i$ is the dominant weight of a fundamental representation of dimension $\binom{2m}{i}$ (when $1 \le i \le l = 2m-1$); see [9, Table 2]. In addition, we put

$$\bar{\omega}_0 := 0 =: \bar{\omega}_{2m}.$$

Notice that the only nontrivial automorphism of (R, B) is the involution

$$\alpha_i \to \alpha_{2m-i} \quad \forall i = 1, \ldots, 2m-1 = l.$$

Hence, each dominant weight $\bar{\omega}_{m+i} + \bar{\omega}_{m-i}$ is $\mathrm{Aut}(R, \mathbf{B})$-invariant for all $i = 0, \ldots, 2m$.

It follows from results of [17, p. 140, Example 9a, last displayed formula] (see also [16, Exercises 6.16 on p. 81 and 15.32 on p. 226]) that the $\bar{\mathfrak{g}} = \mathrm{hdg}_{T,\mathbb{C}}$-module

$$\bar{\mathcal{V}} = \wedge^2_{\mathbb{C}}(\mathbf{V}(\bar{\omega}_m))$$

is isomorphic to a direct sum

$$\oplus_{i \text{ odd},\ 1 \le i \le m} \mathbf{V}(\bar{\omega}_{m+i} + \bar{\omega}_{m-i}). \tag{73}$$

This implies that the $\bar{\mathfrak{g}}$-module $\bar{\mathcal{V}}$ splits into a direct sum of mutually non-isomorphic simple $\bar{\mathfrak{g}}$-modules; one of them is trivial if and only if m is odd (one should take $i = m$ in order to get the summand $\mathbf{V}(0)$.)

Let $\mathcal{W}$ be a simple $\bar{\mathfrak{g}}$-submodule of $\bar{\mathcal{V}}$. Let $\lambda_{\mathcal{W}}$ be the highest weight of $\mathcal{W}$. We know that $\lambda_{\mathcal{W}}$ is $\mathrm{Aut}(R, B)$-invariant. It follows from Theorem 8.6 combined with Proposition 7.9 that the simple $\bar{\mathfrak{g}}$-submodules $\mathcal{W}$ and $\sigma(\mathcal{W})$ have the same highest weight and therefore are isomorphic. This implies that

$$\sigma\mathcal{W} = \mathcal{W}\, \forall \sigma \in \mathrm{Aut}(\mathbb{C}).$$

By Lemma 7.4, $\mathcal{W}$ is defined over $\mathbb{Q}$, i.e., there is a $\mathbb{Q}$-vector subspace $\mathbf{W}$ of $\mathcal{V}$ such that

$$\mathcal{W} = \mathbf{W} \otimes_{\mathbb{Q}} \mathbb{C}.$$

Clearly, such $\mathbf{W}$ is a simple hdg_T-submodule of $\mathcal{V}$. It follows from (73) that the hdg_T-module $\mathcal{V}$ splits into a direct sum

$$\oplus_{i \text{ odd},\ 1 \le i \le m} \mathbf{W}_i \tag{74}$$

of hdg_T-modules such that

$$\mathbf{V}(\bar{\omega}_{m+i} + \bar{\omega}_{m-i}) \cong \mathbf{T}_{\mathbb{Q},\mathbb{C}}(\mathbf{W}_i) = \mathbf{W}_i \otimes_{\mathbb{Q}} \mathbb{C}.$$

This implies that all $\mathbf{W}_i$ are mutually non-isomorphic simple hdg_T-modules. In addition, one of them is trivial if and only if m is odd. (Namely, if m is odd then $\mathbf{W}_m$ is a trivial hdg_T-module of $\mathbb{Q}$-dimension 1.)

Thus, if m is *even*, then the hdg_T-module $\mathcal{V}$ splits into a direct sum of $(m/2)$ simple modules, none of which is trivial. If m is *odd*, then the hdg_T-module $\mathcal{V}$ splits into a direct sum of $(m + 1)/2$ simple modules, and precisely one of them is trivial. It follows that the hdg_T-module $\mathcal{V}$ is simple if and only if $m = 2$. Since $m \ge 3$ (72), we conclude that $\mathcal{V}$ is never simple. On the other hand, it's clear that $\mathcal{V}$ is a direct sum of a simple hdg_T-module and a trivial one if and only if $m = 3$.

Recall that we are actually interested in the dual hdg_T-module

$$\mathrm{H}^2(T, \mathbb{Q}) = \mathrm{Hom}_{\mathbb{Q}}(\mathcal{V}, \mathbb{Q}).$$

By duality, the hdg_T-module is never simple; it is a direct sum of a simple hdg_T-module and a trivial one if and only if $m = 3$. Now the 2-simplicity of T implies that $m = 3$ and therefore

$$2g = \binom{2 \cdot 3}{3} = 20,$$

i.e., $g = 10$. □

Acknowledgements I am grateful to Frédéric Campana and Ekaterina Amerik for interesting stimulating questions. My special thanks go to Grigori Olshanski for a very informative letter about *plethysm*. I am grateful to Tatiana Bandman and the referee, whose thoughtful comments helped to improve the exposition.

I was partially supported by Simons Foundation Collaboration grant # 585711 and the Travel Support for Mathematicians Grant MPS-TSM-00007756 from the Simons Foundation. Most of this work was done in January–May 2022 and December 2023 during my stay at the Max-Planck Institut für Mathematik (Bonn, Germany), whose hospitality and support are gratefully acknowledged.

References

1. E. Amerik, F. Campana, On algebraically coisotropic submanifolds of holomorphic symplectic manifolds, in *Épijournal Géom. Algébrique, Special volume in honour of C. Voisin*, Article No. 11 (2023). arXiv:2205.07958 [math.AG]
2. E. Artin, G. Whaples, Axiomatic characterization of fields by the product formula for valuations. Bull. Am. Math. Soc. **51**(7), 469–492 (1945). Reprinted in: Exposition by Emil Artin: A selection. American Mathematical Society, Providence (2007)
3. T. Bandman, Yu.G. Zarhin, Simple complex tori of algebraic dimension 0. Proc. Steklov Inst. Math. **320**, 21–38 (2023)
4. A. Borel, *Linear Algebraic Groups*, 2nd enlarged edition (Springer, New York, 1991)
5. Z.I. Borevich, I.R. Shafarevich, *Number Theory* (Academic Press, New York, 1986)
6. C. Birkenhake, H. Lange, *Complex Tori* (Birkhäuser, Boston, Basel, Stutgart, 1999)
7. M.V. Borovoi, The Hodge group and the endomorphism algebra of an abelian variety, in *Questions in Group Theorem and Homological Algebra*, ed. by A.L. Onishchik (Yaroslavl', 1981), pp. 124–126. arXiv:1310.5236 [math.AG]
8. N. Bourbaki, Groupes et algébres de Lie. Chapitres 4–6. Actualitiés Sci. Indust., vol. 1337 (Hermann, Paris, 1968)
9. N. Bourbaki, Groupes et algébres de Lie. Chapitres 7, 8. Actualitiés Sci. Indust., vol. 1364 (Hermann, Paris, 1975)
10. F. Campana, Isotrivialité de certaines familes kählériennes de variétés non projectives. Math. Z. **252**, 147–156 (2006)
11. F. Charles, Two results on the Hodge structure of complex tori. Math. Z. **300**, 3623–3643 (2022)
12. C. Chevalley, Théorie des groupes de Lie. Tome II. Groupes algébriques. Actualités Sci. Ind., vol. 1152 (Hermann, Paris, 1951)
13. K. Conrad, Zorn's Lemma and some applications. II. https://kconrad.math.uconn.edu/blurbs/zorn2.pdf
14. P. Deligne, Theorie de Hodge II. Publ. Math. IHES **40**, 5–57 (1971)
15. P. Deligne, Variétés de Shimura: Interprétation modulaire, et techniques de construction de modéles canoniques, in *Automorphic Forms, Representations and L-Functions. Proceedings of Symposia in Pure Mathematics*, vol.33(2) (American Mathematical Society, Providence, 1979), pp. 247–289
16. W. Fulton, J. Harris, *Representation Theory: A First Course. GTM*, vol. 129 (Springer, New York, 1991)
17. I. McDonald, *Symmetric Functions and Hall Polynomials*, 2nd edn. (Oxford University Press, Oxford, 1995)
18. D. Mumford, A note on Shimura's paper "discontinuous groups and Abelian varieties". Math. Ann. **181**, 345–351 (1969)
19. D. Mumford, *Abelian Varieties*, 2nd edn. (Oxford University Press, London, 1974)

20. A.L. Onishchik, *Lectures on Semisimple Real Lie Algebras and Their Representations* (European Mathematical Society Publishing House, 2004)
21. F. Oort, Yu.G. Zarhin, Endomorphism algebras of complex tori. Math. Ann. **303**(1), 11–30 (1995)
22. R.S. Pierce, *Associative Algebras. GTM*, vol. 88 (Springer, New York, Heidelberg, Berlin, 1982)
23. K.A. Ribet, Galois action on division points of Abelian varieties with real multiplications. Am. J. Math. **98**, 751–804 (1976)
24. J.-P. Serre, Representations ℓ-adiques, in *Algebraic Number Theory (International Symposium, Kyoto, 1976)* (Japan Society for the Promotion of Science, Tokyo, 1977), pp. 177–193. Reprinted in Œuvres **III** n 112, pp. 384–400. Springer, Berlin, Heidelberg (1986)
25. J.-P. Serre, Groupes algébriques associés aux modules de Hodge-Tate. Astérisque **65**, 155–188 (1979). Reprinted in Œuvres **III** n 119, pp. 469–502 (Springer, Berlin, Heidelberg, 1986)
26. J.-P. Serre, *Abelian ℓ-Adic Representations and Elliptic Curves*, 2nd edn. (Addison-Wesley, 1989)
27. J. Tits, Représentations linéaires irréductibles d'un groupe réductif sur un corps quelconque. Crelle **247**, 196–220 (1971)
28. E.B. Vinberg, A.L. Onishchik, *Lie Groups and Algebraic Groups* (Springer, Berlin, Heidelberg, 1990)
29. C. Voisin, *Hodge Theory and Complex Algebraic Geometry I* (Cambridge University Press, Cambridge, 2002)
30. V.E. Voskresensky, *Algebraic Tori (Russian)* (Nauka, Moscow, 1977)
31. B.L. van der Waerden, *Algebra*, vol. I (Springer, New York, 1991), xiv+265 pp
32. J. Xue, Yu.G. Zarhin, Centers of Hodge groups of superelliptic jacobians. Transform. Groups **15**, 449–482 (2010)
33. J. Xue, Yu.G. Zarhin, Hodge groups of certain superelliptic jacobians. Math. Res. Lett. **17**, 371–388 (2010)
34. J. Xue, Hodge groups of certain superelliptic jacobians II. Math. Res. Lett. **18**, 579–590 (2011)
35. Yu.G. Zarhin, Weights of simple Lie algebras in the cohomology of algebraic varieties. Izv. Akad Nauk SSSR **48**, 264–304 (1984). Math. USSR Izvestya **24**, 245–281 (1985)
36. Yu.G. Zarhin, Linear semisimple Lie algebras containing an operator with small number of eigenvalues. Arch. Math. **46**, 522–532 (1986)
37. Yu.G. Zarhin, Linear irreducible algebras and Hodge structures, in *Proceedings of the USA - USSR Symposium on Algebraic Geometry, Chicago, June–July 1989*, ed. by S. Bloch, I. Dolgachev, W. Fulton. Springer Lecture Notes in Mathematics, vol. 1479 (1991), pp. 281–297
38. Yu.G. Zarhin, Actions of semisimple Lie groups and orbits of Cartan subgroups. Arch. Math. **56**, 491–496 (1991)
39. Yu.G. Zarhin, Hodge classes on certain hyperelliptic prymians, in: *Arithmetic, Geometry, Cryptography and Coding Theory (AGCT 2011)*, ed. by Y. Aubry, Ch. Ritzenthaler, A. Zykin. Contemporary Mathematics, vol. 574 (2012), pp. 171–183

www.ingramcontent.com/pod-product-compliance
Lightning Source LLC
Chambersburg PA
CBHW070045060125
19957CB00003B/43

9 783031 741333